Animal Behaviour

Animal Behaviour

Mechanism, Development, Function and Evolution

Chris Barnard

Professor of Animal Behaviour, University of Nottingham, UK

PEARSON
Prentice
Hall

Harlow, England • London • New York • Boston • San Francisco • Toronto • Sydney • Singapore • Hong Kong
Tokyo • Seoul • Taipei • New Delhi • Cape Town • Madrid • Mexico City • Amsterdam • Munich • Paris • Milan

For my family

Pearson Education Limited
Edinburgh Gate
Harlow
Essex CM20 2JE
England

and Associated Companies throughout the world

Visit us on the World Wide Web at:
www.pearsoned.co.uk

_____

First published 2004

© Pearson Education Limited 2004

ISBN 0130899364

British Library Cataloguing-in-Publication Data
A catalogue record for this book is available from the British Library

Library of Congress Cataloging-in-Publication Data
Barnard, C. J. (Christopher J.)
    Animal behavior : mechanism, development, function, and evolution / Chris Barnard.
       p.   cm.
   Includes bibliographical references and index.
   ISBN 0-13-089936-4
     1. Animal behavior.   I. Title.

QL751.B18563 2003
591.5–dc22   2003060928

10 9 8 7 6 5 4 3 2 1
09 08 07 06 05 04

Typeset in 9.5/12pt Sabon by 35
Printed and bound by Ashford Colour Press, Gosport, Hants
The publisher's policy is to use paper manufactured from sustainable forests.

Brief contents

Contents

Preface

Animal behaviour has been one of the most exciting and fastest-growing scientific disciplines of recent years. Its impact on the whole way we think about biology has spawned lucid 'best sellers' such as *The Selfish Gene* and widespread scientific and public debate about our view of the natural world and our place in it. This book provides a comprehensive introduction to the study of behaviour, from its basis in the animal's anatomy and physiology to its adaptive value in the environment. It is aimed at undergraduate students in the biological sciences and psychology and is designed to serve as both a detailed introduction and an extensive, up-to-date source of reference enabling students to pursue topics in the primary literature. Animal behaviour is a subject rich in debate, much of it fundamental to our understanding of biology as a whole. The book therefore highlights issues currently occupying researchers and examines their implications for underlying theory and empirical studies. The overall theme of the book is evolutionary, showing how the different levels at which behaviour can be studied ultimately relate to the consequences of behaviour for the animal's reproductive success – *mechanism and development serve function through evolution*, as it says in Chapter 2.

The book begins by asking the most basic questions of all: what do we mean by 'behaviour', and what does the apparent purposefulness of many behaviour patterns imply about the organisms performing them? This leads to a consideration of Tinbergen's four levels of inquiry (the 'Four Whys') as a framework for posing questions about behaviour, and a brief review of the various traditions of behaviour study (ethology, comparative psychology, behavioural ecology and evolutionary psychology) to provide historical anchor points for different approaches and ideas. Chapter 2 then discusses the evolution of behaviour and its main theoretical cornerstones, using these to establish the evolutionary theme running through the remaining chapters. Mechanism is examined in Chapters 3 and 4, first in terms of the animal's physiological systems, then from a motivational perspective. Different models of motivation are compared along with the implications of motivational theory for intentionality and cognition, and ultimately our views on suffering and welfare in other species. Chapter 5 focuses on the effects of genes in behavioural development. Arguments over the shorthand 'genes for behaviour' language of selfish gene theory are examined, along with examples of what it might mean in different behavioural systems, and the relationships between genes at the level of the genome are discussed in the light of recent ideas about genetic conflict. This is followed in Chapter 6 by a consideration of maturation and learning, looking in particular at how developments in learning theory have begun to blur the classical distinctions between different types of learning that still pervade the literature. Chapters 7–11 deal with the behavioural ecology of decision-making in some of the major functional contexts of the animal's life: finding a place to live, foraging for food, avoiding predators, social behaviour, reproduction and communication. The book rounds off by looking at the behaviour of our own species, and our ability to account for it within the theoretical frameworks applied to other organisms. The discussion throughout is wide-ranging, but each chapter contains several features that are designed to make navigating and assimilating information easier. These include:

- an outline header at the start, indicating the main points to be covered;
- boxed features summarising key concepts, theories or debates underlying the main discussion;
- extensive cross-referencing to other chapters;
- a full summary of conclusions at the end;
- a list of recommended further reading to enable issues to be followed up.

What's new about the book?

While some other textbooks provide broad treatments of the subject, and many more focus on the now dominant theme of behavioural ecology, this book develops the evolutionary framework of behavioural ecology but also uses it to reflect on the design of underlying mechanism and development. The study of animal behaviour has certainly revolutionised our understanding of evolutionary biology over the last three decades, much as advances in molecular genetics now promise to do for developmental biology. But, as the revolution comes of age, the focus is broadening once again to take in its implications for the machinery and developmental processes of behaviour and the constraints these in turn impose on functional design. The balance of inquiry advocated so strongly by Tinbergen nearly half a century ago is thus beginning to re-emerge in a new light. This newly broadening perspective is one of the major emphases of the book.

Author's acknowledgements

During the months of writing, I have benefited more than I can adequately express from the comments of colleagues on various chapters, some of them wading through the entire manuscript in unexpurgated first draft. For this heroic effort I am indebted to Mike Ritchie, Ian Hartley, Mike Siva-Jothy, Manfred Gahr, Ted Burk, Carel ten Cate, Sarah Collins, Francis Gilbert and Angela Turner. Given the present distorted priorities of academic life, I am doubly grateful to them all. Andy Higginson, Zoe Ong, Lucy Barnard and Anna Barnard gave me a salutary student's-eye view of the text by reading through much of the same. My intended readership has benefited more than it will know from their judgements on style and clarity. I have paid careful attention to all the comments I received, and can only hope my critics feel their efforts have been worthwhile. Needless to say, any shortcomings that remain are entirely my own. Geoff Parker, Alex Weir, Alex Kacelnik, Nicky Clayton, Marion Petrie, Jane Brockmann, Richard Cowie, Wolfgang and Roswitha Wiltschko, James Osmond, Gerald Wilkinson, Leigh Simmons, Mario de Bono, Barbara Webb, John Manning, Marc Bekoff, Rob Beynon and Jane Hurst kindly gave permission for me to reproduce their excellent photographs and other material, and Mark Spedding's wonderful drawings illuminate the text and remind us what it is we are actually talking about. My warmest thanks to them all.

I have also been extremely fortunate in my editorial and production team at Pearson. Alex Seabrook is one of the most persuasive and committed commissioning editors I have met, and writing for her has always been a pleasure. This venture has been no exception. Of course, agreeing to write something is very different from actually doing it, and certainly different from doing it within an agreed deadline. I therefore have Stuart Hay to thank for the many timely prods that have kept the project to something resembling its original timetable. But Stuart has been much more than a necessary goad; his advice on virtually every aspect of the book has always been spot on, and I am hugely indebted to his experienced judgement. Many other people at Pearson have played a vital role in bringing the book to life, including Nicola Chilvers, Bridget Allen, Amanda Thomas, Tina Cadle-Bowman, Pauline Gillett, Kevin Ancient and Carol Abbott. My grateful thanks to each of them.

In addition to the people who have helped with the book itself, I should like acknowledge a much longer standing debt of gratitude to Rex Knight, Ron Pearson, Geoff Parker, Arthur Cain, David McFarland, Richard Dawkins, John Krebs and Peter Usherwood for inspiration and support at various critical stages in my career, and to Peter McGregor, Francis Gilbert, Jerzy Behnke, Ian Duce, Jane Hurst and Sarah Collins, my past and present colleagues at Nottingham, for the privilege of teaching and researching with them over the years, and for their unfailingly stimulating and enjoyable friendship. Not least, of course, I thank the many hundreds of students who have passed through my various courses during the last 25 years, both for their helpful comments, and for keeping me in touch with much more than would otherwise have been the case with their constantly evolving questions. They are a refreshing reminder of what we in the university system are really here to do. Finally, special thanks to my family, Siân, Anna, Lucy and Matt, for yet again putting up with teetering mounds of books and papers spilling through the house, and months of a near social hermit scribbling in their midst.

Chris Barnard
Nottingham 2003

Acknowledgements

Publisher's acknowledgements

We are grateful to the following for permission to reproduce copyright material:

Table 2.1 from 'A gene mutation which changes a behaviour pattern', *Evolution*, Vol. 10, pp 421–39 (Bastock, M. 1956); Figure 2.4(b) after 'Behaviour genetics of nest clearing honey bees. IV. Responses of F1 and backcross generations to disease-killed brood', *American Zoologist*, Vol. 4, pp 111–23 (Rothenbuhler, N. 1964b); Figure 7.20 after 'Celestial orientation in wild mallards', *Bird Banding*, Vol. 29, pp 75–90 (Bellrose, F.C. 1958); Figure 10.21(a) after 'Parent–offspring conflict', *The American Zoologist*, Vol. 14, pp 249–64 (Trivers, R.L. 1974); Figure 10.12(b, c) adapted from 'Effects of a haematophagous mite on the barn swallow *Hirundo rustica*: a test of the Hamilton and Zuk hypothesis', *Evolution*, Vol. 44, pp 771–84 (Moller, A.P. 1990) reproduced with permission of the Society for Integrative and Comparative Biology. Tables 2.2 and 2.4 from 'The evolution of behavior', *Scientific American*, September, pp 186 and 189 (Maynard Smith, J. 1978); reproduced with permission of Adolph E. Brotman. Table 4.1 and Figure 4.27 from 'Welfare by design: the natural selection of welfare criteria', *Animal Welfare*, Vol. 5, pp 405–33, Figures 1 and 2 and table (Barnard, C.J. and Hurst, J.L. 1996) reproduced with permission of the Universities Federation for Animal Welfare, Wheathampstead. Table 7.1 from *Migration: The Biology of Life on the Move* by Hugh Dingle, copyright © 1996 by Oxford University Press Inc; Figures 4.19 (from), 4.26 (from), 6.22 (after), 6.25 (after), 11.21 (after), 11.22(a) (after), 11.22(b) (after) and 11.23 (after) *Cognition, Evolution and Behavior*, Oxford, Oxford University Press (Shettleworth, S.J. 1998) used by permission of Oxford University Press, Inc. Table 8.1 reprinted from *Animal Behaviour*, Vol. 34, Caro, T.M., 'The functions of stotting: a review of the hypotheses', pp 649–62, copyright 1986a; Table 10.2 reprinted from *Animal Behaviour*, Vol. 59, Westmeat, D.F., Walters, A., McCarthy, T.M., Hatch, M.I. and Hein, W.K., 'Alternative mechanisms of nonindependent mate choice', pp 467–76, copyright 2000; Figure 1.1 reprinted from *Animal Behaviour*, Vol. 28, Brockman, H.J., 'The control of nest depth in a digger wasp (*Sphex ichneumoneus* L.)', pp 426–45, copyright 1980; Figure 1.2 reprinted from *Animal Behaviour*, Vol. 58, Hughes, B.O., Duncan, I.J.H. and Brown M.F., 'The performance of nest building by domestic hens: is it more important than the construction of a nest?', pp 1177–90, copyright 1989; Figure 1.9 reprinted from *Animal Behaviour*, Vol. 58, Nelson, D.A., 'Ecological influences on vocal development in the white-crowned sparrow', pp 21–36, copyright 1999; Figure 1.10 reprinted from *Animal Behaviour*, Vol. 23, Halliday, T.R., 'An observational and experimental study of sexual behaviour in the smooth newt', pp 291–322, copyright 1977; Figure 2.2(a, b, c) reprinted from *Animal Behaviour*, Vol. 51,

Kennedy, M., Spencer, H.G. and Gray, R.D., 'Hop, step and gape: do social displays of the Pelecaniformes reflect phylogeny?', pp 273–91, copyright 1996; Figure 2.3 reprinted from *Animal Behaviour*, Vol. 36, Irwin, R., 'The evolutionary importance of behavioural development: the ontogeny and phylogeny of bird song', pp 814–24, copyright 1988; Figure 2.11(a) reprinted from *Behavioural Biology*, Vol. 11, DeFries, J.C., Hegmann, J.P. and Halcomb, R.A., 'Response to 20 generations of selection of open-field activity in mice', pp 481–95, copyright 1974; Figure 2.17 reprinted from *Journal of Theoretical Biology*, Vol. 77, Brockman, H.J., Grafen, A. and Dawkins, R., 'Evolutionarily stable nesting strategy in digger wasp', pp 473–96, copyright 1979; Figure 3.15 reprinted from Lloyd, D.P.C., 'Synaptic mechanisms', in J.F. Fulton (ed.) *A Textbook of Physiology*, Philadelphia, Saunders, pp 35–47, copyright 1955; Figure 3.32 reprinted from *Psychoneuroendrocrinology*, Vol. 3, Mendoza, S.P., Coe, C.L., Lowe, E.L. and Levine, S., 'The physiological response to group formation in adult squirrel monkeys', pp 221–9, copyright 1979; Figure 3.36(a, b) reprinted from *Animal Behaviour*, Vol. 51, Podos, J., 'Motor constraints on vocal developments in a songbird', pp 1061–70, copyright 1996; Figure 4.6 reprinted from *Animal Behaviour*, Vol. 20, McFarland, D. and Silby, R.M., 'Unitary drives' revisited, pp 548–63, copyright 1972; Figures 4.10, 4.13 and 4.14 reprinted from *Animal Behaviour*, Vol. 23, Silby, R.M., 'How incentive and deficit determine feeding tendency', pp 437–46, copyright 1975; Figure 4.11 reprinted from *Animal Behaviour*, Vol. 25, McCleery, R.H., 'On satiation curves', pp 1005–15, copyright 1977; Figure 4.16 reprinted from *Advances in the Study of Behaviour*, Vol. 5, McFarland, D., 'Time-sharing as a behavioural phenomenon', pp 201–25, copyright 1974; Figure 4.23 reprinted from *Animal Behaviour*, Vol. 53, Povinelli, D.J., Gallup, G.G. Jr, Eddy, T.J., Bierschwale, D.T., Engstrom, M.C., Perilloux, H.K. and Toxopeus, I.B., 'How incentive and deficit determine feeding tendency', pp 437–46, copyright 1975; Figure 5.3 reprinted from *Animal Behaviour*, Vol. 20, Bentley, D. and Hoy, R., 'Genetic control of the neuronal network generating cricket (Teleogryllus) song patterns', pp 478–92, copyright 1972; Figure 6.7 reprinted from *Behavioural Biology*, Vol. 8, Peeke, H.V.S. and Veno, A., 'Stimulus specificity of habituated aggression in three-spined sticklebacks (*Gastrosteus aculeatus*)', pp 427–32, copyright 1973; Figure 6.15 reprinted from Moore, B.R., 'The role of directed Pavlovian reactions in simple instrumental learning in the pigeon', in R.A. Hinde and J. Stevenson-Hinde (eds) *Constraints on Learning and Predispositions*, London, Academic Press, pp 59–88, copyright 1973; Figure 6.21 reprinted from Warren, J.M. 'Primate learning in comparative perspective', in A.M. Schrier, H.F. Harlow and F. Stollnitz (eds) *Behavior of Non-human Primates*, Vol. 1, New York, Academic Press, pp 249–81, copyright 1965; Figure 6.22 reprinted from *Animal Behaviour*, Vol. 31, Galef, B.G. Jr and Wigmore, S.W., 'Transfer of information concerning distant foods: a laboratory investigation of the "information centre" hypothesis', pp 748–58, copyright 1983; Figure 6.24 reprinted from Galef, B.G. Jr, 'Social enhancement of food preferences in Norway rats: a brief review', in C.M. Heyes and B.G. Galef Jr (eds) *Social Learning and Imitation in Animals: The Roots of Culture*, New York, Academic Press, pp 49–64, copyright 1996a; Figure 6.27 reprinted from *Animal Behaviour*, Vol. 27, Bateson, P.P.G., 'How do sensitive periods arise and what are they for?', pp 470–86, copyright 1979; Figure 7.3(a, b) reprinted from *Animal Behaviour*, Vol. 24, Partridge, L., 'Field and laboratory observations on the foraging and feeding techniques of bluetits (*Parus caeruleus*) and coaltits (*Parus ater*) in relation to their habitats', pp 534–44, copyright 1976; Figure 7.5(b) reprinted from *Trends in Evolution*, Vol. 3, Milinski, M., 'Games fish play: making decisions as a social forager', pp 325–50, copyright 1988; Figure 8.6(b) reprinted from *Journal of Experimental Marine Biology and Ecology*, Vol. 25, Kislaliogu, M. and Gibson, R.N., 'Prey "handling time" and its importance in food section by the 15-spined stickleback *Spinachia spinachia*', pp 151–8, copyright 1976; Figure 8.8 reprinted from *Animal Behaviour*, Vol. 32, Thompson, D.B.A. and Barnard, C.J., 'Prey selection by plovers: optimal foraging in mixed-species groups', pp 534–63, copyright 1984; Figure 8.12 reprinted from *Theoretical Population Biology*, Vol. 14, Belovsky, G.E., 'Diet optimisation in a generalist herbivore: the moose', pp 105–34, copyright 1978; Figure 8.15 reprinted from *Animal Behaviour*, Vol. 19, Dawkins, M.S., 'Perceptual changes in chicks: another look at the "search image" concept', pp 566–74, copyright 1971; Figure 8.18(a) reprinted from *Animal Behaviour*, Vol. 33, Lima, S.L., Valone, T.J. and Caraco, T., 'Foraging efficiency–predation risk tradeoff in the grey squirrel', pp 155–65, copyright 1985; Figure 8.21 reprinted from *Animal Behaviour*, Vol. 36, Burghardt, G.M. and Greene, H.W., 'Predator simulation and duration of death feigning in neonate hognose snakes', pp 1842–4, copyright 1988; Figure 8.25 reprinted from *Animal Behaviour*, Vol. 31, Hoogland, J.L., 'Nepotism and alarm calling in the black-tailed prairie dog *Cymomys ludovicianus*', pp 1472–9, copyright 1983; Figure 9.2 reprinted from *Animal Behaviour*, Vol. 30, Jennings, T. and Evans, S.M., 'Influence of position in the flock and clock size on vigilance in the starling *Sturnus vulgaris*', pp 634–5, copyright 1980; Figure 9.3 reprinted from *Animal Behaviour*, Vol. 29, Treherne, J.E. and Foster, W.A., 'Group transmission of predator avoidance in a marine insect: the Trafalgar effect', pp 911–17, copyright 1981; Figure 9.6 reprinted from *Animal Behaviour*, Vol. 29, Elgar, M.A. and Catterall, C.P., 'Flocking and predator surveillance in house sparrows: test of an hypothesis', pp 868–72, copyright 1981; Figure 9.7(a, b) reprinted from *Animal Behaviour*, Vol. 32, Hart, A. and Lendrem, D.W., 'Vigilance and scanning patterns in birds', pp 1216–24, copyright 1984; Figure 9.9 reprinted from *Animal Behaviour*, Vol. 28, Barnard, C.J., 'Flock feeding and time budgets in the house sparrow (*Passer domesticus* L.)', pp 295–309, copyright 1980a; Figure 9.11 reprinted from *Animal Behaviour*, Vol. 28, Barnard, C.J., 'Equilibrium flock size and factors affecting arrival and departure in feeding sparrows', pp 503–11, copyright 1980c; Figure 9.17 reprinted from *Animal Behaviour*, Vol. 22, Barlow, G.W., 'Hexagonal territories', pp 876–8, copyright 1974; Figure 10.3 reprinted from *Animal Behaviour*, Vol. 34, Clutton-Brock, T.H., Albon, S.D. and Guiness, F.E., 'Great expectations: maternal dominance sex rations and offspring reproductive success in red deer', pp 460–71, copyright 1986; Figure 10.6 reprinted from *Animal Behaviour*, Vol. 27, Clutton-Brock, T.H., Albon, S.D., Gibson, R.M. and Guiness, F.E., 'The logical stag: adaptive aspects of fighting in red deer (*Cervus elephus* L.)', pp 211–25, copyright 1979; Figure 10.9(b) reprinted from *Animal Behaviour*, Vol. 37, Baker, R.R. and Bellis, M.A., 'Number of sperm in human ejaculate varies in accordance with sperm competition theory', pp 867–9, copyright 1989; Figure 10.19 reprinted from *Animal Behaviour*, Vol. 45, Medvin, M.B., Stoddard, P.K. and Beecher, M.D., 'Signals for parent–offspring recognition: a comparative analysis of the begging calls of cliff swallows and barn swallows', pp 841–50, copyright 1993; Figure 11.4 reprinted from *Journal of Theoretical Biology*, Vol. 69, Norman, R.F., Taylor, P.D. and Robertson, R.J., 'Stable equilibrium strategies and penalty functions in a game of attrition', pp 571–8, copyright 1977; Figure 11.7 reprinted from *Animal Behaviour*, Vol. 40, Enquist, M., Leimar, O., Ljungberg, T., Mallner, Y. and Segerdahl, M., 'A test of the sequential assessment game: fighting in the cichlid fish *Nannacara anomala*', pp 1–14, copyright 1990; Figure 11.8 reprinted from *Animal Behaviour*, Vol. 35, Moller, A.P., 'Variation in badge size in male house sparrows (*Passer domesticus*): evidence for status signalling', pp 1637–44, copyright 1987; Figure 11.9 reprinted from *Animal Behaviour*, Vol. 28, Dawkins, R. and Brockmann, H.J., 'Do digger wasps commit the Concorde fallacy?', pp 892–6, copyright 1980; Figure 11.10 reprinted from *Animal Behaviour*, Vol. 26, Davis, N.B., 'Territorial defence in the speckled wood butterfly (*Parage aegeria*): the resident always wins', pp 138–47, copyright 1978b; Figure 11.11 reprinted from *Animal Behaviour*, Vol. 31, Austad, S.N., 'A game theoretical interpretation of male combat in the bowl and doily spider, *Frontinella pyramitela*', pp 59–73, copyright 1983; Figure 11.20 reprinted from *Animal Behaviour*, Vol. 28, Seyfarth, R.M., Cheney, D.L. and Marler, P., 'Vervet monkeys alarm calls: semantic communications in free-ranging primate', pp 1070–94, copyright 1980; Figure 11.21 reprinted from *Animal Behaviour*, Vol. 40, Seyfarth, R.M. and Cheney, D.L., 'The assessment of vervet monkeys of their and another species' alarm calls', pp 754–64, copyright 1990; Figure 11.22(b) reprinted from *Language Learning by a Chimpanzee*, Rumbaugh, D.M. (ed.), New York, Academic Press, copyright 1977; Figure 12.16 reprinted from *Personality and Individual Differences*, Vol. 21, Lalumire, M.L. and Quinsey, V.L., 'Sexual deviance, antisociality, mating effort, and the use of sexually coercive behaviors', pp 33–48, copyright 1996; Box 4.2 figures (i) and (ii) reprinted from *Animal Behaviour*, Vol. 26, Larkin, S. and McFarland, D.J., 'The cost of changing from one activity to another', pp 1237–46, copyright 1978; Box 5.3 figure (i) reprinted from *Trends in Genetics*, Vol. 16, Lakin-Thomas, P.L., 'Circadian rhythms: new functions for old genes?', pp 106–14, copyright 2000; Box 7.4 figure (ii) reprinted from *Journal of Theoretical Biology*, Vol. 55, Lundberg, P., 'Partial evolutionarily stable strategies', pp 1216–32, copyright 1987; Box 9.2 figures (i) and (ii) reprinted from *Animal Behaviour*, Vol. 29, Barnard, C.J. and Sibly, R.M., 'Producers and scroungers: a general model and its application to feeding flocks of house sparrows', pp 543–50, copyright 1981; Box 9.4 figure (i) reprinted from *Animal Behaviour*, Vol. 28, Caraco, T., Martindale, S. and Whitham, T., 'An empirical demonstration of risk-sensitive foraging preferences', pp 820–30, copyright 1980a; Box 10.4 figure (i) reprinted from *Animal Behaviour*, Vol. 25, Maynard Smith, J., 'Parental investment – a prospective analysis', pp 1–9, copyright 1977; with permission of Elsevier. Table 9.2 from Woolfenden, G.E. and Fitzpatrick, J.W., *The Florida Scrub Jay*, copyright © 1984 by Princeton University Press; Figure 10.10(a, b) from Anderson, M., *Sexual Selection*, copyright © 1994 by Princeton University Press; Figure 12.1 from Cavalli-Sfoza, L.L. and Feldman, M.W., *Cultural Transmission and Evolution: A Quantitative Approach*, copyright ©1981 by Princeton University Press; Figure 12.5 and Box 12.1 table (i) from Barrett, L., Dunbar, R.I.M. and Lycett, J. *Human Evolutionary Psychology*, copyright © 2002 Palgrave, North American rights copyright © 2002 by Princeton University Press reprinted by permission of Princeton University Press. Table 10.1 and Figure 10.15(a, b) from 'Optimal outbreeding', in *Mate Choice*, P. Bateson (ed.), reproduced with permission of Professor Patrick Bateson (Bateson, P. 1983); Table 11.3 after 'The comparison of vocal communication in animals and man', pp 27–48 in *Non-Verbal Communication*, R.A. Hinde (ed.) (Thorpe, W.H. 1972); Figures 1.15 and 11.12 after *Essentials of Animal Behaviour*, reproduced with permission of Professor Peter Slater (Slater, P.J.B. 1999); Figure 2.19 after 'Sperm competition and its evolutionary consequences in the insects', *Biological Reviews*, Vol. 45, pp 525–67 (Parker, G.A. 1970); Figure 3.5(a) after *The Insects: Structure and Function*, London, English Universities Press (Chapman, R.F. 1971); Figure 4.4 adapted from *The Structural Basis of Behaviour* (Deutsch, J.A. 1960); Figure 6.1 after 'The behaviour and neuro-anatomy of some developing teleost fishes', *Journal of Zoology*, Vol. 149, pp 215–41 (Abu-Ghidieri, Y.B. 1966); Figure 6.5 after *Bird Song* (Thorpe, W.H. 1961); Figures 6.8 and 6.26 after *An Introduction to Animal Behaviour*, 5th Edn (Manning, A. and Dawkins, M.S. 1998); Figure 6.29(a, b, c) after 'Play in common ravens (*Corus corax*)', pp 27–48 in *Animal Play: Evolutionary, Comparative and Ecological Perspectives*, M. Bekoff and J.A. Byers (eds) (Heinrich, B. and Smolker, R. 1998); Figure 7.7 after *The Evolutionary Ecology of Animal Migration*, London, Hodder & Stoughton (Baker, R.R. 1978); Figure 7.8(a–c) after *Migration: Paths Through Time and Space*, London,

Hodder & Stoughton (Baker, R.R. 1982); Figure 7.21(a) after *Bird Navigation: The Solution of a Mystery?* London, Hodder & Stoughton (Baker, R.R. 1984); Figure 9.24(a) and Box 9.7 figure (i) from 'Of mice and kin: the functional significance of kin bias in social behaviour', *Biological Reviews*, Vol. 66, pp 379–430 (Barnard, C.J. *et al.* 1983) reproduced with permission of Cambridge University Press. Table 11.1 after T.A. Sebeok, *Behavioural Science*, Vol. 7, pp 430–42, copyright © 1962 by Sage Publications, Inc., reprinted by permission of Sage Publications, Inc. Table 12.1 adapted from Cartwright, J. *Evolution and Human Behaviour*, 2000, © Macmillan Press; Figure 12.5 and Box 12.1 table (i) from Barrett, L., Dunbar, R.I.M. and Lycett, J. *Human Evolutionary Psychology*, 2002, © Palgrave reproduced with permission of Palgrave Macmillan. Table 12.1 after *Evolution and Human Behaviour*, Basingstoke, Macmillan (Cartwright, J. 2000) reproduced in North America and United States territories by permission of the publisher, The MIT Press. Figure 1.3 based on diagrams from *Darwin's Dangerous Idea: Evolution and the Meanings of Life* by Daniel C. Dennett (Penguin Press, 1995), copyright © Daniel C. Dennett, 1995. Figure 1.3 after Dennett's (1995) *Darwin's Dangerous Idea* by Daniel C. Dennett (NY: Simon & Schuster, 1995, pp 374–8). Figure 1.5 after 'Steering responses of flying crickets to sound and ultrasound: male attraction and predator avoidance', *Proceedings of the National Academy of Sciences of the USA*, Vol. 75, pp 4052–4026 (Mosieff, A. 1978). Figure 1.6 from 'Aerial defense tactics of flying insects', *American Scientist*, Vol. 79, pp 316–29, Figure 14 (May, M. 1991); reproduced with permission of Virge Kask. Figure 1.8 after 'Song "dialects" in three populations of white-crowned sparrows', *Condor*, Vol. 64, pp 368–77 (Marler, P. and Tamura, M. 1962); © Cooper Ornithological Society, reproduced with permission of Cooper Ornithological Society. Figures 1.11 and 1.12 after 'Towards a model of the courtship of the smooth newt, *Triturus vulgalis*, with special emphasis on problems of observability in the simulation of behaviour', *Medical and Biological Engineering and Computing*, Vol. 15, pp 49–61 (Houston, A.I. *et al.* 1977). Figure 1.13 after 'Zur Deutung der phasianidenbalz', *Ornithologische Beobachter*, Vol. 53, p 182 (Schenkel, R. 1956), reproduced with permission of the Swiss Ornithological Institute. Figure 2.4(a) from *Darwin to DNA: Molecules to Humanity* by G. Ledyard Sebbins, 1982 by W.H. Freeman and Company, used with permission; Figures 3.1 and 3.3 from *Animal Physiology: Mechanisms and Adaptations* by David Randall, *et al.*, © 1978, 1983, 1997 by W.H. Freeman and Company; Figure 3.16 from *Cellular Basis Behavior: An Introduction to Behavioral Neurobiology* by Eric R. Kandel, © 1976 by W.H. Freeman and Company, used with permission; Figure 11.19 after *Biology* by Helen Curtis, *et al.* © 1968, 1975, 1979, 1983, 1989 by Worth Publishers, used with permission. Figures 2.7 and 2.8 after 'Clock mutants of *Drosophila melanogaster*', *Proceedings of the National Academy of Science of the USA*, Vol. 68, pp 2112–16 (Konopka, R.J. and Benzer, S. 1971); Figure 5.7(a) after 'Genetic dissection of Drosophila nervous systems by means of mosaics', *Proceedings of the National Academy of Science of the USA*, Vol. 67, pp 1156–63 (Hotta, Y. and Benzer, S. 1970); reproduced with permission of Professor S. Benzer. Figure 2.9 reprinted with permission, and by permission of Professor C.P. Kyriacou, after 'Molecular transfer of species-specific behavior from *Drosophila simulans* to *Drosophila melanogaster*', *Science*, Vol. 251, pp 1082–5 (Wheeler, D.A. *et al.* 1991); Figure 2.11(b) reprinted with permission, and by permission of Professor P. Berthold, after 'Genetic basis of migratory behavior in European warblers', *Science*, Vol. 212, pp 77–9 (Berthold, P. and Querner, U. 1981); Figure 2.18 reprinted with permission, and by permission of Professor M. Hori, after 'Frequency-dependent natural selection in the handedness of scale-eating cichlid fish', *Science*, Vol. 260, pp 216–19 (Hori, M. 1993); Box 2.3 figure (i) reprinted with permission, and by permission of Dr Marcia Barinaga, adapted from 'From fruit flies, rats, mice: evidence of genetic influence',

Science, Vol. 264, pp 1690–3 (Barinaga, M. 1994); Figure 3.34(c) reprinted with permission, and by permission of Professor Dr E. Gwinner, after 'Endogenous reproductive rhythms in a tropical bird', *Science*, Vol. 249, pp 906–8 (Gwinner, E. and Dittami, J. 1990); Figure 6.17(a–c) reprinted with permission, and by permission of Professor Juan D. Delius, from 'Rotational invariance in visual pattern recognition by pigeons and humans', *Science*, Vol. 218, pp 804–6 (Hollard, V.D. and Delius, J.D. 1982); Figure 9.18 reprinted with permission, and by permission of Professor F. Lynn Carpenter, after 'Threshold model of feeding territoriality and test with a Hawaiian honeycreeper', *Science*, Vol. 194, pp 639–42 (Carpenter, F.L. and MacMillen, R.E. 1976); Figure 9.24(c) reprinted with permission, and by permission of Professor L. Greenberg, after 'Genetic component of bee odor in kin recognition', *Science*, Vol. 206, pp 1095–7 (Greenberg, L. 1979); Figure 10.4 reprinted with permission, and by permission of Professor Robert Trivers after 'Haplodiploidy and the evolution of the social insects', *Science*, Vol. 191, pp 249–63 (Trivers, R.L. and Hare, H. 1976); Figure 11.23 reprinted with permission after 'Symbolic communication between two pigeons (*Columba livia domestica*)', *Science*, Vol. 207, pp 543–5 (Epstein, R. *et al.* 1980); copyright 1991, 1981, 1993, 1994, 1990, 1982, 1976, 1979, 1976, 1980, respectively, American Association for the Advancement of Science. Figure 2.12 after Belyaev, D.K., 'Destabilizing selection as a factor in domestication', *Journal of Heredity*, 1979, Vol. 70, pp 301–8; Figure 3.2 after *The Nerve Impulse*, Oxford Biology Readers (Adrian, R.H. 1974); Figure 3.12(a) after 'The neural basis of avian song learning and perception', pp 113–25, in J.J. Bolhuis (ed.) *Brain, Perception, Memory: Advances in Cognitive Neuroscience* (Clayton, D.F. 2000), Figure 3.24(a, b) from *Eye and Brain: The Psychology of Seeing*, 5th Edn (Gregory, R.L. 1998); Figures 4.2(b) and 4.17 after *The Study of Instinct* (Tinbergen, N. 1951); Figure 4.25 after *Machiavellian Intelligence: Social Expertise and the Evolution of Intelligence in Monkeys, Apes and Humans* (Byrne, R.W. and Whiten, A. 1988); Box 5.4 figure (i) after *Asymmetry, Development Biology and Evolution* (Moller, A.P. and Swaddle, J.P. 1997); Figure 8.19 after *Protean Behaviour: The Biology of Unpredictability* (Driver, P.M. and Humphries, D.A. 1988); copyright © 1979, 1974, 2000, 1998, 1951, 1988, 1997, 1988, respectively, Oxford University Press, reprinted by permission of Oxford University Press. Figure 2.14 from 'A latitudinal cline in a Drosophila clock gene', *Proceedings of the Royal Society of London, Series B*, Vol. 250, pp 43–9 (Costa, R. *et al.* 1992); Figure 3.9(c) after 'The nervous anatomy of the body segments of nereid polychaetes', *Proceedings of the Royal Society of London, Series B*, Vol. 240, pp 135–96, figure 4 (Smith, J.E. 1957); Figure 3.12(b) after 'Relations between repertoire size and the volume of the brain nuclei related to song: comparative evolutionary analyses among Oscine birds', *Proceedings of the Royal Society of London, Series B*, Vol. 254, pp 75–82, figures 1 and 3 (DeVoogd, T.J. *et al.* 1993); Figure 5.14 after 'Evolutionary change in a receiver bias: a comparison of female preference functions', *Proceedings of the Royal Society of London, Series B*, Vol. 265, pp 223–8 (Basolo, A.L. 1998); Figure 5.16 after 'Primate brain evolution: genetic and functional considerations', *Proceedings of the Royal Society of London, Series B*, Vol. 262, pp 689–96 (Keverne, E.B. *et al.* 1996); Figure 8.17 after 'Discrimination of flying mimetic, passion-vine butterflies, Heliconius', *Proceedings of the Royal Society of London, Series B*, Vol. 266, pp 2137–40 (Srygley, R.B. and Ellington, C.P. 1999); Figure 10.11 from 'Female choice response to artificial selection on an exaggerated male trait in a stalk-eyed fly', *Proceedings of the Royal Society of London, Series B*, Vol. 255, pp 1–6 (Wilkinson, G.S. and Reillo, P.R. 1994); Figure 10.13(b–d) from 'Endocrine–immune interactions, ornaments and mate choice in red jungle fowl', *Proceedings of the Royal Society of London, Series B*, Vol. 260, pp 205–10 (Zuk, M. *et al.* 1995); Figure 10.17(a, b) after 'Differences across taxa in nuptial gift size correlate with differences in sperm number and ejaculate volume in bushcrickets (Orthoptera: Teggoniidae)', *Proceedings of the Royal Society of London, Series B*, Vol. 263, pp 1257–65 (Vahed, K. and Gibert, F.S. 1996); Figure 10.20 after 'Nestling cuckoos, *Cuculuc canorus*, exploit hosts with begging calls that mimic a brood', *Proceedings of the Royal Society of London, Series B*, Vol. 265, pp 673–8 (Davies, N.B. *et al.* 1998); Figure 10.23 after 'Begging intensity of nestling sibling relatedness', *Proceedings of the Royal Society of London, Series B*, Vol. 258, pp 73–8 (Briskie, J.V. *et al.* 1994); Figure 11.17 after 'Know thine enemy: fighting fish gather information from observing conspecific interactions', *Proceedings of the Royal Society of London, Series B*, Vol. 265, pp 1045–9 (Oliveira, R.F. *et al.* 1998); Figure 12.10(b) after 'Impact of market value on human mate choice decisions', *Proceedings of the Royal Society of London, Series B*, Vol. 266, pp 281–5 (Pawlowski, B. and Dunbar, R.I.M. 1999); Figure 12.15 after 'The relationship between serial monogamy and rape in the United States, 1960–1995', *Proceedings of the Royal Society of London, Series B*, Vol. 267, pp 1259–63; reproduced with permission of the authors, Professor C.P. Kyriacou, Professor Tim DeVoogd, Dr Alexandra Basolo, Professor E.B. Keverne, Professor Bob Srygley, Professor Jerry Wilkinson, Professor Marlene Zuk, Professor Karim Vahed, Professor Nick Davies, Professor Jim V. Briskie, Professor Rui Oliveira, Professor Robin I.M. Dunbar, Professor Philip T.B. Starks, respectively, and the publisher, The Royal Society. Figure 2.15 from 'Parasites and altered behavior', *Scientific American*, Vol. 250, pp 108–5 (Moore, J. 1984); reproduced with permission of Nelson H. Prentiss. Figures 2.20 (after) and 9.10 from *Gulls and Plovers: The Ecology and Behaviour of Mixed-Species Feeding Groups*, London, Croom Helm (Barnard, C.J. and Thompson, D.B.A. 1985); Figures 3.25(b), 8.10 and 8.18(b) (after) from *Animal Behaviour: Ecology and Evolution*, London, Croom Helm (Barnard, C.J. 1983); Figure 4.21, Box 6.2 figure (ia) and Box 6.3 figure (iia) after *Animal Learning and Cognition: An Introduction*, 2nd Edn, London, Psychology Press (Pearce, J.M. 1997); Figure 4.24 after 'Grades of mind-reading', in *Children's Early Understanding of Mind: Origins and Development*, C. Lewis and P. Mitchell (eds), Psychology Press (Whiten, A. 1994); Figure 6.19 from *The Mentality of Apes*, New York, Harcourt Brace (Kohler, W. 1927); Box 8.2 figure (i) after *Modelling in Behavioural Ecology*, London, Croom Helm (Lendrem, D.W. 1986); reproduced with permission of Taylor & Francis. Figure 2.21(b) after 'Prey size selection and competition in the common shrew (*Sorex araneus* L.)', *Behavioural Ecology and Sociobiology*, Vol. 8, pp 239–43 (Barnard, C.J. and Brown, C.A.J. 1981); Figure 3.9(a) after 'Funktionelle Anatomie de dorsalen Riessenfaser-System von *Lumbricus terrestris* L. (annelida, Oligacaeta)', *Zeitschrift für Morphologi der Tiere*, Vol. 70, pp 253–80 (Gunther, J. and Walther, J.B. 1971); Figure 3.18(a) after 'The behavioural hierarchy of the mollusk *Pleurobranchaea*. I. The dominant position of feeding behaviour. II. Hormonal suppression if feeding associated with egg-laying', *Journal of Comparative Psychology*, Vol. 90, pp 207–43 (Davis, W.J. *et al.* 1974); Figure 3.23 from *Neurobiology*, New York Springer-Verlag (Ewert, J.-P. 1980); Figure 4.8(b) after 'Decision making by rats', *Journal of Comparative Psychology*, Vol. 59, pp 1–12 (Logan, F.A. 1965); Figure 4.9(c) after 'The stimulation of territorial singing in house crickets (*Acheta domesticus*)', *Zeitschrift für vergleichende Physiologie*, Vol. 53, pp 437–46; Figure 5.11 after 'Classical conditioning and retention in normal and mutant *Drosophila melanogaster*', *Journal of Comparative Psychology*, Vol. 157, p 127 (Tully, T. and Quinn, W.G. 1985); Figure 7.11 from 'Inheritance of migratory direction in a bird species: a cross-breeding experiment with SE- and SW-migratory blackcaps (*Sylvia atricapilla*)', *Behavioural Ecology and Sociobiology*, Vol. 28, pp 9–12 (Helbig, A.J. 1991); Figure 7.14 after 'The significance of landmarks for path integration in homing honeybee foragers', *Naturwissenschaften*, Vol. 82, pp 635–45 (Chittka, L. *et al.* 1995); Figure 7.15(a–c) after 'Pigeon navigation: effects of wind deflection at home cage on homing behaviour', *Journal of Comparative Psychology*, Vol. 99, pp 177–86 (Baldaccini, N.E. *et al.* 1975); Figure 7.15(d) after 'Pigeon navigation: effects upon homing behaviour by reversing wind deflection at the loft', *Journal of Comparative Psychology*, Vol. 128, pp 285–95 (Ioale, P. *et al.* 1978); Figure 8.5(a, b) after 'Timing mechanisms in optimal foraging: some applications of scalar expectancy theory', pp 61–82, in R.N. Hughes (ed.) *Behavioural Mechanisms of Food Selection*, NATO ASI Series G, Ecological Sciences, Vol. 20, Heidelberg, Springer-Verlag (Kacelnik, A. *et al.* 1990); Figure 8.11 after 'Risk-sensitive foraging in common shrews (*Sorex araneus* L.)', *Behavioural Ecology and Sociobiology*, Vol. 16, pp 162–4 (Barnard, C.J. and Brown, C.A.J. 1985a); Figure 9.28 after 'The adaptive significance of communal nesting in groove-billed anis, *Crotophaga sulcirostris*', *Behavioural Ecology and Sociobiology*, Vol. 38, pp 349–53 (Vehrencamp, S.L. 1978); Figure 10.9(a) after 'Male crickets increase sperm number in relation to competition and female size', *Behavioural Ecology and Sociobiology*, Vol. 38, pp 349–53 (Gage, A.R. and Barnard, C.J. 1996); copyright © 1981, 1971, 1974, 1980, 1965, 1966, 1985, 1991, 1995, 1975, 1978, 1990, 1985, 1978, 1996, respectively, Springer-Verlag, reproduced with permission of Springer-Verlag. Figures 2.23 and 2.24 after 'Sneakers, satellites and parentals: polymorphic mating strategies in North American sunfishes', *Zeitschrift für Tierpsychologie*, Vol. 60, pp 1–26 (Gross, M.R. 1982); Figure 6.4 after 'The development of behaviour in the cichlid fish *Etroplus maculates*', *Zeitschrift für Tierpsychologie*, Vol. 33, pp 461–91 (Wyman, R.L. and Ward, J.A. 1973); Figure 7.1(b, c) after 'Über den Einfluß statischer Magnetfelder auf die Zugorientierung der Rotkenlchen (*Erithacus rubecula*)', *Zeitschrift für Tierpsychologie*, Vol. 25, pp 537–58 (Wiltschko, W. 1968); Figure 9.4 after 'Experiments on the selection by predators against spatial oddity of their prey', *Zeitschrift für Tierpsychologie*, Vol. 43, pp 311–26 (Milimski, M. 1977); reproduced with permission of Blackwell Verlag. Figure 3.4(a) from *Zoology for Intermediate Students*, London, Longman (Chapman G. and Baker, W.B. 1966); Figure 4.18 from *Animal Behaviour*, 3rd Edn, Harlow, Longman (McFarland, D. 1999); Figure 4.28 after *Problems of Animal Behaviour*, Harlow, Longman (McFarland, D. 1989); Figure 8.22 after *Defence in Animals*, Harlow, Longman (Edmunds, M. 1974); reproduced with permission of Pearson Education Ltd. Figures 3.4(b) and 3.5(b) after *Invertebrate Zoology*, 2nd edn, London, Saunders, with kind permission of Kluwer Academic Publishers. Figure 3.7(c) from Jerison, H. 'The evolution of neural and behavioural complexity' in Roth and Wulliman (eds) *Brain Evolution and Cognition* (2001) © Spektrum Akademischer Verlag, Heidelberg, Berlin, and with permission of Professor Harry Jerison. Figure 3.8 after *Biology*, 5th Edn by Campbell, N.A., Reece, J.B. and Mitchell, L.G., copyright © 1999 Addison Wesley, reprinted by permission of Pearson Education Inc. Figure 3.9(b) after 'Annelid giant fibres', *Quarterly Review of Biology*, Vol. 23, pp 291–324 (Nicol, A.C. 1948); Figure 7.4(d) after 'The theory of habitat selection examined and extended using *Pemphigus* aphids', *American Naturalist*, Vol. 115, pp 449–66 (Whitham, T.G. 1980); Figure 9.12 after 'Territory quality and dispersal options in the acorn woodpecker, and a challenge to the habitat saturation model of cooperative breeding', *American Naturalist*, Vol. 130, pp 654–76 (Stacey, P. and Ligon, J.D. 1987); Figure 10.7 after 'On the evolution of mating systems in birds and mammals', *American Naturalist*, Vol. 103, pp 589–603 (Orians, G.H. 1969); Figure 10.13(a) from 'Parasites, bright males and the immunocompetence handicap', *American Naturalist*, Vol. 139, pp 603–22 (Folstad, I. And Karter, A.J. 1992); Figure 10.16 after 'Sexual selection and nuptial feeding behavior in *Bittacus apicalis* (Insecta: Mecoptera)', *American Naturalist*, Vol. 110, pp 529–48 (Thornhill, R. 1976), Box 7.3 figure (i) after 'A theory of partial migration', *American Naturalist*, Vol.142, pp 59–81 (Kaitala, A., Kaitala, V. and Lundberg, P. 1993); Box 11.5 figure (i) after 'The dance-language controversy', *Quarterly Review of Biology*, Vol. 51, pp 211–44 (Gould, J.L. 1976); © 1948, 1980, 1987, 1969, 1992, 1976, 1993, 1976, respectively, by The University of Chicago, all rights reserved. Figure 3.11 after 'Evolution of somatic sensory specialization in otter brains', *Journal of Comparative Neurology*, Vol. 134, pp 495–506, copyright © 1968 Wiley–Liss, Inc., reprinted by permission of Wiley–Liss, Inc., a subsidiary of John Wiley & Sons, Inc. (Radinsky, L.B. 1968). Figure 3.14 Scrub jay, photograph courtesy of Ian Connell and Nicky Clayton; mountain chickadee, photograph courtesy of Daniel Griffiths and Nicky Clayton; marsh tit, photograph courtesy of Andy Bennett and Nicky Clayton. Figure 3.17 from *Journal of Neuroscience* by Katz, P.S. and Frost, W.N., copyright 1995 by the Society for Neuroscience; Figure 6.2 from *Journal of Neuroscience* by Levine, R.B. and Truman, J.W., copyright 1985 by the Society for Neuroscience; reproduced with permission, respectively, of Professor Paul Katz and Professor Richard Levine and with permission of the Society for Neuroscience in the format Textbook via Copyright Clearance Center. Figure 3.18(b) after 'Escape swim network interneurons have diverse roles in behavioural switching and putative arousal in Pleurobranches', *Journal of Neurophysiology*, Vol. 83, pp 1346–55 (Jing, J. and Gillette, R. 2000); reproduced with permission of The American Physiological Society. Figure 3.19 from *Journal of Neuroscience* by Katz, P.S. and Frost, W.N., copyright 1995 by the Society for Neuroscience, reproduced with permission of Professor Paul S. Katz; Figure 9.13 from *Journal of Neuroscience* by Levine, R.B. and Truman, J.W., reproduced with permission of Professor Richard Levine, copyright 1985 by the Society for Neuroscience, reproduced with permission of the Society of Neuroscience in the format textbook via Copyright Clearance Center. Figure 3.20(a, c) after 'March of the robots', *New Scientist*, Vol. 160, No. 2163, pp 26–32, reproduced with permission of the New Scientist (Graham-Rowe, D. 1998). Figure 3.20(b) reproduced by permission of Barbara Webb and Andrew Horchler. Figures 3.21 and 3.22 from 'The neural basis of visually guided behavior', *Scientific American*, Vol. 230, pp 34–42 (Ewert, J.P. 1974); Figures 5.6 and 5.7(b) after 'Genetic dissection of behavior', *Scientific American*, Vol. 229, pp 24–37 (Benzer, S. 1973); reproduced with permission of Donald Garber on behalf of the Estate of Bonji Tagawa. Figures 3.25, 3.27 and 8.24 from *An Introduction to Animal Behaviour*, 5th Edn, Cambridge, Cambridge University Press (Manning, A. and Dawkins, M.S. 1998); reproduced with permission of Dr Nigel Mann. Figure 3.26(a, b) reproduced with permission of Professor R.J. Beynon. Figure 3.31 after 'Differential reactivity of individuals and the response of the male guinea pig to testosterone proportionate', *Endocrinology*, Vol. 51, pp 237–48 (Grunt, J.A. and Young, W.C. 1952); reproduced with permission of The Endocrine Society. Figure 3.34(a) after *Animal Behaviour: Mechanisms, Ecology, Evolution*, 4th Edn (Drickamer, L.C., Vessey, S.H. and Meikle, D. 1966) copyright 1966, Chicago, W.C. Brown, reproduced with permission of The McGraw-Hill Companies. Figure 3.37(a, b, d) adapted from 'Social status and resistance to disease in house mice (*Mus musculus*): status-related modulation of hormonal responses in relation to immunity costs in different social and physical environments', *Ethology*, Vol. 51, pp 1061–70 (Barnard, C.J. *et al.* 1996a); Figure 4.3 after 'Wheel-running activity: a new interpretation', *Mammal Review*, Vol. 11, pp 45–51 (Mather, J.G. 1981); Figure 5.17 after 'Plumage condition affects flight performance in starlings: implications for developmental homeostasis, abrasion and moult', *Journal of Avian Biology*, Vol. 7, pp 103–11 (Swaddle, J.P. *et al.* 1996); Figure 8.4(a, b) after 'Predator ingestion rate and its bearing on feeding time and the theory of optimal diets', *Journal of Animal Ecology*, Vol. 47, pp 529–47 (Cook, R.M. and Cockrell, B.J. 1978); Figure 8.6(a) after 'Energy maximization in the diet of the shore crab, *Carcinus maenas*', *Journal of Animal Ecology*, Vol. 47, pp 103–16 (Elner, R.W. and Hughes, R.N. 1978); Figure 9.5 after 'Hawks and doves: factors affecting success and selection in goshawk attacks on woodpigeons', *Journal of Animal Ecology*, Vol. 47, pp 449–60 (Kenward, R.E. 1978); Figure 9.18 from 'Ecological questions about territorial behaviour', pp 317–20 in J.R. Krebs and N.B. Davies (eds) *Behavioural Ecology: An Evolutionary Approach*, 1st Edn (Carpenter, F.L. and MacMillen, R.E. 1976); Figure 11.14 after 'Geographical variation in the song of the great tit (*Parus major*) in relation to ecological factors', *Journal of Animal Ecology*, Vol. 48, pp 759–85 (Hunter, M.L. and Krebs, J.R. 1979); Figure 12.9 after 'Natural, kin and group selection', pp 62–84 in J.R. Krebs and N.B. Davies (eds) *Behavioural Ecology: An Evolutionary Approach*, 2nd Edn (Grafen, A. 1984); Figure 12.13 after 'Sex differences in jealousy in evolutionary and cultural perspective: tests from The Netherlands, Germany and the United States', *Psychological Science*, Vol. 7, pp 359–'63 (Buunk, B.P. *et al.* 1996); Box 7.3 figure (i) after Models of avian migration: state, time and predation', *Journal of Avian Biology*, Vol. 29, pp 395–404 (Houston, A.I. 1998); reproduced with permission of Blackwell Publishers Ltd. Figure 4.1(b, c) after 'Effects of drugs on motivation: the value of using a variety of measures', *Annals of the New York Academy of Sciences*, Vol. 65, pp 318–33, copyright © 1956 New York Academy of Sciences, USA (Miller, N.E. 1956). Figure 4.2(a) after 'The comparative method in studying innate behaviour patters', *Symposia for the Society of Experimental Biology*, Vol. 4, pp 221–68 (Lorenz, K. 1950). Figure 4.9(b) adapted from 'Ethological studies of *Lebistes reticulates* (Peters): I. An analysis of male courtship pattern', *Behaviour*, Vol. 8, pp 249–34 (Baerends, G.P. *et al.* 1955); Figure 4.15 after 'A model of the functional organization of incubation behaviour', pp 265–310 in G.P. Baerends and R.G. Drent (eds), *The Herring Gull and its Egg, Behaviour Supplement XVIII*, (Baerends, G.P. 1970); Figure 8.23(a) after 'The spines of sticklebacks (Gasterosteus and Pygosteus) as a means of defence against predators (Perca and Esox)', *Behaviour*, Vol. 10, pp 205–36 (Hoogland, R. *et al.* 1957); Figure 8.26 after 'Specific distinctiveness in the communication signals of birds', *Behaviour*, Vol. 11, pp 13–39 (Marler, P. 1957); Figure 9.8 from 'Factors affecting flock size mean and variance in a winter population of house sparrows (*Passer domesticus* L.)', *Behaviour*, Vol. 74, pp 114–27 (Barnard, C.J. 1980b); Figure 11.5 after 'Communication by agonistic displays: what can games theory contribute to ethology?', *Behaviour*, Vol. 68, pp 13–69 (Caryl, P.G. 1979); Figure 12.6(a, b) after 'Conditional mate choice strategies in humans: evidence from lonely hears advertisement', *Behaviour*, Vol. 132, pp 755–79 (Waynforth, D. and Dunbar, R.I.M. 1995) reproduced with permission of E.J. Brille. Figures 4.19 after 'Postconditioning devaluation of a reinforcer affects instrumental responding', *Journal of Experimental Psychology: Animal Behavior Processes*, Vol. 11, pp 120–32, reproduced with permission of Professor Ruth Colwill (Colwill, R.M. and Rescorla, R.A. 1985); Figure 5.1 from 'Genetic influences on the behaviour of mice can be obscured by laboratory rearing', *Journal of Comparative and Physiological Psychology*, Vol. 72, pp 505–11 (Henderson, N.D. 1970); Figure 6.11 after 'Positive and negative relations between a signal and food: approach–withdrawal behavior', *Journal of Experimental Psychology: Animal Behavior Processes*, Vol. 3, pp 37–52 (Hearst, E. and Franklin, S.R. 1977); Figure 6.20 after 'The formation of learning sets', *Psychological Review*, Vol. 56, pp 51–65 (Harlow, H.F. 1949); Box 6.2 figure (i) after 'Effects of conditioned stimulus intensity on the conditioned emotional response', *Journal of Comparative and Physiological Psychology*, Vol. 56, pp 502–7 (Kamin, L.J. and Schaub, R.E. 1963); Box 6.3 figure (ii) from 'The conditioned emotional response as a function of intensity of the US', *Journal of Comparative and Physiological Psychology*, Vol. 54, pp 428–32 (Annau, Z. and Kamin, L.J. 1961); Figures 12.9 and 12.10(a) after 'Sexual strategies theory: an evolutionary perspective on human mating', *Psychological Review*, Vol. 100, pp 204–32, reproduced with permission of Professor David, M. Buss (Buss, D.M. and Schmidt, D.P. 1993); Figure 12.6(c) after 'Integrating evolutionary and social exchange perspectives on relationships: effects of gender, self-approach and involvement level on mate choice', *Journal of Personality and Social Psychology*, Vol. 25, pp 159–67 (Kenrick, D.T. *et al.* 1990); copyright © 1985, 1970, 1977, 1949, 1963, 1961, 1993, 1990, respectively, by the American Psychological Association, reprinted/adapted with permission. Figure 4.20 reproduced with permission of the Behavioural Ecology Research Group, Oxford University. Figure 4.22 after 'Shortcut ability in hamsters (*Mesocricetus auratus*): the role of environmental and kinaesthetic information', *Animal Learning and Behavior*, Vol. 21, pp 255–65 (Chapuis, N. and Scardigli, P. 1993); Figure 6.10 after 'Classical conditioning of the rabbit nictitating membrane response: effects of reinforcement schedule on response maintenance and resistance to extinction', *Animal Learning and Behavior*, Vol. 6, pp 209–15 (Gibbs, C.M. *et al.* 1978); Figure 6.28(a) after 'Relation of cue to consequence in avoidance learning', *Psychonomic Science*, Vol. 4, pp 123–4 (Garcia, J. and Koelling, R.A. 1996); Figure 6.28(b) after 'Trace conditioning with X-rays as the aversive stimulus', *Psychonomic Science*, Vol. 9, pp 11–12 (Smith, J.C. and Roll, D.L. 1967); reproduced with permission of the Psychonomic Society. Figure 5.2 from *Ants: Their Structure, Development and Behavior*, by W.M. Wheeler, © 1910 Columbia University Press, reprinted with permission of the publisher. Figure 5.5 reproduced with permission of Dr Mario de Bono. Figure 5.12 from 'Variations in the morphology of the septo-hippocampal complex and maze learning in rodents: correlations between morphology and behaviour', pp 259–76, Fig. 4, in E. Alleva, A. Fasolo, H.-P. Lipp, L. Nadel and L. Ricceri (eds) *Behavioural Brain Research in Naturalistic and Semi-naturalistic Settings, Vol. 82, Proceedings of the NATO Advanced Science Institute*, Aquafredda di Maratea, Italy, 10–20 September 1994, Dordrecht, Kluwer (Schwegler, H. and Lipp, H.-P. 1995); Figures 6.12 and 6.13 after 'The role of learning in the aggressive and reproductive behavior of blue gouramis, *Trichogastr trichopterus*', *Environmental Biology of Fishes*, Vol. 54, pp 355–69, Figs 2, 3, 9 and 10 (Hollis, K.L. 1999); Figure 10.1 from *Producers and Scroungers: Strategies of Exploitation and Parasitism*, Kluwer, p. 142, reproduced with permission of Professor Chris J. Barnard (Barnard, C.J. 1988); copyright © 1995, 1999 and 1984, respectively, Kluwer Academic Publishers; Figure 12.18 after 'Household compositions and female strategies in a Trinidadian village', pp 206–33, in A.E. Rasa, C. Vogel and E. Voland (eds) *The Sociobiology of Sexual and Reproductive Strategies*, New York, Chapman & Hall, copyright © 1989 Chapman & Hall, reproduced with permission of Professor Mark Flinn (Flinn, M. 1989) with kind permission of Kluwer Academic Publishers. Figure 5.15 after 'Urinary odour preferences in mice', *Nature*, Vol. 409, pp 783–4 reproduced with permission of Professor Michael Baum (Isles, A.R. *et al.* 2001); Figure 6.3(a, b) adapted from 'Development of the brain depends on the visual environment', *Nature*, Vol. 228, pp 477–7, reproduced with permission of Professor Colin Blakemore (Blakemore, C. and Cooper, G.F. 1970); Figure 7.12(a–c) after 'Rapid microevolution of migratory behaviour in a wild bird species', *Nature*, Vol. 360, pp 668–70 (Berthold, P. *et al.* 1992); Figure 7.13(a) after 'A demonstration of navigation by rodents using an orientation cage', *Nature*, Vol. 284, pp 259–62 (Mather, J.G. and Baker, R.R. 1980); Figure 8.2(a–c) after 'Test of optimal sampling by foraging great tits', *Nature*, Vol. 275, pp 27–31, reproduced with permission of Professor Alex Kacelnik (Krebs, J.R. *et al.* 1978); Figure 8.3(c) after 'Optimal foraging in great tits (*Parus major*)', *Nature*, Vol. 268, pp 137–9, reproduced with permission of Professor Richard Cowie (Cowie, R.J. 1977); Figure 8.13 after 'Influence of a predator on the optimal foraging behaviour of sticklebacks (*Gasterosteus aculeatus*)', *Nature*, Vol. 275, pp 642–4, reproduced with permission of Professor Manfred Milinski (Milinski, M. and Heller, R. 1978); Figure 9.1 after 'Evidence for the dilution effect in the selfish herd from fish predation of a marine insect', *Nature*, Vol. 293, pp 466–7, reproduced with permission of Dr William Foster (Foster, W.A. and Treherne, J.E. 1981); Figure 9.23 after 'Jackal helpers and pup survival', *Nature*, Vol. 277, pp 382–3, reproduced with permission of Dr Patricia Moehlman (Moehlman, P.D. 1979); Figure 9.26(b) after 'Tit for tat and the evolution of cooperation in sticklebacks', *Nature*, Vol. 325, pp 433–5, reproduced with permission of Professor Manfred Milinski (Milinski, M. 1987); Figure 10.2 after 'Intrasexual selection in Drosophila', *Heredity*, Vol. 2, pp 349–68 (Bateman, A.J. 1948); Figure 10.12(a) adapted from 'Female choice selects for male sexual tail ornaments in a swallow', *Nature*, Vol. 332, pp 640–2, reproduced with permission of Professor Anders Pape Moller (Moller, A.P. 1988b); Figure 10.14(a) after 'Female swallow preference for symmetrical male sexual ornaments', *Nature*, Vol. 357, pp 238–40, reproduced with permission of Professor Anders Pape Moller (Moller, A.P. 1992); Figure 10.14(b) after 'Preference for symmetric males by female zebras', *Nature*, Vol. 367, pp 165–6, reproduced with permission of Professor Innes C. Cuthill (Swaddle, J.P. and Cuthill, I.C. 1994); Figure 11.6 after 'Deep croaks and fighting assessment in toads, *Bufo bufo*', *Nature*, Vol. 274, pp 683–5, reproduced with permission of Professor Nick Davies (Davies, N.B. and Halliday, T.R. 1978); Box 10.10 figure (i) after 'Relationship between egg size and post-hatching chick mortality in the herring gull (*Larus argentatus*)', *Nature*, Vol. 228, pp 1221–2 (Parsons, J. 1970); copyright © 2001, 1970, 1992, 1980, 1978, 1977, 1978, 1981, 1979, 1987, 1948, 1988, 1992, 1994, 1978, 1970, respectively, Nature Publishing Group. Box 5.2 figure (i) from 'How an instinct is learned', *Scientific American*, Vol. 221, pp 98–108 (Hailman, J. 1967); reproduced with permission of Eric Mose Jr. Figure 6.16 after 'Relative and absolute strength of response as a function of frequency reinforcement', *Journal of the Experimental Analysis of Behavior*, Vol. 4, p 267 (Hernstein, R.J. 1961); Figure 8.2(d) after 'Maximizing and matching on concurrent ration schedules', *Journal of the Experimental Analysis of Behavior*, Vol. 24, pp 107–16 (Hernstein, R.J. and Loveland, D.H. 1975); copyright 1961, 1975, respectively, by the Society for the Experimental Analysis of Behavior, Inc. Figure 6.25 from 'A demonstration of observational learning in rats using a bi-directional control', *Quarterly Journal of Experimental Psychology*, Vol. 45B, pp 229–40 (Heyes, C.M. and Dawson, G.R. 1990) reprinted by permission of Professor Cecilia Heyes and by permission of The Experimental Psychology Society. Figure 6.26 from 'A laboratory approach to the study of imprinting', *Wilson Bulletin*, Vol. 66, pp 196–206 (Ramsay, A.O. and Hess, E.H. 1954). Figure 7.4(a–c) after 'Habitat selection by *Pemphigus* aphids in response to resource limitation and competition', *Ecology*, Vol. 59, pp 1164–76 (Whitham, T.G. 1978); Figure 8.1 after 'Food searching behaviour of titmice in patchy environments', *Ecology*, Vol. 55, pp 1216–32 (Smith, J.N.M. and Sweatman, H.P. 1974); Figure 9.19(b) after 'Influence of economics, interspecific competition and sexual dimorphism on territoriality of migrant rufous hummingbirds', *Ecology*, Vol. 59, pp 285–96 (Kodric-Brown, A. and Brown, J.H. 1978); Box 9.4 Figure (ii) after 'Time budgeting and group size: a test of theory', *Ecology*, Vol. 60, pp 618–27 (Caraco, T. 1979); reproduced with permission of the Ecological Society of America. Figure 7.5(a) vignette (i) reproduced with permission of G.A. Parker. Figure 7.10 after 'Two types of orientation in migrating starlings, *Stuenus vulgaris* L, and chaffinches, *Fringilla coelebs*, as revealed by displacement experiments', *Adrea*, Vol. 46, pp 1–37 (Perdeck, A.C. 1958). Figures 7.13(b) and 7.16(a) reproduced by permission of Roswitha Wiltschko. Figure 7.21(b) after *Perspectives on Animal Behavior*, copyright © 1993 John Wiley & Sons, this material is used by permission of John Wiley & Sons, Inc. (Goodenough, J., McGuire, B. and Wallace, R. 1993). Box 7.3 figure (ii) after Srygley, R.B., 'Sexual differences in tailwind drift compensation in *Phoebis sennae* butterflies (Lepidoptra: Pieridae) migrating over seas', *Behavioural Ecology*, 2001, Vol. 12, pp 607–11, reproduced with permission of Professor Robert

Srygley and by permission of Oxford University Press. Figure 8.3(a), Figure 8.3(b) vignette and Figure 8.3(c) vignette reproduced with permission of Dr Richard Cowie. Figure 8.14(a) after *Journal of Entomology* by DeBach P. and Smith H.S., copyright 1941 by Entomological Society of America, reproduced with permission of the Entomological Society of America in the format textbook via Copyright Clearance Center. Figure 8.20 after *Mimicry in Plants and Animals*, published by Weidenfeld & Nicholson, all attempts to tracing the copyright holder of *Mimicry in Plants and Animals* were unsuccessful (Wickler, W. 1968). Figure 8.21 reproduced with permission of Peter M.C. Davies. Box 8.5 figure (i) from *The Dynamics of Competition and Predation*, Figure 5.1, London, Edward Arnold, © 1976 Edward Arnold (Hassell, M.P. 1976) reproduced by permission of Hodder Arnold. Figure 9.14 reprinted by permission of the publisher from *Communication Among Social Bees*, by Martin Lindauer, p 20, Cambridge, Mass, Harvard University Press, copyright © 1961 by the President and Fellows of Harvard College. Figure 9.15(a) after *The Ants*, Berlin, Springer Verlag (Holldobler, B. and Wilson, E.O. 1990) reproduced with permission of Professor E.O. Wilson. Figure 9.15(b) after 'Behavioural studies of army ants', *University of Kansas Scientific Bulletin*, Vol. 44, pp 281–465 (Rettenmeyer, C.W. 1963). Figure 9.19(a) after 'Weight gain and adjustment of feeding territory size in migrant hummingbirds', *Proceedings of the National Academy of Sciences of the USA*, Vol. 80, pp 7259–63 (Carpenter, F.L. *et al.* 1983) reproduced with permission of Professor Mark Hixon. Figure 9.25(a) reproduced with permission of Professor Marion Petrie. Figure 9.25(b) reproduced with permission of James Osmond (www.jamesosmond.co.uk). Figures 9.27(a, b) reproduced by permission of Gerald Wilkinson. Figure 10.11(a, b) reproduced with permission of Phil Savoie and Gerald Wilkinson. Figure 10.17(a) vignette reproduced with permission of Leigh William Simmons. Figure 10.21(b) reprinted, with permission, from the *American Journal of Orthopsychiatry*, copyright 1961 by the American Orthopsychiatric Association, Inc. and with permission of Professor Jay S. Rosenblatt (Schneirla, T.C. and Rosenblatt, J.S. 1961). Box 10.10 figure (i)b from 'Selection for adult size in Coho salmon', *Canadian Journal of Fisheries and Aquatic Science*, Vol. 43, pp 949–1057 (Holtby, L.B. and Healey, M.C. 1986) reproduced with permission of NRC Research Press. Figure 11.2 courtesy of Marc Bekoff. Figures 11.3 and 11.13 after 'Outdoor sound propagation over ground of finite impedance', *Journal of the Acoustical Society of America*, Vol. 59, pp 267–77 (Embleton, T.F.W. *et al.* 1976) reproduced with permission of the Acoustical Society of America. Figure 11.9 reproduced with permission of H. Jane Brockmann. Figure 12.2(a, b) and Figure 12.3 vignette reproduced by permission of PA Photos. Figures 12.7 and 12.9 from Buss, David M. *Evolutionary Psychology: The New Science of the Mind* © 1999, published by Allyn and Bacon, Boston, MA, copyright © 1999 by Pearson Education, reprinted by permission of the publisher and by permission of Professor David M. Buss. Figure 12.11 reproduced with permission of J.T. Manning. Figure 12.12 after 'Sexual selection for cultural displays', in R.I.M. Dunbar, C. Knight and C. Power (eds), *The Evolution of Culture*, pp 71–91, Edinburgh University Press (Miller, G.F. 1999). Figure 12.17 reprinted with permission from *Homicide*, by Martin Daly and Margo Wilson, p 90, copyright © 1988, by Walter de Gruyter, Inc., New York; and Figure 12.19 reprinted with permission from *Human Nature Vol. 6, Book 3*, edited by Jane Lancaster, p 281, copyright © 1995, by Walter de Gruyter, Inc., New York.

In some instances we have been unable to trace the owners of the copyright material, and we would appreciate any information that would enable us to do so.

1

Questions about behaviour

Introduction

Among the various branches of science, the study of animal behaviour arguably embraces a unique combination of theoretical sophistication and engaging accessibility. It is at one and the same time a rigorous scientific discipline with much to say about ourselves and biology as a whole, and a source of widespread popular fascination with the natural world. But what exactly *is* behaviour? How can we define it within the constant stream of actions performed by the animal? How should we set about asking sensible questions about why animals do particular things, and what does behaviour imply about the mental attributes of the animals performing it? Like most scientific disciplines, animal behaviour has evolved through a number of traditions, each taking different approaches and developing its own theoretical frameworks. Why is this, and to what extent have these different approaches become integrated?

'... we animals are the most complicated and perfectly-designed pieces of machinery in the known universe ... it is hard to see why anyone studies anything else.'

Richard Dawkins (1989)

If one needed an exhortation to study animal behaviour, Richard Dawkins's confident claim certainly provides one. The time and money that television producers and the publishers of books and magazines devote to the subject, however, suggests it needs little in the way of a recruiting sergeant. Perhaps uniquely among the various branches of science, animal behaviour combines philosophical debate and practical relevance with an intrinsic ability to engage and entertain. A cheetah driving down its prey, the dazzling light shows of communication among cuttlefish, a parasitic spider stealthily snipping its booty from the web of an unsuspecting host are living theatre to the casual observer. A moment's reflection, however, turns entertainment into a stream of thought-provoking questions. How does the cheetah choose its victim from the panicking throng around it? Do cuttlefish really talk to each other? Is the thieving spider aware of its hazardous lifestyle? Do other organisms know they exist at all?

The spectacle of behaviour, whether our own or of other species, gives rise to a panoply of such questions. Some, like the existence of consciousness, remain puzzles, as we shall see, but many more are yielding to insights that have revolutionised our perception of behaviour over the last 30 years, and heightened awareness of its relevance to ourselves and the world in which we live.

For such a flourishing science, animal behaviour is surprisingly young. 'Behaviour' as a label for what animals do did not enter regular use until the early years of the twentieth century and the writings of Conway Lloyd Morgan, a disciple of the fledgling comparative psychologist and protégé of Darwin, George Romanes. Psychology itself is not much older, emerging from nineteenth century physiology through the work of Helmholtz and Wundt in the 1860s and 1870s (Plotkin 1999). Before 'behaviour', science imbued animals with 'emotions', 'habits', 'manners', 'customs' and 'instincts' with a distinct focus on mental attributes that might compare with our own (Sparks 1983; Boakes 1984). This anthropocentric approach viewed animals as essentially little furry or feathery people (Shettleworth 1998), model systems for understanding the general principles of human psychology and the evolutionary continuity of animal and human minds. Such a cognitive perspective was eclipsed by the mechanical stimulus–response philosophy of behaviourism (1.3.4) but is now once again a major focus of research and debate (see 4.2).

Of course, the behaviour of other species had been observed for centuries before it was formalised into any kind of science, but this was largely for utilitarian reasons: to hone hunting skills, avoid death or manage domestic stocks. Behaviour is still studied for these reasons. But new utilitarian incentives are also emerging. Anthropologists and social scientists increasingly acknowledge animal behaviour as a framework for understanding human society and social problems, such as child abuse, drug addiction and social discrimination (e.g. Lenington 1981; Buss 1999; Barrett *et al.* 2002). Studies of social behaviour have revealed its effects on reproductive physiology and prenatal mortality, and on the immune system and resistance to disease (e.g. Wingfield *et al.* 1990; Clutton-Brock 1991; Barnard & Behnke 2001). Models simulating the distribution of predators around feeding sites can help predict some of the effects of habitat fragmentation on natural populations (e.g. Sutherland 1996; Frankham *et al.* 2002).

This book provides a comprehensive introduction to animal behaviour by developing examples in the context of their theoretical background and the various different levels at which behaviour can be explained. It emphasises the importance of integrating different approaches to the study of behaviour and of evolution by natural selection as a framework for interpreting behaviour at whichever level it is studied. Indeed, a theme of the book is that no level of explanation makes complete sense except in the context of the others and that adaptation by natural selection (Chapter 2) is the process that ensures this is so. Why will become clear as we progress through the book. To begin with, however, we must establish some terms of reference.

1.1 What is behaviour?

What exactly *is* 'behaviour'? As good a working definition as any is that of the pioneering psychologists B.F. Skinner and D.O. Hebb (see Barnard 1983) who include as behaviour:

all observable processes by which an animal responds to perceived changes in the internal state of its body or in the external world.

While this admits some physiological responses, such as the secretion of a sweat gland or the expansion of a chromatophore, this is not as serious a drawback as it might seem. Such responses may, for example, play a crucial role in communication or provide incidental information about the animal's internal state. What is less obvious about the definition is that it incorporates a variety of different levels of response within the concept of 'behaviour'. This is captured more clearly by the *Oxford English Dictionary* which summarises 'behaviour' as:

> deportment, manners . . . moral conduct . . . way in which a . . . machine, substance etc. acts or works; response to stimulus.

The blinking of an eye, the sudden movement of a limb as it is withdrawn from a sharp object, the cacophonous display of a male blue bird of paradise (*Paradisaea rudolphi*), the care shown by a female chimpanzee (*Pan troglodytes*) to a distressed infant, our own cultural etiquettes; all fall within these definitions of behaviour and our working use of the term. At first sight this seems hopelessly broad. How can we ever ask sensible questions about behaviour if its units range from the twitch of a few small muscles to sophisticated social interaction? There are two kinds of answer to this.

1.1.1 Repeatable measurement

The first hinges on the inherent consistency of many behaviour patterns and the fact that this allows reliable repeated observation. All science progresses by testing hypotheses about the world, but in order to do so it must be able to measure the world reliably. From Darwin onwards, people studying behaviour have recognised that muscular movements and complex sets of actions can be ordered into **repeatedly recognisable**, and therefore **measurable**, units (see Dawkins 1983 and Martin & Bateson 1993 for a good introduction to the problem). Despite differences in complexity and duration between units, individuals of a given species perform them in more or less the same way every time. A dog relieving an itch performs a characteristic series of movements with one of its hind legs. Sometimes these are directed at an ear, sometimes a shoulder, but wherever they are directed the movements have a clear beginning and end and are distinctively different from other kinds of movement involving the legs. So distinctive are they, in fact, that we confidently give the whole sequence a label, 'scratching', to sum up its apparent function. 'Scratching' can then be distinguished from other units of behaviour such as 'walking' and 'cocking' which also involve characteristic movements of the leg but serve different functions (locomotion and urine marking respectively).

Sometimes the units we define incorporate many lower level units such as limb movements and body orientations but nevertheless add up to a self-contained and repeatable pattern that is distinguishable from other patterns. 'Pup retrieval' by a mother rat (*Rattus norvegicus*), 'nest building' by a female blackbird (*Turdus merula*), 'allogrooming' (grooming of other individuals) by a hamadryas baboon (*Papio hamadryas*) are three randomly chosen examples. Each is an identifiable pattern of activity that can be recognised whenever it is performed and differs clearly from other identifiable patterns of activity performed by the species. There is therefore no particular level of complexity that characterises a unit. Units can be as simple or as complex as repeatability and the demands of the study dictate.

1.1.2 The function of behaviour patterns

The second kind of answer is couched in terms of function. Diverse as they are, our examples of behaviour above (and any others we might have come up with) fall into one of two camps. Moving a limb or batting an eyelid each involves a relatively simple set of motor responses resulting in a clear-cut and seemingly isolated action. The display of the bird of paradise and the helpful chimpanzee, on the other hand, employ a complex set of actions that has all the appearance of contriving a desired outcome (attracting a female, comforting an infant). The sequence has a functional outcome, but more than that it seems to have **purpose**. Animals appear to do things *in order* to achieve something. This is reflected in our very descriptions of behaviour. We talk of 'searching for food', 'hiding from predators', 'migrating home' or 'exploring a cage'. Each involves a diversity of actions, perhaps no more complicated than moving a limb or blinking an eye, but actions that are clustered together into functionally organised sequences resulting in the animal achieving an important outcome. What can we infer from this?

1.1.2.1 Purpose and goal-directedness: means to ends

The apparent purposefulness of much behaviour begs two important questions. How has such purposefulness come about in natural systems, and what does it imply about the organisms that show it?

As McFarland (1989) has pointed out, there is a widespread temptation to interpret apparent purposefulness as indicating some kind of internal representation of what has to be achieved. The representation may be a set point (an ideal state of the system), as envisaged in certain homeostatic physiological processes such as temperature regulation, or it may be some kind of mental image, as postulated in many cognitive explanations of behaviour (a hungry rat sets out with an image of food in its head, a blackbird weaves twigs and grass into its notion of a nest) (see 4.2). McFarland uses the term '**goal-directedness**' to describe purposefulness based on these kinds of internal representation and contrasts it with two other ways of appearing to pursue a goal.

In '**goal-achieving**' systems the requirement for some commodity, say food, makes the animal restless; it moves around until it happens to encounter some food, then it becomes quiescent. This implies that the goal is recognised when it is encountered rather than the behaviour resulting in its discovery (restlessness) being driven by a prior internal representation. Template recognition in certain molecular systems, such as 'lock and key' enzyme–substrate complexes, would be an analogous process. In '**goal-seeking**' systems, on the other hand, there is no representation or recognition of the goal at all. The apparent attainment of a goal is due entirely to the physical forces acting on and/or within the system. McFarland uses the example of a marble rolling around a bowl. Depending on where it starts rolling, the marble could take a variety of routes round the bowl but it will inevitably end up sitting in the bottom. There is no internal representation of the bottom of the bowl and the effectiveness of its downward movement in bringing the marble to rest is independent of its environment (unlike a goal-achieving system in which triggered activity such as restlessness will lead to an appropriate outcome, say finding food, only in specific kinds of environment). The marble would head downward and come to rest whether it was rolled around a bowl or dropped from a skyscraper. It arrives at its 'goal' simply by gravity.

McFarland's distinctions, while schematic, show there are several ways of appearing purposeful, each with different implications for underlying mechanism. These kinds of distinction become extremely important, as we shall see later, when we consider the nature of internal mechanisms responsible for behaviour. Some examples make them clearer.

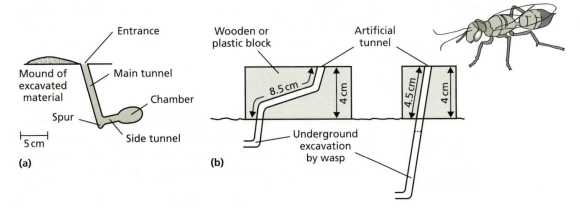

Figure 1.1 (a) Diagram of the nesting burrow of a great golden digger wasp (*Sphex ichneumoneus*); (b) when burrows are artificially lengthened by placing blocks on top of the entrance, wasps reduce their excavation accordingly. After Brockmann (1980).

Locomotion in woodlice

A nice example of goal-achieving behaviour comes from Fraenkel & Gunn's (1961) classic study of locomotion in woodlice (*Porcellio scaber*) (see also Benhamou & Bovet 1989). Woodlice tend to end up in dark, damp places, but they do not get there by knowing where these are and setting out to find them. Instead, individuals that find themselves in places that are too dry or too light simply start moving about randomly. The drier or lighter the area they encounter, the faster they move, the damper or darker the area, the slower they move. When they reach a place that is sufficiently damp and dark, they stop moving altogether. Thus, like the marble in the bottom of the bowl, woodlice end up in damp, dark places simply because that is where they stop moving about.

Burrowing behaviour in digger wasps

Females of the great golden digger wasp (*Sphex ichneumoneus*) dig a burrow in the ground which they provision with paralysed katydids (relatives of grasshoppers) as food for their larvae. Burrows consist of a main shaft with one or more side tunnels branching off. At the end of each side tunnel the female excavates a nesting chamber in which she lays a single egg (Fig. 1.1a). When she has completed one burrow, the female moves off to start digging another. But how does she know when a burrow is complete? Does she have some model she is working to, or does she follow a simple rule such as, say, 'dig downwards for *t* minutes then along for *t'* minutes' which produces the required design without any working blueprint? To find out, Jane Brockmann manipulated the depth of digger wasp burrows during their excavation to see how females responded (Brockmann 1980).

Brockmann found that if she artificially lengthened the main shaft so that it was deeper than the wasps would normally dig, females backed up the shaft to the appropriate level for the side tunnel and filled in the excess depth as they went. If a wooden or plastic extension shaft was added to a burrow in progress, wasps reduced their excavation to compensate. Moreover the degree of reduction could be manipulated by varying the length of extension shaft (Fig. 1.1b). This suggests that females registered and were satisfied with the modified shafts despite not having completed the excavation themselves. When the extension was removed, however, they resumed digging. This combination of digging and filling in response to different experimental manipulations is consistent with the wasps having some criterion for deciding when a burrow is finished and moving on to the next. Such responses are suggestive of goal-directed burrowing behaviour, perhaps based on some set point mechanism for gauging depth (Brockmann 1980).

Nest-building in domestic hens

A completely different picture emerges when we look at nesting behaviour in domestic hens (*Gallus gallus domesticus*). Hens go through a well-defined series of activities during nest-building, from scraping a shallow depression in the ground to placing nesting materials around the edges. Do hens have the finished product of a nest as a goal to head for as the digger wasps seem to with their burrows? A study by Hughes *et al.* (1989) suggests not. Hughes *et al.* presented hens with three different types of nest: (a) a flat litter surface, (b) a pre-formed hollow nest and (c) a pre-formed nest with an egg. If a fully formed nest was the goal of building activity we should expect hens to show less building with nest types (b) and (c) than with nest type (a) because the job is essentially done in the first two cases. Somewhat surprisingly, however, Hughes *et al.* found that the amount of nest-building by hens was in fact *greater* with nest types (b) and (c), with hens taking longer to reach the egg-laying stage after entering the nest area (Fig. 1.2). This suggests that nest-building is driven by factors other than the functional consequences (a complete nest) of the behaviour and that it is the performance of the behaviour itself that matters to the animal. In this and its dependence on an appropriate environment (no pre-formed nest available) for the right outcome, the hens' response is more in keeping with a goal-seeking system than a goal-directed one (the hen behaves

Figure 1.2 Domestic hens (*Gallus gallus domesticus*) spend more time nest-building and thus take longer to lay eggs after entering a nesting area if they are presented with pre-formed nests, especially if these already have an egg. Nest types: A – litter surface only; B – pre-formed nest; C – pre-formed nest with egg. See text. After Hughes *et al.* (1989).

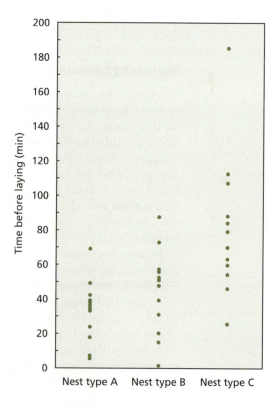

along the lines of 'keep adding bits of nesting material until it's time to lay an egg' rather than 'add bits of nesting material until the structure looks like a nest').

1.1.2.2 Purpose and mind

While there may be several different ways of generating the same purposeful outcome, there is also a hierarchy of potential decision-making processes by which animals might arrive at their choice of behaviour. The hierarchy is one of increasing sophistication in the scope and flexibility of choice, and thus justification for thinking in terms of 'mind' and 'intelligence' as opposed to pre-programmed stimulus–response relationships. Dennett (1995) captures this, again schematically, in his so-called 'Tower of Generate-and-Test' (Fig. 1.3). In Dennett's scheme, organisms face various problems of survival and reproduction set by the environment and have to come up with appropriate solutions from a battery of options available to them. The 'Tower of Generate-and-Test' envisages four levels of complexity by which choices might be made.

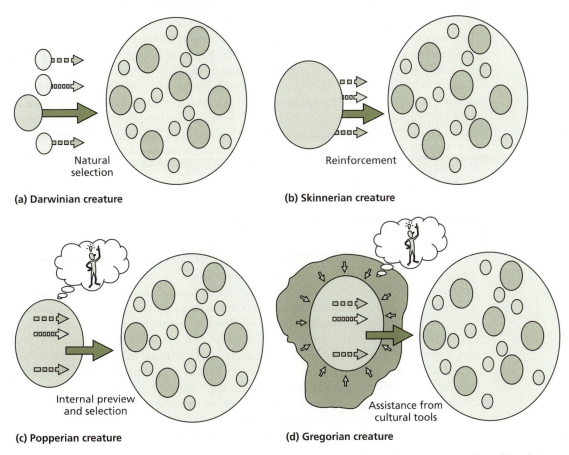

(a) Darwinian creature

Natural selection

(b) Skinnerian creature

Reinforcement

(c) Popperian creature

Internal preview and selection

(d) Gregorian creature

Assistance from cultural tools

Figure 1.3 The levels of Dennett's 'Tower of Generate-and-Test'. In each case there is a range of possible solutions to problems of survival and reproduction set by the environment, but organisms at each level of the hierarchy arrive at one of them by different means. In Darwinian creatures (a), natural selection chooses between different genetically encoded options. In Skinnerian creatures (b), individuals arrive at a choice by trial and error learning. Popperian creatures (c) also choose by a trial and error process, but this is in the form of internalised hypothesis testing. Gregorian creatures (d) are like their Popperian predecessors, but hypothesis testing is now aided by cultural tools. See text. After Dennett (1985) *Darwin's Dangerous Idea* by Daniel C. Dennett (NY: Simon and Schuster, 1995, pp. 374–8).

Hard-wired solutions: 'Darwinian creatures'

At the simplest level, decisions are **genetically hard-wired** (what we shall later [Chapter 5] call 'developmentally fixed'), so that each decision is an inflexible property of a particular genotype. A variety of candidates is generated by mutation and (in sexually reproducing organisms) recombination and outcrossing. Candidate decision-makers compete for resources and the most effective genotypes reproduce to make it down the generations. These are Dennett's '**Darwinian creatures**', products of natural selection between alternative, inflexible, genetically encoded responses. To pursue (now hypothetically) our example of nest-building hens, this would mean that hens inheriting different versions (alleles) of a putative nest-building gene would build nests in the one particular way encoded in their genotype which would be inflexibly different from the way hens with other genotypes built their nests.

Reinforcement and learning: 'Skinnerian creatures'

In time, after many generations and varieties of innovative, hard-wired solution, the Darwinian process throws up a chance novelty that takes things to the next level of Dennett's scheme: **phenotypic plasticity**. Here, choices are still genetically encoded but the coding allows for a variety of responses. There is thus more opportunity to hit on a good solution, but the advantage over pure 'Darwinian creatures' is not very great unless there is a means of biasing choice in favour of the more effective options. The advent of a reinforcer of some kind (see Box 6.1), a mechanism that increases the likelihood of a good solution being tried again next time, does the trick. Now organisms have *conditionable* flexibility and become Dennett's '**Skinnerian creatures**', named after the famous learning theorist B.F. Skinner (see 6.2.1.2). A 'Skinnerian' hen might winnow down her range of nest-building moves according to their contribution to a sense of support. Moves increasing the feeling of support would be repeated next time, and those that did not would be abandoned.

Hypothesis-testing: 'Popperian creatures'

Conditionable flexibility is a useful advance over the rigid single option of the 'Darwinian creature', but it is still a very inefficient way to proceed. It is far better to avoid mistakes altogether, or at least reduce their performance costs, by thinking through available options first and dispensing with those that are obviously useless. To paraphrase the philosopher of science Sir Karl Popper, it is better to let your hypotheses die in your stead (Dennett 1995). '**Popperian creatures**' have the edge over their 'Skinnerian' precursors because they have a better chance of choosing the best option at the outset. However, the 'Popperian' process presupposes some elaborate internal machinery for storing, sifting and integrating information as well as a means of modelling the world and testing possibilities against the model's predictions; in other words it assumes a certain level of cognitive processing. In terms of our nest-building example, a 'Popperian' hen would view the problem in hand, consult its memory of past approaches and outcomes and select the approach most likely to complete the task effectively.

Cultural enhancement: 'Gregorian creatures'

Dennett envisages a stage beyond his 'Popperian creatures', a stage informed and assisted by a cultural environment (see 12.1). The inhabitants of this level he calls '**Gregorian creatures**' after the psychologist Richard Gregory. Gregory's thesis is that cultural artefacts are not just a result of intelligence but in many cases endow intelligence by providing new means of choosing appropriate responses. Gadgets such as lathes and computers potentiate clever options that could never be realised in their absence. But cultural 'tools' are not limited to physical artefacts. As Gregory (1981) and others have argued, language, one of the consummate skills of our own species, can be considered a cultural facilitator of new intellectual possibilities (11.3.2). Indeed, some view language as *the* principal driving force in the evolution of intelligence (Dennett 1995). 'Gregorian creatures' thus have an additional armoury that can be focused on problems and their potential solutions and gain a cultural boost over their 'Popperian' rivals. In our nesting hen, a 'Gregorian' builder might glean some tips by watching the hens around her and incorporate the acquired information into her own repertory of options.

The ascent of Dennett's tower is one of gradual emancipation from rigid pre-programmed responses to a flexible, higher-level choice of options which depends on a host of contingencies and a carefully integrated bank of information. The complex on-board computer of the brain, rather than simple neural circuits, assumes responsibility for translating incoming information into suitable action, aided in its task in 'Gregorian creatures' by the collective intelligence of culture. We, of course, appear to be the most sophisticated 'Gregorian creatures'. Our culture has an evolutionary life of its own, one that is many times faster than the neoDarwinian process from which it arose (12.1). The ability of culture to invent and propagate mind tools that themselves accelerate the pace of change partly explains this. But the positive feedback effect is enhanced by the fact that our on-board computer is an elaborate problem-solving machine honed by natural selection to aid survival in a diverse and changeable environment. One of its key propensities is to set goals and achieve them, but, as we have seen in our discussion of goal directedness, this propensity may be brought to bear inappropriately when reflecting on the problem-solving mechanisms of other species.

Of course, McFarland's and Dennett's distinctions are crude categorisations for the purpose of illustration. Nevertheless they highlight the important point that much may be hidden within an observed behaviour. A variety of mechanisms and processes can produce the same apparent outcome and some deft investigation may be necessary to discover which of the possibilities it is. Which behaviours are goal-directed and which merely goal-achieving? Which are the result of 'Popperian' choice and which of the 'Skinnerian' or first-order 'Darwinian' variety? How sophisticated are models of the world in 'Popperian' and 'Gregorian' heads? We shall see what progress has been made in this respect in later chapters. First we must look at questions about behaviour from a different perspective.

1.2 Questions about behaviour

1.2.1 Tinbergen's Four Whys

Our discussion of purpose has emphasised the need to understand both what a behaviour is designed to achieve and how its performance is supported by the animal's decision-making machinery. Dennett's schema also points to developmental and evolutionary factors; 'Skinnerian creatures' home in on the best option after bitter experience with less useful ones, and each tier of the 'Tower of Generate-and-Test' is an evolutionary progression from the one below. Together, these points capture the four fundamental levels at which any behaviour pattern can be explained. Aristotelian in origin, these were first formalised in a behavioural context by the Nobel Prize-winning ethologist Niko

Underlying theory

Box 1.1   Tinbergen's Four Whys

As Tinbergen (1963) pointed out, there are four different ways of answering the question 'Why?' in biology. We can illustrate them in the context of behaviour by thinking about why a common shrew (*Sorex araneus*) is selective in what it accepts as prey. While taking a broad range of prey species, shrews bias their intake towards energetically the most rewarding items when given a choice (Barnard & Brown 1981). Why do they do this? Answers at Tinbergen's four different levels might be as follows:

☐ **Function** (*what is the behaviour for?*) – answers at this level are concerned with what the system is designed to do, i.e. its role in the life of the organism. Our shrew may thus feed selectively to make sure it takes only the most nutritious prey, so maximising its foraging efficiency.

☐ **Mechanism** (*how is the behaviour achieved?*) – answers here are concerned with how the system operates in terms of underlying mechanism and organisation. So, for example, our shrew's tactile and visual senses may be most responsive to large, active prey (which also turn out to be the most nutritious), thereby biasing the animal's intake towards these items.

☐ **Development** (*how does the behaviour develop?*) – these kinds of answer are concerned with the way the system reflects its embryological, cultural or other developmental influences. Our shrew's early foraging experience may therefore teach it which types of prey are easy to locate and subdue, and so the most efficient to deal with in terms of energy return.

☐ **Evolution** (*where has the behaviour come from?*) – here we are concerned with the ancestral selection pressures and phylogenetic pathways that have shaped and constrained the system. Thus shrews may be selective foragers because less discriminating ancestors foraged inefficiently so were less likely to survive and reproduce. Selective foraging is thus the result of generations of natural selection for increased foraging efficiency.

Tinbergen (Tinbergen 1963) into what are now referred to as **Tinbergen's Four Questions** or **Four Whys** (Box 1.1).

As is evident from Box 1.1, the four kinds of question are *complementary* answers to the question, not rivals for the truth. It makes no sense to claim that the shrews in Box 1.1 forage selectively because active, nutritious prey are the ones they tend to detect *rather* than because they are attempting to forage efficiently. Why? Because efficient foraging is not some abstract property that exists independently of the shrew's physical systems, it is the result of selection shaping those systems so they are likely to hit on nutritious prey rather than less nutritious ones. One effect therefore leads to the other; answers in terms of mechanism and function are not alternatives to each other. Nevertheless it is surprising how often this kind of confusion arises. Functional explanations are sometimes referred to as **ultimate** explanations because they are concerned with the adaptive consequences of a behaviour and thus the reason it has evolved. By the same token, explanations in terms of mechanism are referred to as **proximate** explanations because they are concerned with the physical systems that influence the performance of the behaviour. Since it is these two levels of explanation that are most often confused, it is worth making the distinction clear with a real example.

1.2.1.1 Socially mediated changes in brain and behaviour

Social competition is a harsh fact of life for many species. Natural selection has there-fore given rise to a rich diversity of behavioural solutions for reducing it, some of which we shall encounter later. One widespread solution is territoriality, the aggressive defence of patches of habitat against competitors (9.2). An example comes from a cichlid fish, *Haplochromis burtoni*, that lives in Lake Tanganyika in Africa.

Reproduction among *H. burtoni* males is monopolised by a relatively small proportion (about 10%) of individuals that are able to defend a breeding site (Fernald & Hirata 1977). Territorial males are strikingly coloured and use their dazzling black, yellow, blue and red appearance as a territorial signal both to warn off rival males and to attract females to a nest on the lake floor. They also have much higher androgen (male sex hormone) levels than their non-territorial counterparts. Non-territorial males, along with females and juveniles, are a uniform dull brown colour. They tend to aggregate into schools that drift round the colony area, attempting to feed in the territories of the brighter males. Territorial males chase them off but also try to court the females in the school.

The crucial point, however, is that the dichotomy between territorial and non-territorial males is not fixed and inflexible but depends on the prevailing social environment. When males are reared together, maturation rate is socially regulated; males reared with adults have delayed maturation compared with those reared without adults (Fraley & Fernald 1982). More surprisingly, however, if a territory owner loses an encounter with a non-territorial intruder, he immediately adopts the drab hues of the latter and moves into the school on the fringe of the nesting area. Conversely, a sudden vacancy in territory owner-ship, through ousting or mortality, can lead to a non-territorial male metamorphosing into the brilliant territorial form. How is all this achieved?

The key lies in a set of specialised cells in the pre-optic area of the hypothalamus (a region in the floor of the forebrain [3.2]) which secrete gonadotrophin-releasing hormone (GnRH). Via the pituitary, GnRH stimulates development of the testes which in turn secrete androgens that promote aggressiveness by acting on other clusters of cells in the brain. Francis *et al.* (1993) conducted an elegant series of experiments in which they compared GnRH cells and testes in territorial and non-territorial males and looked for any changes when social roles reversed. The results were extraordinarily clear cut. When established samples of the two types of male were compared, both the GnRH cells and testes of territorial individuals were significantly larger than those from the non-territorial males. This is as might be expected from their behavioural profiles. But when social roles were reversed, by exposing territorial males to communities of larger territorial males and non-territorial males to females and smaller males, the size of GnRH cells and testes changed accordingly (Fig. 1.4). Previously territorial males now exhibited the reduced GnRH cells and testes characteristic of non-territorial males and vice versa. Plasticity of behavioural phenotype was reflected in plasticity of underlying mechanism (see also 3.1.3.1).

The distinction between ultimate and proximate explanations for the socially mediated territorial behaviour male *H. burtoni* is clear. From a functional perspective, large males can capitalise on their competitive advantage by defending a territory and advertising their status to both competitors and females by being brightly coloured. Smaller males are not able to do this, so adopt a non-aggressive, dull coloration instead, perhaps to deceive their way into being mistaken for a female and getting access to food, or even mates, on other males' territories. If a territorial male is ousted it signifies a more powerful male is around, so it pays the ousted male to switch into drab, non-aggressive mode to avoid

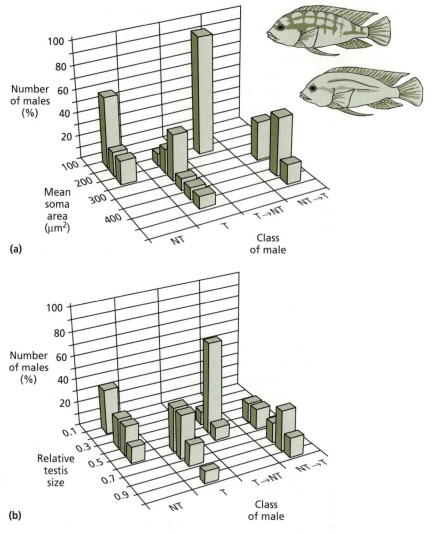

(a)

(b)

Figure 1.4 (a) The percentage of non-territorial (NT) and territorial (T) male *Haplochromis burtoni* with soma regions (containing GnRH cells) of different size. When territorial status changes (NT to T, or T to NT) the size of GnRH cells changes too. (b) Similar changes occur in relative testis size. After Francis *et al.* (1993).

further provocation and possible injury. Conversely, a vacant territory suggests an opportunity to move up the social scale, but a full behavioural and cosmetic conversion is necessary if the takeover is not to invite a challenge.

A mechanistic explanation casts the same scenario in terms of the effect of social signals on hormones regulating aggression and reproductive physiology. The hypothalamic–pituitary–gonadal axis integrates various behavioural and physiological responses relating to reproduction and is subject to external influences (of which social signals are one) via neural and hormonal feedback. As perceived signals change (territory owners are beaten or disappear), so responses within the axis change, triggering or suppressing their associated behaviours (in this case aggression).

While it is obvious that these two explanations are complementary rather than mutually exclusive alternatives, the example of the flexible male *H. burtoni* highlights the fact that natural selection simultaneously shapes the mechanisms underlying behaviour as well as the functional consequences of the behaviour. The two may belong to different categories of Tinbergen's Four Whys but they are both adaptive products of the same evolutionary process.

1.2.2 Addressing the Four Whys

The Four Whys identify complementary levels of explanation. But how do we arrive at the answers in particular cases? In Box 1.1, for example, how could we establish that shrews chose prey to maximise their foraging efficiency, or that their sense organs responded more strongly to large active stimuli? In fact there are several ways we could do it and these apply generally to any aspect of biology.

1.2.2.1 Experiments

One way is to conduct an **experiment**. Scientific research proceeds by distinguishing between competing hypotheses and systematically reducing the number of ways in which the world can satisfactorily be explained. Experiments, at least good ones, work by manipulating variables that are of interest while controlling those that are not. The outcome of each manipulation should allow the experimenter to distinguish between rival explanatory hypotheses. In the shrew example the experimenter might test the hypothesis that shrews select prey to maximise their net rate of energy intake by pitting it against a plausible alternative, such as shrews forage randomly but take mainly the most profitable prey because these just happen to be the most active and easily noticed. Experiments controlling the shrews' encounters with different quality prey or varying the relationship between quality and noticeability would help distinguish between these possibilities. Of course, if it turned out that shrews went for noticeability, this would beg the next question: did they go for it simply because that was what they happened to see, or had their perception been shaped to respond selectively that way *because* noticeability generally correlated with profitability (i.e. noticeability is used as a guide to profitability)? One experiment therefore leads to another and explanations become more refined.

1.2.2.2 Observation

Experiments seek to control the world in order to focus on specific phenomena of interest. In many situations this may not be possible or ethically desirable. However, carefully designed **observational studies** can provide just as powerful a means of distinguishing between competing hypotheses. Indeed, many celebrated field studies of behaviour have been based almost entirely on observation. Observational studies can sometimes be viewed as experiments in retrospect. Whereas experiments control unwanted variables at the outset and watch what happens as a result, many observational studies painstakingly record them as they go along and then control for them in subsequent analyses. As long as the analyses are driven by a priori predictions, and are not just retrospective 'fishing exercises', they can be as powerful as any well-controlled experiment. In our shrew example, careful observation in a vivarium might reveal rejection of some potential prey items despite the shrew having noticed and inspected them, thus increasing our faith in selective foraging as a working hypothesis for its behaviour.

1.2.2.3 Comparison

A powerful way of understanding behaviour at all four levels is to compare performances across different species and environments. The assumption here is that exposure to different selection pressures will have moulded behaviour in predictable ways and that something of the routes of evolutionary change will be discernible in comparisons of closely related species. The **comparative approach** applied across species is a major tool in evolutionary explanation, though one that must be used with care as we shall see in Chapter 2, but it is also effective within species where populations are associated with different physical and social environments. In our shrews, for example, we could compare foraging selectivity in animals from populations in which the range of prey species was different. Perhaps one was richer in molluscs and earthworms and the other in ground-dwelling insects. If shrews chose prey that maximised their foraging efficiency in both cases, despite big differences in the nature and activity of the prey, it would again reinforce our confidence in the foraging efficiency hypothesis. Comparative studies may be experimental or observational so are not in any sense alternatives to the first two categories above. The key point is that existing ecological and putative evolutionary differences between species or populations provide a ready-made set of 'treatments' for comparison.

1.2.2.4 Theoretical models

An entirely different approach is to take things back to first principles and construct a **model**. A model is a distillation of the essential elements of a system (as perceived by the modeller) that explores their potential for accounting for the observed world. Models can take a variety of forms, from mathematical equations and computer programs to physical constructs such as robots (3.1.3.4). Many, of course, are adjuncts to experiments or other kinds of investigation, helping to set up precise predictions for testing, but they can also be used to simulate outcomes and see whether our assumptions about a system accord with the way it performs. Models that simulate different degrees of competition for resources and look at their effects on overall population dynamics are a good example. To return once again to foraging shrews, we might develop a model of long-term time and energy budgeting and explore how well taking different ranges of prey satisfied the animals' requirements for survival and reproduction. The predictions from different hypothetical diets could then be compared with what shrews actually take. We shall now look at some real examples to see how different levels of explanation can combine with different modes of investigation.

Crickets and bats: an experimental test of mechanism

Like several flying insects that fall prey to foraging bats, the Polynesian cricket *Teleogryllus oceanicus* possesses a sound detection system honed to respond to the telltale signals of approaching death. When they detect an approaching bat, crickets can take evasive action by steering away from it. An ultimate explanation for this is obvious: crickets that move out of the way are likely to survive, those that do not are not. But what is the proximate explanation in terms of underlying mechanism?

A key component of the system is a set of receptors in the cricket's ears (situated on its forelegs) that are sensitive to ultrasound and thus to approaching bats. Sensory messages from these receptors are relayed to cells in the central nervous system which include a pair of sensory interneurons, *int-1* cells, located one on either side of the body. *Int-1* cells become excited when the cricket's ears receive ultrasound; the more intense the sound the stronger the cells' response and the shorter their latency to respond, properties they share with analogous cells in certain night-flying moths (see Chapter 3). When a flying cricket (tethered for experimental purposes) is exposed to ultrasound it characteristically bends its abdomen away from the source of the sound (Fig. 1.5). The response can be elicited experimentally by stimulating the *int-1* cells with an electrode (Nolan & Hoy 1984). The converse is also true: when *int-1* cells are temporarily inactivated the bending response to ultrasound is abolished. Thus stimulation of these interneurons appears to cause the bending response and steer the cricket away from potential danger. However, the *int-1* cells are *sensory* neurons; they do not issue motor commands to the cricket's muscles. These are relayed from the cerebral ganglia (brain) after information from diverse other receptors has been integrated. So what is the mechanism for changing the direction of flight?

A clue comes from the beating of the wings during flight. Observations by Michael May suggested that, when tethered crickets were exposed to ultrasound, the beating rate

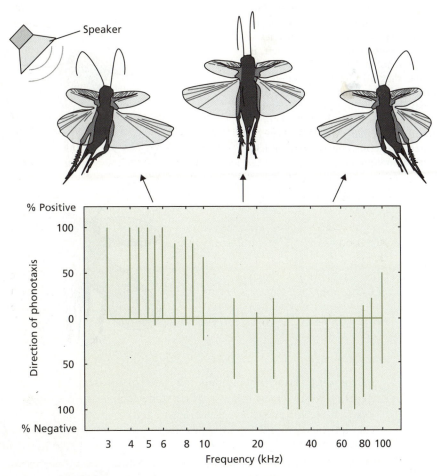

Figure 1.5 Veering response (percentage of positive [towards the sound] or negative [away from the sound] turns) of tethered flying crickets (*Teleogryllus oceanicus*) to different sound frequencies. Crickets tend to veer away from high-frequency sounds that might indicate an approaching bat. The tendency to turn towards low frequencies may reflect a response to calls by other males. After Moiseff *et al.* (1978).

of the hindwing further from the source was reduced (May 1991). Since only the hindwings power flight in crickets (the forewings are specialised to protect the hindwings from damage), this unilateral slowing would make the cricket yaw away from the sound source. But how was the slowing brought about? A possibility that occurred to May was that the hind leg might interfere with the beating of the wing by being lifted into its path. May tested this in a simple experiment in which he removed the hind legs of some crickets and took high-speed photographs of them in tethered flight. When he compared these with similar photographs of intact animals, it was clear that the wingbeat responses of the two groups to ultrasound were very different. In the absence of a hind leg, both hindwings of amputee crickets continued to beat as normal when ultrasound was played. In the unmanipulated crickets the hind leg on the side opposite the sound source could be seen being lifted into the downward sweep of the wing, disturbing its beat pattern (Fig. 1.6). While amputee crickets could eventually make a turn, it took them roughly 40% longer

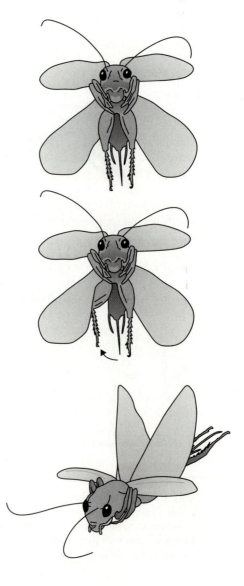

Figure 1.6 How a flying *Teleogryllus oceanicus* responds to ultrasound that might indicate an approaching bat. As the cricket detects ultrasound coming from its left, it swings its right hindleg into the path of its right wing, thus reducing the thrust on that side. Tilting the forewings at the same time combines the resulting yaw with a power dive and rapid escape from the potential flight path of the bat. After May (1991). Reprinted by permission of Virge Kask, Scientific Illustrator.

than their intact counterparts to begin to do it, a margin likely to be fatal. Through simple but revealing experiments, therefore, the proximate mechanism for rapid turning proves to be a command loop from a set of sensory interneurons via the cerebral ganglia and appropriate motor neurones to the musculature of the appropriate hind leg.

Clutch size in magpies: an experimental test of function

How many eggs should a bird lay? The obvious answer might seem to be as many as possible since lots of eggs turn into lots of offspring that will later reproduce. A moment's thought, however, shows that things are not as simple as that. Offspring have to be fed and protected and parental care is not an infinite resource (10.2). We might thus expect some kind of compromise between maximising the number of offspring produced and maximising their chances of being reared to independence. Lots of offspring are no good if they all die before maturity, and lavishing everything on a singleton might waste investment that could have sustained two or three more. In other words we might expect some kind of optimum number that maximises the overall reproductive output of the clutch. This is an example of a general argument about decision-making by organisms that we shall come to in more detail in Chapter 2.

Like many other bird species, European magpies (*Pica pica*) show some variation in clutch size. We might predict this if individuals differ in their physical condition, or food supplies vary between seasons and locations. But this assumes that magpies adjust their clutch size according to their ability to raise chicks on different occasions. Before we can accept this we need to test it. Goran Högstedt did so in a classic study of magpies in Sweden. Högstedt (1980) noticed that clutch sizes varied between five and eight eggs in the birds in his study population and that this variation was associated with differences in food availability on territories. Was this due to females making adaptive decisions based on their chances of rearing different numbers of offspring? To find out, Högstedt carried out a simple field experiment in which he selected nests with different clutch sizes over three breeding seasons and either added or removed nestlings or left broods unchanged. For each starting clutch size he thus had some nests in which brood size had been reduced, some in which it had been increased and some where it remained at its initial size. He then recorded the number of young still surviving in the nest just prior to fledging. The results showed that survival was best in broods that were the same size as the original clutch (Fig. 1.7), good evidence that magpies were adjusting their clutch sizes according to their ability to rear chicks.

Song learning in sparrows: a comparative test of development

The males of songbird species usually learn their songs by imitating those of adult males around them, often during some critical period in their development (6.2.1.5). As a result, local song 'dialects' (Marler & Tamura 1962) can form in which clusters of neighbouring males sing similar songs that are different from those of males further away (Fig. 1.8). However, local dialects are not a ubiquitous feature of songbird species. One reason for this might be that differences in dispersal and breeding seasons between species affect the acquisition of songs. This is borne out by a recent study of white-crowned sparrows (*Zonotrichia leucophrys*), a North American bunting, in which differences between subspecies in the tendency to form dialects appear to relate to differences in the length of the available breeding period. Douglas Nelson compared the songs of four subspecies of *Z. leucophrys*: a sedentary race, *Z. l. nuttalli*, two migratory races, *Z. l. oriantha* and

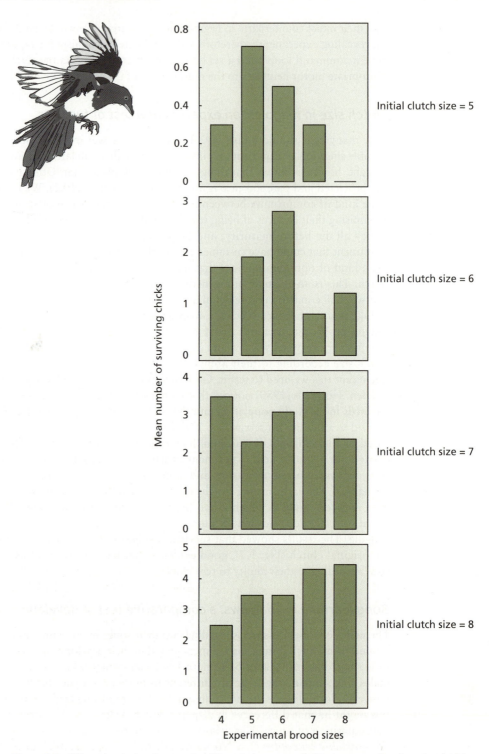

Figure 1.7 When clutch sizes in magpies (*Pica pica*) were made bigger or smaller than that laid by the female, the number of chicks surviving to fledge declined, suggesting magpies optimise clutch size in relation to available resources. Plotted from data in Högstedt (1980).

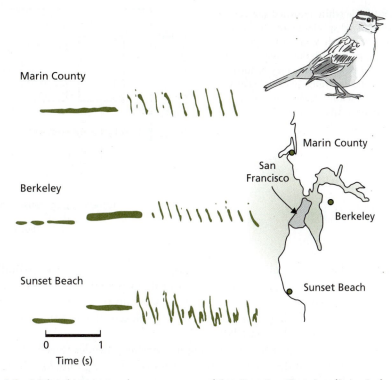

Figure 1.8 Male white-crowned sparrows around San Francisco Bay sing distinctively different songs (dialects) in different areas. These can be visualised in sound frequency plots above. After Marler, P. and Tamura, M. (1962) Song 'dialects' in three populations of white-crowned sparrows, *Condor*, 64, 368–77. Reproduced with permission of the Cooper Ornithological Society, © Cooper Ornithological Society.

Z. l. pugetensis, and an Arctic/sub-Arctic race, *Z. l. gambelli* (Nelson 1999). Males of each subspecies sing one song type which they maintain for life. However, the first three subspecies, *nuttalli*, *oriantha* and *pugetensis*, all form local dialects whereas *gambelli* does not. Nelson speculated that the lack of dialects in *gambelli* was due to the subspecies' northerly breeding grounds in which the breeding season starts late and is very short compared with those of the other three subspecies. This compression of the breeding season, Nelson suggested, may have influenced the song learning process in *gambelli*.

To test his idea, Nelson compared the development of song in *Z. l. gambelli* with that of the other subspecies both in the field and in controlled laboratory conditions. The first thing Nelson noticed was that, in the field, *gambelli* males arrived on their breeding grounds already singing only a single song type. Males of the other subspecies arrived with several song types which were then winnowed down to one, possibly, as suggested by responses to songs played in the laboratory, by a process of matching to the songs of neighbouring males. Unlike males in the field, hand-reared *gambelli* males in the laboratory did start by singing more than one type of song, but when they were played tutor songs, only 2 out of 11 selected their eventual song to match that of the tutor. In contrast to those of the other subspecies, therefore, *gambelli* males appeared to select song types randomly with respect to those they could hear around them.

A second aspect of song learning in *gambelli* that appeared to have been influenced by the abbreviated breeding season, and which might help to explain the differences in

Figure 1.9 Male white-crowned sparrows of the *gambelli* subspecies learn almost all their songs before they are 50 days old. Other subspecies acquire their songs over a longer period. *Gambelli* males may be forced to truncate the learning period because of the short breeding season at northern latitudes. See text. After Nelson (1999).

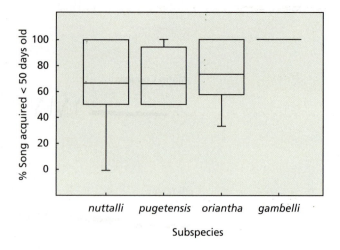

the song selection process above, related to the sensitive period for song acquisition. Experiments with hand-reared birds in the laboratory showed that *gambelli* had a dramatically shorter sensitive period than any of the other subspecies (Fig. 1.9), suggesting that chicks had been selected to be most sensitive to song stimulation at a significantly earlier age (Nelson 1999).

The exigencies of breeding in extreme latitudes thus seem to have imposed a constraint on the song development process in the *gambelli* subspecies that has caused a departure from the pattern of song production at the population level seen in other races of the white-crowned sparrow.

Courtship in newts: a modelling approach to mechanism

Courtship in the smooth newt (*Triturus vulgaris*) is an elaborate and time-consuming affair that takes place on the floor of ponds. The process has been studied extensively by Tim Halliday (e.g. Halliday 1975; Halliday & Sweatman 1976; Houston *et al.* 1977). Three phases can be distinguished in which the male orientates and manoeuvres in various ways close to the female in an attempt to gain and maintain her attention (Fig. 1.10). At some point during the third (Retreat) phase, the male switches to a behaviour called Creep which involves turning away from the female (Fig. 1.10) and creeping forward until the female nudges his tail (Tail-touch). When this happens the male stops and deposits a spermatophore (a package of sperm) on the substrate before creeping slowly on to encourage the female to follow and take the spermatophore up into her cloaca. The female can lose interest at any point in this sequence in which case the male has to go back to Retreat and start again.

Houston *et al.* (1977) were interested in how various internal and external factors interacted to determine the sequence of courtship behaviours shown by a male on any given occasion. On the basis of Halliday's detailed investigations they made some educated guesses at what these would be and how they might interact. They then simulated the result using a control systems model (Fig. 1.11). Their study is a nice example of the development of a simulation model hand-in-hand with studies of the behavioural system in question, but we shall be concerned only with the final version of the model here.

In the model, the variable *Hope* represents the male's assessment of the readiness of the female to accept his spermatophore. The model assumes that ongoing events in the newt's internal and external environment induce changes in *Hope* and cause the male

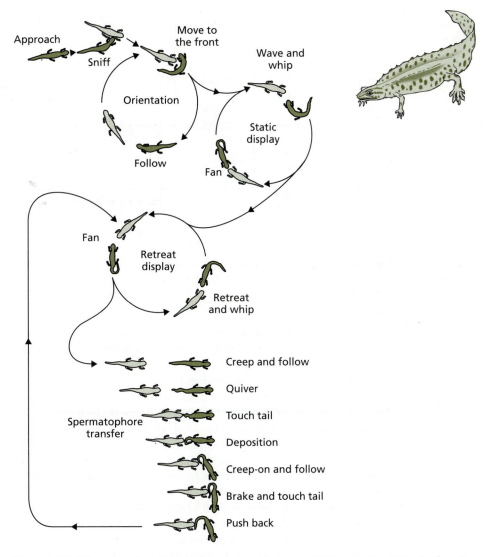

Figure 1.10 The courtship sequence of the male smooth newt (*Triturus vulgaris*). In the flow diagram the male is dark and the female pale. From Halliday (1975).

to adjust his responses accordingly. These changes are effected by integrators in the circuitry of the model (boxes i in Fig. 1.11). To begin with, *Hope* depends on the state of readiness of the female (F) and the sperm supply (S) of the male (males may have a number of spermatophores ready to transfer). Since the model is concerned only with the latter (Retreat and Spermatophore Transfer) phases of courtship, the male begins by performing the Retreat display. During Retreat, *Hope* rises (via boxes 1 and 4) until it reaches a threshold value (T_2) whereupon the male switches to Creep. (The threshold boxes T_1–T_4 operate in such a way that when the variable represented by the arrow [in this case *Hope*] leading into the box exceeds a certain value, the arrow leading out of the box is activated.) During Creep, *Hope* decreases (by negative feedback via box 2) because the male can no longer see the female and thus does not know whether she is responding. In the model, the male proceeds to spermatophore deposition once a certain

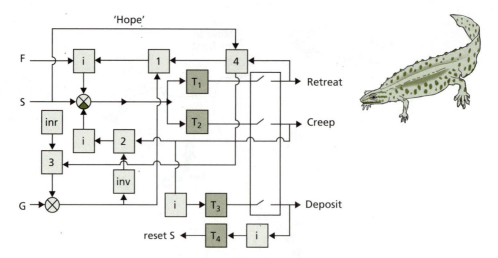

Figure 1.11 A control systems model of the spermatophore transfer phase of courtship in the male smooth newt (see text and Fig. 1.10): F, the state of readiness of the female; S, the sperm supply of the male; G, the male's oxygen debt; i, integrator; inr, increase in oxygen debt with time; inv, invertor. After Houston *et al.* (1977).

amount of Creep has been performed (T_3). In reality, however, he would have to receive a Tail-touch from the female before depositing his spermatophore. If *Hope* decays too quickly so that it drops below threshold T_1 (before the supposed Tail-touch), the male reverts back to the Retreat display (Fig. 1.10).

However, there is another complicating factor to take into account: oxygen debt. Although newts are able to respire to a certain extent through their skin and buccal cavity, they are dependent while in water on air obtained at the water surface. The frequency with which newts have to ascend to the surface depends on their level of activity, the temperature and gas content of the water and the relative concentration of atmospheric gases above the water surface. This requirement for air places a serious constraint on the length of time for which a male can display (Halliday & Sweatman 1976). Effectively, breathing behaviour and courtship behaviour compete with each other. Oxygen requirement is therefore built into the model as a factor (G) competing with courtship. The magnitude of the oxygen debt affects the rate at which *Hope* changes as a result of Retreat display (via box 1) and Creep (via box 2). As oxygen debt will increase through the sequence, its effect will be to make the male more likely to get through to spermatophore deposition as time goes on because he behaves at an increasingly greater rate.

So much for outlining the model: does it capture the behaviour of real males courting real females? Figure 1.12(a) shows mean durations of Retreat, Creep and Spermatophore Transfer for courting males observed in the laboratory. Figure 1.12(b) shows output from the simulation model in Fig. 1.11 for values of F, S and G deemed appropriate for the sequences observed. The match in terms of the temporal patterning of the sequence is remarkable, suggesting that the model captures some essential elements of the decision-making process in male newts. The faithfulness of the model can also be tested experimentally by manipulating some of the factors that are represented by its components. For example, Halliday (1977) reduced the amount of oxygen available in the air above the water surface by adding nitrogen and found that courtship sequences were performed faster. The converse was true when the oxygen content was increased. This concurs with the output of the model (courtship finishes earlier or later) when the value of G is altered.

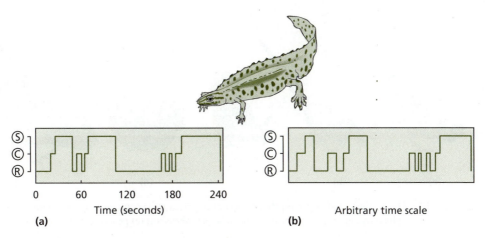

Figure 1.12 (a) The sequence of Retreat (R), Creep (C) and Spermatophore deposition (S) (see Fig. 1.10) observed in the laboratory. (b) The sequence predicted by computer simulation of the model in Fig. 1.11. After Houston *et al.* (1977).

Ritualised displays: a comparative study of evolution

Displays during courtship and aggressive confrontations provide some of the most striking spectacles of the animal world. A conspicuous feature of many displays is their highly stylised and stereotyped performance. This has traditionally been explained in terms of 'ritualisation' whereby movements and morphological features become modified during evolution to enhance their impact as signals (11.1.1.4). The displays of various male birds of paradise and ducks are well-known examples. There are various adaptive explanations for ritualisation which we shall return to in Chapter 11, but it is generally agreed that ritualised displays evolved from various non-stereotyped, functional behaviours such as intention movements (e.g. preparing to take off or run away) and autonomic responses (e.g. sweating and urinating) (see Hinde 1970 and Box 4.3). The question is how? Apart from indirect clues such as footprints and burrows, behaviour does not leave much of a fossil record. There is thus little evidence with which to reconstruct the evolutionary history of a display. However, some insight into likely pathways can be gained by comparing displays in closely related species.

An old but classic example is Schenkel's (1956, 1958) comparison of courtship displays in phasianid birds (pheasants and their relatives) (Fig. 1.13). Schenkel traced the evolution of the highly ritualised displays of peacocks (*Pavo cristatus*) and male pheasants from food enticing activities in species with less elaborate displays. The sequence charts the progressive elaboration of various movements involved in food enticement and their ultimate divorce from their original function.

At the least stylised end of the spectrum is the domestic fowl in which males simply scratch the ground with their feet, then step back and peck at it while calling to attract a female (Fig. 1.13a). Food may or may not be present when the display takes place. If food is not present, the male may instead pick up small stones as if they were food objects. Either way a female usually approaches and searches for food in vicinity of the male's display, thereby making herself available for mating (see also 11.2.1.1).

Male ring-necked pheasants (*Phasianus colchicus*) behave in a similar way (Fig. 1.13b). Male impeyan pheasants (*Lophophorus impejanus*), however, do not scratch for food but only peck at it. During courtship, the male bows in front of the hen, spreads his tail

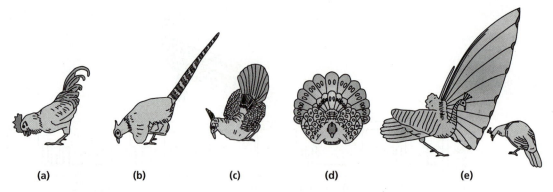

Figure 1.13 The evolution of courtship from food enticement behaviour in male phasianids: (a) domestic fowl; (b) ring-necked pheasant; (c) impeyan pheasant; (d) peacock pheasant; (e) peacock. See text. After Schenkel (1956).

slightly and pecks energetically at the ground. While the hen approaches and searches in the vicinity of the pecking, he spreads his wings and tail to the full and, maintaining a head-down position, gently fans his fully spread tail up and down thus exaggerating the body movements normally involved in pecking (Fig. 1.13c).

After scratching at the ground, the peacock pheasant (*Polyplectron bicalcaratum*) also bows with raised wings and spread tail (Fig. 1.13d). When a female approaches, he moves his head backwards and forwards in a conspicuous fashion. This is usually done in the absence of food, but if food is given to the male, he will offer it to the female, thus betraying the origin of the bowing and stylised head movements in feeding behaviour.

In the peacock, the degree of ritualisation is such that the origins of the courtship display in feeding enticement are all but unrecognisable. Only by surmising the intermediate steps from the displays of related species do its affinities become clear. The male spreads and shakes his tail feathers, then arches the spread tail forward and down over his upright head (Fig. 1.13e). While there is no attempt by the male to peck the ground, the female nevertheless searches in front of him, guided by the concave fan of his tail. The male's display has thus become completely emancipated from the act of finding food. Interestingly, however, juvenile males show classic food enticing behaviour involving both scratching and pecking at the ground. Ritualisation of the behaviour into a courtship display therefore seems to occur on a developmental, as well as an evolutionary, timescale.

1.3 Approaches to the study of behaviour

We have looked at what we mean by behaviour and the different ways of studying it. While all behaviour is, in principle, amenable to each of Tinbergen's four approaches, it is perhaps not surprising that various traditions have developed that tend to emphasise some approaches over others. Animal behaviour is a rich and varied subject. It is divided into many subdisciplines with philosophies, terminology and perspectives of their own. The same behavioural problem is often approached from completely different angles by people in different fields, making it difficult for them not to talk past each other. Learning is a good example. Both psychologists and zoologists are interested in learning, but psychologists have traditionally focused on the underlying principles and mechanisms of learning while zoologists have been more concerned with how learning fits the animal for the

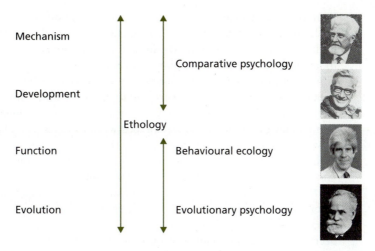

Figure 1.14 Historical subdisciplines in the study of behaviour as they map onto Tinbergen's Four Whys. See text. Photographs from top to bottom: Konrad Lorenz (Photograph: Hermann Kacher), Niko Tinbergen (Photograph: B. Tschanz), William Hamilton, Ivan Pavlov.

environment in which it has to function. Until relatively recently these two approaches had little to say to each other. In fact, for a while, they were symptoms of a fundamental rift of opinion about the very nature of behaviour itself, the so-called **'nature–nurture' debate** (see Slater 1999 for a good summary). As it turns out, however, the two offer complementary insights into learning which now mutually inform the way it is studied (e.g. Staddon 1983; Giraldeau 1997). We shall see more of this in Chapter 6.

This is not the place for a detailed history of the study of behaviour. However, discipline labels abound in the literature and it is worth briefly clarifying some of them. We shall look at four whose approximate mapping onto the Four Whys is shown in Fig. 1.14.

1.3.1 Ethology

Tinbergen's Four Whys originate in the mid-twentieth century European tradition of behaviour study known as **ethology**. Its founding practitioners, foreshadowed by the German ornithologist Oscar Heinroth, were Konrad Lorenz, Karl von Frisch and Tinbergen (Fig. 1.14) who shared the Nobel Prize for Physiology in 1972 for their contributions to the study of behaviour. Tinbergen defined ethology as the 'biological study of behaviour', indicating its all-encompassing scope in explaining *how* animals achieved things as well as *why* they did (Fig. 1.14). While people from many disciplines – psychologists, physiologists, veterinary scientists and psychiatrists among them – were (and still are) interested in behaviour, the defining feature of ethologists was that they were biologists, scientists with an awareness of the diversity of species and the importance of natural context in interpreting behaviour (Hinde 1982). Even though many ethologists worked in the laboratory, most attempted to relate their work to the natural environment in which behaviour was assumed to be adaptive. Indeed, as Slater (1999) has pointed out, this driving principle of ethology lent animal behaviour a reputation as a branch of natural history.

The ethologists' approach was based on detailed description of behaviour, allowing categories to be identified and subsequently investigated. Tinbergen's emphasis on the

Four Whys encouraged people to study behaviour at different levels. Only by this process could the nature of behaviour be thoroughly understood. As a result, ethology fostered a tradition of hypothesis-driven experimental investigation, often in the field, which flourishes today in the burgeoning discipline of **behavioural ecology** (see below). It also spawned grand theories about the organisation and control of behaviour which eventually contributed to its demise. These were mainly theories about internal control mechanisms and the extent to which they were genetically hard-wired ('**innate**' or '**instinctive**' in the terminology of the time) and stereotyped within species (see Chapters 3–5). While these proved to be hostages to fortune and withered for lack of experimental support, they highlighted an important dichotomy in concepts of causal mechanism which still informs our view of behaviour. This is the distinction between so-called 'nuts and bolts' or 'hardware' explanations of mechanism (e.g. which neurones fire in what sequence to produce behaviour A), and motivational or 'software' explanations, which view the animal's physical (hardware) systems simply as a 'black box' and focus on how the decision-making rules (software) use the 'black box' to translate incoming and stored information into adaptive motor output. We have already encountered the two approaches in our worked examples above. The neuronal loop causing crickets to veer away from ultrasound is explained at the 'nuts and bolts' level, while the control of courtship sequences in smooth newts is couched in terms of motivation. We shall look at the relative merits of these approaches in Chapters 3 and 4.

1.3.2 Comparative psychology

While unsubstantiated grand theories was one factor in the decline of the European school of ethology, another was its failure to gain a foothold in North America. In part this was due to the conflict of perspectives between ethology and the American school of **comparative psychology** (epitomised at the time by T.C. Schneirla and N.R.F. Maier) referred to above. Ethologists, as we have seen, were zoologists interested in a wide range of species and the species-typical ways in which they behaved. Comparative psychologists, on the other hand, were not so much interested in *differences* between species as in *similarities*, seeking **general laws** of behaviour, especially relating to development, that could be applied to all species, including humans. The focus was on rigorously controlled laboratory experiments involving a handful of species, mainly rats and pigeons (*Columba livia*), on the assumption that if behaviour was underpinned by general laws it made no difference which species you studied; a convenient laboratory vehicle such as the rat could serve for the lot. This, of course, was in stark contrast to the necessarily less tightly controlled field observations of many ethologists and their concentration on the adaptiveness of species-specific behaviours in their natural environment (Fig. 1.15). Indeed the two traditions have been characterised as the study of general processes (psychology) versus the study of adaptive specialisation (ethology) (Riley & Langley 1993; Shettleworth 1998).

Conflict between ethology and comparative psychology over common behavioural territory was inevitable and reached its peak in the pre-1960s 'nature–nurture' debate (1.3.1; Chapter 6). The ethologists' concept of 'fixed action patterns', rigidly structured, innate blocks of behaviour performed in their entirety on each occasion (3.1.3.2), provided a particular focus here. Other differences of opinion centred on ethological models of motivation in which behaviour resulted from the discharge of pent up action-specific energies in the central nervous system (CNS) (4.1.1.1). While Tinbergen in particular

Figure 1.15 A traditional caricature of ethologists and psychologists saw the latter as putting their animals in a box and peering in to see what they did, while the former put themselves in a box and peered out to see what animals were up to. After Slater (1999).

attempted to translate the concept into respectable neuroanatomical language, the psychologists rightly regarded it as unrealistic and incompatible with anything that was understood about the workings of the CNS.

Of course there were many other areas of disagreement but a detailed history is not appropriate here. After years of mutual hostility, meaningful dialogue and ultimately rapprochement followed a famous critique of Lorenzian ethology in 1953 by the psychologist, and student of Schneirla, D.S. Lehrman. Ethologists came to accept the flexibility of behaviour patterns and that learning and other environmental influences probably had a role to play in every case. They also came to recognise the value of carefully controlled laboratory experiments which, while divorced from the natural environment, could provide insights unattainable in the field. For their part, psychologists conceded that evolution in the context of a species' ecology placed constraints on what could be learned and how, and that the pursuit of general laws was probably naïve.

While it is fair to highlight the historical differences between ethologists and comparative psychologists, they should not be mythologised into more than they were. In this respect, Dewsbury (1984) sounds a healthy note of caution about retrospective finger-wagging in discussions about comparative psychology. Today, animal behaviour is a considerably more integrated discipline within which old subject boundaries have become blurred or disappeared altogether. Many comparative psychologists and 'ethological' animal behaviourists work on similar problems and publish side by side in the same journals.

1.3.3 Behavioural ecology

While 'ethology' as a label has largely faded from use, its guiding principle of hypothesis-driven experimentation in the context of natural history is very much alive. Behavioural ecology, the overwhelmingly dominant force in animal behaviour for the past 30 years,

Underlying theory

Box 1.2 Cornerstones of behavioural ecology

The explanatory power, and successful rise, of behavioural ecology stems from three core theoretical frameworks emerging in the 1960s and 1970s:

1. Hamilton's (1964a,b; see Fig. 1.14) notion of **inclusive fitness**, which recognised related-ness as an indirect route through which alleles affecting social interaction could influence their transmission to subsequent generations, thereby revolutionising evolu-tionary explanations of social behaviour and phenotypic altruism.

2. The concept of **evolutionarily stable strategies (ESSs)** (Maynard Smith 1972), which recognised that adaptive decision-making was often frequency-dependent (the best thing to do depends on what everyone else is doing).

3. The development of economic models of decision-making (of which ESS models are a special kind) that used cost/benefit analysis and **optimality theory** as a framework for predicting evolutionary outcomes (Parker 1970; Schoener 1971; Charnov 1976a,b).

The rationale behind each of these is discussed in detail in Chapter 2.

is the direct intellectual descendant of Tinbergen's pioneering approach. Concentrating on questions of function and evolution, behavioural ecology emerged in the 1960s as a fusion of ethology, ecology and evolutionary biology. Its spectacular rise was based on three mould-breaking developments in evolutionary thinking (Box 1.2 and 2.4.4), which between them fostered a rigorous, quantitative approach to behaviour often based on explicit genetic models. To paraphrase Krebs & Davies (1993), it allowed us to ask not simply why sunbirds (*Nectarinia reichnowi*) defend feeding territories or male dungflies (*Scatophaga stercoraria*) copulate for long periods of time, but why sunbird territories comprise around 1600 flowers and male dungflies copulate for an average of 41 minutes.

The application of behavioural ecology to social behaviour, particularly phenotypic altruism and cooperation, is sometimes referred to as 'sociobiology', after the seminal volume of the name by E.O. Wilson (Wilson 1975a). Wilson's quite reasonable attempt to extend this kind of thinking to human behaviour led to a celebrated fracas with anthropologists and a handful of politically enthusiastic biologists that has turned 'sociobiology' into something of a red rag term. Since it is also unnecessary (it is simply a subset of behavioural ecology), the term is probably best avoided, along with its knee-jerk opponents.

As predictive science, behavioural ecology has been a stunning success, not only yielding new insights into behaviour but influencing the parent subjects of ecology and evolution from which it partly emerged. However, success has not come without criticism. The enthusiasm for behavioural ecology has often been blamed for an unhealthy bias towards questions of function, ignoring underlying mechanisms and the balanced approach of Tinbergen's Four Whys. This bias is now being redressed as mechanism and develop-ment play an increasing role in tests of functional hypotheses and speculations about evolutionary origins, but the rise of behavioural ecology undoubtedly contributed to the decline of interest in mechanism and development during its early years.

Behavioural ecology has also been accused of Panglossism, after the character Dr Pangloss in Voltaire's *Candide*, whose philosophy was that 'all was for the best in the best of all possible worlds', meaning everything in life was as good as it could possibly be. In a landmark paper, Gould & Lewontin (1979) used Pangloss to parody the kind of adaptive explanation promulgated by behavioural ecology and its subdiscipline sociobiology, suggesting that it uncritically viewed every aspect of an organism as an adaptation to something and encouraged a naïve perfectionist view of the design of organisms. However, historical (phylogenetic) constraints, and those imposed by alternative demands on the animal, have long been recognised as factors limiting the evolution of perfect adaptation in the sense implied by Gould & Lewontin (Dawkins 1999, and see Chapter 2).

The bogey of genetic determinism has been another problem (see 5.1). The use of 'genes for' language (Chapter 5) in explaining the evolution of behaviour has led to misunderstanding about the role of development and the environment in adaptive decision-making (Lewontin *et al.* 1984; Lewontin 1991; Dawkins 1995). The approach is often caricatured as implying rigidly encoded behaviour patterns that are always performed in a particular way like an automaton, or as somehow justifying antisocial behaviour and social stereotyping in our own species because it offers a plausible evolutionary explanation for them (the crux of much of the opposition to sociobiology). Dawkins (1989, 1999) once again provides a readable antidote to this kind of muddled (or in some cases mischievous) argument, to which we shall return in Chapter 5.

1.3.4 Evolutionary psychology

Psychology is an older science than animal behaviour. Yet it is the younger discipline that captures the wider imagination. Why should this be? The most likely reason is that animal behaviour has its roots in natural history and evolutionary biology, which together can tell us something about why we are what we are and how we came to be. Evolutionary thinking has been remarkably lacking, though not entirely absent, from contemporary psychology and we have already seen a consequence of this in the contretemps with the ethologists. There are good historical reasons for this odd impoverishment (Buss 1999; Plotkin 1999). One is that comparative psychology pre-dated the temporary demise of Darwin's theory of natural selection in the early years of the last century (see Nordenskiöld 1928). While the onslaught on Darwinism came largely from the early geneticists, psychology suffered by association because Darwin, and later his student Romanes, speculated about behaviour and its evolutionary origins. This was compounded by the early post-Darwinian rush to ascribe human behavioural traits, indeed the perceived behavioural traits of entire nations, to simplistic inbuilt instincts inherited from our animal past. William James in particular saw instinct in everything, from biting, sleeping, sitting up and walking in infants to imitation, fear of certain objects, shyness, sociability and parenting in adults (James 1890). Flying in the face of contemporary opinion, James claimed that humans possessed more instincts than any other species, but his long, atomised lists of putative examples soon came to be ridiculed. Concepts based on a crumbling theory were thus easy prey and their vulnerability paved the way for a school of thought that banished evolutionary thinking from psychology for the best part of half a century: **behaviourism**.

The beginnings of behaviourism can be traced to John B. Watson's seminal paper in the *Psychological Review* in 1913. Behaviourism rejected any explanation of behaviour that depended on unobservable causes. Thus mental states and concepts such as instinct were not the province of scientific explanation because they had no tangible reality; they were hypothetical constructs and thus immune to scrutiny. Evolutionary explanation was given short shrift precisely because it relied on inference and hypothesis about history rather than direct measurement of the physical present. The cause célèbre of behaviourism was the study of learning which emerged from the early experiments of Pavlov (Fig. 1.14) and Thorndike and, from the 1920s onward, provided psychology with its main claim to be an empirical science (Plotkin 1999). This emphasis on learning, and the quest for the general laws that underpinned it, did two things. It encouraged an inflated confidence in 'nurture' as the major determinant of behaviour and behavioural abnormalities, and ensured a mechanistic, physiological focus in the causal explanation of behaviour.

With minor perturbations, and reinforced by the blossoming discipline of cultural anthropology, this perspective held sway until the 1960s when certain empirical findings began seriously to undermine the concept of general laws of learning. Chief among these were Harlow's studies of mother–infant relationships in monkeys and Garcia's experiments on associative learning in rats (see 6.2.2).

Harlow's studies demonstrated the limited power of reinforcement in shaping mother object preferences in young monkeys; infants could not be persuaded to prefer a wire mesh 'mother' over a soft cloth version, no matter how strongly contact with the former was rewarded with food (e.g. Harlow & Zimmerman 1959). Garcia's work showed that associative learning could be highly dependent on the form of stimulus–response relationship and that basic tenets of temporal association between stimulus and response could be violated without jeopardising the formation of an association (Garcia & Koelling 1966). In short, animals found some things easy to associate and some things difficult or impossible. Moreover, things that were easy to associate could be linked even when presentation and consequence were separated by several hours. Clearly, learning was not an homogeneous, generalised property of living organisms that could be accounted for solely by reinforcement. Instead it was honed to respond selectively to things that were likely to be important in the day-to-day survival (and thus ultimately the reproductive success) of the individual.

These findings helped pave the way for a renaissance of evolutionary thinking in psychology. They led to a new perspective on information processing and the integration of sensory input and motor output that could be generalised to human cognitive attributes. Perhaps humans too were designed to respond to certain kinds of information but not others in order to further their reproductive interests. Indeed, might not the entire architecture of the brain and its legion capabilities have been shaped by the forces of natural selection? If so, a science of **evolutionary psychology** would seem an essential starting point for any understanding of brain and mind. A new and flourishing literature shows that many agree, and evolutionary psychology is now emerging as a productive, and (as ever with human behaviour) contentious, new field of enquiry (12.2).

While it is often defined in terms of an evolutionarily informed study of human mental attributes, Daly & Wilson (1999) rightly point out that evolutionary psychology draws ideas from a wide range of intellectual and taxonomic sources. Many contributors are neither psychologists nor do they work on humans; they are interested in the commonality of cognitive processes in humans and other animals as we might expect natural selection to have shaped them. In the main, evolutionary psychology has reflected an extension of behavioural ecology. Thus ideas align along the characteristic subject divisions of behavioural ecology: sexual selection, social behaviour, kinship, reciprocity and cooperation, parental investment, conflict and aggression (Buss 1999; Daly & Wilson 1999; Barrett *et al.* 2002). They are sometimes based on the concept of an **environment of evolutionary adaptedness (EEA)** (Bowlby 1969; Tooby & Cosmides 1990; Daly & Wilson 1999), a reflection of the critical selection pressures that have shaped mental processes in the past (Tooby & Cosmides 1990) and which are thus essential to understanding responses to environments in the present. Perhaps not surprisingly, the nature of any such EEA, indeed its usefulness as a concept at all, is a matter of some debate (Chapter 12).

Summary

1. 'Behaviour' includes a wide range of responses, from simple movements of a limb to complex social interaction. In order to study it we must break it down into measurable units that can be recognised reliably each time they are performed. Sometimes units are simple actions such as batting an eyelid, but more often they are sequences of actions that produce a functional outcome, such as finding food or escaping from a predator.

2. The fact that behaviour patterns frequently have a functional outcome gives them an air of purpose. It is tempting to attribute apparent purposefulness to a pre-conceived goal on the part of the animal. However, several different mechanisms could give rise to the appearance of purpose, each assuming very different things about the internal processes of the animal and its perception of the external environment.

3. The same functional outcomes can be achieved through very different mechanisms of decision-making. Decisions may be 'hard-wired' into the animal so that it always does the same thing when a response is required, or they may be devolved to a higher-level centre that integrates incoming and stored information and responds flexibly to the demands of the moment. Deciding which kind of decision-making process is responsible for behaviour can be difficult.

4. Questions about behaviour can be answered at four different but complementary levels, generally known as Tinbergen's Four Whys. Answers can be in terms of function (how the behaviour helps the animal survive and reproduce), mechanism (the underlying mechanisms producing the behaviour), development (the embryological or other developmental factors influencing the behaviour) or evolution (the historical selection pressures and phylogenetic constraints that have shaped the behaviour).

5. Questions at any of these four levels can be investigated by a number of methods: experimentation, observation, comparison and modelling. In many cases combinations of these methods are used.

6. Various traditions of behaviour study have given different emphasis to the Four Whys. While ethology maintained a broad basis in all four, comparative psychology has tended to emphasise mechanism and development while behavioural ecology has emphasized function and evolution. However, these kinds of traditional biases are now disappearing, both within existing approaches and in emergent new fields such as evolutionary psychology.

Further reading

Dawkins (1983) and Martin & Bateson (1993) discuss some of the problems of defining and measuring units of behaviour, and McFarland (1989), Kennedy (1992) and Dennett (1995) deal, in somewhat different ways, with various issues surrounding purposefulness. Tinbergen's (1963) classic paper is still well worth reading as an introduction to the 'Four Whys'. Several books introduce and/or review some of the different fields of behaviour study. Good starting points are Boakes (1984) (for a historical overview of early thinking), Manning & Dawkins (1998), McFarland (1999) and Slater (1999) (for broadly ethological approaches), Krebs & Davies (1993) (by far the best introduction to behavioural ecology), and Buss (1999) and Barrett *et al.* (2002) (for evolutionary psychology).

2 Evolution and behaviour

Introduction

Evolutionary questions ask where behaviour patterns have come from, and how and why they have become what they are. But behaviour leaves little in the way of a fossil record, so how can we deduce its evolutionary history? Can suggested evolutionary pathways ever be more than uncorroborated speculation? Adaptive behaviour patterns are assumed to have evolved through natural selection. This presupposes that variation in behaviour has a heritable genetic basis. Is there any evidence for this? Since we do not yet know the genetic basis for most behaviours, can we model their evolution by natural selection with any accuracy? Natural selection may be one way in which behaviour can evolve, but are there others? The fact that behaviour patterns can be acquired through observation and learning suggests there might be. What is the relationship between such cultural evolution and evolution by natural selection?

'Nothing in biology makes sense except in the light of evolution'

Theodosius Dobzhansky (1973)

Dobzhansky's dictum has been quoted almost to the point of cliché. But it is especially pertinent to the study of behaviour and its outing again here needs no justification. In the previous chapter we were careful to distinguish between the different levels at which behaviour can be investigated, but at the same time we stressed that these different levels are complementary and united in having been shaped by the same evolutionary processes – *mechanism and development serve function through evolution*.

The neoDarwinian theory of evolution is the one, all-important, conceptual framework uniting the biological sciences. As a coordinated product of the animal's morphology and physiology, behaviour is a reflection of, but also a driving force for, the evolution of morphology and physiology. The veering response of Polynesian crickets to ultrasound in Fig. 1.5, for example, reflects the sensitivity of the *int-1* interneurons to ultrasound (behaviour *reflects* physiology), but the fact that these cells are sensitive to ultrasonic frequencies in the first place is due to the evolutionary pressure for crickets to avoid bats (behaviour *drives* physiology). In this chapter, we shall be looking at the evolution of behaviour and the process of evolutionary explanation. Many of the examples and arguments relate to functional questions because the contribution of behaviour patterns to individual reproductive success is the principal engine of evolutionary change. Thus we shall encompass two of Tinbergen's Four Whys in Box 1.1. However, we shall also see that evolutionary changes at the functional level reflect underlying changes in mechanism and development.

2.1 NeoDarwinism and the modern synthesis

The modern theory of evolution is based on Charles Darwin's theory of 'descent with modification' (Darwin 1859). Darwin's phrase captures the twin essences of the theory: (a) an unbroken line of descent of modern forms from a common ancestor and (b) departure from ancestral forms through a process of adaptive modification. Evolutionary studies are therefore all about history and change. Where have modern forms come from, and how have they ended up as they are? The generally accepted view is that Darwin's process of **natural selection** provides most of the answers. But before we see how, we must expand a little on the modern view of evolution.

2.1.1 The neoDarwinian synthesis

Current evolutionary theory is not strictly Darwin's theory as he originally conceived it. There are two main reasons for this. First, Darwin hung on to a number of old ideas, such as the inheritance of acquired characters and direct effects of the environment, which he saw operating alongside his new process of natural selection but which are now largely disregarded as mechanisms of evolutionary change (but see some remarkable evidence that defensive responses to predators induced in one generation can affect offspring in the next – so-called **maternally induced defence** [Agrawal *et al.* 1999]). Second, and more importantly, he understood nothing of genetics. He appreciated that many characteristics were passed on from parent to offspring, i.e. were heritable, but had no inkling as to the mechanism of inheritance.

Darwin's failure to appreciate the principles of heredity was a major factor in the temporary eclipse of his ideas during the early years of the twentieth century. Ironically, the very discovery that was to be their salvation, Mendel's theory of particulate inheritance, initially contributed to their demise. Mendel's own experiments, and those of the early Mendelian geneticists such as Bateson and Morgan, had focused on relatively large-scale variation segregating into clear-cut characters, such as round versus wrinkled peas, or different eye colour in fruit flies. There seemed to be little evidence of the ubiquitous small-scale variation required by Darwin's natural selection. As a result, a theory of evolution based on large-scale mutation (macromutationism) held sway, with natural selection, if it was a force to be reckoned with at all, being relegated to a minor tinkering role. (Interestingly, an earlier theory of particulate inheritance [Maupertuis 1753 – see Boas 1966], foreshadowing Mendel to an astonishing degree, had also led to a mutational theory of evolution.)

Two things led to the rehabilitation of Darwin's ideas. Continuing experimental work by the early Mendelians began to undermine confidence in the segregation of clear-cut characters as the general rule. For instance, many characters that were discrete in the parental generation yielded a profusion of intermediates when the parents were crossed, suggesting variation on a much finer scale than superficial appearances might indicate. Secondly, evolutionary biologists with a sound grasp of quantitative genetics, such as Ronald Fisher, Sewall Wright and Sergei Chetverikov, realised that variation in natural, sexually reproducing populations was very different from that in the restricted laboratory populations studied by many of the early Mendelians, and that it was more in accord with the abundant small-scale variation envisaged by Darwin. The combined effect of these two insights was to rationalise the conflicting fields of Mendelian genetics and Darwinian evolution into a single framework, the **Modern Synthesis** as it

Underlying theory

Box 2.1 Assumptions of the neoDarwinian synthesis

Six key assumptions underlie the neoDarwinian synthesis (see text):

1. *Heritable variation*. Populations of sexually reproducing, cross-breeding organisms contain large pools of heritable genetic variation.

2. *Genetic recombination*. The amount of variation is increased by recombination and segregation at meoisis.

3. *Population genetics*. Evolution results from the differential spread (fitness) of alleles (alternative versions of individual genes) through the population gene pool. The fitness of alleles is not absolute, but relative to that of other alleles in the gene pool.

4. *Constancy of gene frequency (the Hardy–Weinberg law)*. The frequency of alleles within a gene pool remains constant unless altered by mutation, natural selection or genetic drift (random effects).

5. *Natural selection*. Mutation rates are too low to have a significant effect on allele frequencies except over long periods, genetic drift is important only under special conditions, so changes in allele frequency (i.e. evolution) is due mainly to natural selection.

6. *Adaptive complexes*. Most adaptive characters are coded for by many different genes and can be altered only by the occurrence and establishment of changes in several genes. Mutations of single genes that produce large effects are usually detrimental.

Based on discussion in Stebbins (1982).

was dubbed by Huxley (1942), in which Darwin's central idea was underpinned by a firm understanding of variation and inheritance. **NeoDarwinism**, the basis of modern evolutionary understanding, was born.

Stebbins (1982) identifies six sets of assumptions that characterise the neoDarwinian synthesis (Box 2.1). The two key assumptions are that (a) natural populations contain an abundance of heritable variation in anatomical, physiological and other characters and (b) natural selection is the main driving force for evolutionary change in these characters. Darwin was clear that behaviour had been shaped by the same evolutionary processes as other characters (Darwin 1872), a theme taken up again later by the early ethologists (Chapter 1). We shall see how well the assumptions of the neoDarwinian synthesis apply to behaviour shortly. Since Darwin's theory of descent with modification is about continuity as well as change, however, we shall first look at how the evolutionary history of behaviour patterns can be traced.

2.2 Phylogeny and behaviour

We have little trouble convincing ourselves that organisms have evolved along various lines of descent from a common ancestor. Species simply share too many features of their biochemistry, anatomy and other characteristics for this not to have been the case. The fossil record even allows us to chart some of the pathways of descent, at least for

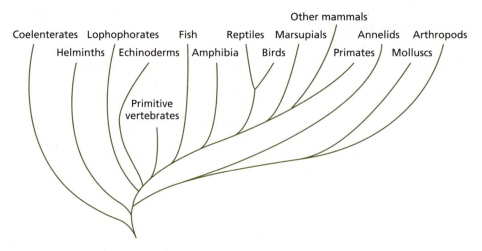

Figure 2.1 An example of a phylogenetic tree, in this case for some major groups of animals. After Ridley (1995).

those features, such as skeletal morphology, that have been preserved. The problem with behaviour is that we are largely stuck in the present. It seems highly unlikely that behaviour has not evolved along with the features preserved in the fossil record – indeed, many of them such as beaks, horns and limbs beg interpretation in terms of behaviour. Unlike anatomy and morphology, however, behaviour leaves little or nothing behind in the way of fossil evidence. We are left having to infer evolutionary history from the behaviour of modern organisms. How can we do it?

Schenkel's pheasants in the last chapter (Fig. 1.13) give us a clue: we can compare behaviours across groups of related species. In Schenkel's case, extreme courtship displays in peacocks and male pheasants were seen as elaborations of an ancestral food pecking response, still shown in a basic form by present-day jungle fowl and domestic chickens. Whether or not we are convinced by Schenkel's scenario, however, it unfortunately amounts to little more than imaginative story-telling as it stands. Why? Because it lacks a crucial ingredient: an independent **phylogeny**. In principle (and with sufficient imagination), we could posit any number of evolutionary routes through such a set of behaviours. To make a convincing case, we must show that any proposed sequence squares with other, independent, evidence of evolutionary change. A phylogenetic tree based on morphological or molecular evidence provides an effective comparator.

The essentials of constructing a phylogenetic tree can be gleaned from Fig. 2.1. The figure shows a phylogeny, based on morphology and molecular evidence, of some major groups of animals. Groups close together in the tree, e.g. birds, reptiles and amphibia, have more in common in terms of their morphology and molecular structure than any of them do with more remote groups, such as echinoderms (starfish, sea urchins and their relatives) or helminths (tapeworms, threadworms, etc.). We can infer that birds, reptiles and amphibians shared a common ancestor with each other more recently than with echinoderms and helminths because the arrangement in the tree minimises the number of evolutionary steps needed to account for their shared characters (the **principle of parsimony**). If we separated the three groups in different parts of the tree, perhaps lumping birds with helminths and reptiles with echinoderms, we would have to explain why basic shared characters such as a dorsal nerve cord and backbone evolved on three different occasions. (The principle of parsimony, it is worth emphasising, applies throughout scientific explanation, not just to deriving phylogenies.)

Once we have a phylogeny, we are in a good position to look at evolutionary relationships between behaviours. Behaviours shared by species close together in the tree are likely to be shared through recent common ancestry. Those shared by widely separated species, where species in between show very different behaviours, are likely to be shared by **convergence** (the independent acquisition of similar characters – e.g. flippers in penguins and porpoises – under similar selection pressures). To increase our faith in Schenkel's model for the evolution of phasianid courtship displays, we should need to map the different behaviours onto an independent phylogeny of the relevant species. While this has not yet been done for phasianids, it has been done for courtship displays in another group of birds, the Pelecaniformes (pelicans, tropicbirds, frigatebirds, gannets, cormorants and their relatives) (van Tets 1965; Kennedy *et al.* 1996).

2.2.1 Phylogeny and courtship displays

G.W. van Tets was interested in the evolution of social displays in pelecaniforms, a taxonomically complex group in which comparative studies of behaviour have been used to try to unravel the evolutionary relationships between different species. In particular, he looked at courtship displays, such as wing-fluttering and sky-pointing (raising the bill vertically to the sky) and throwing the head back, that seemed have their roots in flight intention movements (see Box 4.3).

Cormorants (*Phalacrocorax* spp.) engage in various forms of wing-fluttering and waving display. In the pelagic (*P. pelagicus*) and red-faced (*P. urile*) shags, for instance, the fluttering is rapid and vigorous, while other cormorants, such as the black cormorant (*P. carbo*), wave their wings more slowly. Anhingas (or darters) (*Anhinga anhinga*), which are thought to be closely related to cormorants, have a unique alternating wing-waving display, first waving one wing then the other. Gannets and boobies (*Morus* and *Sula* spp.), on the other hand, do not wave their wings, but instead show variations of pre-takeoff and sky-pointing behaviour.

As Schenkel had for pheasants, van Tets postulated an evolutionary sequence for these displays (Fig. 2.2a). His sequence moved from basic flight intention movements in pelicans, frigatebirds and tropicbirds (sharing the most distant common ancestor with cormorants in the group), through pre-takeoff and sky-pointing displays in gannets and boobies, to slow wing-waving in cormorants and, finally, rapid wing-fluttering in shags. The implication is that the rapid-fluttering modern shags had a recent ancestor that waved its wings more slowly, a more distant ancestor that showed slightly exaggerated wing movements associated with pre-takeoff behaviour, and a more distant ancestor still that just showed flight intention movements. All very plausible, but is there any *independent* evidence that it is true? A study by Kennedy *et al.* (1996) suggests there is.

While van Tets compared his behavioural relationships with phylogenetic trees (based on anatomy and egg-white proteins) proposed for the group at the time, these comparisons were not based on an attempt to find the best (i.e. most parsimonious, see above) tree. Kennedy *et al.* compared the suggested evolutionary sequence of displays with different phylogenies of pelecaniform birds based on DNA analysis, skeletal morphology and bone anatomy (Fig. 2.2b). All of the independently derived phylogenies matched van Tets's proposed relationships more closely than expected by chance. Kennedy *et al.* then combined the information from the trees to create the most parsimonious best-estimate tree, and mapped van Tets's proposed behavioural transitions onto it (Fig. 2.2c). Comparisons with alternative possible mappings suggested that van Tets's original proposal was

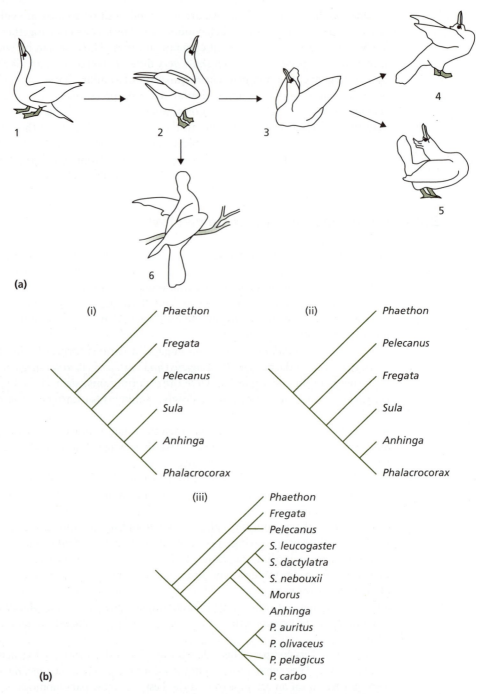

Figure 2.2 (a) van Tets's proposed transitions of courtship display characters form initial flight intention movements to (1) pre-takeoff (gannets, *Morus* spp.) to (2) sky-pointing (boobies, *Sula* spp.) to (3) slow wing-waving (some cormorants, *Phalocrocorax* spp.) and (6) alternating wing-waving (darters, *Anhinga*). Slow wing-waving then develops into (4) rapid wing-waving (some shags, *Phalocrocorax* spp.) and (5) throwback (European shag, *P. aristotelis*). From Kennedy *et al.* (1996) after van Tets (1965). (b) Independently derived phylogenies of the Pelecaniformes based on skeletal morphology (i), osteology (ii) and DNA analysis (iii). Taxa have been pruned in different ways to accommodate differences in available information in the various trees. After Kennedy *et al.* (1996).

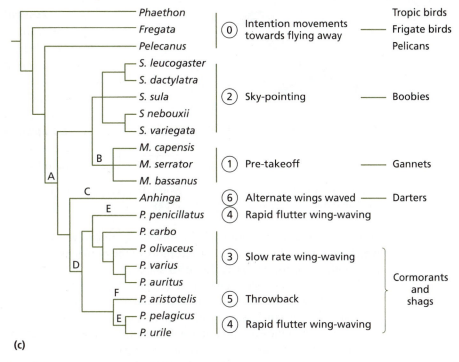

(c)

Figure 2.2 (c) The mapping of behavioural characters onto a best-estimate tree (see text), and the junctures (A–F) at which transitions can be proposed on the basis of van Tets's (1965) suggested sequence (see a). After Kennedy *et al.* (1996).

indeed the most plausibly parsimonious in terms of the number of independent character transitions and losses required.

2.2.2 Phylogeny and song development

Phylogenetic analysis helps us determine evolutionary pathways. But such pathways do more than just tell us what might have evolved from what. They can also indicate constraints on evolutionary change, features that are conserved from branch to branch in the tree despite changes in other features. Development can be one major source of constraint.

Development may limit novel evolutionary changes either because the same embryological or postnatal pathways serve a number of different but equally important traits, or because evolution occurs early on in the developmental process, thus making radical change, with its potential for disruption, unlikely. However, novel changes could arise by adding new stages onto the end of an existing developmental pathway (**terminal addition**), in which case the more derived species (those with the new addition) pass through the developmental and adult stages of more primitive species in their own development ('ontogeny recapitulates phylogeny' in the famous aphorism of Haeckel's 'biogenetic law'). One prediction arising from terminal addition and conservation of early development is that early developmental stages of traits will be generalised and widely distributed across species, while specialised features will come later and be more narrowly distributed (von Baer's law). Conversely, novel changes might be also brought

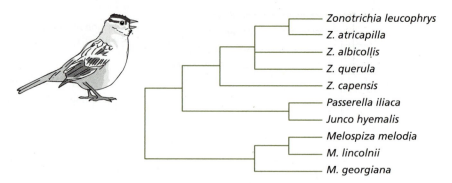

Figure 2.3 A phylogenetic tree for some North American sparrows based on a parsimony analysis of allozyme data. See text. After Irwin (1988).

about by truncating developmental pathways at an earlier stage (**paedomorphosis**), in which case derived species do not recapitulate the ontogeny of more primitive species, but simply reach adulthood at the earlier stage.

Theories based on such changes in development are familiar in studies of the evolution of morphological novelty, but their role in the origin of behavioural novelty is less well understood. Nevertheless, evidence is emerging that developmental constraints have been important in the evolution of new behaviour patterns. Rebecca Irwin's (1988) study of song development in sparrows is a nice example.

Bird song is a good vehicle for looking at developmental constraints because the development of song is well understood in certain groups of closely related species (Marler & Peters 1982), and there is evidence that various aspects of song remain stable (are conserved) through evolution (Payne 1986). Irwin looked at patterns of song development in oscine birds (a group of passerine [perching] birds with a complex syrinx [vocal apparatus]). In particular, she looked at a group of North American sparrows of the genera *Zonotrichia* (white-crowned sparrows), *Passerella* (fox sparrow), *Melospiza* (song and swamp sparrows) and *Junco* (juncos) for which she had worked out a parsimonious phylogeny based on enzyme electrophoresis (Fig. 2.3).

Among oscines in general, the pattern of song development was remarkably similar across species up to the point where species-specific song emerged. Songs in all species were longer and more continuously variable earlier in development, and progressed by increasing the repetition of particular syllables (sound elements) and reducing overall song length. Song development in some species, such as chaffinches (*Fringilla coelebs*) and song sparrows (*M. melodia*), stopped at the point where they sang many different songs, each consisting of more than one syllable type. In others, such as white-crowned sparrows (*Z. leucophrys*) and swamp sparrows (*M. georgiana*), change occurred after this stage, resulting in a reduction in song repertoire size to one (or rarely two) in *Z. leucophrys*, and in syllable types per song to one in *M. georgiana*. The more generally distributed stages of song development therefore occurred earlier in ontogeny, and specialisations occurred later. Song development in oscines thus appears to accord with von Baer's law. But does this square with independent phylogenetic evidence?

Irwin found out for the sparrow group by using the molecular phylogeny in Fig. 2.3. She found that assuming single song repertoires to be derived necessitated only one transformation of the tree (within the *Zonotrichia* group). Assuming it to be primitive, however, required two independent derivations of large song repertoires (in *Passarella/Junco* and *Melospiza*). The hypothesis that large repertoires are the primitive state was therefore more parsimonious. By the same token, assuming single-syllable songs to be derived required independent origins in two lineages (*Melospiza* and *Junco*), while assuming it to be primitive required independent origins in three (*Zonotrichia*, *Melospiza* and *Passarella*). Multi-syllabic song as the primitive state was also, therefore, more parsimonious. On the basis of these results, Irwin concluded that most modification of song in sparrows occurs at the end of the developmental process, so that songs with derived features, such as single-song repertoires and single-syllable songs, must first pass through stages similar to more primitive song. Developmental pathways for song among sparrows thus appear to constrain the possibilities for novel changes, either through inertia of the developmental process itself or through selection against major changes.

2.3 Variation and heredity

Having satisfied ourselves that we can say something about pathways of evolutionary change in behaviour, we can now ask *how* behaviours are passed down these pathways. While Darwin knew nothing about the basis of heritable variation, we now, of course, know that it broadly follows the principles laid down by Mendel. Mendel's principles are crucial to understanding how natural selection works, so we must be satisfied that they apply to behaviour. We can then see whether natural selection has indeed been the major shaping force in the evolution of behaviour.

2.3.1 Mendelian genetics and behaviour

While Mendel's classic experiments, and many of those that followed the rediscovery of his work in 1900, were concerned with morphological characters, it is reasonable to suppose that the same principles of inheritance work for behaviour. We shall look at evidence for this in a moment, but first we must introduce some basic terminology.

2.3.1.1 Genes and Mendelian inheritance

The basis of inheritance is the **gene**, a sequence of nucleotide bases in a molecule of DNA (deoxyribosenucleic acid) that codes for a particular protein. In most organisms, the long DNA molecule, comprising many thousands of nucleotide bases and therefore genes, is twisted and folded into a **chromosome**. Each gene occupies a particular position, or **locus** (plural **loci**), along the chromosome. Different species have different numbers of chromosomes and, in most cases, more than one version (most often two) of *each* chromosome. Organisms that have two copies of each chromosome are known as **diploid** organisms, those with three copies are known as triploids, those with four, tetraploids and so on. Organisms that have only one copy, such as the males of some social insects, are referred to as **haploid**. The copies of each chromosome are **homologous**; that is they carry the same sequence of genes. Importantly, however, they do not necessarily, or even usually, carry the same *version* of a gene at any given locus. Different versions of the same gene are known as **alleles** and each allele codes for a slightly different form of the protein produced by the gene. If homologous chromosomes carry the same allele, the individual is said to be **homozygous** at that locus; if they carry different alleles, the individual is **heterozygous**. But if homologous chromosomes carry different alleles, which one is going to be expressed and exert its effects on the organism (i.e. at the level of the **phenotype**)? One of Mendel's early discoveries (also pre-empted by Maupertuis) was that, when an individual carries two different alleles, one of them (the **dominant** allele) is usually expressed at the expense of the other (the **recessive** allele). Dominance interactions between alleles are a form of **epistasis** (effects of alleles on the expression of others in the same **genome**). A variety of epistatic interactions may occur between loci, each modifying the phenotypic expression of the genes involved. Since genes may have many different alleles, the effect of any given gene on an organism will vary according to which versions the organism happens to be carrying and the nature of their interactions with their own and other alleles.

Mendelian inheritance and behaviour in bees

A study of brood maintenance in honey bees (*Apis mellifera*) by Rothenbuhler (1964a,b) provides a nice illustration of simple Mendelian inheritance in behaviour. Figure 2.4 summarises Rothenbuhler's study alongside some of Mendel's classic results from peas for comparison.

Figure 2.4(a) shows the outcome of Mendel's crosses between two pure breeding lines of pea plants, one producing smooth round seeds, the other wrinkled seeds. When he crossed the two pure lines Mendel found that all the progeny (F1 generation) produced round seeds, but when he crossed F1 plants he found that the next generation (F2) produced both round and wrinkled seeds in the ratio of 3 round : 1 wrinkled. Mendel's results can be explained in terms of two alleles of a gene affecting the appearance of the seeds (Fig. 2.4a). We can denote the allele for roundness '*R*' and that for wrinkliness '*r*'. In the F1 generation all progeny inherit a copy of *R* from their 'round' parent and a copy of *r* from their 'wrinkly' parent (peas are diploid, so have two copies of the gene for seed appearance). The **genotypes** of these individuals are therefore all *Rr*. But all the F1 generation have round seeds, so the allele for roundness must be dominant (hence we represent it with a capital letter) and that for wrinkliness recessive. In the F2 generation, however, each parent had contributed both alleles instead of just one, resulting in a mixture of *RR*, *Rr/rR* and *rr* genotypes among the progeny (Fig. 2.4a). Because *R* is dominant to *r*, only the homozygous *rr* genotype expressed the wrinkled phenotype, thus yielding a ratio of 3 : 1 (in a sufficiently large sample) in favour of round seeds.

Figure 2.4(b) shows the segregation of 'hygienic' behaviours in Rothenbuhler's honey bees. Honey bee larvae are susceptible to various diseases, among them American foul brood caused by the bacterium *Bacillus larvae*. Some strains of honey bee are called *hygienic* because when larvae die inside their cells, workers uncap the cells and remove them. Dead larvae in *unhygienic* strains are left in their cells to decompose, thus risking the spread of disease. If *unhygienic* and *hygienic* strains are crossed, the F1 progeny are all *unhygienic*. *Unhygienic* behaviour is thus dominant and *hygienic* behaviour recessive. In his experiment, Rothenbuhler backcrossed his F1 bees (mated them with the homozygous recessive parental strain) with the following outcomes:

☐ Nine of the resultant colonies uncapped cells in which larvae had died but did not remove the corpses.

☐ Six colonies would remove dead larvae from uncapped cells, but would not uncap the cells themselves.

☐ Eight colonies were *unhygienic* and neither uncapped cells nor removed dead larvae if the cells were uncapped for them.

☐ Six colonies were *hygienic* and did both.

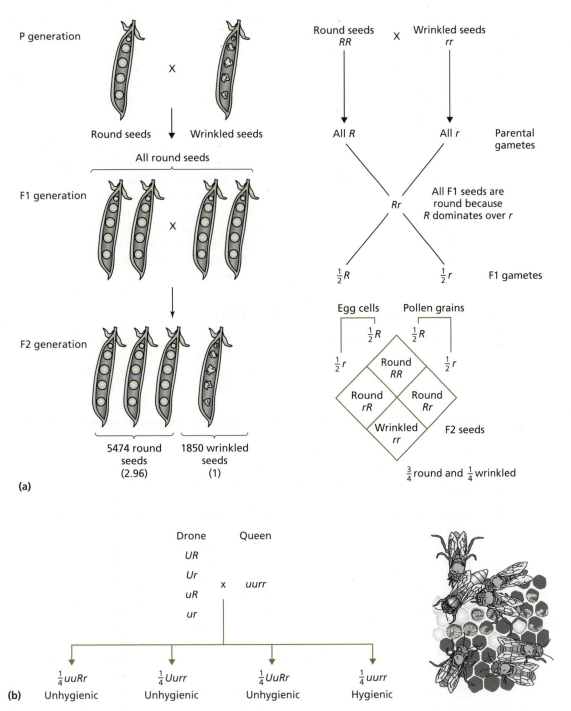

Figure 2.4 (a) The results of Mendel's experiment crossing true-breeding lines of peas, one producing round seeds, the other wrinkled seeds (after Stebbins 1982). The pattern of segregation suggests two alleles at a single locus, with the allele for round seeds (*R*) being dominant. (b) *Hygienic* behaviour in honey bees can be explained in terms of two alleles at each of two loci, one locus controlling uncapping cells, the other removal of dead larvae. *Unhygienic* behaviour is dominant at both loci. The figure shows Rothenbuhler's backcross experiment (see text) between drones of different genotype and recessive *hygienic* queens (after Rothenbuhler 1964b).

Statistically, the proportions of these four types of progeny did not differ from equality. The *hygienic* trait can thus be explained in terms of two independently segregating genes each with two alleles. One gene codes for uncapping behaviour as dominant *unhygienic* (*U*) and recessive *hygienic* (*u*) alleles, the other codes for removing the larva, again with *unhygienic* alleles (*R*) being dominant and *hygienic* (*r*) alleles recessive. Since honey bee workers are diploid, they possess two copies of each of the uncapping and removal genes. Workers of the pure breeding *hygienic* and *unhygienic* strains in Rothenbuhler's study therefore had the genotypes *uurr* and *UURR* respectively. To achieve the backcross, Rothenbuhler had to take into account the fact that male honey bees (drones) are haploid and have only one copy of each gene. Rothenbuhler had to cross 29 different F1 drones with *uurr* queens to produce the full range of backcross colonies. Among them were the four possible male genotypes for *hygienic* behaviour: *UR*, *ur*, *Ur* and *uR* (Fig. 2.4b). While the pattern of inheritance in the backcrosses suggests two independently segregating loci, the *U/u* and *R/r* genes may not actually code for the behaviours themselves. *Unhygienic* workers do in fact perform *hygienic* activities but at a very low frequency, one possibility being that the *U/u* and *R/r* loci act as switches controlling the expression of a number of other loci in an almost all-or-nothing way.

2.3.1.2 Quantitative characters and heritability

We can therefore demonstrate Mendelian patterns of inheritance for a particular behaviour. A problem, however, is that many phenotypic characteristics of organisms do not segregate in the neat, clear-cut fashion of Mendel's peas and Rothenbuhler's bees. Many, like body weight or the size of appendages, do not occur in two discrete states but vary more or less continuously. The same is true for most behaviours. Does this mean that Mendel's principles do not apply here? No. What it implies is that such characters are controlled by a large number of genes (and/or environmental factors) and not just by alleles at one or two loci, though, in some cases, clusters of loci form **linkage groups** and segregate as if they were a single locus.

The problem with such quantitative, or **polygenic**, characters is that many different combinations of alleles can give rise to similar phenotypes, so that it is virtually impossible to determine the precise genotype of an individual simply from breeding experiments such as those carried out by Mendel. We can still probe the genetic basis of polygenic characters using breeding experiments, but what we seek is the **heritability** of a character rather than the identity of the alleles coding for it.

Heritability is calculated as the proportion of variation in the character that is attributable to genes, and thus takes a value between zero and one. Heritability is zero if none of the variation is genetic. The fact that different individual blackbirds (*Turdus merula*) in a population build their nests out of different mixtures of plant material, for instance, is likely to be due to local differences in the availability of materials rather than genetic differences between blackbirds; thus choice of material will have zero heritability. At the other extreme, the heritability of albinism in blackbirds is likely to be one: differences in plumage colour between albino and normal blackbirds are entirely genetic. In between are characters such as body size and the number of eggs laid in a clutch, in which both genetic and environmental factors are likely to affect the final outcome.

The usual way to measure heritability is to breed selectively from parents that differ in the character in question, then either look for a correlation in the character between parent and offspring (Fig. 2.5a), or see whether mating parents with extremes of the

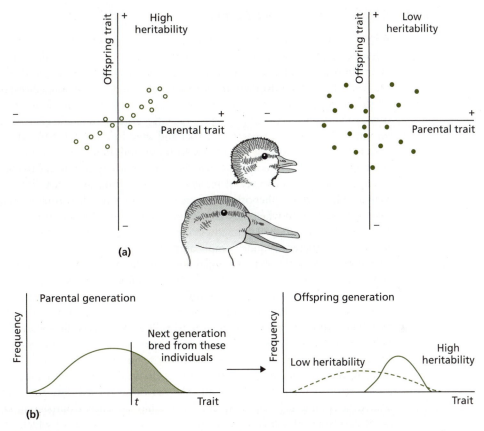

Figure 2.5 (a) Heritability of a given trait can be demonstrated as a correlation between the trait in a parent and the same trait in its offspring. A close correlation (left) indicates high heritability, while a poor correlation (right) indicates little or no heritability. (b) Heritability can also be demonstrated as a shift in the mean value of a trait as a result of selective breeding. After Ridley (1995).

character over several generations shifts its mean value in the population in the same direction (Fig. 2.5b). The closeness of the correlation between parent and offspring values, or the extent to which the population responds to repeated selective breeding, indicates the degree of heritability in the character in question.

Having introduced the basic principles of Mendelian genetics and seen that they can be applied to behaviour, we now look at the types of evidence that suggest differences in behaviour between individuals more generally have a genetic basis.

2.3.2 Evidence that differences in behaviour are due to genes

The evidence for a genetic basis to behavioural variation is diverse, and much of it, it has to be said, holds little interest for the evolutionary biologist since it emerges from gross disruptions to normal behaviour or other dysfunctional effects (see e.g. van der Steen 1998). However, other lines of evidence have considerable functional/evolutionary interest, and dysfunctional effects can, as we shall see in Chapter 5, have much to say about the cellular developmental processes that lead to changes in behaviour.

2.3.2.1 Single genes and behaviour

While no behaviour is encoded at a single locus in a literal sense, there is a wealth of evidence that changes at single loci can affect behaviour through the cascade of developmental processes influenced by their expression. Indeed, we have already seen an example for two behaviours in Rothenbuhler's study of brood care in honey bees, though the role of the loci in question in controlling the expression of the behaviours in Rothenbuhler's bees remains unclear. Until relatively recently, studies of single gene effects have relied on mutations arising at particular loci and then looking for differences in behaviour between mutant and non-mutant (wild-type) individuals. Useful mutations can arise spontaneously, or, more usually, artificially by exposing organisms to chemical or other mutagens. A useful property of many single gene mutations affecting behaviour is that they also have other phenotypic effects (i.e. they show **pleiotropy**) so that carriers are easy to identify even when they are not performing the behaviour in question. However, as we shall see in Chapter 5, such pleiotropic effects can also make the interpretation of behavioural differences quite complicated.

A classic example of a single gene mutational effect comes from Bastock's (1956) study of courtship in male fruitflies (*Drosophila melanogaster*). Bastock investigated effects on mating success of a sex-linked (carried on one of the so-called X or Y sex chromosomes) recessive mutation, the most obvious effect of which was to change the fly's body colour from wild-type grey to yellow. Wild-type flies were crossed with the *yellow* line for seven generations to make sure flies were genetically similar except at the *yellow* locus. When males from wild-type and *yellow* lines were allowed to court females, Bastock found that males carrying the *yellow* allele were much slower to mate and sometimes failed to mate at all. The reason, it turned out, was that *yellow* males were deficient in crucial elements of the courtship sequence. Courtship in *D. melanogaster* involves a series of displays and responses by the male, three of which are illustrated in Fig. 2.6. Failure to perform adequately at any stage could compromise a male's chances of mating. When Bastock analysed the courtship sequence in her flies, she found that the 'licking' (contact between the male's proboscis and the female's genitalia) and 'vibration' (wing movement) phases were longer or more frequent in wild-type males (Fig. 2.6b). Unfortunately for *yellow* males, it seems that both 'licking' and 'vibration' serve to stimulate the female and are a necessary prelude to mounting, so courtship tends to abort if they are not performed properly.

As well as demonstrating a behavioural effect of the *yellow* mutation, Bastock's study also showed that the effects of the *yellow* locus depend on the genetic background. Table 2.1 compares the percentage mating success of wild-type × *yellow* crosses in stocks where wild-type and *yellow* lines had not been crossed at all and stocks that had been crossed for seven generations (thus producing similar genetic backgrounds). In the uncrossed stock there was a significant effect of female genotype (62% *vs* 87% and 34% *vs* 78% in the reciprocal sex/line pairings), whereas no such effect occurred in the crossed stock. This illustrates an important point: the effects of particular alleles on individual reproductive success (and therefore their own fitness [see 2.4.3]) are not absolute but depend on the genetic environment in which they are being expressed. We shall come back to this when we consider the process of adaptation.

Various other single gene mutations in *Drosophila* have been shown to affect behaviour. *Coitus interruptus* (in which the male dismounts from the female after only half the normal copulation time) and *stuck* (where the opposite happens – the male fails to dismount even after the normal copulation time) are two notable ones affecting

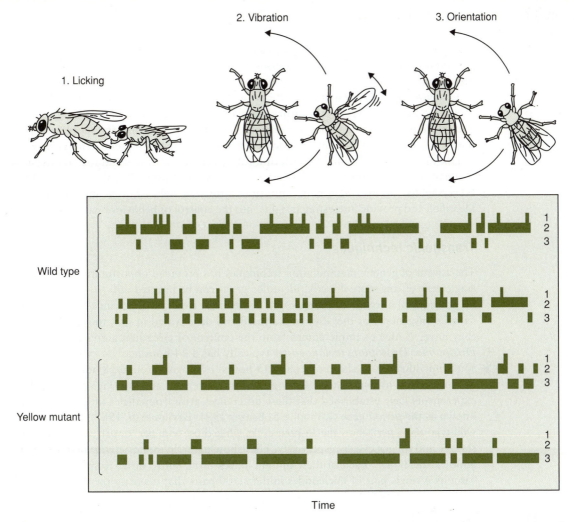

Figure 2.6 The temporal patterning of licking (1), vibration (2) and orientation (3) in the courtship sequence of male *D. melanogaster* of wild-type and *yellow* stocks. Note the shorter bouts of licking and vibration in *yellow* males (see text). After Bastock (1967).

Table 2.1 The percentage mating success of wild-type and *yellow* mutant male *Drosophila melanogaster* in within and between line crosses and before or after lines had been crossed to generate similar genetic backgrounds. See text. From Bastock (1956).

Before crossing wild stock	After crossing wild stock	
Matings	with *yellow* stock for seven generations	with *yellow* stock for seven generations
Wild male × wild female	62	75
Yellow male × wild female	34	47
Wild male × *yellow* female	87	81
Yellow male × *yellow* female	78	59

sexual behaviour. Many others, such as *hyperkinetic, easily-shocked, non-phototactic* and *dunce*, affect movement, responses to environmental stimuli, learning or other aspects of behaviour.

Of course, single gene mutation effects are not limited to *Drosophila*, but in the main they have been studied in laboratory stalwarts such as fruitflies, laboratory rodents and, more recently, nematodes (e.g. Thomas 1990). One reason is the simple practical one of generating and screening sufficient mutations to identify useful genotypes. Even in laboratory mice, attainable mutation rates, generation times and the colony sizes required can be an enormous disincentive for this kind of study (Takahashi *et al.* 1994). Nevertheless, many single gene mutations affecting behaviour in mice have been identified and provide interesting insights into the control of behaviour patterns.

Transgenic techniques

The advent of genetic manipulation techniques has provided opportunities for probing single gene effects more directly. Identified genes can be inserted into (so-called 'knock-in' procedures) or removed ('knock-out' procedures) from the organism's cells, a process called **transgenesis**, or they can be activated and deactivated *in situ*, to see what effect they have. A nice example comes from the control of circadian activity rhythms in *Drosophila*. *Drosophila melanogaster* typically has a 24-hour activity cycle. However, some individuals have shorter (around 19 hours) or longer (around 28 hours) cycles, or do not cycle at all but instead show a random pattern of activity (Fig. 2.7). Breeding experiments have established that these differences arise from alleles of the same gene, known as the *period* gene (Konopka & Benzer 1971; Baylies *et al.* 1987). Flies with the wild-type allele, *per*⁺, show the 24-hour cycle, while those with *per*ˢ or *per*ᴸ alleles show the shorter and longer cycles respectively. Flies showing arrhythmic activity patterns carry another allele, *per*ᵒ. Each of these mutant alleles turns out to differ from the wild-type form by a single pair of nucleotides in the 3,500 pairs that make up the gene (Yu *et al.*

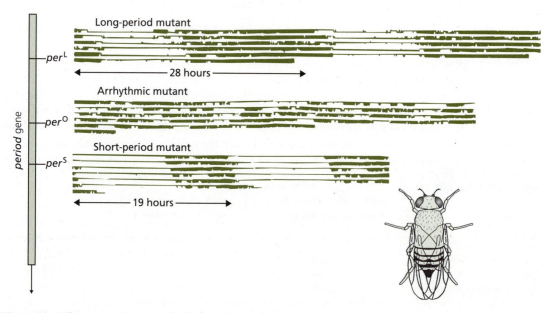

Figure 2.7 The temporal pattern of activity in *Drosophila melanogaster* with different mutations of the *period* gene (see text). After Konopka & Benzer (1971) and Baylies *et al.* (1987).

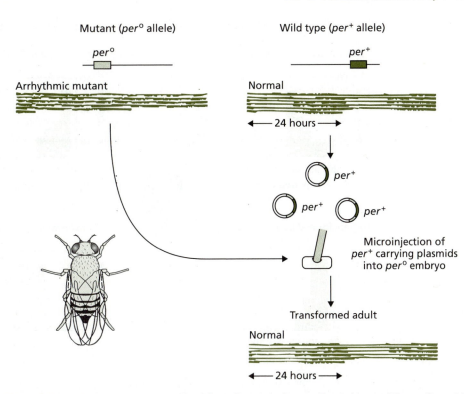

Figure 2.8 Arrhythmic *per°* mutant fruit fly embryos can be transformed into wild-type flies with a normal activity pattern by microinjecting plasmids containing the wild-type *per+* gene. After Konopka & Benzer (1971) and Alcock (1998).

1987). By transplanting the fragment of the wild-type fly's chromosome that contains the *per+* allele into a virus-like vector known as a plasmid, it is possible to transfer the wild-type allele into the cells of a developing mutant fly. When Zehring *et al.* (1984) did this with mutant *per°* (arrhythmic) flies, they found that the adults subsequently exhibited a normal 24-hour activity pattern (Fig. 2.8). The arrhythmic mutants had therefore been **genetically transformed** into wild-type flies.

The *period* gene story also illustrates the pleiotropic nature of many mutations affecting behaviour. In *Drosophila* the gene affects not only activity cycles but also the timing of eclosion (hatching from the pupa) and male courtship 'song' (pulses of sound produced during the wing vibration phase of courtship; see Fig. 2.6a). Variation in song can be detected as differences in the intervals between pulses of sound, usually measured as the inter-pulse interval (IPI), the gap between successive bursts of sound, and the IPI period, the cycle of increasing and decreasing IPIs as the song progresses. *Drosophila melanogaster*, for example, has IPIs of around 30 milliseconds and an IPI period of about 60 seconds, while *D. simulans* has IPIs in the region of 50 milliseconds and an IPI period of 35 seconds. Both inter- and intraspecific differences in song pattern appear to be under the control of the *period* gene. Thus male *D. melanogaster* with the wild-type *per+* genotype sing normal *D. melanogaster* song (similarly in *D. simulans*), while those with the mutant *per°* sing arrhythmic song, where the normal pattern of ISIs and ISI periods is broken. Gene transfer experiments have shown that, as with activity patterns, song can be altered predictably by manipulating the *per* genotype (Wheeler *et al.* 1991). Figure 2.9 shows a reciprocal transformation of *per°* *D. melanogaster* and *D. simulans* males by transferring the *per+* allele of their own or the other species. Thus *per°* male

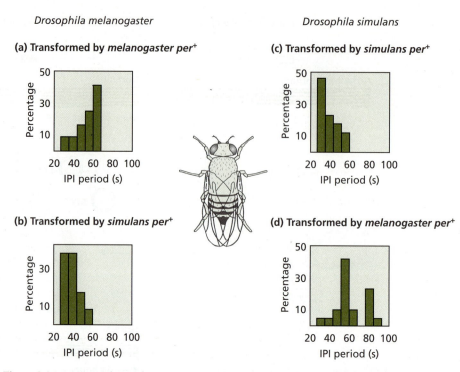

Figure 2.9 Reciprocal transformation of song pattern in male *Drosophila melanogaster* and *D. simulans*. Arrhythmic *per°* flies of each species can be transformed to sing the wild-type song of their own or the other species by transferring the appropriate *per+* gene. (a) and (b) show results for *D. melanogaster* and (c) and (d) for *D. simulans*. Reprinted with permission after 'Molecular transfer of species-specific behaviour from *Drosophila simulans* to *Drosophila melanogaster*', *Science*, Vol. 251, pp. 1082–5 (Wheeler, D. A. *et al.* 1991). Copyright 1991 American Association for the Advancement of Science.

D. melanogaster could be induced to sing normal conspecific song (Fig. 2.9a), or the normal song of male *D. simulans* (Fig. 2.9b), depending which allele was transferred. The same was true for *D. simulans* (Fig. 2.9c,d).

'Knock-in' and 'knock-out' procedures are also being used extensively in mice to probe the genetic underpinnings of behaviour, particularly sexual behaviour, aggression and learning (e.g. Rousse *et al.* 1997; Gimenez-Llort *et al.* 2002; Scordalakes *et al.* 2002), though questions have been raised as to whether behavioural differences can always safely be attributed to the engineered mutations concerned (e.g. Cook *et al.* 2002; Rodgers *et al.* 2002). Despite some fringe doubts, however, genetic manipulations are allowing refined dissection of some of the physiological pathways influencing behaviour, an issue to which we shall return in Chapter 5.

2.3.2.2 Multiple genes

While single gene mutations show that small differences in genotype can have pronounced effects on behaviour, many, if not most, behavioural differences are likely to be underpinned by many different loci with complex epistatic relationships between them. The evidence for multiple gene effects is diverse, and much of it long established in the behaviour genetics literature. Some of the key approaches are summarised in Box 2.2, and many other examples can be found in textbooks of the field such as

Supporting evidence

Box 2.2   Evidence for multiple genes encoding behaviour

Differences between strains

Genetic strains within species usually differ at many loci and provide a wealth of evidence for a genetic basis to different behaviour patterns. The genetic 'purity' of strains depends on the level at which they are defined. Thus strains in terms of different populations of the same species will differ more widely in their genotypes than those derived from controlled laboratory breeding programmes. Because of this, most work has been done with inbred laboratory strains. Serial inbreeding increases homozygosity and thus genetic uniformity within lines, but also, because the particular alleles that go to fixation at any given locus is a matter of chance, differences between lines. Extensive work with laboratory mice have shown differences between strains in, among other things, aggressiveness (Southwick 1968), mating behaviour (McGill 1970) and maze learning (Upchurch & Wehner 1988; Rousse *et al.* 1997), as well as various behavioural pathologies such as audiogenic seizure (seizures induced by loud sound; Fuller & Thompson 1978). Such differences are evident even when post- and prenatal experience is accounted for (e.g. by cross-fostering [Southwick 1968] or ovarian transplant [DeFries *et al.* 1967; Robertoux & Carlier 1988]). Local population-level strains among various invertebrates have similarly revealed differences in, e.g., mating behaviour (the mosquito *Aedes atropalus*; Gwadz 1970) and foraging communication (honey bees, *Apis mellifera*; Dyer & Seeley 1991).

Effects of chromosomal changes

Differences in behaviour can sometimes be traced to chromosomal abnormalities. Many of these are deleterious and, if not lethal, give rise to grossly dysfunctional phenotypes. Abnormalities take three basic forms: (a) a change in the entire complement of chromosomes (**euploidy**) from, say, diploid (2*n*) to haploid (*n*) or triploid (3*n*) (multiplication of the complement is known as **polyploidy**), (b) the addition or loss of a single chromosome (**aneuploidy**) and (c) chromosome breakage. Evidence for relationships between polyploid events and behaviour is thin, but there are suggestive associations with dietary preferences in diploid and triploid whiptail lizards (*Cnemidophorus tesselatus*) (Paulissen *et al* 1993) and call structure in polyploid tree frogs (*Hyla versicolor*) (Keller & Gerhardt 2001). In contrast, aneuploid changes are associated with a wide range of behavioural effects, mostly pathological, such as the cognitive, motor and sexual retardation accompanying an extra copy of chromosome 21 (Trisomy 21 or Down's syndrome) in humans and chimpanzees (*Pan troglodytes*), and the personality and spatial orientation problems associated with changes in the complement of sex chromosomes (e.g. Turner's and Klinefelter's syndromes). Changes through chromosome breakage can occur in four ways (Fig. (i)). Inversions are of some significance because the reversal of genetic material prevents normal crossing over during meiosis so that inverted sequences are transmitted as single units (linkage groups). Inversion karyotypes appear to be maintained in various natural populations of flies through their effects on mating speed (Brncic & Koref-Santibañez 1964; Speiss & Langer 1964, 1966; Crean *et al*. 2000).

Hybridisation and polygenic traits

While differences at the level of chromosomes usually involve multiple loci, the behavioural changes associated with them may, of course, reflect the addition, subtraction or spatial

Box 2.2 continued

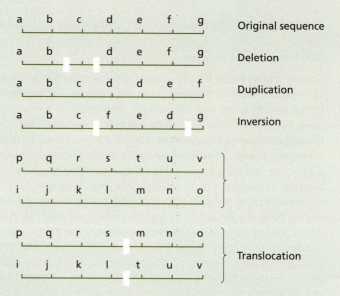

Figure (i)   The four major changes in gene sequence arising from chromosome breakage: deletion, duplication, inversion and translocation. After Ehrman & Parsons (1981).

rearrangement of one particular gene. In principle, a single gene difference could also underlie a behavioural difference between inbred strains. Other evidence, however, demonstrates very clearly that many behaviours are under the influence of several different loci (i.e. are polygenic). Hybridisation is one approach that has been used widely to look at such traits. When two closely related species with different forms of a particular behaviour are crossed, the pattern of segregation of the behaviour in the resulting offspring tells us something about the number of loci influencing it. In Dilger's (1962) classic study of nest-building in lovebirds (*Agapornis* spp.), for example, hybrid crosses between peach-faced (*A. roseicollis*) and Fischer's (*A. fischeri*) lovebirds yielded offspring with a variety of (usually dysfunctional) combinations of nest-building behaviours characteristic of the two parental species, implying that nest-building in the two species was controlled by several independently segregating loci. The same was true for other behaviours in the birds, such as courtship (Dilger 1962). Crosses between inbred strains of mice have shown that various strain differences in behaviour have similarly scrambled patterns of inheritance, again implying polygenic control (see Ehrman & Parsons 1981).

Ehrman & Parsons (1981). But while approaches such as those in Box 2.2 can tell us that many genes underlie behaviour, they do not tell us how many genes are involved or where they are on the animal's chromosomes. Some steps towards this have been made by so-called **cosegregation** studies, in which the pattern of segregation of behaviours and genetic markers yield some clues as to the loci influencing behaviour (e.g. Hunt *et al.* 1995), but recent developments in molecular genetic techniques may now allow us to say much more.

Quantitative trait loci

Quantitative trait locus (QTL) analysis, developed in its present form in the late 1980s, is one of a number of techniques that allows us to trace the different genes, or at least their position on particular chromosomes, underlying polygenic traits. QTL analysis can be done on any species for which there are inbred strains. The basic procedure is summarised, using maze-learning in mice as a hypothetical example, in Box 2.3. The process uses cross-breeding between strains and strain-specific genetic markers to seek associations between genotype at particular points along an organism's chromosomes and the possession of a given phenotypic trait. QTLs emerge as a series of locations, often on several chromosomes, that contribute to the trait. Several behavioural traits have been traced to particular chromosomal positions using the approach. Fear conditioning is one example.

Inbred strains of mice differ in their tendency to show conditioned (trained to be expressed in response to particular stimuli; see Chapter 6) fear, suggesting a degree of genetic influence. Wehner and colleagues have looked at differences in fear responses under different contextual and stimulus conditions in a number of inbred strains. In BXD recombinant mice, correlations between strain genetic markers and conditioning responses suggested multiple QTLs. The strongest associations were on chromosomes 1 and 17, for freezing to context. However, associations also emerged on chromosome 12 for freezing to an altered context, and again on chromosome 1 for responses to an auditory stimulus (Owen *et al.* 1997). Using C57BL/6J and DBA/2J mice, QTLs for contextual conditioning mapped to chromosomes 10 and 16, with further suggestive sites on chromosomes 1, 2 and 3 (Wehner *et al.* 1997).

Tafti *et al.* (1997) traced QTLs associated with patterns of sleep in CXB recombinant mice. Sleep during the light period was associated with loci on chromosome 7, while that in the dark period mapped to chromosome 5, near the locus controlling activity cycles (the mouse equivalent of the *period* gene in *Drosophila* [see above]). The periodicity of sleep was influenced by yet other sites on chromosomes 2, 17 and 19.

More recently, QTL analysis has been used in *Drosophila* itself, and has indicated a large phenotypic effect of three chromosomal locations on courtship song (Gleason *et al.* 2002). Interestingly, these mostly differ from candidate 'song' genes that had been proposed by earlier work demonstrating underlying polygenic control of song.

QTL analysis is thus beginning to yield some insights into the distribution of loci influencing complex behaviour patterns, but also to throw up more questions. Depending on the complexity of the trait, however, the technique can be very demanding of time and resources. In particular, screening programmes may need to be enormous in order to achieve satisfactory statistical rigour in associating loci with traits. Furthermore, once putative QTLs have been identified, further tests are needed to pin down the genes involved.

2.4 Natural selection and 'selfish genes'

So, the evolutionary history of behaviour can be deduced and we can demonstrate the heritable genetic basis on which such a history depends. But how has this heritable variation produced the dramatic diversity of behaviours we see today? The prevailing consensus is that most evolutionary change, certainly all *adaptive* evolutionary change, is driven by the process first recognised by Darwin and Wallace as natural selection. The process is easy to understand and, following Krebs & Davies (1993), can be boiled down to a handful of simple assumptions. In each case the familiar summary in terms

Underlying theory

Box 2.3   Quantitative trait locus (QTL) analysis

Two inbred strains are chosen which differ in a trait of interest, say speed of learning a maze task (Fig. (i)). The two strains, 'fast' (F) and 'slow' (S), are crossed so that the F1 progeny possess a homologous chromosome from each parental strain for each pair of chromosomes (1). F1 mice are then backcrossed to the parental strains to produce F2 individuals

Figure (i)

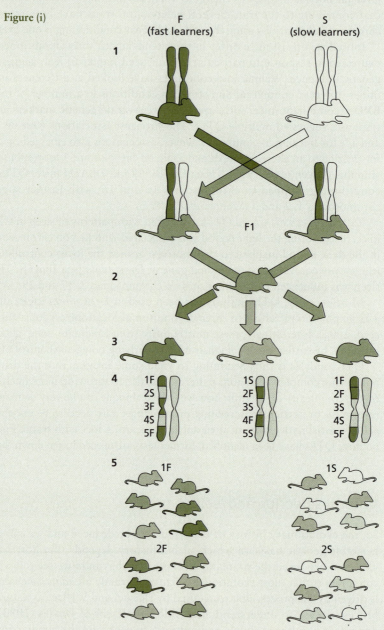

with one homologue of each pair that combines material from both parental strains (through recombination during meiosis), and one homologue which is from only one of the parental strains (2). F2 mice are then tested in the maze task and ranked according to their performance (degree of shading in 3). To find the QTLs that contribute to maze-learning, the genomes of the F2 mice are screened and sorted for genetic markers specific to the two parental strains (4) and the occurrence of F or S strain loci at each marker noted. Maze performance is then checked (5) and if F-type material at a particular marked spot is associated with fast learning, that particular section of the chromosome represents a QTL that may contain genes contributing to maze learning.

of individuals and populations (which Darwin would still recognise) is recast in the modern language of genetics:

☐ Individuals within species populations differ in morphology, physiology, behaviour and other characteristics. Populations thus show *variation* in these characteristics.

   ■ *Organisms possess genes which code for protein synthesis and thus regulate the structure, physiology and, through these, behaviour of the organism. Many genes occur as two or more alleles within the population, which code for slightly different versions of the same protein. These cause differences in development, so individuals within the population will differ in various phenotypic characters.*

☐ This variation (or at least some of it) is *heritable*, so that offspring by and large resemble their parents more than other members of their population.

   ■ *Copies of each allele are passed from parent to offspring, which thus tend to share the developmental and phenotypic characteristics of their parents.*

☐ Resources such as food, water and mates are limited, and organisms tend to produce far more offspring than will eventually reproduce. There is therefore *competition* for resources (a 'struggle for existence' in Darwin's terminology) and many individuals die or fail to reproduce.

   ■ *The likelihood of any given allele being passed on to the next generation depends on whether its bearer manages to reproduce.*

☐ Some individuals (variants within the population) happen to be better at coping with competition (are better *adapted*) than others and will therefore leave more offspring. These will inherit the characteristics of their parents, which thus become more common in the population (*natural selection* has occurred).

   ■ *Alleles within the population gene pool 'compete' for representation at their locus. Alleles that improve the chances of their bearer reproducing pass more copies to the next generation and may replace other, competing, alleles at their locus. Natural selection is thus the differential survival of alternative alleles.*

☐ As a result of natural selection and ongoing variation, organisms will gradually become adapted to their environment.

   ■ *Mutation is likely to generate new alternative alleles from time to time, so continuing selection among alternatives will hone the reproductive performance of their bearers in their particular environment.*

Natural selection thus describes changes in allele frequencies within population gene pools through the effect of different alleles on the reproductive success of their bearer individuals. Dawkins (1989) has characterised the process in his immortal metaphor of the 'selfish gene'. Because alleles are selected through their effects on individual phenotypes, individuals will come to possess those characteristics that promote the propagation of their alleles; i.e. individuals will appear to act in the best interests of their genes (hence the 'selfish' metaphor). Individuals are thus best regarded as the devices through which alleles engineer their transmission from one generation to the next. Individuals reproduce *within* generations, but alleles replicate, multiply and decline *across* generations (the distinction between 'vehicles' and 'replicators' in Dawkins's terminology). This is the essence of the **gene selection** view of evolution. While Dawkins's innocuous metaphor still generates misunderstanding in some quarters, its gene-centred view has spawned enormous leaps of understanding in evolutionary biology over the last three decades, many of them stemming from studies of behaviour. Before we discuss some of these, however, we must first make a further distinction, then see whether there is any evidence that genetic variation in behaviour responds to selection as neoDarwinian theory assumes.

2.4.1 Natural selection and sexual selection

Natural selection happens when different alleles have different consequences for the reproductive success of their bearers. Darwin (1859, 1871) recognised that individuals, at least in sexually reproducing species, could improve their chances of reproducing in one or both of two ways: first by competing successfully in the 'struggle for existence', thus reproducing by dint of surviving and acquiring resources, and, second, by competing successfully for mating opportunities. Darwin saw the latter as giving rise to a special form of selection which he called **sexual selection**. While selection acting through the 'struggle for existence' led to adaptations that aided survival, such as efficient feeding mechanisms, means of escaping predators, insulation against cold and so on, sexual selection produced characters that (a) aided competition within one sex (usually males) for access to the other (**intrasexual selection**), or (b) enhanced the attractiveness of individuals of one sex (again, usually males) to members of the other (**intersexual** or **epigamic selection**). Thus, intrasexual selection tends to favour characteristics such as large size, weaponry (antlers, horns, etc.), mate-guarding behaviour and other features that give a competitive edge, while intersexual selection leads to the elaborate adornments and displays that appear to be the basis of mating preferences in many species (Fig. 2.10). However, these distinctions are far from absolute, and the same characteristics can serve either or both functions (see Chapter 10).

Because sexually selected characters often appear to conflict with the survival interests of their bearers – antlers squander expensive resources, bright plumage is conspicuous to predators, etc. – some people regard natural and sexual selection as fundamentally different processes. However, from an evolutionary perspective, survival is valuable only to the extent that it contributes to reproduction. Since both processes act through differential reproductive success, there seems little at a fundamental level to distinguish them. How much an individual should invest in the 'struggle for existence' versus the 'struggle to mate' will depend on the relative contribution each makes to the individual's lifetime reproductive output. We shall return to this later when we discuss life history strategies. Sexual selection is discussed more fully in Chapter 10.

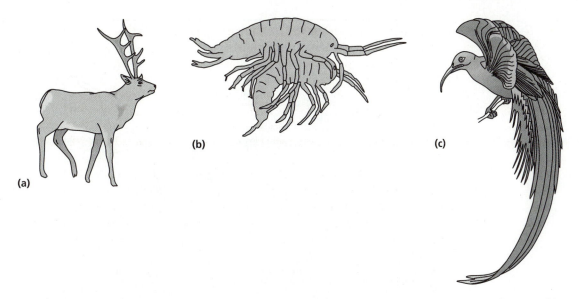

Figure 2.10 Some typical sexually selected characters. (a) Antlers in many large herbivorous mammals are used by males to fight for access to females, while various mate-guarding behaviours, such as that of the freshwater shrimp *Gammarus pulex* (b), are designed to reduce post-copulatory competition from the sperm of rival males. Elaborate adornments, such as the long tail of male birds of paradise (c) are generally assumed to have evolved through female choice. See text and Chapter 10. After Gould & Gould (1989) and Krebs & Davies (1993).

2.4.2 Responses to selection

Is there any evidence that variation in behaviour responds to selection? We shall look at two lines of evidence that suggest it does: the response of behavioural traits to artificial selection, and heritable differences in behaviour associated with presumed differences in selection pressure in the natural environment.

2.4.2.1 Responses to artificial selection

Artificial selection usually involves taking a genetically variable population, testing individuals for the behavioural characteristic of interest, say aggressiveness or learning speed, then pairing high-scoring males with high-scoring females and vice versa to create opposing selected lines. The procedure is repeated over a number of generations to see whether the average score of the population increases or decreases accordingly (see Fig. 2.5b). It is also good practice to create a third line of randomly mated pairs to monitor any underlying drift in the selected character over the period of the experiment. Such experiments rarely fail to show changes in the selected direction, implying that much of the continuous variation in behaviour within populations is genetic and able to respond to selection. Some examples are shown in Fig. 2.11.

Of course, many of the behaviours selected for or against in these kinds of experiment are complex characters that are the end product of many underlying processes, quite possibly different in different individuals. High learning ability, for example, may reflect good memory, attentiveness, boldness in exploration or a number of other attributes. An initial trawl of good learners and poor learners from a starting population could include any or all of these causes, and it can be instructive to see what is selected

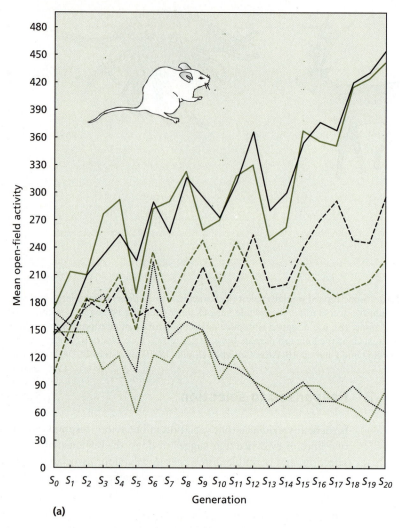

(a)

Figure 2.11 Some examples of artificial selection for behavioural attributes. (a) Results of selection for increased (solid lines) and reduced (dotted lines) open field activity (locomotory activity in an open arena) in laboratory mice. The dashed lines are randomly mated controls. After DeFries *et al.* (1974).

along with the trait of interest as the selection process proceeds. A nice example comes from a study of tameness in silver foxes (*Vulpes vulpes*) by Dmitry Belyaev (1979).

Belyaev was interested in the evolution of domestication in dogs. As predators and scavengers, often hunting in packs, dogs are not the most obvious candidates for companionship with humans, and indeed the behaviour of wild dogs towards people is vastly different from that of the familiar domestic animal. The word we use to sum up the difference is 'tameness'. Generally, the tamer an animal the less likely it is to attack us or run off, and the more likely it is to appear friendly. But these are complex traits, and only a handful of those we might want to incorporate into our estimate of tameness in any particular case (Price 2002). So what *does* constitute tameness and what changes as tameness evolves? To find out, Belyaev conducted an artificial selection experiment with his foxes.

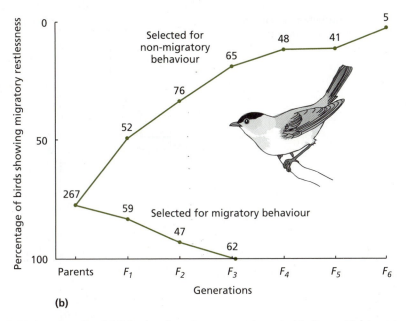

(b)

Figure 2.11 *(continued)*   (b) Selection for migratory tendency in blackcaps (*Sylvia atricapilla*). Starting from a parental population in which about 75% of birds migrate, lines of non-migrators could be produced in six generations of selection, and lines of 100% migrators in three. Numbers indicate how many birds were reared in each generation. Note the reversed scaling of the *y*-axis. Reprinted with permission after 'Genetic basis of migratory behavior in European warblers', *Science*, Vol. 212, pp. 77–9 (Berthold, P. and Querner, U. 1981). Copyright 1981 American Association for the Advancement of Science.

Silver foxes are close enough to the domestic dog's jackal ancestor to provide a reasonable starting point for a selection experiment. They are also bred commercially for their fur, so were readily available for Belyaev's experiment. Belyaev's first problem, however, was to decide what to select. Since tameness is a complex character, his approach was to select on the basis of a broad range of aggressive, fearful, friendly and inquisitive responses towards the experimenters. He then selectively mated the tamest foxes by these criteria each generation and continued until the process had been carried out for 18 generations. Figure 2.12 summarises the outcome. The *x*-axis in the figure represents Belyaev's composite 'tameness' score, with tameness increasing to the right and decreasing to the left. As the generations go by, the distribution of the tameness score shifts to the right, from the aggressive end of the scale in the original, unselected population towards tamer values by generation 18. Thus variation in tameness among the foxes had a genetic basis and responded to selection. What was especially interesting, however, was that, along with the behavioural characteristics actively selected by Belyaev, came a range of other behaviours which echoed those of domestic dogs, for example approaching people and licking their hands and faces, or barking and tail-wagging when a person came into view. How did these associated changes come about?

As well as tracking changes in behaviour in his fox populations, Belyaev also measured changes in various hormones, among them serotonin, which is known to inhibit certain kinds of aggression and play an important role in the central regulation of stress and sex hormone secretion. When he compared hormone levels in his selected and unselected

Figure 2.12 The distribution of 'tameness' scores in silver foxes after different generations of artificial selection. Modified from Belyaev (1979).

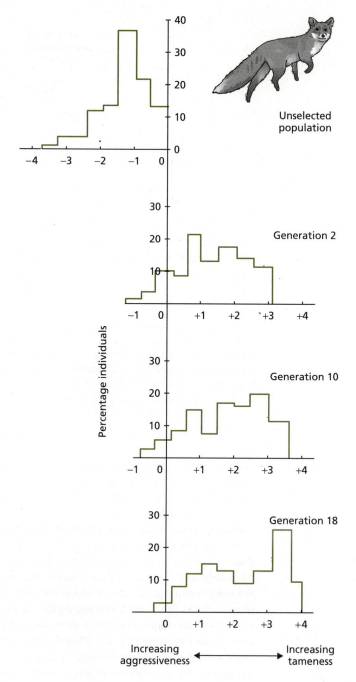

lines, Belyaev found higher concentrations of serotonin and (in females over the first few days of pregnancy) the reproductive hormones oestradiol and progesterone in animals that had been selected for tameness He also found reduced levels of corticosteroids and associated changes in adrenal gland morphology in his tame foxes. Selection for tameness therefore appeared to be associated with changes in hormones associated with the neuroendocrine control of metabolism and development. Thus what Belyaev may have done by using his composite behavioural selection criterion was actually select for changes in the mechanisms regulating early ontogeny. Since these affected a wider range of behaviours than those chosen by Belyaev, the additional tameness characteristics were incidentally affected too. These changes in physiological profile were also accompanied by changes in the reproductive cycle (towards two oestrous cycles a year [dioestrousness]) and external appearance (drooping ears, turned-up tails and variegated coat colours began to appear, for example) towards features familiar in domestic dog breeds. The experiment thus illustrates nicely what we stressed at the beginning of this chapter: behaviour patterns evolve in concert with their underlying mechanisms. Function and mechanism are two sides of the same coin.

2.4.2.2 Adaptive differences between species populations

If selection has shaped behaviour in response to the demands of the local environment, then changes in environmental selection pressures should result in different adaptive behaviours. Populations of a given species living in different environments should thus show appropriate differences in behaviour. Several studies suggest this is the case. Prey choice in gartersnakes (*Thamnophis elegans*) is a good example.

Slugs and gartersnakes

Arnold (1980, 1981) studied gartersnakes from two regions of California: the low, wet, coastal region and the drier interior uplands. Coastal and inland snakes differed markedly in their diets. Coastal snakes, hunting in a warm, humid environment, took mainly slugs, while inland snakes, coming from a dry environment that did not support slugs, took mainly fish and frogs, which they caught in lakes and streams. The interesting question, however, is whether inland snakes would take slugs if given the opportunity. When Arnold offered slugs to wild-caught inland snakes he found they refused to eat them. However, this may not be very informative, because their prior experience of fish and frogs may cause wild-caught snakes to reject slugs simply because they are novel. Arnold therefore used isolated captive-born snakes from the two regions to eliminate any effects of different feeding experience or social influence. Once again, he found that most inland snakes refused to take slugs, while coastal snakes took them quite happily (Fig. 2.13).

Arnold then went a stage further and tested the response of isolated newborn snakes that had never fed on anything to the odour of slugs presented on a cottonwool swab. He counted the number of tongue flicks directed by snakes from the two environments at swabs soaked in different prey fluids. While there was a greater spread of response this time, there was still a marked difference in the acceptability of slug extract between inland and coastal snakes. By comparing the tongue flick rate of siblings within each population, Arnold was able to determine the percentage variation in response that was due to genetic as opposed to environmental differences between them. It turned out that only about 17% of the difference in response had a genetic basis, implying that genetic variation in responsiveness to slug odour had all but disappeared in the two populations. Thus most coastal snakes have the allele(s) that allow them to detect and respond to slug odour, while most of the inland population have different alleles that do not. Crossing snakes from the two populations produced a variety of responses (see Box 2.2), but most individuals refused to take slugs. The difference in slug acceptance between the populations thus has a strong genetic basis with the allele(s) for slug rejection being dominant to that (those) for acceptance.

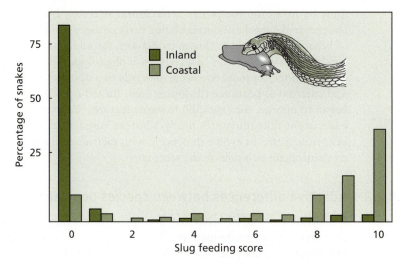

Figure 2.13 The distribution of responses to cubes of slug by naïve laboratory-reared garter-snakes from inland and coastal populations in California. Snakes derived from coastal populations showed high feeding scores, while those from inland populations generally refused to feed. From Alcock (1998) after Arnold (1980).

But why has this happened? While it is easy to envisage a selective advantage to taking slugs where they are an abundant source of food, as they are on the coast, it is not obvious why snakes should *reject* slugs if they live inland. The answer seems to be to avoid ingesting leeches. Arnold has shown that the tendency to accept slugs predisposes snakes to accept leeches. Eating leeches is a bad idea because they can live on in the gut and potentially cause serious damage. This is not a problem along the coast because there are no leeches. Inland, however, they abound in the lakes in which the snakes hunt. A taste for soft, mucus-laden invertebrates here could prove fatal.

Temperature and rhythmicity in fruit flies

A different example comes from a geographical survey of the *period* gene in *Drosophila* (see 2.3.2.1). Costa *et al.* (1992) looked at geographical variation in the protein encoded by the *period* gene in *D. melanogaster*. The protein is characterised by an alternating series of threonine–glycine (amino acid) pairs, and the region of the gene encoding the repeat varies in length across populations of *D. melanogaster*. Costa *et al.* sampled populations from 18 locations across Europe and North Africa and looked at the frequency of different threonine–glycine alleles at each one. What they found was a marked latitudinal trend (a north–south cline) in the frequency of some of the alleles (Fig. 2.14). Why should this be?

A clue came from some genetic transformation experiments in which the efficacy of inducing arrhythmia using a *per⁰* transformation (see 2.3.2.1) with the threonine–glycine element removed was affected by temperature. Recovery of the arrhythmic phenotype was good at 25 °C but much weaker at 29 °C. The threonine–glycine repeat therefore appeared to be important for the thermal stability of the circadian rhythm. Subsequent behavioural experiments have convincingly confirmed this (Sawyer *et al.* 1997). Under different temperature regimes, the periodicity of the clock differs subtly between the major threonine–glycine variants, with the common southern variant (17 repeats) having

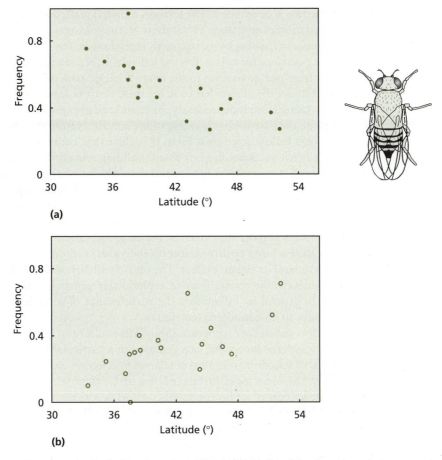

Figure 2.14 Latitudinal clines in variants 17 (a) and 20 (b) of the threonine–glycine repeat in the protein encoded by the *period* gene in *Drosophila*. See text. From Costa *et al.* (1992).

a period close to 24 hours under hotter conditions, and the most abundant northern variant (20 repeats) having a period that is better buffered against temperature swings (i.e. is better temperature compensated) (Costa & Kyriacou 1998). Each variant is thus adapted to its particular climatic environment. Furthermore, studies of the molecular conformation of the region have shown that the variants predominating in natural populations (14, 17, 20 and 23) have significantly better responses to temperature perturbations than other variants (e.g. 15 and 21) (Sawyer *et al.* 1997; Fig. 2.14b). The significance of the north–south cline thus appears to be that different threonine–glycine alleles confer different adaptive circadian responses to geographical variation in temperature.

2.4.3 Fitness and adaptation

2.4.3.1 What is fitness?

As we saw at the beginning of the section, natural selection can be summarised at different levels. We can describe it in terms of changes in the characteristics of individuals and populations, or we can describe it in terms of changes in allele frequency. The connection between the two is heritability. The features of individuals, and their relative abundance within populations, can change only if those features are passed on between generations. The mechanism of 'passing on' is, of course, reproduction. But while it is undoubtedly individuals that reproduce (or fail to), it is not individuals, or even their particular features, that pass intact from one generation to the next. What do pass intact (and are thus selected for or against) are the alleles that code for these features, each (in sexually reproducing organisms) segregating independently, or in linkage groups, and mixing in subsequent offspring with the independently segregating equivalents from their bearer's mate.

Two things follow from this. First, it is clear that contributing to the features of individuals (such as camouflage or slender gliding wings) that improve their chances of reproducing (adaptations) is a crucial determinant of allele transmission and thus selection. Second, the abundance and longevity of an allele in its population gene pool will depend on its *average* contribution to individual reproductive success in different combinations of genes (each allele is likely to be passed on by several different individuals per generation, and more than once by the same individual). In some combinations, an allele may have a large positive effect on individual reproductive success, in others it may have a neutral or negative effect. The transmissibility, and thus rate of spread, of an allele, its **fitness** in the terminology of evolutionary genetics, is therefore both relative and likely to depend on influencing the performance of some higher-order entity, of which we have so far considered individuals.

Population geneticists recognise the relative nature of genetic fitness in a simple formula in which the fitness (W) of a genotype at a particular locus is defined in terms of the strength of selection (s) against it relative to selection against a standard alternative genotype at the same locus (arbitrarily set at 1); thus $W = 1 - s$ (see also Box 2.1). Unfortunately, 'fitness' is also commonly used to denote the survival-enhancing qualities of individuals, such as size, speed or strength ('survival of the fittest'), or individual reproductive success (reproductive 'fitness'). However, as Dawkins (1999) has forcefully pointed out, its use in this sense confuses the distinction between replicators (what is inherited, i.e. beneficial alleles) and vehicles (how inheritance is brought about, via the reproductive success of individuals). For this reason, and another that will become clearer when we discuss relatedness, we shall use 'fitness' to refer to the spread of alleles rather than any quality of individuals.

2.4.3.2 Outlaw genes and extended phenotypes

While adaptations at the individual level reflect the fitness of different alleles over past generations, they are not the only way in which alleles can engineer their spread through the gene pool. An obvious alternative is to subvert the cell division process so that the distribution of alleles to daughter nuclei is biased in favour of the subverting allele. Examples of such **segregation distorter** genes are well known. They belong to a class of genes dubbed **outlaw genes** by Alexander & Borgia (1978), or **self-promoting** or **selfish genetic elements** by others (Hatcher 2000), because, while acting in their own selfish interest, they are likely to conflict with the interests of genes at other loci within the genome (see Dawkins 1999 and Hatcher 2000 for detailed discussions). This intragenomic conflict of interests favours alleles at other loci that dampen or negate the effects of their outlaw companions. Some fundamental epistatic relationships between loci, such as dominance and recessiveness, may be the result of selection for such **modifier** effects. The point of all this is that segregation distortion, modifier effects and other manifestations of outlaw genes are as much adaptive phenotypes of genes as the individual-level

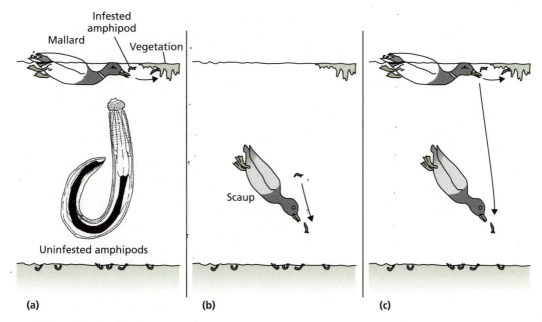

Figure 2.15 Extended phenotypes in the form of manipulation of host behaviour by parasitic acanthocephalan (spiny-headed) worms. Three species of acanthocephalan, *Polymorphus paradoxus*, *P. marilis* and *Corynosoma constrictum*, use freshwater 'shrimps' (*Gammarus lacustris*) as intermediate hosts, relying on them being eaten by ducks (definitive hosts) to complete their life cycle. Uninfected amphipods avoid light and burrow into the mud when disturbed. Amphipods infected with *P. paradoxus* (a), however, move *towards* light, and, when threatened, cling to floating vegetation rather than burying themselves. As a result they are likely to be eaten by their dabbling duck hosts such as mallard (*Anas platyrhynchos*). Crustaceans infected with *P. marilis* (b) are also attracted to light but do not go to the water surface. Instead they swim in mid-water where they are vulnerable to their diving duck hosts such as scaup (*Aythya marila marila*). Animals infected with *C. constrictum* do go to the surface, but some dive when disturbed. The definitive hosts of these worms are both dabbling and diving ducks. The behaviour of the worms' intermediate hosts are therefore altered in ways that subtly reflect the transmission needs of the parasite (e.g. Bethel & Holmes 1974). From Moore (1984).

adaptations we normally think of as phenotypes. They just happen to be adaptations that work (increase the fitness of the alleles coding for them) at a sub-individual level. In addition, while their primary selective advantage is intragenomic, it is clear that outlaw genes may have important implications for behaviour, in particular choosing mates for genetic quality (Wilkinson *et al.* 1998; Hatcher 2000; Chapter 10).

We can also think of adaptations extending *beyond* the physical boundary of the individual. For example, some parasites cause changes to the behaviour and physiology of their hosts that, while not in the host's reproductive interests, enhance the reproductive interests of the parasite (Barnard & Behnke 1990; Moore 2002; but see Poulin 2000) (Fig. 2.15). Such parasitic manipulation is a phenotype of the parasite's genes in just the same way as the physical features of its body; the genes responsible simply use the host's metabolic pathways to express themselves rather than those of the parasite. Parasitic manipulation is thus an example of what Dawkins (1978, 1999) calls an **extended phenotype**, a phenotype expressed in the world outside the bearer of the gene(s) in question. Dawkins (1999) discusses many examples of extended phenotypes, from parasitic manipulation to animal artefacts and cross-generational control of shell-coiling in snails. The concept has a general importance in the study of behaviour because much of what animals do can be construed as manipulating the external world, especially the other living components of it, to their own advantage. Signalling behaviour is one obvious example (see Chapter 11). The question as to whether the behaviour of an organism at a particular time reflects its own adaptive phenotype or the adaptive extended phenotype of another organism clearly has important implications for understanding how behaviours have evolved, and we shall come back to it again in later discussions.

2.4.4 Levels of selection

While we can define adaptation in terms of what genes do to enhance their fitness, it is clear that it is often expressed as the improved performance of some higher-order entity. Individual-level adaptations, such as the specialised digging forefeet of a mole (*Talpa europaea*), undoubtedly exist, and there is a sense in which the genes responsible for them benefit the individual in benefiting themselves. Improved performance by individuals may in turn result in the growth of their population and an expansion of their species' range. Natural selection can thus have consequences at different levels of organisation. As a result, it is easy to imagine that it acts for the *benefit* of one of these other levels of organisation, i.e. attributes that are favoured because they improve the vigour of the population or species, for example. Superficially appealing though this view can be, it is seriously misleading as a framework for understanding evolution. We have already indicated why (in the case of individuals) in the distinction between replicators and vehicles. However, the problem of the 'unit of selection' can be a subtle one and still leads to polarised debate (Keller 1999; Johnson & Boerlijst 2002). It is therefore worth pursuing a little further.

2.4.4.1 Group selection

Until the mid 1970s, it was common to read or hear that animals behaved in the interests of their species or their social group. Buck antelope competed with each other for females to ensure only the strongest perpetuated the species. Worker honey bees toiled selflessly for the good of the colony, even sacrificing their ability to reproduce (worker females are sterile) for the greater cause. Such explanations are based on **group selection**, the idea that natural selection chooses between groups, populations or species on the basis of how well their individuals serve the collective unit's interests. Group selection reached its apotheosis (and ultimately met its Waterloo) in V.C. Wynne-Edwards historic book *Animal Dispersion in Relation to Social Behaviour* (Wynne-Edwards 1962, 1986).

Wynne-Edwards argued that animal populations regulated their size to track fluctuations in limited resources. That is, in times of plenty, populations were allowed to expand, but in times of dearth, reproduction was held in check to ensure populations did not exceed the capacity of the environment to support them. Wynne-Edwards's observation that populations by and large did tend to track resource levels was uncontroversial. What created all the fuss was his assertion that animals adopted social systems specifically to bring this about. The core of his argument (see Box 2.4) was that, instead of competing for resources in a selfish free-for-all, animals adopted social conventions, such as territory ownership or dominance hierarchies, that regulated access to resources and thus breeding opportunity. Individuals selflessly obeyed the convention, breeding or not according to their status in the social system. Selection then chose between social systems on the basis of how efficiently they regulated the size of the population.

Underlying theory

Box 2.4 Wynne-Edwards's (1962) hypothesis for the evolution of social behaviour by group selection

Wynne-Edwards suggested that various aspects of social behaviour in animals were geared to regulating population sizes to the available food supply. The main assumptions of his argument were as follows (see text):

☐ Animals are variously adapted to control their own population densities.

☐ The mechanisms of control work homeostatically to adjust population densities to changing levels of resource.

☐ The mechanisms depend partly on substituting conventional 'prizes', such as territories or dominance status, for resources as the proximate objects of competition.

☐ Groups of animals adopting such conventional rules of competition constitute a society.

☐ Selection chooses between societies on the basis of how well their rules of competition regulate population size.

Based on discussion in Wynne-Edwards (1962).

Wynne-Edwards's theory caused consternation because it was difficult to see how selection acting at the level of social systems could be powerful enough to constitute a major evolutionary force. Social systems that operate in the interests of the population as a whole are vulnerable to invasion by individuals who ignore the conventions and act in their own selfish interests. Such mavericks are likely to arise in a population from time to time through immigration, mutation or recombination and outcrossing, and their selfish characteristics would spread rapidly because of their immediate reproductive advantage. A group-selected social system would thus collapse in the face of competition from selfish alternatives.

This is not to say that group selection is completely implausible as a mechanism of evolution. It *can* work and can even be demonstrated in the laboratory. Michael Wade (1977) carried out a series of experiments with flour beetles (*Tribolium castaneum*) in which he manipulated laboratory populations so as to distinguish between the effects of group and individual selection on the number of adults produced per generation. He found that group selection contributed significantly to the total variance in adult numbers when it was acting both with and against individual selection. The objection is thus not that group selection *cannot* work but that it is unlikely to be a major force *under natural conditions*. One of the main problems is that group selection requires unrealistically low levels of gene flow between groups, and low mutation rates within groups. Groups would effectively have to be genetically isolated with little chance of a selfish renegade cropping up in their ranks. These conditions can, of course, be met in theory, and Maynard Smith's (1964) 'haystack model' of group-selected altruism (Fig. 2.16) is a well-known example. However, Wade's experiments suggest that isolation need not be as complete as was at first thought, and other models, notably by D.S. Wilson (1980), have pointed to effects of population structure on the potential for group selection (see also Johnson & Boerlijst 2002).

Figure 2.16 Maynard Smith's 'haystack' model of group selection. Hypothetical altruistic mice (genotype *aa*) living in discrete haystacks can be maintained in the population in the face of competition from selfish individuals (*AA/Aa*) as long as there is little migration between colonies. Heterozygous (*Aa*) colonies drift towards selfish homozygosity. From discussion in Maynard Smith (1964).

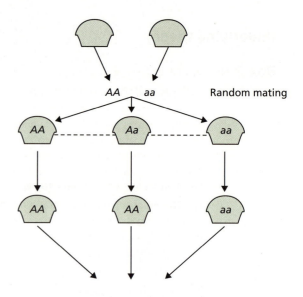

Trait group selection

In Wilson's models, populations are randomly structured into small groups (called 'trait groups' by Wilson) within which selection for or against a particular characteristic takes place. Trait groups may have different levels of aggression, tendencies to give predator alarm calls, or other characteristics likely to affect the reproductive performance of the group. After selection has occurred, the population mixes then re-segregates into new groups (again randomly) for the next round of selection. When this happens individuals from erstwhile trait groups that conferred greater benefit (e.g. by being socially more benign) fare better than those from other groups. But this works only if trait groups are reasonably small. Consider Wilson's argument for the spread of altruism. The evolution of altruism is problematic because there appears to be an immediate disadvantage to individuals showing it. Within a Wilsonian trait group, therefore, individual altruists will be at a disadvantage because of their self-sacrifice. However, trait groups containing an altruist are more likely to contribute to the next generation than those without. The reason, argues Wilson, is that splitting the population up into small groups reduces the margin of advantage to the altruist (compared with non-altruistic individuals) necessary to overcome the initial disadvantage of imposing a cost on itself. If the population comprised just one large group, altruism would spread only if the reproductive benefit to the altruist was greater than that of *every* other individual in the population. But if the population is divided into small groups, the number of other individuals the altruist has to better is considerably reduced. Indeed, Wilson argues that the relative reproductive success of other trait group members under these conditions can effectively be ignored, so making it much easier for the altruistic trait to spread.

2.4.4.2 Evolutionarily stable strategies

While models like Wilson's have continued to breathe some life into the group selection debate, there is no escaping the fact that group selection requires special conditions to work. In the majority of natural situations it is likely to be swamped by selection at other levels. In Maynard Smith's terminology, therefore, group-selected traits are unlikely to be **evolutionarily stable**. Nevertheless, traits that tempt interpretation in terms of group selection, such as altruism and selflessly obeying social conventions, demand an explanation, and one that is evolutionarily robust. How can we account for them without falling into the group selection trap? Games theory, a branch of mathematics long used in the study of economics, provides a way forward.

Games theory models

The use of games theory in an evolutionary context was pioneered by John Maynard Smith in the early 1970s (see Maynard Smith 1972, 1974, 1982; Dugatkin & Reeve 1998). Its starting point is the assumption that any problem can be approached in a number of different ways. We can think of these as different **strategies**, courses of action that provide different potential solutions to the problem. In economics, strategies may be different ways of tackling inflation or imposing taxes; in biology they may be different ways of arranging leaves up a stem, choosing a mate or timing the secretion of a hormone. A crucial difference, of course, is that economic strategies are guided by (accurate or otherwise) foresight and design, the result of sizing up a problem and estimating the effect of different approaches to it. No intelligent foresight is involved in arranging leaves up the stem of a plant, or setting daily and seasonal patterns of hormone secretion. So what exactly do we mean by a 'strategy' here?

When we talk about strategies in an evolutionary context, we mean the various options that have been available to natural selection during the evolutionary history of the organism. That is, the set of genotypes that have variously affected the character in question and been selected for or against depending on their phenotypic consequences. In any given generation, mutation, recombination and migration are likely to present selection with new stategies to choose between, favouring the ones that best serve the reproductive interests of its player. As the generations pass, more and more of the feasible genetic alternatives will have arisen and been played against each other, so that, eventually, one or more (equivalently effective) strategies will emerge triumphant and be resistant to invasion by any alternatives. Such strategies are referred to as **evolutionarily stable strategies** or **ESSs** (Maynard Smith 1972). An ESS can be defined as: *a strategy that, if most members of a population adopt it, cannot be bettered by any feasible alternative strategy*. We can use the concept of an ESS to explore the evolution of the kind of behaviour that led Wynne-Edwards into group selection.

Hawks and doves   Imagine, like Wynne-Edwards, a population of individuals competing for limited food supplies. Imagine also that contests between individuals are the restrained and formalised kind common in such populations and envisaged by group selectionists as restraint for the good of the group or species. We have already seen that such group-selected restraint is unlikely to be evolutionarily stable, yet it clearly exists. So how come? To help us think about this, we can consider a simple evolutionary scenario of just two competing strategies (Maynard Smith 1972).

The first strategy, which we will call *Dove*, corresponds to our restrained competitors above. *Dove* strategists simply display at one another and never engage in serious fighting. *Doves* therefore fight **conventionally**. The second strategy, *Hawk*, however, is overtly aggressive, fighting to injure or kill its opponent. *Hawks* thus fight in an **escalated** fashion. As we have already implied, a *Hawk* individual popping up in a population of *Doves* will be at an immediate advantage since any individual it encounters beats a retreat as

Table 2.2 A payoff matrix for a contest between *Hawks* and *Doves* (see text). Payoffs are arbitrary fitness units. If neither *Hawk* nor *Dove* is a pure evolutionarily stable strategy (ESS), the mixed ESS is given by the equation under the matrix. After Maynard Smith (1978b).

Serious injury = −20
Victory = +10
Long contest = −3

$E(H, H) = \frac{1}{2}(+10) + \frac{1}{2}(-20) = -5$
$E(H, D) = +10$
$E(D, H) = 0$
$E(D, D) = \frac{1}{2}(+10) + (-3) = +2$

		Against:	
		Hawk (H)	Dove (D)
Payoff to:	Hawk (H)	−5	+10
	Dove (D)	0	+2

The ESS is where:

$$p[E(H, H)] + (1 - p)[E(H, D)] = p[E(D, H)] + (1 - p)[E(D, D)]$$

soon as *Hawk* become aggressive. Genes for *Hawk* will therefore spread rapidly through the population. As *Hawk* spreads, however, the law of diminishing returns starts to bite, because, instead of just encountering *Doves*, it now begins to encounter other *Hawks* and experience serious aggression in return. To appreciate the consequences of this for the two strategies, we need to model the process more formally.

Let us assume for the sake of argument that the fitness payoff to either the *Hawk* or *Dove* genotype from winning a contest (getting the resource) is +10 units. Let us also assume that the fitness cost of injury during a contest is −20 units and the cost of being involved in a long display is −3 units. We can see how these payoffs accrue to each strategy in Table 2.2. When *Dove* meets *Dove*, there is no predictable winner at the outset. Either contestant could win. On average, therefore, each gets half the value of the resource $(0.5 \times 10 = 5)$ but pays the time cost of display (-3), and so experiences a net payoff of +2. When *Dove* meets *Hawk*, however, *Dove* always loses because it retreats, but, at the same time, it does not waste any time displaying. The payoff to *Dove* is therefore 0, and the payoff to *Hawk* is +10 (the total value of the resource). When *Hawk* meets *Hawk*, the winner is again unpredictable, so each contestant gets half the value of the resource (5). But, since there is an equal probability of losing, each pays half the cost of injury $(0.5 \times (-20) = -10)$ as well. The net payoff under these conditions is thus $-10 + 5 = -5$. We can now see that, in a population of *Doves*, *Hawk* has a runaway advantage of +10 fitness units versus +2 for *Doves*. On the assumption that payoffs translate directly into copies of the genes for each strategy in the next generation, *Hawk* will begin to spread through the population. But we can also see that, in a population consisting entirely of *Hawks*, *Doves* have the advantage. They may get nothing, but nothing is better than losing 5 units. Neither **pure strategy** is therefore an ESS. Each can be invaded by the other. The benefit accruing to each strategy thus depends on what everyone else is doing; in other words it is **frequency-dependent**.

If neither *Hawk* nor *Dove* are ESSs in their own right, and there is a frequency-dependent advantage to each strategy when it is rare, the evolutionarily stable solution must be the combination of the two strategies where both do equally well. We can find this combination by solving the simple equation in Table 2.2. For the arbitrary fitness values in the table, the critical combination turns out to be 8/13 *Hawk* and 5/13 *Dove*.

This means the ESS is a **mixed strategy** in which both *Hawk* and *Dove* are played in these stable proportions. If the proportions drift in either direction for any reason, frequency-dependent selection will tend to push them back again. But how exactly is the mixed ESS expressed? There are two possibilities. It could be expressed as a stable proportion of *individuals*, 8/13ths of them playing *Hawk* all the time and 5/13ths *Dove*, or all individuals could play both strategies, but play *Hawk* for 8/13ths of the time and *Dove* for the other 5/13ths. Genetically speaking, therefore, mixed ESSs can be polymorphic or monomorphic. In the first case, convention would regard *Hawk* and *Dove* as **alternative strategies** within the population. In the second, it would regard them as a single mixed strategy with **alternative tactics** (see Gross 1996; Brockmann 2001). However, as Brockmann (2001) points out, this distinction between strategies and tactics demands a knowledge of underlying genetics that is simply not available in most cases. Thus, she suggests, it is more useful to consider alternative strategies and tactics under the broader heading of 'allocation strategies' (Lloyd 1987), which simply assumes that: time or metabolic resources are divided between two or more mutually exclusive solutions to the same problem regardless of their genetic underpinings. We shall therefore continue to use the term 'strategy' in the remaining discussion.

The 8/13ths : 5/13ths ratio in the *Hawk/Dove* example above is, of course, entirely a product of the arbitrary fitness values in Table 2.2. But it makes the point that restrained contest behaviour can be accounted for without recourse to group selection. The point is made more emphatically if we change the fitness values, for example by increasing the cost of injury from –20 to –90 units (Table 2.3). Now the stable proportion of *Hawk* drops to 3% and the population consists almost entirely of *Doves*. Conversely, dropping the cost of injury to –5 leads to *Hawk* becoming a pure ESS. A second point worth noting from Table 2.2 is that the average payoff to the mixed strategy at equilibrium works out at 1.9, a value less than the payoff of 2 if everyone played *Dove*. In an ideal world, therefore, it would pay everyone to agree, in true group selection style, to play *Dove*. As we have seen, however, this simply would not be stable. The temptation to cheat and play *Hawk* is too great and the cosy conspiracy would be wide open to exploitation.

Table 2.3 Varying the magnitude of the various fitness costs and benefits in the *Hawk/Dove* game affects the stable frequency of *Hawks* in the population.

Fitness parameters	Stable proportion of *Hawk*
Serious injury = –20 Victory = +10 Long contest = –3	0.62 (mixed ESS)
Serious injury = –5 Victory = +10 Long contest = –3	1.00 (pure ESS)
Serious injury = –90 Victory = +5 Long contest = –3	0.03 (mixed ESS)

Table 2.4 Adding a conditional strategy, *Bourgeois*, to the game in Table 2.2 generates an entirely different ESS, even though the fitness costs and benefits stay the same. After Maynard Smith (1978b).

Serious injury = −20		$E(H, B) = \frac{1}{2}E(H, H) + \frac{1}{2}E(H, D) = -\frac{5}{2} + \frac{10}{2} = +2.5$			
Victory = +10		$E(D, B) = \frac{1}{2}E(D, H) + \frac{1}{2}E(D, D) = 0 + \frac{2}{2} = +1$			
Long contest = −3		$E(B, B) = \frac{1}{2}E(H, D) + \frac{1}{2}E(D, H) = +\frac{10}{2} + 0 = +5$			

		Against:		
		Hawk (H)	Dove (D)	Bourgeois (B)
Payoff to:	Hawk (H)	−5	+10	+2.5
	Dove (D)	0	+2	+1
	Bourgeois (B)	−2.5	+6	+5

In Table 2.2, individuals play *Hawk* or *Dove* all the time, and nothing other than the respective strategies of the contestants influences what is played. In the real world, the strategies adopted are likely to depend on all sorts of contingencies, for example relative body size or whether females are present. The ESS is thus likely to be a **conditional strategy**, one that takes the form: 'if faced with situation x, do A; if faced with y, do B'. We can illustrate this by adding a simple conditional strategy to our *Hawk/Dove* game. *Bourgeois* strategists act as both *Hawks* and *Doves*, but according to the rule: 'escalate aggressively if already the owner of the resource, but display conventionally if not'. Assuming it owns the resource on 50% of occasions, *Bourgeois*' conditional choice of approach risks only half the injury cost of unconditionally playing *Hawk*. Keeping the same fitness payoffs as before (Table 2.2), we can see that, in a population of *Hawk*, *Dove* and *Bourgeois*, *Bourgeois* does better against itself than do either of the other two strategies (Table 2.4). This is because it pays neither the cost of injury, nor the time cost of display – one player is always the owner, so the other player always retreats. Thus, when everyone is playing *Bourgeois*, neither *Hawk* nor *Dove* can invade (but see Mesterton-Gibbons 1992 for further discussion).

Waiting games and the 'War of Attrition' Of course, alternative strategies compete in many ways, but success in competition is always measured ultimately in terms of the spread of genes for the strategies within each competing set. In the *Hawk/Dove* game, phenotypic competition happens to be literal and aggressive, but many competing strategies we shall encounter later on have nothing to do with aggression. Nevertheless precisely the same principles of analysis apply. Before we look at some real ESS analyses, however, we shall introduce one further general model, the so-called 'War of Attrition'.

In many situations, rewards depend on persistence rather than overt aggression ('all things come to he who waits'). Gaining a mate, for example, or winning a contest for food, may depend on displaying for longer than your rivals. However, the longer you display or wait around, the more costly it is likely to become, either metabolically or in lost opportunities for doing other things. The problem is therefore to choose a persistence time that gets the reward but does not cost more than the reward is worth. Assuming that choice of persistence time shows genetic variation and offspring inherit the choices of their parents, how is selection likely to work? Obviously choosing a fixed time, say t minutes, is no good because it will always pay someone else to go for slightly longer, say $t + 1$ minutes. But, equally, there cannot be a never-ending escalation of persistence times because the cost will quickly become too great. It is clear, therefore, that no pure strategy of persistence time can be an ESS. In fact, the evolutionarily stable solution to the problem, as Maynard Smith (1974) has shown, is to play a mixed strategy of *random* persistence times. As in the *Hawk/Dove* model, the mixed ESS could be a polymorphism, with individuals playing different fixed persistence times within a random frequency distribution, or each individual could play the same variable strategy, choosing shorter or longer persistence times on a random schedule. We shall return to the War of Attrition model in the context of signalling in Chapter 11.

The *Hawk/Dove* game and the War of Attrition introduce the principle of ESS analysis. However, it is necessary to sound a note of caution. While populations may reach an equilibrium for particular strategies where each strategy fares equally well, the equilibrium is not necessarily stable. Dawkins (1989), for example, presents a games theory model of sexual strategies in which stable mixtures of *Coy* and *Fast* female strategies and *Faithful* and *Philanderer* male strategies are sought. An equilibrium mixture of the four strategies can be found, but it is unstable, like a pencil balanced on its point. A drift away from the equilibrium point is not corrected by frequency-dependent selection, as in the *Hawk/Dove* model, but *accelerated*. With this caveat in mind, can we apply ESS theory to behaviour in real animals?

Nesting strategies in great golden digger wasps

One of the best field applications of ESS theory comes from Jane Brockmann's study of burrowing behaviour of great golden digger wasps (*Sphex ichneumoneus*). As we saw in Chapter 1 (Fig. 1.1), female golden digger wasps lay their eggs in burrows dug in the ground which they provision with katydids as food for the developing larvae. Digging a burrow is hard work, taking over an hour and a half on average. Perhaps not surprisingly, therefore, Brockmann discovered that females sometime use burrows that have already been dug by another female (Fig. 2.17). Entering an existing burrow seems a sensible idea, but there is a catch. Females do not seem to be able to distinguish between burrows that are vacant and those that are still in use. If the burrow is vacant, the female can proceed to provision and lay as normal, but if it is still occupied, there will come a time when the two females using the burrow meet and fight, leaving only one of them in possession. Even if a female finds an unoccupied burrow, or has gone to the effort of digging her own,

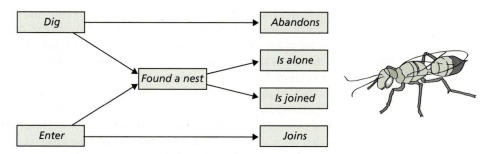

Figure 2.17 A female digger wasp can either dig a nesting burrow for herself (*Dig*) or move into (*Enter*) one that has already been dug (or begun) by another female. A decision to dig for herself may result in owning her own burrow, being joined by another female, or abandoning it as a bad job. Similarly, taking over another female's burrow may mean she has either exclusive or shared ownership. The various possible outcomes have different implications for the female's expected reproductive success. See text. After Brockmann *et al.* (1979).

she may herself be joined by another female. Clearly, then, there are various costs and benefits to digging your own burrow or entering someone else's. Can ESS theory help us to understand how the two approaches, which we can think of as two alternative strategies, *Digging* and *Entering*, are maintained in the wasp population?

Like *Hawk* and *Dove*, the advantages of *Digging* and *Entering* are likely to be frequency-dependent. If everyone digs, there will be lots of vacant burrows from past breeding attempts, and therefore a temptation to save time and effort by *Entering*. If everyone *Enters*, however, aggressive competition is likely to be fierce and it will pay a female to go off and dig for herself. Neither *Digging* nor *Entering* are therefore likely to be pure ESSs. Instead, we might expect a stable mixture of the two strategies as we found in the *Hawk/Dove* game.

The stable mixture should be the one in which the two strategies do equally well. In the *Hawk/Dove* game this would mean the mixture in which the average arbitrary fitness payoffs were the same. But what is the measure going to be in the digger wasps' case? As we might suspect from the above, individual wasps play both *Digging* and *Entering* at different times, rather than each wasp playing one of the strategies all the time, and neither strategy is associated with any particular phenotype, such as large size, or environment. Any mixed ESS will therefore be expressed within individuals rather than as a polymorphism at the population level. This means the fitness of the two strategies must be estimated in terms of digging and entering *decisions* rather than digging and entering *individuals*. As the most practicable measure of fitness, Brockmann *et al.* (1979) chose to count the number of eggs laid per unit time as a result of *Digging* or *Entering*. This meant observing all digging and entering behaviour for a number of wasps and calculating the number of eggs laid in each attempt, taking account of the range of possible outcomes in Fig. 2.17. From over 1500 hours of observation, and a sample of some 410 burrows, Brockmann *et al.* found that the success rates of the two strategies were not significantly different. Wasps showed *Digging* on 59% of occasions, laying 0.96 eggs per 100 hours as a result, and *Entering* on 41% of occasions, laying 0.84 eggs per 100 hours. The results are therefore consistent with *Digging* and *Entering* being a mixed ESS, with frequency-dependent selection setting the stable equilibrium at 59% *Digging* : 41% *Entering*.

Jaw-handedness in cichlid fish

Another apparently frequency-dependent mixed ESS, this time for a morphological trait linked to behaviour, comes from a species of cichlid, *Perissodus microlepis*, in Lake Tanganyika in Africa. The trait in question, which relates to the unusual feeding behaviour of the fish, is the orientation of the jaw. *Perissodus microlepis* makes its living by tearing scales from the flanks of other fish. To aid this, the jaw is twisted slightly to one side. However, in some individuals the twist is to the right, while in others it is to the left (Fig. 2.18a). Michio Hori, who has studied the trait in detail, has shown that the direction of twisting is heritable (Hori 1993). An obvious explanation for this polymorphism is that there is some frequency-dependent feeding advantage to each morph when rare. From his studies, Hori found that in years when one morph made up more than 50% of the population, the opposite morph seemed to have more success, as judged by the number of scales missing from different sides of their prey (left-jawed individuals attack the right flank and right-jawed individuals the left). Increased feeding success of the rarer jaw type results in an increase in the frequency of the morph in the next generation, leading, over a number of years, to an oscillation in the frequency

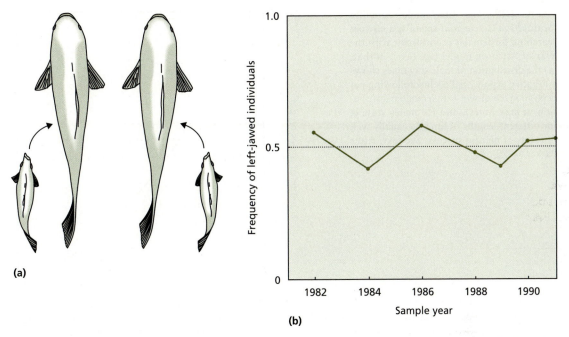

Figure 2.18 The cichlid *Perissodus microlepis* feeds by tearing scales from the flanks of other fish. To do this, it has evolved laterally displaced jaws. However, there is a polymorphism, with some individuals having jaws displaced to the left and others with jaws displaced to the right (a). Over successive years (b), the two morphs seem to be maintained in a ratio of 1 : 1. See text. From Alcock (1998) reprinted with permission, after 'Frequency-dependent natural selection in the handedness of scale-eating cichlid fish, *Science*, Vol. 260, pp. 216–19 (Hori, M. 1993). Copyright 1993 American Association for the Advancement of Science.

of the two morphs about 50% (Fig. 2.18b). Hori's interpretation of this frequency-dependent polymorphism is that, if one morph comes to predominate in the population, prey fish are attacked more often on the relevant flank. They therefore learn to guard against attacks on that flank, but in doing so expose the other, creating an advantage for the opposite morph. A less interesting alternative, however, might be that the balance simply reflects competition for two-sided prey!

Intermale competition in dungflies

A third example illustrates a War of Attrition-style ESS. It concerns waiting strategies among male dungflies (*Scatophaga stercoraria*) around a fresh cowpat, as studied by Geoff Parker (1970).

Soon after a cowpat has been deposited, female dungflies begin arriving to lay their eggs. Males accumulate at the pat to intercept females and copulate with them. They then guard the females from other males until her eggs have been laid, since competition in the early stages is extremely fierce and aggressive (Fig. 2.19). As time goes on, however, the pat starts to dry up and becomes less suitable for egg-laying, so the arrival rate of females declines. A problem facing males is to decide how long to stay at the pat. The answer depends on what other males are doing. If most males stay for only a short time, it might pay a male to stay a bit longer and copulate with some of the few late-arriving females. If most males stay longer, it will pay to move on to a new pat and mate with females arriving early there. As in the War of Attrition model, therefore, no single, fixed

Figure 2.19 The time male dungflies (*Scatophaga stercoraria*) spend at a cowpat searching for females is consistent with the War of Attrition model (see text), in that: (a) the distribution of male stay times shows a negative exponential decline following pat deposition (consistent with a random distribution of stay times), and (b) males staying for different lengths of time do equally well in terms of mating success. See text. After Parker (1970).

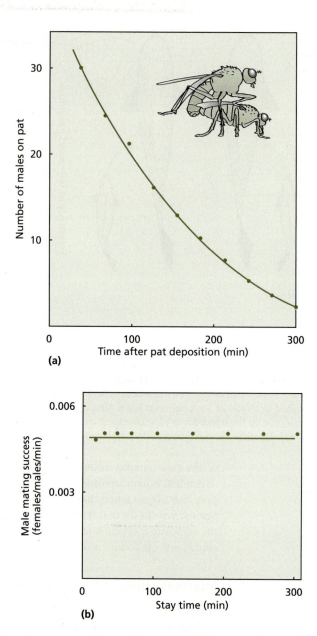

waiting time is likely to be stable. But what will be stable, and what do male dungflies do? The results of some of Parker's observations are shown in Fig. 2.19. Figure 2.19(a) shows the number of males at a pat with time since it was deposited. The graph shows a curvilinear decline in the number of males which approximates to a negative exponential (one form of random) distribution. This suggests that males stay for randomly distributed lengths of time, exactly as predicted by the War of Attrition model. Moreover, Fig. 2.19(b) shows that the mating success of males, calculated on the basis of male-minutes at the pat, is more or less equal for all stay times, again as predicted by the War of Attrition. Male dungflies thus appear to play a mixed ESS of randomly distributed stay times, though it is not clear whether this is due to different individuals adopting different fixed stay times or to all individuals adopting a random mix of stay times.

2.4.4.3   Optimality theory

ESS models seek stable strategies in terms of fitness cost–benefit payoffs. The ESS is the strategy whose payoff (ratio of benefits to costs) when played against itself exceeds that of other competing strategies in the set. ESS theory is therefore a special case of **optimality theory**. An ESS is an optimal outcome, but one that depends on what everyone else is doing. However, the solutions to many problems do not depend on what everyone else is doing. Maynard Smith & Parker (1990) therefore distinguish between competitive (frequency-dependent) and simple (frequency-independent) optimality. A solitary bird foraging for prey may be able to take the largest, most profitable items because there is no risk of another bird stealing them before they can be swallowed (simple optimum). In a flock, however, large items may often be lost to thieves, so pushing the optimal prey size down (competitive optimum) (Fig. 2.20). Frequency-dependent competition is thus a **constraint** on free choice that is built in to the optimisation process (see below and Box 2.5). Dawkins (1995) uses a somewhat similar argument to distinguish between **short-** and **long-term optimality**. If we look at each problem an organism has to solve in isolation, we can come up with an optimal solution for it (such as take the largest prey if you are a single bird). This is short-term optimisation *sensu* Dawkins because it focuses on an immediate solution to a problem in the absence of other considerations. However,

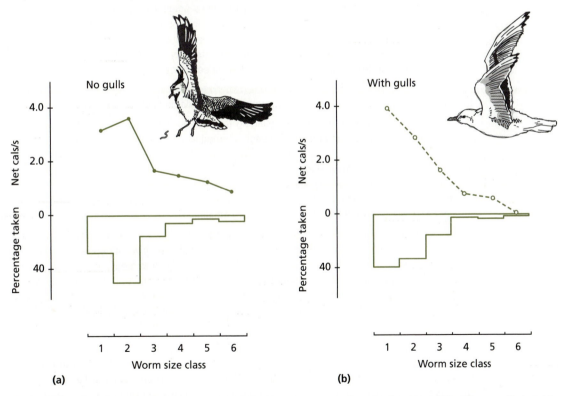

Figure 2.20   The principle of 'simple' versus 'competitive' optima in foraging lapwings (*Vanellus vanellus*). (a) In the absence of kleptoparasitic (food-stealing) gulls, lapwings get most net energy by concentrating on the second smallest size class of earthworm (top), and bias their intake towards these worms (bottom). (b) When gulls are present, however, they do best by taking the smallest worms, and shift their intake accordingly. After Barnard & Thompson (1985).

Underlying theory

Box 2.5 Assumptions of optimality theory models

Like any theoretical framework, optimality theory makes certain assumptions that are crucial to the design of rigorous tests. Optimisation models of behaviour assume we can:

☐ Identify the problem to be solved (the *decision* variable).

■ What decision is the organism trying to make? Is it trying to decide how long to scan for predators, whether to take prey type A rather than prey type B, when to moult, or what? The problem has to be correctly identified before we can ask which solution the organism should adopt.

☐ Choose the right currency (the *currency* variable).

■ What is the decision seeking to achieve? Is it trying to maximise the rate of food energy intake, minimise the risk of predation, maximize the rate of fertilization? If we do not know the currency, we cannot measure costs and benefits.

☐ Identify the available alternative solutions and constraints (the *constraint* variable).

■ As we have stressed repeatedly, adaptation is relative. Whether a particular decision is best depends entirely on the alternatives against which it is compared. It is therefore crucial to identify the alternative possibilities that selection is choosing (or has chosen) between. In addition, we must identify any constraints on these alternatives; for example, aerodynamic requirements may limit the elaboration of wing feathers for sexual display, or the need for specific trace elements may limit flexibility in choice of diet.

☐ Quantify the costs and benefits accruing from the available alternatives.

■ In order to identify the decision yielding the highest net return, we must be able to quantify the costs and benefits of each alternative in the relevant currency.

In addition to all this, we must also:

☐ Assume that appropriate genetic variation has arisen in the organism's evolutionary history and that the population is at evolutionary equilibrium.

■ Optimality theory assumes selection acts on genotypes coding for different solutions to problems. If a genotype appropriate to a particular solution happens not to have arisen, then that alternative cannot have been part of the strategy sets in which the existing strategy competed. Optimality theory also assumes that we are looking at the end product of the evolutionary process, that all feasible alternatives have been played off and the population has reached a stable evolutionary equilibrium.

optimisation is not simply about solving individual problems; it is about solutions that maximise reproductive success and the transmission of alleles coding for them. It is therefore the contribution of short-term decisions to reproductive success – long-term optimisation in Dawkins's terminology – that matters. It is then not difficult to see that all manner of concerns, such as risk of predation, maintaining a territory, or feeding a brood, will influence a given decision such as choosing prey, since they each contribute to the organism's lifetime reproductive output.

Assumptions and caveats

Optimality theory itself applies broadly in evolutionary biology (McNeill Alexander 1996). Why? Because selection can be thought of as a natural optimising agent. Ongoing variation and selection over generations (see 2.4) will tend to hone phenotypic solutions to problems as more and more alternative genetic strategies are played off against each other. Once evolutionary equilibrium is reached, the gamut will have been run and the current strategy (genotype) will have triumphed against all the options in the strategy sets (gene pools) in which it competed. We cannot know what these historical options were, but if we assume our population has reached equilibrium, we can make some educated guesses and see whether our existing strategy is likely to yield the highest benefit : cost ratio.

The process is a little like a business person making an investment. Any investment requires capital to start with, but stands to make a return at some point in the future. Many opportunities for investment are likely to be on offer, each demanding a different capital outlay and offering different prospective returns. The trick is to choose the investment that maximises the net return. Business people who are good at this tend to flourish; those who are not go to the wall. By analogy, we can think of any decision an organism makes as incurring an initial downpayment, not in terms of dollars or euros, but in time, risk and metabolic resources – in short, future reproductive potential. The reproductive cost, of course, is offset by a resulting benefit that enhances future reproductive potential, such as obtaining food or successfully completing cell division. As with the business-person's investment, organisms are likely to be faced with a range of options at each decision point. The trick now is to choose the option that maximises net reproductive gain. As in the case of ESSs, there is no suggestion that organisms consciously weigh up the reproductive consequences of the various options; the 'choice' is made by natural selection acting through differential reproductive success. We have already encountered the optimality approach in Högstedt's study of clutch size in magpies (Fig. 1.7) and shall encounter it again in many other contexts in later chapters.

Applying optimality theory involves a number of critical assumptions, summarised in Box 2.5. If these assumptions are not met for any particular optimisation model, then the model cannot fairly be tested. Failure of results to conform to the model's predictions could be due to a violation of one or more of the assumptions and not to inappropriateness of the optimality approach itself.

Evolutionary lags

In addition to the factors in Box 2.5 there may be other reasons why the predictions of optimisation models fail. Sudden changes in the environment, for example, may render current adaptations inappropriate until selection has had time to correct the mismatch. Hosts that accept the eggs of brood-parasitic cowbirds, rather than ejecting them from the nest, may be examples of just such an **evolutionary lag**. Field observations and egg manipulation experiments suggest that many non-rejecting hosts would do better by rejecting, even though occasionally it might mean throwing out one of their own eggs by mistake. However, it turns out that many of these non-rejectors are relatively new cowbird hosts in which selection may not yet have had a chance to shape a counter-strategy. Longer-established hosts are much more likely to reject, presumably because they are further advanced in their **arms race** with the cowbirds (Davies 1999). Coevolutionary arms races (Box 2.6), such as those between parasites and their hosts, or predators and prey, are likely to be a major cause of evolutionary lags, as selection adapts one lineage to new

Underlying theory

Box 2.6 Coevolutionary arms races

If we think of a predator chasing its prey, as, say, a hunting dog (*Lycaon pictus*) might chase a zebra (*Equus* spp.), we can imagine a race occurring on two different timescales. First, there is a race in the immediate sense of individual dogs chasing an individual zebra. Second, there is a race on an evolutionary timescale in which hunting dog and zebra lineages adapt and counter-adapt to each other. This second, evolutionary, type of race is analogous to the mutually progressive advances of battleship and submarine designs during a war, and can be thought of as a coevolutionary **arms race**. While we talk of arms races as 'progressing' and their competing lineages as 'improving', however, this does not assume that predators end up catching more prey, or parasites, say, infecting more hosts. Prey and hosts 'improve' too. As Dawkins & Krebs (1979) put it, Recent predators might massacre Eocene prey, but Recent prey would almost certainly outrun Eocene predators. It is also important to stress that the mere existence of anti-predator adaptations in prey does not by itself indicate a mutually counter-adaptive arms race (see Endler 1991). Although coevolutionary arms races occur in a wide variety of contexts, they can be classified according to whether the adaptive currencies of the competing lineages are the same (*symmetric* arms races) or different (*asymmetric* arms races):

A simple classification of coevolutionary arms races

	Asymmetric **(e.g. attack/defence)**	**Symmetric** **(e.g. competition to be bigger)**
Arms analogy:	swords sharper shields thicker swords sharper still	2 megatonne bombs by A 3 megatonne bombs by B 4 megatonne bombs by A
Biological example:	predator/prey host/parasite assessor/cheat	intermale trials of strength Batesian models/mimics
Distribution:	mainly interspecific	mainly intraspecific

By imposing strong directional selection on coevolving lineages, arms races can have several important evolutionary consequences, including sudden acceleration in the rate of evolutionary change (leading to apparent 'jumps' in the fossil record), evolutionary lags while lineages wait for appropriate genetic variation to allow them to catch up with their rivals (the basis of the Red Queen effect – see text), and extinction. Various factors are likely to dictate who stays ahead in, or even wins, an arms race. The most important are (a) differences in the strength of selection for counteradaptation (the so-called *life–dinner principle* – selection will generally act more strongly on prey to avoid predators, and thus losing their life, than on predators to catch prey, where failure means only a missed meal), and (b) conflicting demands from different selection pressures, which limit the resources available for counterresponse.

Based on discussion in Dawkins & Krebs (1979) and Endler (1991).

changes in the other. Indeed, continual coevolutionary change in the biotic environment may mean that adaptation is in a more or less permanent state of catching up – the so-called **Red Queen effect** (van Valen 1973), after the character in Lewis Carroll's *Through the Looking Glass* who is always having to run to stay in the same place.

Rules of thumb

Another reason organisms may appear to respond sub-optimally is because they are using approximate **rules of thumb** to solve their problems. One way of deciding which is the most profitable of a range of prey types, for example, is to do some calorimetry and work out the energetic value of each type. By judging this against how long it takes to subdue and eat the prey (handling time), the profitability of each prey type can be calculated (see 8.1.1.3). This is indeed the approach *we* might take to test whether an animal is optimising its choice of prey. The animal itself, however, uses a very different yardstick, perhaps going for the largest prey, or the ones that move about most, or those that are the deepest green. As long as size, movement or greenness correlate closely with profitability such rules of thumb do the job very well. They may result in the occasional mistake, but if mistakes are infrequent and not too costly, there may be little pressure to evolve a more elaborate method of choice. Figure 2.21 shows just such an apparent rule of thumb in action in foraging shrews.

Rules of thumb may be only approximations of perfect decisions, but in terms of fitness costs and benefits and the raw material selection has to work on, they may provide the most cost-effective means of decision-making. The critical point, however, is that they are specific to the environment in which they evolved. Size, movement

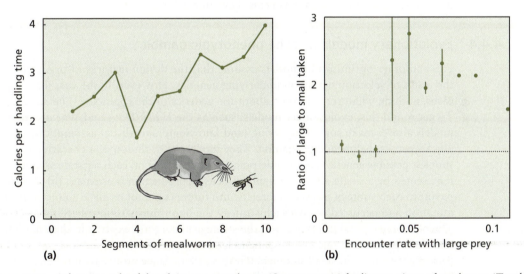

(a)

(b)

Figure 2.21 A foraging rule of thumb in common shrews (*Sorex araneus*) feeding on pieces of mealworm (*Tenebrio molitor*). (a) Shrews generally do best by taking larger pieces, because their net intake of energy increases with size. However, when mealworm pieces become bigger than three segments, net energy intake drops sharply before increasing again. Thus four segment pieces yielded *less* net energy than two or three segment pieces. This is because there is a sudden increase in the time necessary to pick up and chew the prey between three and four segments. If shrews optimised their choice of prey, therefore, they should have reversed their normal preference for larger pieces when presented with both two and four segment options. Instead, they still preferred large pieces (b) when they encountered enough to make specialisation worthwhile, suggesting they were using an approximate 'go for the biggest because it is usually best' rule of thumb. From Barnard & Brown (1981).

or greenness may not be universal indications of prey profitability, and moving the animal to a novel environment may cause its rule of thumb to misfire badly. This is a particular problem for laboratory environments, where things may be very different from the situation in the field. Tests of optimisation models can thus easily fall foul of rules of thumb being expressed in inappropriate environments.

Local versus global optima

Finally, it is important to emphasise that optimality theory does not test whether organisms are optimal in an absolute sense; that is, adopt the best of all possible solutions to a problem. All it tests is whether the particular costs, benefits and constraints identified provide a good account of what the organism does. There are several reasons why it is not sensible to test for global optimality. An obvious one is that evolution is a continuous process and natural selection can work only on the raw material handed down by history. Nothing can be invented from scratch; innovations must be wrought from existing designs, and this automatically imposes constraints. Likewise, phenotypes are often the product of complex developmental pathways, which also limit the opportunity for novelty (see 2.2.2 and Chapter 5). Yet another problem is that solutions to the same problem may differ from one place to another, not because some solutions are optimal and others not, but because of chance historical differences. For example, driving on the right-hand side of the road is just as good a strategy for avoiding accidents as driving on the left, but some countries have opted for one rule, some for the other. Similarly, the presence of one horn in the Indian rhinoceros (*Rhinoceros unicornis*) but two in its African counterpart (*Diceros bicornis*) may simply reflect chance differences in development rather than optimal solutions to different problems (Lewontin 1978).

2.4.4.4 Evolutionary models and the phenotypic gambit

ESS and other optimisation models assume that the design features of organisms are the result of selection acting on underlying genetics. However, in the vast majority of cases, nothing whatever is known about the genetics of the character(s) being studied. To get round this, evolutionary models, such as the *Hawk/Dove* and War of Attrition models above, make some simplifying (and knowingly unrealistic) assumptions, generally known as the **phenotypic gambit**. They model the evolution of a character as if the simplest possible genetic system underpinned it; that is as if (a) each separate strategy was represented by a single allele at a haploid locus (reproduction is asexual), (b) the fitness payoff to each strategy translated directly into future copies of its allele and (c) sufficient mutation had occurred to give each strategy the opportunity to invade (Grafen 1984). The phenotypic gambit implies that the strategies currently existing in the population are all as successful as one another, and at least as successful as any feasible new strategy that might turn up in small numbers (Grafen 1984). Since no behavioural character is likely to be controlled by such a simple system, is the phenotypic gambit reasonable?

There are certainly genetic systems where the gambit would be seriously misleading. Grafen (1984) points to heterozygous advantage, such as the resistance to malaria conferred by the sickle cell anaemia trait, as an obvious case where the gambit could not apply because only the heterozygous contingent of the population would be able to show the optimal phenotype. However, such special cases are relatively rare. Maynard Smith (1982) has explored the robustness of the gambit in more general genetic systems and found it to be broadly satisfactory, partly because the polygenic nature of many phenotypic (especially behavioural; see Box 2.2) characters reduces the opportunity for large fitness differences, such as those arising from heterozygous advantage. Attempts to square population genetic models of allele fitness with the simplified assumptions of the phenotypic gambit have also been encouraging (Moore & Boake 1994; Marrow & Johnston 1996). Explicit genetic models have suggested that the long-term outcome of selection is the kind of evolutionary equilibrium predicted by ESS models (Eshel & Feldman 1984). However, short-term evolutionary change is better characterised by population genetic models because it is more dependent on the range of alleles present in the gene pool at the time (Marrow & Johnston 1996).

The problem of evolutionary equilibrium

One serious question raised by population genetic models, however, concerns the assumption that populations are generally at evolutionary equilibrium, a critical plank in the optimality approach (Box 2.5). This is rarely tested, and where it has been, has often been found wanting (Moore & Boake 1994). How serious this is for the predictions of optimality models has yet to be ascertained, but it is clear that it can account for departures from predicted outcomes. A good example comes from Susan Riechert's study of aggression in web-spinning spiders (*Agelenopsis aperta*) (Riechert 1993).

Riechert and coworkers studied aggressive behaviour in two populations of *A. aperta* in the south-western United States, one a desert-grassland population, the other riparian. Aggression in both populations was related to territorial defence, the payoff being a tradeoff between obtaining more prey and the risk of mortality from fighting and predation. Levels of aggression were more or less the same in the two populations, a surprising finding because food abundance in the two cases was strikingly different. The disparity was reinforced when ESS models of the behaviour were tested and showed that, while the desert-grassland population, with its poor food supply, fitted the predictions very well, the well-provided riparian population did not. From an analysis of the genes underlying aggression in the species, Riechert was able to rule out insufficient variation or evolutionary lags as explanations for the mismatch. Extensive exploration of the model, and tests of genetic crosses, also suggested there were no flaws in the model. Electrophoretic and experimental evidence, however, strongly suggested there was gene flow from the desert-grassland population to the riparian population, and that this was preventing the latter from reaching an evolutionary equilibrium. If so, this could account for the riparian spiders failing to achieve the predicted optimal state for their habitat.

2.4.4.5 Inclusive fitness

Optimisation models base their predictions on the fitness of putative genotypes coding for phenotypic strategies. In most cases, genes stand or fall (their fitness increases or decreases) on the basis of their contribution to the reproductive success of their bearer. However, as W.D. Hamilton (1964a,b) pointed out in a now legendary pair of papers, this is too narrow a view of genetic fitness.

An individual possessing a particular allele shares it by common descent with others in the population. The probability that any two individuals share copies of the allele depends on their degree of relatedness. An individual has a probability of 0.5 of sharing the allele with its parent, offspring or sibling, a probability of 0.25 of sharing it with a grandparent or grandchild, of 0.125 of sharing it with a first cousin and so on. This probability is referred to as the **coefficient of relatedness** (*r*). The effect an allele has on its fitness, by causing its bearer to develop or respond in a particular way, depends not only on changes in that individual's reproductive success, but also on the effect of its phenotype on the success of others sharing the allele. An allele's effect on these cobearers, weighted by r, thus provides an additional, indirect route by which it can influence its transmission into the next generation (so-called **indirect fitness** effects). Hamilton referred to this extended concept of fitness as **inclusive fitness**, and selection acting via indirect fitness has been dubbed **kin selection** by Maynard Smith (1964).

Crucially, inclusive fitness does *not* simply mean all the individual's offspring plus all those of its relatives weighted by r, which is one of many erroneous definitions scattered through the literature (see Grafen 1984 and Dawkins 1979 for common misunderstandings of inclusive fitness and kin selection). We have to discount the proportion of the individual's own offspring that were produced only because of 'help' from neighbours and not by its own effort alone, then include the *additional* offspring produced by relatives (devalued by the appropriate coefficient of relatedness) as a direct result of the individual's helpfulness. The purpose of this tortuous 'stripping and augmenting' (as Hamilton 1964a put it) is to ensure that offspring are not double-counted in calculating inclusive fitness benefits. For instance, if an individual had been helped by his brother, his male offspring could, if we were not careful, be counted once as his father's son and again as his helpful uncle's nephew.

Self-evidently, Hamilton's stripping and augmenting process is a complicated one; impossible, in fact, for many practical purposes. Happily, there is an alternative. We can simply count the number of offspring produced by individuals that help and those that don't! Why? Because we are interested in the fitness of the allele for helping behaviour. If the sum total of individuals carrying the putative allele produce more offspring than the sum total of those that do not (because the extra offspring produced by help-receivers so far outnumber those lost by their related help-givers in helping), then the allele will spread. Counting the offspring of helpers and non-helpers and comparing them thus estimates the fitness of the allele for helping (Grafen 1984). The equivalence between the two accounting procedures is demonstrated mathematically in Box 2.7.

Underlying theory

Box 2.7 Measuring inclusive fitness

Alan Grafen (1984) nicely demonstrates how, in terms of the fitness of an allele for phenotypic altruism, inclusive fitness and the number of offspring produced by the bearer add up to the same thing. The calculations are summarised in the following table.

Two ways of calculating reproductive success – as the number of offspring and as inclusive fitness. The advantage to a heterozygous individual carrying an allele for helping is the same in both cases. Modified from Grafen (1984).

Measure of reproductive success: Genotype:	Number of offspring		Inclusive fitness	
	aa	*Aa*	*aa*	*Aa*
Basic (non-social) fitness	1	1	1	1
Cost of act		c		c
Benefit of act	$pb/2$	$(1+p)b/2$		$b/2$
Total	$1+pb/2$	$1-c+(1+p)b/2$	1	$1-c+b/2$
Advantage to *Aa*		$b/2-c$		$b/2-c$

Counting the number of offspring

The table assumes a population in which every breeding male has one brother of the same breeding age nearby. Each male's response to his brother is controlled by a single locus and the population is currently at fixation (all individuals have the same allele) for allele *a* at the locus. We now consider a mutant allele *A* that causes its (homozygous and heterozygous) bearer to help its sibling rear *b* more offspring than it would otherwise have done, at cost *c* to the bearer. Will *A* spread when rare?

The number of offspring a mated pair produces depends only on the genotypes of the male and his brother. If the male and his brother are both *aa*, the pair produces 1 offspring. If an *aa* male's brother is *Aa*, then the pair produces $(1 + b)$ offspring. If the proportion of *Aa* individuals in the population is *p*, the probability that a male has an *Aa* brother is $p/2$ and the average number of offpring he produces will be $(1 + pb/2)$. The probability that an *Aa* male has an *Aa* brother is $(1 + p)/2$. However, the *Aa* male loses *c* offspring through carrying *A*, so the average number of offspring accruing to an *Aa* male becomes $1 - c + (1 + p)b/2$. If *Aa* males have more offspring than *aa* males, the allele *A* will spread. The conditions for this are therefore:

$$1 - c + (1 + p)b/2 > 1 + pb/2 \qquad (2.7.1)$$

which reduces to:

$$b/2 - c > 0 \qquad (2.7.2)$$

Calculating inclusive fitness

We have just derived the conditions for an allele causing phenotypic altruism to spread by accounting for the effect of everybody's actions on males of different genotype. Can we arrive at the same conclusion by looking at the effect of a given individual's actions on everyone else, in other words via Hamilton's process of 'stripping and augmenting' (see text)?

Here, *aa* males have an inclusive fitness of 1. This is because the extra *b* they receive from males carrying *A* has to be discounted as Hamilton's 'help from the social environment'. *Aa* males get $(1 + b/2 - c)$ because their relatedness to their brother is 1/2 (so the benefit *b* is halved) and they pay the cost of helping *c*. For *Aa* to have a greater inclusive fitness than *aa*, therefore, the condition that must be met is:

$$1 + b/2 - c > 1 \qquad (2.7.3)$$

which reduces to:

$$b/2 - c > 0 \qquad (2.7.4)$$

exactly as before.

While these calculations derive the conditions for the spread of an allele for helping, a more useful expression from a practical viewpoint is **Hamilton's rule**. Hamilton's rule specifies that an animal will be selected to perform an action for which $rb - c > 0$, where *r* is the coefficient of relatedness between actor and recipient (see text). Several studies of phenotypic altruism have attempted to apply the rule (Chapters 9 and 10).

Based on discussion in Grafen (1984).

The importance of inclusive fitness is that it provides an evolutionarily robust explanation for phenotypic altruism, the actions by one individual that detract from its own reproductive success while enhancing the reproductive success of others. Altruism takes many forms, from parental care and helping at the nest, to self-sacrifice and sterility (Box 2.8; Chapters 9 and 10). With the exception of parental care, such traits fly in the face of an individual selection view of evolution. Before Hamilton, therefore, most explanations of altruism relied on some form of group selection (see 2.4.4.1). By accounting for phenotypic altruism in terms of genetic selfishness, inclusive fitness removed the need to resort to group selection. At least as importantly, it also emphasised the necessity for modelling evolutionary outcomes on genetic fitness rather than individual reproductive success. Inclusive fitness and kin selection make sense *only* in a selfish gene perspective, even in cases such as parental care where genetic fitness and individual 'fitness' arguably coincide.

Kin selection is not the only explanation for phenotypic altruism. Apparent altruism can arise between unrelated individuals too. However, the main contender to account for this, **reciprocal altruism** (Trivers 1971), works on the same principle as inclusive fitness and depends on a similar genetic fitness argument, as we shall see later (9.3.2).

2.4.5 Life history strategies and behaviour

Genetic fitness is usually a consequence of the reproductive success of individuals. Reproductive success, however, depends on the allocation of resources not only to reproduction itself, but to the growth and survival necessary to achieve it. How this allocation is best achieved will in turn depend on the kind of environment in which the organism lives, and how reproductive opportunities change with age, size, social status and a host of other factors. The outcome of the allocation constitutes the organism's **life history strategy**.

Life history strategies are concerned with optimising investment of metabolic resources over the organism's lifetime. Since reproduction is the arbiter of gene transmission, the earlier an organism reproduces, the greater the potential long-term spread of alleles coding for early investment. However, many things conspire to delay the time when it is best to reproduce. If the environment provides *ad libitum* breeding opportunities, and/or adult life expectancy is short, early and copious reproduction is likely to be favoured. If, on the other hand, offspring must compete or disperse for limited breeding opportunities, selection will favour later reproduction and an extensive period of growth and parental care. Since there is only so much resource to go round, our expectation is therefore that life history components will be **traded off** against one another in order to maximise the organism's lifetime reproductive success.

Adaptive adjustments in the allocation may be made in different ways: physiological, genetic, developmental or behavioural (Horn & Rubenstein 1984; Daan & Tinbergen 1997; Ricklefs & Wikelski 2002). How and when they are made, however, depends not only on selection pressures from the environment, but also on changing **reproductive value**. The opportunities for reproduction, and their consequences for survival and reproduction in the future, are likely to vary with a number of things, including age, sex, social status and physical condition. Optimal investment policy will therefore vary with them. Rhesus macaques (*Macaca mulatta*) nicely illustrate a relationship between reproductive value and social rank. Here, rank, and thus breeding opportunity, among females depends on sexual maturity and age. Mothers have a higher rank than their daughters, but among sexually mature daughters, younger individuals rank above their older sisters, a position

Underlying theory

Box 2.8 Inclusive fitness and sterile castes in hymenopteran insects

Several species of hymenopteran insect, such as honey bees (*Apis mellifera*), are eusocial; that is they have a reproductive division of labour based on a system of castes, some of which are sterile. Sterile castes appear to have given up their own reproductive potential in order to help the colony queen to produce offspring. Why should they do this? Part of the answer may lie in the unusual sex determination system in these species. Males hatch from unfertilised eggs produced by the diploid queen and are therefore haploid, while females hatch from fertilised eggs and are diploid, a system known as **haplodiploidy**. The coefficient of relatedness, *r*, between females and their mother is the usual 0.5 for diploid individuals, but because males are haploid, females are sure to share all their father's genes, so *r* for father and daughter is 1. If we now work out the coefficient of relatedness between *sisters*, i.e. worker females in the colony, the consequences of this asymmetry become clear. Assuming the queen has been fertilised by a single male, sisters share half their genome via their father, so the probability of a female sharing a paternal allele with a sister is $0.5 \times 1 = 0.5$. Females share the other half of their genome with their mother, but because both are diploid, the probability of sharing a maternal allele with a sister is $0.5 \times 0.5 = 0.25$. The overall coefficient of relatedness between sisters is therefore $r = 0.5 + 0.25 = 0.75$. If a female were to produce offspring of her own, she would be related to them by only 0.5. Haplodiploidy therefore means that it pays a female to help the queen produce reproductive sisters (i.e. new queens, *not* more sterile workers), rather than reproduce independently. It should be noted, however, that the argument hinges critically on the queen mating with only a single male, and that factors other than haplodiploidy predispose species to eusociality and the production of sterile castes (see Andersson 1984a; Bourke 1997; and 9.3.1).

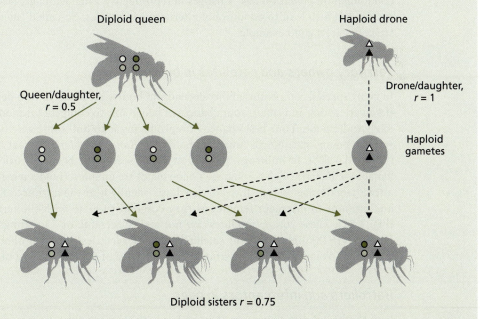

Figure (i)   Potential coefficients of relatedness between queens, drones and workers in colonies of honey bees, see above.

they retain until a still younger sister matures and usurps them. The reason that reaching sexual maturity leads to a rise in status is that mothers increase their support for daughters in aggressive contests just as they become mature, the point at which their reproductive value reaches its peak (Schulman & Chapais 1980). Their older sisters already have less of their reproductive life before them, so are worth correspondingly less to the mother.

A nice example of life history response to environmental selection pressures comes from guppies (*Poecilia reticulata*) in Trinidad. Field and laboratory studies by Reznick & Endler (1982) have shown that guppies tend to mature and breed earlier in habitats where predators of the adult fish are common and the prospect of long-term survival slim. In high-risk habitats, therefore, guppies adjust their life history strategy in favour of growth and reproduction and away from long-term survival. Similarly, differences in predation pressure in lake-dwelling sticklebacks (*Gasterosteus aculeatus*) affect the tendency for males to develop the characteristic red throat during the breeding season. Bright red throats are attractive to females, but they also make males more conspicuous to predators. Moodie (1972) found that males in shallower waters, where red throats were especially conspicuous, had much duller throats than those in more dimly lit deeper waters. In this case, therefore, investment in reproduction (secondary sexual ornamentation) appeared to be traded-off in favour of survival (future breeding opportunity) when predation risk was high.

2.4.5.1 Life histories and alternative strategies

In the stickleback example, environmental selection pressures have resulted in different patterns of life history investment within populations. Male sticklebacks have **alternative strategies** (see 2.4.4.2) of secondary sexual ornamentation – 'bright' and 'dull' – that are conditional on predation risk. Changes in reproductive value with age, health and other factors can also lead to conditional alternative strategies. So-called 'sneaky breeding' strategies are a good example.

Territory owners and satellites in bullfrogs

In many species, competitive ability depends on body size, which in turn depends on age. If you are young and small, your chances of being able to compete for food or mates are likely to be poor. Your best chance of reproducing is to find some way of bucking the system. Young male North American bullfrogs (*Rana catesbiana*) appear to do just this (Howard 1978). In bullfrogs, the largest males secure the best territories and attract females to the territory by croaking. Young males are unable to hold a territory so cannot attract females themselves. Instead, they loiter silently near the territory of a larger male and attempt to intercept females as they come in to mate with the territory holder (Fig. 2.22). These young 'satellites' are not very successful; of 73 matings observed by Howard (1978), only 2 were by satellite males. Nevertheless, this is probably as good as it gets for a young male, whose strategy has to be one of **making the best of a bad job**.

Patrollers and interceptors in bees

In the bullfrogs' case, young, small males eventually become older, larger males, so the best of a bad job satellite strategy is only a temporary expedient. In many insects, adult body size is determined by larval feeding conditions and is fixed. Small adults may therefore

Figure 2.22 Small male bullfrogs (*Rana catesbiana*) (centre) sometimes try to sneak matings by taking up a 'satellite' role (centre) around territorial males (left) and intercepting females (right) that are attracted to their calls. See text. After Krebs & Davies (1993) from Howard (1978).

be stuck with making the best of a bad job for the whole of their reproductive life. An example is the solitary bee *Centris pallida* (Alcock *et al.* 1977). Large male *C. pallida* patrol the ground where virgin females are about to emerge from their burrows. When they encounter an emerging female, they dig her out and attempt to mate with her. Because it takes a while to dig the female out, however, other males are attracted by the activity and there is often intense aggression over possession of the female. As a result, it is only the larger males that adopt a patrol and dig strategy. Small males hover around the general area where females are emerging and attempt to pursue any females that manage to escape the mêlée. Intermediate-sized males may show both patrolling and hovering. Alcock *et al.*'s observations showed that large, patrolling males had by far the greater mating success, so hovering/intercepting males appeared to be making the best of the bad job inflicted by their small size.

Parental and cuckolder males in fish

In some cases, alternative strategies appear to be equally successful. Unlike the best of a bad job strategies above, they may be maintained in evolutionary equilibrium like the alternative burrowing strategies in digger wasps earlier. Mart Gross's (1982) study of bluegill sunfish (*Lepomis machrochirus*) is a good example. Bluegill sunfish are common North American freshwater fish that breed colonially in ponds and lakes. Gross discovered two distinct breeding strategies among males that are reminiscent of the territorial/satellite male system in bullfrogs (Fig. 2.23). He called the two strategies *Parental* and *Cuckolder*. *Parental* males are large and construct breeding nests on the pond/lake floor to which they attract females for mating (Fig. 2.23a). When successfully mated females have spawned, the males then provide all subsequent parental care. *Parental* males become reproductively active late in their life cycle, reaching maturity at about 7 years. *Cuckolder* males, on the other hand, mature when they are still small (about 2 years old). *Cuckolders* adopt one of two age-dependent sneaky breeder strategies. When they are still small, they behave as *Sneakers*, lurking in the vegetation close to a *Parental* male until the *Parental* male attracts a female and the moment of spawning and fertilisation approaches. At this point the *Sneaker* rushes out from cover and darts between the *Parental* male and the female shedding its own sperm on the eggs (Fig. 2.23b). These sneaky fertilisations are accomplished at high speed and the *Parental* male has little opportunity to prevent them.

Figure 2.23 *Parental* and *Cuckolder* males in bluegill sunfish (*Lepomis machrochirus*). (a)–(c) as in the text. After Gross (1982).

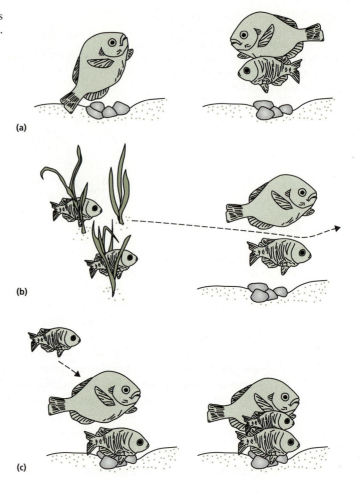

(a)

(b)

(c)

When they are larger and reach the size of adult females, *Sneaker* males become *Satellites* and adopt female-like striped markings. Now when *Parental* male are shedding sperm of the eggs of a spawning female, the *Satellite* male moves in slowly and insinuates himself between the mating male and female and sheds his sperm (Fig. 2.23c). Presumably because the *Satellite* closely resembles a female in size and appearance, the *Parental* male appears not to notice it is being cuckolded and allows the *Satellite* to proceed. *Parental* and *Cuckolder* males represent different developmental pathways with no opportunity for switching between the two (Fig. 2.24; Gross 1982). When Gross looked at the relative reproductive success of the two strategies, he found that about one-fifth of the males entered the *Cuckolder* pathway and appeared to fertilise about one-fifth of the eggs produced. The average fertilisation success of the two strategies thus appears to be the same, suggesting a polymorphism maintained at evolutionary equilibrium within the population. The equilibrium is likely to be maintained by frequency-dependent selection because *Cuckolders* will be at an advantage when they are rare (and there are plenty of *Parental* males to go round), but at a disadvantage when they are common (and have to compete for a small number of *Parentals*).

Figure 2.24 *Parental* and *Cuckolder* males represent an irreversible dichotomy in sunfish life history strategy. About one-fifth (21%) of males become *Cuckolders*, maturing at a much earlier stage than *Parental* males. See text. After Gross (1982).

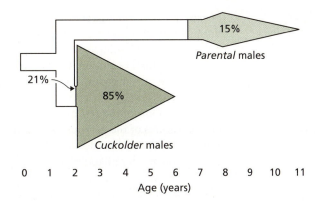

2.5   Cultural evolution

So far we have discussed the evolution of behaviour as it occurs through the differential transmission of genes (natural selection). As Dawkins (1989) has stressed (2.4), the special thing about genes is that they are replicators, entities capable of being copied and transmitted from one individual to another. What genes transmit is information: information to reproduce the phenotypic qualities that facilitated their transmission in the first place. In the case of behaviour, however, information can also be transmitted and propagated in a different way, through observation and learning (so-called horizontal transmission within generations, as opposed to vertical transmission between generations in the case of genes [though cultural information can be transmitted vertically too]; see Chapter 6). As with genetic information, if the copying process is sometimes imperfect, this culturally transmitted information changes, allowing a process of **cultural evolution** (12.1.1). Dawkins (1989) has coined the term **meme** to describe the cultural analogue of the gene, the unit of replication that passes from one individual to another (12.1.1.2).

Memes can be ideas, beliefs, customs, skills, fashions, anything that constitutes an identifiable trait copied between individuals. In much the same way that we can view organisms as the vehicles of selfish genes, created by accumulated copying errors that happened to enhance the transmission of their various mutant genes, so we can look on the brain as a repository and transmission station designed by memes in their battle for cultural supremacy (Dawkins 1976; Blackmore 1999). Dawkins's analogy is persuasive, and Susan Blackmore (1999), in her book *The Meme Machine*, develops it into a general theory of the human mind, an issue to which we shall return in Chapter 12. However, the concept of memes also finds broader application in animal communication. Burnell (1998) applied it to patterns of cultural variation in song in populations of savannah sparrows (*Passerculus sandwichensis*). In this case memes comprised different numbers of syllables in the song. Differences in the degree of similarity between populations for memes of different size suggested that differences in the mutation rate of memes, and their relative susceptibility to random loss (memetic drift), were sufficient to account for the observed patterns of similarity (and difference) in overall song structure across populations. A comparative study of chimpazees (*Pan troglodytes*), while not explicitly evoking memes, has revealed multi-behavioural 'customs', including activities such as tool use, grooming and courtship, that are performed in distinctively different ways across groups (Whiten *et al.* 1999). The characteristic repertoires of behaviour in different groups are very reminiscent of cultural differences between human populations, indeed uniquely so (Whiten *et al.* 1999), and strongly suggest local traditions based on true imitative social learning (see 6.2.1.4 and 12.1.1).

While cultural evolution has its own characteristics and momentum (see Chapter 12), it is dependent on genes in that it requires sophisticated central nervous systems to propagate information. Central nervous systems are built by genes and depend on the life support systems of the rest of the body, which are also built by genes. But as one of the design features of the nervous system, the ability to propagate and acquire cultural traits can also be seen as a product of natural selection, adaptations serving the selfish genes that code for them. This adaptationist view of gene–culture coevolution has been championed in different ways by Lumsden & Wilson (1981) and Cavalli-Sforza & Feldman (e.g. 1973, 1981) (see Feldman & Laland 1996 for a recent review). But different evolutionary relationships between genes and culture, where cultural traits are independent of genotype rather than adaptive products of it, can also be envisaged (Boyd & Richerson 1985; Blackmore 1999). We look at these ideas again and discuss the relationship between cultural and biological evolution more fully in Chapter 12. While the focus of gene–culture models has been on human culture, the approach has begun to influence thinking in relation to other species. The interaction between the cultural transmission of song characteristics and natural selection for song learning in birds, for instance, is explored in the context of gene–culture coevolution by Lachlan & Slater (2000), who conclude that cultural evolution could maintain song learning against naturally selected trends.

Summary

1. Evolutionary studies are concerned with where behaviour patterns have come from and how and why they have changed over time. Although behaviour leaves little in the way of a fossil record, comparative studies of related species can shed light on possible evolutionary pathways. However, such pathways remain speculative unless they can be corroborated using an independent phylogeny. While phylogenetic analyses can verify suggested evolutionary pathways, they do not explain *why* behaviours have evolved the way they have. The prevailing view is that most change is adaptive and driven by Darwin's process of natural selection.

2. Arguments based on natural selection assume that differences in behaviour have a heritable genetic basis. Several lines of evidence support this assumption, and both artificial selection experiments and ecological studies confirm that behaviour patterns respond to selection. Selection experiments also show that evolutionary change at the functional level is accompanied by corresponding change in underlying mechanism, thus emphasising the complementarity of Tinbergen's different levels of explanation (1.2.1).

3. Natural selection describes changes in allele frequencies within population gene pools, usually (but not always) through the effect of different alleles on the reproductive success of their bearers. Individuals thus come to possess those phenotypic characteristics that promote the spread of their encoding alleles, and so appear to act in the best interests of their genes (the basis of Dawkins's 'selfish gene' metaphor).

4. While adaptation is what genes do to enhance their fitness, it is often interpreted in terms of the improved performance of some higher-order entity, such as an individual, population or even species. This leads to the confused view that selection acts for the benefit of one of these higher levels of organisation. It is easy to show, however, that strategies based on this assumption are unlikely to be evolutionarily stable.

5. Optimality theory provides a general mathematical framework for testing evolutionary stability. However, it is based on a set of simplifying assumptions about underlying genetics known as the phenotypic gambit. Despite this, studies comparing the predictions of optimality theory with those of population genetic models suggest that the two accord well when applied to long-term evolutionary outcomes.

6. In many cases, the fitness of an allele depends on its contribution to the reproductive success of its bearer. However, as Hamilton first pointed out, this is too narrow a view of genetic fitness. An individual possessing a particular allele is likely to share it by common descent with others in the population. The allele's effect on these cobearers, weighted by their coefficient of relatedness to the bearer, provides an additional, indirect, route by which it can influence its transmission into the next generation. This extended concept of fitness is known as inclusive fitness, and selection acting via inclusive fitness effects is known as kin selection. Inclusive fitness and kin selection have had a profound effect on our understanding of social behaviour generally and phenotypic altruism in particular.

7. While genetic fitness usually relies on the reproductive success of individuals, reproductive success itself depends on the allocation of resources to the three major components of life history: growth, survival and reproduction. How this allocation is best achieved in turn depends on the environment and how reproductive opportunities change with age, size, social status and other factors. Behavioural polymorphisms within populations sometime reflect adaptive differences in life history strategy.

8. Behaviour patterns can spread not only via genes and natural selection, but also through observation and learning. Once again, the copying process is sometimes imperfect, this time leading to a process of cultural evolution. Dawkins (1989) has coined the term meme to describe the cultural analogue of the gene, the unit of replication that passes from one individual to another. While cultural evolution has been discussed mainly in the context of human societies, there are parallels in other species.

Further reading

Dawkins's (1989) *The Selfish Gene* and (1999) *The Extended Phenotype* are mandatory reading for a grasp of the 'selfish gene' approach. Williams (1992), Keller (1999) and Freeman & Herron (2001) provide thorough discussions of evolutionary theory and analysis generally, and Maynard Smith (1982) and the volume edited by Dugatkin & Reeve (1998) good accounts of games theory in an evolutionary context. Moore (2002) reviews the evidence for manipulation of host behaviour by parasites. The paper by Brockmann (2001) takes a critical look at the idea of alternative strategies and tactics, while Hamilton's (1996) *Narrow Roads of Gene Land*, volume 1 is an unmissable opportunity to appreciate his landmark thinking on altruism through the original papers.

Physiological mechanisms and behaviour

Introduction

Organised behaviour patterns are the result of sensory and motor integration in the organism. In most animals, this is the province of the nervous system. To what extent have the anatomy and organisation of the nervous system been shaped by the behavioural needs of the animal? Are there specific centres and pathways in the nervous system dedicated to the control of particular behaviours? Are adaptive behaviours 'hard-wired' in this sense? Nervous systems coordinate the animal's response to events in its environment using a range of sensory modalities. But at any one time the animal's sense organs are bombarded with information, most of which is irrelevant to its immediate needs. How is important information sifted from all the 'noise'? The nervous system is not the only means of coordinating behaviour. Hormones can also play an important role. But how do hormones affect behaviour, and how do they work alongside the nervous system? Physiological mechanisms serve functional outcomes, but function and mechanism can mutually constrain one another. What is the evidence for such constraints?

Behaviour is the tool with which the animal uses its environment. Through behaviour, an animal manoeuvres itself in an organised and directed way and manipulates objects in the environment to suit its requirements. In order to do this, it must act as an integrated and coordinated unit. It must juggle a bewildering array of stimuli from inside and outside its body and organise the information so as to generate appropriate commands to its muscles or other effector systems. When we ask what causes an animal to behave in a particular way at this level, we are asking about proximate causation or mechanism, the third of Tinbergen's Four Whys (Box 1.1). As we have noted (1.3.1), attempts to understand mechanisms of behaviour divide into two kinds: the 'nuts and bolts' approach, which tries to account for behaviour in terms of the animal's physiological mechanisms, and the motivational approach, which regards physiological mechanisms as a 'black box' and instead seeks general rules to explain the translation of internal and external stimuli into behavioural output. We shall look at the second of these in Chapter 4. In this chapter, we shall be concerned with 'nuts and bolts' explanations – accounts of behaviour in terms of physiological mechanism (Fig. 3.1). In particular, we shall look at the way underlying mechanisms reflect the adaptiveness of the behaviours they control. We shall thus be concerned once again with the complementary relationship between proximate and ultimate explanation. Before we begin, however, we need to know something about the physiological systems that constitute the animal's internal mechanisms and the general trends in their evolution.

Multicellular animals have evolved complex systems of cells and chemicals whose task it is to detect, transmit, integrate and store information supplied by the animal's internal and external environment for use in making decisions. They consist of: (a) various types of sensory cell, which detect different changes in the environment, (b) a more or less complex

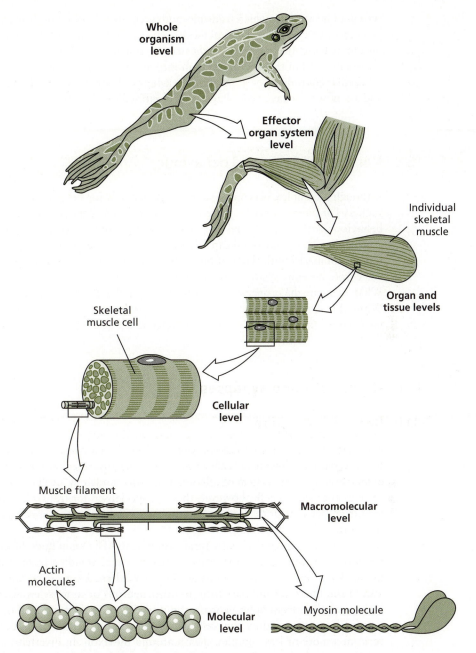

Whole organism level

Effector organ system level

Individual skeletal muscle

Organ and tissue levels

Skeletal muscle cell

Cellular level

Muscle filament

Macromolecular level

Actin molecules

Molecular level

Myosin molecule

Figure 3.1 In principle, the coordinated behaviour of the whole organism can be explained at the level of its effector organ systems, and onward down to the cellular and molecular substructure of each effector system (exemplified here in terms of skeletal muscle). 'Nuts and bolts' explanations of behaviour are usually pitched at the level of effector organ systems and their respective neural and hormonal control mechanisms. See text. After Randall *et al.* (1997).

system of nerve cells, which transmits and integrates information from sensory receptors and other nerve cells, (c) chemical messengers, which transmit information round the body on a more leisurely timescale than the nervous system, and (d) muscle cells, which translate information from the nervous system into actions. Many single-celled organisms have subcellular equivalents of some of these elements. We shall look at various of them in turn and ask how they have been shaped and organised to allow animals to behave adaptively.

3.1 Nervous systems and behaviour

Although a complex nervous system is not essential for behaviour – motile protozoans such as *Paramecium* get by quite nicely without one – the scope and sophistication of behaviour within the animal kingdom is broadly linked with the evolution of neural complexity. The behavioural capabilities of sea anemones and earthworms are extremely limited compared with those of birds and mammals, though it is important to note that vertebrate nervous systems are not always more complex than those of invertebrates. What, then, are the properties of a nervous system that make complex behaviour possible? While this is not the place to discuss neural anatomy and physiology in detail, some basic understanding is necessary to appreciate the role of neural circuits in controlling behaviour patterns.

3.1.1 Nerve cells and synapses

3.1.1.1 Nerve cell structure

True nervous systems are found only in multicellular animals. Here they form a tissue of discrete, self-contained nerve cells or **neurons**. Like any other type of animal cell, neurons comprise an intricate system of cell organelles surrounded by a cell membrane (Fig. 3.2a). Unlike other animal cells, however, they are specialised for transmitting electrical messages from one part of the body to another, a specialisation that is reflected in both their structure and their physiology.

A neuron has three basic structural components. The main body of the cell, or **soma**, is a broad, expanded structure housing the nucleus. Extending from it are two types of cytoplasm-filled process called **axons** and **dendrites**. Axons carry electrical impulses away from the soma and pass them to other neurons or to muscle cells, while dendrites receive impulses from other neurons and direct them to the soma. All three components are usually surrounded by **glial cells** which, though not derived from nerve tissue, come to form a more or less complex sheath around the axon. In invertebrates, the glial cell membranes may form a loose, multilayered sheath in which there is still room for cytoplasm between the layers (*tunicated* axon). In vertebrates, the sheath is bound more tightly so that no gaps are left. The glial cells are now known as **Schwann cells** and are arranged along the axon in a characteristic way. Each Schwann cell covers about 2 mm of axon. Between neighbouring cells, there is a small gap, known as a node of Ranvier, where the membrane of the axon is exposed to the extracellular medium. Axons with this punctuated Schwann cell sheath arrangement are called **myelinated** or medullated axons, and the formation of the myelin sheath greatly enhances the speed and quality of impulse conduction.

Figure 3.2 (a) A motor neuron connecting with a muscle fibre (see text). (b) The relationship between ion flow and membrane potential during an action potential in a non-myelinated axon. In squid neurons, the concentration of potassium (K^+) ions is some 20 times greater inside the cell than outside, whereas sodium (Na^+) ions are about 10 times more concentrated outside. Diffusion of ions towards equilibrium is counteracted both by the low permeability of the cell membrane to Na^+ and by a metabolic pump which transfers the ions against their gradients. The net result of this ionic imbalance is a negative **resting potential** across the membrane of some −60 to −70 mV, depending on the neuron. When the neuron is stimulated, the membrane suddenly becomes highly permeable to Na^+ at the site of stimulation and there is a massive influx of Na^+ which results in a sharp **depolarisation** to around +40 mV known as an **action potential**. The formation of an action potential at one part of the membrane stimulates an increase in Na^+ permeability in the adjacent part, and a wave of depolarisation courses down the axon. As soon as the action potential has passed a given point, the resting potential is then restored by a metabolic pump. Modified from Barnard (1983) after Adrian (1974).

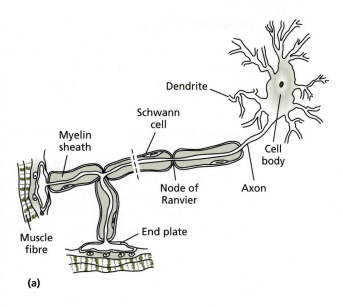

(a)

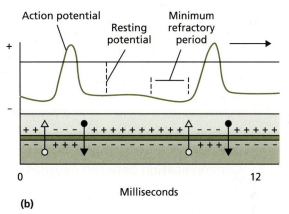

(b)

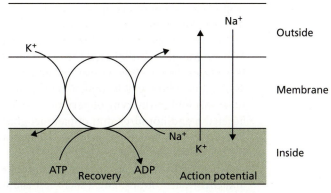

(c)

3.1.1.2 Nerve cell function

The basis of the neuron's ability to conduct electrical impulses (action potentials) is the distribution of electrically charged ions inside and outside the cell (see Fig. 3.2b). The rate of conduction of the action potential depends on how good a conductor the axon is. One way to increase conductivity is to make the diameter of the axon larger, as has occurred in the 'giant' axons of squid and earthworms (Annelida, Oligochaeta). However, increasing the size of axons is costly in both materials and space. Insects and vertebrates have solved the conduction problem by insulating their axons instead. Their myelinated (see above) axons allow the flow of current only through certain areas (the nodes of Ranvier [Fig. 3.2a]). Consequently, action potentials are conducted from node to node extremely rapidly.

In most neurons, action potentials are all-or-nothing events, though in some arthropod sensory receptors and certain other specialised cells, where short-distance communication is required, information can be coded as graded potentials. Neurons are limited in the kind of information they can transmit because they have only two states: 'on' or 'off'. Transmitted information is therefore digital. Different grades of information can be transmitted by neurons, but only through changes in the frequency of depolarisations (Fig. 3.2b), not by changes in their intensity. Increasing the strength of the stimulus will simply generate more depolarisations, not bigger ones. The mechanical reason for this comes back to the nature of depolarisation, which depends on a fixed influx of sodium ions before the neuron reverts to its negative resting potential. The functional reason seems to be to preserve the integrity of the information as it is transmitted down sometimes lengthy axons. The simple signal 'on' can be transmitted reliably from one end of the axon to the other, almost regardless of length. A graded signal, however, is susceptible to the vagaries of signal decay over long distances: what starts out as 90% 'on' at one end of the axon might dwindle to 30% 'on' by the time it gets to the other end.

3.1.1.3 Communication between neurons

The membrane boundaries of neurons are complete, and the contact between cells that is essential to the transmission of impulses is accomplished by close juxtaposition rather than the formation of a continuous syncytium (collection of cells without cell walls). The region of juxtaposition is known as a **synapse,** a term that is also applied to neuronal junctions with sensory receptors and muscle fibres. In most cases, transmission across the synapse occurs through the medium of transmitter substances known as **neurotransmitters,** although electrical communication is known where the juxtaposition is very close (e.g. in invertebrate 'giant' axons [3.1.3.1]). Various neurotransmitters are known, including acetylcholine (an excitatory transmitter at vertebrate neuromuscular junctions and in the central nervous systems of both vertebrates and invertebrates), adrenalin and noradrenalin, dopamine, glutamate, γ-aminobutyric acid (GABA) and 5-hydroxytryptomine (serotonin). Neurotransmitters have important consequences for behaviour and show some differences between vertebrate and invertebrate nervous systems. Apart from the general excitatory or inhibitory properties of acetylcholine and GABA, neurotransmitters such as β-endorphin (one of the opioid group), dopamine and serotonin have important effects on many things, including mood, pain perception, sleep, attention and learning. Synapses facilitate complex cross-connections between neural pathways and are a crucial feature in the evolution of behavioural coordination and integration.

However, not all communication between neurons is by chemical transmission across synapses. Non-synaptic interactions are also recognised. Recent evidence, for example, suggests that nitric oxide can establish non-synaptic communication between glutamatergic neurons and surrounding cells where transmission is mediated by monoamine. These interactions appear to be important in the function of regions of the brain concerned with learning, memory and the coordination of movement (Kiss & Vizi 2001).

3.1.2 Evolutionary trends in nervous systems and behaviour

As we ascend the evolutionary scale from simple unicellular organisms to vertebrates, the organisation and complexity of sensory and motor responses, and, in multicellular animals, nervous systems, change in two major ways: first towards greater **differentiation**, and second towards greater **centralisation** (see Guthrie 1980).

3.1.2.1 Unicellular organisms

Even in unicellular organisms, without any nervous system at all, there can be spatial differentiation between sensory and motor functions. *Paramecium*, for example, is an aquatic protozoan covered with motile cilia (hair-like processes) that propel it in a helical fashion through the water. When *Paramecium* collides with an obstacle, a mechanoreceptor at the anterior end is stimulated, causing the direction of beating of the cilia to be reversed and the organism to back away and move off in a different direction (Fig. 3.3a). If it is hit from the rear, a posterior mechanoreceptor triggers a forward thrust. But how do the mechanoreceptors manage to communicate information to the cilia in the absence of nerve fibres? The answer seems to be by causing changes in the electrical potential of the cell membrane, which are then picked up by each of the cilia in turn. In short, the entire organism acts like a single nerve cell. As in a neuron, there is a negative resting potential across the membrane. When the anterior mechanoreceptor fires, it causes a drop in the potential which spreads over the whole cell and is detected by voltage-sensitive channels around each of the cilia (Fig. 3.3b). The channels open to allow calcium to flow into the cell down its concentration gradient. The calcium then interacts with the cilia to reverse their beat before being pumped out of the cell again. Stimulation of the rear receptor increases potassium permeability, which raises the membrane potential and increases the rate of beating of the cilia and thus forward propulsion. Thus the location of the mechanoreceptors determines the sign of the change in the membrane potential and, through that, the direction of beating of the cilia and the subsequent movement of the organism.

3.1.2.2 Nervous systems and behaviour in invertebrates

Although *Paramecium* can respond in a directed way to stimuli from its environment, control is limited because the means of linking receptors to effectors is relatively unrefined. In multicellular animals, this role is taken over by the axons and dendrites of neurons and the various synapses between them. In an advanced nervous system, there are five major components linking stimulus perception and motor response: (a) a cell or group of cells acting as a sensory receptor, (b) an **afferent** or **sensory neuron** carrying impulses from the sense cells, (c) an **efferent** or **motor neuron** carrying impulses to effector cells, (d) an internuncial neuron or **interneuron** linking sensory and motor neurons, and (e) an effector organ (that performs the motor task).

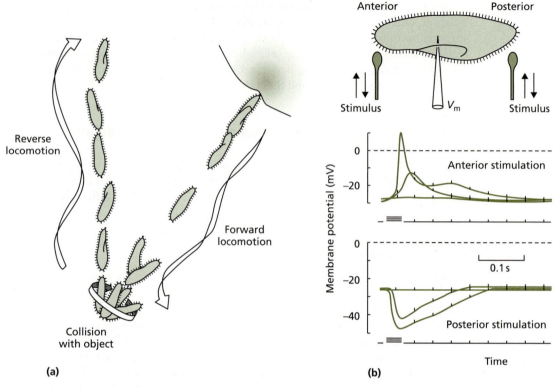

Figure 3.3 The avoidance response of *Paramecium*. (a) After colliding with an object, *Paramecium* reverses its direction of locomotion and backs away. (b) Forward and reverse movements are driven by different changes in membrane potential (curves show responses to three different stimulus intensities). See text. After Collett (1983) and Randall *et al.* (1997).

Nerve nets

Examples of this link-up system at its simplest are found in the nerve nets of cnidarians (sea anemones, jellyfish and their relatives) and echinoderms (starfish, sea urchins, etc.). The simplest kind of net is the type found in the sedentary freshwater cnidarian *Hydra* (Fig. 3.4a), which lies just under the epidermis and consists of a series of synaptically linked bipolar and tripolar (two and three connections respectively) cells. Transmission is slow because impulses have to traverse large numbers of synapses and lack directionality, thus dissipating in several different directions. Trends in other cnidarians, which still possess nerve nets, are towards the differentiation of nerve cells into fast-conduction tracts (e.g. Fig. 3.4b), mainly through the lengthening and thickening of individual axons. Impulses can then be channelled in particular directions to bring effector organs into play more rapidly. The rapid withdrawal reaction of anemones such as *Actinia* and *Metridium* is a good example.

The behaviour of animals relying on nerve nets is characteristically stereotyped (unvarying). Most exhibit simple reflexes (see 3.1.3.2), stereotyped motor sequences and rhythmic locomotory activities. Even more elaborate behaviour patterns, such as shell-climbing in epizooic (living on other animals) anemones, consist of only three or four elements. There is relatively poor stimulus discrimination, and such learning as exists consists of habituation (6.2.1.1) and reflex (3.1.3.2) facilitation rather than any

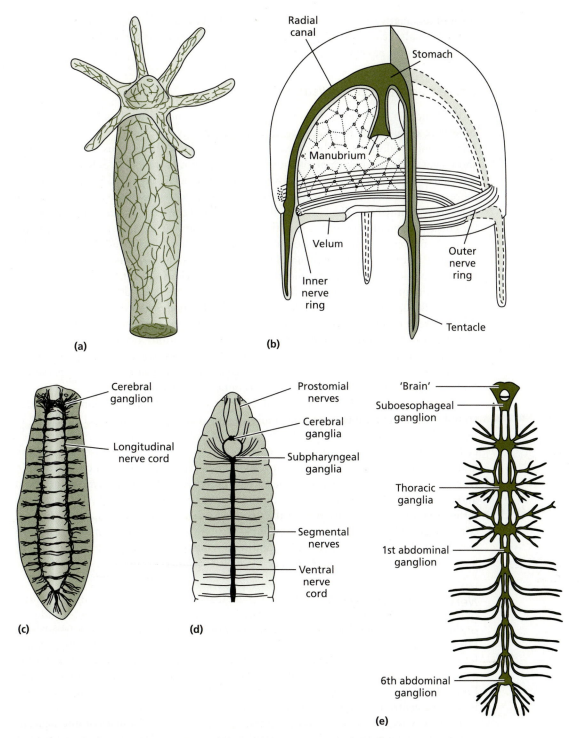

Figure 3.4 (a) The nerve net of *Hydra*; (b) the concentration of neural tissue into nerve rings in a hydroid medusa; (c) the CNS of a turbellarian flatworm; (d) and (e) the segmental arrangement of ganglia and nerves in the CNS of an oligochaete worm (*Lumbricus terrestris*) and a cockroach (*Periplaneta americana*). After Chapman & Barker (1966) and Barnes (1968).

form of associative learning (6.2.1.2). However, these simple responses function well in the animals' relatively stable aquatic environment with its ample food supply and opportunities for passive dispersal.

Nerve tracts and centralisation

In the Cnidaria, through-conduction tracts are little more than local channels in an otherwise diffuse net. In relatively more advanced invertebrates, such as flatworms (Platyhelminthes), the tracts become more pronounced and the nervous system begins to show signs of the second major evolutionary trend, *centralisation* (Fig. 3.4c). Even within the Cnidaria there is a trend towards more deeply seated nerve rings (Fig. 3.4b) and tracts in more mobile species, but flatworms are the first level of life to possess a recognisable **central nervous system (CNS)**. They are also the first to show the early stages of **cephalisation**, the concentration of nervous tissue in the head region into an anterior ganglion or simple brain. In bilaterally symmetrical organisms (those with distinct 'head' and 'tail' ends), incoming information arrives mainly at the front as the organism moves forward through its environment. Consequently there is an obvious advantage in concentrating sensory and integration centres at the anterior end. Two nerve cords, linked by nerves in a ladder-like arrangement (Fig. 3.4c), extend down the body from the anterior ganglion. Nerve fibres extend from the cords to all regions of the body in a network arrangement, constituting the **peripheral nervous system**. The division into a central and peripheral nervous system is common to most invertebrates and all vertebrates, but is seen in its simplest form in the flatworms. In general, the CNS houses most of the motor nerve cell bodies, while the peripheral nervous system contains the sensory receptors.

Sensory cells in the head region of flatworms such as *Planaria* respond to various stimuli, including temperature, touch and chemical changes in the water. Changes in light intensity can also be registered by a pair of eyespots (clusters of photoreceptor cells). Impulses from the various sensory receptors are routed to the anterior ganglion and from there to the appropriate muscles. The nerve cords allow much more rapid transmission of impulses than nerve nets, with a consequent enhancement in the speed and variety of behavioural responses to different environmental stimuli. The increased differentiation and centralisation of the nervous system in flatworms is also associated with a degree of learning ability, for example learning which way to turn in a T-maze to avoid a noxious mechanical stimulus, and with relatively sophisticated mechanisms of assessment during mate choice (Vreys & Michiels 1997; see Chapter 10).

Nerve cords and ganglia

In the higher invertebrates, which include the **metamerically segmented** (animals with a serially segmented body plan) annelids (earthworms, ragworms, etc.) and arthropods (crustaceans, insects, spiders, etc.), and the non-metamerically segmented (it is generally assumed) molluscs, the nervous system has become differentiated into a series of ganglia linked by nerve cords lying near the ventral surface of the body (Fig. 3.3c–e). This increasing centralisation has produced a kind of neural 'switchboard'. Afferent fibres from sensory receptors plug into the central switchboard where a mass of interneurons is ready to connect them with a variety of motor neurons. Depending on the type of input from a sensory receptor, different motor neurons are brought into play so that the animal can respond appropriately.

Ganglia (other than cerebral ganglia) in the CNS may contain anything from 400 cells (in leeches) to over 1500 (in the mollusc *Aplysia*). Well-defined tracts and **glomeruli** (aggregations of neuron terminals) can also be distinguished, especially in the cerebral ganglia which may form an elaborate brain-like structure. The primary function of each ganglion is the regulation of local reflex arcs, providing for a degree of local control of movement impossible with a diffuse nerve net. Ganglia also exercise longer-range control via long interneurons extending along the nerve cords, thus facilitating the coordinated operation of different parts of the body. A number of behavioural advances are associated with these developments. In particular, elaboration of appendages and musculature, aided by the emergence of a fluid-filled body cavity (the **coelom**), makes subtle movements and complex, manipulative tasks possible, as, for example, in the web-building activities of spiders and the elaborate courtship songs and ornamented nest constructions of some insect species. Stimulus discrimination and learning also show advances over species with less structured nervous systems. However, learned responses are seldom retained for long, probably because of the small capacity of the cerebral ganglia, itself perhaps a reflection of the short-generation life cycles of many invertebrate species – where there is little time for sophisticated learning, simple pre-programmed responses may be more economical (see Chapter 5).

Among the invertebrates, there is a trend towards enlargement of the 'brain' by amalgamation of somatic ganglia, and an increase in the brain's control over regional centres. Despite the increasing importance of the brain, however, the somatic ganglia still retain considerable independence of control. Earthworms, for instance, can crawl normally, feed, copulate and burrow after removal of the cerebral ganglia, though they are hyperactive and their movements phrenetic. Nereid species (ragworms) are able to learn certain tasks, or persist with previously learned tasks, after disconnection of the cerebral ganglia from the rest of the CNS. The cerebral ganglia thus appear to be just one of several memory storage sites.

Independent control of behaviour by somatic ganglia is particularly well developed in arthropods. If still connected to its ganglion, the isolated leg of a cockroach (*Periplaneta americana*) will continue to show stepping movements when stimulated appropriately by pressing on the trochanter (articulating joint in the upper leg). Indeed, even if the ventral nerve cord is completely severed, coordinated walking can be elicited by stimulating a single leg. Movement of one leg exerts a traction force on the leg behind, stimulating specialised proprioceptors (cells sensitive to mechanical stimulation) called campaniform organs and eliciting a reflex response (Zill & Moran 1981). Even in cephalopods (squid and octopuses), where cephalisation has reached its peak within the invertebrates, many responses are still under the control of somatic ganglia. A constraint on the octopus's (*Octopus vulgaris*) otherwise impressive object learning prowess is its inability to distinguish objects by weight. This is because the movements of each tentacle are regulated by local axial ganglia, so that information from proprioceptors in the tentacle are processed in the ganglia rather than being passed to the brain where it could become available for learning.

Evolutionary trends in invertebrate brains

The brains of invertebrates vary considerably in structure and complexity. At the lower end of the scale, flatworms brains contain some 2000 cells. Insect brains are intermediate, with around 340 000, whereas those of cephalopods contain up to 170 million, almost a tenth of the number found in the human brain. Despite its outwardly advanced appearance, however, the anatomy of the cephalopod brain reveals its derivation from amalgamated somatic ganglia. The cephalopod CNS contrasts sharply with the loose string of ganglia (comprising fewer than 50 000 cells in total) found in the slower moving gastropod (slugs,

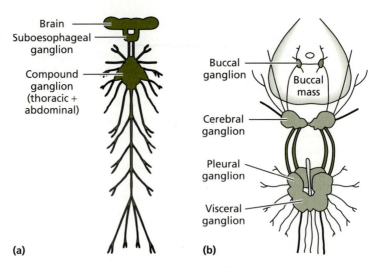

Figure 3.5 The trend towards fusion of ganglia in the CNS of higher invertebrates: (a) the house fly (*Musca domestica*) and (b) the Roman snail (*Helix pomatia*). After Chapman (1971) and Barnes (1968).

snails) and lamellibranch (bivalve) molluscs. There are separate visual and tactile learning centres within the lobes of the brain, and centres for the integration of visual information from the high-resolution eye, and tactile information from the mobile tentacles.

Among arthropods, brain structure is remarkably conservative. In common with annelids, it develops as three main regions: (1) the *protocerebrum*, the main components in arthropods being the paired optic lobes, the median body and the *corpora pedunculata*, each functioning in the integration of information from the anterior sense organs and the control of subsequent behaviour; (2) the *deuterocerebrum*, containing association centres for the first antennae; and (3) the *tritocerebrum*, nerves from which extend to the upper alimentary canal, second antennae (where they exist) and the upper 'lip'. The same functional zones can be recognised throughout most of the phylum and are derived from homologous zones in the polychaete annelids (ragworms). Within broad taxonomic groups, the relative size of different centres in the brain reflects differences in predominant sensory modalities and lifestyle (see 3.1.3).

The gross plan of the CNS in arthropods, with its central nerve cord and ganglia in each segment, is broadly similar to that in annelids. The segmental arrangement is clearly discernible in primitive arthropods, but is obscured in advanced forms by extensive fusion of ganglia (Fig. 3.5a), a development that also characterises more advanced nervous systems in molluscs (Fig. 3.5b). The general tendency towards fusion of ganglia in invertebrate nervous systems goes hand in hand with the evolution of more sophisticated sensory systems and behaviour.

3.1.2.3 Nervous systems and behaviour in vertebrates

In vertebrates, the nervous system develops from dorsal tissue and as a tube rather than as a solid structure. Nevertheless, traces of the ancestral segmental pattern remain in the distribution of sensory and motor zones within the system. Centralisation, cephalisation and the functional differentiation of the nervous system reach their peak in the vertebrates. Indeed, while structural centralisation occurs at various levels of the evolutionary scale, true centralisation of *function* is the exclusive property of the vertebrate nervous system.

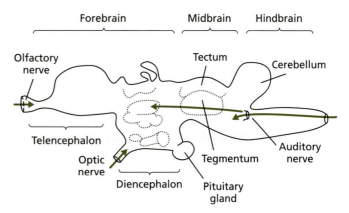

Figure 3.6 A generalised scheme of the vertebrate brain, showing the major regions and pathways of sensory input (arrows). After Guthrie (1980).

Structure and function in the CNS

The vertebrate CNS consists more clearly than that of invertebrates of two principal components: the brain and the spinal cord. The spinal cord comprises an outer region of mostly myelinated tracts (so-called 'white matter') connecting the brain with spinal control centres, and an inner region of neuron cell bodies ('grey matter'). The *dorsal horns* of the inner region accept afferent, sensory neurons entering the CNS, while the *ventral horns* send out efferent motor fibres. The arrangement of axon and cell body material in the vertebrate spinal cord is the opposite of that in the invertebrate CNS, where the open circulatory system requires cell bodies to be exposed in the haemocoel (body cavity).

The brain develops as three major regions: the forebrain (*prosencephalon*), the **midbrain** (*mesencephalon*) and the **hindbrain** (*rhombencephalon*) (Fig. 3.6). The forebrain is further divided into two parts: the anterior *telencephalon*, associated with olfaction and giving rise to the cerebral cortex, and the posterior *diencephalon* with its pathways connecting with the pituitary gland. The roof of the midbrain houses the optic lobes, or optic tecta in lower vertebrates, while a dorsal projection of the hindbrain, the *cerebellum*, acts a major coordination centre for movement by integrating messages from receptors in joints and muscles, though it also plays a role in higher brain functions and cognition in advanced vertebrates (Riva 2000). The hindbrain is also divided into two parts: the *metencephalon*, containing the anterior part of the *medulla oblongata*, the cerebellum and, in mammals, the *pons* and the *myelencephalon*, which contains the posterior medulla oblongata. The medulla oblongata is the centre of control for certain vital functions, including breathing and blood vessel tone. It contains a number of nuclei for the cranial nerves, including the vagus nerve, which is concerned with heart and gastro-intestinal reflexes, and large numbers of fibres pass through it on their way between the spinal cord and the forebrain cortex.

The relatively simple arrangement of 'white' and 'grey matter' in the spinal cord is extensively modified in the brain. Here it is the central regions surrounding the fluid-filled ventricles that originate in the outer regions of the spinal cord. The outer areas of the brain are specialised tissues consisting of dense masses of cell bodies homologous with the 'grey matter' of the spinal cord. The regions of the brain other than the cerebral cortex, and sometimes other dorsal lobes, are often referred to as the **brain stem**. The brain stem is the most primitive part of the brain. As well as acting as a conduit for information from the body flowing into the brain, various centres in the brain stem determine general alertness and regulate, as we have seen, the automatic maintenance processes of the body such as breathing and circulation.

Evolutionary trends in vertebrate brains

Vertebrate brains show two main evolutionary trends. The first, exemplified by the bony fish (Actinopterygii), is an elaboration of the midbrain in which the optic tectum becomes thickened and stratified and acts as a major integration centre for information from other parts of the brain. The diencephalon lies underneath it and differentiates into the central *thalamus* and the ventral *hypothalamus* and pituitary. The second trend, shown in mammals, involves the elaboration of the cerebral hemispheres of the forebrain, which now become the major association centres (Fig. 3.7a,b). The forebrain of the lower vertebrates remains in the form of the *hippocampus* and some other ventrolateral

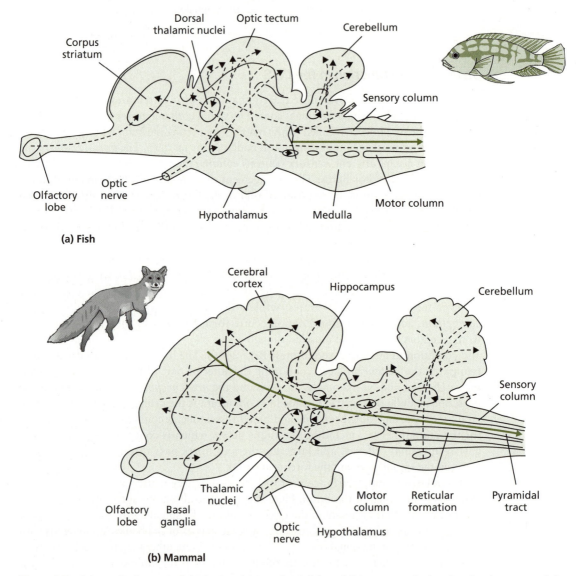

(a) Fish

(b) Mammal

Figure 3.7 Schematic diagram of the brain of (a) a teleost fish and (b) a mammal, showing developments of the major regions (see Fig. 3.6) and connecting pathways (arrows). Note the elaboration of the optic tectum in fish and the cerebral cortex in mammals (see text). After Guthrie (1980).

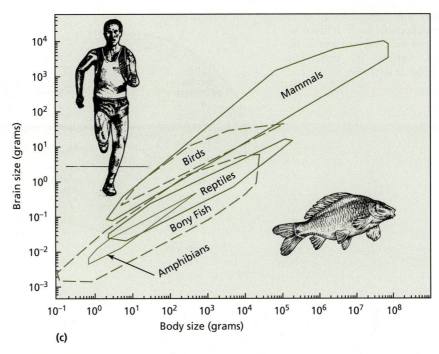

Figure 3.7 (*continued*) (c) The relationship between brain and body mass ('encephalisation quotient') in extant higher and lower vertebrates. The plot shows minimum convex polygons enclosing points for 647 mammal, 180 bird, 59 reptile, 42 amphibian and 1027 bony fish species After Jerison (2001).

elements, but the cerebral hemispheres themselves consist of new material, the **neocortex**, which, in humans, has extended to cover the rest of the brain. A plot of the relationship between brain and body size in vertebrates emphasises the discontinuity caused by the elaboration of the forebrain cortex in higher vertebrates (Fig. 3.7c). In lower vertebrates, the brain accounts for less than 0.1% of the body mass, while in birds and mammals it often exceeds 0.5% with a high extreme of 2.1% in humans.

The neocortex has several externally visible divisions, many of which reflect different functional areas (Fig. 3.8a). In advanced mammals, voluntary motor control areas lie in front of those for somatic sensory functions and are separated from them by the deep *central fissure*. A proportion of the motor cortex cells communicate directly with the spinal cord via a large through-conduction pathway, the *pyramidal tract* (Fig. 3.7b). The two halves of the cortex are connected by another large tract, the *corpus callosum*. Severance of the corpus callosum in humans has revealed pronounced differences in the functional dominance of the left and right hemispheres, for instance in the control of speech (usually a left hemisphere job) and verbal comprehension (right hemisphere).

Further motor control occurs subcortically in various centres, an important one being the *corpus striatum*. In birds, the corpus striatum is an association centre controlling the performance of stereotyped behaviour patterns. In mammals, it houses the *basal ganglia* which coordinate motor control by acting as 'switches' for impulses from different motor systems. Damage to the basal ganglia, or the cells immediately communicating with them (as in Parkinson's disease) results in passive immobility as the nuclei can no longer send motor messages to the muscles. A second set of subcortical nuclei, the limbic system,

Figure 3.8 Lateral views of the human brain showing (a) functional areas of the cerebrum and (b) the limbic system. See text. After Campbell *et al.* (1999).

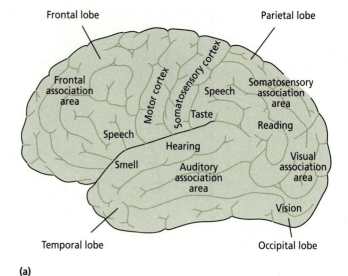

constitutes the hippocampus, *cingulate gyrus, septum* and *amygdala*. The limbic system (Fig. 3.8b) connects with the hypothalamus and the *reticular formation* (a system of branching cells traversing the three main regions of the brain [Fig. 3.7b] and controlling, among other things, the strength and specificity of arousal) and plays a role in associative learning (hippocampus), agonistic behaviour (amygdala) and decision-making.

The thalamus (Fig. 3.8b) contains a number of important nuclei which relay information to the cortex from the retina (*lateral geniculate nucleus*), the ear (*medial geniculate nucleus*), the cerebellum (*ventrolateral nucleus*) and the tectum (*posterolateral nucleus*), and also appears to function in the appreciation of temperature, pain and pleasure. A complex array of nuclei in the hypothalamus regulate both the production of behaviourally important hormones (3.4), via connections with the pituitary gland, and the activity of the **autonomic nervous system**, which is important in, among other things, emotional arousal. The hypothalamus also controls complex behaviours such as sleep, feeding (via the *ventromedial nucleus* [3.1.3.2]) and aggression, and hypothalamic osmoreceptors (cells sensitive to blood sodium concentration) appear to play a role in the initiation of drinking-orientated behaviour.

3.1.3 Nervous systems and the adaptive organisation of behaviour

Clearly, at a gross level, there is an association between organisation within the nervous system and the range and complexity of behaviour shown by different taxonomic groups. While reassuring from an evolutionary point of view, however, the association is hardly a great surprise. Since behaviour results from the coordinated neural control of effector systems, it must trivially reflect something of the underlying organisation of the nervous system. A more interesting question concerns the extent to which variation in organisation reflects adaptive specialisation, both between and within species. We shall look at this in two ways: first by asking whether the gross anatomy of the nervous system, particularly the CNS, reflects different adaptive behaviour patterns, and second by looking at the extent to which adaptive behaviours are 'hard-wired' into the nervous system, that is have identifiable neural circuits dedicated to their control.

3.1.3.1 Neuroanatomy and adaptive behaviour

Can we infer anything about an animal's behavioural specialisations from the anatomy of its nervous system? Given its overarching evolutionary trends towards centralisation and cephalisation (3.2), the obvious place to look is the CNS.

Comparative studies of invertebrates

Lifestyle As we have seen, an early development in the structural organisation of the nervous system was the appearance of rapid through-conduction pathways in the nerve nets of cnidarians (Fig. 3.4). Their purpose is to facilitate a quick response to noxious stimuli in an otherwise slow and inefficient nervous system. The principle of fast through-conduction tracts persists in the nerve cords of annelids and arthropods in the form of the large axons known as 'giant fibres'. But these advanced groups are diverse and show a wide range of adaptive behavioural specialisations. Are some of these reflected in the anatomical arrangement of the giant fibres?

In annelids, a comparison of oligochaete (earthworms) and polychaete (ragworms and their relatives) species reveals differences in both the anatomy and conduction physiology of the giant fibres associated with different lifestyles. In *Lumbricus*, a typical burrowing oligochaete, three myelinated longitudinal giant axons make up some 10% of the cross-sectional area of the nerve cord (Fig. 3.9a). The fibres are syncytial (the neurons are fused without cell walls) to reduce synaptic delay in impulse transmission. The central fibre conducts impulses from head to tail and the two (thinner and slower) lateral fibres conduct in the reverse direction. While there are various segmental connections, the primary function of the giant fibres appears to be to generate fast contraction in the longitudinal muscles and a prompt forward or reverse withdrawal response to mechanical stimulation, important emergency responses in a habitually burrowing organism. Among polychaetes, the even larger (25–70% of the nerve cord, Fig. 3.9b) single or paired fibres in the sedentary fan-worms (e.g. *Sabella*, *Myxicola* and

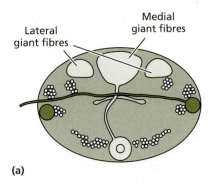

(a)

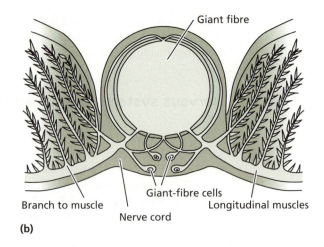

(b)

Figure 3.9 Giant fibres providing fast conduction pathways in the nerve cord of annelids. The size and conduction properties of the fibres differ between species, reflecting differences in their lifestyle and patterns of movement. (a) Transverse section of the median and lateral giant fibres mediating forward and backward withdrawal responses of the earthworm *Lumbricus terrestris*. After Günther & Walther (1971). (b) Section through the nerve cord of the sabellid polychaete *Myxicola*, a syncytium of motor neurons and interneurons which avoids synaptic delay. After Nicol, A.C. (1948) Annelid giant fibres. *Quarterly Review of Biology* **23**: 291–324, reprinted by permission of the University of Chicago Press, © 1948 by The University of Chicago. All rights reserved. (c) Stereogram of the giant fibre system in *Nereis diversicolor*. After Smith (1957). See text.

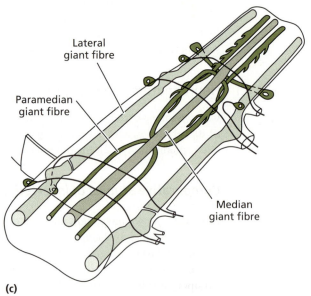

(c)

Branchiomma) serve a similar function, protecting the all-important feeding mechanism, the branchial crown, by withdrawing it rapidly into the burrow. In contrast, the errant polychaetes, such as the ragworm *Nereis*, are active surface predators and occupy burrows only temporarily. They have complex eyes and a battery of sensory tentacles at the anterior end, a large and well-differentiated brain, and a greater locomotory repertoire, ranging from withdrawal and creeping to side-to-side swimming movements with a rotary motion of the parapodia ('legs'). The giant fibres in *Nereis* comprise the same three as in *Lumbricus* but with the addition of two, much thicker, lateral fibres (Fig. 3.9c). They also have more extensive connections and closer association with motor axons than in *Lumbricus*. This has resulted in the central three fibres (the median and paramedians) having control over the parapodia, while control of the longitudinal muscles of the body has devolved to the lateral fibres. The division of labour between the giant fibres, and their various segmental connections, allows the rapid but diverse locomotory control important to a mobile predator.

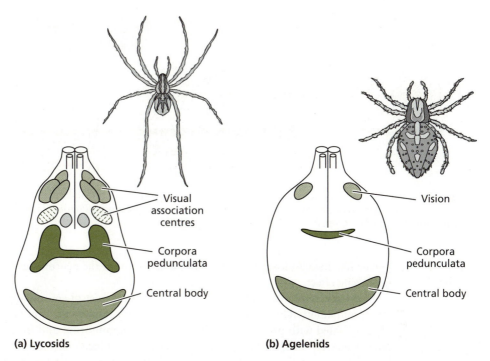

(a) Lycosids

(b) Agelenids

Figure 3.10 A schematic comparison of the major glomeruli in the brains of two groups of spiders: (a) the active, surface-hunting lycosids (wolf spiders) and (b) the web-spinning agelenids, which are sit-and-wait predators. After Hanström (1928).

The brains of annelids and arthropods share a broadly similar glomerular (3.1.2.2) structure (arrangement of neuronal fields analogous to the nuclei of vertebrate brains). Interspecific comparisons, however, show that the relative sizes of different glomeruli can be related to differences in predominant behavioural modalities.

Hanström (1928) compared the major glomeruli in the brains of two groups of spiders: the actively hunting lycosids (wolf spiders) and the web-spinning agelenids (Fig. 3.10). Compared with web-spinning species, lycosids had much more extensively developed optic centres and corpora pedunculata, glomeruli involved in sensory association and visual memory. Web-spinners, on the other hand, had a somewhat larger central body, a region associated with the integration of pre-programmed behaviour patterns. The internal anatomy of the CNS thus reflected the different lifestyles of the two groups and their different demands on sensory integration and motor skills. A comparison across different groups of insects reveals a more general association between the size and structure of the corpora pedunculata and behaviour, particularly in relation to social organisation and the spatial complexity of foraging behaviour (Howse 1974).

Life history strategies Differences in neural organisation also relate to sex and life history strategy (2.4.5). The fruit fly *Drosophila melanogaster*, for example, shows a sexual dimorphism (difference) in locomotory activity: males exhibit a steady, stereo-typed walking pace, while females show greater variability. Using transgenic techniques, Gatti *et al.* (2000) mapped the control of the dimorphism to a small cluster of neurons in the *pars intercerebralis*, a region associated with the cerebral ganglia. Adult males transformed to express a dominant feminising transgene in this region showed variable

(a)   (b)   (c)

Figure 3.11 Somatic sensory projections from the face (medium shading) and forelimbs (dark shading) in the cortex of different species of otter: (a) the river otter *Lutra canadensis*; (b) the giant otter *Pteroneura braziliensis*; (c) the sea otter *Enhydra lutris*. After Radinsky (1968).

female-like locomotion, while ablation (destruction) of the neuron cluster prevented the feminising effect.

Many invertebrates, and some lower vertebrates, exist in entirely different forms as larvae and adults. The changes from one life history stage to the other (**metamorphosis**) are associated with profound changes in neural organisation and function as muscles take on entirely different tasks. Studies of the effects of metamorphosis at the neural level have shown dramatic increases in the dendritic branching of neurons as the system re-wires to control new functional groups of muscles (Truman & Reiss 1995; see Fig. 6.2).

Comparative studies of vertebrates

Lifestyle Elaboration of different parts of the nervous system in vertebrates, but particularly the brain, also correlates well with the broad behavioural characteristics of the animal. However, such specialisations occur in the context of more general long-term evolutionary trends within the CNS, as shown in Fig. 3.8(b). It is thus important to take account of other factors that affect the relative size of different parts of the brain when invoking adaptation to particular ways of life as an explanation for differences in nervous system structure.

Comparative studies between closely related species overcome some of the problems of phylogenetic 'noise'. Otters provide a nice example (Fig. 3.11). As a group, otters show a range of different foraging skills, including a tendency to manipulate food with the forepaws. These manual skills, however, vary among species. At one end of the spectrum, the clawless sea otter, *Enhydra lutris*, has dexterous forepaws that it uses to break shellfish against a stone anvil balanced on its chest. Its clawed relatives, such as the river otter *Lutra canadensis* and the South American giant otter *Pteroneura braziliensis*, on the other hand, use their forepaws in a less specialised way and place more emphasis on sensory information from the face and vibrissae. When he compared the brains of different species, Radinsky (1968) found that the forelimb projection in the cortex was much bigger in sophisticated handlers such as *Enhydra* than in manually less dexterous species such as *Lutra* and *Pteroneura*, while the projection field from the face and vibrissae was more extensive in the latter species (Fig. 3.11).

Song is a conspicuous feature of many bird species and can range from a handful of simple notes to constructions of stunning complexity and beauty (Chapter 11). The control system for song consists of several discrete nuclei in the forebrain hemispheres that

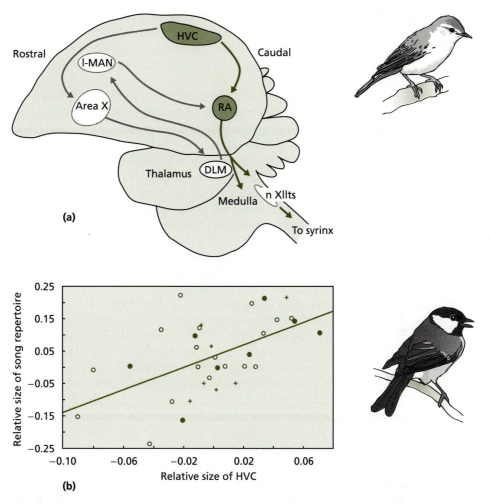

Figure 3.12 (a) Schematic diagram of the major nuclei making up the song system in the cerebral hemispheres of birds (see text). All nuclei occur in both hemispheres, but with no projections between them in most cases. DCM, medial dorsolateral (thalamic relay) nucleus; n XIIts, the tracheosyringeal part of the hypoglossal motor neuron controlling the syrinx. From *Brain, Perception, Memory: Advances in Cognitive Neuroscience* by J.J. Bolhuis (2000). Reprinted by permission of Oxford University Press. (b) The correlation between the size of the song repertoire and the volume of the HVC in birds. Based on analysis at the level of species (solid circles), genera (open circles) and families (+). After DeVoogd *et al.* (1993).

ultimately project to the syrinx, the vocal organ in birds (Nottebohm *et al.* 1976, 1990; DeVoogd *et al.* 1993; Fig. 3.12a). These nuclei fall into two groups: those such as the so-called 'higher vocal centre' (HVC) and the *robustus archistriatalis* (RA) that play a role in song production, and others such as Area X and the *lateral magnocellular nucleus* (l-MAN) that are concerned with song acquisition. DeVoogd *et al.* (1993) compared the volume of two key nuclei, the HVC and Area X, across 45 species of songbird which differed in two measures of song complexity: repertoire size (the number of different songs sung) and the number of syllables per song. In doing so, they used a DNA phylogeny (2.2) of the sample to control for statistical non-independence between related species. When this was done, a significant positive correlation emerged between the size of the song repertoire and the volume of the HVC (Fig. 3.12b). Species with larger repertoires had larger nuclei associated with the control of song production. While no correlations emerged with the number of syllables in a song or with the size of Area X across the sample as a whole, syllable number did correlate positively with HVC size among warblers (Sylviidae) where song complexity in males has been under strong sexual selection (2.4.1) (Szekèly *et al.* 1996).

Sex differences in brain structure   Features of the vertebrate nervous system differ between the sexes in relation to the different reproductive roles, and often ecology, of males and females. Many of these differences are driven by hormonal events during development (3.3.1.4), some of which may be mediated by sex differences in the distribution of hormone receptors in the brain (Gahr 2001). In humans, there are three main differences in the structural architecture of the brains of males and females (Carter 1998).

First, differences in sexual behaviour appear to be associated with a dimorphism in the size of a nucleus, called INAH-3, in the medial preoptic area of the hypothalamus. INAH-3 is generally two to three times larger in males than in females, is packed with androgen (male sex hormone)-sensitive cells, and appears to be responsible for male-typical sexual behaviour. Variation in the size or hormonal stimulation of the nucleus within one or other sex appears to be associated with a drift in sexual behaviour towards that more typical of the opposite sex (LeVay 1994). Thus high circulating levels of androgen in women are broadly associated with male-like assertive sexual behaviour, small breasts, low vocal pitch and hirsuteness. More famously (and controversially), LeVay (1991) has also linked male homosexuality to the INAH-3 nucleus. Comparing the brains of deceased homosexual AIDS victims with those of (presumed) heterosexual men who had not died of AIDS, LeVay discovered that the INAH-3 nucleus in his homosexual group was smaller than in the heterosexual group and roughly the same size as in women. While this is suggestive of a neural basis for homosexuality, however, it is certainly not conclusive, partly because sexual orientation in the study is confounded with having died of AIDS (which may itself have affected the size of the nucleus [but see Swaab & Hofman 1990]), and partly because reduced nucleus size may be a *result*, rather than a cause, of homosexual orientation. However, similar differences in the preoptic hypothalamus have recently been found in domestic rams that show homosexual behaviour, and genetic studies subsequent to LeVay's findings have supported the idea that physical developmental causes underlie different sexual orientations (Hamer & Copeland 1994).

A second difference between the sexes lies in the connection between the two cerebral hemispheres. The corpus callosum and anterior commissure, the bundles of nerve fibres connecting the hemispheres (3.1.2.3), are relatively larger in women than in men. The greater connection between the hemispheres in women may be responsible for sex differences in empathy and emotional sensitivity, as more information is able to be passed from the 'emotional' right hemisphere (see 5.3.3) to the 'analytical' left and become incorporated into thoughts and verbal expression. Women also show greater connectivity between the two halves of the thalamus, an important relay centre for sensory information to the cortex (3.1.2.3).

Third, the ageing process of the brain shows some important differences between the sexes. As a rule, men tend to lose more brain tissue as they age and they tend to lose it earlier in life. Moreover, tissue tends to be lost from different areas of the brain in the two sexes: from the frontal and temporal lobes in men, but the hippocampus and parietal areas in women. This may account for some differences in personality and behavioural changes with age: for instance, increased irritability among men and reduced memory and visual spatial skills in women.

Ecology and sexual dimorphism in brain structure While the sexes may differ in general features of their brain anatomy, more specific differences can arise as a result of ecological selection pressures. Selection for spatial memory and navigational skills is a good example.

In many species, it is males that have the greater requirement for spatial awareness and navigational ability, though this depends on the mating system (see Chapter 10). Males frequently maintain larger home ranges or territories and are often more peripatetic than females. Presumably as a result, they also tend to perform better in spatial learning tasks (Williams *et al.* 1990; Astur *et al.* 1998), though the sex difference in spatial learning can depend on the extent to which males are polygynous (mate with more than one female) and thus move around (Gaulin & Fitzgerald 1986). If any part of the brain was particularly associated with spatial memory, we should thus expect it to be better developed in males. Just such a region is the hippocampus, a phylogenetically old part of the brain (see 3.1.2.3) which appears to play a role in spatial awareness and navigation in a wide range of species. Where comparisons have been made, it turns out that males do indeed have larger hippocampi than females, a difference that holds true in both birds and mammals (Healy & Krebs 1992; Roof & Havens 1992). As with many generalisations, however, there is an exception that proves the rule.

Brown-headed cowbirds (*Molothrus ater*) are brood parasites. Female cowbirds search for host nests and, having located them, return later to lay a single egg in each nest. A female may lay up to 40 eggs in the course of a breeding season. Male cowbirds play no part in the location of nests. On this basis, David Sherry and coworkers predicted that the spatial abilities required to locate and return to host nests would have produced a sex difference in the size of the hippocampus in favour of females, the reverse of the normal sex difference (Sherry *et al.* 1993). They compared the size of the hippocampus relative to the non-hippocampal telencephalon (3.1.2.3) in males and females and compared the difference with that in the closely related, but non-parasitic, red-winged blackbird (*Agelaius phoeniceus*) and common grackle (*Quiscalus quiscula*). As predicted, female cowbirds emerged with significantly larger hippocampi than males, while no sex difference was found in either of the two control species (Fig. 3.13). Nevertheless, it is important to note that sex differences in brain structures do not always reflect differences in associated behaviour. Gahr *et al.* (1998), for example, found that the HVC and RA nuclei (see Fig. 3.12a) in the vocal control systems of male and female African bush shrikes (*Laniarius funebris*) (in which the two sexes sing duets) were larger in males, but song repertoire sizes were the same in the two sexes.

Figure 3.13 Mean volume of the hippocampus in male and female brown-headed cowbirds (*Molothrus ater*). Females have a significantly larger hippocampus than males, in contrast to the situation in the closely related red-winged blackbird and common grackle. See text. After Sherry *et al.* (1993).

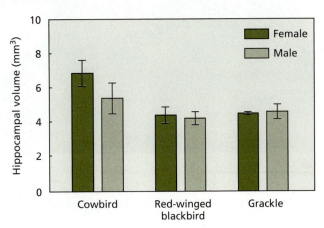

Brain development and experience The differences in brain structure above are the result of long-term selection pressures moulding different areas of the brain to serve the animal's behavioural requirements. However, as a clincher to the cowbird story, later studies of South American cowbird species showed that the hippocampus was bigger, and the sex difference in size more pronounced, during the breeding season, when birds were actively looking for hosts (Clayton *et al.* 1997). The size of the forebrain nuclei controlling song in canaries (*Serinus canaria*) (see above) also shows cyclical variation with season and thus song production (Nottebohm 1981; see also 3.3.1.5). These findings suggest a degree of **adaptive plasticity** in brain development. It turns out that cowbirds and canaries are not alone in showing such plasticity.

Several species of bird store food in the environment for later retrieval. The striking thing about the behaviour is the extraordinary feat of spatial memory required to retrieve items. Black-capped chickadees (*Parus atricapillus*), North American members of the tit family, for example, may scatter-hoard several hundred small seeds or insects in bark or clumps of moss over a considerable area. Items are hidden individually and each location is used only once. Nevertheless, chickadees can remember where items were stored up to a month later (Hitchcock & Sherry 1990). Laboratory experiments with chickadees and other food-hoarding species (such as Clark's nutcracker (*Nucifraga columbiana*), which can cache as many as 9000 items and still find them nine months later) have shown that birds really do remember specific sites and are not using proximate

(a)

Figure 3.14 (a) Scrub jays (*Aphelocoma coerulescens*) can remember what kind of food items they cached in different places, and can time the recovery of items according to how quickly the items decay. Photograph courtesy of Ian Connell and Nicky Clayton.

Figure 3.14 (*continued*) (b) Marsh tits (*Parus palustris*) allowed to cache seeds had a larger hippocampus than those prevented from caching. The volume of the hippocampus also decreased with time in the latter birds. After Clayton & Krebs (1994). (c) The volume of the posterior hippocampus in London taxi drivers increases with the amount of time they have been operating. After Maguire *et al.* (2000).

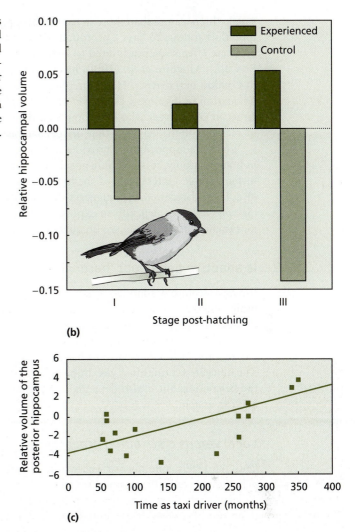

(b)

(c)

visual or odour cues to detect food, or general rules of thumb (2.4.4.3) about the kind of areas in which to search (Sherry 1984; Balda & Kamil 1992). More recent work suggests some species can even remember *what* they cached at each site and when they cached it (so-called **episodic memory**), and adapt their recovery times according to how quickly different items decay (Clayton & Dickinson 1998; Fig. 3.14a).

Once again, the region of the brain involved in this prodigious feat of memory is the hippocampus. We might thus expect the hippocampus to be bigger in food-hoarding species than in closely related species that do not hoard. This largely turns out to be the case (Krebs *et al.* 1989, but see Volman *et al.* 1997). More surprisingly, however, the size of the hippocampus appears to be affected by the hoarding experience of *individual* birds. Nicky Clayton and John Krebs (1994) reared some marsh tits (*Parus palustris*), food-hoarding relatives of the black-capped chickadee, in the laboratory. They divided the birds into two groups. Birds in the first group were given the opportunity to cache sunflower seeds at three different time points after hatching, while birds in the second group were fed powdered seeds which they were unable to cache. When their brains were examined after the experiment, birds from the second group had dramatically smaller hippocampi than those in the first group (Fig. 3.14b). The experience of hoarding food thus appears to have a profound effect on the development of the hippocampus.

These effects are not limited to birds. Using magnetic resonance imaging (MRI; see Box 3.3), Maguire *et al.* (2000) have found similar consequences of spatial learning in London taxi drivers. Here, the posterior hippocampus (the part involved in spatial navigation in mammals [Moser *et al.* 1993; Hock & Bunsey 1998]) was significantly larger in taxi drivers who had completed their training than in control individuals, and increased with the length of time drivers had been operating (Fig. 3.14c).

While the hippocampus provides a particularly good example of adaptive plasticity in brain anatomy, it is certainly not the only case. We have already encountered another in the flexible GnRH cells of the hypothalamus in the cichlid *Haplochromis burtoni* (1.2.1), and recent work suggests that auditory forebrain development in starlings (*Sturnus vulgaris*) depends on exposure to particular kinds of song during development (Sockman *et al.* 2002, but see Brenowitz *et al.* 1995).

3.1.3.2 Is adaptive behaviour 'hard-wired'?

We have seen that adaptive behaviour patterns can be linked to particular areas of the brain, and that the size of these areas can reflect the relative importance of the behaviour in different species or individuals. But to what extent can we think of a particular component of the nervous system as the underlying mechanism controlling a behaviour? Do adaptive behaviour patterns have identifiable neural circuits associated with them? The answer is a qualified yes. In part it depends on the definition and complexity of the behaviour, but it also depends on the nature of information processing within the nervous system.

Local versus distributed processing

Until the 1990s, a number of well-worked examples suggested that the control of certain behaviours could be mapped to relatively simple, self-contained neuronal circuits. Unfortunately there were persistent features of some of the responses that were difficult to account for with the proposed circuitry. Part of the problem seemed to be that the circuits ignored the extensive connections of their constituent cells to the mass of neurons outside them, many associated with putative control circuits for other behaviours. Control of a given response thus might not be a local, self-contained process at all, but the result of a **neural network** functioning in different ways at different times, a **distributed processing** system with ramifications well beyond its immediate site of expression. The implications for this **connectionist** view are particularly profound for the functioning of the vertebrate brain, where even relatively simple tasks may engage 100 million cells or more (John *et al.* 1986). To what extent, then, can models based on relatively simple local circuits provide a convincing account of behaviour?

Reflexes   In terms of the coordinated movement of the whole animal or part of the animal, **reflexes** represent most people's idea of the simplest form of behaviour. A reflex is an automatic, stereotyped unit of behaviour, usually in response to a simple stimulus, whose occurrence may vary with context and habituation, but whose *form* does not. Reflexes allow the animal to respond automatically, and usually quickly, to important internal and external events. Because they are relatively simple, reflexes provide a good opportunity to examine the relationship between behaviour and the functioning of

Figure 3.15 Diagram of adjacent sections of the human spinal cord showing connections between sensory neurons, interneurons and motor neurons making up the *monosynaptic* (direct connection between sensory and motor neurons) reflex arc of the knee-jerk response (components A, B and C) and the *polysynaptic* (sensory and motor neurons connected via an interneuron) arc of the limb withdrawal reflex (components D–H). See text. From Barnard (1983) after Lloyd (1955).

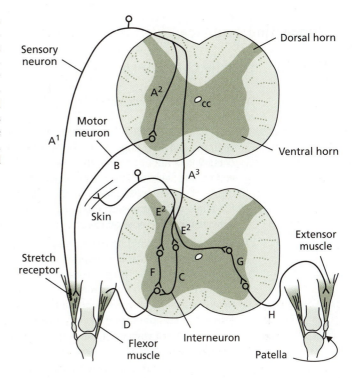

specific neural circuits. The classic human knee-jerk response to a tap on the patellar tendon, and the withdrawal of a limb from a painful stimulus, are well-known examples. Figure 3.15 traces the neural pathways (**reflex arcs**) for these two reflexes.

The apparent simplicity of reflex arcs often belies a complex pattern of muscle control, as well as some physiological features that set them apart from other kinds of neuronal activity and ally them to more complex behaviour patterns (Box 3.1). The limb withdrawal reflex, for example, requires the animal simultaneously to contract the flexor muscles and relax the antagonistic extensor muscles of the stimulated limb so that the limb is pulled towards the body. There is first of all, then, a *flexion reflex*. As the affected limb is flexed, the animal may need to use its other limbs to steady itself. Similar flexion and extension responses in these limbs are controlled by *crossed-extension reflexes*. In combination, these two reflexes control not only rapid emergency actions, but also organise limb movement for locomotion. To do so smoothly, *stretch reflexes* operate via stretch receptors in the muscles to grade flexion and extension in antagonistic pairs so that the limb moves in controlled stages rather than in one violent action.

Although invertebrates do not possess such an elaborate CNS as vertebrates, sensory receptors and effector organs still communicate through the switchboard of the central nerve cord and ganglia, and reflex arcs are very much part of the armoury of response. The gill withdrawal response of the sea hare *Aplysia*, a tectibranch mollusc, is a much-studied example.

The gill in *Aplysia* is a delicate and sensitive respiratory organ which is retracted into the mantle cavity in response to weak mechanical stimulation of the siphon, an extensible aperture opening into the cavity. The reflex arc controlling the response involves a group of excitatory and inhibitory interneurons relaying sensory information from the siphon to a battery of motor neurons in the abdominal ganglion which then effect the withdrawal of

Supporting evidence

Box 3.1 Special characteristics of reflex arcs

While superficially simple, reflex arcs show a number of unusual features related to their adaptive function:

☐ *Latency.* When a dog withdraws its paw from a painful stimulus the response should occur within about 27 ms if the only limiting factor is the rate of transmission through appropriate neurons. In fact it does not appear for about 60–200 ms because of synaptic delay. However, unlike more complex behaviours, the latency of reflexes decreases as the stimulus becomes stronger, an important property given that many reflexes are designed to act in emergencies.

☐ *Summation.* Among its other integrating properties, the CNS is able to accumulate repeated stimuli over time (*temporal summation*) and from different parts of the body (*spatial summation*). The scratch reflex of the dog is a good example. A dog scratches with its hind leg if an irritating stimulus is applied to its back. However, if the stimulus is weak, scratching may not occur until the stimulus has been applied 20 or more times. This is due to more neurons being brought into play with successive stimulation (*motor recruitment*) and leads to a characteristic *warm-up effect* in the expression of the reflex (the first few strokes of the paw do not have such a broad sweep as later ones).

☐ *Fatigue.* Normally a muscle stimulated to contract remains responsive for several hours. If a muscle is stimulated via a reflex arc, however, its response declines very rapidly. In some cases, like the scratch reflex in the dog, responses last only about 20 s. Despite the persistence of an irritating stimulus, the dog eventually stops scratching. What seems to happen is that, with repeated stimulation, interneurons begin to block impulse transmission by increasing the resistance of their synaptic junctions. A stronger, or novel, stimulus, however, will quickly re-establish a fatigued reflex.

the gill (Fig. 3.16). The stimulation thresholds of the sensory and motor neurons involved are similar to that for the reflex as a whole (about 0.25 g), and the firing of both types of neuron, and the magnitude of gill retraction are linearly related to the intensity of the stimulus. Together, these relationships make for a smooth withdrawal of the gill.

While early studies explained the withdrawal reflex in terms of a relatively simple neural circuit (Fig. 3.16), the response is capable of several simple forms of learning, suggesting involvement of neurons elsewhere (Cohen *et al.* 1991; Hawkins *et al.* 1993). Thus the reflex may be better characterised in terms of distributed neural processing rather than a dedicated local circuit. Later work by Cohen *et al.* (1997), has shown that, while habituation (the decrease in response with repeated stimulation, Chapter 6) was due to depression at local sensory synapses, dishabituation and sensitisation (the heightening of responsiveness to one stimulus by previous responses to another) involved several different sensory and interneurons at other locations in the nervous system. Moreover, these different elements came into play at different times after the initial training of the response, showing that information for the reflex was distributed in time as well as space. Despite this, however, much of the reflex (about 84% of the response strength) turns out to be mediated through the single LD_{g1} motor neuron (Fig. 3.16), changes in the firing of which can account for most of the variation in behaviour (Cohen

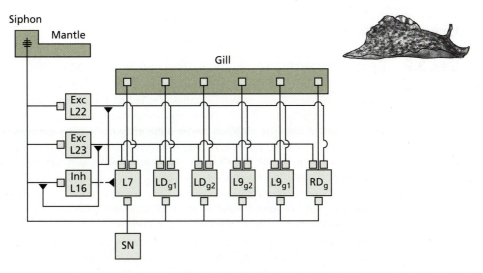

Figure 3.16 Neural circuit of the gill withdrawal reflex of *Aplysia* following weak stimulation of the siphon. Input from receptors to the motor neurons L7, LD_{g1}, LD_{g2}, $L9_{g1}$, $L9_{g2}$ and RD_g is mediated by two excitatory interneurons, L22 and L23. SN, sensory neuron; □ excitatory synapses; ▼ inhibitory synapses. After Kandel (1976).

et al. 1997). Thus, while the reflex is controlled by a distributed network of neurons, it is not a highly distributed system, and much of its plasticity can be explained in terms of a limited number of neurons making a disproportionate contribution to the response. Similar localised distribution appears to underlie the bending reflex in leeches (Lockery & Kristan 1990; Lockery & Sejnowski 1992).

More complex behaviours   The distinction between reflexes and more complex behaviours hinges on the number of different actions involved and the number of factors affecting their expression. In a reflex, there is often little more than a simple neural pathway and a clear-cut, often momentary, response. Most other behaviours are subject to a wide range of influences from the animal's internal and external environment and vary in their performance accordingly. Nevertheless, complex behaviours can be stereotyped, with elements of the behaviour occurring in a predictable and inflexible sequence once triggered. Ethologists, such as Lorenz and Tinbergen, (1.3.1) referred to such stereotyped behaviours as **fixed action patterns** (FAPs). They saw the elicitation of a FAP as being dependent on a specific stimulus, variously called a **key stimulus**, **sign stimulus** or **releaser**, according to context. These stimuli triggered the performance of a FAP via an **innate releasing mechanism** (IRM), a hypothetical neural centre distinct from the receptor initially detecting the stimulus. Once a FAP had been triggered it was always performed in exactly the same way and in its entirety.

While attractive as an idea, FAPs *sensu* Lorenz and Tinbergen run into a number of difficulties when applied to behaviour at large (Box 3.2) and have long fallen from general use. However, certain behaviours do appear to fit some of the basic assumptions behind them and suggest a degree of 'hard-wiring' in their underlying control. The egg-retrieval response of nesting greylag geese (*Anser anser*), studied by Lorenz, is a classic example. When an egg rolls out of the nest, as occasionally happens, the goose stretches out its neck and rolls it back with the underside of its bill, moving its head from side to side as it goes to prevent the egg from rolling off to the side. Sometimes the manoeuvre fails and the egg rolls away

Underlying theory

Box 3.2   Problems with the concept of fixed action patterns

The idea that stereotyped behaviour patterns reflected 'hard-wired' responses which were performed invariably and in their entirety each time they were elicited was one of the defining concepts of classical ethology. While apparently convincing examples of such 'fixed action patterns' (FAPs), such as the egg-rolling response of the greylag goose (see text), were forthcoming, the concept rapidly ran into problems and has now fallen from use. Difficulties included the following:

☐   Most behaviours were simply too variable to warrant the term FAP, which thus applied only to a small minority of responses showing apparent stereotypy.

☐   Most apparently stereotyped behaviours showing endogenous control also showed some influence of environmental feedback. Even in greylag geese, the side-to-side movement of the head during egg-rolling depended on the vicissitudes of the egg's trajectory.

☐   With careful study, many putative FAPs turned out to vary in their performance from one occasion to another.

☐   Many FAPs showing apparent endogenous control also turned out to depend on experience for their full development.

again, but instead of retrieving it, the goose continues its neck and head movements back to the nest as if the egg was still there. The same thing happens if the egg is removed mid-retrieval by an experimenter. Once triggered, it seems, the egg-retrieval response is seen through to completion regardless of changes in the environment. However, is there any evidence that such behaviour patterns are stamped into the animal's nervous system as identifiable, 'hard-wired' circuits? Studies of some invertebrate species suggest there is. The escape response of the sea slug (opisthobranch mollusc) *Tritonia* is an example.

When it encounters chemicals emanating from the tentacles of a starfish, its main predator, *Tritonia* initiates a highly stereotyped, undulating escape response (Fig. 3.17). The response is accomplished by alternate flexing of dorsal and ventral longitudinal muscles in the body wall and lasts for about 30 s, enough to remove the animal from the vicinity of the starfish. While it would be easy to account for the response as a series of reflexes, through reciprocal stimulation of stretch receptors in the two sets of muscles, studies of the underlying neural circuitry have revealed a very different story. Contractions in the antagonistic sheets of muscle are regulated by two kinds of motor cell, the dorsal (DFN) and ventral (VFN) flexion neurons. Early work (Willows & Hoyle 1969), based on electrode stimulation, assumed mutually inhibitory connections between the DFN and VFN (thus causing alternating flexion), but with both cells connected to a separate general excitatory neuron (GEN) which was activated when the DFN fired in response to detecting a starfish. Later work, however, has established a more complex control system (Katz & Frost 1995; Fig. 3.17). It now appears that there is a cerebral neuron (C2), with excitatory and inhibitory connections to three dorsal and two ventral swim interneurons (DSI and VSI respectively). This cluster of neurons in the CNS acts as a central pattern generator, regulating the activity of two dorsal flexion neurons (DFN-A and DFN-B) and a ventral flexion neuron (Fig. 3.17). The flexion neurons in turn regulate

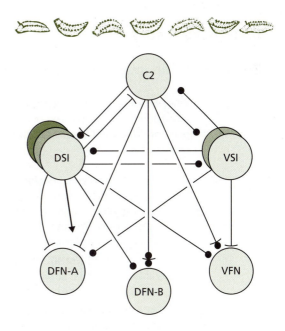

Figure 3.17 Schematic representation of the neural network controlling the escape swimming response in *Tritonia*. A central interneuron (C2), three dorsal swim inter-neurons (DSI) and two ventral swim inter-neurons (VSI) comprise a central pattern generator regulating the activity of two dor-sal (DFN) and one ventral (VFN) flexion neurons that have no direct communication with each other. See text. After Katz & Frost (1995).

the muscles but do not communicate directly with each other. The interesting property of the system from the point of view of behavioural control, however, is that the nerve impulses that control muscle flexion are issued even when connections with the muscles are severed and there is no peripheral feedback. *Tritonia*'s escape response thus appears to be programmed into its CNS, to be performed in its stereotyped entirety whenever triggered – exactly as envisaged by Lorenz in his concept of a FAP. Feeding in snails, flight in locusts and swimming in lampreys are other examples of behaviours that have turned out to be governed by neural pattern generators of the kind demonstrated in *Tritonia* (Robertson & Pearson 1982; Camhi 1984; Grillner & Wallen 1985).

Brain centres and the control of behaviour The twin evolutionary processes of centralisa-tion and cephalisation have meant that behavioural control has devolved more and more to the centre as nervous systems have become more structured and organised. Thus, in vertebrates, attention has focused on control centres and pathways within the CNS, initi-ally using lesion (cutting through selected areas), ablation (selectively destroying particular areas) or electrical stimulation techniques to elucidate pathways of control. Studies of the ventromedial nucleus (VMN) in the mammalian hypothalamus provide an example.

Among other things the VMN in mammals is involved in the control of feeding behavi-our. In conjunction with the *lateral nucleus* (LN) and receptors outside the brain, the VMN monitors levels of glucose in the blood. An early model of the control of feeding in rats (*Rattus norvegicus*) suggested a central role for the relationship between the VMN and LN (Teitelbaum 1955; Teitelbaum & Epstein 1962). Here, when glucose levels drop, the VMN releases its inhibitory control over the LN which then stimulates food searching and ingestion. Once the animal is satiated, sensory feedback re-establishes the VMN's control and feeding ceases. If the VMN is ablated, the animal will gorge itself to distension (show hyperphagy). If the LN is ablated, it will starve to death even in the presence of preferred food. Conversely, experimental stimulation of the VMN inhibits feeding, while stimulation of the LN causes it to be elicited (Teitelbaum & Epstein 1962).

Underlying theory

Box 3.3   Brain imaging techniques

While still expensive and used largely for clinical purposes, the development of brain scanning technology is revolutionising our ability to track behaviour patterns to events in particular parts of the central nervous system. The following are some of the techniques and their applications:

☐ **Magnetic resonance imaging (MRI).** MRI (sometimes called nuclear magnetic resonance imaging or NMR) aligns atomic particles in the body using magnetism then bombards them with radio waves so that they give off radio signals that differ according to the type of tissue. Computerised tomography (CT) then converts the information into a three-dimensional image of the target tissue.

☐ **Functional MRI (fMRI).** fMRI builds on MRI by using local variation in oxygen concentration (neuronal activity is fuelled by glucose and oxygen) to highlight the areas of greatest activity in the brain at the time of scanning.

☐ **Positron emission tomography (PET).** PET operates in a similar way to fMRI, highlighting areas of brain activity through their energy consumption. However, unlike fMRI, it requires injection into the bloodstream of a radioactive marker. For safety reasons, therefore, PET scans need careful regulation.

☐ **Near-infrared spectroscopy (NIRS).** This technique also produces an image based on energy consumption but works by measuring the reflection of low-level light beams from different parts of the brain.

☐ **Electroencephalography (EEG).** A long-established technique that uses electrodes to detect changes in the electrical activity of neurons. Modern EEG takes readings from several different locations to map variation in activity across the brain. It often uses event-related potentials (ERPs), i.e. electrical changes associated with a particular stimulus, such as a noise or image, in the construction of activity maps.

☐ **Magnetoencephalography (MEG).** MEG works like EEG except that it uses changes in magnetic, rather than electrical, activity to map brain activity.

From discussion in Carter (1998).

Neat though this story is, however, the VMN and LN are not quite the feeding control centres it suggests. Later work by Winn (Winn *et al.* 1984; Winn 1995) has shown that damage to pathways simply passing through the LN inhibits feeding, suggesting that, in fact, areas of the brain beside the hypothalamus are involved in the control of feeding, including the cortex. Indeed, as understanding of the central nervous control of behaviour grows, it is becoming clear that the performance of even simple actions frequently engages diverse areas of the brain. Recent advances in brain imaging techniques in humans (Box 3.3) have revealed complex, dynamic interactions between different neural centres as individuals respond to sensory input, attend to tasks and organise thoughts and actions (Carter 1998). Attending to a simple external stimulus such as a rustle in the bushes (which might signify danger) illustrates the point.

On detecting such a stimulus, the reticular formation in the limbic system (3.1.2.3) puts the brain into an alert state by releasing adrenalin. This stimulates neurons to fire throughout the brain and shuts down other unnecessary activity. Firing of dopaminergic and noradrenergic neurons in the reticular formation generates the alpha brainwaves (electrical oscillations at 20–40 Hz, detected using electroencephalography) characteristic of arousal. Orientation towards the source of the stimulus is achieved by neurons in the *superior colliculus* (thalamus) and *parietal* (top rear) *cortex* (see Fig. 3.8a,b), the former directing the eyes to the stimulus, the latter disengaging attention to the preceding activity. Finally, the brain focuses on the stimulus via the *lateral pulvinar nucleus* in the thalamus, which passes information forward to the frontal lobes. These in turn lock on to the stimulus and maintain attention (Carter 1998).

While brain imaging has largely been used to study the human brain, adaptation of imaging techniques has provided comparable information for some other species. Blaizot *et al.* (2000), for instance, used positron emission tomography (PET) scanning (Box 3.3) to map brain activity associated with a matching-to-sample task (here matching arbitrary geometric shapes to previous exemplars) in baboons (*Papio hamadryas*). Brain activity during the task was distributed through occipital and temporal regions, including the hippocampus, and in the frontal cortex, a pattern consistent with imaging studies of humans and lesioning in monkeys. The study also indicated other local areas of activity and a degree of left hemispheric dominance in the control of object matching.

Although many aspects of sensorimotor control appear to be mediated by distributed activity in the brain, there can be a surprising degree of focus in attributes that, a priori, might be expected to show distributed control. Cognitive performance is a case in point. Widespread positive correlations in performance on different kinds of cognitive test have led to the concept of 'general intelligence' or **Spearman's g**. Duncan *et al.* (2000) used PET scanning to compare brain activity during spatial, verbal and perceptual tasks associated with high *g* scores, with that during matched control tasks associated with low *g*. While *g* is generally thought to reflect a broad range of major cognitive attributes, Duncan *et al.* found that their high *g* tasks elicited remarkably focused activity in the lateral frontal cortex. The pattern of lateral cortex activity was very similar in the three kinds of task, despite their different demands. The study thus suggests that 'general intelligence' derives from (or is at least coordinated by) a specific frontal cortex system involved in the control of a diverse range of behaviours. A similar cortical focus appears to underlie the integration of multiple sensory information relayed from other centres, particularly the superior colliculus in the thalamus, in cats (Stein *et al.* 2000).

3.1.3.3 Command centres and neural hierarchies

A point that emerges clearly from both vertebrates and invertebrates is that there are apparent 'chains of command' within nervous systems. Behaviour patterns arise via a cascade of information through various CNS centres and peripheral circuits. This has led to the idea of behavioural '**command centres**', regions of neuronal activity exercising high level control over behaviour through a hierarchy of lower level centres or local circuits. The song control centre in birds (Fig. 3.12; and see Yu & Margoliash 1996), and *Tritonia*'s central pattern generator (Fig. 3.17) discussed above are examples. Of course, as we have seen, such command centres need not be discrete single locations. Rather, they may arise through the concerted activity of several different locations distributed through the nervous system. However they are arranged physically within the nervous system, hierarchies of command are likely to have several evolutionary advantages in terms of organisational efficiency (Box 3.4).

One such advantage is that they allow potentially conflicting demands on the animal to be coordinated and prioritised through inhibitory relationships between different command centres, such as that postulated between the VMN and LN above. In this way,

Underlying theory

Box 3.4   Advantages of behavioural hierarchies

The control of behaviour is frequently hierarchical in its organisation, both at the level of priorities between different activities and in underlying neural mechanisms (see Figs 3.18 and 4.15). Why should this be? Dawkins (1976) points to a number of potential advantages of hierarchical control.

The evolutionary rate advantage

Imagine two watchmakers, A and B. A's watches are just as good as B's but he takes 4000 times as long to make them. The reason is that B first assembles the 1000 components of his watches into 100 sub-assemblies of 10 components each, and then assembles these into 10 larger sub-assemblies before finally putting the whole thing together. A, in contrast, puts the 1000 pieces of his watch together in one go. If anything goes wrong he has to dismantle the whole thing to put it right. By analogy, complex and thermodynamically unlikely systems, such as the nervous system and complex adaptive behaviour patterns, might evolve more rapidly if they are put together as a series of functional subcomponents. We might thus *expect* complex biological systems to have hierarchical structures rather than be surprised by them.

The local administration advantage

The analogy here is with central and local modes of government. Central government is important for all-round coordination of its different regions, but is likely to be inefficient in terms of policy-making at a local level. It is more effective to have local administrations managing the latter. In the same way, we might expect different activities such as feeding and nesting, each of which involves the coordination of many different sub-activities, to be controlled by different centres in the CNS, but for these centres to be controlled by a higher-level coordinator (e.g. Fig. 3.17).

The redundancy reduction advantage

The number of possible states of a complex system is likely to be enormous. This is as true for physiological systems as any other. For instance, the human retina contains some four million light-sensitive cells. If each signals the presence or otherwise of light, the number of possible states of the retina is an unimaginable $2^{4\,000\,000}$. To cope with this the cubic capacity of the brain would have to astronomical. An important point, however, is that a very large proportion of these states will be *redundant* as far as the animal is concerned because they are biologically meaningless or duplicate the information of other states. To reduce this redundancy, lateral inhibition between retinal cells limits firing to cells bearing most information (such as those scanning the edges of objects). The contraction and relaxation of muscle fibres in coordinated groups, and the correlated responses of different muscle systems, illustrate the same principle.

From discussion in Dawkins (1976).

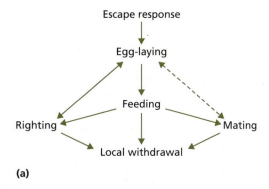

(a)

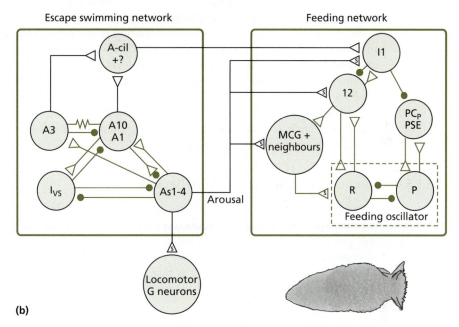

(b)

Figure 3.18 (a) The proposed hierarchy of escape, mating, egg-laying, feeding, righting and with-drawal behaviours in the predatory opisthobranch mollusc *Pleurobranchaea californica*. After Mpitsos & Pinneo (1974a,b). (b) The proposed circuitry for part of the hierarchy (the inhibition of feeding by the escape response): the swim central pattern generator comprises a set of cells (A1, A3, A10 and I_{vs}) that generate the swimming pattern and have inhibitory connections (via A-cil) to the cells PC_p, PSE and I2 in the feeding network. MCG, metacerebral giant neuron; R, populations of retractor neurons and P, populations of protractor neurons for the feeding apparatus. Open triangles, excitatory synapses; closed circles, inhibitory synapses; partly shaded triangles, variably excitatory synapses; open triangles with closed circles, excitatory and inhibitory synapses. After Jing & Gillette (2000).

the animal can avoid paralysing indecision. In many cases, such relationships between putative command centres and their associated hierarchies of control are inferred purely from behavioural output (see, for example, Fig. 4.15). In some, however, it has been possible to trace the neural components of the hierarchy.

The predatory marine snail *Pleurobranchaea californica* shows an apparent hierarchy of control between a number of behaviours, the main aspects of which are shown in Fig. 3.18a. Feeding behaviour dominates mating, withdrawal from a tactile stimulus and righting behaviour, but is itself dominated by egg-laying. Not surprisingly, the emergency escape swimming response (like that of *Tritonia*) takes priority over all other activities. Beginning with carefully controlled behavioural experiments by Davis *et al.* (1974a,b), successive studies have been able to elucidate some of the neural components of the hierarchy (e.g. Jing & Gillette 2000; Fig. 3.18b). The collective results suggest that feeding dominates righting because chemoreceptors on the oral veil (a fleshy structure around the mouth) respond to chemical stimuli in food and suppress impulses in the neurons controlling righting behaviour. Withdrawal behaviour, however, is inhibited directly by feeding because activity in the complex set of neurons controlling feeding inhibits activity in the neurons controlling withdrawal. Withdrawal can thus occur only when the feeding neurons are not activated. The suppression of feeding by egg-laying appears to be hormonally based, with a single hormone both inducing egg-laying and suppressing feeding, perhaps by competing for messenger receptor sites on the feeding neurons.

Recent work also suggests that sensory pathways mediating feeding and withdrawal from food stimuli have dual access to the neural networks controlling the two behaviours and that their effects are modulated by the level of satiation (Gillette *et al.* 2000). Switching between feeding and withdrawal thus appears to be regulated by the animal's need for nutrients weighed against energetic cost and the risk of predation associated with attacking prey. The behavioural control mechanism thus seems to incorporate a cost–benefit analysis that allows the animal to optimise (see 2.4.4.3) its choice of response – function meets mechanism once again.

3.1.3.4 Deducing mechanism from behaviour

Although underlying control mechanisms can be surmised purely from observing behaviour, they remain in the realm of speculation unless they are tested. In *Pleurobranchaea*, it was possible to identify the neuronal pathways contributing to a hierarchical control mechanism, but in many cases, like the gulls in Fig. 4.15, such an approach is likely to be impracticable. Does this mean it is pointless to speculate about unseen mechanism? An entertaining model system suggests not.

Toy elephants and behaviour sequences

Hailman & Sustare (1973) used a talking toy elephant, called Horton, to demonstrate the deduction of mechanism from behaviour to a group of students. When his string was pulled, Horton uttered one of a number of pre-recorded sentences. The sentences were effectively fixed action patterns (3.1.3.2), since, once triggered, they were always uttered in their entirety and were not influenced by events in the environment. Students in the class began the exercise by pulling Horton's string 10 000 times and noting what he said each time. When they looked at the accumulated data, it was clear there was some pattern in his responses. While there was a roughly equal chance he would say any of his repertoire of 10 sentences when his string was first pulled, the first sentence uttered was more likely to be followed by certain other sentences. From this, the students hypothesised that the speaking mechanism consisted of some sort of pointer moving round a circle with each pull of the string. Wherever the pointer stopped, the sentence at that point was uttered. By looking at the tendency for one sentence to be followed by another, the students were able to map the sequence of sentences around the circle and deduce some of the rules underlying transitions from one sentence to another (Fig. 3.19). Thus sentences were highly unlikely to be repeated or followed by those immediately next to them, but were more likely to be followed by particular sentences

Figure 3.19 A model of Horton's talking mechanism, showing the probability of different utterances following each other (see text). Arrows indicate a low probability of two utterances occurring in sequence (e.g. G is unlikely to be followed by either C or E). Utterances are also unlikely to follow themselves (recurrent arrows). After Hailman & Sustare (1973).

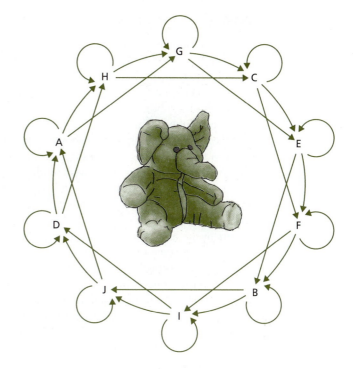

further away. Cross-checking outcomes from different starting points showed that the proposed arrangement had high internal consistency. So what was the actual mechanism? Favourites of candidates were a recording disc with sound grooves spiralling into the centre, or a cylinder with grooves spiralling round its surface. In either case, pulling and releasing the string would cause the recorded surface to spin and a needle to drop into a groove at whatever point was determined by the sequence. When Horton was finally sacrificed and opened up, it turned out that the mechanism was indeed a disc with spiral grooves, to the great satisfaction of the 'disc' proponents among the students.

Even though it did not involve a real animal or real behaviour, the Horton story shows that much can be gleaned about internal mechanism by careful, systematic observation of behaviour. Apart from deducing a disc and needle system, the students also revealed the probabilistic (or **stochastic**) nature of Horton's sequences of sentences. Sentence E was often followed by sentence A, but not always. As Dawkins (1983) points out, this is a feature of behavioural sequences in real animals. A male Siamese fighting fish (*Betta splendens*) which has just raised its gill coverts to an opponent, for example, is very likely to follow it up with another kind of aggressive display, but it is not possible to say with certainty whether this will be an approach towards the opponent or a display of fin-spreading (Simpson 1968). However, while realistic in some ways, Horton is, not surprisingly, unrealistic in many others. For example, the predictability of behavioural sequences in many animals is time-dependent: it works in the short term but not over longer periods. Thus, the sequence of courtship behaviours in many male fish, or of notes in the songs of various bird species, are predictable over periods of a few seconds, but not if consecutive actions are separated by longer than this (Dawkins 1983). Unlike in Horton, therefore, where the sequence of sentences is entirely independent of the time between successive pulls of the string, pattern in the activity of animals may be dependent on the timescale of sampling. Dawkins (1983) discusses a number of other disparities between Horton and the behaviour of real animals.

Animals and animats

A different modelling approach to the relationship between mechanism and behaviour is literally to build a mechanism and see whether it generates the responses expected. 'Building' may involve simulation in the form of a computer program, as in some of the tests of neural network models above, or it may involve constructing a physical mechanism in the form of a robot. In the inevitable jargon, robots that attempt to simulate animals have been dubbed '**animats**'.

One of the advantages of a robotics approach is that it allows the experimenter to go back to first principles and create alternative options in order to test hypotheses about the control of behaviour. If we wish to identify the neural circuitry controlling behaviour in a real animal, we must resort to invasive surgery, preparations isolated from their normal companion circuits, sedation or other drastic measures. Often this is not practicable, and even when it is, there are doubts as to whether it tells us anything credible about events in the intact, freely responding animal. Animat roboticists (or 'biroboticists') aspire to overcome these difficulties by avoiding them altogether. Recent results suggest they may be making headway (see Holland & McFarland 2001).

One of the pioneers of the approach is Barbara Webb of Edinburgh University (Webb 2000, 2002). Webb has used animat 'crickets' to probe the mechanisms underlying phonotaxis (orientation towards sound – in this case the call of a male) in real female crickets (*Gryllus bimaculatus*). Male *G. bimaculatus* call (stridulate) in the night by rubbing their specially modified forewings (tegmina) together. Females home in on the sound by means of a rather extraordinary auditory system. *G. bimaculatus* has an ear on each of its forelegs that is connected, via internal tubes, to an aperture in its body (Fig. 3.20a). The system generates phase differences in front of and behind each ear so that the cricket can tell where sound is coming from. Crucially, however, the arrangement works only if the sound is pitched at the carrier frequency of the species (see stimulus filtering [3.2]).

To try to simulate this process, Webb devised an electronic neural network fed by microphone 'ears' mounted on a wheeled robot (Fig. 3.20b). The network was based on two key neurons known to be involved in phonotaxis, and Webb tuned their model counterparts so that they had the same firing pattern as the real things. A key question was how complex the system had to become to show naturalistic phonotaxis.

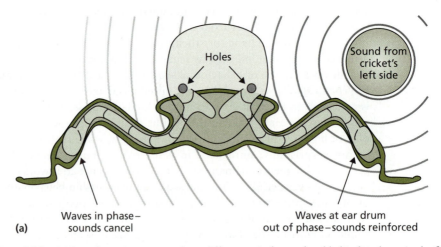

(a) Waves in phase – sounds cancel   Waves at ear drum out of phase – sounds reinforced

Figure 3.20 (a) Sound waves generate phase differences in front of and behind each ear in the field cricket *Gryllus bimaculatus*, so females can tell where a male is calling from. After Graham-Rowe (1998).

(b)

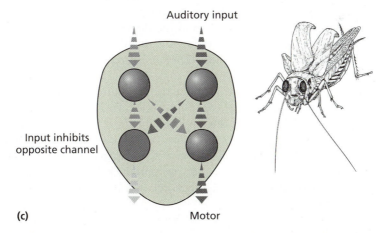

(c)

Figure 3.20 (*continued*) (b) The process can be simulated using robots with an electronic neural network fed by microphone 'ears'. Photograph courtesy of Barbara Webb and Andrew Horchler. (c) A series of robots, progressively incorporating more biological detail, showed that only four components were needed: matching left and right input/output nodes to control the robot's left and right motors, and an inhibitory connection between the input node on one side and the output node on the other. See text. After Graham-Rowe (1998).

Neuroanatomical investigations of phonotaxis in various gryllid crickets had identified several neurons that were activated during the behaviour, but had not established the connectivity between them or the means by which the behaviour was controlled. By building a series of robots that progressively incorporated more biological detail (Fig. 3.20b), Webb and her coworkers (e.g. Lund *et al.* 1998; Webb & Scutt 2000) found that only four components were needed: matching left and right input/output nodes to control the robot's left and right motors, and an inhibitory connection between the input node on one side and the output node on the other (Fig. 3.20c). Neuroethologists have speculated that phonotaxis required two neural control systems: one to recognise the male's call, the other to locate it. However, when Webb played calls to her robot system, it not only recognised the sound, but also moved towards it. The two control system model was therefore redundant; one simple system could control *both* components. But is this really what the cricket's system is like? To find out, Webb explored other response features of the robot to see if they reflected behaviour in the real animal. One thing she tested was the robot's response to different syllable rates in the call. Female *G. bimaculatus* prefer calls with faster syllable rates. When given a choice, Webb's robot did too. Suggestions that such preferences indicate a special control centre in the cricket's brain are therefore questionable. The simple auditory system and neural network mimicked in the robot are sufficient in themselves to generate the preference.

Webb and coworkers' study is not alone in being able to replicate surprising aspects of behaviour using simple robots. Lambrinos *et al.* (1997, 2000) have achieved something similar with the navigation system of desert ants of the genus *Cataglyphis*. *Cataglyphis* forages up to 200 m from its nest, but although its outward foraging path may meander in several different directions, the ant is able to head back home in an almost unerring straight line. How does it do it? The key seems to lie in specialised receptors in the compound eye that are sensitive to different orientations of polarised light. These send information to three neurons, called POL neurons, which each produce a peak output at a different orientation of polarised light relative to the long axis of the ant's body. Initial theory assumed that each POL neuron acted independently in rotation, allowing the ant to scan the environment, find the direction of brightest light and choose a bearing home. To test this, Lambrinos *et al.* built a robot ant (the Sahabot) with three sets of polarised light sensors linked to amplifiers, thus simulating the POL neurons and their incremental response to the angle of polarised light. When they released the robot in the desert, it behaved very like the ant, but did least well when programmed according to the independent scanning model. To get round this, Lambrinos *et al.* allowed the amplifiers of the three polarised light sensors to combine their output, a development that considerably sharpened the robot's response. While theory required only one POL neuron to account for the ant's behaviour, empirical testing with a robot suggested that all three were necessary to reproduce the response correctly. The robot experiment thus provided a novel insight into why the orientation system of *Cataglyphis* involves three neurons.

While animat approaches such as these can be illuminating, and may indeed be a powerful tool in unravelling physical mechanisms controlling behaviour, they are not without their critics. One problem is that the actual mechanisms underlying a behaviour are often very much more complicated than the minimum necessary mechanism suggested by robotics. There is thus a danger of cleverly reproducing the output of a behavioural system but by a means entirely different from that used by the animal. Nevertheless, the ability of some robot mechanisms to show counterintuitive emergent properties that are akin to real behavioural systems is beginning to increase faith in the approach.

3.2   Sensory mechanisms, perception and behaviour

Neural control mechanisms coordinate incoming and outgoing information to ensure the animal responds appropriately to events in its environment. However, the nervous system is not a passive observer of such events. Stimuli from the environment are filtered through the animal's sense organs and processed by its nervous system so that it obtains a very selective view of the world around it. Exactly what this view comprises depends on what natural selection has shaped the animal to do. The environment is likely to contain an enormous amount of information, only a small proportion of which is important to the animal at any given time. Getting rid of this redundancy, while at the same time maintaining adequate sensory input, is the task of specialised sensory receptors and organs and the CNS. This often results in the animal responding to specific components of objects or events confronting them (the basis of key stimuli, sign stimuli and releasers in ethology [3.1.3.2]). In this section, we shall look at the way different kinds of sensory information are sifted to meet behavioural needs. The structure and physiology of sensory systems *per se* are beyond the scope of this book, but a good account can be found in Randall *et al.* (1997).

3.2.1   Sensory mechanisms, stimulus filtering and perception

For an animal to behave appropriately, its nervous system must receive the right kind of information. It must be able to register relevant changes in the environment. Several types of change may be important to the animal's survival and reproductive success. Changes in light intensity, temperature, sound, tactile stimuli, odours, barometric pressure or other factors may signal the approach of a predator, a potential mate or prey item, or a critical change in weather conditions. All therefore need careful monitoring. How have the various sensory mechanisms evolved to cope with this monitoring task?

3.2.1.1   Visual stimulus filtering

At one time, the eye was thought of as little more than a means of translating images of the environment into electrical impulses. It was only in the brain that information was sorted and interpreted. A ground-breaking study of retinal function in the leopard frog (*Rana pipiens*) by Lettvin *et al.* (1959), however, changed all that. Earlier studies of the leopard frog had looked at its responses to points of light and dark, and concluded that the frog's eye simply registered changes in the tone of the viewed object. Lettvin *et al.*, however, stimulated the retina with images possessing some of the basic features of natural objects that were salient to the frog in its day-to-day life. By this means they discovered a remarkable apparent functional diversity among the ganglion cells of the retina (Table 3.1). The qualification 'apparent' is important because later work showed that, while retinal ganglion cells undoubtedly send important perceptual information to the brain, they do not quite fulfil the role of specific stimulus detectors that Lettvin *et al.*'s classification might suggest (Ewert 1997).

Studies of a variety of frog and toad species have shown that visual information from the retina is sorted by different classes of ganglion cell according to their different sensitivities to the size, contrast, motion, colour and edge characteristics of stimuli. This

Table 3.1 Lettvin *et al.*'s (1959) functional classification of retinal ganglion cells in the leopard frog (*Rana pipiens*). See text

---

1. *Sustained-edge detectors* showed the greatest response when a small, moving edge entered and remained in their receptive field. Immobile or long edges did not evoke a response.
2. *Convex-edge detectors* were stimulated mainly by small, dark objects with a convex outline.
3. *Moving-edge detectors* were most responsive to edges moving in and out of their receptive field.
4. *Dimming detectors* responded most to decreases in light intensity.
5. *Light-intensity detectors*: the responsiveness of these cells was inversely proportional to light intensity. They were most responsive in dim light.

---

structured information is received by neurons in the optic tectum and pretectal thalamus and coordinated to generate appropriate motor output. Jörg-Peter Ewert's studies of the common toad, *Bufo bufo*, show how the relationship between peripheral and central processing works.

In a series of classic experiments, Ewert (1974, 1980) probed the events in the eye and brain of the toad that determined its response to different kinds of visual stimuli. Using microelectrodes to record from single cells in the tectum of freely moving toads, he showed that visual perception depended on a combination of **peripheral** and **central stimulus filtering**. Peripheral filtering occurs via the structured receptive fields of the retinal ganglion cells (whose axons extend down the optic nerve to the brain), each of which receives input from a small elliptical area of the retina. Stimulation of the ellipse sends impulses to the appropriate ganglion cell via bipolar cells which determine whether or not the ganglion cell is likely to fire. However, the receptive field of each ganglion cell is divided into two regions: a central *excitatory* region, which is stimulated to respond when it receives input from bipolar cells, and a peripheral *inhibitory* region, which reduces the likelihood of the ganglion cell firing when stimulated. The extent to which objects passing across the toad's visual field stimulate the central and peripheral receptive fields therefore determines whether the cells pass information back to the brain. A small beetle moving across the toad's field of view is likely to stimulate the central fields of the ganglion cells but leave the peripheral fields unmoved, thus eliciting a stream of messages back to the tectum. A large object looming close to its eye, on the other hand, will stimulate both central and peripheral fields, thereby inhibiting a response. In effect, a toad sees only those things that change the light intensity falling on its ganglion cells: in the main, small moving objects, such as beetles, flies or distant predators, that impact on its chances of survival. Large, stationary images are more likely to be rocks or tree stumps, or some other inconsequential object unworthy of a response.

While peripheral filtering goes on in the retina, neurons within the optic tectum receive information from clusters of neighbouring ganglion cells. Thus, each tectal neuron has its own receptive field based on the area of the retina serving its ganglion cells. Pathways also extend from the retina to cells in the thalamus, with additional connections between the thalamus and the tectum. Ewert investigated the receptive fields of individual cells in the tectum and thalamus by implanting an electrode and recording the responses of the cells when different objects were passed in front of the toad's eyes. The results

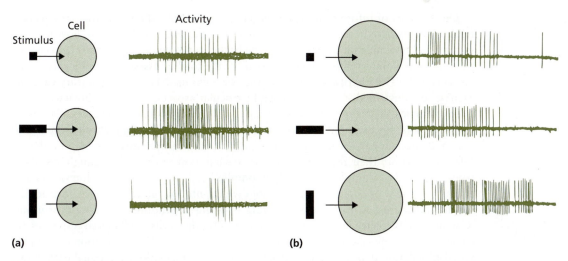

Figure 3.21 Visual stimulus filtering in cells of the tectum and thalamus of the toad *Bufo bufo*. Electrical recordings show that cells in the tectum are stimulated most strongly by a moving object extended in the direction of movement (a). Cells in the thalamus, however, respond most to objects extended perpendicularly to the direction of movement (b). From Ewert (1974).

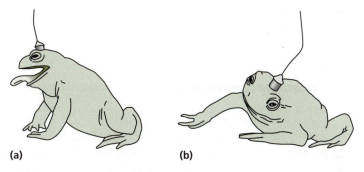

Figure 3.22 (a) Electrical stimulation of the optic tectum (a visual centre in the brain) in toads elicits a sequence of prey-catching behaviours such as snapping, while (b) stimulation of the thalamus elicits defensive movements. After Ewert (1974).

showed a variety of responses to the different objects, some very strong, others weak or non-existent. These variations create a further, central, tier of stimulus filtering, with different cells responding more strongly to objects of different shape or orientation. Some cells in the tectum, for instance, are most responsive to long, thin objects moving horizontally across the toad's field of view (Fig. 3.21a). Others in the thalamus respond most to objects moving through in a vertical orientation (Fig. 3.21b). Electrical stimulation of tectal and thalamic cells has shown that excitation in the tectum results in the toad orientating and leaning towards the perceived object, opening its mouth and snapping the object up with its tongue, and performing actions associated with cleaning its mouth (Ewert 1974; Fig. 3.22a). Excitation in the thalamus, on the other hand, elicits defensive crouching, rising up or avoidance behaviours (Fig. 3.22b). Ewert characterises these opposing sets of responses in terms of potential 'prey' and 'enemy' (respectively) stimuli. However, excitatory and inhibitory connections between the thalamus and tectum are also important in modulating the toad's response.

On the basis of information from retinal ganglion cells, therefore, the optic tectum tells the toad where in the visual field an object is located, how large it is, how fast it is moving and how much it contrasts with the background, while connections between the tectum and thalamus refine perception to allow the toad to assess the significance of the visual signals. The filtering process can thus be envisaged as a discriminatory cascade, each tier of which analyses and amplifies a particular aspect of the object in view.

In mammals, central processing of visual stimuli follows similar principles. The visual cortex of cats, for example, contains two main types of cell. 'Simple' cells respond to lines and edges in particular orientations or locations via excitatory 'on' and inhibitory 'off' zones within their receptive fields, somewhat akin to the retinal ganglion cells in frogs and toads. 'Complex' and 'hypercomplex' cells also respond to lines, slits and edges in different orientations, but are not divided into 'on' and 'off' zones. Instead, the whole unit increases or decreases its rate of firing depending on the kind of input. 'Complex' cells receive information from several 'simple' cells, and filtered responses are passed on from the visual cortex to other parts of the brain. The recognition of complex visual stimuli thus depends on neural activity at many stages along the visual pathway.

Although cats rely more on central rather than a peripheral filtering process, there is not a simple evolutionary progression towards centralisation from lower to higher organisms. Pigeons, for example, show even more retinal differentiation than the leopard frog, while many invertebrates, such as crabs (*Podophthalamus vigil*) and locusts (*Locusta migratoria*), have sophisticated central processing. The selective elicitation of behaviour as a result of central stimulus filtering in part prompted the concept of innate releasing mechanisms (IRMs) in ethology (see 3.1.3.2).

Perceptual rules of thumb

Visual stimulus filtering provides common toads with a rough but workable guide to what is edible and what should best be avoided. We can thus think of the toad having perceptual rules of thumb (see 2.4.4.3). Like all rules of thumb, the toad's 'prey' and 'enemy' rules work well enough in the world in which the animal normally operates, but can easily be fooled by novel cue configurations or experimental manipulation. Thus the toad's 'enemy' response to a snake with a raised head (Fig. 3.23a) can be elicited by an abstract pattern with a raised element (Fig. 3.23b) or even a leech, which is normally regarded as prey, if its front suckers are raised off the ground (Figs 3.23c,d) (Ewert & Traud 1979).

Our own visual perception is also heavily dependent on rules of thumb. Our propensity for visual illusions tells us a great deal about the filtering processes and rules of interpretation that determine the model our brains build of the world. As Richard Dawkins (1998) puts it, the brain is a natural onboard virtual reality computer, constructing images of the world according to rules honed by natural selection. Information is amplified, integrated, suppressed or synthesised to generate the best working hypothesis for functioning in the environment. Thus, a glimpse of fur in the undergrowth is extrapolated into a predator, images of different size translate into perspective, subtle changes in facial expression become beacons of social information. Several excellent illustrations can be found in Gregory (1998).

The brain as hypothesis generator is exemplified particularly clearly when it dithers between equivalent alternatives. A good example is the Necker cube (Fig. 3.24a), a simple two-dimensional set of lines of paper which the brain interprets as a three-dimensional

Figure 3.23 Visual rules of thumb in the toad *Bufo bufo*. 'Enemy' responses are provoked by images of a snake (a), a head-rump dummy (b) and a leech with a raised front sucker (c). If the leech's sucker lies in the plane of movement (d), however, the toad responds as if it is prey. From Ewert (1980).

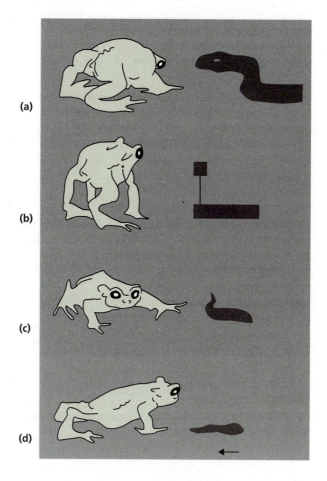

cube. Stare at the apparent cube for a few seconds, however, and it flips in depth perspective between two alternative forms, each interpreting a different 'end' facet as being to the fore. Being faced with two equally valid alternatives, the brain flips backwards and forwards between them rather than plumping arbitrarily for one. Familiar ambiguous images illustrate the same principle (Fig. 3.24b).

The capacity for interpretation is illustrated by Fig. 3.24(c), in which we perceive a white shape lying across the triangles where in fact none exists. The broken lines of the triangles, and the white segments of the circles, are suggestive enough for the brain to invent the rest. Our ability to recognise faces or familiar scenes in simple line sketches relies on the same inventive filling-in. But why does the brain do this? Because we inhabit a world of objects, and objects have boundaries that distinguish one from another. Anything that suggests the presence of an object is worth noting, and if necessary extrapolating, because many objects have salience: they are food, predators, companions or dangerous obstacles. We overlook them at our peril.

The adaptive value of these perceptual tricks is that they allow the brain to rationalise the environment according to well-tried rules of operation, often on the basis of partial or ephemeral information and with limited time to waste. Like all rules of thumb (2.4.4.3), they are prone to error, but as long as errors are sufficiently rare, and/or are not too costly, they provide an economical but effective means of interpreting the world.

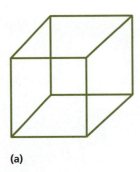
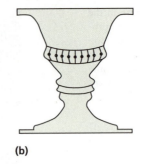
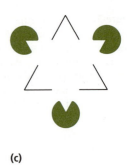

(a) (b) (c)

Figure 3.24 Visual illusions. (a) A Necker cube: a two-dimensional apparent cube that appears to flip between downward left and upward right orientations. (b) A classic ambiguous image alternating between a chalice and two opposing faces in profile. From *Eye and Brain: The Psychology of Seeing* by R.L. Gregory (Fifth Edition, 1998). Reprinted by permission of Oxford University Press. (c) An illusion of a second triangle overlying the one outlined. From Marr (1982).

3.2.1.2 Auditory stimulus filtering

Like vision, hearing has become specialised in different species according to the demands of their way of life, a process that also frequently relies on stimulus filtering. In some cases, filtering is achieved by adjusting the mechanics of the hearing apparatus to focus on specific components of the sound environment. This is particularly clear in nocturnal hunters such as bats and owls, some of which show extraordinary sensitivity to the sounds generated by their prey. Pioneering work by Payne (1971), for example, has shown that barn owls (*Tyto alba*) can home in on the faint noises made by mice as they move through the litter or gnaw their food. Even barely audible (to humans) sounds such as a leaf being pulled across a floor can be pinpointed. Barn owls achieve this remarkable accuracy by means of the positional asymmetry of the ears on either side of the head (the left ear is higher than the mid-point of the eye, the right lower) and the arrangement of feathers around the face, which form the facial 'disc' (Fig. 3.25a). The facial disc helps to channel incoming sound into the auditory meatus of the two ears and filter out sounds that are not arriving along the line of vision. Information about the prey's position in the horizontal plane can then be gleaned from the relative stimulation of each ear (Fig. 3.25b), while the asymmetric positioning of the ears provides information in the vertical plane (Knudsen & Konishi 1979).

Auditory stimulus filtering in barn owls involves mainly the removal of directional redundancy. Among insects, however, there are many examples of 'ears' which respond solely to a limited range of sound frequencies. One of the best known comes from noctuid moths. These moths are heavily preyed upon by night-flying bats that use echolocation to identify and home in on their airborne prey. Echolocation in bats works like human sonar in that it relies on the animal emitting pulses of sound and listening to their echoes as they bounce back from the environment. The pattern of returning echoes enables the bats to create a 'sound topography' of the environment by which they can orientate and navigate their way around. Objects distort the returning sound waves in different ways, allowing bats to judge their size, shape and texture as well as, in the case of moving objects, their speed and direction of movement (Simmons & Stein 1980). Whether or not an object will generate an echo, however, depends on its size and the wavelength of the sound. The wavelength must be roughly equal to the diameter of the object to produce an

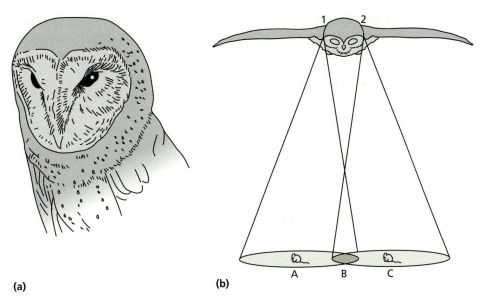

(a)

(b)

Figure 3.25   (a) The facial 'disc' of the barn owl (*Tyto alba*) is a fan arrangement of feathers around the face that helps to channel sound to the asymmetrically positioned ears. (b) Barn owls match the sound input to each ear to locate and catch prey in the dark. A mouse at A will stimulate ear 1, but not ear 2, and vice versa at C. A mouse at B, however, will stimulate both ears. From Barnard (1983) after Alcock (1975).

echo. Since the aerial prey of most bats are small insects, bats are constrained to use very high frequency (ultrasonic) sound, somewhere in the region of 50–80 kHz in fact, well above the upper limit (around 20 kHz) of human hearing. While humans may not be able to hear bats, however, some of their insect prey certainly can, noctuid moths among them. In fact, the ears of noctuids are almost entirely geared to listening out for bats.

The moths possess a pair of ears, one either side of the thorax. Each ear consists of just two sensory neurons connected to a tympanum, one sensitive to low-intensity sound, the other to high-intensity sound (Roeder 1970). The low-intensity (A1) neuron also responds more to intermittent pulses of sound rather than continuous bursts. Neither low- nor high-intensity (A2) neurons, however, respond to different frequencies of sound, only their intensity and temporal pattern of emission. The A1 neuron responds to the faint ultrasonic emissions of bats up to 10 m away, and fires as the emissions increase in intensity. The moth can therefore tell whether the bat is getting closer. By comparing the relative stimulation of the A1 neuron in its two ears, the moth can also pinpoint the position (above, below, to the side) of the bat and move away. The A2 neuron comes into play only when the bat flies close to the moth and the neuron is exposed to very high-intensity emissions. When the A2 neuron fires, impulses are transmitted to the cerebral ganglia and inhibit the centre controlling activity in the thoracic ganglion. The thoracic ganglion controls the pattern of wing beat, so, when it is inhibited, the wings beat asynchronously or stop beating altogether. As a result, the moth either flies erratically or drops like a stone out of the flight path of the bat.

Noctuid moths therefore show an extreme degree of auditory stimulus filtering. In the myriad noises of the night, their hearing system is tuned exclusively to the sound signatures of their predators. While motor responses are mediated by neurons in the

CNS, filtering is achieved peripherally via the highly selective neurons of the ear. An intriguing example of auditory peripheral filtering in vertebrates comes from the tree frog *Eleutherodactylus coqui*, whose specific name derives onomatopoeically from the characteristic 'co-qui' call of the male. The call is particularly interesting because its two components ('co' and 'qui') appear to be directed at different recipients – males and females respectively. Selective targeting could be achieved in a number of ways. However, Navins & Capranica (1976) discovered that the tympanic membranes of the two sexes are in fact tuned differently, so that each sex hears *only* the relevant part of the call. Males hear the 'co', females the 'qui', and neither hears the other component. In mammals, centrally mediated selective attention appears to be the more important mechanism for biasing responses to auditory stimuli.

3.2.1.3 Other mechanoperception

Hearing works through high-frequency stimulation of **mechanoreceptors**, structures (such as the tympanum of the insect and mammalian ear in the case of hearing) that are sensitive to vibrations caused by pressure changes. Many kinds of mechanical stimuli are important to animals and they have evolved a wide range of receptors and organs to detect them. Most of these are low-frequency stimuli which the animal senses as touch or pressure.

Many species, including insects and mammals, possess touch-sensitive hairs over the body surface. These can be crucial for orientation and coordinated movement. The intricate foraging dance of the honey bee (Chapter 11), for example, depends on such hairs. Bees usually dance on the vertical surface of the honeycomb, where information about orientation is essential to ensure the directional components of the dance are correct. Touch-sensitive hairs between the head and thorax provide this information. Because the top half of the bee's head is lighter than the bottom half, the bee's orientation on a vertical surface causes the head to press on the hairs in different ways. If the bee is facing vertically up, the bottom of the head presses against the ventral set of hairs. If it moves to the left or right, the left or right ventral hairs are stimulated. If it faces down, the lower part of the head falls forward and the top of the head presses on the dorsal set of hairs. Central processing of this tactile information tells the bee how it is orientated on the comb. If the hairs are denervated, the animal wanders about chaotically. If weights are glued to different parts of the head, the bee's movements are altered predictably depending on the resulting pattern of stimulation of the hairs (Lindauer 1961).

The sensitivity of mechanoreceptive hairs can be extraordinary. Camhi *et al.* (1978) showed that cockroaches were able to escape from toads by detecting tiny changes in air pressure caused as the animal manoeuvered to strike. The minute draughts are detected by hairs on the appendages at the tip of the abdomen known as cerci. If these are clogged with glue, cockroaches are far less able to escape from toads. On much the same plane of incredibility, some caterpillars are able to detect approaching parasitic wasps up to half a metre away as a result of pressure changes caused by the wasp's wings (Markl 1977).

As with vision, perception based on mechanoreception sometimes works on crude rules of thumb. Web-spinning spiders can be fooled into attacking a tuning fork because its vibrations are reminiscent of those generated by prey as they struggle to break free of the web. Vibrations transmitted through the web provide useful information about type and whereabouts of prey, and, since tuning forks have not been a regular hazard in the spiders' evolutionary history, vibrations are a sufficiently reliable cue on which to act.

3.2.1.4 Olfactory and gustatory perception

Chemical cues provide information about many important aspects of life: food, predators, mates, companions, toxins, disease, even direction. Not surprisingly, therefore, animals have evolved diverse and often remarkable sensitivities to chemical information, some of which have given rise to specialised chemical signals known as **pheromones** (see Whyatt 2003 for an excellent account). The **chemoreceptors** involved are broadly divided into those concerned with detecting odours (olfactory receptors) and those concerned with taste (gustatory receptors). As usually conceived, smell is concerned with detecting relatively low concentrations of airborne chemicals, whereas taste involves direct contact with chemicals at higher concentrations. Since in both cases the chemical(s) concerned reach the receptor in solution, however, it is difficult to maintain a hard and fast distinction between the two senses, especially for animals that live in water. Indeed in humans, and quite possibly other mammals as well, the flavour of food *depends* on the sense of smell, as can be demonstrated easily by pinching the nostrils closed when trying to discern flavours. Nevertheless, the two kinds of receptor are served by separate neural pathways in many species.

Rather little is known about the molecular events that translate chemical information within receptors, and it seems unlikely that there is a simple mapping of specific taste or odour molecules to particular types of receptor. Instead, individual receptor cells probably possess a combination of receptors that vary in their proportions across different types of cell. The diversity of olfactory and gustatory discriminations made by many animals is therefore likely to result largely from central processing of the enhanced and depressed impulses emanating from various populations of cells.

Olfactory sensitivity varies considerably across taxonomic groups and, like some of the senses already mentioned, shows remarkable acuity in some species. One of the best known examples among invertebrates is the mate attraction pheromone of the female silk moth (*Bombyx mori*). The sensitivity of the male's antennae to the pheromone (a polyalcohol known as 'bombykol') is such that a female has simply to emit a small quantity of the substance into the air and sit tight in order to secure a mate. The molecules drift downwind where they are detected by a wandering male who needs only 200 molecules to strike his elaborate, feathery antennae within a second for him to be able to orientate towards the female and home in on her. Recent experiments using robotics (3.1.3.4) have begun to tease apart some of the neural and behavioural mechanisms by which males achieve homing accuracy (Kanzaki 1996; Ishida *et al.* 1999). The female's altogether simpler antennae are not designed to detect airborne pheromones; instead her olfactory sensitivities are geared to detecting good oviposition sites.

A similarly remarkable olfactory homing feat appears to underlie the spawning migration of north Pacific coho salmon (*Oncorhynchus kisutch*). After spending up to five years feeding at sea, salmon migrate back to the river, indeed the very tributary, in which they hatched in order to mate. Migration to the river itself appears to be accomplished using a combination of sensory cues, including a sun compass, magnetism, polarised light and odour (see Chapter 7). Once in the river, however, odour seems to be the principal cue used to locate the spawning ground (Brannon 1982). Experiments involving sensory deprivation and exposure to artificial odours in the natal stream, as well as observations of the choice behaviour of migrating fish, have shown beyond reasonable doubt that odour is involved (Scholz *et al.* 1976; Johnson & Hasler 1980; Brannon 1982). There is still debate as to the nature of the odour cue denoting home; contenders are cues emanating from local rocks, soil and plants, and/or those from conspecifics. Fish could

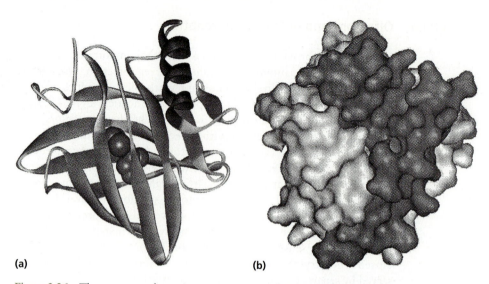

(a)   **(b)**

Figure 3.26   The structure of a major urinary protein (MUP) molecule of the house mouse (*Mus domesticus*). (a) The male pheromone inside the cavity of the molecule; (b) the surface location (dark-shaded area) of most of the polymorphic variation between MUPs. See text. From Beynon *et al.* (2002). Photographs courtesy of Rob Beynon and Jane Hurst.

learn either kind of cue during their early sojourn in the stream; they could even learn the sequence of odours from different streams as they encounter them on their outward journey (Brannon & Quinn 1990). Whichever sources are used, however, fish are, like the male silk moths above, responding to minute concentrations of odour molecules, less than one part per million in fact.

Fine-tuned olfactory discrimination underlies the social organisation of many mammal species. Jane Hurst's ground-breaking studies of house mice (*Mus domesticus*) provide a nice illustration. Commensal populations of house mice are structured into group territories in which social and reproductive relationships are maintained by a complex olfactory communication system based on urinary odours (e.g. Hurst 1989, 1993; see also 9.3.1.3). Competitive relationships between males are partly regulated by a system of countermarking (adding own urine to) the marks of other males on the substrate. By analysing the biochemical composition of the marks, Hurst and coworkers were able to show that countermarking depends on two levels of odour discrimination. Initial attraction to a mark is stimulated by volatile ligands (thiazole and brevicomin) bound to larger protein molecules (major urinary proteins, or MUPs; Fig. 3.26), while countermarking itself appears to be a response to the non-volatile protein component of the mark (Humphries *et al.* 1999). In fact, variation in the protein–ligand complex and its ageing on the substrate appear to provide several tiers of social and sexual information that prime interactions when male or female odour donors are eventually encountered (Hurst 1993; Hurst *et al.* 2001; Beynon & Hurst 2003).

3.2.1.5   Electromagnetic perception

As well as the familiar modalities above, animals have evolved others to complement, or even replace, them when conditions become too difficult. Sensitivity to natural electric and magnetic fields are two such modalities.

Figure 3.27 The duck-billed platypus (*Ornithorhynchus anatinus*) detects prey in the mud at the bottom of a stream using electroreceptors in its 'bill'. Drawing after Manning & Dawkins (1998).

Various organisms from bacteria to humans have been shown to respond to variation in natural or artificial magnetic fields. Magnetic sensitivity in bacteria, honey bees and pigeons appears to be mediated by sensory structures containing crystals of magnetite whose anatomical location varies across groups; in honey bees they are found in the abdomen, in pigeons, the head. An ability to respond to variation in the Earth's magnetic field plays an important role in orientation and navigation, for instance in calibrating celestial compasses and providing backup cues when celestial compasses are not available. We discuss this in more detail in Chapter 7.

Sensitivity to electric fields has evolved most conspicuously among fish that inhabit murky waters, but it is also present, uniquely among mammals, in the monotremes (duck-billed platypus [*Ornithorhynchus anatinus*] and echidnas [*Tachyglossus aculeatus* and *Zaglossus bruijnii*]) (Proske *et al.* 1998; Pettigrew 1999; Fig. 3.27).

Three broad groups of fish use electricity, but only two appear to possess electrical sensitivity. Elasmobranchs (sharks and rays) and some other non-teleost (non-bony) fish such as lungfish (Dipnoi) are sensitive to local distortions of the Earth's electric field and use this to detect prey, even, in the case of dogfish (*Scyliorhinus canicula*), when prey are buried in the substrate. Such sensitivity has recently been exploited to produce shark repellents for divers. A second group, the so-called 'weakly electric' fish (which are, in fact, two independent groups: the African Mormyridae and the South American Gymnotidae), can generate their own electric fields but are also sensitive to changes in the electric field around them. Objects of different conductivity in the environment cause different patterns of distortion in the fishes' electric field, allowing them to orientate and navigate, as well as detect predators and prey.

'Weakly electric' fish have two kinds of electrosensitive receptor: *ampulla* receptors and *tuberous* receptors, though some species have only one or the other. Ampulla receptors respond to slow changes in the electric field, while tuberous receptors respond to more rapid changes. Electrical discharges are pulsed at rates of up to 300 s^{-1} by modified muscles or axons comprising the *electric organ*. Variable pulse rates allow fish to communicate and reduce interference from each others' electrical fields. They may even be able to exploit individual differences in signal characteristics to 'eavesdrop' on one another and size up potential competitors (McGregor 1993; Scudamore 1995; see also 11.2.1.1).

Lungfish have electrosensitive organs analogous to the ampulla receptors of 'weakly electric' fish. In monotremes, however, this role is performed by sensory mucus glands in the skin which are innervated by large-diameter nerve fibres. There may be up to 40 000 mucus sensory glands in the upper and lower bill of the platypus, which lives in a similar murky water environment to the 'weakly electric' fish, a sharp contrast with the 100 or so in the snout of the terrestrial echidna. It is not clear why the echidna is sensitive to electric fields, but it may help detect electrical activity from the muscles of prey moving in moist soil (Proske *et al.* 1998).

The third group of fish that use electricity are the 'strongly electric' fish, and include the electric eel (*Electrophorus electricus*) and electric ray (*Torpedo* spp.). Unlike the low-voltage (a few millivolts) field of the 'weakly electric' species, the electric organs of these fish produce up to 900 volts and are used to paralyse or kill prey. In these cases, electricity is used purely as a weapon and 'strongly electric' species seem not to possess an electroreceptive modality.

3.3 Hormones and behaviour

Working in conjunction with the nervous system in many cases, and providing another means of communication within the animal's body, is a specialised group of organs called the **endocrine glands**. The endocrine glands secrete chemical messengers, or **hormones**, into the circulatory system in response to various internal and external stimuli. Hormones are also secreted by special neurons (neurosecretory cells) within the nervous system itself, where they are transmitted along axons or, again, into the blood. When they are transmitted via the circulatory system, hormonal messages are much slower than the electrical messages of the nervous system. Thus, they are particularly suited to regulating functions, physiological and behavioural, that are sustained over minutes, days or even months (see Randall *et al.* 1997 for a good summary). The fact that they are transported in the bloodstream also means that hormones can travel everywhere in the body, in contrast to the more focused transmission of nerve impulses to particular muscles. Together, the rapid responses of the nervous system and the more sustained influences of hormones complement one another in controlling the animal's actions.

3.3.1 How hormones affect behaviour

The effects of hormones on behaviour can be traced to four broad areas of influence: the nervous system, sensory perception, effector systems and development. Techniques for studying them range from surgery (glandectomy) followed by hormone replacement to manipulation of circulating hormone concentrations (e.g. by injection, implants or cross-transfusion) and correlational studies (e.g. Figs 3.28 and 3.31). We discuss the effects of hormones on the development of behaviour in more detail in Chapter 6 (see 6.1.2); here we focus mainly on their role in mechanism.

3.3.1.1 Effects on the nervous system

Hormones affect many aspects of the nervous system, including anatomy, biochemistry and impulse transmission. In some cases, they may be responsible for basic structural and functional changes within the CNS. Reflex connections (3.1.3.2), for example, are accelerated by high levels of thyroxin, a hormone secreted by the thyroid gland. Some sex differences in behaviour in rats are associated with sexual dimorphism in the anatomy of cells in the neuropile of the hypothalamus, a dimorphism that appears to be

Figure 3.28 Correlational evidence for a relationship between circulating testosterone concentration and aggression in male Egyptian spiny mice (*Acomys cahirinus dimidiatus*). The vertical axis scale is deviations from the sample mean. From Barnard *et al.* (2003).

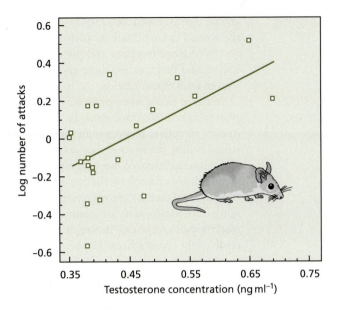

mediated by neonatal levels of androgen (male steroid sex hormones secreted mainly by the testes) (Raisman & Field 1973). However, care is needed in interpreting the neuronal effects of hormones. For instance, progesterone and oestrogen (female steroid sex hormones) both enhance the responsiveness of neurons in the rat CNS to stimulation of the vaginal cervix. However, while the effect of progesterone is rapid and associated with a general increase in arousal of the animal, behavioural effects of oestrogen are usually not apparent for several days, implying longer-term changes in the state of neurons.

A given hormone can have very different effects on behaviour depending on the region of the CNS on which it acts. Nyby *et al.* (1992) looked at the effects of intracranial implants of testosterone (an androgen) in male house mice on various aspects of social and sexual behaviour, including ultrasonic vocalisations, urine marking, mounting and aggression. They implanted the hormone in one of four regions of the brain: the septum, the medial preoptic area, the anterior hypothalamus or the ventromedial hypothalamus. Control animals were given subcutaneous implants of testosterone or empty implants in the appropriate region of the brain. The testosterone controls performed all the behaviours at the normal level for sexually responsive males, while the empty implant controls showed no response. When it came to the site-specific implants, however, Nyby *et al.* found a range of responses. Implants in the median preoptic area showed increased ultrasonic vocalisation, while those in other regions had little effect. Those in the medial preoptic or either hypothalamic regions resulted in more urine marking than empty implant controls, but less than testosterone controls. Mounting occurred in testosterone controls and males implanted in the median preoptic area, while aggression was rare in all males given brain implants. Together, the results thus suggest complex functional interactions between different areas of the brain containing testosterone receptors.

Rather than influencing behaviour directly through their effects on the nervous system, hormones may act as 'primers' facilitating the action of other hormones. In female hamsters (*Mesocricetus auratus*) primed with oestrogen, behaviour typical of oestrus can be induced by injecting small doses of progesterone into the brain ventricles. If injected subcutaneously, or in the absence of oestrogen, such doses fail to induce the response. Similar priming effects, but in relation to electrical stimulation, have been shown in rats. Electrodes implanted in a rat's brain can be connected in such a way that the animal can simulate its own brain by pressing a pedal. Stimulation of some areas results in more frequent and persistent pedal-pushing than others. Olds (1961), however, found that treating these areas with androgen could enhance the rat's responses to a given level of stimulation.

In other cases, hormones may not so much exert a positive effect on the nervous system as remove inhibition. In neonatal guinea pigs (*Cavia porcellus*), the lordosis (female receptive) posture is an integral part of excretory behaviour. At first, excretion is stimulated by the mother, but, as the young mature, micturation and defaecation come under internal control mechanisms located in the spine. These spinal centres are subject to inhibitory control by the brain, and, as females become adult, inhibition of spinal control is relaxed by the secretion of ovarian hormones. Hormones may also *establish* inhibition. Oestrogen, for example, inhibits aggressive behaviour in female hamsters, while sexual receptivity in female grasshoppers is inhibited by hormones produced as a result of the spermatheca (female sperm storage organ) filling with sperm.

3.3.1.2 Effects on sensory perception

Many studies suggest that hormones affect an animal's sensory capabilities. In doing so, they alter the animal's perception of its environment and therefore the way it responds to particular stimuli. The seasonal migration of three-spined sticklebacks (*Gasterosteus aculeatus*) provides a good example. In spring, male sticklebacks migrate from the sea to their freshwater breeding grounds. In doing so, they move through water that gradually changes in salinity. To facilitate this transition, hormones secreted by the pituitary and thyroid glands, but particularly thyroxin, alter the fishes' salinity preference from salt water to freshwater (Baggerman 1962).

In many female mammals, sensory perception is influenced by the oestrous cycle. Female rats fluctuate in their ability to detect certain odours according to circulating levels of oestrogen and progesterone. Similarly, visual sensitivity in women varies with the stage of their menstrual cycle (and thus relative oestrogen and progesterone levels), being most acute around the time of ovulation, and least acute during menstruation, a difference that is abolished when taking oral contraceptives. In female rats, oestrogen also has the effect of extending the sensory field of the perineal nerve, which innervates the genital tract (Komisaruk *et al.* 1972). During oestrus, she is thus more responsive to the tactile stimulus of intromission, and orientates her body to facilitate penetration (lordosis; Fig. 3.29). In birds, oestrogen causes the formation of a 'brood patch'. Feathers are lost from part of the ventral body surface in females and the exposed skin becomes more heavily vascularised, thus increasing the sensitivity of the female to the nest cup and influencing her nest maintenance and incubatory behaviour.

Figure 3.29 Female rats take up the characteristic copulatory posture known as lordosis partly in response to the effects of oestrogen on the tactile sensitivity of the genital tract.

Sensory perception in males is also influenced by hormones. Sexually experienced male rats prefer the odour of urine produced by females in oestrus compared with dioestrus, but the preference disappears if males are castrated. Androgens produced by the testes thus appear to modify the animals' response to urinary odours, though the effect is partly confounded with those of experience. As in females, genital sensitivity in male rats is increased by sex steroids. Testosterone causes the skin of the glans penis to become thinner so that underlying sensory cells receive more stimulation and the male can respond more effectively to the copulatory movements of the female. Conversely, castration results in loss of the surface papillae of the glans, with a concomitant reduction in sensitivity and copulatory behaviour.

3.3.1.3 Effects on effector systems

Animals use a range of appendages and other external structures in performing different behaviours. Various hormones affect the development of such structures and thus their efficacy in performing whatever behaviours depend on them. In some cases, hormones induce appropriate muscle development, as in the hypertrophied brachial musculature used by male frogs in amplexus (coupling). The aminergic neurotransmitters serotonin and octopamine in lobsters (*Homarus vulgaris*) prime receptors in muscles of the exoskeleton to respond appropriately to particular stimuli. Serotonin primes the postural muscles for flexion, for example when another lobster comes into view, while octopamine inhibits flexion. Together, they help coordinate dominant and subordinate responses to social stimuli (Kravitz 1988). Sex differences in call characteristics in the clawed toad, *Xenopus laevis*, are also due to hormonal effects on muscle development. In this case, androgens increase the number of muscle fibres in the larynx of males as they mature, and stimulate the development of more so-called 'fast twitch' fibres that produce the characteristic rapid trill of the male call (Kelley & Gorlick 1990). In women, changing oestrogen and progesterone levels during the menstrual cycle affect the strength of several muscle systems, causing fluctuations in physical performance at different stages of the cycle (Reilly 2000).

Secondary sexual adornments provide other examples of hormonally induced effector systems. In newts, prolactin appears to be important in the development of the enlarged tail fin in males, which is used to fan a stream of water at the female during courtship. While prolactin also influences the vigour of tail fanning, there is a synergistic effect with growth hormone in enhancing the overall performance of the response (Toyoda *et al.* 1992). Sexually selected plumage characteristics in birds provide other examples. These range from minor markings, such as the orange patch in the cap of male goldcrests (*Regulus regulus*), to elaborate, gaudy structures such as the tails of male birds of paradise, or the combs and wattles of male fowl. In many species, the development and maintenance of secondary sexual characters depend on sex steroids, in some cases the presence of androgen, in others the absence of oestrogen (Owens & Short 1995; Kimball & Ligon 1999; Fig. 3.30). For example, comb and wattle size in newly hatched male and female domestic chicks can be increased to proportions normally found in sexually mature males by injecting testosterone, while castration of mature males results in a pronounced reduction of comb and wattle size. Plumage dimorphism in birds of this group (galliforms), as in ratites (ostriches, rheas, etc.) and anseriforms (ducks, geese, etc.), on the other hand, depends on the presence or absence of oestrogen. Secretion of oestrogen leads to dull plumage typical of females, while an absence of the hormone leads to bright male-like plumage. Removal of the gonads in either sex results in

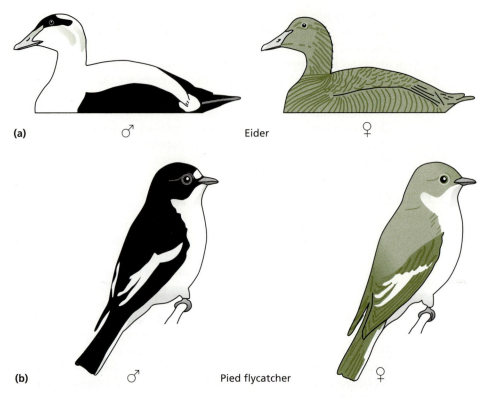

Figure 3.30 Plumage dimorphism in eider ducks (*Somateria mollissima*) and pied flycatchers (*Ficedula hypoleuca*). Although ducks and passerines often show similar degrees of dimorphism, bright plumage in males of the two groups depends on very different hormonal influences. See text.

the bright plumage typical of males. Among passerines (perching birds such as sparrows and finches), bright male plumage depends on secretion of luteinising hormone, and gonadectomy has no effect on the development of plumage colouration (Kimball & Ligon 1999; Fig. 3.30).

Horn and antler size in artiodactyls (cloven-hoofed mammals) such as deer and gazelle, often an important determinant of success in intermale disputes and mate attraction (see Chapter 10), appears to be at least partly determined by androgen levels. In addition, sexual behaviour in mammals may be mediated by hormonal effects on odour production. Oestrogen administered intravaginally to female rhesus monkeys (*Macaca mulatta*) increases the tendency for males to mount or press a lever to gain access to a female, the latter effect disappearing when males are rendered anosmic (Michael & Keverne 1968; Michael & Saayman 1968).

3.3.1.4 Effects on development

Hormones have a profound effect on the development of young animals and impart characteristic features to their behaviour as adults. While hormonal effects on behaviour and associated physiological processes in adults are usually reversible and independent of age (once adult), those affecting development are permanent and irreversible, often limited to clearly defined periods in the developmental process, and usually manifested later in life rather than at the time of effect (Beach 1975). For example, hormone-mediated anatomical and physiological changes are important in the development of sexual behaviour. In guinea pigs, testosterone levels influence the development of the genitalia, so that female offspring born of females treated with testosterone proprionate during pregnancy have male-like genitals. If they are then ovariectomised and treated with gonadal hormones, these offspring develop more male-like behaviour than controls whose mothers were not treated with testosterone (Young 1965).

Male and female rats are born with a CNS that is largely undifferentiated with respect to sexual behaviour, though with a tendency towards a characteristically female pattern (Harris & Levine 1965). Differentiation of male behaviour is brought about by the later action of testosterone. If female rats are given testosterone at around 4 days of age, their oestrous cycle and sexual behaviour as adults are suppressed and remain unresponsive to ovariectomy and oestrogen treatment. Similarly, treatment of 4-day-old males with oestrogen results in some loss of sexual responsiveness through partial functional castration and impaired development of the penis.

Studies of a range of species suggest that hormonal effects on sexual responses such as those above occur during more or less clear-cut **critical periods** in an animal's development. Sometimes the critical period is prenatal, sometimes postnatal. While guinea pigs and rats exemplify these respective conditions, the critical period in guinea pigs, which are more precocial (born at a later stage of development), in fact occurs at the same *developmental* stage as in rats. However, there is evidence that hormonal influences are at work rather earlier than the critical period suggested by experimental manipulations. In both rats and mice, for example, the sex of neighbouring foetuses *in utero* may have a marked effect on female genital morphology and sexual behaviour as a result of testicular androgens being synthesised and released late in gestation (Clemens 1974; vom Saal 1989). Other evidence from birds suggests that hormonal influences on development may not always be restricted to a sharply defined critical period, and that behavioural effects can occur at different times during development (Schumacher *et al.* 1989).

Prenatal hormonal influences in mammals can have more profound effects on neural development and behaviour. Thyroxin deficiency in human mothers is known to be associated with deficits in motor and cognitive functions in subsequent offspring. Rats born to thyroid-deficient dams also show marked behavioural deficits, including reduced learning ability and responsiveness to emotionality and open field tests. These effects are associated with reduced levels of the cortical neurotransmitters influencing activity, mood and learning in animals with normal thyroid function (Friedhoff *et al.* 2000).

3.3.1.5 Factors influencing relationships between hormones and behaviour

Hormones, then, have diverse effects on behaviour, both directly and indirectly. Not surprisingly, therefore, these effects can be conditional on a host of internal and external factors.

Individual genotype

Differences in individual genotype are one source of variation in response. Experiments with different genetic strains (2.3.2.2) have indicated various strain-specific effects of hormone treatment. Strains of domestic fowl, which had been selected for different tendencies to complete mating, showed very different responses to castration and subsequent androgen treatment: in both cases hormone treatment caused birds to resume the precastration mating behaviour typical of their strain (McCollom *et al.* 1971).

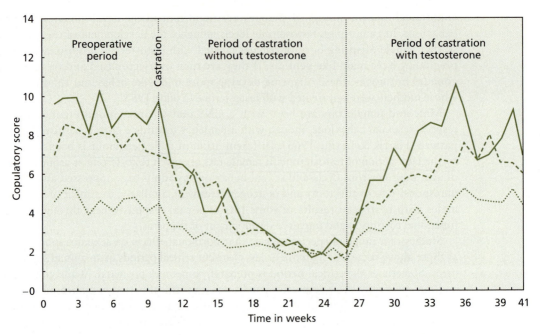

Figure 3.31 Individual male guinea pigs differ in their levels of sexual activity and these differences are maintained in their response to castration and hormone replacement. Solid line, 'high drive'; dashed line, 'medium drive'; dotted line, 'low drive' individuals. After Grunt & Young (1952).

Similarly, perinatal administration of androgen to female mice increased aggressive behaviour if females came from a strain in which males were typically aggressive, but had little effect if males of the strain were non-aggressive (Vale *et al.* 1972). Even within strains individuals show consistent differences in hormone-related behaviour. Individual differences in copulatory behaviour among male guinea pigs, for example, were resumed after castration when animals were given the same dose of testosterone (Fig. 3.31). On a broader scale, effects of sex steroids on the development of sex-typical behaviour appears to depend on which sex is homogametic (has two sex chromosomes the same). In mammals this is the female (females XX, males XY), in birds the male (males ZZ, females ZW). As a result, the development of sexuality in the absence of steroid influence defaults to female in mammals, but male in birds. Recent evidence suggests that interactions between sex steroids and disease resistance genes may also account for some sex-typical relationships between behaviour and immune function (Klein 2000; see 3.5.2).

Seasonal variation

Seasonal effects are also important in determining behavioural responses to hormones. In red deer (*Cervus elephus*), the administration of testosterone to stags in winter brings about full rutting behaviour, whereas in late spring it has no effect until the normal rutting period (Lincoln *et al.* 1972). Similar seasonal influences are evident in the receptivity of female anoles (*Anolis carolinesis*), a desert lizard, to approaches by males following ovariectomomy and oestrogen and progesterone treatment (Wu *et al.* 1985). The seasonal cycling in the size of song control centre nuclei in the forebrain of canaries

(3.1.3.1, Fig. 3.12a) appears to be under the control of testosterone. Goldman & Nottebohm (1983) showed that increase in size (recrudescence) of the HVC (Fig. 3.12a) in male canaries each spring was due to the influence of testosterone on the formation of new nerve cells. Interestingly, and in contrast to the testosterone-induced sexual dimorphism in rats (3.3.1.4), testosterone treatment induces the same pattern of development in female canaries, which then also sing. Seasonal changes in hormone levels and associated behaviour like these are widespread in animal species and can persist even when external cues to seasonality are removed under controlled laboratory conditions (see 3.4).

Effects of experience

Past experience is another factor that can have a profound influence on the behavioural effects of hormones. The maintenance of copulatory behaviour following castration in male cats, for example, is more protracted if males have previously had experience of mating. In young male chickens, there appears to be an interaction between prior copulatory experience and treatment with male and female sex steroids in the lateralisation of copulatory control in different cerebral hemispheres. Using eye patches to restrict vision to one eye, and treating birds with 5-alpha-dihydrotestosterone (5-alpha-DHT) or oestradiol, Bullock & Rogers (1992) found that different hemispheres played a role in copulatory activity depending on which eye was covered. However, once patches had been swapped between eyes, treatment with 5-alpha-DHT led to high copulation scores regardless of which eye was covered, suggesting the hormone facilitated either interocular transfer of behavioural control (i.e. transfer from the part of the brain served by one eye to the part served by the other), or equal access of both eyes to the regions of the brain which control copulation (Bullock & Rogers 1992).

Experience can itself lead to changes in hormonal state, emphasising the frequent bidirectionality of relationships between behaviour and hormonal changes. In male squirrel monkeys (*Siamiri sciureus*), testosterone levels prior to grouping did not predict the social rank a male subsequently adopted (Fig. 3.32). Once rank relationships were established, however, there was a close correlation between rank and testosterone concentration, with the alpha (top ranking) male having the highest testosterone concentration and the gamma (lowest ranking) the lowest (Mendoza *et al.* 1979). The

Figure 3.32 Mean levels of testosterone in male squirrel monkeys (*Siamiri sciureus*) in relation to their social status in different social environments. See text. After Mendoza *et al.* (1979).

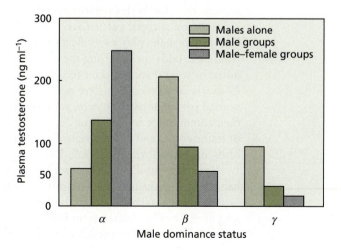

disparity in testosterone levels between ranks was particularly pronounced if females were also present in the group (Fig. 3.32). Thus, the social relationships between males, and the nature of the social environment, appear to determine hormone levels rather than the other way round. Similar effects of social environment are evident in red deer, where hinds exposed to roaring by vasectomised stags, or played tapes of stags roaring, are quicker to come into oestrus (as calculated from calving dates following contemporaneous matings with intact stags) than hinds not exposed to roaring (McComb 1987).

Ecological influences

The secretion of hormones affecting behaviour may vary with ecological conditions and the impact of hormonal effects on individual reproductive success. In male red-winged blackbirds (*Agelaius phoeniceus*), a territorial North American species, testosterone levels peak during the roughly two-week period when males are defending breeding territories and guarding their mates from rivals. In a study by Beletsky *et al.* (1992), testosterone levels were higher in males defending territories in area of high population density, where competition was stiffer, and higher in territorial males than in so-called 'floaters' (itinerant non-territorial males). Elevated testosterone thus appeared to reflect the need for aggressive defence of breeding resources, and it is therefore not surprising that it also correlated with the number of females on a male's territory and the eventual number of offspring fledged. That testosterone levels were associated with a male's ecological circumstances, rather than intrinsic individual qualities, is evident in the lack of correlation in levels within males across years (Beletsky *et al.* 1992).

While evidence from a wide variety of vertebrate species suggests that testosterone increases aggressiveness, especially in males, it is clear that it does not always do so. It seems that the hormone is associated most closely with aggression related to reproduction, such as the territorial and mate-guarding behaviour of the red-winged blackbirds above, and less with aggression in other contexts, such as anti-predator responses, or non-sexual social intolerance (Wingfield *et al.* 1990). Thus, increased reproductive aggression as a result of elevated levels of testosterone is most evident during periods of social instability, and declines when social relationships are stable. Wingfield *et al.* (1990) refer to this social instability hypothesis of testosterone-driven aggression as the **challenge hypothesis.**

While reproductive aggression is ultimately a device for increasing reproductive success, its value in this regard depends on potential conflict with other components of reproductive success, particularly, argue Wingfield *et al.*, caring for offspring. Thus in birds where males play a significant role in caring for the young, testosterone levels drop from their initial peak during the competitive phase of breeding, a trend that has been noted in several such species (e.g. Wingfield & Moore 1987; Vleck & Brown 1999). Converse observations, where males have been treated with testosterone during their normal period of paternal care, support this interpretation in that treated males reduce their commitment to caring for dependent young (Cawthorn *et al.* 1998). The secretion of testosterone, and its behavioural consequences, thus appear to be constrained by the birds' life history strategy (2.4.5, 3.5).

While many studies have focused on the effects of particular hormones, the control of behaviour is frequently influenced by several different hormones acting simultaneously or in sequence, each subject to the kinds of modulating factors discussed above. Some idea of this can be gleaned from Fig. 3.33, which summarises some classic studies of the hormonal control of reproductive behaviour in the canary.

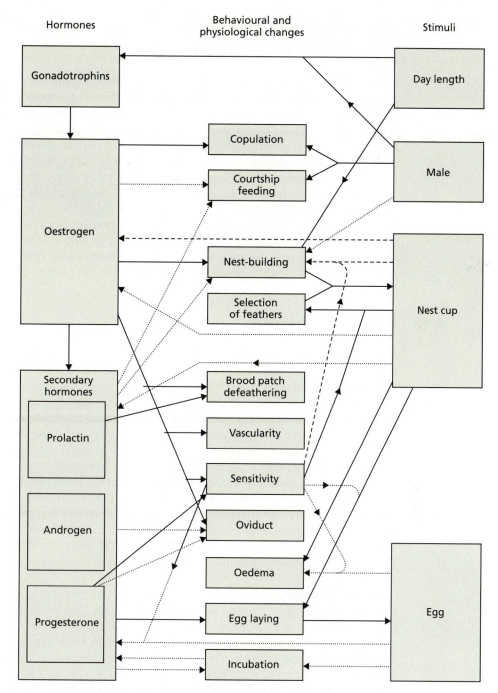

Figure 3.33 Summary of interrelationships between hormones, external stimuli, physiological function and behaviour in the reproductive cycle of the canary (*Serinus canarius*). Solid lines, positive effects; dashed lines, negative effects; dotted lines, putative positive effects. After Hinde (1965).

3.4 Neural and hormonal mechanisms and the long-term control of behaviour: biological rhythms

Between them, the nervous and endocrine systems allow a combination of immediate, short-term control of behaviour, and longer-term patterns of activity such as seasonal cycles. The interplay is particularly well illustrated by the various forms of rhythmic behaviour shown by different species. Life has evolved under many cyclical influences in the environment, from the daily light–dark cycle and familiar annual seasons to the rhythms imposed by the tides and moon. During their evolution, animals have acquired a variety of endogenous physiological and behavioural rhythms whose periodicity is matched to the relevant rhythm(s) in the environment. The remarkable thing about these rhythms is that they persist even when the environmental cycles to which they are tuned are removed. They thus reflect an endogenous time base or 'clock' within the animal that regulates its activities. The genetic and developmental basis for these internal clocks has been subject to intensive study and we have encountered some of the results already in Chapter 2 (2.3.2.1). These aspects are developed further in Chapter 5.

3.4.1 Rhythmic behaviour

Box 3.5 summarises the main kinds of rhythmicity in animals, and Fig. 3.34 shows some examples. While the timescale of these different classes varies, they share a number of basic features (Drickamer et al. 1996).

Rhythmic behaviour is characterised by a series of repeating units or **cycles**, the duration of which constitutes the rhythm's **period** (Fig. 3.34a). The degree of change in response between the peak and trough of the cycle is the **amplitude**, while any particular part of the cycle is generally referred to as a **phase**. A crucial feature of behavioural and other biological rhythms is that they can be **entrained** to cues (**Zeitgebers**) in the external environment, so that the pacemaker of the clock can be set or adjusted appropriately. Moreover, the rhythm of the clock is generally robust to disturbance by factors that normally affect cell physiology or the rate of chemical processes, such as metabolic poisons and temperature variation, though chemical manipulation experiments, particularly involving protein synthesis, have elucidated some of the biochemical underpinnings to the clock mechanism (e.g. Hastings et al. 1991; see also Chapter 5).

While endogenous clocks can be entrained to environmental cues, however, they are not dependent on them for their subsequent expression. External light–dark, seasonal or tidal cycles are not responsible for the timing of the behaviour in the way that, say, temperature variation drives the daily pattern of sun- and shade-seeking in thermo-regulating lizards. Nevertheless, the entraining properties of cyclical environmental cues keep the clock running to time. When animals are kept in a constant laboratory environment, with no cyclical external cues, their clocks continue to run, but slightly out of phase with the cue to which they were originally entrained (Fig. 3.34b,c). Thus the flying squirrel and stonechat in Fig. 3.34(b,c), maintained under constant light and temperature conditions in the laboratory, show a systematic drift in their normal circadian activity and circannual testicular and moult cycles respectively. This drift is the reason biological rhythms are given the prefix 'circa' (from the Latin for 'around' or 'about'). The approximate free-running periodicity of many circadian clocks studied

Supporting evidence

Box 3.5 Behavioural rhythmicity

Several environmental influences on behaviour are rhythmic in nature. We might therefore expect organisms to have adapted to these and entrained their patterns of behaviour accordingly. But what are these influences, and how has behaviour become tuned to them?

Circadian rhythms

The most obvious source of environmental rhythmicity is the daily light/dark cycle, and almost all taxonomic groups have evolved complementary circadian (from the Latin *circa* [about] *dies* [a day]) rhythms of behaviour to accommodate it, if only in alternating phases of activity and rest. **Circadian** clocks are so-called because their periodicity is generally not exactly 24 h, and, under constant light conditions, tend to drift. Nevertheless, they provide the underlying mechanism for some other, longer-term, behavioural rhythms (see text).

Circannual rhythms

Behaviour in many species is entrained to the annual seasonal cycle. This is reflected particularly in seasonal reproductive behaviour and dormancy. Like circadian rhythms, the periodicity of **circannual** rhythms is seldom exactly a year. In hibernating ground squirrels, for instance, they vary from 229 to 445 days.

Circatidal and circasyzygic rhythms

Any organism living in the intertidal zone of the seashore is alternately submerged by water and exposed to air. Given the number of environmental factors affected by these cyclical inundations and exposures, it is not surprising that organisms show some associated behavioural rhythms. The **circatidal** vertical migration cycles of sand-dwelling platyhelminths, polychaetes and diatoms, and cycles of contraction in anemones, filtration in bivalves and swimming activity in fish are good examples. Some seashore organisms, such as the periwinkle *Littorina rudis* and some crabs, show **semilunar** or **circasyzygic** activity rhythms synchronised to the fortnightly cycles of spring and neap tides.

Circalunar rhythms

In some animals, behaviour is entrained to the 29-day lunar month, thus showing **circalunar** or **circasynodic** rhythmicity. The Mediterranean polychaete *Platynereis dumerlii* transforms into its sexual phase and swarms at the water surface in synchrony with the full moon. A similar endogenous rhythm underlies the famous reproductive swarming of palolo worms (*Eunice viridis*) in the Pacific. Among terrestrial organisms, antlion (*Myrmeleon obscurus*) show patterns of pit-building activity that peak at around the full moon.

Ultradian rhythms

Some behavioural cycles are of very short duration and are called **epicycles** or **ultradian rhythms**. For example, lugworms (*Arenicola marina*) in their intertidal burrows feed with a periodicity of roughly six to eight minutes, while predominantly diurnal meadow voles (*Microtus pennsylvanicus*) intersperse activity with bouts of resting that last from a few minutes to two hours. The significance of these short-term rhythms is not always clear, and their sometimes high variability can make them difficult to identify rigorously.

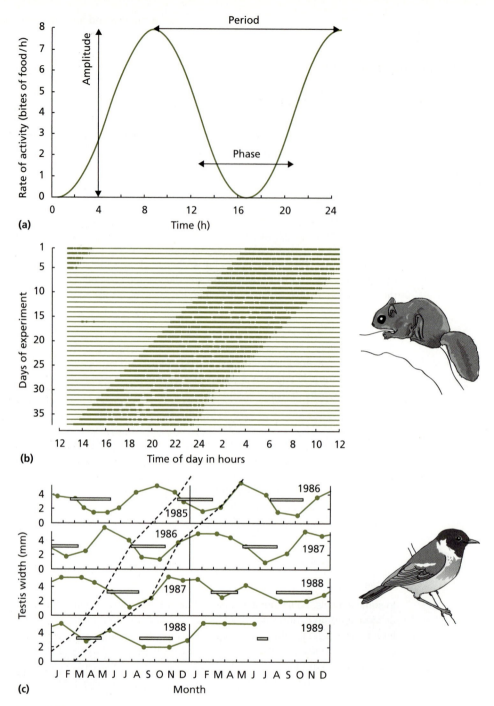

Figure 3.34 Many species show rhythmic patterns of behaviour entrained to natural cycles in the environment (see Box 3.5). (a) The basic characteristics of rhythmic behaviour (see text). After Drickamer, L.C., Vessey, S.H. and Meikle, D. *Animal Behaviour: Mechanisms, Ecology, Evolution*, 4th Edn, copyright 1966, Chicago, W.C. Brown, reproduced with permission of The McGraw-Hill Companies. (b) A circadian rhythm of wheel-running in a flying squirrel (*Glaucomys volans*) under conditions of constant darkness. After DeCoursey (1960). (c) Male stonechats (*Saxicola torquata*) maintain a circannual rhythm of testicular growth (solid lines), feather moults (bars) and reproductive behaviour under constant (non-seasonal) conditions in the laboratory. The dashed lines indicate the characteristic drift in the rhythm. From Alcock (1998) after Gwinner & Dittami (1990). Reprinted with permission after 'Endogenous reproductive rhythms in a tropical bird', *Science*, Vol. 249, pp. 906–8 (Gwinner, E. and Dittami, J. 1990). Copyright 1990 American Association for the Advancement of Science.

follow **Aschoff's rule** (e.g. Aschoff 1979), which states that the rate and direction of drift away from 24 h depends on light intensity and whether the animal is nocturnal or diurnal in its pattern of activity. Thus in the nocturnal flying squirrel kept in constant darkness, for example, the free-running period of activity starts slightly earlier each 'night' (Fig. 3.34b), while in male crickets (*Teleogryllus commodus*) kept in constant light, normally nocturnal singing behaviour starts slightly later (Loher 1972). However, there are several exceptions to the rule.

3.4.1.1 Interactions between clocks

While there is widespread evidence for the clocks of different periodicity in Box 3.5, it is also apparent that clocks of one kind may underlie those of others. Circadian clocks, in particular, appear to drive some longer-term behavioural rhythms. The northward spring breeding migration of the white-crowned sparrow (*Zonotrichia leucophrys*) is a good example (Farner & Lewis 1971).

White-crowned sparrows show cyclical changes in the light sensitivity of particular cells in the brain (mainly in the pineal gland and hypothalamus) over the 24-h period. The cycle is set at first light each day after which the cells remain insensitive to light for around 16–20 hours. Sensitivity then increases before declining again after 24 hours. If daylength is short (11–12 h), no light reaches the brain during its period of photo-sensitivity. As days get longer (14–15 h), however, the hours of light extend into the sensitive period and a number of light-activated systems are brought into play, stimulating the pituitary gland to secrete prolactin and gonadotrophic hormone. In response to these, the gonads increase rapidly in size (*recrudesce*) to 300–400 times their winter resting size, and birds start to feed more, increase their fat deposition and become more active, particularly at night ('migratory restlessness' or **Zugunruhe**). After the spring migration and breeding, the gonads atrophy again and birds enter a period of photo-refractoriness. It is not until the following spring that they once again become sensitive to changing daylength and the process repeats itself. Thus, the combination of changing daylength and a circadian pattern of light sensitivity ultimately produces an annual cycle of migration and reproduction.

Circadian rhythms play a similar role in some circatidal, circasyzygic and circalunar activity cycles (e.g. Naylor 1958; Youthed & Moran 1969). Indeed, different levels of rhythmicity may be subsumed within the same control mechanism. In the sea hare *Aplysia*, for example, the R15 (so-called 'parabolic burster') cell in the parieto-visceral ganglion emits a rhythmic 12–13 impulses every 30 s. Superimposed on this background rhythm are circadian fluctuations with a marked increase in spike frequency an hour before dawn. This pattern persists in a modified form even when the ganglion is isolated (see 3.4.1.2). In addition, there is seasonal variation in the R15 spiking pattern, so the same mechanism appears to provide for both short- and long-term rhythmicity.

3.4.1.2 The physiological basis of endogenous clocks

Endogenous clocks are a widespread feature of organisms of all levels, even single cells, which shows that nervous and endocrine systems are not essential prerequisites for rhythmic control. In fact, it is not even essential to have an intact cell. In *Acetabularia*, a large, unicellular alga, the circadian clock continues to function even if the nucleus is removed or destroyed (Woolum 1991). Since individual cells can possess a clock, it is not surprising that tissues and organs also show clock activity. Isolated adrenal and pineal glands, for example, can continue their rhythmic secretion of hormones (corticosteroids and melatonin respectively) for several days (e.g. Takahashi *et al.* 1980). This being the case, it stands to reason that entire organisms must actually have several clocks, not just one, an assumption that can easily be corroborated by manipulating different clocks into running independently at different rates. This happens to some extent when animals are kept in constant conditions. For example, humans living in underground shelters without any time cues begin to show disparities in the periodicity of sleeping–waking and body temperature cycles (Aschoff 1965). Such asynchronous phase shifts as different rhythms try to readjust underlie some of the unpleasant effects of 'jet lag' after long-distance travel. In some cases, however, the rhythmicity of different functions is governed by the same clock, as, for example, in the control of ultradian locomotory and cell division cycles in *Paramecium* (Kippert 1996).

Clearly, these multiple clocks require coordinating if the animal is to function coherently in relation to cyclical environmental events. How does this happen? The answer seems to be through one or more 'master' clocks that regulate others via the nervous and endocrine systems.

Clock regulation in invertebrates

In cockroaches (*Leucophaena maderae* and *Periplaneta americana*), the circadian clock appears to be regulated by large ocellar fibres extending from cells in the optic lobes of the brain to neurosecretory cells in the thoracic nerves. In fact, cockroaches have two regulatory clocks, since each of the two optic lobes keeps time independently (Page 1985). In some other insects, such as certain species of moth, brain transplant experiments have shown that communication between the cerebral clock and the mechanism controlling rhythmic activity is hormonal rather than neural. Both hormones and neural connections are important in crustacean clocks, with neurons in the eyestalk regulating the basic activity rhythm and hormones from the X-organ (a group of neurosecretory cells in the optic ganglia) determining its persistence. In some molluscs, seasonal rhythmicity is associated with changes in the ionic status of the haemolymph and nerve cells. After hibernation, the resting potential of neurons (3.1.1.2) is lowered from 60 mV to around 40–50 mV, probably through increased permeability of the membrane to sodium and calcium. This brings the membrane potential nearer to the impulse threshold and may even elicit autorhythmic spiking. This galvanising effect on neural activity is reversed in the summer when activity levels drop and animals prepare to aestivate. Thus the seasonal cycle of activity and torpor in these species appears to be regulated by the chemical environment of the nervous system.

Clock regulation in vertebrates

The pacemaker of the clock and its mechanism of entrainment appear to be one and the same thing in invertebrates. In vertebrates, the picture is more complicated. The circadian system in birds is based on three interacting clocks: the eyes, pineal gland and suprachiasmatic nucleus (SCN) of the hypothalamus, the relative importance of each differing among species. Between them, these clocks control others in the body through the rhythmic secretion of the hormone melatonin by the pineal gland. Transplant experiments, in which the pineals of sparrows entrained to various 24-h light–dark cycles are surgically implanted into arrhythmic pinealectomised birds, have shown that the pineal is the circadian pacemaker in these species, transplant recipients showing the circadian rhythmicity of the donors. However, in some other bird species, such as the Japanese quail (*Coturnix japonica*), the pineal is less important in this respect than the eyes and SCN (Underwood *et al.* 1990; Hastings *et al.* 1991). Blinding, but not pinealectomy, abolishes rhythmicity in quail, as it does in pigeons (*Columba livia*), but has no such effect in sparrows or chickens. While the eyes undoubtedly provide light information to the clock mechanism, however, entrainment to light cycles in some species appears to be directly via photoreceptors in the brain, a route demonstrated by manipulating the opacity of the scalp and roof of the skull (Menaker 1968).

Interaction between the three clocks is mediated by neural connections and melatonin secretion. The SCN communicates with the pineal via nerve connections and appears to receive feedback hormonally via melatonin receptors. The eyes also secrete melatonin, but experimental evidence suggests their influence on free-running activity rhythms is driven by neural connections rather than hormone secretion (Underwood *et al.* 1990). However, melatonin secretion, especially by the pineal, appears to be the route by which the three-way clock exerts its influence over of the rest of the circadian system.

Clock mechanisms among mammals differ in two important ways from those of birds. First, entrainment to light cycles in mammals does not appear to operate via extraretinal photoreceptors as it does in birds, and, second, the master clock influences the rest of the system through nerve connections rather than hormones. Circadian clock mechanisms have been particularly well studied in rodents.

The SCN appears to be the master clock in rodents and regulates several other clocks via neural connections. Surgical lesions isolating the SCN result in a loss of rhythmicity in all regions of the brain other than the SCN itself. As with the pineal in birds, transplanting the SCN (using foetal or newborn tissue to allow new nerve connections to establish) restores rhythmicity in arrhythmic animals whose own SCN has been destroyed (Lehman *et al.* 1987). Photoreception for entrainment occurs via the eyes, with information reaching the SCN through the retinohypothalamic tract (a nerve tract connecting the retina with the hypothalamus). A complicating factor is that there are two SCNs, one in each half of the brain, each capable of acting as an independent oscillator. Normally, these act in concert, producing a coordinated phase of activity during each cycle. However, when nocturnal animals, such as hamsters, are kept in constant light, their activity pattern sometimes splits into two phases with different free-running periodicities. Ablation experiments suggest the split is due to the usually coordinated activity of the two oscillators becoming uncoupled (Pickard & Turek 1982).

3.5   Mechanism and constraint

So far we have discussed the relationship between function and mechanism in terms of adaptive features of mechanism and the extent to which identifiable mechanisms underpin adaptive behaviour. We have also stressed here, and in Chapter 2, that function and mechanism both potentiate and constrain each other: function provides the evolutionary incentive for mechanism, but conflicting requirements may limit developments in any particular direction, while mechanism provides the wherewithall for behaviour, but also places limits on its possibilities. To complete our discussion of mechanism, we shall look at some examples of the constraints function and mechanism can impose on each other.

3.5.1 Mechanism as a constraint on function

3.5.1.1 Perceptual constraints

How the animal perceives the world inevitably shapes how it can respond to objects and events around it. We have already seen how perception can constrain response in our examples of perceptual rules of thumb (3.2.1.1, 3.2.1.4). Crude approximations, such as the 'prey' and 'enemy' distinction in toads, or the more sophisticated inventiveness of human visual perception, can provide economical mechanisms for decision-making, but also expose the limitations of the system.

Perceptual rules of thumb predispose the animal to notice and respond to particular cue configurations in the environment. Recent studies suggest that such perceptual or **sensory biases** may constrain later evolutionary developments and play a key role in the evolution of animal signalling systems, including those involved in mate choice (e.g. Basolo 1990; Ryan *et al.* 1990; Chapters 10 and 11). The exploitation of sensory biases (dubbed **sensory exploitation** by Ryan 1990) in mating leads to individuals, usually females (Chapter 10), preferring mates with certain characteristics simply because they have inherited a sensory/perceptual system that is responsive to some aspect of them, not because the characteristics themselves confer any reproductive advantage. Thus, for example, female platyfish (*Xiphophorus maculatus*) prefer conspecific males with artificial extensions to their tail fins (Basolo 1990), perhaps because they suggest the males are larger than they really are. The extensions resemble those sported naturally by males of the closely related swordtail (*X. helleri*; Fig. 3.35) to attract females of their own species, but do not occur naturally in platyfish. However, a molecular phylogeny (2.2) suggests that platyfish and swordtails inherited their preference from a common ancestor, so the preference for a tail extension among female platyfish appears to be a retained, but now functionless, ancestral trait (Basolo 1995). A potential hazard of this ancestral baggage is that females might end up mating with males of the wrong species. In the *Xiphophorus* case this is a small risk because platyfish and swordtails seldom co-occur, but there are several other groups of closely related species, for example in birds, where new zones of overlap may well lead to hybridisation through sensory bias.

Pre-existing sensory biases may reflect other features that act as attention-grabbers. In some water mites and jumping spiders, for example, males entice females by exploiting

Figure 3.35 Male swordtails (*Xiphophorus helleri*) possess an exaggerated extension to their caudal fin that appears to have been sexually selected through female choice. However, females of the closely related platyfish (*X. maculatus*) also prefer males with tail extensions, even though males of their own species do not possess them. See text. After Manning & Dawkins (1998).

their responsiveness to cues suggesting food, rather like the courtship displays of pheasants and peafowl in Figure 1.13. In the wax moth *Achroia grisella*, male signals seem to be derived from the ultrasonic emissions of predatory bats, to which the female's ears are particularly attuned (3.2.1.3) (Greenfield & Weber 2000). In these species, therefore, courtship signals have been channelled by pre-existing response characteristics of females honed in other contexts. Intuitively, one might expect such biases to be widespread, and Phelps & Ryan (2000) have recently shown how important an influence they might be using a neural network approach (3.1.3.2), in this case modelling the evolution of phonotactic cues in female túngara frogs (*Physalaemus pustulosus*) (10.1.3.2).

Neural network models have also been used to explore another kind of perceptual bias in animals, commonly revealed as enhanced responsiveness to **supernormal stimuli**. In many cases where they have been tested, animals have shown greater responsiveness to stimuli that exaggerate those of the normal target stimuli. Thus, oystercatchers (*Haematopus ostralegus*) will show a stronger incubation response to an artificial egg several times the normal size. Similarly, male sticklebacks will preferentially court model females with bellies distended well beyond those of normal gravid females. Supernormal stimuli are a reflection of the approximate rules of thumb used to make decisions in a world of normal variation. Such rules are, as we have noted, easily fooled by artificial stimuli. Nevertheless, neural network models of visual preferences suggest that heightened responsiveness to supernormal stimuli may be a basic property of perceptual systems (Arak & Enquist 1993).

3.5.1.2 Motor constraints

Perceptual mechanisms impose some constraints and biases on behaviour. Effector systems impose others. Motor constraints form the basis for some forms of 'honest' signalling (Chapter 11), where, for instance, vocal status signals depend on the pitch of the sound, and the latter depends on the size and resonating qualities of the vocal apparatus. Here, only large, competitive individuals can emit deep, penetrating calls (Davies & Halliday 1978). In some cases, however, the vocal apparatus imposes more basic constraints on vocal behaviour. A study of song development in swamp sparrows (*Melospiza georgiana*) by Jeffrey Podos provides a nice example.

While it is well known that the development of elaborate songs in birds can be limited by constraints on learning and memory, Podos (1996) tested the idea that the physical performance of song may also impose a mechanical constraint. Swamp sparrows have a complex song structure in which syllables (groups of notes) are repeated in a series of trills (Fig. 3.36a). Such songs are likely to be demanding to produce because of the rapid breathing and muscular modulations required, but may nevertheless be under strong sexual selection (2.4.1, Chapter 10) to become more elaborate. If vocal mechanism is a factor constraining further elaboration, then pushing the system beyond its normal performance limits should result in characteristic production errors, such as dropped syllables or breaks in song performance. Podos challenged young birds by exposing them during their song learning phase to songs with trill rates artificially accelerated by between 26% and 92% (Fig. 3.36a). When he examined their later songs, he found precisely the kind of production errors expected if performance was subject to motor constraint (Fig. 3.36b). Errors included trill rate reductions (but never augmentations), note omissions, pauses in trill structure and deceleration within trills, all features consistent with motor constraint rather than limitations of learning or memory. Constraints in the context of learning are discussed further in Chapter 6.

Figure 3.36 (a) Sonograms of three advertisement songs of swamp sparrows (*Melospiza georgiana*) (i, iii, v), and the manipulated versions of each (ii, iv, vi) that were used to train young birds. Note the structuring of the songs into syllables that are repeated as trills, and the compression of the trills in the manipulated versions. See text. (b) Imitations of manipulated training songs (ii, iv, vi in Fig. 3.36a) by five different individuals. Song 1 has a reduced trill rate compared with its model (ii in Fig. 3.36a), song 2 omits some of the notes in its model (iv in Fig. 3.36a), while songs 3, 4 and 5 have pauses interspersed among their syllables – all changes consistent with a motor constraint on song learning. See text. After Podos (1996).

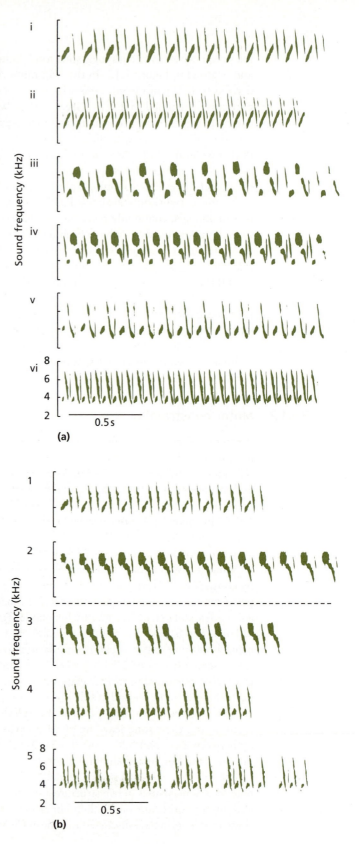

3.5.2 Function as a constraint on mechanism

An animal's resources are in demand for several competing activities. No activity, therefore, can command total investment. Instead, it is assumed that tradeoffs occur between conflicting priorities so that the animal maximises its lifetime reproductive success within the constraint of its limited resources (see 2.4.5). This applies as much to the anatomical and physiological machinery of behaviour as to any other characteristic (Ricklefs & Wikelski 2002). In addition, a particular mechanism, such as a neural network or hormone, may be involved in the control of a number of different activities whose demands also conflict. There are thus different ways in which functional demands may constrain mechanism. We shall illustrate the point by considering potential conflicts between functions regulated by testosterone.

3.5.2.1 Testosterone, immunity and behaviour

As we have seen, testosterone and other sex steroids are important mediators of a wide range of behaviours. Apart from their effects on behaviour and reproductive physiology, however, sex steroids also affect the immune system in ways that may conflict with, and thus cause downregulation of, their other activities. Hormone-mediated interactions between the immune system and the CNS have been known for a long time, and provide mechanisms by which psychological state and behaviour can influence immunocompetence and vice versa – the field of **psychoneuroimmunology** (Maier & Watkins 1999; Ader *et al.* 2001). More recently, however, interest has centred on the significance of this bidirectional relationship from a functional and evolutionary perspective (Folstad & Karter 1992; Sheldon & Verhulst 1996; Barnard & Behnke 2001). Evolutionary biologists interested in the role of parasites and disease in sexual selection have focused on the immunomodulatory effects of steroid hormones, particularly testosterone, involved in the development of some secondary sexual characters and behaviours (10.1.3.2). While this has spawned a research industry of its own, it has also fuelled a more general interest in hormone-mediated tradeoffs between immunity and behaviour (Sheldon & Verhulst 1996; Beckage 1997; Barnard & Behnke 2001). Barnard and Behnke (2001) have studied such tradeoffs in relation to social status in laboratory mice.

Barnard and Behnke argue that dominant and subordinate male mice represent different life history strategies centring on competitive ability. Competitive dominant males can command access to limited resources such as food and nesting sites and invest heavily in the reproductive opportunities this brings. Less competitive subordinate males cannot compete for resources so have to make do with sneaking opportunistic matings as and when they can (see 'best of a bad job' strategies in 2.4.5.1). A prediction that follows from this is that dominant males will be more likely to trade off future survival for short-term reproductive opportunity than subordinates, who would do better to safeguard survival and maximise the likelihood of chance matings. One way of regulating the tradeoff between reproduction and future survival might be to link the secretion of immunodepressive sex steroids to current immunocompetence – i.e. secrete testosterone only if your immune system is robust. Barnard *et al.* (1996a) found that subordinate males did exactly this by regulating the secretion of testosterone in relation to their current circulating antibody levels (one measure of immunocompetence) (Fig. 3.37a). As a result, their resistance to a subsequent infection was unaffected by testosterone (Fig. 3.37b). Dominants, on the other hand, did *not* regulate their secretion of testosterone in relating to antibody levels, with the result that testosterone reduced their subsequent resistance to infection (Fig. 3.37c,d).

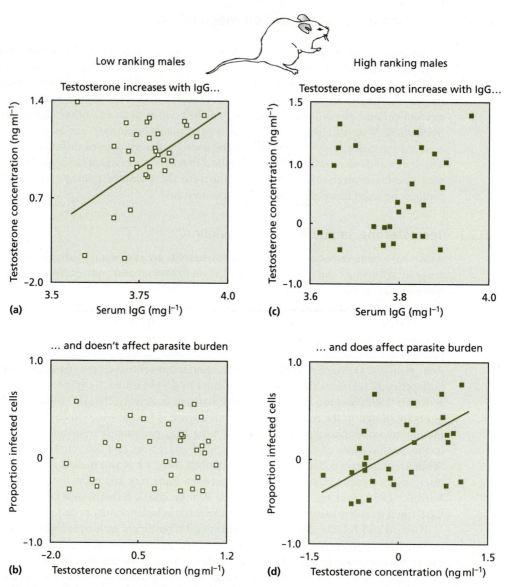

Figure 3.37 Secretion of testosterone in male laboratory mice depends on both social status and current immuno-competence. Subordinate males tend to secrete testosterone only when their current antibody titre is high (a), thus avoiding a testosterone-dependent reduction in resistance (% red blood cells infected) to an experimental infection of the blood protozoan parasite *Babesia microti* (b). Dominant males, in contrast, secrete testosterone regardless of current immunocompetence (c) and do suffer a testosterone-dependent reduction in resistance (d). Vertical axis scales are deviations from sample means. Modified from Barnard *et al.* (1996a).

Further support for the adaptive immunity tradeoff idea came from experimentally immunodepressed males. In these, both testosterone levels and behaviours such as aggression and locomotory activity, which tend to depress immunity, were downregulated, while sleep, which tends to enhance immunity, went up, regardless of social status (Barnard *et al.* 1997a). The really telling point, however, was that the regulation of both testosterone and behaviours affecting immunity in depressed males was abolished when female odours (suggesting reproductive opportunity) were present in the environment (Barnard *et al.* 1997b).

Taken together, therefore, these results suggest that males bring costly physiological mechanisms into action only if it is reproductively worthwhile to do so. Thus dominants are more prepared to risk their future survival by secreting testosterone than subordinates because they are in a better position to secure matings in the short term. However, both classes of male are prepared to risk it if current reproductive opportunity seems high enough (there is lots of female odour about). In this case, therefore, functional considerations appear to limit the deployment of a physiological mechanism.

Summary

1. In most animals, behaviour patterns reflect the complexity and functional organisation of the nervous system. As we ascend the evolutionary scale, nervous systems show two major trends: towards greater differentiation and greater centralisation. Comparative studies show that many adaptive differences in behaviour are reflected in the gross anatomy of the central nervous system, some aspects of which are influenced by the behavioural history of the individual.

2. Some adaptive behaviour patterns can be traced to specific neural pathways, with eventual motor responses sometimes under the control of a hierarchy of 'command centres'. In many cases, however, the control of behaviour is not a local, self-contained process, but the result of a distributed neural network functioning in different ways at different times. Computer and robotic simulations are providing revealing new ways of modelling the neural control of behaviour.

3. Neural control mechanisms coordinate the animal's response to events in its environment. However, the nervous system is not a passive observer of such events. Stimuli from the environment are filtered through the animal's sense organs and processed by its nervous system so that the animal obtains a biased view of the world shaped by natural selection. We can thus think of animals having perceptual rules of thumb which reflect the reproductive costs and benefits of perceiving the world in different ways.

4. Hormones transmit information around the body more slowly than nerve cells. Together, the rapid responses of the nervous system and more sustained influences of hormones complement one another in controlling the animal's actions. Hormones influence behaviour through a variety of routes, including the nervous system, sensory perception, effector systems and development. However, their effects are themselves influenced by a wide range of factors, from individual genotype and experience to seasonal, ecological and life history factors.

5. The interaction between neural and hormonal influences is shown particularly clearly in the entrainment and control of rhythmic behaviour patterns associated with environmental cycles. Together they provide animals with an adaptive internal clock which is regulated to relevant external cyclical events.

6. Function and mechanism can mutually constrain one another. Examples of constraints on function can be seen in perceptual biases and mechanical constraints, while hormone-mediated immunity tradeoffs show how functional considerations can limit the deployment of hormonal control mechanisms.

Further reading

Ewert (1980), Guthrie (1980), Huber (1990) and Young (1996) provide good introductions to the neurobiology of behaviour, while Randall *et al.* (1997) is an accessible source of general background physiology. Carter (1998) and Greenfield (2000) are readable general discussions of brain structure and function in humans, and Barton & Harvey (2000) and Gil & Gahr (2002) focus on developmental interrelationships between brain centres in various species from an evolutionary perspective. Gregory (1998) is unrivalled as an introduction to visual perception, and Bolhuis (2000) is a good collection of more specialised reviews of neurobiology and perception. Webb (2000) and Holland & McFarland (2001) provide good introductions to robotics in the study of behaviour. The extensive work of Crews and Wingfield and their respective coworkers is an excellent source for relationships between hormones and behaviour, while Ricklefs & Wikelski (2002) review the mediating role of physiological mechanisms in life history tradeoffs. Finally, Whyatt (2003) provides an up-to-date review of pheromones and behaviour.

4 Motivation and cognition

Introduction

We have seen that it is possible to account for behaviour in terms of underlying physiological mechanism. But such explanations are not without problems. An alternative approach is to regard mechanisms as a 'black box' and focus on the rules that translate environmental stimuli into behavioural action. How can the same stimuli elicit different responses at different times, or by different individuals, for example? The term often used to refer to this modulation of stimulus-response relationships is 'motivation'. But what *is* motivation, and how can we measure it? The term might suggest purposefulness, and imply that animals initiate behaviour with some sort of goal in mind. Is there any evidence that they do? Indeed, is there any evidence that behaviour is driven by cognitive attributes such as awareness and intentionality at all? Answers to these questions would be interesting in themselves, but they also have important implications for our view of welfare in other species.

While we can account for the proximate causation of behaviour in terms of its 'nuts and bolts' mechanisms, it is clear from the last chapter that the usefulness of such explanations is limited. The most obvious problem is that they are unwieldy. Except for relatively simple actions, such as reflexes, the sheer quantity of information required to trace a behaviour to its underlying physiological events is likely to prove defeating, certainly in terms of any simple summary. Indeed, whole tiers of complexity are already avoided in many 'nuts and bolts' explanations. Accounts of responses in terms of pathways within the CNS, for example, gloss over the multitude of cellular events and neuromuscular pathways that actually produce the behaviour. More than this, however, 'nuts and bolts' explanations say little about why mechanisms are designed the way they are. Why *do* they respond to particular kinds of stimuli to produce particular kinds of behaviour? What mediates the translation of stimulus into response? How do physiological systems allow the establishment of different behavioural priorities? Does the animal express behaviour like an automaton, as an unthinking product of its physical structure and chemistry, or does it consciously reflect on possibilities and outcomes before acting?

4.1 Causation and motivation

These kinds of question lead to a rather different concept of proximate causation, no longer dealing in local circuits and explicit physiological systems, but taking a more holistic view. To be sure, the 'nuts and bolts' components of the organism are recognised as the hardware ultimately responsible for behaviour, but they are simply taken as read and not the focus of analysis in themselves. To borrow an engineering term, the organism's physiological mechanisms are instead treated as a 'black box', a system of as yet undetermined design that translates input into output according to certain rules (just as the 'black box' electronics of our computer turns our key-tapping into desired output according to the rules of the program we are using).

In a behavioural context, the input is internal and external stimuli and the output the behavioural response. The rules converting one into the other are assumed to be shaped by natural selection, so that the organism performs adaptively in the face of prevailing stimuli. In some cases, such as the limb withdrawal reflex in 3.1.3.2, the adaptive behavioural response is a simple stereotyped action: a stabbing pain triggers rapid withdrawal of the limb from the noxious stimulus. The rules of the 'black box' thus appear to be simple, and the translation of stimulus into response can be traced to a particular neural circuit (Fig. 3.15). In most other cases, the translation is not nearly so straightforward.

Take a predator such as a lioness (*Panthera leo*), for example. As she roams around her pride's home range she regularly encounters antelope, which are high on her list of favourite prey. Frequently she attempts to stalk and capture the antelope, as we might expect. But just as often she completely ignores them and walks slowly by. Assuming nothing about the antelope has changed significantly, why should the lioness behave in this contrary way? The only explanation is that something about the lioness has changed, causing her to respond differently on different occasions. But what has changed? Nothing about her outward appearance apparently, so it must be something internal. An obvious possibility is that the lioness has recently fed elsewhere when she appears uninterested in the antelope. If so, then we might say that it was hunger that determined whether or not she attacked. We have no idea what this means in terms of the activity of specific nerve cells or the subjective feelings, if any, of the lioness, it is just something that alters her response to spotting an antelope. The term widely used to describe this kind of modulation of stimulus–response relationships is **motivation**. Animals are motivated in different ways on different occasions, so the same stimulus can elicit a variety of responses. Even reflexes can be influenced by motivation (Toates 1986). Lordosis in female rats (Fig. 3.29), for example, has all the hallmarks of a reflex response, but its expression depends on the female's hormonal state and social environment. The label 'motivation', however, begs two important questions. First, does it do anything other than hang a name on the phenomenon we wish to explain? And, if it does, how exactly does it influence the animal's response?

4.1.1 Models of motivation

4.1.1.1 Intervening variables, drive and energy models of motivation

'Motivation' is shorthand for the internal decision-making process by which the animal chooses to perform a particular behaviour. At any given time, the animal is likely to be faced with several possibilities. It could go for the food over there in the middle distance, or it could court the female that has just come into view, or perhaps that rustle in the bushes is an approaching predator and it should dive down its burrow instead. The three choices are mutually incompatible, so the animal must prioritise. The animal's motivation at the moment it decides is inferred from the option it selects.

Figure 4.1 (a) Motivation as an intervening variable translating various antecedent conditions, such as water or food deprivation and detecting a territorial intruder, into appropriate behavioural output. (b) Being deprived of water, fed dry food or injected with saline all lead to behavioural responses associated with water intake. But rather than express these as separate cause and effect relationships, it is simpler to say that the antecedent conditions induce a state of 'thirst' in the animal (c) and the animal responds by working to obtain water. After Miller (1956).

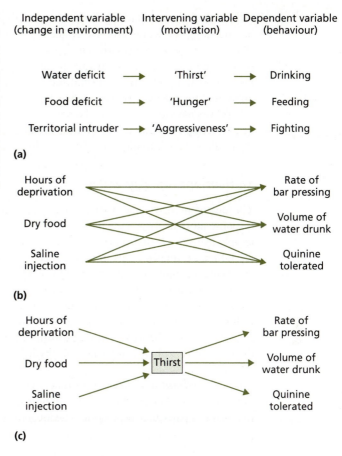

Motivation as an intervening variable

Ethologists and early experimental psychologists referred to motivation as an **intervening variable** because it seems to mediate the response of the animal to internal and external stimuli. More particularly, it predisposes the animal to behave in ways that fulfil important needs, thereby giving the animal the appearance of purposefulness. Thus in Fig. 4.1(a), thirst, hunger and aggressiveness, induced by the prevailing conditions on the left of the figure, cause the animal to drink, feed and fight, activities that are designed to meet its needs in each case. However, as we have already seen in Chapter 1, purposiveness in behaviour is a tricky issue and we shall come back to it in the context of motivation shortly.

Motivational terminology also provides a more economical way of describing the cause and effect relationship between stimulus and response. As Fig. 4.1(b) shows, it is far easier to attribute the various drinking responses on the right to the fact that the conditions on the left make the animal thirsty, than to deal in the cat's cradle of relationships between each condition and response separately. There are dangers in this economy, however. For one thing, calling the effects of all the antecedent conditions in Fig. 4.1(b) 'thirst' implies they all change the animal's internal state in the same way. Classic experiments by Miller (1956), Burke *et al.* (1972) and others, however, suggest this may not be so. For example, injecting saline into rats results in apparently different 'thirst' motivated responses depending on the measure used (Miller 1956). The volume of water drunk shows relatively little change with time since injection, while the amount of bar-pressing for water and the concentration of bitter-tasting quinine tolerated rise sharply. If these responses all reflect a single state of thirst, why don't they all show the same temporal pattern? A further problem is that labels such as 'thirst' and 'hunger' may conceal a plethora of specialised and very different requirements. For example, an animal may incur a specific deficit for protein, or salt, or a vitamin, each requiring a different behavioural solution. To call them all 'hunger' would confound any possibility of predicting the animal's behaviour accurately. Our economy of terms may thus be a misleading over-simplification. There is also a second danger.

Motivational energy and drive

The idea of an intervening variable, especially one that makes an appropriate behavioural response more likely as time passes or the relevant (**causal**) stimulus increases in strength, tempts us to think that motivation has some physical reality in the animal. Ethologists, via Lorenz, formalised this in their concept of **motivational energy**, a hypothetical force paralleling the motivational **drives** of the comparative psychologists. The idea of energising drives was a distant descendant of Cartesian dualism, the belief that mind and body, while jointly controlling the actions of the individual, are themselves separate entities. In his theory of instinct (see 1.3.1, 1.3.2), Lorenz, echoing Descartes, saw the energy for a given action as being something distinct from the machinery that executed it. This is demonstrated explicitly in his famous 'hydraulic model' of motivation (Fig. 4.2a).

Lorenz's hydraulic model In Lorenz's view, motivation was akin to the build-up of fluid in a tank. The fluid represents **action-specific energy**, chemical energy in the CNS that drives a particular behaviour. The longer the animal has gone without performing the

Figure 4.2 (a) Lorenz's hydraulic model sees motivation as the result of action-specific energy accumulating in the CNS like fluid in a reservoir. Behavioural output is determined jointly by the pressure of fluid on a release valve and the pull of an external incentive (weight in the pan) against the valve. See text. After Lorenz (1950).

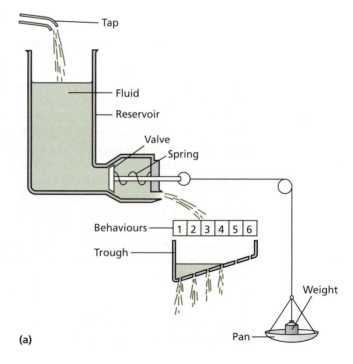

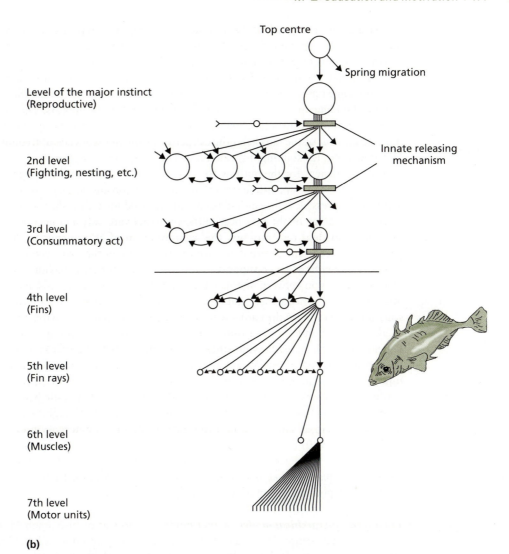

Top centre

Spring migration

Level of the major instinct
(Reproductive)

Innate releasing
mechanism

2nd level
(Fighting, nesting, etc.)

3rd level
(Consummatory act)

4th level
(Fins)

5th level
(Fin rays)

6th level
(Muscles)

7th level
(Motor units)

(b)

Figure 4.2 (*continued*)   (b) Tinbergen's hierarchical model of motivation, in which motivational energy cascades down a hierarchy of behavioural control centres. Here, the model relates to reproductive behaviour in male three-spined sticklebacks. See text. From *The Study of Instinct* by N. Tinbergen (1951). Reprinted by permission of Oxford University Press.

behaviour, the more fluid (action-specific energy) accumulates in the tank (the animal's CNS). The animal's drive to perform the behaviour consequently increases. The behaviour can be activated when the fluid is released from the tank. This can happen in two ways. Either enough fluid builds up for the pressure to open a valve at the base of the tank, in which case the behaviour is performed spontaneously, or external weights, analogous to appropriate stimuli in the environment, pull against the valve and release the fluid (Fig. 4.2a). In most cases, the internal and external forces of the fluid and weights combine to open the valve. How much the valve is opened, and therefore how far the fluid gushes out, depends on the combined strength of the forces. The outflow may be the same if a small weight is combined with a full tank or a large weight with a tank only a quarter full.

Depending on the force of the outflow, different outlets in the trough in Fig. 4.2a come into play. These represent the component behaviours activated by the drive, the activation threshold increasing with distance from the tank. Thus a weak dribble from the tank might activate outlet 1, but outlet 5 would require a strong torrent. Imagine a cat that has not eaten for some time. In Lorenz's model, action-specific energy for feeding (fluid in the tank) has been building up in the cat's CNS when suddenly a mouse appears. The mouse presents a strong stimulus to feed (a heavy weight on the pan), so the pent-up energy is released (fluid gushes out of the tank) and the cat performs the full sequence of feeding behaviours: (1) orientating, (2) stalking, (3) catching, (4) killing and (5) eating. It is as if the head of fluid was sufficient for the outflow to activate all the outlets in the trough, allowing the whole repertoire of **appetitive** (searching out) and **consummatory** (consuming) feeding acts to be expressed. If the cat had been without food for only a short time, however, the reservoir of energy might have been sufficient to activate only a momentary orientation.

In motivational terminology, the accumulation of action-specific energy and the presence of a relevant external cue are **causal factors** in the expression of a behaviour. Lorenz's model finds some support in the real world in that behaviour (a) often results from an interaction between internal and external causal factors, (b) is sometimes regulated only by its own performance, and (c) can occur in the absence of an appropriate external stimulus (no weight on the pan) – what Lorenz called a **vacuum activity**. Thus Vestergaard *et al.* (1999) found that the timing and intensity of dustbathing in domestic chicks depended on the time elapsing since dustbathing was last performed and how much dustbathing was performed on the previous occasion. Performance was unaffected by its functional consequences (e.g. it was performed similarly by featherless chicks on sand and feathered chicks on sand covered by a sheet of glass). Domestic hens kept in a wire cage also show characteristic dustbathing movements even though there is no suitable material present at all, a response reminiscent of vacuum activity and attributed by Vestergaard (1980) to the hens' strong motivation to perform the frustrated activity. In other contexts, however, its failure to take account of feedback from the consequences of behaviour (eating food affects the internal causal factors for feeding) makes Lorenz's model a very poor reflection of real behaviour.

Tinbergen's hierarchical model   Other models of motivation also assumed the existence of motivational energy. Tinbergen (1951) viewed the organisation of functionally related behaviours in terms of a hierarchy of control centres in the CNS. Centres were activated by motivational energy cascading down the hierarchy as appropriate stimuli (**releasers** or **key stimuli**, see 3.1.2.3) removed the inhibitory block of **innate releasing mechanisms** (**IRMs**) in each centre. Thus, rather than the specific drive of Lorenz's model, motivational energy was a general driving force, activating lots of different behaviours in carefully controlled sequences. Figure 4.2(b) illustrates the idea using reproductive behaviour in male three-spined sticklebacks as an example.

At the top of the hierarchy is the migration centre, which is stimulated by hormones and changing environmental conditions and causes the fish to migrate into shallower water. Migration continues until the fish encounters habitat suitable for a territory. This excites the IRM of the reproductive centre and motivational energy passes through the centre. Energy can pass down to the next centres in the hierarchy only when appropriate stimuli cause their IRMs to fire. Thus, for example, it will flow to the fighting centre when a rival male is perceived. Activation of the various behavioural control centres involves tiers of lower level centres, including fins, muscles and individual motor units, so that the hierarchy connects the behaviour of the whole animal with its component mechanisms (Fig. 4.2b).

Difficulties and debates

Box 4.1 Some problems with energy/drive models of motivation

While superficially attractive, explanations of motivation in terms of motivational energy or drive run into a number of problems:

☐ There is no physiological evidence for energy associated with particular motivational states in the CNS.

☐ 'Drive' is often just a label with little explanatory power. Thus, to say an animal feeds because of a hunger drive is circular and does not get us any further in understanding why feeding happens.

☐ 'Drives' can be subdivided almost endlessly. A 'reproductive drive' in a female bird, for instance, is arguably made up of 'nest material-finding drive', 'nest-building drive', 'egg-laying drive', 'incubation drive' and so on. Each of these can be further subdivided: e.g. 'egg-turning drive', 'repositioning drive', etc. within 'incubation drive'. We thus end up with as many drives as there are behaviours, and again lose explanatory value.

☐ Unitary drives such as 'hunger' and 'thirst' can incorporate a wide range of associations between antecedent conditions and the animal's behaviour, thus undermining confidence in the drive as a single state.

☐ Classical energy/drive models omit feedback from the consequences of behaviour.

☐ Responses to a given 'drive' (e.g. feeding in response to hunger) affect other aspects of internal state (e.g. water deficit and thirst), thus rendering one-to-one relationships between internal state and behaviour unlikely on a priori grounds.

Problems with energy/drive models of motivation

While energy models such as those of Lorenz and Tinbergen, and the idea of drive as an energiser of behaviour, have intuitive appeal, critical analysis quickly reveals difficulties. Some of these are summarised in Box 4.1. Apart from the absence of any physiological evidence for a build-up or circulation of motivational energy within the nervous system, and the fact that many apparently unitary drives, such as hunger and thirst, may be anything but unitary (see above), the main problem is that the energy/drive concept lacks general explanatory power. In the case of drives for specific behaviours, for example, it is easy to envisage a hunger or thirst drive as the underlying cause of apparent appetitive searching, but it is less easy to see exploratory behaviour or sitting gazing idly around in these terms. Regarding drive as a mechanism of general (rather than specific) arousal, does not get us off the hook either. Certainly several experiments have shown that specific deficits can lead to generalised increases in activity (e.g. Teghtsoonian & Campbell 1960; Dethier 1982), but whether or not they do can depend greatly on context (e.g. Fig. 4.3).

For these and other reasons, motivational energy and drive have long been abandoned as a general framework for thinking about motivation, although recent reconsiderations of some behaviours, such as dustbathing in chickens (see above), sleep and some aspects of aggression, have begun to rekindle comparisons with Lorenz's model (Hogan 1997; Vestergaard et al. 1999). Despite their fall from grace, however, some of the terminology associated with energy models still lingers in the literature (see Dawkins 1986 and Box 4.3).

Figure 4.3 Activity in rats in three different environments (a running wheel [solid line], stabilimeter cage [dashed line ○]) and Dashiell maze [line ●]) after food and water deprivation. Only in the running wheel was there an increase in activity with degree of deprivation. After Mather (1981).

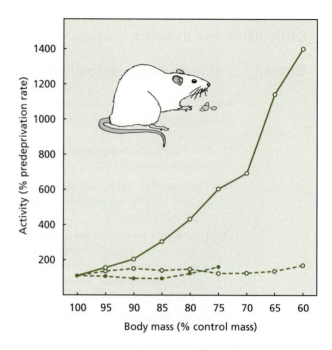

4.1.1.2 Homeostatic models of motivation

A different approach, though not necessarily a rival to models based on motivational energy and drive (Toates 1986), sees some form of homeostatic regulation as the basis for motivation (but see Colgan 1989). The physiological parameters of the body, such as blood sugar concentration, temperature and oxygen levels, tend to be maintained around a norm. Departure in any particular direction sets in train various corrective measures that restore the norm. Where corrective measures involve behaviour, the behaviour can be thought of as motivated by the need to restore normality. Models based on this kind of process introduce two important elements: (a) norms or **set points** as 'goals' for motivational systems, and (b) **feedback** from the internal environment telling the animal when the set point has been restored. The latter, as we have noted, is a significant omission from some of the classical energy-based models of motivation.

Figure 4.4 shows a simple homeostatic model of motivation derived from a broader model of behavioural control proposed by Deutsch (1960). A **deficit** or other imbalance in the animal's physiology is detected by, and excites, a neural centre in the CNS ('link' in the model), the degree of excitation depending on the magnitude of the imbalance. The excited 'link' activates an appropriate motor system which sets corrective behaviour in motion. As a result of the behaviour, the animal's internal environment changes – it has eaten or drunk, say – and the change is registered by an 'analyser' which inhibits the 'link' so that it no longer responds. This inhibition then gradually decays so that the 'link' once again becomes susceptible to excitation. We have already seen a more sophisticated form of this kind of model in action in the control of courtship in male newts (1.2.2, Fig. 1.11).

A simple experiment compares the ability of homeostatic and energy/drive models to explain behaviour triggered by accumulated internal change. Janowitz & Grossman (1949) placed oesophageal fistulas in a number of dogs so that, as they ate, food passed out through the fistula instead of entering the stomach (so-called 'sham-eating'). Food

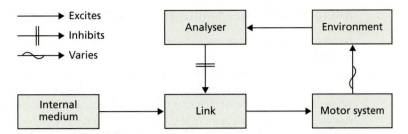

Figure 4.4 A simple homeostatic model of motivation. A feedback loop monitors changes in the animal's internal and external environment and regulates behaviour to maintain the animal in its preferred state. Modified after Deutsch (1960).

could also be placed directly into the stomach through the body wall, without the animal having to eat. If a fistulated dog is deprived of food, but has food placed directly in its stomach, Lorenz's type of energy model predicts that the dog will still eat if presented with food because the accumulated action-specific energy for eating has not been used up. A homeostatic model, however, predicts that the dog will not eat because the full stomach is feeding back the message that it has already eaten. In fact, the dogs in Janowitz & Grossman's experiment did *not* eat, suggesting that the homeostatic model was a better analogue of their behaviour. In general, homeostatic models not only predict maintenance behaviours such as feeding and drinking more accurately, but we can also find physiological support for their assumptions. Thus stretch receptors in the stomach are stimulated by distention due to food and send impulses to the CNS to inhibit feeding, hormones control drinking behaviour via the hypothalamus, and so on. This contrasts with the complete absence of evidence for motivational energy (see Box 4.1).

However, homeostasis cannot comfortably account for all aspects of behaviour. As Bindra (1978) notes, even eating, drinking and other maintenance activities, which fit the precepts of homeostasis quite well, often occur in the absence of a deficit and are thus difficult to reconcile with departure from a set point. Furthermore, displacement from an internal set point cannot, by itself, guide the animal to an appropriate corrective outcome. Additional information is needed to tell the animal how this might be achieved. The important ingredient missing here is, of course, stimuli from the external environment. In some cases, such as exploration, behaviour may appear to be driven entirely by external cues, and the concept of 'displacement' of internal state hardly seems to make sense at all. External cues provide the **incentive** for behaviour; that is, information that the required outcome is achievable. Bindra proposed two categories of incentive: *hedonistic* and *neutral*, depending on whether a cue stimulates directed activity (is *affectively potent* in the jargon). Affective potency may be positive (the animal is attracted to the stimulus) or negative (the animal avoids or retreats from it). While neutral incentives may not stimulate active directed responses, they may be important in organising activity in a more open-ended way, for instance through exploration and latent learning (see Chapter 6). However, as Bindra (1978) emphasises, motivation arises not just from changes in internal state, or the incentive value of external stimuli, but from an interaction between the two. This interaction defines the animal's **motivational state** at any given moment. But what guides the interaction? What determines the relative weighting of internal and external factors in triggering behaviour? How does this interaction work to coordinate adaptive sequences of behaviour when several things need to be done at once? We shall see how these questions can be addressed shortly. First we need to consider one further complication.

Interactions between motivational states and 'feedforward' effects

In responding to one motivational state an animal is likely to alter another. Thus, for example, eating food tends to make it thirsty. Motivational state is thus dynamic, constantly changing as the animal responds to internal and external imperatives. This poses yet another problem for models based on unitary drives since the animal is likely to be driven to do a number of different things at the same time. Interactions between motivational states may be predictable, like those between hunger and thirst above. If so, animals may evolve to anticipate them in their choices of behaviour. Thus they may drink *in advance* of feeding, in order to offset the dehydrating effects of food intake (Fitzsimons & LeMagnen 1969). Just such an anticipatory response to temperature-induced thirst has been shown in barbary doves (*Streptopelia risoria*). Budgell (1970) recorded the amount of water drunk by birds at different ambient temperatures after having been deprived of water for two days at a constant temperature of 20 °C. Despite undergoing the same degree of water stress, birds tested at higher temperatures drank more water. The difference in intake arose before there could have been any temperature-induced change in hydration, so it would appear that the doves drank in anticipation of the future effect of temperature on their water requirement.

Such anticipatory responses can be surprisingly sophisticated. Rats, for example, increase their fluid intake when given the poison lithium chloride (so-called 'antidotal thirst'). This is not because the toxin causes dehydration directly, but because the animal increases its urinary output to void it from its bloodstream, thus risking dehydration later on (Smith *et al.* 1970). Other anticipatory responses occur in relation to predictable time courses of requirement. Thus many animals show a distinctive diurnal pattern of food intake, with meals occurring at similar times each day. Physiological responses may become entrained to this pattern. In humans, blood sugar levels may fall just before an expected meal because the liver stops releasing stored glycogen. The person feels hungry and eats to restore their blood sugar. These kinds of anticipatory responses constitute **feedforward** effects, pre-emptory actions offsetting future potential changes in motivational state (McFarland 1971).

4.1.1.3 The state-space approach

Homeostatic models of motivation introduce two important concepts: (a) motivational state, the combined physiological and perceptual state of the animal as determined by its internal and external environments, and (b) the idea that behaviour acts to maintain the animal's internal milieu at some preferred, or **optimal**, state. As we have seen, however, simple homeostatic models do not cater for some important aspects of motivation such as feedforward effects and interactions between different motivational states. In the 1970s, therefore, McFarland and coworkers developed a new approach which, while retaining the notions of motivational and optimal states, explicitly catered for complex interrelationships between the causal factors for different behaviours. For reasons that will become obvious shortly, this became known as the **state-space approach**. As well as offering a more sophisticated, but manageable, framework for thinking about motivational decision-making, the state-space approach also viewed the optimisation of motivational state as a result of cost–benefit decisions shaped by natural selection (see 2.4.4.3). In doing so, it explicitly linked mechanism and function, two of Tinbergen's complementary Four Whys (see 1.2.1), in its predictions about behaviour.

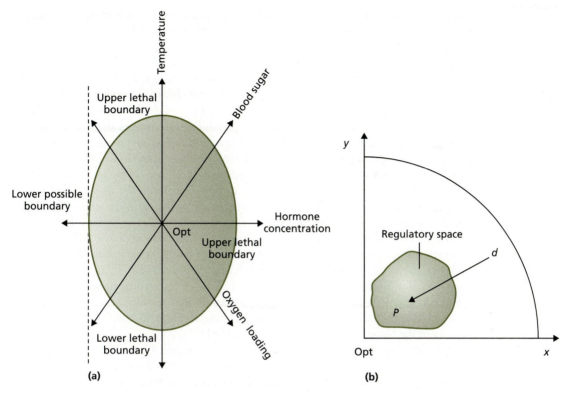

Figure 4.5 (a) A representation of physiological space as defined by state variable axes for blood oxygen loading and sugar levels, concentration of a given hormone and body temperature. A large number of other axes are not represented. Physiological space is bounded by lethal or otherwise biologically impossible states. In an ideal world, there is an optimal state (Opt) for the animal that represents the optimal value for each state variable axis. From discussion in Sibly & McFarland (1974). (b) In reality, the animal will not be able to optimise along every state variable axis, so instead it will adopt an acclimatised optimum (*P*) that reflects conflicting demands. The animal's physiological state is constantly displaced from its acclimatised optimum (e.g. to *d*) and regulatory processes attempt to restore it. Conflicting demands from different displacements result in physiological state being maintained in a subspace of physiological space around the acclimatised optimum. This subspace is referred to as regulatory space, represented here in relation to two state variable axes. From discussion in Sibly & McFarland (1974).

Physiological and regulatory space

It is helpful to begin by considering what determines the animal's internal state. The state-space approach views the internal environment of the animal as a system of interacting variables (hormones, nerve impulses, sugar levels, blood osmolarity, etc.). The finite set of variables that completely characterises the animal's physiology makes up its *physiological state variables*, the value of each state variable at any given moment determining the animal's **physiological state**. Each of these physiological state variables can be represented as an independent axis of a multidimensional hyperspace which we can call **physiological space** (Fig. 4.5). This allows us to pinpoint the animal's current physiological state as a function of all its physiological state variables. As the value of any of these variables changes, the position of the animal's current state changes along the respective axis. Physiological space is bounded by what is biologically possible and what the animal is able to survive (Fig. 4.5a). Thus, for example, negative hormone levels are impossible and there are upper and lower lethal boundaries to temperature. In principle, we can envisage an optimal state that represents the ideal for the animal along each of its physiological state variable axes (Fig. 4.5a). In practice, of course, this is impossible; achieving the optimal state for any one axis will compromise the animal's ability to optimise for others. Moreover, the optimal position with respect to each axis is likely to differ between environments where these impose different selection pressures on the animal. In reality, therefore, the animal will seek an **acclimatised optimum**, a compromise between conflicting requirements adjusted to local environmental conditions (Fig. 4.5b).

Let us now imagine that the animal's physiological state is displaced from its acclimatised optimum, perhaps because the animal has been deprived of water and is dehydrated. We can represent this in Fig. 4.5(b) as a displacement to point *d*. In order to correct this displacement, appropriate regulatory processes come into operation. In this case, the regulatory processes are likely to be behavioural: the water deficit makes the animal thirsty, so it drinks. In other cases, such as temperature stress, regulation might be wholly or partly physiological (e.g. vasodilation or sweating). If exposure to extreme conditions was prolonged, or a regular occurrence, physiological changes might start to acclimatise the animal to its now more stringent environment, perhaps by modifying the action of crucial enzymes. In this case the animal shifts to a new acclimatised optimum. All three processes may act together. Thus in responding to oxygen depletion at high altitude, for example, physiological regulation may initially adjust breathing rate, behavioural regulation the amount of physical exertion, and acclimatisation the oxygen capacity of the blood (McFarland & Houston 1981).

Acclimatisation will be favoured on cost/benefit grounds if it reduces sufficiently the amount of regulatory effort required to maintain an optimal internal state. However, regulatory processes are likely to provide the corrective mechanisms for most day-to-day displacements of the animal's physiological state, such as sugar depletion, water loss, changes in body temperature, etc. As we have already noted, though, regulatory changes in response to one kind of displacement (feeding) are likely to cause displacements elsewhere (e.g. a water deficit). The process of regulation back to the acclimatised optimum is thus likely to oscillate in different directions as the animal deals with the knock-on consequences of each course of action. The result is that, rather than achieving the acclimatised optimum *per se*, the animal's physiological state is maintained within a subspace of physiological space called **regulatory space** (Sibly & McFarland 1974; Fig. 4.5b).

An advantage of thinking about regulation within a multidimensional space is that it is easy to envisage the components of complex motivation like hunger. Hunger state can be mapped in relation to the various component axes (fat, protein, carbohydrate, etc.) making it up (Fig. 4.6). What exactly the animal seeks in order to relieve hunger will depend on its state relative to these different axes: sometimes mainly fat, sometimes mainly protein, sometimes combinations of dietary constituents. Another advantage is the ability to model direct and indirect trajectories of corrective response. For example, a water deficit (say) can be reduced directly by drinking, or indirectly by reducing intake of dry food. In the latter case, water conservation will result in increased hunger.

Motivational space and optimal trajectories of recovery

Where regulation is brought about by behaviour, regulatory space is equivalent to **motivational space**, with the acclimatised optimum as its origin. The axes of motivational space can now be thought of as **causal factor axes**. Just as displacement from the origin of physiological regulatory space results in compensatory physiological regulation, so displacement from the origin of motivational space elicits compensatory

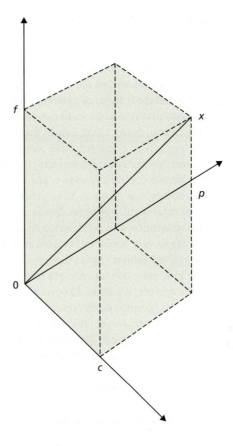

Figure 4.6 'Hunger' (x) can be represented as a multidimensional state determined by deficits of different commodities, e.g. fat (f), protein (p) and carbohydrate (c). After McFarland & Sibly (1972).

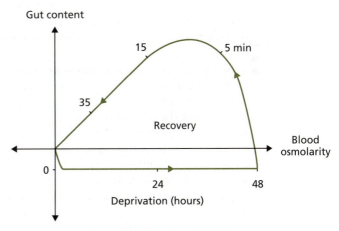

Figure 4.7 A hypothetical trajectory of recovery to the optimal state in motivational space following a period of water deprivation. See text. After McFarland (1970).

behaviour. Such displacements are sometimes called **commands** because they set behaviour in motion.

Of course, once a corrective behaviour has been elicited, the animal must choose how to perform it. But what criterion should it use? Imagine the situation in Fig. 4.7, which shows a simplified two-dimensional motivational space for drinking. The axes represent two internal causal factors: blood osmolarity and the water content of the gut. These tell the animal something about its requirement for water by determining its degree of thirst and telling it how much water it has drunk. Let us assume the animal has been deprived of water for 48 hours, so that blood osmolarity has increased away from the optimum. The animal is then given access to water and drinks. As it does so, its motivational state recovers as indicated in the figure. First, the water content of the gut increases sharply as water is imbibed, then, as water is absorbed through the gut wall, both gut content and blood osmolarity decline until the animal regains its optimal state.

Of course, many other causal factors are likely to come into play in the real world, including positive and negative external incentives, but our attention for the moment is focused on the recovery trajectory in Fig. 4.7. As it stands this is entirely arbitrary. In principle, we could have drawn any number of recovery curves to the optimum. Each of these, however, is likely to have different consequences for the efficiency with which the animal regains the optimum. Precisely how they differ will depend on the physiology of water absorption and transport, and time, energy and risk implications and so on, which are likely to vary across individuals and circumstances. Whatever the differences might be, however, we should expect them to be acted upon by natural selection, since differences in the efficiency of recovery are likely to affect survival and reproductive potential. In short, therefore, we should expect the animal to *optimise* (see 2.4) its return trajectory.

In order to optimise the return of a displaced internal state by means of behaviour, the animal must decide: (a) which behaviour is most appropriate, and (b) how the behaviour should be performed. The first decision depends on the combination of internal and external causal factors acting at the time; that is on the nature of the displacement and the external incentives. The second depends on the cost of performing the behaviour relative to the cost of the animal's current state. We shall look at choosing a behaviour first.

Motivational isoclines   That internal and external causal factors combine to determine behaviour is intuitively obvious. *How* they combine, however, is considerably less obvious. Yet to predict whether an animal will perform a particular behaviour requires a quantitative understanding of their combined effect. One approach is to use a form of **indifference curve** called a **motivational isocline**. An indifference curve joins points of equal probability that something will happen as a function of the values of its axes. Thus, the curve in Fig. 4.8(a) joins points of equal probability that an animal will go for food as a function of how much food it can see and the risk involved in going to get it. The animal will go for a small amount of food if the environment is safe, but will only risk danger if the reward is large. The **utility** (the consequences for the animal's reproductive potential) of a large amount of food in a dangerous place is the same as that for a little food in a safe place, and the animal is just as likely to feed in the two circumstances. Indifference curves can be derived experimentally. Figure 4.8(b) shows some curves from a study by Logan (1965) in which rats experienced different delays before being able to get to food. Logan found that rats were prepared to tolerate a long delay as long as the expected food reward was large. They would only put up with a short delay if the reward was small.

Motivational isoclines are indifference curves joining combinations of causal factor strengths that make a behaviour equally likely to be performed. Thus Fig. 4.9(a) shows hypothetical isoclines for feeding in relation to the strength of both the internal causal factor of food deficit (hunger) and the external causal factor of food availability (incentive). Figure 4.9(a) is therefore a simple two-dimensional motivational space defined by these two causal factor axes. The important point, however, is that the *shape* of the isocline tells us something about the quantitative relationship between the two sets of causal factors.

Figure 4.8 (a) A hypothetical indifference curve for feeding behaviour, trading off food availability and predation risk. Modified after McCleery (1978). (b) Indifference curves in rats, trading off delay against reward size. See text. Modified from Logan (1965).

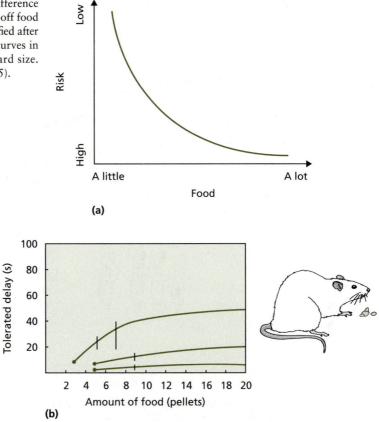

(a)

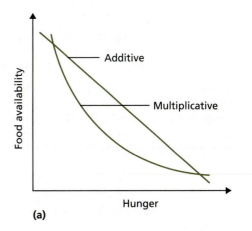

(b)

A smooth curve indicates a *multiplicative* relationship, so whether or not the animal feeds depends on the product 'internal causal factor strength × external causal factor strength'. A straight line, on the other hand, indicates an *additive* relationship; feeding thus takes place according to the rule 'internal causal factor strength + external causal factor strength'. This is all well and good in theory, but is there any evidence for these different kinds of relationship?

Figure 4.9 (a) Motivational isoclines join combinations of causal factor strengths (in this case, hunger and food availability) that are equally likely to elicit a given behaviour (here, feeding). The shape of the isocline indicates the quantitative relationship between causal factor strengths. See text. Based on discussion in McFarland (1993).

(a)

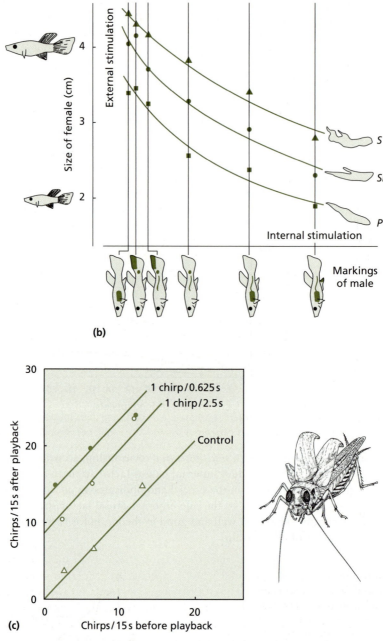

(b)

(c)

Figure 4.9 (*continued*)   (b) Apparently multiplicative motivational isoclines for components of courtship display (*S*, *Si* and *P*) in male guppies, *Poecilia reticulata*, as a function of male sexual state (*x*-axis) and attractiveness of the female (*y*-axis). See text. Modified after Baerends *et al.* (1955). (c) Apparently additive effects of external stimulation (playback of chirps) on chirping rate in male house crickets (*Acheta domesticus*). Different rates of playback had the same magnitude of effect whatever the prior chirping rate (internal state) of the subject. After Heiligenberg (1966).

Courtship in male guppies: Baerends *et al.* (1955) investigated courtship behaviour in male guppies (*Poecilia reticulata*). Here, the most important external causal factor influencing the male's behaviour was the size of the female: males preferred large females. Measuring external causal factor strength was therefore easy. Determining the males' internal sexual state was less straightforward. However, Baerends *et al.* used the fact that sexual activity in males correlated with changes in various bodily markings, which thus provided a guide to their internal state. Using this, they drew up a calibration curve relating a male's marking pattern to the frequency with which he performed various courtship activities (Fig. 4.9b). They then established the size of female necessary to elicit a given display from a male in a known sexual state (determined by his markings). From the combinations of female size and male internal state in their experiment, Baerends *et al.* were able to plot motivational isoclines connecting points of equal likelihood that a male would perform various components of courtship. The results (Fig. 4.9b) described smooth curves, suggesting that male sexual state and female size combined multiplicatively to produce courtship behaviour. However, the arbitrary scaling of the male internal state axis (a rank order based on markings) means that the precise form of the isoclines must be interpreted with some caution.

Territorial song in crickets: An example of an additive relationship comes from a study of territorial song (stridulation) in house crickets (*Acheta domesticus*) by Walter Heiligenberg (1966). Stridulation is an important means of advertising territory ownership. Males stridulate spontaneously, but increase their rate in response to stridulation from other males. Using captive crickets, Heiligenberg monitored spontaneous stridulation before 'challenging' subjects with chirps from another male. Subjects were challenged with two different rates of stridulation: a chirp every 0.625 seconds, or one every 2.5 seconds. Figure 4.9(c) shows the relationship between the subject's chirp rate just before a challenge (a measure of its current internal calling state) and the effects of the two different challenges (external incentive). As the figure shows, chirp rate increased by more or less the same amount regardless of current internal state when subjects were challenged, and increased more in response to the higher rate of stridulation. These results thus imply an additive effect of external stimulation.

Costs, benefits and the performance of behaviour   We have shown that something can be gleaned about the quantitative relationship between internal and external causal factors, and we shall see shortly how this can help us predict which of a range of behaviours an animal should perform. Before that, however, we must consider another important factor in the choice of behavioural response: the cost to the animal of its current state.

By definition, displacement from an optimal state imposes some kind of penalty. If displacement occurs in the animal's physiological state, the penalty is likely to be in terms of reduced function and thus, ultimately, survival and reproduction. The reproductive cost of a small displacement may be negligible; a slight drop in blood sugar, for instance, is unlikely to have serious consequences. As the displacement becomes larger, however, the cost of any further departure is likely to increase. Thus a unit increase in food or water deficit when the deficit is already large is likely to be more costly than the same increase when the deficit is small. In the jargon, the **cost function** of displacement is non-linear (Fig. 4.10). This has important implications for the performance of corrective behaviour.

Reducing a single deficit: While a displaced state may be costly, so too is the performance of corrective behaviour. At the very least the behaviour costs time and energy. Moreover, the cost is likely to rise the more vigorously the behaviour is performed. If the animal is in a seriously displaced state, however, it may be worth incurring a high

Figure 4.10 A hypothetical cost function for food or water deprivation (deficit). The cost of prolonged deprivation is disproportionately higher than that for a short period of deprivation. After Sibly (1975).

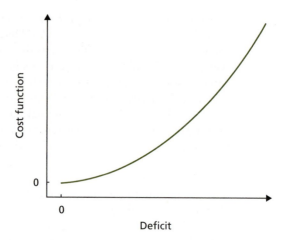

performance cost in order to shift its state out of the danger zone. But, as the displacement reduces, the cost becomes less and less worth paying. If the cost function for displacement is convex as in Fig. 4.10, therefore, the animal's behavioural response curve should start off steep and become shallower as displacement reduces to zero. To predict exactly how it should decelerate, however, we need to know the precise form of the cost function.

A plausible guess from the mathematics of homeostatic control systems might be that it is quadratic; that is the cost (C) is proportional to the square of the displacement ($C \propto x^2$). Robin McCleery (1977) tested this by looking at feeding satiation curves in food-deprived rats. If the cost function for food deprivation is quadratic, then it can be shown mathematically that the optimal, decelerating, feeding response is exponential in form. When McCleery plotted the satiation curves of his rats he found that they were indeed decelerating (Fig. 4.11), but were they best described by an exponential curve? To see, McCleery compared the fit to an exponential curve with that to two other kinds of decelerating curve: a rectangular hyperbola and a parabola (Fig. 4.11). Although similar to an exponential, these two curves are in fact predicted by very different cost functions (McCleery 1977), so they provide a good test of the quadratic cost function hypothesis. Although all three curves seem to fit the data quite well to the naked eye (Fig. 4.11), statistically, the exponential curve came out the clear winner. Thus, the rats appeared to optimise their feeding response according to the presumed cost of their displaced physiological state.

Reducing more than one deficit: But what happens if an animal has more than one deficit to reduce? We have seen that this must often be the case and that, indeed, corrective behaviour itself can induce new deficits. How does the animal choose between the necessary, and probably conflicting, corrective behaviours? It cannot perform them all simultaneously, so can it prioritise and sequence behaviours to optimise the recovery of its internal state?

Sibly & McFarland (1976) argued that which behaviour is chosen at any given time should depend on three things: (a) the position of the animal's internal state relative to the optimum (i.e. the degree of deficit), (b) the maximum rate at which a behaviour can be performed (because this constrains its usefulness in reducing a deficit) and (c) the availability of the relevant commodity (the incentive). To test this, they used barbary doves (*Streptopelia risoria*) that had been deprived of both food and water. Sibly & McFarland were able to show that the motivational isoclines for feeding and drinking are likely to be multiplicative (see Fig. 4.9a). If doves optimised their choice of behaviour, *and the cost functions of their displaced food and water states are quadratic* as above, therefore,

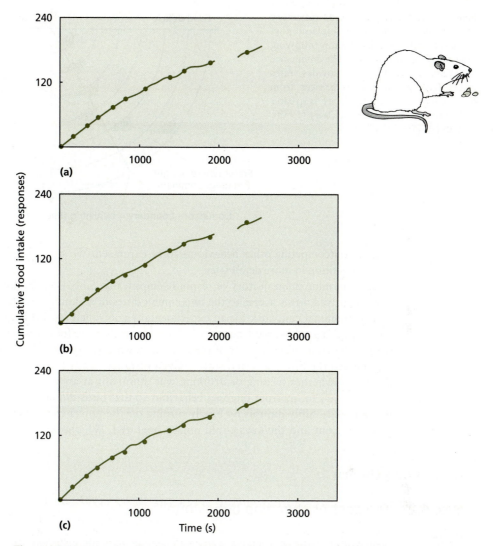

Figure 4.11 Cumulative food intake rate in food-deprived rats with fitted curves appropriate to three different cost functions (see Fig. 4.10): (a) exponential, (b) parabola, (c) hyperbola. The exponential curve in (a) provides a significantly better fit to the data. See text. From McCleery (1977).

they should perform that behaviour at any given time for which the product deficit × incentive is greatest (feeding and drinking behaviour are assumed to be equivalently effective in reducing their respective deficits). The product is thus equivalent to the *tendency* to perform the behaviour.

Figure 4.12 represents a deficit space bounded by the food and water deficit axes, and shows how the deficit × incentive rule should move the doves' state towards the optimum from point *P*, the birds' displaced internal state following food and water deprivation. During the first phase, internal state moves towards the line dividing states in which deficit × incentive is greater for one corrective behaviour (say feeding) from those in which it is greater for the other (drinking). This **switching line** is then tracked back to the origin (satiation) (Phase 2). The switching line is also sometimes referred to as a **dominance boundary** because it divides states in which different behaviours are dominant in the

Figure 4.12 The predicted trajectory (dashed line) for returning a displaced physiological state to the optimum following food and water deprivation. The dominance boundary (≡ switching line) divides states where feeding tendency is greater from those where drinking tendency is greater (solid line). See text. After McCleery (1978).

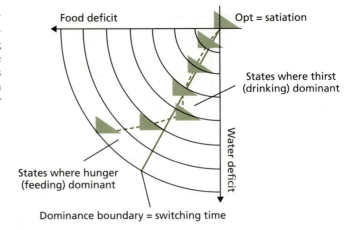

Food deficit

Opt = satiation

States where thirst (drinking) dominant

Water deficit

States where hunger (feeding) dominant

Dominance boundary = switching time

sense of outcompeting other behaviours to be expressed. We shall discuss the nature of this competition in more detail later.

To determine the trajectory of response empirically, Sibly (1975) allowed food- and water-deprived doves access to the two commodities in a Skinner box (an apparatus in which an animal can work for an experimentally controlled reward by, e.g., pressing a bar or pecking a key; see Chapter 6). A peck at either the food key or the water key delivered a controlled amount of the relevant reward to the bird. Each delivery therefore moved the bird's internal state a given distance towards the satiation point. To determine whether feeding or drinking was dominant at any given time, however, it was necessary to interrupt ongoing behaviour so that birds did not 'lock on' to one key (see Box 4.2). Sibly did this by briefly turning off the power to the Skinner box so the light went out and the keys could not be operated. Whichever key was pecked first

Underlying theory

Box 4.2 The cost of changing behaviour

Sequences of behaviour require animals to change from one behaviour to another. Such changes may be costly, leading to a reduced incentive to switch. A bird eating seeds in a field, for example, will build up a water deficit but may have to fly several kilometres to find water. In addition to the energetic cost of the flight, the bird will lose valuable time in the transition and may run a serious risk of predation. The cost of changing might thus be expected to affect the animal's decision to switch. But how should the cost be allocated? Should it militate against the current behaviour, or against the behaviour to which the animal should switch? Larkin & McFarland (1978) tried to find out.

Using barbary doves, they observed birds switching between feeding and drinking in Skinner boxes and an experimental room. The cost of switching was manipulated by making the birds negotiate a partition or fly up a step in order to get from one commodity to the other. Larkin & McFarland predicted that the cost of switching would act as a disincentive to change to the next behaviour in the sequence. In the Skinner boxes, therefore, birds should become more reluctant to change as the length of the dividing partition increases (in control theory jargon, the birds should 'lock on' to current behaviour). Figure (i) shows that the tendency to lock on to current behaviour did indeed increase with the length of the partition.

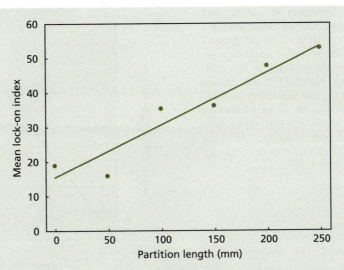

Figure (i) The tendency for doves to keep pecking the current key in a Skinner box instead of switching (mean 'lock on' index) increased significantly with the length of an opaque partition separating the keys. After Larkin & McFarland (1978).

In the experimental room, they predicted that having to fly up a step would reduce the incentive for behaviour appropriate to the elevated commodity and cause the dominance boundary between feeding and drinking to rotate towards the deficit axis for that commodity. Thus if water is elevated, birds should continue to feed at greater values of water deficit (the boundary will rotate towards the water deficit axis) and vice versa. Once again, these predictions were borne out (Fig. (ii)). The cost of switching therefore appears to be allocated to the behaviour to which the animal is about to change.

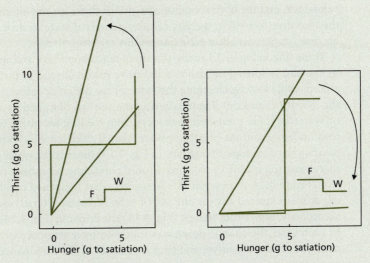

Figure (ii) When food (F) or water (W) were elevated relative to the other commodity, so that doves had to fly up to switch between feeding and drinking, the dominance boundary rotated (arrows) towards the deficit axis with the reduced incentive. After Larkin & McFarland (1978).

Figure 4.13 Sequences of feeding and drinking (pecking food and water keys in a Skinner box) by a barbary dove (*Streptopelia risoria*) following food and water deprivation conform to the prediction in Fig. 4.12. The dominance boundary is constructed empirically as the line separating pecks to the food (*H*) and water (*T*) keys following brief interruptions. *S* represents the empirically determined point of satiation and *P* the bird's state of deprivation prior to testing. See text. After Sibly (1975).

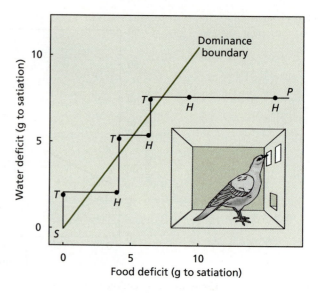

when the power went back on indicated the behaviour that was dominant at the time. The sequence of dominance changes could then be plotted on a graph until the bird ceased to peck at either key for 10 minutes and was regarded as satiated. Figure 4.13 shows such a plot. A straight line can now be drawn dividing points where feeding was dominant (*H*) from those where drinking was dominant (*T*). This is the empirically determined dominance boundary. Note that the bird's responses extend a considerable distance either side of the boundary; that is, they do not switch behaviour the instant they cross it. An obvious reason for this is that it would result in 'dithering', rapid alternations of behaviour, each performed for tiny amounts of time. Such sequences are likely to be grossly inefficient in time and energy expenditure, so the bird's decision-making mechanisms are designed to avoid it. But how? One possibility is that the presence of external cues for current behaviour, and the corresponding absence of those for the alternative behaviour, reduce the incentive to switch, thereby requiring internal state to drift further before a switch becomes imperative. Box 4.2 explores this problem further.

If the line in Fig. 4.13 really is a dominance boundary and its position is determined by the bird following the deficit × incentive rule, it should be possible to manipulate it in a predictable way by changing the value of the deficit or incentive. Sibly (1975) changed the incentive, and did this by altering the reward obtained by pecking either the food or water key. He predicted that increasing feeding incentive would cause the bird to feed under conditions of greater water deficit than it would otherwise have done; thus rotating the dominance boundary in Fig. 4.13 towards the water deficit axis. Conversely, increasing the incentive to drink should rotate the boundary towards the food deficit axis. Figure 4.14 shows that this is exactly what happened in Sibly's doves. Birds in the experiment started off on one set of reward rates, alternating feeding and drinking accordingly, until either feeding (Fig. 4.14a) or drinking (Fig. 4.14b) reward was increased. In each case, the change point was followed by a sequence of the behaviour with the newly increased incentive, and thus a rotation of the dominance boundary towards the predicted axis.

Simply showing that the dominance boundary rotates in the predicted direction, however, is only a qualitative test of the deficit × incentive rule. A much stronger, quantitative, test would be to show that the *degree* of rotation was as predicted by the

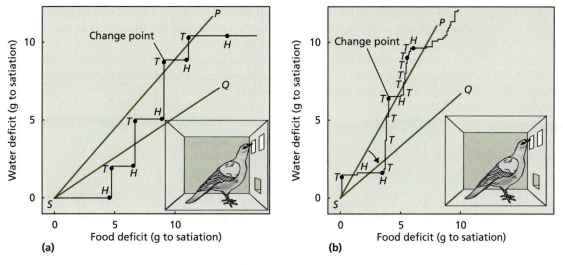

Figure 4.14   (a) When food availability is increased (change point) in experiments with doves, the dominance boundary rotates towards the water deficit axis. The boundary rotates in the opposite direction when water availability is increased (b). After Sibly (1975).

rule. Sibly therefore compared the ratio of slopes of the dominance boundary before and after the change in incentive with that predicted by the ratio of reward rates *if* birds were using the deficit × incentive rule. The result showed that the ratio of the slopes did not differ significantly from that predicted by the rule.

The dominance boundary rotation method can be used to test other predictions about sequences of behaviour. One of these concerns the cost of switching between activities. When an animal changes from one behaviour to another, such as the doves changing from feeding to drinking above, the change itself is likely to cost time and energy. This **switching cost** may then act as a disincentive to change. Box 4.2 looks at some predictions and experimental tests relating to the effects of switching cost on behaviour.

4.1.1.4   Motivational systems and sequences of behaviour

So far we have seen how we can model motivational state in order to predict the performance of behaviour. We have also seen how an animal might be expected to cope when motivated to behave in different ways at the same time. In thinking about the motivational control of behaviour, ethologists and psychologists have often resorted to the notion of *motivational systems*.

The animal's behavioural repertoire is often composed of a large number of discrete behaviours. These behaviours are not performed in a random jumble but occur in functionally related clusters. Thus behaviours relating to feeding occur together in organised sequences to ensure food is obtained and ingested. Each of these functionally related clusters can be thought of as being controlled by a higher-order motivational system. We can therefore envisage a 'feeding' system, a 'drinking' system, a 'reproduction' system and so on. At any particular time, several such hypothetical systems will be active (the animal will be hungry, thirsty, sexually aroused, etc. simultaneously) and must compete for expression, or for access to what Tinbergen referred to as the **behavioural final common path** (a term that has fallen from use with the demise of energy models of motivation, but that reflected the last link in Tinbergen's hierarchy of motivational control centres

[Fig. 4.4]). However, the level of causal factors for some systems will be greater than that for others, so these systems will gain expression and the animal can be said to have *motivational priorities*. Feeding might be top, grooming second and sleep third, for instance. Systems that have top priority at any given time can be thought of as *dominant* systems, and the others as *subordinate* systems. In general, dominant systems are seen as inhibiting subordinate systems, thus allowing their own behaviours to be expressed.

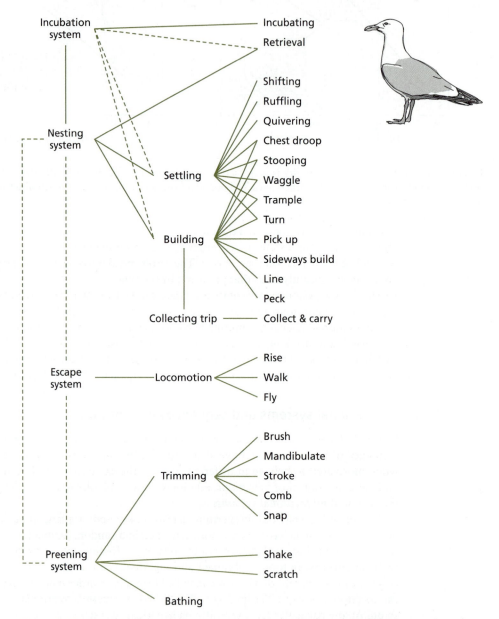

Figure 4.15 By carefully analysing sequences of behaviour, it is possible to speculate about underlying control systems and their interrelationships. The figure arranges behaviours related to feather maintenance, escape from predators, nesting and incubation in herring gulls (*Larus argentatus*) under putative, mutually inhibitory (dashed lines) or excitatory (solid lines) control systems. After Baerends (1970).

Baerends's (1970) model of the motivational control of behaviour in herring gulls (*Larus argentatus*) in Fig. 4.15 is a descriptive model of this kind. On the basis of an enormous amount of field work, Baerends proposed four motivational systems – a preening system, a predator escape system, a nesting system and an incubation system – each controlling its own repertoire of behaviours. As well as organising the behaviours under their individual control, however, Baerends suggested a variety of excitatory and inhibitory relationships between the systems. One major system, the nesting system, controls all nest-building and maintenance and incubatory behaviours (via an excitatory link, see Fig. 4.15) and at the same time inhibits the activity of the other systems. This allows the bird to spend long uninterrupted periods keeping its eggs at the optimum temperature. However, if a predator suddenly appears, or the bird's feathers become ruffled and wet, the escape or preening system will assume temporary priority and inhibit nesting and incubatory behaviour via their inhibitory links with the nesting system (Fig. 4.15). Importantly, the strength of the inhibitory relationships varies with circumstances. Thus a female with five eggs that has been off the nest feeding for some time will be highly motivated to incubate on her return and inhibition of escape and preening is likely to be strong. A female that has been sitting on a single egg for several hours, on the other hand, will easily be distracted into other behaviours if their causal factors increase.

Conflict behaviour

Of course, it is conceivable that two or more motivational systems will become equally aroused. Perhaps a female has just returned to a large clutch when a predator appears on the horizon and is torn between incubation and escape. What should she do? In practice what she might do is neither; instead she preens! We have already encountered this phenomenon in our discussion of motivational energy. It is one example of what ethologists dubbed **conflict behaviour**, responses that not only appeared to say something about interactions between motivational systems but arguably provided the basis for various kinds of signal (see Chapter 11). Box 4.3 presents a summary.

Interactions between motivational systems

When we ask which motivational system is to have priority, the obvious answer seems to be the system with the highest level of causal factors at the time. This system is said to gain priority through **motivational competition** with other systems and thus suppress the expression of these systems by **inhibition** (Fig. 4.16a). Under motivational competition, sequences of behaviour arise because the levels of causal factors for different systems fluctuate and different behaviours assume priority at different times (as, for example, in Fig. 4.13). Ethologists and psychologists long considered this the principal mechanism governing sequences of different behaviours. McFarland (1974), however, points out that sequences could arise through a number of kinds of interaction between motivational systems (Fig. 4.16).

In competition, the causal factors for the second-in-priority behaviour are ultimately responsible for the removal of its inhibition. Thus sequences arise through a cascade of inhibitions (Fig. 4.16b). Alternatively, causal factors for the second-in-priority behaviour may not be responsible for removing inhibition. Instead the top priority behaviour may terminate for some other reason and allow the next behaviour to be expressed. The second-in-priority behaviour is then said to appear by **disinhibition** (Fig. 4.16a). A simple experiment illustrates the distinction.

Underlying theory

Box 4.3 Conflict behaviour

Conflict behaviours are actions performed when two or motivational systems are assumed to be aroused and cannot be expressed at the same time. They are characterised by their seeming inappropriateness to the situation in hand and fall into a number of categories.

Displacement activities

'Displacement activity' is a term for apparently irrelevant behaviours performed when motivational systems conflict. Because the behaviours controlled by the two systems were incompatible and could not be expressed, the pent-up nervous energy was seen as 'sparking over', or being 'displaced', into a third, irrelevant, system. 'Displacement activity', while still in use, is thus a vestige of the old energy-based models of motivation. While such behaviours may appear to be out of context, however, they may not always be irrelevant at all. In some cases, such as displacement cooling behaviour by male buntings (Aves: Emberizidae) during sexual chases, they may be autonomic responses stimulated by exertion or fear. Alternatively, they may be relevant, but lower priority, activities that normally follow one of the current behaviours but are disinhibited early when the ongoing behaviour is curtailed by motivational conflict (e.g. van Iersel & Bol 1958).

Redirected behaviour

Sometimes behaviour patterns appropriate to one of the conflicting systems are shown, but are directed to an irrelevant object. A male herring gull confronted on the boundary of its territory by an aggressive neighbour may suddenly tear at the grass. Aggressive behaviour thus seems to be redirected away from the stimulus that elicited it. Again, however, there can be problems with 'irrelevance'. For instance, jungle fowl show similar redirected pecking to the herring gull, but redirected 'aggressive' ground pecks in this case sometimes result in food being taken. Moreover, if a bird is made aggressive towards a coloured stick, it pecks more at food particles that are similar in colour to the stick. There thus seems to be some interaction between aggressive and feeding states, perhaps relating to energy reserves in anticipation of future conflict.

Intention movements

It may be that inhibition of one motivational system by another is incomplete, so that behaviour appropriate to the inhibited system is not suppressed altogether but is merely reduced in intensity or frequency, or appears in an incomplete form. In some cases, this may consist of the initial phases of some movement or sequence of movements. Such truncated movement patterns are then referred to as *intention movements*. For example, the takeoff leap of a bird consists of two phases: first the bird crouches and withdraws its head and tail, then it reverses these movements as it springs off. If the bird is in motivational conflict, the first phase may be repeated several times. Gulls and gannets (*Morus bassanus*) torn between remaining with their eggs and flying away from a predator, show repeated 'wing-lifting'. Wing-lifting is normally a prelude to flight, but here it is performed several times without the bird taking off.

Ambivalence

Sometimes, intention movements appropriate to two conflicting motivational systems combine into a single motor pattern containing elements of both. Ambivalence is often shown in approach–retreat conflict, where, for example, animals may combine approach with a sidling posture, or repeatedly approach and retreat.

Figure 4.16 Alternating sequences of behaviours A and B can result from the three different combinations of inhibition and disinhibition (a) shown in (b)–(d). See text. After McFarland (1974).

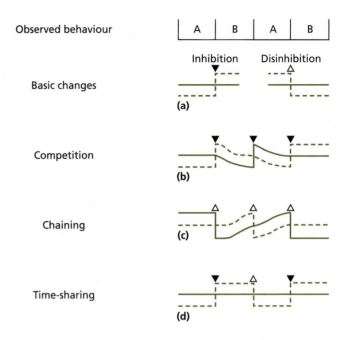

McFarland deprived barbary doves of food and allowed them to work for a food reward in a Skinner box. After 5–10 minutes feeding was usually interrupted by preening and then resumed. To see whether preening occurred by competition (inhibition) or disinhibition, McFarland repeated the experiment under the same conditions, except that he increased the causal factors for preening by fastening paper clips to the primary feathers of the birds' wings. If preening was appearing by competition, it should now appear sooner in the behavioural sequence. Instead the timing of preening remained as before. What changed was the *intensity* with which it was performed. Preening thus appeared to occur by disinhibition when feeding ceased temporarily, rather than by an increase in its motivational status relative to feeding. Sequences based on a series of disinhibitions are sometimes known as **chaining** (Fig. 4.16c), and one famous example may be the courtship dance of the male three-spined stickleback (Fig. 4.17). Here, the sequence of behaviour depends on a succession of stimuli which each fish presents to the other. Careful analysis, however, shows that these are **consummatory stimuli** terminating the ongoing behaviour (like a full stomach after feeding), rather than causal factors for the next behaviour in the sequence. In other words, each behaviour self-terminates on presentation of the right stimulus, thus disinhibiting the next.

While competition and chaining are two ways in which sequences of behaviour could arise, McFarland (1974) recognised a third, more interesting, possibility. Sequences could come about entirely under the control of one (dominant) motivational system. Dominant behaviour expresses itself by inhibition of preceding behaviour, self-terminates after a time, thus disinhibiting a second behaviour, but then regains expression by reinhibition (Fig. 4.16d). The second, *subdominant*, behaviour therefore appears 'by permission' of the dominant system, an arrangement that McFarland called **time-sharing**, since the dominant and subdominant motivational systems effectively share the behavioural final common path. Time-sharing may be adaptive in a number of situations. For instance, time-shared bouts of drinking during feeding may anticipate, and thus offset, the thirst-inducing effects of dry food (see above). Several experiments have provided evidence for

Figure 4.17 The courtship dance of the male three-spined stickleback appears to be an example of 'chaining' because each behaviour in the sequence self-terminates and disinhibits the next (see text and Fig. 4.16). After Tinbergen (1951).

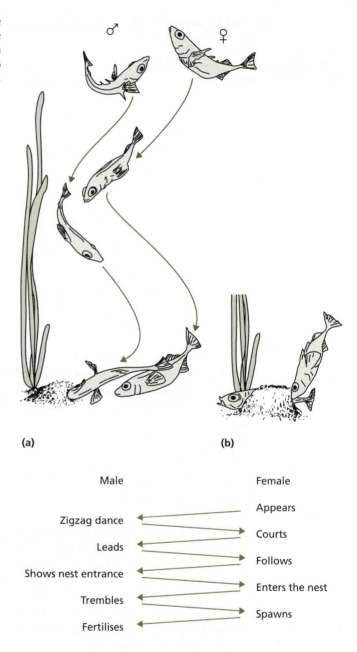

(a) (b)

Male		Female
		Appears
Zigzag dance		Courts
Leads		Follows
Shows nest entrance		Enters the nest
Trembles		Spawns
Fertilises		

time-shared sequences of behaviour, for instance in alternating bouts of copulation and feeding in male rats (Brown & McFarland 1979), and sequences of courtship and nest maintenance in male three-spined sticklebacks (McFarland 1974; Cohen & McFarland 1979). In these cases, one behaviour in the alternating sequence was 'permitted' for only a limited time before the other resumed, and/or its time of appearance in the sequence (but not its intensity) was independent of the strength of its own causal factors (like preening in the barbary doves above), criteria that fit well with the definition of dominant and subdominant systems in McFarland's scheme (but see Houston 1982 for a critique).

4.1.2 Motivation, intentionality and anthropomorphism

So far, 'motivation' has acted as a metaphor for adaptive rules of decision-making. Models of motivation have, by and large, been concerned with capturing the criteria by which animals prioritise and regulate the performance of behaviour. However, there have been pitfalls along the way, not least thinking of motivation as some kind of physical driving force in the animal's nervous system. While many of these have been overcome, a fundamental difficulty remains: the inherent anthropomorphism of the 'motivation' metaphor itself, and what it implies about proximate causal mechanisms of behaviour.

As the late J.S. Kennedy (1992) eloquently pointed out, unwarranted anthropomorphism – the attribution of human qualities to (in this case) animals – is a hazard as widespread as it is usually unintentional in the study of behaviour. Motivation is a paradigm of the problem for a number of reasons. First the term derives directly from our own subjective feeling that we do things because of some inner urge, a desire to make something happen. This itself implies that our actions have some purpose or goal, and reflect an intention to achieve it. But motivation is not necessarily recognisable at the outset; it usually requires an outcome to identify it. Thus the fact that our urgent locomotory behaviour takes us straight to a chip shop tells an observer that it was probably prompted by hunger rather than a need to urinate. While the assumption of purposefulness may be reasonable (though by no means indisputable [see McFarland 1989, Ch. 6, for a clear discussion]) when applied to ourselves, it is seriously problematical when applied to other species. Whether other organisms are purposeful and goal-directed, and whether they have conscious intentionality are important questions, but the mere fact that they behave as if they might, cannot by itself provide the answer. We have already seen why in Chapter 1.

For Kennedy (1992) the essential danger of the motivation metaphor is that functions become causes. An animal ceasing to wander about when it comes across food is assumed to have been wandering about *in order* to find the food. It is then just a small step to assuming it was *searching for food* because it was *hungry*. Such reasoning, of course, is entirely circular. That the animal's behaviour resulted in it finding food in fact tells us nothing about why it started moving about in the first place. To find out whether the animal really did set off intending to find food requires some careful experimentation. In the second part of the chapter, we shall look at progress in this direction. This takes us into the contentious realms of animal **cognition**.

4.2 Causation and cognition: the intentional animal

Cognition is difficult to define unambiguously. It centres on the ways animals acquire, integrate, store and use information in deciding how to behave, but is the collective upshot of all these things rather than a property of any one of them. Thus it may involve learning in some species, individuals or contexts, but not in others, so learning ability by itself is not a measure of cognition, even though it may figure prominently in an assessment of cognitive ability (Chapter 6). More than this, however, cognition usually denotes a capacity for flexible, adaptive responses that are based on a conceptual appreciation of the world rather than a set of hard-wired responses or blind learning rules. The difference between Dennett's Popperian creatures and his Darwinian or Skinnerian creatures in

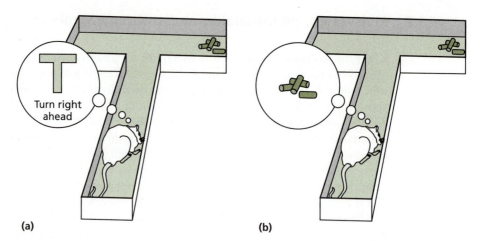

(a)   **(b)**

Figure 4.18 The difference between (a) procedural and (b) explicit responses to obtaining food in a T-maze. In (a) the rat follows a simple rule, such as 'turn right ahead', in (b) it has a mental image of the goal. After McFarland (1999).

Fig. 1.3 captures something of the distinction. In this vein, McFarland (1991) distinguishes between two kinds of knowledge: (a) **procedural** and (b) **explicit** (Fig. 4.18), the difference between following blind rules of response that work in the current situation, and knowing that the world works in a particular way and being able to exploit this knowledge in different circumstances. 'Cognitive' is generally reserved for the latter. Thus, in Fig. 4.18(a), the procedural rat turns right in the T-maze because turning right has been rewarded (and thus reinforced) more often than turning left; the rat therefore adopts 'turn right' as its local rule of response. The cognitive rat (Fig. 4.18b), on the other hand, knows that food is 'over there' and can therefore be got by turning right ahead. Cognition is thus proximate causation that works through some kind of understanding of the world rather than blind, mechanical responses.

4.2.1 Goal-directedness, intentionality and consciousness

Cognition leads naturally to considerations of goal-directedness (see 1.1.2.1), intentionality and consciousness, each a source of considerable controversy in its own right. As a result, and as Dawkins (1995) emphasises, traditions of cognitive theory vary in the extent to which they embrace or avoid these sequelae. In some cases, cognition is simply taken to imply sophisticated information processing (e.g. McClean & Rhodes 1991; Tooby & Cosmides 1992), something that computers and some other machines can do as well. By this yardstick, the study of cognition is about how brains use information and model the world, regardless of whether this is done consciously or unconsciously. However, since human cognition has provided much of the basis for comparison when it comes to other species, the subject has inevitably become enmeshed in the debate about consciousness. Indeed, as Dawkins points out, cognition and consciousness are virtually synonymous in many people's eyes.

Regardless of how inclusively we choose to define cognition, the main problem with goals, intentions and awareness is that they belong to the private experience of the individual. We do not, and probably never will, have access to the private experiences of others, even those of our own species, so we are left having to infer their properties indirectly from what their owners do. And hereby lies a second major problem: it is difficult to envisage an observational or experimental test that unequivocally distinguishes cognitive from procedural causes of behavioural outcomes. We have touched on this already in the context of goal-directedness and intentionality (1.1.2.1), but the problem is just as acute when it comes to awareness (see 4.2.1.4). There seems to be nothing that consciousness would allow an animal to do that a specified, if sophisticated, set of blind procedural rules could not also achieve. Indeed, much of the argument for consciousness in other species rests simply on giving the 'benefit of the doubt', a belief that evolutionary continuity with ourselves makes consciousness more, rather than less, parsimonious as an assumption (e.g. Griffin 1981; McFarland 1989). Despite these difficulties, however, some progress has been made in charting the apparent mental attributes of other species, as we shall now see.

4.2.1.1 Intentionality and learning

As Shettleworth (1998) points out, to say that an act is intentional implies that the actor is motivated by two things: a *belief* that the act will achieve something (a goal), and a *desire* to achieve whatever that is. On this basis, rats pressing a bar for food in a Skinner box appear to show intentional behaviour because the tendency to press the bar is contingent on there generally being a reward and on the desirability of the reward. Rats that have not experienced an association between bar-pressing and food reward do not press the bar, neither do rats that used to but have since been exposed to so-called extinction conditions (where bar-pressing no longer yields food). Rats thus do not press the bar without evidence that doing so produces food – in short, bar-pressing requires 'belief'. Furthermore, bar-pressing also depends on the desirability of the food. Satiated rats, or those that have been made ill as a result of eating the food, do not press the bar either. Figure 4.19, for instance, shows the results of 'revaluing' sucrose or pellet rewards by adding a mild poison. During extinction, poisoned rats showed a drop in their rate of

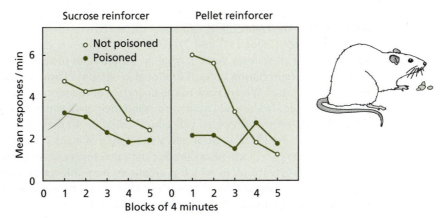

Figure 4.19 After experiencing mild poisoning associated with either sucrose or pellets, poisoned rats showed a reduction in response to the commodity under extinction. See text. From Shettleworth (1998) after Colwill, R.M. and Rescorla, R.A. (1985) Postconditioning devaluation of a reinforcer affects instrumental responding, *Journal of Experimental Psychology: Animal Behavior Processes*, 11, 120–32. Copyright © 1985 by the American Psychological Association. Adapted with permission.

bar-pressing compared with controls (Colwill & Rescorla 1985). These, and a host of similar results, make it difficult to avoid interpretations in terms of intentionality (e.g. Colwill 1994; Balleine & Dickinson 1998).

While bar-pressing in rats may provide evidence of intentionality, this does not follow for all such learned tasks. Unlike rats, pigeons pecking at lighted keys in a Skinner box (as did the doves in Sibly's experiments above) do not appear to peck on the basis of belief and desire. One line of evidence for this comes from experiments in which birds trained to obtain food on pecking a lighted key are faced with a reward each time the key is lit but *not* pecked (a so-called **omission procedure**). Under these conditions pigeons persist in pecking, though at a reduced rate (Williams & Williams 1969), thus showing none of the response extinction expected under an intentional model and demonstrated by bar-pressing rats. Pigeons also show little evidence of intentionality in choosing between keys in various delayed response procedures (Urcuioli & DeMarse 1997).

However, some other bird species show extremely convincing evidence of intentionality, albeit in a different kind of learning task. A stunning set of observations by Weir *et al.* (2002) on New Caledonian crows (*Corvus moneduloides*) revealed that a captive female called Betty had spontaneously learned to shape a straight piece of wire into a hook in order to lift a small bucket of food out of a pipe (Fig. 4.20). New Caledonian crows are known to use tools in the wild (Hunt 2000) and to select tools that are appropriate to a given task from the range available (Chappell & Kacelnik 2002). The remarkable thing about Betty, however, was that she fashioned hooks without any prior experience of the apparatus or of wire as a pliant material. To make the hook, she wedged one end of the wire in the sticky tape round the base of the pipe, or held on to it with her foot, then bent the opposite end with her bill. She bent the wire and succeeded in pulling the food out on nine out of ten experimental trials, in all but one after having first failed to get the food with the straight wire (Weir *et al.* 2002).

4.2.1.2 Cognitive maps

Clearly, finding its way about is an important capability in any animal setting off intentionally to find something, and many species have evolved impressive capabilities in this respect (see Chapters 3 and 7). As always, however, there are different ways of accounting for these feats, some relying on simple rules of response, others asserting a sophisticated appreciation of local or global geography. How can we tell which it is in any particular case? We are back to the quandary of the procedural versus cognitive rat (Fig. 4.18) again. Do animals navigate around their environment by means of naturally selected or currently reinforced procedural rules, or do they consult a **cognitive map** (O'Keefe & Nadel 1978) as we would a street map of a city? The problem is that, most of the time, the two mechanisms predict the same outcome, so to distinguish between them requires situations in which a cognitive map allows an outcome beyond the capability of procedural rules. Two such situations are the selection of a novel route to a goal and making a detour round an obstacle (Pearce 1997).

Choosing a novel route

Perhaps surprisingly, one of the more convincing examples of a cognitive map comes from an insect. During their foraging phase, worker honey bees make extensive use of directional information to find sources of nectar and pollen and provision the hive.

(a)

(b)

Figure 4.20 Betty, a female New Caledonian crow, spontaneously learned to bend a piece of wire into a hook (a) in order to pull food out of a pipe (b). See text. Photographs courtesy of Alex Weir and Alex Kacelnik.

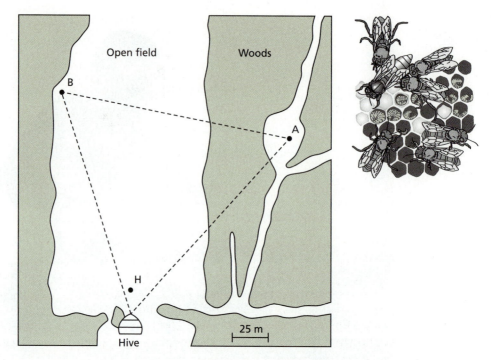

Figure 4.21 The layout of the area and experimental points used by Gould (1986) to test for the use of cognitive maps by honey bees. See text. After Pearce (1997).

In a now well-known experiment, Gould (1986) trained bees to fly from their hive to a feeding station (A in Fig. 4.21). Bees were then captured as they left the hive, removed in a dark container to a novel point B and released. All the bees tested flew off in the direction of A and all were noted arriving at A. The time taken to get to A suggested that they had flown directly there and not spent time meandering about. It is also important to note that none of the features around A was visible from B because of the sloping topography of the land. Gould therefore concluded that the bees must have had a cognitive map of their foraging area and used it to plot a course to the feeder.

While Gould's claim has been challenged (e.g. Menzel 1990; Dyer 1991), the alternative explanations proffered (based on extrapolation from landmarks) demand highly complicated integration of spatial coordinates, and at least as much directional information as would be contained in a cognitive map. They are thus no more parsimonious in accounting for the behaviour. Moreover, in further support of the cognitive map interpretation, Gould & Gould (1988) cite one of Dyer's own findings in which bees were trained to collect food from a boat moored in the middle of a lake. While the bees could be trained to the station easily enough, Dyer found that when they returned to the hive and tried to recruit new foragers, the naïve bees would not respond. When the boat was moved to a location on the shore, however, the bees had no problem recruiting. These results strongly suggest the naïve bees had a cognitive map of their foraging area, and were not attracted to the lake station because the directional information from the trained forager made no sense – food is not found in the middle of lakes. When the information pointed to a terrestrial station, however, it made sense to respond to it (Gould & Gould 1988).

Figure 4.22 A diagram of the circular maze used by Chapuis & Scardigli (1993) to study the response of hamsters to a detour problem. See text. After Chapuis & Scardigli (1993).

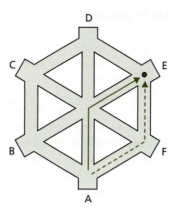

Detour experiments

Another approach to testing for cognitive maps is to train an animal to a goal and then block its normal route of access to the goal and see whether it can choose the best alternative route. A variety of techniques have been used to do this. Chapuis & Scardigli (1993) used the simple circular maze in Fig. 4.22 to see whether hamsters (*Mesocricetus auratus*) were able to solve a detour problem. A hamster was placed in chamber A and had to pass along the route indicated by the dashed line to get food from chamber E. Access to all other routes in the maze was barred by locked doors. Once the hamster had completed several sessions running between A and E, the route was blocked but the doors to all the other routes in the maze were opened. What would the hamster do now? If it had formed some kind of spatial map of where the food was relative to its own starting position, it should be able to choose the shortest new route to the food. If not, it should blunder about randomly until it happens upon it by chance. In Chapuis and Scardigli's experiment, the hamsters chose the shortest new route significantly more often than would be expected by chance. That the hamsters based their choice on a map of the maze itself was made even more likely by the fact that, during the training phase, the maze was rotated within the room between trials so that extraneous cues in the room itself could not provide additional directional information. The starting chamber was also varied between trials. The experiment thus provides strong suggestive evidence for a mental spatial map in the hamsters.

4.2.1.3 Concepts of 'self'

Moving around the environment in order to achieve some outcome, and particularly using a spatial map to do so, is likely to benefit from an awareness of self and one's current location and orientation. Indeed, one theory for the evolution of the concept of self in humans suggests that clambering through the complex three-dimensional world of the tree canopy provided the selection pressure for just such an awareness in our primate ancestors (Povinelli & Cant 1995). However, as Dawkins (1993a) and Shettleworth (1998) point out, many other situations may require some form of self-reference, such as getting out of the way of objects that 'self' has set in motion (Johnson-Laird 1988), or pouncing on prey. Thus there may be more general properties of living in a world of cause and effect that favour the evolution of self-awareness. While there are good a priori grounds for expecting self-awareness in other animals, however, the evidence for it is highly contentious.

'Me in the mirror' experiments

The studies that have contributed most to the argument involve the responses of animals to their own image in a mirror. The approach was pioneered by Gallup in the 1970s. In the first experiments, Gallup (1970) separated young chimpanzees (*Pan troglodytes*) and placed a mirror outside each of their cages. He then recorded their changing responses to it over a number of days. None of the chimpanzees had seen a mirror before and their initial response to seeing their image was to react as if it was another chimpanzee. Thus they showed a variety of social responses, such as threat and greeting. Fish, birds and other mammals also react to their mirror image as a social cue. However, Gallup noticed that, as time went on, the chimpanzees began to show fewer social responses and more self-directed behaviour, using the image in the mirror to guide their own activity. For example, they used the mirror to groom areas of their body they otherwise could not see, or pick pieces of food from between their teeth. This suggested that the animals recognised the image in the mirror as being themselves rather than another individual.

Gallup then devised his famous mark test, which has since become a standard in the field of self-awareness experimentation. Chimpanzees were anaesthetised and marked on one eyebrow with an odourless, non-irritating red dye. When they had recovered, they were observed for a period without a mirror, during which virtually no behavioural responses were directed towards the mark. As soon as the mirror was reintroduced, however, the chimpanzees began touching and picking at the marks and inspecting their fingers between touches. Chimpanzees which had not seen a mirror before, but which were otherwise treated in the same way, showed no such change in behaviour. More recent experiments have extended the controls within the experimental design, for instance by including 'unmarked' manipulations to matching areas of the face (Povinelli *et al.* 1997; Fig. 4.23), thus controlling for the effects of generally elevated levels of grooming following anaesthesia (Heyes 1994). Similar tests with bonobos (pygmy chimpanzees) (*Pan paniscus*), gorillas (*Gorilla gorilla*) and various species of monkey suggested that the tendency was more or less limited to chimpanzees and bonobos (e.g. Povinelli & Cant 1995; Tomasello & Call 1997). Thus it was tempting to conclude that the mark test really did reveal something about the emergence of self-awareness among primates. Unfortunately, a number of issues render this conclusion questionable.

One problem is that these kinds of mirror experiments may not provide a fair test for many species. Hauser *et al.* (1995), for example, point out that for many monkeys a staring image like that in a mirror is threatening, causing them to look away and thus potentially interfering with the acquisition of a self-image response. A fairly dramatic manipulation of appearance may be needed to overcome this. When Hauser *et al.* tried this with cotton-top tamarins (*Saguinus oedipus oedipus*), by dyeing the normally white tufts of fur on their heads garish colours, they indeed found that tamarins with three or four weeks of prior experience with mirrors stared into the mirror longer than undyed or naïve controls and were the only individuals to touch their marked fur during trials. There is also evidence that some non-mammalian species can respond in a broadly comparable fashion. Pigeons, for instance, can be trained to peck at a spot on their breast that can be seen only in a mirror (Epstein *et al.* 1981). However, many other species, including parrots (Pepperberg *et al.* 1995) and elephants (Povinelli 1989), can be trained to use mirrors to respond to objects in the environment without showing any evidence of a sense of self, thus raising questions about the significance of apparent self-reference in some of these experiments.

Figure 4.23 The amount of time chimpanzees spent touching marked (solid rectangles in vignette) and unmarked (shaded rectangles) areas of their face that had been manipulated by an experimenter before and after access to a mirror (dashed line). See text. After Povinelli *et al.* (1997).

By themselves, therefore, mark tests are limited in what they can tell us about the cognitive nature of mirror-directed self-referencing. Our sense of self is a profound one, consisting of much more than knowing our physical position in the world and that the reflection in the mirror is us. We also have a philosophical notion of self, embracing an awareness of our uniqueness and mortality and our emotional relationship with others. As Shettleworth (1998) says, mirror experiments provide evidence that chimpanzees have self-*perception* but not that they have a *concept* of self in the deeper sense above. Indeed, poor correlations between spontaneous self-exploratory behaviour in front of a mirror and later responses to a mark test in chimpanzees (Povinelli *et al.* 1993) suggest little in the way of a unified concept of self. Taken together, the experimental evidence to date suggests that some level of self-referencing is present in a number of species, but the jury is still out as to what this tells us about the cognitive status of the sense of self, even in our closest living relative.

4.2.1.4 Higher-order intentionality and 'theory of mind'

Awareness of self and a set of intentions about what to do are two useful attributes in a cognitive, goal-directed world. Such a world, however, is likely to be populated by other intentional beings whose goals may complement or conflict with those of our questing individual. Knowing something about the intentions of others could thus be extremely useful. Have animals evolved such an ability? This question is equivalent to asking whether animals have a **'theory of mind'** (Premack & Woodruff 1978) about those around them; i.e. an ability to impute mental states to others as well as oneself. Rat A would have a theory of mind if it decided not to go for a concealed source of food because it can see rat B and believes rat B would want the food and steal it if it knew its location. If rat A avoids the food simply because it associates rat B with aggression and theft in the past, there is a simple stimulus–response relationship and no basis for invoking a theory of mind.

A theory of mind opens up new, sophisticated possibilities for behaviour. Information about the environment can be garnered if the beliefs of others can be read. If individual A is looking keenly into the middle distance there may be something there that is worth having, especially if you can get it first. The beliefs of others can be manipulated, perhaps allowing you to deceive other individuals into behaving in your interests rather than their own. Communication offers rich possibilities here, as we shall see in Chapter 11. A theory of mind thus implies a higher order of intentionality, the imputation of intentions and beliefs in others and an intention on the part of self to exploit them. As ever, of course, the problem is to distinguish a behavioural outcome driven by a theory of mind from one driven by blind procedural rules and stimulus–response relationships. All we see in our rat example above, for instance, is rat A not taking food in the presence of rat B. Rat A's theory of mind about rat B is a hypothetical construct on our part to help explain it. This is analogous to inferring in our discussion of motivation that a rat drinks because it is thirsty. The state of thirst is a hypothetical construct in the same mould. A theory of mind is thus a putative intervening variable in exactly the same way as motivational state (4.1.1.1; Shettleworth 1998; Fig. 4.24). We have seen that rats and crows at least appear to base some learned responses on 'desire' and 'belief' (4.2.1.1), so is there any evidence that a non-human species has evolved the ability to attribute these properties to other individuals?

Figure 4.24 A theory of mind as an intervening variable (cf. Fig. 4.1b,c) accounting for responses based on various social experiences. See text. After Whiten (1994).

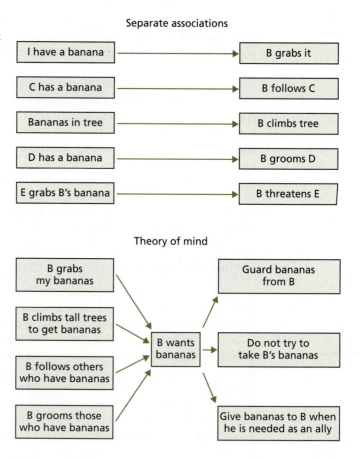

Knowing that others know

At the heart of a theory of mind hypothesis is the suggestion that animals can infer the state of understanding and knowledge in others. Considerable effort has therefore gone into experiments designed to test whether responses are guided by such imputations of knowledge. One sort of experiment asked animals to choose between potential reward locations indicated by individuals in a position to know where the reward was and those indicated by individuals who were not in a position to know. Povinelli *et al.* (1990), for example, allowed chimpanzees to watch a person putting food in one of a number of containers hidden behind a screen. In some experiments a second person (the 'knower') was present while the container was baited, while a third (the 'guesser') was either out of the room or present but with a paper bag over their head so that they could not see where the food went. When the screen was removed and both the 'knower' and the 'guesser' pointed to different containers, the chimpanzees were more likely to choose the container indicated by the 'knower'. This was so whether the 'guesser' had previously been absent from the room or present but worn a bag. At first sight, therefore, the results seem to suggest the chimpanzees had judged the reliability of the two indicators on the basis of an assumption about their knowledge of the food. However, the outcome emerged over a number of trials, so it is possible that the chimpanzees had simply learned that they were rewarded when they chose the person who had been in the room or not worn a bag, but not when they chose the other person. The fact that discrimination improved over successive trials is also consistent with the latter rather than with choice based on a theory of mind. These somewhat negative conclusions are borne out by work on gaze interpretation in chimpanzees, which is similarly failing to provide strong evidence for theory of mind (Povinelli *et al.* 2002). One cautionary note, however, is that experiments with students making analogous 'knower–guesser' discriminations from video footage have shown that subjects who indisputably do possess a theory of mind can sometimes turn in results consistent with simple learning (Gagliardi *et al.* 1995). Once again, therefore, the evidence for a theory of mind is suggestive but not conclusive. Interestingly, similar experiments with rhesus monkeys (*Macaca mulatta*) failed to find evidence for either theory of mind or learning (Povinelli *et al.* 1991), a surprising outcome given the monkeys' propensity for learning other tasks.

At the level of mechanism, reading the actions of others may have some foundation in so-called 'mirror' neurons, a class of nerve cell first discovered in the pre-motor cortex of macaques (Barrett *et al.* 2002). These neurons fire when an individual performs a particular motor action, but also when it sees another perform the same action. Rizzolatti & Arbib (1998) thus suggested that mirror neurons might provide a link between the performance of behaviour and its interpretation when performed by others. In this way, they would support a theory of mind by allowing animals to simulate the behaviour and experiences of other individuals in their heads and guess their feelings and intentions (Gallese & Goldman 1998).

Deception

The natural world is full of examples of deception in a purely functional sense. Stick insects (Insecta: Phasmidae) 'deceive' predators into mistaking them for twigs, grass snakes (*Natrix abstreptophora*) 'feign' death when alarmed, angler fish (*Lophius piscatorius*) snare prey with a deceptive lure. In none of these cases, however, is one tempted to invoke intentionality. But do animals ever intentionally deceive others? Do

Figure 4.25 The mental representations implied if a subordinate baboon intentionally conceals its activities with a female from a dominant male. Here, the subordinate wants the dominant to believe there is no other baboon behind the rock. From *Machiavellian Intelligence: Social Expertise and the Evolution of Intelligence in Monkeys, Apes and Humans* edited by R.W. Byrne and A. Whiten (1988). Reprinted by permission of Oxford University Press.

they deceitfully exploit the perception and understanding of other individuals to their own advantage? Does rat A, for example, hide some food from rat B to make B believe she doesn't have any food? If she does, this would be an example of what Byrne & Whiten (1988) dub 'tactical deception', a higher order of deception than looking like a twig or fooling another fish into regarding part of your anatomy as a worm. Perhaps not surprisingly, evidence for tactical deception has been sought mainly among primates.

While various lines of evidence have been adduced (Byrne & Whiten 1988; Quiatt & Reynolds 1993; Byrne 1995; Whiten & Byrne 1997), principal support has been anecdotal, mainly because opportunities to observe apparently deceitful acts are rare and unpredictable. In some of Byrne & Whiten's examples (Fig. 4.25), a subordinate male baboon appears to conceal the sexual or other contacts he is having with a female from a nearby dominant male by carrying out the activity behind a large rock or other obstacle. Were he to be seen with the female, the subordinate would almost certainly incur the wrath of the dominant, so concealing his activities makes obvious sense. Another apparently deceptive tactic is to use gaze to distract the attention of other individuals. One amusing example cited by Byrne & Whiten (1988) involved a young male baboon which had attacked a younger animal, causing it to scream. Several adults immediately came running to the scene, vocalising aggressively and making to attack the young assailant. On seeing the adults, the young male suddenly and conspicuously stared fixedly into the distance, a posture usually associated with having spotted an approaching predator, though, on this occasion, there was nothing there. The adults halted their advance and followed his line of gaze, thus allowing the young male to make his escape.

While superficially convincing, these examples, and the many others like them, suffer the usual problems of anecdotes (Heyes 1993; Quiatt & Reynolds 1993). In some cases, for instance, they may simply reflect chance coincidence. The young male 'sneakily' consorting with a female behind a rock may just happen to have encountered the female there and seized his opportunity. The 'manipulatively' gazing male, on the other hand, may have interpreted the commotion of the approaching adults as alarm and gazed into the distance to see what was coming. The relatively stable social groups of long-lived species such as baboons also provide ample opportunity for stimulus–response learning in relation to particular individuals and contexts, and 'clever' behaviour may reflect this rather than sophisticated attribution of mental states. Without knowing the detailed social history of the individuals involved it is usually impossible to say. The obvious thing to do is carry out controlled experiments in which individual knowledge and the outcome of applying it are manipulated to distinguish between different kinds of explanation. Several such experiments have been done, looking, for example, at whether chimpanzees and monkeys can exploit the gaze of others (Povinelli & Eddy 1996; Emery *et al.* 1997; Povinelli *et al.* 2002), or monkeys can use assumed knowledge in others to modulate their own responses to desirable commodities (Coussi-Korbel 1994; Gygax 1995; Kummer *et al.* 1996). Although various responses have been consistent with deception based on theory of mind, no outcome has provided unequivocal support; it has always been possible to invoke a more parsimonious explanation.

4.2.1.5 Consciousness and unpredictability

Self-reference and theory of mind beg the question of whether animals are consciously aware of their existence and actions. The more complex and sophisticated a decision process seems to be, the more readily we ascribe it to conscious intention. This, of course is the age-old anthropomorphic trap. Consciousness may suggest itself as a plausible, even favoured, hypothesis to account for a behaviour, but hypothesis is what it must remain until we find some way of testing for it. Is there such a way?

If animals have evolved consciousness, as we think we have, it presumably confers some kind of reproductive advantage; the problem is that it is not yet clear what that advantage is (Dawkins 1993a). For everything that might suggest conscious action, our imagination can invent a consciousness-free alternative explanation. Distinguishing between these possibilities is all the more difficult now that we realise hard-wired procedural rules do not necessarily imply rigidly fixed stereotyped behaviour, but can lead to highly flexible and adaptable responses (see 5.1).

This difficulty, combined with the hard-line empirical legacy of behaviourism (1.3.4), kept considerations of consciousness out of cognitive psychology all through the 1960s and 1970s. The focus was squarely on the processing of information and its influence on behavioural output (Shettleworth 1998). In the 1980s, however, Weiskrantz's (1986) discovery of 'blindsight' in humans, and its subsequent demonstration in brain-lesioned monkeys (Cowey & Stoerig 1995), appeared to open up new possibilities for dissociating awareness from the mechanics of information processing and behaviour. Individuals with 'blindsight' deny (verbally in humans and using carefully controlled response procedures in monkeys) seeing or remembering something (i.e. they are apparently unaware of it), but show behavioural evidence that they actually *do* see or remember it. It thus appears that behaviour can be emancipated from the awareness that is normally assumed to accompany it. The problem when it comes to non-human species, of course, is the reliability of the non-verbal reporting system; which brings us hard up against the privacy of subjective states again. So do we despair, or *is* there a way forward in our quest for a litmus test of consciousness?

Well, there are as yet no decisive empirical data to say for sure, but there is a promising line of argument that could lead to them. A readable summary can be found in Dawkins (1993a). The argument hinges on uncertainty and the somewhat paradoxical claim that much of what we as humans do works better if we are not conscious of doing it (Baars 1988). Playing a complex piece of piano music, for instance, can flow effortlessly if the player is relaxed and allows their fingers to follow their well-trained patterns of movement. As soon as the player thinks about what they are doing, however, mistakes come thick and fast. Juggling and riding a bicycle are other skills where the intrusion of conscious thought can be disastrous. Situations that are predictable and routine, even if they are complicated, may thus be better handled unconsciously. Conscious attention just increases the scope for error.

Unfortunately, much of life is not predictable and routine. Even such seemingly straightforward acts as going out to look for food can be fraught with unpredictable contingencies. The food is not where it should have been, a novel alternative food has appeared in the environment, a predator threatens at the usual feeding site, the food smells of decay. A routine feeding trip suddenly requires some attention and decision-making. This need to react to novelty and unpredictability, to drag attention into focus at a moment's notice, may be what makes the difference in terms of selective advantage between conscious awareness and unconscious procedural rules (Baars 1988; Dawkins 1993a). Animals whose activities are governed by procedural rules, and which thus behave like automata, are notoriously vulnerable to unexpected departures from the normal chain of events (5.1.1.3). A development of this argument suggests that the radical contribution of consciousness to dealing with contingency lies in allowing the animal to restructure its mental model of the world when it suddenly has to change its mind (Oatley 1988). Departures from expectation may require a complete reappraisal of goals and priorities, something that is unlikely to be effective if the animal has to rely on hard-wired rules of response. Dennett (1991), in a lucid and highly recommendable book, has developed this putative 'updating' role of consciousness into a novel view of consciousness itself. Here, the central 'I', the sense of an integrated self, is an illusion created by a constant stream of updated drafts of reality. These drafts are multiple and produced in parallel by different parts of the brain, each draft as it arises becoming a temporary candidate on which to base the next decision. Such a view is beginning to find support from brain imaging studies (see Box 3.3), which report multiple and widely distributed centres of neuronal activity in response to various tasks (Carter 1998; Greenfield 2000; see 3.1.3.2).

'Machiavellian intelligence' and consciousness

While novelty and unpredictability are likely to pervade many aspects of an animal's life, they come to the fore in one particular context: the social environment. As we have seen in 4.2.1.4, inferring states of mind in other individuals may be crucial to achieving important outcomes and avoiding unwelcome penalties. But many social environments are highly dynamic; associations between different individuals ebb and flow, and both states of mind and the consequences of acting on their deduction change with them. Moreover, mind-reading is a two-way activity. Individuals may have to guess not only what is going on in the heads of others around them, but what these individuals think is going on in theirs'. Thus there is intelligence and counter-intelligence, a world in which it is as important to disguise your own attentions as it is to glean the intentions of others. Escalating social dynamism in the primate lineages leading to humans has been suggested by several people as the key force driving the evolution of intelligence. Byrne & Whiten (1988) have labelled this socially driven intelligence **'Machiavellian intelligence'** (see Chapter 6), after the Renaissance political theorist Niccolò Machiavelli, whose name has become a byword for scheming and intrigue. Humphrey (1983) and others have therefore argued that consciousness is most likely among animals requiring a high degree of social 'cleverness' (Dawkins 1993a). This focuses attention on primates, particularly higher primates, and their evolutionary juxtaposition to humans. Others, as we shall now see, use the evolutionary continuity argument in a rather different way.

4.2.1.6 Cognitive ecology, adaptation and modularity

Historically, cognition has been studied mainly at the levels of mechanism and development, especially perception and learning. However, as with consciousness, cognitive skills beg explanation in terms of their adaptive significance (e.g. Sherry & Shacter 1987; Rozin & Schull 1988; Cosmides & Tooby 1995; but see Macphail & Bolhuis 2001). If cognition has evolved through natural selection to solve ecologically relevant problems, then it is unlikely to be a single, generalised property across the animal kingdom. In much the same way as perceptual mechanisms emphasise different modalities (vision, hearing, electroperception, etc.) in different species and environments, cognition should also reflect the different specialist requirements of organisms (Shettleworth 1998, 2001). Fodor (1983) therefore argues that cognitive skills have evolved in a **modular** fashion, with different cognitive mechanisms (or modules) arising to cope with functionally incompatible computations (Sherry & Shacter 1987; Shettleworth 1998). In the jargon, these modules are *domain-specific*, that is they deal with specific kinds of information and response. Thus, prodigious feats of spatial memory are limited to species such as scatter-hoarding birds (see 3.1.3.1), an ability to keep track of fluid social relationships is found in species that live in long-term social groups, and so on. However, some cognitive modules will be widespread among species because they reflect skills common to many tasks. Learning stimulus–response relationships and some aspects of memory, such as its tendency to decay with time, are good examples. Nevertheless, even here, we might expect specialisation. As Shettleworth (1998) points out, animal nervous systems are not blank slates; they are organised in species-specific ways and these are as likely to affect basic processes of learning as they are specialist perceptual abilities (Roper 1983; but see Macphail & Bolhuis 2001). We shall see some examples of this in Chapter 6.

The same argument can be applied to rationality, the capacity for logical inference and deduction. Evolutionary psychologists, such as Cosmides & Tooby (1994, 1995), have argued that rationality is also unlikely to be a blanket property, but should instead reflect the selection pressures acting on the organism. Since social behaviour has provided much of the focus for speculation about the evolution of consciousness and intelligence, it is reasonable to ask whether cognitive skills relating to social information show evidence of such adaptive specialisation. One problem with testing this, however, is that it is difficult to control for all the factors that may affect social and non-social cognition differently, in particular the effects of social experience on the motivation to perform tasks (Shettleworth 1998). A neat way round this is to test humans with abstract problems that reflect the different kinds of task (Cosmides & Tooby 1992). The so-called Wason test (named after its inventor, the psychologist Peter Wason) is an example (Fig. 4.26).

The Wason test asks subjects to detect violations of logical rules. In one version of the test, subjects are asked which of four cards needs to be turned over to see whether the rule 'If a card has **p** on one side, it has **q** on the other' has been violated (Fig. 4.26a). The majority of subjects simply turn over the card showing **p** to see if it has **q** on the back. Very few also turn over the necessary additional card showing **not q**. Much the same logical error emerges when the test is repeated with other more or less abstract tasks. However, when the same task is given social salience, such as the version in Fig. 4.26(b), performance improves dramatically, with as many as 75% of subjects correctly turning over just the two end cards. Cosmides & Tooby (1992) and others argue that rationality in humans evolved to detect social cheats in early hominid groups, and that 'offences'

Figure 4.26 Three versions of the Wason test in which the cards at each end of a row, and only those cards, need to be inspected to verify the statement below them. From Shettleworth (1998).

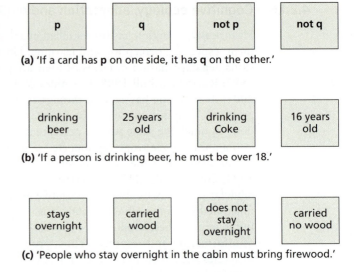

(a) 'If a card has **p** on one side, it has **q** on the other.'

(b) 'If a person is drinking beer, he must be over 18.'

(c) 'People who stay overnight in the cabin must bring firewood.'

such as drinking under age are forms of social cheating; thus we are quick to detect them. Sophistications of the Wason test ask subjects to perform the task in different imaginary social contexts. Gigerenzer & Hug (1992), for instance, charged subjects with solving the problem in Fig. 4.26(c) under one of the following conditions: (a) subjects were German mountain climbers visiting a Swiss mountaintop to see whether the Swiss had the same rule about bringing firewood as them, or (b) they were members of a Swiss mountaineering club checking whether lodgers were obeying the firewood rule. Subjects showed the correct choices 50% of the time under (a), but 90% of the time under (b). Logically, they should have performed to the same level in both cases, but the 'detect cheats' context of (b) elicited a markedly sharper response.

Heuristics: cognitive rules of thumb

Like any other adaptation, cognitive skills are subject to constraints. The computations necessary to solve all problems facing the animal perfectly and rationally may demand more neural resources and time than their reproductive benefits merit. Thus, as well as focusing investment in rationality on problems with most reproductive salience, like detecting cheats in the Wason test above, we might also expect animals to adopt *cognitive rules of thumb*. Like the perceptual and other rules of thumb discussed earlier (2.4.4.3), these would be workable approximations of perfect solutions balancing the cost of investment against proneness to error. We are not talking here about rules of thumb that might superficially mimic cognitive behaviour, like the rule 'choose prey that are mainly green' might lead to a predator taking the most profitable prey *as if* it was calculating the ratio of energy ingested and expended (see Dawkins 1993a). Instead, we are *assuming* cognition and asking how sophisticated it really is. Again, humans provide the best testing ground, since our cognitive world is directly reportable.

In Chapter 3, we saw that optical illusions give away some of the perceptual rules of thumb by which the brain models the world. A now substantial body of work, pioneered largely by Amos Tversky and Daniel Kahneman at the Massachusetts Institute of Technology, suggests that human rationality works on similar approximate rules (or

heuristics as they are referred to by cognitive psychologists). Despite our confidence in our logical ability, careful analysis reveals we are subject to many illusions and fallacies in our reasoning (Kahneman *et al.* 1982; Piattelli-Palmarini 1994). These 'errors', however, often make sense if viewed in terms of adaptive rules of thumb.

One line of evidence comes from so-called 'mental economies' in calculation (Piattelli-Palmarini 1994). For example, if asked to perform the following multiplications in their head:

$$2 \times 3 \times 4 \times 5 \times 6 \times 7 \times 8 = ?$$
$$8 \times 7 \times 6 \times 5 \times 4 \times 3 \times 2 = ?$$

people routinely arrive at different answers for the two products. They also inflate the product of the second multiplication relative to the first and dramatically under-estimate the actual product in both cases. In one large-scale experiment, the average estimate for the first calculation was 512, and for the second, 2250 (Piattelli-Palmarini 1994). The correct answer in both cases, of course, is 40 320. So why the gross and systematic error? The reason seems to be that the first numbers in each series create an immediate, and unconscious, assumption about the rough magnitude of the outcome, a phenomenon known as 'anchoring'. Far from carefully calculating the real product (though that is what they think they are doing), people in fact use a crude rule to quickly guesstimate it. In an artificial situation such as serial mental multiplication, such a rule leads to serious inaccuracy. The natural world, however, is unlikely to demand quite such feats of quantitative juggling. Instead the premium will be on quick decisions about things such as size, distance and time, when partial information may indeed provide a reasonable guide for projection. Estimates of probability betray similar, and equally misleading, mental shortcuts (Piattelli-Palmarini 1994; Dawkins 1998).

Our assessments in other contexts are also heavily influenced by seemingly irrational rules. For example, our emotional response to a given event is greater if we have caused the event than if the same event occurs independently of our own actions. Thus we are more likely to regret losing £20 000 worth of investments if we had just moved shares into a company that suddenly failed than if we lost the same amount by indolently leaving the shares where they were. Why the difference in response? Financially, it makes no sense; the outcome is exactly the same in both cases. From an evolutionary point of view, however, it may make perfect sense. If emotional reactions such as regret serve a useful purpose, it is likely to be through their influence on future actions. Thus perceptions of individual causation become crucial in judging what should be done. Good decisions are positively reinforced (we feel good about them) and we are more likely to make them again; poor decisions are negatively reinforced (we feel bad about them) and they are therefore not repeated. Things that happen independently of our own actions have little to tell us about what we should do in the future, so why worry about them?

Piattelli-Palmarini (1994) discusses many more examples of these and other cognitive rules. While the examples are limited to humans, cognitive processes in other species are likely to employ similar kinds of rules. Once again, therefore, this highlights the practical difficulty in testing for different underlying causal mechanisms. Distinguishing between a procedural rule of thumb that mimics a cognitive outcome (e.g. eat green rather than red objects because these were quicker to eat in the past) and a cognitive rule of thumb that provides a mental shortcut (prey are more often green than red so concentrate on green) might be quite a challenge.

4.3   Motivation, suffering and welfare

The mental worlds of other species are of interest because they should affect how different species behave. But there is another reason they demand our attention. If animals are sentient and have subjective experiences similar to ourselves, they may have the capacity to **suffer**. Thus how we treat them takes on a moral dimension: we become concerned for their well-being or **welfare**.

4.3.1   What is welfare?

There is little agreement as to how welfare should be defined and measured (e.g. Dawkins 1980; Rushen & de Passillé 1992; Mason & Mendl 1993; Barnard & Hurst 1996). Most attempts rely on anthropomorphism, emphasising comfort (adequate space, food and other resources), health (freedom from injury, disease and stress), 'normal' behaviour (the repertoire of behaviours performed in the wild) or philosophical stances on ethics and animal rights (e.g. Thorpe 1965; Moberg 1983; Regan 1984; Rollin 1989; Wolfensohn & Lloyd 1994). Within these, there is a broad division of emphasis between emotional states of suffering (Griffin 1981; Rollin 1989; Duncan 1993) and impairment of biological functioning (Broom 1991; Rushen & de Passillé 1992). Of course, these are not strictly alternatives; because suffering is a private experience, invoking it necessarily requires support from measures of function or performance (Dawkins 1993a; Fraser 1995; Table 4.1). But what is suffering and when might it occur?

Table 4.1 Examples of measures used to assess an organism's welfare and likelihood of suffering. These are divided into measures relating to (a) decision-making priorities, (b) the behavioural and/or physiological consequences of failure to achieve priority outcomes, and (c) the reproductive impact of supposed poor welfare. Even though they are used to infer subjective mental states, all measures relate to functional performance in the animal. See text. Modified after Barnard & Hurst (1996).

Welfare ≡ consequences for ability to prioritise and cope		Welfare ≡ consequences for physical function and reproductive potential
(a) *Assessing priority* e.g.	(b) *Assessing impact* e.g.	(c) *Assessing impact* e.g.
behavioural resilience[1]	physiological stress responses	growth
rebound effects[2]	stereotypies	longevity
tolerance of cost	displacement activities	offspring production
(e.g. work effort)	behavioural repertoire	immunodepression
preference tests		disease
		injury
		fluctuating asymmetry[3]

1 See 4.3.2.1.
2 The extent to which a response is exaggerated following a period of enforced non-performance.
3 See 5.3.3.

4.3.1.1 Suffering as a special state

As a negative subjective state, suffering is presumed to be a causal mechanism that triggers adaptive aversive responses in the animal. However, not just any negative subjective state qualifies as suffering. As Barnard & Hurst (1996) argue, negative states, where (if) they exist, are likely to have arisen as a proximate mechanism for gauging the reproductive cost of the animal's actions. Since all decisions are likely to have both costs and benefits for its reproductive success, an animal must judge the appropriate benefit : cost ratio for each decision it makes (2.4). Choices trade off the animal's time, metabolic resources, health and survival against resulting offspring. A few crumbs of food may not be worth risking life and limb for, but a harem of females may well be. Thus costs such as reduced growth, depressed immunity, disease and injury reflect a price the animal is designed to pay for its reproductive reward. Any negative subjective states associated with these costs are simply part of the animal's adaptive regulatory machinery. They remain so, however, only as long as the animal is operating in an environment appropriate to its decision-making rules of thumb (2.4.4.3). Step outside this, or frustrate the animal's decision rules in other ways, and the system of regulatory cost-gauging is likely to break down. Under these conditions, argue McFarland (1989) and Barnard & Hurst (1996), it is reasonable to think of negative subjective states as suffering.

Suffering is thus an unpleasant generalised state of emergency that arises when the animal incurs or risks costs it has no rules of thumb to regulate. Panic is a good example. In humans, panic is an unpleasant and extreme subjective state that can arise in a crisis. It can combine different negative emotions, such as terror and despair, and the usual upshot is some form of precipitate action that may be frenzied and irrational, but has some chance of extricating the individual from their situation when reasoned action has failed. Pain, where it is not 'willingly' entered into (e.g. by fighting for a valuable territory regardless of injury), can have a similar galvanising effect.

4.3.2 Welfare and adaptive expendability

On this basis, Barnard & Hurst (1996) consider two contrasting views of welfare (Fig. 4.27). The traditional view (Fig. 4.27a) emphasises the maintenance and well-being of the individual. Here, fitness (see Chapter 2) is equated to measures of growth, longevity, offspring production, health and so on, and the organism is seen as having coping mechanisms that maintain normal bodily function through maintenance behaviours (feeding, grooming, etc.), immunological defences and body repair systems. Coping thus acts homeostatically to maintain the physical integrity of the individual, and welfare can be defined in terms of the organism's 'state as regards its attempts to cope' (Broom & Johnson 1993). Failure to cope, or having to work excessively hard to cope, leads to reduced welfare and some form of suffering, and may result in physical impairment (reduced growth, longevity, etc.) (Fig. 4.27a).

By focusing on the well-being of the individual, however, the traditional view commits what Barnard & Hurst (1996) call the 'fallacy of individual preservationism', the assumption that the organism's priority is to ensure its integrity and survival. As we have already seen, well-being and survival are tradeable commodities. Their reproductive value depends on the organism's life history strategy, its chosen tradeoff between growth, survival and reproduction (2.4.5). From an evolutionary perspective,

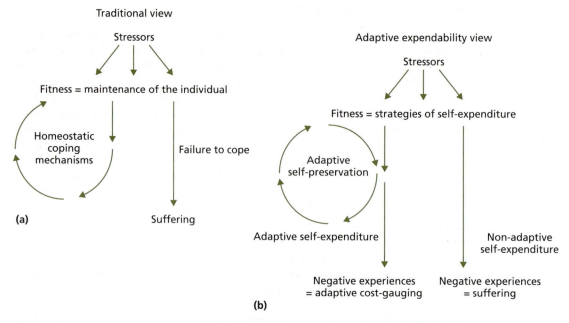

Figure 4.27 (a) The traditional concept of welfare in which fitness is equated with individual maintenance and well-being and homeostatic coping mechanisms. (b) The adaptive expendability concept of welfare, in which fitness is equated with strategies of self-expenditure, and suffering arises when expenditure becomes non-adaptive. From Barnard & Hurst (1996).

individuals are expendable; they 'use themselves up' in the pursuit of reproduction. Fitness thus becomes a property of alleles coding for different strategies of self-expenditure, rather than something reflected by the physical condition of the individual (Fig. 4.27b). Humans place a high priority on individual well-being because we are a long-lived species with a long period of parental care. Health and survival thus contribute greatly to reproductive success. The very notion of welfare may thus be little more than an anthropomorphic conceit.

In the adaptive expendability view in Fig. 4.27(b), negative functional consequences for (and thus negative subjective states in) the individual arise in two ways: (a) via adaptive self-expenditure, where negative subjective states reflect adaptive cost-gauging (e.g. fatigue as a guide to energy expenditure while foraging, or pain as a guide to injury sustained in a territorial contest, see above), or (b) through non-adaptive self-expenditure (e.g. hunting fruitlessly in an inappropriate environment, mounting an ineffective immune response against a novel parasite). On the 'state of emergency' argument above, only the negative effects in (b) qualify as suffering and give rise to concerns about welfare. Indeed, frustrating the organism's attempts to 'spend itself' adaptively in (a), on the mistaken assumption that such expenditure reflected an inability to cope, could itself *create* a welfare problem. A possible example centres on the choice of group versus solitary housing in laboratory rats. Rats are often housed singly to prevent problems with aggression or other forms of social stress. That this reduces stress is suggested by a number of behavioural and pathophysiological differences between solitary animals and those in groups (Hurst *et al.* 1997). However, if given the opportunity, singly housed rats will work to regain contact with other individuals, even though they would be more stressed if they succeeded. From a functional perspective, the response makes sense in that the rat's life history strategies and decision-making priorities have evolved in a social environment. Indeed, solitary rats showed a marked increase in self-directed behaviour, especially tail manipulation, which correlated negatively with organ pathology and may have been a tactic for avoiding an activity 'limbo' (Hurst *et al.* 1997; see 4.3.2.1).

4.3.2.1 Measuring welfare

It is easy to talk in a hand-waving way about adaptive self-expenditure, but proposed yardsticks of welfare are no good unless they can be implemented. We must thus be able to assess the adaptiveness of self-expenditure and recognise departures from it. As we have argued above, the basis for this is the organism's decision-making rules of thumb and the appropriateness of the environment in which they have to be expressed. Decision rules can be grouped into two categories: (a) *rules of time and energy budgeting*, which determine how the organism spends its limited time and metabolic resources on different activities and components of life history, and (b) *rules of response*, which determine how it reacts behaviourally and physiologically to environmental contingencies (e.g. mating opportunities, predators, infection, changes in temperature) (Barnard & Hurst 1996). Of course, there is an interaction between the two categories. If a predator appears, for instance, the organism's rules of time and energy budgeting allocate a high priority to escape behaviour, but the form of escape (bolting down a hole, freezing, fleeing, etc.) is determined by its rules of response.

Rules of time and energy budgeting

The work of McFarland and Houston (e.g. McFarland & Houston 1981; McFarland 1989; Houston & McNamara 1999) provides the clearest analysis of time and energy budgeting and its relevance to welfare, and the following adheres closely to the approaches developed there. We shall consider two kinds of constraint on time and energy budgeting that have implications for welfare as defined by the adaptive expendability argument.

Resilience and squashing   It has long been argued that patterns of time and energy budgeting reflect the relative importance of different activities for individual reproductive success (Houston & McFarland 1980; Shaffery *et al.* 1985; Dawkins 1988). When time and energy are short, or priorities are distorted by, for instance, persistent attempts to escape, allocation to reproductively important activities is conserved at the expense of less important activities. By analogy with economic demand theory, conserved activities are said to be *resilient*, and demand for their outcome *inelastic*, while the remainder are *squashable* and demand for them *elastic*. The severity of time and energy constraint can thus be measured in terms of its impact on normally resilient activities. If these begin to be squashed, the organism is seriously constrained. Deciding when this is the case, however, may not be straightforward. A persistent change in the environment may result in the organism acclimatising physiologically to the new conditions, so that previously resilient behaviours become less important and thus squashable. Changes in time budgets may also be misleading as an index of welfare for other reasons. For example, some organisms may budget their time to minimise energy expenditure, so that as much time as possible is spent sitting around doing nothing (Pyke 1979; Herbers 1981). In a well-provisioned captive environment, therefore, inactivity may reflect precisely what the organism is designed to do. Indeed, in some situations, such as caged groups of laboratory rats, a lot of activity, or a wide range of behaviours, may be a serious *negative* indicator of welfare (Hurst *et al.* 1999).

Figure 4.28 An animal may enter states of 'limbo' and 'purgatory' as it approaches the optimum and outer boundary respectively of its physiological space (defined here by just two hypothetical state variable axes [see 4.1.1.3]). 'Limbo' and 'purgatory' arise where the animal's adaptive rules of response or time and energy budgeting break down. See text. After McFarland (1989).

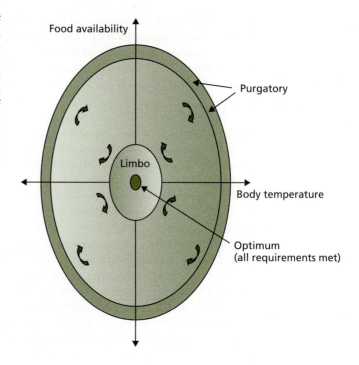

Limbo While inertia may be adaptive in some cases, it is likely to be highly maladaptive in others. Another view of the well-provisioned captive environment above is that it leaves the organism with time and energy on its hands that it has no rules for spending. Under normal circumstances, once it had found food, water, shelter and other vital requirements, the organism would switch to next-in-priority behaviours, perhaps looking for a mate or patrolling a territory boundary. In the captive environment, such opportunities are unlikely to exist. The organism then risks entering a state of 'limbo' (McFarland 1989) in which it is effectively paralysed by a lack of options. 'Limbo' can be represented as a region of motivational state space close to the origin and thus organism's acclimatised optimum (Fig. 4.28). Normally, the organism will be kept away from its optimum, and therefore 'limbo', by interactions between causal factors for different behaviours which tend to pull its state away again (see 4.1.1.3). Whether or not this happens may depend on the role of resource availability in the instigation of behaviour. If the decision rule is to instigate a behaviour on perceiving an appropriate resource, then providing the resource will permit switching to the behaviour. Thus domestic sows (*Sus scrofa*) show a greater tendency to seek out nest material if such material is made available (Baxter 1983). Failure to provide the material reduces the sows' options for searching once other behaviours have been completed, so searching does not occur. A caged migratory bird, such as a starling (*Sturnus vulgaris*), on the other hand, initiates winter migratory activity at the appropriate time of year *regardless* of the presence of food, the resource for which migration is normally a seasonal necessity (Kramer 1957). The risk of entering 'limbo' is thus arguably greater for the nesting sow than it is for the migratory starling.

Rules of response

As with time and energy budgeting, we can envisage a number of factors that are likely to impact on an organism's rules of response.

Frustration Lack of opportunity may frustrate decisions both in terms of prioritising activities (budgeting rule) and performing the chosen activity (response rule). For example, if a hungry animal is presented with food in an inappropriate fashion (e.g. a container it does not recognise), it may switch to its next-in-priority behaviour without feeding, or it may attempt to feed and fail. In either case, failure to feed is the result of a frustrated rule of response. The failure of escape responses by subordinate laboratory mice to protect them from attack by dominants when groups are confined in a cage is a familiar example of such a frustrated rule (e.g. Barnard *et al.* 1996b).

Inappropriate outcome In other situations, the organism may achieve an outcome from its response, but the outcome is inappropriate in some way. Thus, it is easy to imprint (Chapter 6) the young of various precocial bird species on artificial objects, such as boxes and balloons (Hess 1973; Bateson 1979). This is because the rule of response leading to attachment works crudely; it can usually afford to because it did not evolve in an environment where mistaken attachment to boxes and balloons was a problem. Inappropriate outcomes can sometimes arise because relevant cues in the environment become corrupted. The urinary cues used by house mice to prime social interactions (see 3.2.1.4), for example, become disrupted in laboratory animals by regular cage cleaning. In particular, subordinate mice can no longer reliably identify dominant individuals by the predominance of their odours. Cleaning thus results in dominants having substrate odour templates indistinguishable from those of lower ranking mice, leading to inappropriate challenges when they are encountered (Gray & Hurst 1995). Well-intentioned environmental 'enrichment' programmes can also have unexpected negative consequences. Providing caged groups of male laboratory mice with shelters and platforms, for example, can *increase*, rather than reduce, levels of aggression, because of the opportunity it creates for dominant males to defend spatially discrete resources. The upshot is greater stress with a consequent reduction in immunocompetence and resistance to infection (Barnard *et al.* 1996b).

Purgatory Finally, rules of response may lead to appropriate behaviour but be overwhelmed by the circumstances that triggered them. A simple example would be the overwhelming of panting as a cooling response in dogs by exposure to excessively high temperature. Panting is the correct response, and is normally effective in regulating body temperature, but excessive temperatures are outside its range of competence. Under these conditions, we can envisage an organism entering a state of 'purgatory' (McFarland 1989; Barnard & Hurst 1996). 'Purgatory' can be represented in state space as a narrow region close to the boundary of possible states (Fig. 4.28). The disruption of social odour cues in the mice above might lead to 'purgatory', for example, if inappropriate aggressive challenges create a state of persistent social flux, with different individuals temporarily assuming unstable dominant roles.

Corruption of the animal's rules of time and energy budgeting and rules of response can result in various behavioural pathologies, of which some forms of stereotypy are a familiar example (Mason 1991; Garner & Mason 2002; see also 6.1.1.1). Stereotypies are repetitive actions that become fixed in form, intensity and orientation, and constitute one of the commonest behavioural pathologies in confined animals. In an elegant series of studies, Weidenmayer (1997) showed that caged gerbils (*Meriones unguiculatus*) develop stereotyped digging when denied access to a suitable shelter. Gerbils start to dig at bedding as soon as they open their eyes, but quickly concentrate their digging in the corners of the cage because corners provide some of the cues that would guide burrow construction in the wild. Stereotypical digging occurs even when gerbils are provided with sand-filled arenas, indicating that the performance of digging itself fails to satisfy the underlying motivation. In contrast, stereotypical digging does not occur if gerbils are kept without a suitable substrate and given access to a dark chamber. However, the entrance to the chamber has to be shaped like a tunnel; if not, gerbils once again show stereotypy. The gerbils' rules of response in this case therefore seem to depend on very specific stimulus conditions (Würbel 2001). Since it is not difficult to violate them, artificial husbandry conditions can often lead to pathology.

Organisational complexity and welfare

While the temptation to treat more advanced organisms with anthropomorphic empathy must be resisted, their (generally) greater sophistication in terms of sensory perception, information processing, learning, motor skills and so on have a number of obvious implications from a welfare point of view.

First, greater sensory and information processing capacity may increase the sensitivity with which circumstances are distinguished as desirable or aversive. Thus preferences may become more refined and vulnerable to frustration, making infringements of welfare in novel environments more likely. Second, increasing emancipation of decision-making from hard-wired rules to flexible experience may make it difficult to identify underlying rules of time and energy budgeting or response, and thus to distinguish adaptive plasticity of response from imposed constraint – the problem of inertia above is a good example. In addition, as we shall see in Chapter 6, advanced brains, such as those of mammals, may depend for their proper development on appropriate environmental cues. Impoverished environments, such as bleak laboratory cages, can interfere with brain development and behaviour, thwarting decision rules and the adaptive shaping of behaviour, and resulting in aberrant brain function (Würbel 2001; Garner & Mason 2002).

Summary

1. Motivation is a concept of proximate causation that adopts a 'black box' approach to underlying physiological mechanisms. While acknowledging that physiological mechanisms ultimately control behaviour, motivational theory seeks the rules by which factors in the organism's internal and external environments integrate to predict behaviour. Motivation has thus sometimes been referred to as an intervening variable, mediating the organism's response in relation to changes in its environment.

2. Motivational terminology describes behavioural causation more economically than 'nuts and bolts' explanations, but there is a danger of oversimplifying and a temptation to think of motivation as having some kind of separate physiological existence in the organism. Unitary drives and energy-based models of motivation are historical examples of these problems.

3. Homeostatic models of motivation are based on the regulation of motivational state about a set (or optimal) point through feedback from behaviour. The state-space approach combines the assumptions of homeostatic models with an explicitly functional approach to the design of regulatory systems. Motivational control is thus assumed to

have been shaped by natural selection, so that organisms optimise the performance and sequencing of behaviour.

4. The concept of motivation is inherently anthropomorphic, implying the purposefulness we believe guides human behaviour. While this is a problem from a number of stand-points, motivation nevertheless leads naturally to a consideration of cognition generally, and intentionality, goal-directedness and consciousness in particular.

5. A problem with testing for cognition as a proximate causal mechanism is that it is usually possible to imagine a set of procedural rules that could account for the same outcome. Nevertheless, experimental approaches have provided suggestive evidence for intentionality, self-awareness and an awareness of the mental worlds of other individuals in various species.

6. Cognitive skills are likely to incur costs and wells as reap benefits. We might thus expect such skills to be specialised for different ways of life and to be based on approximate rules of thumb where these provide economical shortcuts to making decisions. Ideas based on the modularity of cognitive systems and the widespread use of heuristics in human cognition support these assumptions.

7. If animals are sentient and have subjective experiences similar to ourselves, they may have the capacity to *suffer* and we become concerned about their well-being and welfare. However, life history theory leads us to question the priority given by many species to individual well-being and survival, and to propose an 'adaptive expendability' view of impaired biological functioning and reduced well-being. Here, negative sub-jective states largely reflect adaptive cost-gauging, and suffering becomes a special state of emergency limited to circumstances beyond the competence of the organism's decision-making rules of thumb.

Further reading

Toates (1986) and Colgan (1989) provide excellent and highly readable summaries of motivational theory. McCleery's (1978) chapter in Krebs & Davies's edited volume is by far the best potted introduction to the state-space approach, while McFarland (1971, 1989) and McFarland & Houston (1981) provide deeper treatments. Shettleworth (1998) brings func-tional and evolutionary considerations to bear on cognition in a thorough and readable account, while the volume edited by Halpern (2000) focuses on sex differences in cognition. As a popular account of animal cognition, Hauser (2000) is a lucid and entertaining read. Broom & Johnson (1993) provide a review of traditional approaches to animal welfare, while Barnard & Hurst (1996) take an explicitly evolutionary perspective. The target article by Dawkins (1990) and accompanying commentaries sample various opinions on motivation and welfare from an evolutionary standpoint.

Development is the final level in Tinbergen's 'Four Whys' and has proved to be a hotbed of controversy in the study of behaviour. Much of the disagreement has centred on the role of genes, particularly the gene's-eye view of evolution promulgated by 'selfish gene' theory. But what does the 'genes for . . .' language of selfish gene theory really mean when it comes to behavioural development? Is it suggesting animals are effectively automata, with their behaviour determined in some fixed and inflexible way by their genes? Or can it accommodate the many ways in which the environment and experience appear to play a role? How does learning, for instance, fit into a gene's-eye view of behaviour? The embryological chain of events from gene expression to behaviour is usually long and complicated. Can we ever trace it? Can we identify the developmental route by which a given gene affects a given behaviour? The phenotypic effect(s) of a gene can also depend on the other genes with which it shares the genome. How do these interrelationships influence behaviour?

So far we have talked about behaviour as if it is a hard and fast feature of an organism, as steadfastly recognisable as its species-typical morphology. Of course, the intensity and duration of a behaviour, and the precise form it takes, vary with the nature and strength of its different causal factors, but aggression remains recognisably aggression, copulation recognisably copulation and so on. Or does it? Our definitions of behaviour (Chapter 1) and discussion of its evolution and underlying mechanisms have thus far skirted round a third, and equally fundamental, aspect of behavioural expression: **development**. As the animal grows and matures, its patterns of behaviour change, both in terms of the performance of particular activities and the animal's overall behavioural repertoire. These changes can be profound, leading to wholly different patterns of behaviour in immature and adult individuals of the same species and sex. Indeed, as Bateson & Martin (1999) point out, an organism at different stages of development may be so different that one might almost be forgiven for classifying them as different species.

Development (or, as it is often called **ontogeny**) is the next of Tinbergen's Four Whys (Chapter 1), providing another complementary, and often controversial, perspective on why animals behave as they do. The controversy has arisen largely from arguments about the relative importance of genetic and environment influences on behaviour, particularly as they relate to cognition and culture. While some of the historical divisions in the genes versus environment ('nature–nurture', see 1.3.2) debate have been overcome (Box 5.1), the issue continues to inflame passions, both scientific and emotional, as we shall see shortly. In addition, disagreement has been fuelled by the very complexity of many developmental processes and the convenience of taking them as read in functional arguments about behaviour. In the next two chapters, we shall look at some of the influences at work in the development of behaviour, beginning with the role of genes and then seeing how behaviour is moulded by physical development, experience and learning.

Difficulties and debates

Box 5.1 Genes versus environment

The study of development has experienced some of the most heated controversies in the entire field of animal behaviour, owing largely to polarised opinion about the relative importance of genetic versus environmental influences. For the most part, advocates of genetic determinism (see 5.1) came from biology, while environmentalists came from psychology, thus leading to the so-called 'nature–nurture' debate between the two camps in the middle of the last century (see 1.3).

Bit by bit, both sides have come to agree that what determines a behaviour pattern cannot be addressed in such an either/or fashion. To do so is like asking, as the psychologist Donald Hebb put it, how much of the area of a field is due to its length and how much to its width. Biologists now agree that the expression of a given gene depends on environmental factors, whether in the cell, elsewhere in the organism or in the outside world. Psychologists, for their part, agree that environmental effects act through phenotypes that are heavily influenced by their respective genes.

The ethologist Patrick Bateson likens the process to baking a cake. Like an organism, a cake is made according to a set of instructions (a recipe) for building blocks (ingredients) that become integrated in a particular environment (baking conditions). The final product, however, cannot be predicted from any one of these elements (recipe, ingredients or baking conditions) alone; it is a consequence of all three combined. While there is general agreement on the above, arguments about genetic determinism still continue. This is partly because some old 'nature–nurture' terminology, such as 'innateness' and 'instinct', continues to cause misunderstanding (5.2; Dawkins 1995), but it also reflects a degree of ideological resistance to 'selfish gene' arguments about evolution (1.3.3), often based on a caricature of their assumptions (see Dawkins 1999).

5.1 Innateness and genetic determinism

We have already seen that there are genetic differences in behaviour and that these can be quantified and subjected to selection (Chapter 2). Indeed, the very idea that behaviour evolves hinges on the assumption that such differences exist and are sufficiently consistent across generations to allow adaptive traits to spread through populations. This has spawned the fallacious view among some that genetic variation in behaviour is equivalent to genes *determining* behaviour in some fixed and ineradicable way. Two simple examples show why **genetic determinism** in this literal sense is a straw man (Dawkins 1995).

The first concerns a serious metabolic disease called phenylketonuria. The disease is inherited as a recessive trait at a single locus, and causes a damaging build-up of the amino acid phenylalanine in the brain. People expressing the disease suffer severe mental impairment as a result. However, while phenylketonuria indisputably has a genetic origin, its effects are far from inevitable. If the disease is recognised early enough, affected children can be fed a diet free of phenylalanine, with the obvious result that it does not accumulate in the brain. Such children show normal or virtually normal mental development (Hsia 1970). The phenotypic effects of a clear-cut genetic difference between normal

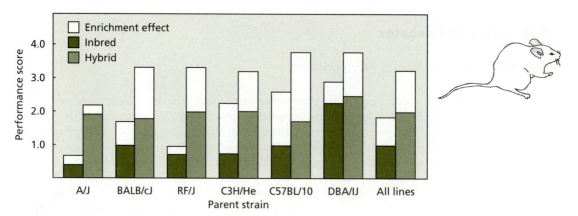

Figure 5.1 The mean performance of different strains of mice (pure strains and hybrids) in a food location task. Performance is scored on an integer scale of 0 (poor) to 5 (good), depending on how long it took to find food. Mice were reared either in non-enriched cages (pale and dark shaded bars) or in cages enriched with tubes, rocks, ramps and other furniture (open bars). In some comparisons, e.g. between RF/J and C3H/He mice, enrichment was required before strain differences in performance became evident. From Henderson, N.D. (1970) Genetic influences on the behaviour of mice can be obscured by laboratory rearing, *Journal of Comparative and Physiological Psychology*, 72, 505–11. Copyright © 1970 by the American Psychological Association. Reprinted with permission.

and phenylketonuric individuals can thus be overcome by appropriate manipulation of the environment. Many other studies have shown that genetic effects on behaviour can be prevented or reversed by (sometimes very minor) changes to the environment (e.g. Crabbe *et al.* 1999; Cabib *et al.* 2000).

The second example concerns the ability of mice to find food in a maze. Henderson (1970) compared the speed with which mice of different strains learned to negotiate a maze of ladders, tightropes and other challenges to get to the food. When mice were all reared in the same standard laboratory cage, there was little difference in the speed of learning between some of the strains (for instance, RF/J and C3H/He; Fig. 5.1), so genetic differences appeared to contribute little to differences in learning ability. However, when mice were reared in cages enriched with toys and climbing surfaces, differences between these previously similar strains began to emerge (Fig. 5.1). While learning speed was enhanced in all mice reared in enriched environments, mice of some strains were very much quicker than those of others, a difference that was attributable to the genetic difference between them. Thus, a particular kind of environment was necessary for the genetic difference in learning ability to be expressed.

Of course, some traits, such as blood group and eye colour, *are* fixed by the individual's genes, but many others – especially behavioural traits – are subject to considerable modification by the environment. Some extreme examples of how powerful environmental influences can be come from the caste systems of eusocial insects. For example, depending on how they are fed as larvae, closely related sisters of the haplodiploid (see Box 2.8) ant species *Pheidole kingi* develop into very different kinds of adult, queens or various forms of worker, not only performing different suites of behaviour but also looking very different morphologically (Fig. 5.2). Thus genetic differences do not equate to genetic determinism in any meaningful sense of the phrase. However, arguments about genetic determinism in this narrow sense are only part of the picture. The nature–nurture debate had much deeper roots in the relationship between genes and development. At its core were the twin notions of **innateness** and **instinct**.

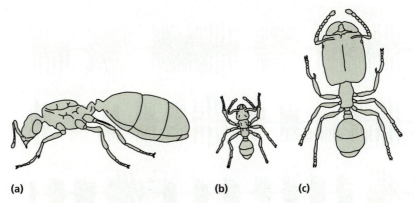

Figure 5.2 Different castes of the ant *Pheidole kingi*: (a) queen, (b) minor worker, (c) soldier. The ants may be sisters, but they differ greatly in their morphology and behaviour as a result of different rearing conditions as larvae. After Wheeler (1910).

5.1.1 Innateness, instinct and 'genes for' behaviour

As Dawkins (1995) points out, the label 'innate' has been applied to behaviour in two different ways. The first means simply that behavioural differences are genetic and thus relates to the discussion above: 'innate differences' means 'genetic differences'. But 'innate' has been used in a second, rather different, way, relating not to differences *between* individuals but to the nature of development *within* individuals. This is the sense in which the pioneering ethologist Konrad Lorenz used it in his theory of instinct (see also 1.3). Lorenz saw an animal's behaviour in terms of instinctive behaviour patterns based on inherited neural pathways. Instincts were thus rather like reflexes only more complicated. They were 'innate' or 'inborn'; the animal did not have to learn them and they required no insight or awareness. In Lorenz's conception, therefore, innate came to refer to behaviour that was not learned. The acid test was to rear a young animal in isolation, so that it had no opportunity to copy others and was prevented from learning on its own. If, when tested later, the animal managed to perform adult behaviour normally, the behaviour was innate, the animal's genes 'told it what to do'.

Of course, in its naïve form, this simple dichotomy (innate versus learned) reaped the same criticism as other kinds of genes versus environment argument. But, while accepting that both genes and environment have a role to play in all cases, it is still possible to retain useful meaning for the term 'innate' (Dawkins 1995), as we shall now see.

5.1.1.1 Innate meaning developmentally fixed

A good example is what Dawkins (1995) has called **developmentally fixed** behaviour. Here, the animal performs the behaviour in typical form irrespective of rearing conditions or current environment. Dawkins uses the song of male crickets of the genus *Teleogryllus* as a good example. Male *Teleogryllus* attract females by singing, which involves scraping one wing quickly over the other. Males of different species sing different species-specific forms of the song (Fig. 5.3), the genetic basis of which can be demonstrated by hybridisation to produce offspring with songs intermediate in character to their two parent species (Bentley & Hoy 1972; Fig. 5.3; see also Box 2.2). The point, however, is that whatever song the crickets inherit it is performed in typical fashion regardless of how individuals

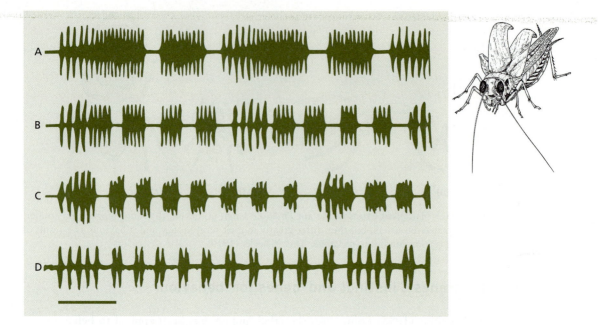

Figure 5.3 Oscillograms of the calling songs of males of the crickets, *Teleogryllus oceanicus* (A) and *T. commodus* (D). The calls of the two species have very different structures. If they are crossed, the hybrid offspring have calls intermediate between the two parental types, but the structure depends on which species is the mother or father (B, *T. oceanicus* ♂ × *T. commodus* ♀; C, *T. oceanicus* ♀ × *T. commodus* ♂). After Bentley & Hoy (1972).

are reared. Rearing in isolation, or with other species of cricket singing different songs, has no effect. At least with respect to the rearing environments tested, therefore, *Teleogryllus* song seems to meet Dawkins's criteria for being developmentally fixed.

The development of *Teleogryllus* song is about development within individuals. It has nothing to do with differences between individuals. While we can reveal genetic differences in song between species (Fig. 5.3), individuals within species all sing their species-typical song. Indeed, where traits have been subject to intense selection, we might expect little in the way of genetic differences because only the most successful variant is likely to remain (the allele(s) coding for the trait have gone to **fixation**; see 10.1.3.2). Thus, in principle, we can have traits with a strong genetic influence in their development but no genetic basis at a population level worth speaking of because there are few if any individual differences attributable to genes (though song in crickets is encoded polygenically [2.3.1.2], so there is always *some* variation within species). 'Developmentally fixed' is therefore not synonymous with 'genetic' in the naïve deterministic sense (Dawkins 1995). To put it in the language of selfish gene theory, we cannot say there are '*genes for*' a trait that has gone to fixation *within* a population; we can say that only at the level of interspecific or inter-population comparisons.

5.1.1.2 Innate meaning inbuilt adaptiveness

There is another sense in which 'innate' retains useful meaning, one that relates to Lorenz's own defence of the concept in the face of criticism from the psychologists. Rather than focusing on the developmental process itself, it is concerned with its *adaptiveness* to the animal (Dawkins 1995). Innate behaviours are those that equip the animal to respond adaptively to the world around it from the outset, without the need for experience and learning. Thus a male three-spined stickleback emerges ready-primed to treat red-throated conspecifics as rivals. It does not have to experience other males fighting or see its own reflection to get the idea. The *Teleogryllus* song above could also be defined as innate in this second sense, in that the song comes ready-tuned to attract the right females. However, caution is still needed here. Innate behaviours in this 'born adapted' sense *can* subsequently be modified by experience, as Jack Hailman (1967) showed in his classic study of the parental bill-pecking response in young laughing gulls (*Larus atricilla*; Box 5.2). Moreover, as we shall see in the next chapter, learning itself is far from an open-ended alternative to innate prescription. Instead, it is often moulded in adaptive, species-specific ways.

Difficulties and debates

Box 5.2   Innate behaviour can be influenced by experience

One of the classic demonstrations of innate behaviour by European ethologists concerned the food begging behaviour of herring gull (*Larus argentatus*) chicks, which consists of pecking at the parent's bill until it regurgitates food. Tinbergen & Perdeck (1950) described the pecking behaviour as innate because it was performed in the same stereotyped way by all chicks soon after hatching. Jack Hailman (1967), however, studied the same response in North American laughing gull (*L. atricilla*) chicks and found that what at first sight appeared stereotyped and thus innate was in fact modified by experience (Fig. (i)).

Figure (i)   The accuracy with which laughing gull (*Larus atricilla*) chicks peck at the image of an adult bill on a piece of card as a function of age. After Hailman (1967).

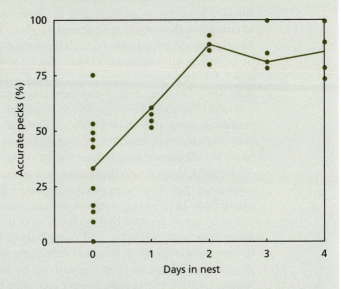

A week-old laughing gull chick begs food from its parent in much the same way as its herring gull counterpart: opening and closing its bill, moving towards the parent, then rotating its head to grasp the parent's bill, causing the parent to regurgitate. Newly hatched chicks, however, do not rotate their heads and thus cannot grasp the parent's bill. Instead they just peck at it, often missing at first. By raising some chicks in the dark and feeding them so that they did not have to beg from a parent, Hailman showed that acquiring the bill-grasping response required learning. Hand-reared chicks were unable to show it at the age their wild counterparts started to grasp. Further experiments, using various model heads and bills, also showed that chicks refined their image of a parent as they gained experience from begging. Chicks kept in the dark for their first 24 hours and then shown only models continued to peck at poor imitations of the parental image when wild chicks had begun to lessen their response to them.

5.1.1.3   Innateness and flexibility

Innateness, the possession of hard-wired adaptive responses, can be taken to imply inflexibility, rigid adherence to a pre-determined set of moves. In *The Selfish Gene*, Richard Dawkins (1989) likened organisms to '. . . robot vehicles blindly programmed to preserve the selfish molecules known as genes', a piece of imagery that took on a life of its own for knee-jerk critics of the selfish gene approach (Dawkins 1999). As Dawkins himself has emphasised, however, much of the objection to the analogy assumes a fallacious equivalence between robots and inflexibility, that robot-like organisms will be slaves to hard-wired mechanical routines acted out regardless of events around them. This begs two questions. First, do organisms ever conform to this narrow concept of robotic behaviour, because, if they do, objections to comparisons with it clearly begin to evaporate? Second, is this really what is implied by likening organisms to robots?

Organisms as robots?

We have already seen that aspects of the behaviour of real organisms can be simulated with robots (3.1.3.4), but this is a far cry from testing the analogy between organisms and robots *per se*. To do that, we must first decide what we mean by robotic behaviour.

The behaviour of robots can be divided into two basic types depending on their responsiveness to current circumstances (McFarland 1999). Primitive robots, such as clockwork automata, or those used in simple factory assembly line tasks, are insensitive to changes in their working environment. Once started, their routines are completed in a fixed, preprogrammed pattern, and their behaviour is said to be **unsituated** (independent of prevailing circumstances) (McFarland 1999). The fixed action patterns discussed in 3.1.3.1 are potential behavioural examples (though see Box 3.2), and the stereotyped behaviour of nesting digger wasps provides a nice illustration.

Gerard Baerends (1941) studied nesting behaviour in the wasp *Ammophila campestris*. Female *A. campestris* show a clear and almost unvarying routine when it comes to the construction and provisioning of nesting burrows. First, the female excavates a burrow, much like the female golden digger wasp in Fig. 1.1. She then hunts down and paralyses a caterpillar, drags it into the burrow and lays an egg on it, repeating the process for her subsequent eggs. As the eggs hatch in her earlier burrows, she returns to them with further caterpillars. In this way, a female may maintain up to five active burrows simultaneously. Baerends noticed that wasps always inspected their burrows first thing in the morning before flying off to find caterpillars. If he added or removed caterpillars *before* this morning inspection, he could induce the wasps to alter their rate of provisioning accordingly. Adding or removing caterpillars afterwards, however, had no effect. Thus female *A. campestris* appeared to act according to a fixed set of simple routines. There is a standard routine for digging and provisioning a burrow and laying an egg, another for inspecting burrows to decide on further provisioning, and another for sealing burrows once they have been adequately provisioned. Once initiated, each routine is seen through to completion; for example, if caterpillars are systematically removed as the wasp brings them to a burrow, she will repeat the hunting and provisioning loop indefinitely, locked into her rigidly programmed sequence.

The behaviour of Baerends's digger wasps thus has something in common with the blind mechanical responses of a simple robot such as a clockwork automaton. The analogy between such stereotyped responses and the behaviour of simple mechanical automata may therefore be reasonable. In other cases, however, it is clear that the behaviour of organisms is highly dependent on prevailing circumstances. The navigation skills of the desert ant *Cataglyphis* in Chapter 3 are a good example. There we saw that ants were able to navigate directly home from any point in their meandering foraging path by integrating patterns of polarised light. The system depends on the external pattern of light available to the ant, which changes with the elevation of the sun above the horizon. Predictable errors in navigation can be induced if the pattern of light available to a foraging ant is manipulated experimentally. The behaviour of the ants is thus wholly dependent on monitoring the environment and responding accordingly; in robotics terms, their behaviour is therefore said to be **situated** (McFarland 1999).

What does this mean for comparisons between animals and robots? Well, as we have seen in Chapter 3, it is perfectly possible to mimic the navigational abilities of *Cataglyphis* with a robot programmed with relatively simple rules. Apparently flexible, skilful behaviour can therefore arise from a modest set of instructions. Dawkins (1999) takes this point further and stresses the fact that modern robots and electronic systems are equipped with conditional rules that allow all manner of contingencies to be taken into account. Computer chess games and space exploration robots are familiar examples. Such systems can respond strategically, with foresight and flexibility – in short, intelligently. In other words, robotics has matured into a byword for sophisticated flexibility rather than limited mechanical response. Nevertheless, however sophisticated, the behaviour of robots is still programmed by relatively simple instructions, so the analogy with genes, and the 'genes for' language of selfish gene theory, is arguably perfectly apposite.

5.2 The meaning of 'genes for' behaviour

The meaning of 'genes for' behaviour *a*, *b* or *c* is obvious enough when used as a heuristic convenience in models, as in 'genes for' altruism or *Hawk* and *Dove* strategies in Chapter 2. But what does it mean in terms of the developmental effects of genes on real phenotypes? How exactly does a 'gene for' altruism *cause* an individual to be altruistic?

5.2.1 Tracing the effects of genes

At the moment, the answer to the last question is that we haven't the faintest idea. Indeed, it may be difficult to decide which developmental effects should count as a contribution of particular genes to behaviour in the first place. As Barinaga (1994) notes, a gene causing blindness would have profound effects on several behaviours, but we should be reluctant to claim it as a 'gene for' any of the affected activities. Nevertheless, as we shall see shortly, some genetic effects we might want to claim in this way occur early on in embryology and do have a broad impact on the organism's physiology. And there are further problems. As we noted in Chapter 2, many different genes can act in concert to influence behaviour. Our definition of a 'gene for' a behaviour may thus have to be operational, encompassing however many loci are involved in making the difference we see (see also Dawkins 1989). A single mutation may also have several phenotypic effects (pleiotropy, see Chapter 2), making it difficult to pin down those that are causally important. Responsiveness to air movement in the crickets *Acheta domesticus* and *Teleogryllus oceanicus* provides a nice example.

These species possess wind-sensitive hairs on a pair of sensory abdominal appendages called cerci. The cerci are innervated by sensory neurons which project into the abdominal ganglion and synapse with the medial giant interneuron (MGI) of the ventral nerve cord. Some mutant individuals are insensitive to air movement over the cerci. Investigation of these mutants revealed that the dendrites of their MGIs were stunted compared with those of normal crickets (Bentley 1975). While this might suggest that an abnormal MGI was responsible for the insensitivity, the mutants also lacked the wind-sensitive hairs on their cerci. Thus there may be an alternative or additional explanation for their lack of response – the lack of an appropriate sensory mechanism. Furthermore, the two morphological consequences of the mutation may have interdependent effects on the behavioural phenotype. MGI dendrites may require appropriate stimulation from the sensitive hairs to develop properly. If so, the absence of hairs in the wind-insensitive mutants will starve MGIs of the necessary impulses and stunt their development. Surgical removal of cerci in normal crickets lends some support to this idea (Murphey *et al.* 1975).

There are also other problems. As Wilcock (1969) pointed out, it is easy to confuse cause and effect in attributing behavioural changes in mutants to associated physiological differences. A mouse with the *obese* mutation, for example, grossly overeats and shows abnormal enzyme activity. The biochemical change, however, turns out to be a consequence of the *behaviour*, not the mutation; normal mice made to overeat by chemical means or surgery show the same changes in enzyme function (Hay 1985). Careful experimental manipulations may therefore be necessary to identify effects unambiguously.

With these caveats in mind, we can look at some approaches to tracing the developmental effects of genes on behaviour. Ultimately, most effects will have their origin in the consequences of gene expression for cellular development and communication, but our ability to trace this will depend on the complexity of the embryological chain of events leading to the behaviour. We are more likely to be successful with relatively simple organisms and behaviours than we are with complex, advanced organisms. Moreover, as with much of the evidence for genetic differences in behaviour in Chapter 2, examples often rely on induced mutations or gross defects in behaviour as a result of naturally occurring mutation. Not all genetic effects work through complex embryology, however. In at least one well-studied example – rhythmic behaviour (3.4.1) – effects on behaviour appear to be set by the pattern of protein production during gene transcription, though exactly how these events at the cellular level translate into the behaviour patterns of organisms is still uncertain (Box 5.3).

5.2.1.1   Sensory and motor effects

Perhaps the most direct impact of genes on behaviour is via the development of the animal's sensory and motor systems. Not surprisingly, therefore, many studies have focused on the physical developmental effects of genes. We shall look at four examples.

Locomotion in Paramecium

A series of natural and chemically induced single-gene mutations in *Paramecium aurelia* affect the ability of the organism to move backwards (see Fig. 3.3). Affected paramecia are known as *pawn* mutants because of the same constraint on the movement of their chess-piece namesake (Chang & Kung 1973). As a single-celled organism, *Paramecium* provides a good opportunity to identify the cellular events producing mutant behaviour. From careful electrophysiological studies, it turns out that *pawn* mutations affect the

Underlying theory

Box 5.3 Gene expression and rhythmic behaviour in fruit flies

In common with many other natural circadian and other oscillators (3.4.1), the control of rhythmic behaviour in *Drosophila* appears to depend on a feedback loop between gene transcription, producing rhythmic levels of 'clock' ribonucleic acids (RNAs), and the translation of RNAs into 'clock' proteins. The 'clock' proteins eventually inhibit the transcription of their own genes, perhaps by interfering with factors that activate transcription. As 'clock' RNA and protein levels subsequently fall, however, gene transcription becomes reactivated. In *Drosophila*, the mechanism depends on two 'clock' genes, *per* (*period* – see 2.3.2.1) and *tim* (*timeless*), which are rhythmically transcribed and translated into their proteins PER and TIM (Fig. (i)). The two proteins combine (dimerise) and enter the nucleus of the cell where they inhibit transcription of the *per* and *tim* genes by interfering with its activation by two other dimerised proteins, CLK and CYC. However, while the model in Fig. (i) accounts for many of the features of the system, there are some aspects, such as the mapping of the time course of the transcription–translation process onto the periodicity of rhythmic behaviours, that suggest the picture may be more complicated.

Figure (i) A model of the genetic control of a circadian oscillator in *Drosophila*. See above. After Lakin-Thomas (2000).

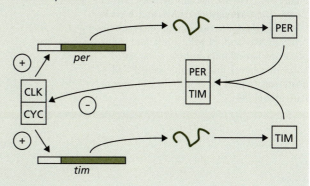

Based on discussion in Lakin-Thomas (2000).

voltage-sensitive membrane that bears the organism's cilia (see 3.1.2.1). However, different mutations do this in different ways. The first *pawn* mutants showed a deficiency in calcium cation (Ca²⁺) conduction in the ciliated membrane. If the membrane is disrupted using detergent, the organism can swim backwards provided enough Ca²⁺ and adenosine triphosphate (ATP, a source of energy for cellular processes) is added to the medium. The mutation is thus specific to the membrane and does not impair the cilia themselves. In other cases, *pawn* mutants turned out to be defective in potassium (K⁺), rather than calcium, ions, so while the phenotypic effect at the level of membrane conductance is similar, the underlying cause is different. In other cases, mutant *pawn* behaviour is conditional on temperature for its expression. In one such strain, paramecia back away from toxic salt solution normally at 23 °C, but fail to back away at 35 °C. Most spontaneously arising temperature-sensitive *pawn* mutations occur at the same loci as temperature-independent mutations, with at least three independently segregating loci appearing to be involved. In all cases, however, it is conduction physiology in the membrane that is affected.

Touch-insensitivity in nematodes

Pawn mutants in *Paramecium* show that, even in a single-celled organism, the same mutant behavioural phenotype can have a variety of underlying causes. This becomes even more evident when we look at multicellular organisms, such as the much-studied nematode *Caenorhabditis elegans*.

Caenorhabditis elegans offers rich opportunities for tracing behavioural changes to their roots in cellular development. It is transparent throughout its brief (3–4 days) life cycle, develops through well-defined, easily traceable cell lineages, and produces an adult with a known number (959) and arrangement of somatic cells, 302 of which are neurons. A wide range of mutations has been induced in *C. elegans* over the years, many affecting behaviour, and it has been possible, using the worm's unusually tractable development, to relate some of these to specific changes in cell development. A group of mutants known as *touch-insensitive* provides a nice example (Chalfie & Sulston 1981).

Touch-insensitive mutations have a very specific effect on behaviour. While affected worms can move normally and respond to a range of mechanical stimuli, they fail to respond to gentle mechanical stimulation, such as being touched with a hair. It turns

Figure 5.4 The six mechanosensory cells in *Caenorhabditis elegans*. ALML and AMLR – left and right anterior lateral microtubule cells; PLML and PLMR – left and right posterior lateral microtubule cells; AVM – anterior ventral microtubule cell; PVM – posterior ventral microtubule cell. From Partridge (1983).

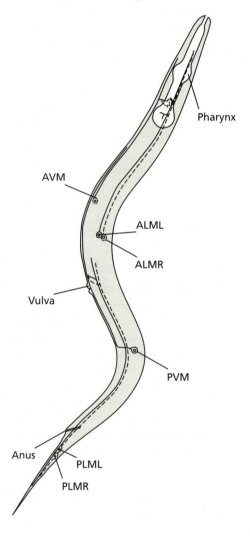

Figure 5.5 Schematic effect of various *touch-insensitive* mutations on the mechanosensory cells of *Caenorhabditis elegans*: (a) normal wild-type cell with cell body, cell process, microtubules and mantle; (b) the *mec-7* mutant, lacking microtubules; (c) *mec-1*, lacking a mantle; (d) *unc-86*, in which no cell body develops; (e) *e1611*, in which the cell body degenerates once formed; (f) *mec-3*, lacking the cell process. After Partridge (1983). (g) Mutations affecting nociceptive neurons seem to mediate social (right) versus solitary (left) feeding in *C. elegans*. See text. Photographs courtesy of Mario de Bono.

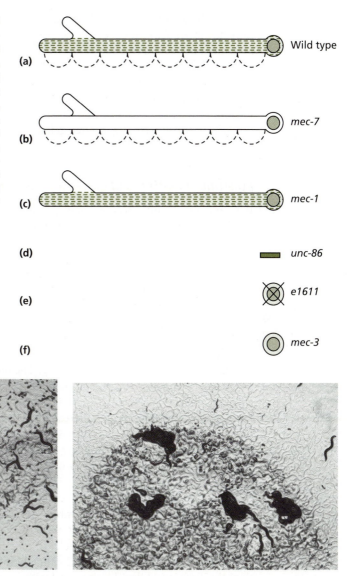

out that the effect is brought about by a variety of abnormalities in the development of specialised touch receptor neurons (Fig. 5.4). There are six touch receptors, each with a long cell process extending anteriorly and cemented to the hypodermis by a mantle. The cells are characterised by unusually large microtubules which appear to be important in impulse conduction. *Touch-insensitive* mutations occur at 12 different loci, seven of which have identifiable effects on the touch receptor cells (Fig. 5.5). In some cases, the cells develop normally in terms of producing a cell body and extended process, but show abnormal function owing to the absence of microtubules or a mantle (Fig. 5.5b,c). In others, growth or development of the receptor cell itself is impaired, so that cells fail to be produced at cell division, are produced but immediately atrophy, or fail to develop a cell process (Fig. 5.5d–f).

The cellular consequences of several other mutations affecting behaviour in *C. elegans* have now been mapped to specific neurons. A striking recent example involves mutations affecting nociceptive (noxious chemical-detecting) neurons that seem to mediate social versus solitary feeding in the worm (de Bono *et al.* 2002; de Bono 2003; Fig. 5.5g). In an organism cellularly and genetically as well characterised as *C. elegans*, it is therefore possible to use mutations to unravel some of the fundamental processes in the development of the nervous system and the control of behavioural responses. Even here, though, the picture is still partial and often reliant on gross abnormalities to provide a lead. As we move to more complex organisms, the leads become ever less direct.

Locating the effects of genes in Drosophila

If a complex machine becomes faulty, one way to find out what is wrong is to replace different components until the source of the problem is located. This is easy enough with an engine or a hi-fi system, but what about an organism? Can we selectively replace components here? The answer, surprisingly, is yes – at least in a sense. Clearly we cannot remove and replace components as we might a spark plug or a bulb, but we can use an ingenious genetic technique to achieve an analogous effect. The approach was pioneered by Seymour Benzer in the 1970s and involves individuals that are **genetic mosaics**.

Mosaics are organisms whose bodies are made up of more than one genotype. Mosaics are known, or can be generated, in a number of species, including mammals. Benzer and his coworkers used the fruit fly *Drosophila melanogaster*, a species in which genetic mosaics can be generated easily in the laboratory. The mechanism Benzer used was the loss of an abnormal, ring-shaped X chromosome during cell division in early embryogenesis. Because chromosomal sex determination in flies is similar to that in mammals, in that females are XX and males XY (though there are some important differences in the chromosomal sex determination of the two groups), females carrying a ring-X chromosome can end up with some of their tissues being female (XX) and some male (X0), depending on which cells lose the ring-X chromosome during development. Such female/male mosaics are known as **gynandromorphs**.

Since, if it is lost (roughly a 35% chance), the ring-X chromosome disappears early in development, roughly equal numbers of XX and X0 nuclei are produced. Migration of nuclei to the surface of the egg to form the *blastula* (a monolayer of cells surrounding the yolk), and the tendency for nuclei to remain near their neighbours, leads to broad areas of the blastoderm (the surface of the blastula) being male or female. Furthermore, since the direction of first division is arbitrary in relation to the egg, there are innumerable ways in which the blastoderm can be divided into male and female parts, so that adult gynandromorphs can end up with a variety of arrangements of male and female tissue (Fig. 5.6a). In general, the division into male and female parts follows the intersegmental (see 3.1.2.2) boundaries and the longitudinal midline of the exoskeleton (Fig. 5.6b). This is because each part forms independently during metamorphosis from a specific region (imaginal disc) in the larva, which is itself derived from a particular area of the blastoderm (see Fig. 5.7b). The final upshot provides a golden opportunity to locate the effects of mutant genes on behaviour, as Benzer and his team realised.

In order to trace which parts of a gynandromorph fly were normal (female) and which mutant (male), Benzer used marker mutations which imparted a different colour to the mutant tissue (e.g. yellow body, white eye; Benzer 1973). By crossing flies bearing surface markers with those showing different sex-linked (carried on a sex chromosome) mutant behaviours, Benzer obtained flies with X chromosomes carrying *both* surface marker and

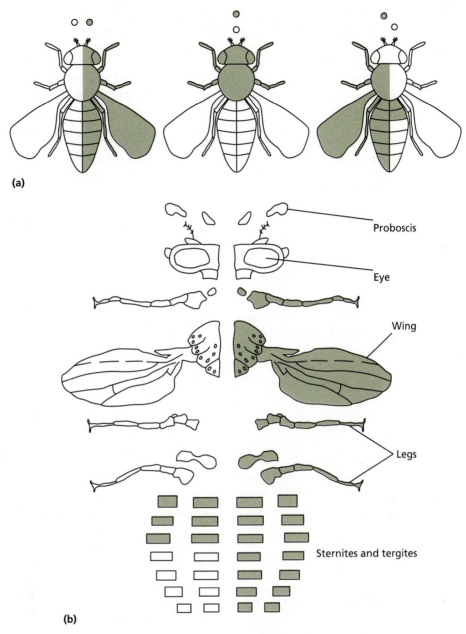

Figure 5.6 (a) Depending on the orientation of single (●) and double (○) X chromosome nuclei at their first division, gynandromorph *Drosophila* develop with different distributions of male (shaded) and female (unshaded) tissue. (b) Male and female tissues tend to follow lines of division between discrete body parts. After Benzer (1973).

behavioural mutations. Males with a mutation-laden X chromosome were then mated with ring-X bearing females. Among the offspring of such matings will be some in which the ring-X chromosome carries normal (wild-type) genes and the normal X chromosome mutant genes. Any cells of these individuals in which the ring-X chromosome is subsequently lost will develop into mutant body parts because the recessive mutations on the remaining X chromosome are no longer masked by their dominant counterparts. Finding out which part of a fly's body needs to be mutant in order to express a behavioural mutation then becomes a relatively simple matter of inspecting the distribution of mutant tissue and testing for the behaviour.

Hotta & Benzer (1969, 1970) used the approach to investigate different mutant behaviours in *D. melanogaster*. In some cases, direct associations emerged between behaviour and particular components of the body. For example, in the strain *nonphototactic*, in which the ability to respond to directional sources of light was impaired, flies showed the mutant behaviour only if one or both eyes had mutant surface characteristics. Subsequent investigation showed that mutant eyes had a defective electroretinogram (pattern of electrical activity in the retina). The cause of the *nonphototactic* behaviour was therefore local impairment of the key sense organ, rather than, say, an enzyme deficiency or abnormal brain development. In many other cases, associations were far less direct and more sophisticated methods were needed to trace the site of effect. Hotta & Benzer used one of these, a technique known as **fate mapping**, but another way is to use internal tissue **marker enzymes**, such as acid phosphatase or succinate dehydrogenase, to distinguish mutant tissue directly (Kankel & Hall 1976; Lawrence 1981; Strauss *et al.* 1992). Sometimes the two techniques can be used together (Hall 1977, 1979).

Hotta & Benzer used fate mapping to trace the cause of the mutant behaviour *hyperkinetic*, in which flies react to anaesthesia by shaking their legs instead of lying still. Gynandromorph *hyperkinetics* shake only some of their legs. While there is a good correlation between the surface genotype of a leg and whether or not it shakes, the correlation is not perfect, suggesting the abnormal behaviour is caused by changes close to the leg rather than in the leg itself, the most likely candidate being the nearby CNS. Fate mapping allows the effect to be located more precisely by using a form of three-dimensional triangulation. By taking a series of 'landmarks', such as the antennae, bristles on the body surface and the legs themselves, and doing sufficient crosses to generate a wide range of mutant/wild-type tissue distributions, it is possible to work out the number of times mutant behaviour co-occurs with a mutant surface genotype at each landmark. The more often they co-occur, the closer together their origins must be in the blastoderm. The number of individuals in which a given landmark and the behaviour are both mutant or both wild type, expressed as a ratio of the total number of individuals in the sample, thus gives the relative distance between the landmark and the site of effect on behaviour in the blastoderm (Fig. 5.7a). The distance is measured in 'sturts', after the geneticist A.H. Sturtevant who pioneered the technique of fate mapping. 'Triangulating' the calculated distances between different landmarks and the site of effect on behaviour suggests a location for the latter which can then be tested experimentally (Fig. 5.7b). When Hotta & Benzer did this for *hyperkinetic* behaviour, they discovered that each leg had an independent shaking focus. The focus was close to the affected leg, as indicated by the correlation with surface genotype, but below it in a region of the blastoderm giving rise to the ventral nerve cord and ganglia (Fig. 5.7a,b). Electrophysiological studies of *hyperkinetic* flies later revealed abnormal activity in neurons of the thoracic ganglion, which is important in coordinating the movement of the legs.

Hyperkinetic provides a nice illustration of the fate mapping technique. However, it is unquestionably an aberrant response with little interest from a functional point of view. Can we use the approach to probe behaviours of more consequence in the life of normal flies? Happily, the answer is yes. Hall and coworkers, for example, used a combination of fate mapping and enzyme tissue markers to see which parts of gynandromorph flies had to be male for them to show the 'following' and 'wing vibration' phases of

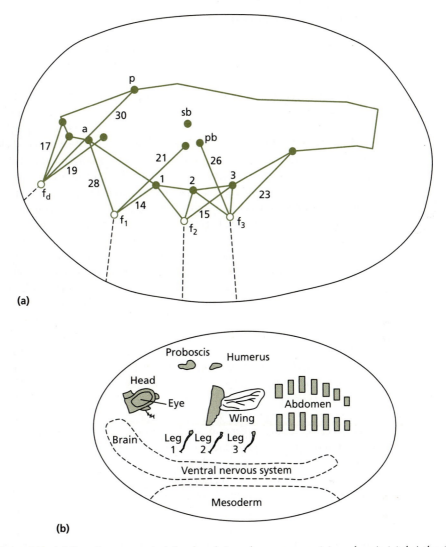

Figure 5.7 (a) By using anatomical 'landmarks', such as antennae (a), proboscis (p), bristles (sb and pb) and the legs themselves (1–3 each side), and seeing how often their genotype (mutant or wild type) coincided with the behaviour of the legs (mutant shaking or wild type nonshaking), Hotta & Benzer (1970) were able to calculate the distance of each landmark from the focus of effect (f_1–f_3) of the *hyperkinetic* mutation in *Drosophila melanogaster* (f_d shows the focus of effect for another behavioural mutant, *drop dead*). After Hotta & Benzer (1969). (b) Combining the distances of the different landmarks enabled the site of effect in *hyperkinetic* mutants to be mapped to a particular region of the blastoderm giving rise to the ventral nerve cord and ganglia (in the case of *drop dead*, the focus maps to the cerebral ganglia [brain]). From Goodenough *et al.* (1993) after Benzer (1973).

courtship (see Fig. 2.6). Hall (1977, 1979) found that tissue in the right or left side of the dorsal brain had to be male for the two behaviours to be expressed, but that additional male tissue was required in the thoracic ganglion for copulation to follow. However, courtship in gynadromorphs was always less vigorous than in normal males, suggesting that tissues elsewhere also need to be male for courtship activity to be normal.

Locating the effects of genes in mammals

Mosaic techniques have also been used to trace the developmental effects of genes on behaviour in mice. However, mutant and laboratory wild-type cells are much more finely mixed than in *Drosophila*, where sizeable regions of the body are one genotype or the other (Partridge 1983). As a result, effects often have to be sought on a finer scale and using behaviours where mutations have pronounced and easily recognised consequences. Locomotory behaviour is a good example (Partridge 1983).

Mutations producing aberrant locomotion mainly affect the cerebellum, the region of the hindbrain responsible for the fine coordination of movement (3.1.2.3). In normal mice, the cerebellum comprises several different cell types in a characteristic arrangement. Ascending neurons forming the input to the cerebellum either (climbing fibres) connect directly with Purkinje cells, or (in the case of mossy fibres) terminate in so-called granule cells, which synapse with the Purkinje cells via a T-shaped axon known as the parallel fibre. Various of these cells are affected in mutants showing locomotory deficiencies. In *staggerer* mutants, for example, in which the animal has tremors and the gait is grossly abnormal, the granule cells are absent. They develop normally initially and establish connections with the Purkinje cells, but then they atrophy along with their mossy fibres. The Purkinje cells also have an abnormal appearance, being reduced in number and lacking the characteristic spiny processes on the end of their dendrites. The question is, does the *staggerer* condition arise because of the lack of granule cells, because of the abnormal Purkinje cells or for a combination of reasons? Mosaic analysis has helped to provide an answer.

In *staggerer*/normal mosaics, generated by mixing the cells of dissociated early embryos together, Mullen & Herrup (1979) found that the cerebellum had an appearance intermediate between normal and mutant. However, the effect was due almost entirely to a mix of Purkinje cell morphologies, some having normal dendrites, some having dendrites that lack processes (as in *staggerer* mice). The granule cell layer appeared almost normal. It thus appeared that the *staggerer* mutation was having its primary impact on the Purkinje cells, with the atrophy of granule cells being a secondary consequence. Rather like the MGI neurons in the crickets above, normal postsynaptic (Purkinje) cells seem to be necessary for the proper development of presynaptic (granule) cells in the mouse cerebellum.

Like the behavioural mutations in *C. elegans* and *Drosophila*, the effects of *staggerer* and other mutations in mice can be traced to critical components of the nervous system. While this is likely to be true for many behavioural mutations, the developmental chain of events can sometimes be surprising. One last example illustrates the point.

Visual development in Siamese cats   The mammalian visual system involves a complex neural link between the retina of each eye and the visual cortex in the brain. The link is shown for a cat in Fig. 5.8. The field of view received by the eyes is relayed by the optic nerve first to the lateral geniculate nucleus (LGN) on each side of the brain and then to the visual cortex. Information from each eye travels to both LGNs as a result of decussation (crossing over) of retinal ganglion cell axons in the optic nerve. The crucial point is that, at each stage, overlapping parts of the visual field in the two eyes are matched up (kept in register) so that an accurate field of view is represented in the visual cortex (Fig. 5.8).

In Siamese cats, and some other mammals with reduced melanin pigmentation, this visual wiring system is corrupted, with the result that representation of the visual field in the brain becomes partially out of register (Fig. 5.9). The problem arises because, instead of some retinal ganglion cell axons from each eye innervating the LGN on the same side of the brain (ipsilaterally), as in normal cats (Fig. 5.8), all the axons cross over to the

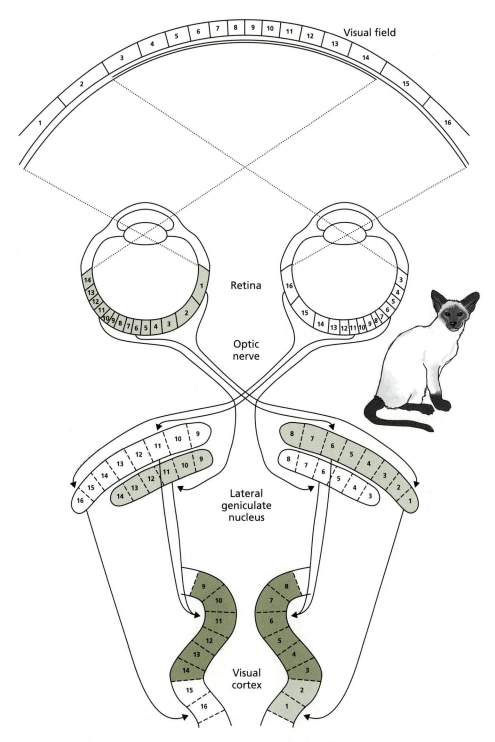

Figure 5.8 The visual circuitry in a normal cat, in which nerve fibres from each eye decussate to connect with the lateral geniculate nucleus (LGN) on each side of the brain. Segments of the visual field are kept in register in the LGN and the two sets of information are then brought together in the visual cortex. After Partridge (1983).

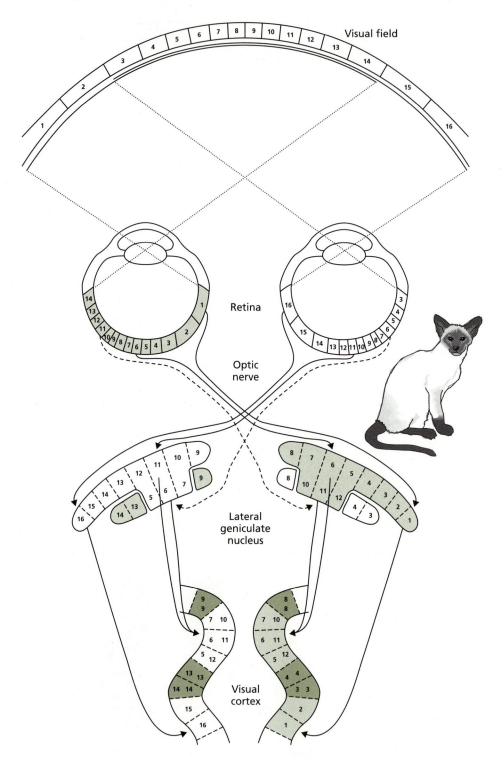

Figure 5.9 In Siamese cats, decussation is abnormal, resulting in complete cross-over of visual information from the eyes. As a result, part of the visual field in the LGN is out of register and confused information is passed to the visual cortex. After Partridge (1983).

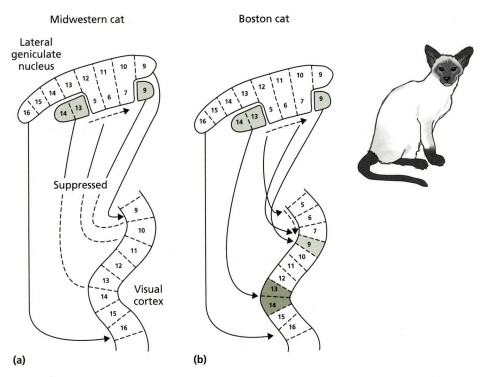

Figure 5.10 Two solutions adopted by Siamese cats to deal with the effect of abnormal decussation: (a) the 'midwestern cat' solution in which information that is out of register is suppressed, and (b) the 'Boston cat' solution, the abnormally arranged segments are reversed and inserted at a different point in the cortex. See text. After Partridge (1983).

LGN on the opposite side (contralaterally) (Fig. 5.9). Interestingly, Siamese cats appear to adopt two different approaches in trying to compensate for the resulting distortion to their visual field (Guillery *et al.* 1974). In one, the so-called 'midwestern cat' solution (Fig. 5.10a), the corrupted part of the field in the LGN is suppressed as information is passed back to the visual cortex. This means the cat loses part of its visual field, something that is easily confirmed in behavioural tests. The second, 'Boston cat', solution reorganises the projection to the cortex so that the segments of the visual field are unscrambled (Fig. 5.10b). One consequence of this, however, is that each side of the cortex receives input from a larger section of the visual field than usual (cf. Fig. 5.8). As in the 'midwestern cat', therefore, though for a different reason, binocular vision is once again disturbed and the cat shows characteristic deficiencies in visual spatial perception. This second solution also appears to be associated with the severe cross-eyes seen in some Siamese cats.

But what does all this tell us about the role of genes in visual development? The key lies in the pigment melanin. Siamese cats carry a temperature-sensitive allele of the gene that codes for tyrosinase, an enzyme important in the production of melanin. The pigment is thus produced only at temperatures below 37 °C in Siamese cats, hence its restriction to the body extremities, such as ears and paws. Abnormal decussation occurs in other albino mammals too, including rodents, mustelids (ferrets, mink, etc.), marsupials and primates (e.g. Giebel *et al.* 1991; Zhang & Hoffmann 1993; Guillery *et al.* 1999), but here it is sometimes not tyrosinase that is aberrant, but another enzyme involved in producing melanin. Thus it is the absence of the pigment, rather than the normal enzyme, that seems to be important. But why?

The answer appears to lie in the pigment epithelium, a layer of cells containing melanin that underlies the retina. The extent to which pigment-deficient mammals show abnormal decussation is closely associated with the amount of melanin in the pigment epithelium. It turns out that the retinal ganglion cells, which send axons back along the optic nerve to the brain, develop next to the pigment epithelium *after* melanin has been laid down. The prior deposition of melanin appears to be essential for the normal development of the ganglion cells and their axons, and thus the establishment of normal decussation. Any deficiency in laying down the pigment is reflected in the development of the visual wiring system, though the cellular processes involved are still unclear (Jeffery 1997).

The examples discussed above show that 'genes for' novel behavioural phenotypes can be traced to specific aspects of cell development and their consequences for sensory and motor function. The behaviours in question all arise through defects in sensory or motor systems caused by particular mutations. Similar associations between behaviour and anatomical/physiological changes can be identified in selected lines, another widely used method of generating genetic differences between individuals (see 2.3.2). In these cases, of course, animals are likely to differ at several loci, so genetic contributions to behaviour patterns are more diffuse and complex. For example, Frans Sluyter and coworkers selected lines of wild house mice (*Mus domesticus*) for either short (SAL) or long (LAL) attack latencies (time taken to initiate an attack). Although latency is only one measure of aggression, it correlated very closely with other measures, and was a good index of general aggressiveness. By means of reciprocal crosses between SAL and LAL lines, and mice identical in genetic background (congenic) but differing in selected chromosomal regions, Sluyter *et al.* (1996) were able to identify contributions of both the Y sex chromosome and autosomes (non-sex chromosomes) to aggressiveness in their two lines (see also 2.3.3.2 and Box 2.3). These hinged on the timing of testosterone surges around birth and at puberty (30–50 days of age), hormonal changes well known to affect aggression, and effects on dopaminergic pathways and hippocampal projections to the hypothalamus in the brain, regions known to be associated with aggression in laboratory strains.

5.2.1.2 Higher-order effects

As well as influencing sensory and motor capability, genetic changes can also affect behaviour via higher nervous system function, such as learning and emotionality. In the case of learning, this underlines the fact that effects of experience depend as much on genotype as any other phenotypic character (see Chapter 6); genes and learning cannot be treated as mutually exclusive influences.

Genes and learning

Perhaps not surprisingly, much of the evidence for developmental genetic effects on learning comes from laboratory species, particularly *Drosophila* and rodents, though significant advances have also been made in understanding the gene–enzyme pathways underpinning song learning in some birds (e.g. Denisenko-Nehrbass *et al.* 2000).

Several mutations affect learning in *Drosophila*. The best known occur at the so-called *dunce* locus, which is located on the X chromosome in *D. melanogaster*. *Dunce* mutants

Figure 5.11 *Drosophila melanogaster* of the *dunce* strain show reduced ability to learn and retain an associative olfactory task compared with wild-type (normal) flies. After Tully & Quinn (1985).

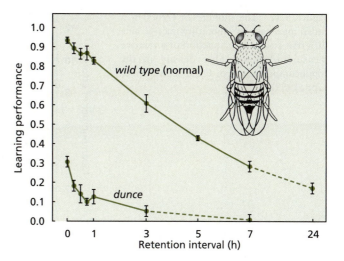

show poor retention of learned avoidance responses to odour cues. Flies can be trained to avoid particular odours by pairing them with electric shock, but while normal flies retain the response in the absence of shock for several hours, *dunce* mutants forget it very quickly (Tully & Quinn 1985; Fig. 5.11). They are also poor at other learning tasks, suggesting a general impairment of learning ability. But what causes this impairment?

Davis & Dauwalder (1991) looked at the biochemistry underlying learning in *D. melanogaster* and discovered that *dunce* mutants were deficient for a particular phosphodiesterase enzyme encoded at the *dunce* locus. Phosphodiesterase breaks down the compound cyclic AMP (adenosine monophosphate), which is important in the transmission of nerve impulses. Its defective activity in *dunce* mutants means that levels of cyclic AMP tend to build up rather than being metabolised, so impairing nerve cell function. Phosphodiesterase activity is normally high in parts of the brain where nerve cell connections and synaptic processes are concentrated, especially in the structures known as 'mushroom bodies' where information is believed to be processed and retained (Hoffmann 1994). Deficiencies here could thus account for the poor learning performance of *dunce* flies.

The biochemistry of neurotransmission has also been implicated in strain differences in learning ability in rodents (Will 1977; Peng *et al.* 1996). Strains of mice showing poor learning ability, such as C57Bl/6, have been found to have lower levels of the enzymes acetylcholinesterase (AChE) and choline acetyltransferase (ChA) in their temporal lobes than those that are good learners, such as DBA/2 and SEC/Re1. The two enzymes modulate activity in cholinergic pathways (those involving the neurotransmitter acetylcholine [ACh]) in the brain, and ChA, among others, plays a role in habituation to novelty and suppressing exploratory behaviour during learning tasks. Similar differences in AChE levels have been found in strains of rat that are good (e.g. TMB strain) or poor (e.g. TMD strain) learners, but further experiments show once again why caution is needed in interpreting these differences. When lines were selected for AChE levels and compared in learning tasks, the *low* AChE line performed better than the high line, the opposite of what the original TMB/TMD strain difference would predict. One reason for these apparently conflicting results may be that it is the ratio of neurotransmitter (ACh) and enzyme that matters in regulating cholinergic pathways, rather than absolute levels of the enzyme itself (Hay 1985).

Figure 5.12 The number of errors in a radial maze task by laboratory mice of nine different strains as a function of a relative measure of mossy fibre density in the hippocampus. See text. From Schwegler & Lipp (1995).

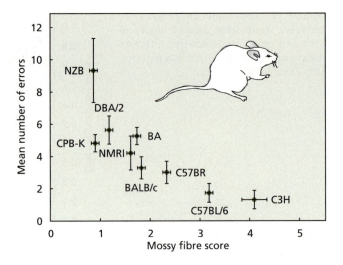

Some inbred strains of mice show pronounced differences in spatial learning ability. In these cases, performance in spatial maze tasks appears to be associated with the length and size of mossy fibre projections to the hippocampus. Schwegler & Lipp (1995) found that the number of errors in a radial arm maze correlated closely with the size of these projections across nine strains of mice (Fig. 5.12). Experimental manipulations of the projections support a cause and effect interpretation of the correlation. In the strain DBA/2, for example, which has short mossy fibre projections and poor spatial learning ability, thyroxin treatment causes an increase in the size of the projection field. When thyroxine-treated animals were tested in a radial maze, performance once again correlated with the size of the (now manipulated) projection field (Schwegler & Lipp 1995). Transgenic experiments (see 2.3.2.1) with other strains have also been able to link poor maze performance with neurological changes. Rousse *et al.* (1997), for instance, genetically transformed mice of the B6C/3F1 strain to produce individuals with defective brain glucocorticoid (steroid hormone) receptors. Glucocorticoids play an important role in memory retention, and transformed B6C/3F1 mice performed poorly compared with the untransformed strain in both water maze and radial maze learning tasks.

Such genetic manipulations in rodents are beginning to shed light on the cellular basis of memory. Work by Stephen Rose and his group at the Open University suggests that the conversion of learned experiences into long-term memory (Chapter 6) depends in part on cell adhesion molecules which 'stick' neurons in the brain together to form lasting connections (Rose 1998). Initial work with domestic chicks showed that blocking the action of these molecules impaired retention of avoidance learning tasks. Mice in which the genes coding for various cell adhesion molecules have been 'knocked out' also show a range of learning and memory deficits depending on which family of adhesion molecules they lack. One molecule in particular, amyloid precursor protein (APP), may provide insights into the development of Alzheimer's disease in humans. Alzheimer's disease results in degenerative memory loss, and is caused by plaques of insoluble beta amyloid protein building up between neurons and interfering with communication between cells. The beta amyloid fragments are formed by biochemical errors in the breakdown of APP. Mice engineered to express the mutant human APP molecule show many of the pathological symptoms of Alzheimer's as they age. Intriguingly, on the other side of the coin, there is suggestive evidence that effects of steroid hormones and so-called 'cognitive enhancer' drugs on learning and memory performance also work via cell adhesion molecules (Rose 1998).

Genes and emotionality

A rapidly developing body of evidence suggests that complex personality and emotional traits in humans are associated with identifiable genetic influences (see Plomin *et al.* 1994; Hamer & Copeland 1998). Individual differences in behavioural predispositions in people, such as aggressiveness and responses to stress, tend to be stable and largely heritable (Costa & McCrae 1988; Loehlin 1992), implying a strong genetic element. Given their role in behavioural differences associated with mutations and inbred strains (see above), neurotransmitters in the brain seem to be a logical focus for investigation here. In particular, increasing evidence suggests that the monoamine neurotransmitter serotonin (5-HT) plays a major role in modulating emotional characteristics in both humans and other animals (Westerberg *et al.* 1996), largely through its regulation by the serotonin transporter 5-HTT. 5-HTT thus plays a pivotal role in regulating neurotransmission throughout the brain, including the cortical and limbic regions important in emotional experience. It also has an important influence on the physical development of the brain, so that differences in 5-HTT function could have profound and lasting effects on personality traits. Recent work by Dean Hamer and his group at NIH in Maryland, suggests that indeed it does, and that differences in personality traits are associated with alleles linked to the 5-HTT coding locus on chromosome 17 (Esterling *et al.* 1997).

The alleles in question are functional length variations of the so-called 5-HTT-linked-polymorphic region (or 5-HTTLPR) which occur in two common forms: *long* and *short*. The dominant *short* allele is associated with reduced 5-HTT gene transcription and thus 5-HT uptake. Using established personality questionnaires, Hamer's team has shown that men or women carrying the *short* allele (either homozygously or heterozygously) are significantly more neurotic (prone to anxiety, hostility and depression), less agreeable and shier than those homozygous for the *long* allele (Lesch *et al.* 1996; Greenberg *et al.* 2000; Osher *et al.* 2000; Fig. 5.13). Animal models confirm aspects of these associations. Similar *long* and *short* 5-HTTLPR alleles are known in rhesus monkeys (*Macaca mulatta*), for example, where the short allele is associated with greater behavioural reactivity to stress (Champoux *et al.* 1999). In addition, experimental disruption of 5-HTT activity in mice enhances anxiety-related responses in light–dark boxes and maze procedures designed to impose stress (Wichems *et al.* 1998).

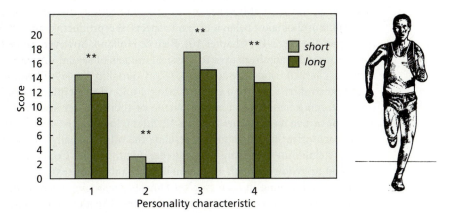

Figure 5.13 Mean scores for various personality traits in people with either *short* or *long* alleles of the serotonin transporter regulatory region (5-HTTP) gene. Subjects carrying the short allele are generally more anxious and depressed, traits that may be related to differences in the activity of the neurotransmitter serotonin. 1, harm avoidance; 2, shyness; 3, anxiety; 4, depression. See text. ** Difference significant at the 1% level. Plotted from data in Osher *et al.* (2000).

5.3   Genomes, development and behaviour

So far, we have discussed the role of genes in behavioural development in terms of genetic determinism and what individual genes do at a biochemical or cellular level to influence behaviour. The expression of individual genes, however, and their consequences for the viability and reproductive success of their bearer individual, frequently depend on other genes with which they share their genome (see also 2.3.1.1 and 2.4.3.2), and, in some cases, the parental genome from which they have been inherited. Such conditional gene expression plays an important role in several aspects of behaviour.

5.3.1   Intergenomic conflict and behaviour

Given genes may express themselves differently in different individuals, not just through variation in the genetic company they keep, but because of fundamental asymmetries in the reproductive interests of different classes of individual. The reproductive strategies of males are very different from those of females, parental strategies differ from those of offspring and so on. But at different times the same allele may find itself in each of these different kinds of bearer.

In sexually reproducing species, recombination and outcrossing mean that alleles at any given locus have only a transient association with those occupying other loci. As a result, natural selection is likely to act quasi-independently at each locus, leading to potential conflicts of interest and antagonistic coevolution (see 2.4.4.3 and Box 2.6) between loci (Rice & Holland 1997). We have already encountered *intra*genomic conflict among loci in the context of outlaw genes and modifiers (2.4.3.2), but conflict might also arise between genomes, expressing itself as conflict between individuals. Rice & Holland (1997) see such intergenomic conflict driving a form of intraspecific arms race (Box 2.6) that has a profound effect on rates of genetic change and accompanying behavioural phenotypes. Conflict between the sexes is a good example.

As we shall see in more detail in Chapter 10, the sexes are likely to disagree over investment in reproduction. With some exceptions, males seek to maximise the number of fertilisations, while females concentrate investment in individual offspring. In the traditional view, this basic difference drives sexual selection, thereby producing many of the different morphologies, behaviours and aspects of reproductive physiology that typify the two sexes (see Fig. 2.10; Chapter 10). Rice & Holland, however, argue that intergenomic conflict provides an alternative to classical models of sexual selection as an explanation for these characteristics (see also 10.1.3.2). Courtship displays illustrate the point.

While conflict over the rate of mating could in principle be resolved by coercion (males forcing females to mate by virtue of their greater strength), this seems to be absent or rare in most species (but see 12.2.1.1). Instead, the sex (usually male) investing less in each offspring attempts to persuade the other (usually female) to mate, often by means of elaborate courtship displays (e.g. Figure 1.13). Courtship has variously been suggested to evolve through all the main mechanisms of sexual selection (Chapter 10), the outcome hingeing in each case on females *preferring* (even if sometimes passively) the displays of certain males over those of others. The idea of an intergenomic arms race, however, suggests a different interpretation.

Taking sensory exploitation (the tuning of signals to pre-existing sensory capabilities in the receiver – see 3.5.1.1 and 10.1.3.2), and the assumption that males and females have different optimum mating rates (see above), as their starting points, Rice & Holland envisage an arms race developing between male persuasion and female resistance. Alleles coding for signals that stimulate female sensory predispositions, and thus cause females to mate at a rate beyond their optimum, are favoured in male genomes, while alleles that modify the female's sensory system to resist such persuasion are favoured in female genomes. While a sexual selection model of courtship signalling predicts a mutually escalating coevolution of male trait and female preference, the intergenomic arms race model predicts the opposite; as male signals evolve to manipulate females, females become *less* responsive to them. Is there any evidence for this kind of persuasion/resistance arms race? The swordtail fish studied by Alexandra Basolo (see 3.5.1.1) suggest there is.

Basolo's studies of poecilid fish (*Xiphophorus* and *Priapella* species) suggested that female preference for males with exaggerated extensions to their caudal fins ('swordtails' – see 3.5.1.1) originated as an incidental consequence of visual sensory bias. The remarkable existence of the preference in species lacking fin extensions altogether (3.5.1.1) offers strong support for this idea. But does this sensory exploitation push females beyond their optimal mating rate? If it does, then female sensitivity, and thus responsiveness, to the stimulating male trait should *diminish* after the trait has evolved. Evidence from Basolo's studies suggest this has indeed happened. When she compared the strength of preference for sworded males in *Xiphophorus* and *Priapella* females, she found that the preference was some three times weaker in the sworded *Xiphophorus* than in the swordless *Priapella* (Basolo 1998; Fig. 5.14). Studies of other swordtail species, in which females show cross-species preferences for larger males, show a similar reduction in preference once the preferred trait (here large size) has become exaggerated (Morris *et al.* 1996).

Intergenomic conflict between the sexes may explain some fundamental features of sexual development, for example why, in mammals, its hormonal control by androgens and oestrogens is not completely differentiated by sex. Manning *et al.* (2000) have speculated on the problem in humans. On the face of it, it would pay males to have sons that developed in a high androgen uterine environment, because high androgen levels boost development of the testes and sperm production, and the differentiation and maintenance of the male vascular system. High androgen levels are also associated with good spatial ability, cognitive skills and various measures of metabolic competence, such as good blood glucose control (Manning *et al.* 2000). Sons with such attributes would thus be healthy, skilled and highly fertile. By the same token, prenatal oestrogen is important in females for differentiation of the ovaries and breasts, while low androgen may reduce the risk of cardiovascular disease, diabetes and other metabolic disorders. So why are both sexes exposed to a mix of hormones during gestation? The answer may lie in the frequency-dependent reproductive costs of investing hormonally in one or other sex. If fathers bias the intra-uterine environment in favour of sons, they stand to reduce the fertility of their daughters, and vice versa for mothers. High androgen-producing fathers thus experience greater variance in reproductive success and females will be selected to mate polygynously (several with each of the best males, none with the poorest males). As polygyny spreads, however, the variance in male reproductive success declines until eventually it pays females to mate monogamously. These conditions then favour a high oestrogen uterine environment and the pendulum swings back in the opposite direction. Alternating cycles of high prenatal androgen and high prenatal oestrogen may thus evolve, driven by the conflict of interest between the sexes and the fact that genes having harmful effects in one sex are maintained in the population by their beneficial effects in the other. Apparently deleterious hormonal effects in each sex are maintained because the genes coding for them experience a fitness advantage when in the other sex (Manning *et al.* 2000).

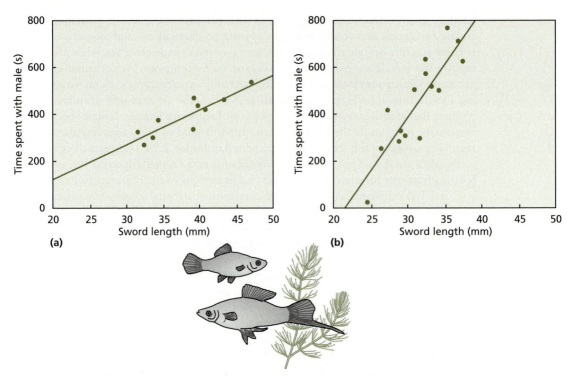

Figure 5.14 The strength of preference for conspecific males with different-sized artificial 'sword' extensions to their tail by female *Xiphophorus helleri* (a) and *Priapella olmeceae* (b). Preference for longer 'swords' is stronger in *P. olmeceae*, in which males do not normally possess 'swords'. The difference in strength of preference between the species may be the result of an intraspecific arms race between male persuasion and female resistance in *X. helleri*. See text. After Basolo (1998).

Rice & Holland discuss other contexts in which intergenomic conflict appears to influence the conditional expression of genes and the physiological and behavioural phenotypes for which they code. The point underlying all of them, however, is that genetic influences on behaviour are subject to conflicting selection pressures, not just within individuals as a result of life history constraints, but because their temporary residence in different individuals leads to a range of phenotypic demands. Development, like mechanism, cannot be divorced from its evolutionary context.

5.3.2 Genomic imprinting

Conflict between the sexes can lead not only to **sex-limited** expression, where genes are silenced or expressed differently depending on the sex of the bearer (note that this has nothing to do with **sex-*linked*** characters, which relate to genes on the sex chromosomes, but to the silencing of one allele of a gene according to its parental origin – i.e. whether it has been inherited from the animal's mother or its father [Barlow 1995]). This is often referred to as **genomic imprinting**. The developmental mechanisms behind genomic imprinting are not fully understood, but it is known that it occurs during the production of gametes (sperm and eggs). This selective silencing means that some traits effectively pass down only the maternal line and some down only the paternal line, unlike normal biparental inheritance where gene expression is indifferent to the parent of origin. Genomic imprinting is particularly prevalent in placental mammals, where parental conflict over investment in the foetus is focused on the period of growth in the uterus (Moore & Haig 1991). Generally, paternal genes would be expected to increase, and maternal genes to decrease, resource allocation to the foetus in the face of potential multiple paternity and developmental competition (Isles & Wilkinson 2000). As might be expected from this, imprinting is known to affect patterns of growth in mammals, but increasing evidence suggests it also has profound effects on behaviour.

Parent-of-origin effects on behaviour have been known for some time (Gray 1972), and careful experiments strongly suggest genomic imprinting has an important role to play. Isles *et al.* (2001) looked at parent-of-origin effects on urinary odour preferences (an important means of social communication [3.2.1.4]) in CBA/Ca and C57Bl/6 mice. To avoid any effects of family cues, Isles *et al.* used mice derived by embryo transfer to genetically unrelated foster mothers. They then tested F1 mice from reciprocal crosses to see whether they had any preference for urine from maternal or paternal strain females compared with urine from unrelated (BALB/c) controls. Both male and female F1 mice from CBA/Ca♀ × C57Bl/6♂ crosses showed a significant preference for control urine over urine from the maternal strain, but no preference for control urine over that from the paternal strain (Fig. 5.15a). Mice from the reciprocal cross (C57Bl/6♀ × CBA/Ca♂), however, showed exactly the same pattern with respect to parental strain, but the opposite preference with respect to genotype (Fig. 5.15b). Since the mice were able to distinguish urine from males of the two strains perfectly well, the lack of preference for

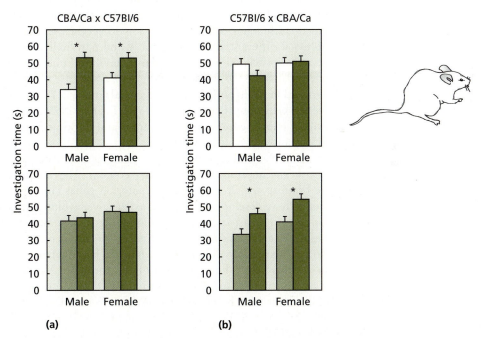

(a)   (b)

Figure 5.15 Urinary odour preferences of F1 males and females produced by reciprocal crosses of CBA/Ca and C57Bl/6 mice. Mice were tested with urine from females of each parental strain (open bars, CBA/Ca; pale fill bars, C57Bl/6) and from the non-parental BALB/c (dark fill bars) strain. (a) Males and females from CBA/Ca♀ × C57Bl/6♂ crosses preferred BALB/c female urine over CBA/Ca urine, but showed no preference when tested with BALB/c and C57Bl/6 urine. (b) Mice from CBA/Ca♂ × C57Bl/6♀ crosses, however, showed the opposite tendency. * Mean values differ significantly. The results suggest genomic imprinting has an influence in the development of urinary preferences. See text. After Isles *et al.* (2001) © Nature Publishing Group (http://www.nature.com/), reprinted by permission.

paternal strain urine was not due to any impairment of olfactory function. Neither was it due to odour characteristics imparted by the genetic parent – BALB/c mice could distinguish between the odours of the two parental strains, but not between those of the reciprocal F1 animals. Isles *et al.*'s findings thus strongly suggested an imprinting effect leading to differences in the functioning of the olfactory system in the reciprocal F1 mice, either via perception or during higher-level processing.

5.3.2.1 Genomic imprinting and brain development

Studies of imprinting effects on brain development imply that consequences for behaviour are likely to be fundamental and profound. The most striking examples come from studies of development in mice that possess only maternal (parthenogenetic) or paternal (androgenetic) chromosomes. While strictly uniparental mice are not viable, and die at around 7–10 days gestation, it is possible to produce chimaeric (mosaic) animals comprising a mixture of uniparental and normal cell types. Azim Surani and Barry Keverne at the University of Cambridge have exploited this to look at brain development in mice with different combinations of maternal-only/wild-type or paternal-only/wild-type tissue (Allen *et al.* 1995; Keverne *et al.* 1996a). Using a genetic marker to identify the different tissue, they found that cells with maternally and paternally expressed genes had distinctly different distributions in the brain. Cells that expressed maternal genes were concentrated in the cortex and striatum, but not in the hypothalamus, septum or preoptic areas, whereas cells expressing paternal genes had the opposite distribution. Using 'knockout' genetic techniques, Surani and his coworkers then set about isolating genes that were expressed differentially in maternally and paternally derived tissue to see whether these were associated with particular behavioural phenotypes. A number of such genes emerged (Isles & Wilkinson 2000). *Peg1* and *Peg3*, for example, which are paternally expressed in the hypothalamus and amygdala (see 3.1), appear to have major effects on maternal behaviour (Lefebvre *et al.* 1998; Li *et al.* 1999), while others, such as *Ube3a*, maternally expressed in the hippocampus and Purkinje fibres, affect learning and motor performance (Jiang *et al.* 1998).

Keverne *et al.* (1996b) have looked at the distribution of maternal and paternal gene expression in the brain in an evolutionary context. They note that, in primates, regions of the brain to which maternal and paternal genes contribute differentially through imprinting have developed to different extents over evolutionary time. The major expansion has been in areas, such as the forebrain neocortex and striatum, in which maternal genes make a substantial contribution to development. In contrast, areas in which paternal gene expression dominates, such as the hypothalamus and septum, have contracted in relative size. The expansion of the forebrain has been important in the evolution of the complex social organisation and decision-making found in many primates, and its domination by maternal genes may be significant in that the maintenance of group cohesion, within and between generations, is heavily dependent on the female line (high-ranking females tend to produce high-ranking daughters that stay in the group [Keverne *et al.* 1996b]). The regions of relative contraction in the brain are those that are target areas for gonadal hormones, and concerned with sexual and parental behaviour. Their regression is consistent with the reduced role of these hormones in regulating primate reproductive behaviour. Keverne *et al.* (1996b) characterise these changes in terms of a shifting balance between the 'executive' and 'emotional' components of the brain as primates have evolved towards greater higher-order control in decision-making (Fig. 5.16).

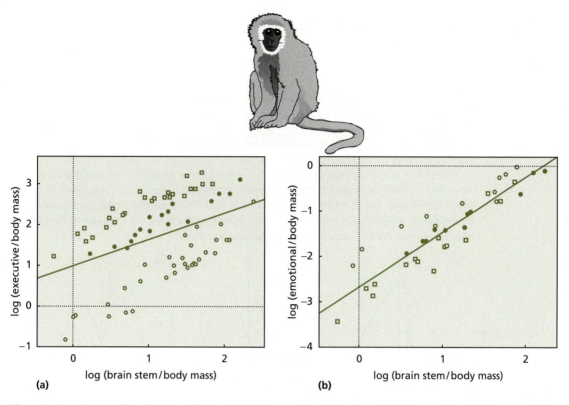

Figure 5.16 Maternally and paternally imprinted genes make different contributions to the development of the brain in primates. (a) Structures in which the maternal genome has a substantial influence (the 'executive' brain – see text) correlate positively with taxonomic groupings based on brain : body mass ratio, and increase from insectivores (○) through prosimians (●) to the more advanced simians (□). (b) Structures associated with the paternal genome (the 'emotional' brain) also correlate positively with brain : body mass ratio, but show the opposite trend across taxonomic groups. See text. After Keverne *et al.* (1996b).

5.3.3 Developmental instability and behaviour

We might imagine that a given genotype being expressed under constant environmental conditions would always produce the same developmental outcome. This is often not the case. For reasons that are only partly understood, certain combinations of genes can undermine the stability of developmental processes and produce randomly varying phenotypes. One symptom of this can be random departures in the symmetry of morphological features, such as appendages, plumage markings or sense organs. In bilaterally symmetrical organisms, these can be measured as the difference between right and left sides in the length, area or other measure of paired features. Such random variation in symmetry is known as **fluctuating asymmetry (FA)** (van Valen 1962; Palmer & Strobeck 1986; Møller & Swaddle 1997), and is distinguished from other forms of asymmetry in a number of ways (Box 5.4). FA has become something of a bandwagon among evolutionary biologists in recent years, largely because, as a potential measure of genetic quality (at least in relation to the environment in which an individual has developed), it could provide a basis for mating preferences and therefore drive sexual selection (Chapter 10). While this is an area of conflicting evidence and lively disagreement (see e.g. Houle 1997; Palmer 1999;

Underlying theory

Box 5.4 Asymmetry and developmental instability

Fluctuating asymmetry (FA) has become widely regarded as a measure of 'phenodeviance', the departure of phenotypes from an assumed optimum (here perfect symmetry). As such it is a potential measure of developmental instability, reflecting the robustness of a given genotype (its ability to cope with nutritional stress, disease and other insults) in the environment in which its phenotype has developed. Two things should be carefully noted, however.

First, FA is a special case of asymmetry, which can be distinguished statistically from other forms in having an approximately normal distribution of *signed* right minus left, or vice versa, differences with a mean value of zero (Fig. (i)a). The two other common asymmetric distributions, *directional asymmetry* and *antisymmetry*, which are not considered to reflect phenodeviance, depart from normality in characteristic ways.

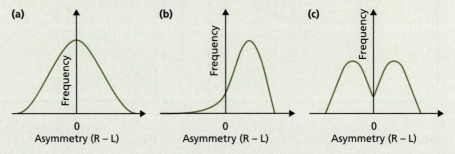

Figure (i) Different kinds of bilateral (right–left) asymmetry. (a) Fluctuating asymmetry, characterised by a normal distribution and a mean of zero. (b) Directional asymmetry, in which there is a consistent bias in favour of one side. (c) Antisymmetry, where there is a bias to one side, but whether left or right varies across the population. From *Asymmetry, Development Stability and Evolution* by A.P. Møller and J.P. Swaddle (1997). Reprinted by permission of Oxford University Press.

Directional asymmetry (Fig. (i)b) occurs when there is a tendency for one side of a trait to develop more than the other. The side that develops more is predictable so, at the population level, there is either a right-sided or left-sided bias (often referred to as a '**handedness**' bias). Examples of directional asymmetry include testis size in humans, the position of the eyes in adult flatfish and wing morphology in species of cricket that use them to stridulate (call). In antisymmetry, there is also a right or left bias, but the direction is not consistent. Antisymmetry typically shows either a flattened (platykurtic) or bimodal (Fig. (i)c) distribution. Examples include the large signalling claws of male fiddler crabs (*Uca* spp.), in which males start off with two large claws but usually end up losing one of them. The lost claw regenerates, but only in a small 'female-type' form. Since the chances of losing the right or left claw are more or less equal, there are roughly equal numbers of left-side large and right-side large males in the population.

The second point to note is that, although much attention has focused on FA as an index of developmental instability, it is only a particular example of putative phenodeviance. Phenodeviance can be measured in relation to any assumed standard, not just right–left symmetry. Recent approaches using fractal analysis (comparisons with repeated identical patterns) have applied the concept to a wide range of biological contexts, from the pattern of sutures in skulls and branching in blood vessels (Graham *et al.* 1993) to anti-predator behaviour in gazelles (Escós *et al.* 1995).

Based on discussion in Palmer & Strobeck (1986) and Møller & Swaddle (1997).

Simmons *et al.* 1999; Thornhill *et al.* 1999; Bjorksten *et al.* 2000; 10.1.3.2), the implications of developmental instability for behaviour, and other functional characteristics of organisms, extend well beyond mate choice and sexual selection.

5.3.3.1 Fluctuating asymmetry and behaviour

Perhaps the most wide-ranging study of FA and behaviour has been carried out by John Manning and his group at the University of Liverpool. Working mainly with human subjects, Manning has revealed associations between behavioural attributes and FA in various morphological features, particularly of the hands and face. On the physical performance side, for example, asymmetric male middle-distance runners recorded slower times over 800- and 1500-m races than their more symmetrical counterparts (Manning & Pickup 1998), a negative effect that also emerged for FA and handicap ratings (an inverse measure of performance) in race horses (Manning & Ockenden 1994). Starlings (*Sturnus vulgaris*) showing greater asymmetry in their primary feathers appear to be less manoeuverable in the air, colliding more often with obstacles in a flight aviary (Swaddle *et al.* 1996; Fig. 5.17). Other studies of locomotory performance in a range of species generally suggest a negative impact of FA, sometimes indicating increased susceptibility to injury or disease through correlations with lameness (Møller & Swaddle 1997). However, not all measures of performance correlate negatively with FA. Swaddle & Witter (1994), for instance, found that dominance in male starlings was *positively* related to stress-induced asymmetry in wing feathers, though this may have reflected the cost of being dominant rather than a pre-determinant of dominance.

Asymmetry also appears to predict more subtle aspects of behaviour, for example the tendency for left-side cradling (the head near the mother's left breast) of infants in women (Manning *et al.* 1997). Left-side cradling is a cross-cultural tendency in humans that is also shared by chimpanzees and gorillas. One explanation for it is that it facilitates the flow of auditory and visual information from the infant via the left

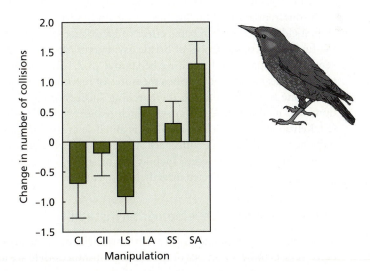

Figure 5.17 Starlings (*Sturnus vulgaris*) in which the left–right asymmetry of the primary feathers had been experimentally increased (long [LA] and short [SA] asymmetric columns) collided more frequently with poles suspended in an aviary than control birds (CI, CII) or those whose manipulated feathers were symmetrical (LS, SS). After Swaddle *et al.* (1996).

ear and eye to the emotional centres of the right hemisphere of the mother's brain. Occasionally, however, women cradle on the right-hand side. Manning *et al.* found that this tendency was significantly greater in women with asymmetric ears, an outcome they suggested might reflect instability in the development of the ear and disruption to normal left–right information flow. Other associations between FA and emotional well-being have emerged for depression and other psychiatric disorders (Lane *et al.* 1996; Martin *et al.* 1999).

These examples relate to correlations between morphological asymmetry and behaviour within individuals. However, asymmetry can also influence behaviour through the perceived quality of external stimuli. In Chapters 10 and 11, we shall see that symmetry in visual signals can be an important determinant of the response of receivers. For the most part, attention here has concentrated on sexually selected signals within species, but stimulus symmetry appears to have widespread effects on behavioural responses. Horridge (1996), for example, trained honey bees to use visual cues to find food in a Y-maze. Using patterns to mimic flowers, Horridge's experiments showed faster learning when patterns were bilaterally symmetrical about a vertical plane of symmetry. Interestingly, when given a choice between a bilaterally symmetrical pattern and the same pattern rotated through 90°, bees preferred the vertical bilateral pattern, but the preference developed during the experiment rather than being expressed at the outset. Giurfa *et al.* (1996) have demonstrated a similar learned preference for symmetry in bees.

5.3.3.2 Developmental instability and modifiers

Given that developmental instability appears to impact on several aspects of performance, we might expect genes that cause it to select for modifier effects at other loci (see 2.4.3.2) in order to correct the disruption. There is evidence that this has happened. Genes conferring insecticide resistance in sheep blowflies (*Lucilia cuprina*) provide an example (Clarke & McKenzie 1987). These genes, known as *Rop-1* and *Rmal*, affect pesticide resistance via carboxylesterase enzymes, which also happen to be instrumental in the differentiation of the bristles on the body surface of the fly. As a result, resistance to pesticides has the pleiotropic effect of increasing FA in the number of bristles. While the functional consequences of bristle asymmetry are not clear, bristles are important in sensory perception and resulting behaviour (3.2.1.3), and asymmetry in their number correlates negatively with various measures of reproductive potential, including development time to adulthood and the hatching success of eggs. Apparently in response, a dominant modifier gene has arisen which reduces this side-effect of the resistance genes (McKenzie *et al.* 1990). The modifier is an allele at the *Scl* locus and appears to interact with the resistance genes to affect cell adhesion properties, and thus the eventual phenotype of morphological characters such as the bristles (Lauder 1993), though whether the latter effect has been selected directly is not yet clear.

5.3.3.3 Genetic causes of developmental instability

While developmental impairment can arise through various environmental insults, there appear to be a number of genetic circumstances that are associated with it. Strong directional selection appears to be one of them, and may be instrumental in associations between elaborate secondary sexual characters, behaviour and FA (but see Simmons *et al.* 1999). We shall return to this in Chapter 10. Since some of the others are also associated with behavioural deficits, they are worth discussing briefly here.

Inbreeding

Inbreeding, and a consequently increased level of homozygosity, risks expression of recessive, and thus normally silenced, deleterious alleles. We might thus expect inbreeding to lead to developmental instability. Extensive studies of inbred laboratory strains of rodents and insects certainly suggest that inbreeding leads to greater FA (e.g. Mather 1953; Leamy & Atchley 1985), and there is similar evidence from populations of other species (Leary *et al.* 1985; Clarke *et al.* 1986). However, this may be only a temporary effect until differential mortality weeds out unstable genotypes, as may have happened, for example, in inbred populations of cheetah (*Acinonyx jubatus*) (Kieser & Groeneveld 1991). In addition, while increased homozygosity associated with inbreeding may correlate with greater FA, it is not clear that the opposite holds, i.e. that increased heterozygosity *reduces* FA (Møller & Swaddle 1997).

Hybridisation

Hybridisation, with its potential for disrupting co-adapted combinations of genes, has also been shown to correlate with increased FA (Tebb & Thoday 1954; Graham & Felley 1985; Ross & Robertson 1990). These effects are likely to reflect the dependence of symmetry on polygenic complexes, with hybridisation fragmenting some of these in much the same way as polygenically inherited behaviours (see Figs 2.17 and 5.3). Such effects may lead to a general association between FA and behavioural deficits in hybrids.

Mutation

While hybridisation may disrupt polygenic traits, it might also influence developmental stability via mutation rates, where, for example, new transposons (genetic elements that 'jump' between locations within genomes) are introduced into a genome that step up the rate of genetic change (Gvozdev 1986). Mutation is a third potential cause of developmental instability, for the obvious reason that it introduces new gene products into the developmental process. In some cases, such as the sheep blowfly example above, spontaneous mutations have been shown to affect stability (at least as indicated by FA), but transgenesis can have similar effects, though, not surprisingly, these often centre on the phenotypic characters influenced by the transferred gene. For example, by inserting a mutant gene for muscular dysgenesis (curtailing muscle development and causing skeletal degeneration) into mice, Atchley *et al.* (1984) produced an increase in FA in bilateral measures of mandibular development, but no increase relating to non-skeletal traits. This suggested that only particular developmental pathways were destabilised by the mutation and that there was not a generalised reduction in developmental stability.

Summary

1. Ultimately all behaviour develops through a combination of genetic and environmental influences. Behaviour is a function of the organism's physical structure and physiology and thus the products of gene expression. Gene expression, however, requires an appropriate biochemical environment in which to take place, and this in turn is affected by factors elsewhere within, and beyond, the organism.

2. Despite this, the idea that genes influence behaviour has a long history of controversy, of which the 'nature–nurture' debate between the ethologists and comparative psychologists is an enduring example. In part the controversy has been fuelled by terminological confusion, especially over the meanings of 'innateness' and 'instinct', but in part it is due to arguments based on naïve genetic determinism and ideological opposition to 'selfish gene' thinking.

3. While it is relatively straightforward to demonstrate a genetic influence on behaviour, identifying the developmental pathway by which it occurs is considerably less so. Nevertheless, it has been possible to identify the cellular, biochemical and other developmental routes by which identified genes affect behaviour in a wide range of taxonomic groups. Many of these affect the development of sensorimotor systems, but others affect higher-order aspects of behaviour such as learning and emotionality.

4. The expression and phenotypic consequences of individual genes can depend on other genes with which they share a genome. There may be intra- and intergenomic conflicts between loci which influence both developmental and evolutionary aspects of behaviour. Several characteristic features of sexual and parent–offspring conflict may be phenotypic reflections of such interlocus conflict. Gene expression, and resulting behavioural phenotypes, may also show parent-of-origin effects (genomic imprinting), where genes inherited from the mother or father are selectively silenced in different tissues. Genomic imprinting may have played a central role in the evolution of brain anatomy, cognition and social behaviour in primates.

5. Different combinations of genes may affect the stability of phenotypic development, leading to physical and behavioural impairments. Fluctuating asymmetry is one potential measure of developmental instability. Increased asymmetry has been associated with various behavioural deficits, from locomotion to sexual behaviour and parental care.

Further reading

Dawkins's (1995) little book provides a characteristically lucid summary of the problems surrounding innateness and instinct. Benzer (1973), Partridge (1983), Hay (1985) and Hoffmann (1994) usefully summarise some developmental aspects of behaviour genetics, and Weiner (1999) gives an engaging account of life and events in Benzer's laboratory. Hamer & Copeland (1998) look at the genetic underpinnings of cognitive and emotional aspects of human behaviour. Rice & Holland (1997) provide a brief but clear introduction to intergenomic conflict. Markow (1994) and Møller & Swaddle (1997) give good accounts of the functional/evolutionary implications of developmental instability and fluctuating asymmetry, and Surani et al. (1987) and Moore & Haig (1991) discuss some developmental consequences of genomic imprinting.

Maturation and learning

Introduction

The complex interaction between genes and environment in Chapter 5 continues to affect the development of behaviour throughout the life of the animal, partly through maturation of sensory, motor and other physical systems, and partly through learning. Can we tease these two routes apart? Are there changes in behaviour that simply reflect physical maturation? If so, how are they brought about? If changes are instead due to experience, what processes underlie them? Animals seem to learn in many different ways, but are they really different or just modifications of the same basic process? Does an ability to learn quickly, or to learn complex tasks, equate to intelligence? Animals vary in their ability to learn. Why? Is there any evidence that such differences reflect adaptive specialisation? Play behaviour is a puzzling feature of development in many species. Explanations have suggested various roles in maturation and learning, but is there any consensus view?

The functional movements we call behaviour are the product of intricately coordinated sensory and motor systems (Chapter 3). In Chapter 5, we saw how the development of these systems, and therefore behavior, is influenced by genes through the cellular and biochemical events triggered by gene expression. Development, however, is an ongoing process throughout the organism's life, with genes continuing to play a fundamental, shaping role. Because gene expression depends on environmental conditions, given genes give rise to different phenotypes according to prevailing resources and constraints. The development of behaviour is thus a complex and changeable interaction between organism and environment. This interaction is reflected in two ways: through maturation of the animal's morphology and physiology, and through learning.

6.1 Maturation and behaviour

When animals emerge into the world, their physical systems may not be fully developed. This is especially true of **altricial** species (such as passerine birds and many small mammals) where the young are born or hatch at a relatively early stage of development. Even in **precocial** species such as guinea pigs and gazelles, however, where the young emerge at a relatively advanced stage, much may still need to be done before the animal's physiology and behaviour achieve adult performance. Progress to maturity can affect behaviour in many ways.

6.1.1 Nervous systems, development and behaviour

For understandable reasons, developmental changes in the nervous system are often associated with changes in behaviour. Such associations are apparent even in the prenatal life of many species. Movement in embryonic Atlantic salmon (*Salmo salar*) is a good example (Abu-Gidieri 1966; Huntingford 1986; Fig. 6.1).

Movement at very early stages of development in salmon is limited to heart beats and twitches of the dorsal muscles. These are entirely muscular in origin (**myogenic**) and occur before the nervous system has formed. It is not until about half way through embryogenesis that the major motor systems differentiate in the spinal cord, and motor neurons establish contact with anterior muscles, allowing the embryo to flex its body. As neural connections extend down the sides of the body, the embryo becomes capable of the undulatory movements associated with swimming. Development of the sensory system and skin connections along the trunk follow soon after this, allowing responses to mechanical stimulation, while completion of neural circuits to the fins and jaws leads to their coordinated articulation and movement. Development of the nervous system

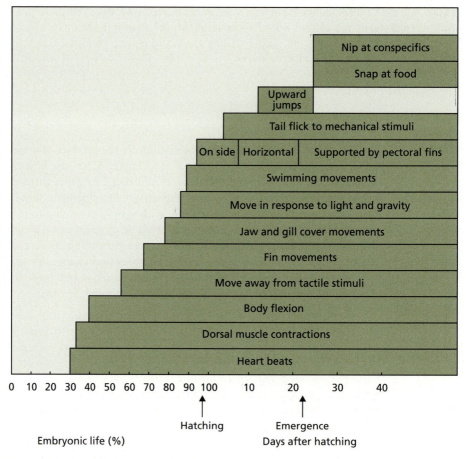

Figure 6.1 The sequence in which behaviour develops in embryonic Atlantic salmon (*Salmo salar*) reflects the development of neural circuitry necessary for different patterns of movement. See text. From Goodenough *et al.* (1993), after Abu-Ghidieri (1966).

continues beyond hatching and is associated with the emergence of an increasing repertoire of behaviours (Fig. 6.1). The sequence in Fig. 6.1 may reflect an unsurprising relationship between neural connection, muscle function and movement of the affected part of the body, but it illustrates the progressive wiring up that shapes the animal's behavioural capabilities, preparing it for the environment with which it will eventually have to cope.

Development, however, is not necessarily progressive. It is important to bear in mind that earlier stages of an organism's life cycle are not some kind of imperfect protoadult, waiting to mature into their proper functional form. They are fully functional organisms in their own right, adapted to the particular circumstances of their time of life. This may entail specialisations that are transient, fulfilling a vital role then disappearing. The larvae of most metamorphosing insects provide clear, and sometimes extreme, examples, such as the cave-dwelling larvae of the fly *Arachnocampa luminosa*, which dangle luminous threads from an elevated web to catch prey, or the caterpillar of the butterfly *Aethria carnicauda*, building 'fences' of hairs along twigs to protect itself from predators during the pupal stage. Young vertebrates also show adaptive specialisations tailored to their early needs: the egg-eviction response of cuckoo chicks, and rhythmic head movements of many young mammals searching for their mother's teat are examples. By definition, these early specialisations come and go as the animal matures. But how complete are the transitions; do they, for example, extend to underlying physiological mechanisms? Behaviours involved in hatching provide some indications.

Hatching is something an animal does only once, but it can require highly specialised anatomy and behaviour to be successful. After that, the specialisations are redundant and are not employed again. So what happens to the underlying machinery that coordinates them? Does it atrophy? Does it become used for something else? An innovative experiment by Anne Bekoff and Julie Kauer (1984) provides answers for hatching behaviour in domestic chicks. During hatching, the chick breaks free from the confines of its shell using a highly stereotyped sequence of body rotation and thrusts of the head and limbs. Since these actions are never seen again in the chick's behavioural repertoire, Bekoff & Kauer wondered about the fate of the neural circuits controlling them, particularly the leg movements. To see whether capacity for the responses disappeared irreversibly, implying a loss of mechanism, they introduced chicks of up to 60 days of age into experimental glass 'eggs', carefully folding them into the hatching position, then recorded their behaviour and muscle movements. To their surprise, reincarcerated chicks quickly began to perform typical hatching movements, qualitatively and quantitatively matching the stereotyped patterns of true hatching. Thus, the machinery for producing the movements remained intact and functional long after hatching, even though the behaviour it controlled was not normally called upon again.

In other cases, the neural connections associated with particular behaviours disappear or are modified for different purposes. Obvious cases are organisms, such as butterflies and frogs, that metamorphose from larvae to a very differently organised adult form. Studies of changes in the nervous system in insects with complete metamorphosis have revealed some intriguing 'conversion' strategies in their neural circuitry. Levine & Truman (1985), for example, have shown that, in the tobacco hornworm (the caterpillar of the hawk moth *Manduca sexta*), all the motor neurons innervating the abdominal muscles are used again in the adult. Some continue to innervate the same muscles as before, which, surprisingly in view of the very different bodily appearance of the adult, retain broadly similar roles, but others lose their previous function in the caterpillar and grow new processes to serve newly formed muscles in the adult (Fig. 6.2). On the sensory side, most larval neurons innervating the sensory hairs atrophy when the caterpillar pupates,

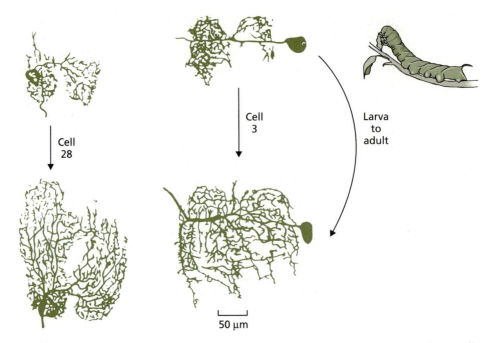

Figure 6.2 Motor neurons in the tobacco hornworm moth *Manduca sexta* are substantially 're-modelled' as the moth metamorphoses from caterpillar to adult. Examples for two cells show expansion of the dendritic field as axons lose their larval connections and innervate new muscles in the adult. After Levine & Truman (1985).

but one group remains to serve specialised trigger hairs on the abdomen of the pupa. These hairs trigger a reflex – existing only during the pupal phase – which causes sudden flexion and serves as an anti-predator device. Thus neural and behavioural development across the metamorphic divide is a mixture of conservatism and innovation.

6.1.1.1 Environmental effects

As we might suspect from Chapter 5, physical maturation can be heavily influenced by the environment. This is certainly true for aspects of neural and sensorimotor development, as the acquisition of visual responsiveness in mammals demonstrates.

The eyes of many mammals develop their basic structure without any exposure to light, but normal function does not develop unless the eyes are 'primed' with certain kinds of visual experience. Blakemore & Cooper (1970) demonstrated this very clearly by exposing kittens to controlled visual environments during their early postnatal development. From the time their eyes opened at two weeks of age, kittens were kept in darkness except for brief exposure to an environment where they saw either vertical or horizontal stripes (Fig. 6.3a). After five months of this, the kittens were brought out into the real world where they generally appeared to cope normally. However, when Blakemore and Cooper tested their visual response to objects in different orientations, they found a dramatic difference between animals raised in the two environments. When presented with a rod held vertically, kittens raised with vertical stripes immediately paid attention to it and approached. Kittens exposed to horizontal stripes, however, failed to show any response. When the rod was held horizontally, the situation was reversed: the kittens exposed to the horizontal stripes responded, but not those shown vertical stripes.

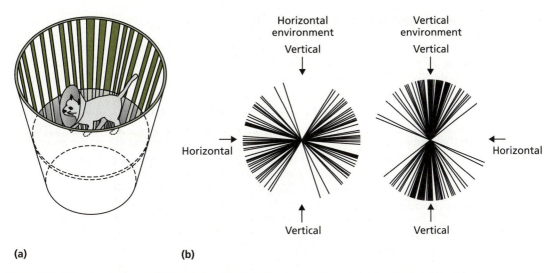

Figure 6.3 The visual cortex of kittens reared in an environment of vertical (a) or horizontal stripes became sensitive only to lines in the orientation experienced (b). The lines in (b) represent the preferred response orientation of cells in the cortex of kittens reared in each of the two environments. Modified from Blakemore & Cooper (1970). © Nature Publishing Group (http://www.nature.com/), reprinted by permission.

To see why the kittens showed these restricted responses, Blakemore & Cooper looked at the arrangement of neurons in the cats' visual cortex, the part of the brain eventually receiving information from the eyes (see Fig. 5.8). Each of the cortical neurons responds to objects in a particular orientation, so they fire in response to lines or edges presenting in that orientation (see 3.2.1.1). In the cortex of normally reared cats, there is a spectrum of responsiveness to different orientations (Fig. 6.3b), but in the kittens reared with vertical or horizontal stripes, the cortex appeared to be dominated by neurons sensitive to the orientation of the stripes. Thus kittens experiencing horizontal stripes had no neurons responding to vertically aligned objects or to lines or edges within 20° of vertical. In this case, therefore, early environmental influences appear to be critical in determining the functional development of visual perception, and thus the later response of the animal to objects in its world.

While the external environment may provide important cues during the development of the nervous system, internal cues may also be important, and illustrate the sometimes flexible nature of development at the cellular level. Bentley & Keshishian (1982) studied the development of neurons from the sensory hairs on the legs of developing grasshoppers. Like those of the cricket *Acheta domesticus* in 5.2.1, the sensory neurons grow out from cells in the hairs and establish connections with the CNS. Bentley & Keshishian, however, showed that the first axons to develop, the so-called 'pioneers', are guided in their quest by special cells that act as way-markers along the route. But, as the legs develop and grow, the intervals between these way-markers become greater and greater. Perhaps as a result, later axons do not appear to use them, but instead grow along the route already taken by the 'pioneers'. Although the final result looks as if all the sensory axons developed in the same way, their developmental histories are in fact quite different. They also show a degree of flexibility, in that the way the synaptic connections with the CNS are formed as the sensory organs develop and their neurons reach their destination depends on which other cells are nearby at the time and what information is flowing from the sense organs. As we stressed in Chapter 5, therefore, the genetic prescription for development can establish **contingency rules** rather than laying down a hard-and-fast blueprint.

6.1.1.2 Brain development and behaviour

Needless to say, much of the behavioural change taking place during an animal's lifetime relates to events in the brain. Much of this in turn involves learning, which we shall discuss shortly. Some changes, however, are associated with the ongoing physical development of the brain.

The direction of neural development in vertebrates is generally caudal–rostral (rear to fore). Thus lower brain structures develop before the higher regions. This is reflected in the presence and subsequent disappearance of neonatal reflexes in mammals (e.g. the grasp and nipple rooting reflexes in human babies and the lordosis response to maternal licking in neonatal rodents [Dewsbury 1978]), for which the excitatory centres are located caudally and inhibitory centres rostrally. The responses are therefore shown early on in development, before the inhibitory centres have matured, after which they are suppressed. Interestingly, some of the reflexes in humans can return in old age as a result of the deterioration of brain tissue and loss of higher-order control. Changes in other types of behaviour, such as locomotory activity, are also consistent with this pattern of brain development (Campbell & Mabry 1972, but see Randall & Campbell 1976).

Environmental factors are also important in shaping the brain and behaviour. The *complexity* of the environment in particular seems to have widespread effects, at least in mammals. For example, it is well known that so-called environmental enrichment (e.g. providing caged animals with toys or other objects) leads to structural changes in the brains of rats, including increased numbers of nerve cells, synapses and dendrites, particularly in the hippocampus and cortex (see 3.1.3.1). These changes can result in improved learning and memory (Rosenzweig & Bennett 1996; van Praag *et al.* 2000; Würbel 2001). Recent studies suggest that impoverished housing, especially involving social deprivation, can affect dopaminergic pathways between the cortex and striatum, leading to attentional deficits. Somewhat soberingly, the symptoms of this deficiency are the same as those used to model key attributes of schizophrenia in humans (Würbel 2001).

Impoverished environments can have other negative effects, including inducing stereotypical behaviour (see 4.3.2.1). Once fully developed, stereotypies persist, even under conditions where they would not normally be induced, suggesting chronic changes in underlying physiology (Mason 1991; Cooper *et al.* 1996). Indeed, behavioural and neurophysiological evidence associates stereotypy with changes in the basal ganglia, a region of the brain important in the initiation and sequencing of movements (Ridley 1994; Hauber 1998; Garner & Mason 2002; see 3.1.2.3). Once again, the changes involve dopamine, with stereotypies apparently developing in response to stress-induced sensitisation of dopamine target cells (Cabib 1993; Steiner & Gerfen 1998). Perhaps not surprisingly, therefore, stereotypies are often associated with other behavioural symptoms of basal ganglia disorders (Würbel 2001; Garner & Mason 2002).

6.1.2 Morphology, maturation and behaviour

Morphological change is an obvious feature of maturation. Size, physique, secondary sexual characters, genitalia and a host of other attributes change as the animal ages, and along with this come new opportunities and limitations for behaviour. The development of a particular behaviour pattern may track that of a specific structure or appendage which is critical for its performance. In the paddlefish (*Polyodon spathula*), a bizarre inhabitant of the Mississippi and Ohio rivers of North America, feeding behaviour changes with the development of the gill rakers, comb-like bony structures projecting into the mouth from the gill arches (Rosen & Hales 1981). While juvenile paddlefish feed selectively on individual zooplankton, adults feed indiscriminately by simply opening their jaws and swimming through the water. The gill rakers strain food particles from the resulting inflow. Paddlefish can grow to over two metres in length, but the gill rakers start to appear as buds along the midline of the gill arches when the fish are about 10 cm long. At this stage, however, the fish are still selective feeders. By the time they reach 12–13 cm the buds line all the gill arches and have begun to increase in length. Nevertheless, this is not enough to allow effective filter feeding and the young fish remain selective. It is not until they reach around 30 cm that the gill rakers are sufficiently well developed for the fish to adopt the indiscriminate adult mode of feeding. Changes in feeding strategy thus track the development of the necessary anatomical structures. By way of contrast, the larvae of some insects sometimes carry out actions that are ineffectually premature; larval grasshoppers, for instance, silently perform the leg movements of song production long before the development of the necessary rasping mechanism (Weih 1951).

6.1.3 Maturation and motivation

The performance of a behaviour by a young animal may occur in a completely different context from that in an adult, suggesting differences in underlying motivation (Chapter 4). In young orange chromides (*Etroplus maculatus*), a species of cichlid fish, approach and 'glancing' behaviour (Fig. 6.4) are expressions of feeding (the fish skim mucus from the side of a parent), and approaches are made only to larger individuals. Overtly similar behaviour in adults is part of pairing or intrasexual competition, and is directed towards individuals of similar size (Ward & Barlow 1967).

Feeding behaviour in mammals provides other examples where underlying causal factors appear to change with age. Feeding actions in neonatal mammals, for instance, are unrelated to the degree of food deficit. Puppies fed through artificial nipples with large apertures, which thus obtain their required amount of milk quickly, suck more on objects unconnected with feeding than puppies given access to nipples with small apertures. Sucking in kittens is similarly independent of the rate of satiation over the first three weeks of life, and human babies actually suck more when satiated or aroused other than by hunger.

6.1.4 Life history strategies and foetal programming

Growth and maturation are not simply a function of an animal's age. They also reflect its life history strategy, the adaptive allocation of resources to growth, survival and reproduction (2.4.5). Relative investment in different components of life history, and thus the timing of maturation and reproduction, vary between individuals in relation to a wide range of factors (see 2.4.5). Accordingly, behaviour is likely to develop along different lines and on different timescales in different individuals. A major factor influencing life history variation is maternal investment. Provisioning of eggs or foetuses *in utero* can have a profound effect on the competitiveness and reproductive potential of resulting offspring, and therefore their strategy for investing metabolic resources.

Figure 6.4 Various forms of adult approach behaviour (b–f) in the orange chromide cichlid *Etroplus maculatus* appear to derive from 'glancing' in juveniles (a). The targets and motivation of the behaviours, however, differ between juveniles and adults (see text). The dark fish is the performer and the pale fish the target in each case. After Wyman & Ward (1973).

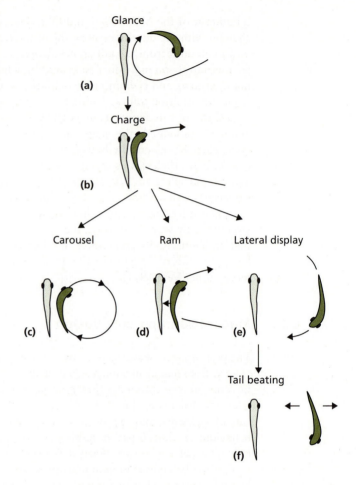

Extensive work in humans and rodents (e.g. Barker 1995; Phillips 1996; Rao 1996) has identified a crucial role of nutritional constraints and other mother/foetus conflicts *in utero* (Haig 1993; see also 5.3.1) in determining a suite of life history attributes in offspring. Patterns of growth and organ development, immune function, menopause and longevity all appear to be predictable from nutritional history in the womb (Barker 1995; Cresswell *et al.* 1997; Hales 1997; Langley-Evans 1997), the effects often extending down later generations (see Lummaa & Clutton-Brock 2002). This is sometimes referred to as **foetal programming** (Barker 1995). The effects are underpinned by various endocrine changes involving many different hormones, but particularly glucocorticoids, insulin and growth hormone.

Such shifts in development and metabolism, mediated by maternal condition, could account for early differences in apparent life history tradeoffs in some species (Lummaa & Clutton-Brock 2002). For example, Barnard *et al.* (1998) found that social status in randomly constituted groups of male outbred laboratory mice was predictable from the sex ratio *in utero* of their natal litters and its consequences for early suckling behaviour and rate of weight gain. High-ranking males tended to be those born of litters with a low male : female ratio that had suckled on more anterior teats and gained weight more quickly while with their mother. Heavier mothers (i.e. those in better condition) also lavished more attention on their pups. Although males did not develop aggressive dominance relationships in their natal litters, eventual high and low rankers showed the same difference in apparent immunity tradeoff as Barnard *et al.* had found in randomised groups in earlier studies (see 3.5.2.1): that is, future low rankers regulated immunodepressive hormone concentrations in relation to current immunocompetence (antibody levels), while high rankers did not (cf. Fig. 3.34). Since Barnard *et al.* (1998) had standardised postnatal litters to four males, their findings suggest that maternal condition influences investment in male offspring prenatally, and that this is reflected in the priority given by well-resourced pups to early reproduction (androgen-driven attributes) over long-term survival (immune responsiveness). Other studies have shown that the odours of male mice reared by well-nourished mothers are more attractive to females than those of males from undernourished females (Meikle *et al.* 1995), and that relationships between maternal investment, hormones and behaviour can programme later responses to stress (Würbel 2001). Key behavioural and physiological characteristics may thus be set environmentally before the animal ever ventures into the world.

6.2 Experience and learning

While many behaviours appear in their fully functional form without any previous practice by the animal, or develop in parallel with the animal's morphology and physiology as it matures, many develop through accumulating experience and therefore the animal's ability to acquire and retain information. The development of feeding skills provide some good examples.

Many small passerine birds, such as sparrows and tits, hold down food items with one or both feet while pecking and tearing at them. In great tits (*Parus major*), young birds attempt this at their first encounter, but their early efforts are clumsy and ineffective. The speed with which they acquire adult competence, however, depends on how much experience they have with suitable objects (Vince 1964). Improvement is thus not simply a consequence of developing musculature and grip. Similarly, red squirrels (*Sciurus vulgaris*) require experience in order to open hazel nuts efficiently. Young squirrels possess all the necessary gnawing and prising movements but cannot deploy them effectively (Eibl-Eibesfeldt 1963). Experienced adults usually gnaw a vertical furrow down the broad side of the nut, then drive their incisors into the aperture and prise the nut open, though some animals develop more idiosyncratic, but nevertheless effective, techniques. Naïve individuals gnaw haphazardly, sometimes creating several furrows until the nut happens to break. Improvement seems to involve learning that the nut breaks more easily when the furrow is gnawed parallel to the grain.

Some of the most detailed investigations of experience in behavioural development have involved bird song. Classic work with young male chaffinches (*Fringilla coelebs*) by the ethologist W.H. Thorpe and others (e.g. Marler 1956; Thorpe 1961; Nottebohm 1967) has shown that birds isolated from other conspecifics, so they hear no vocalisations by other birds, develop far simpler songs than normal adult males (Fig. 6.5a,c). Males reared with another of the same age end up with a song intermediate in complexity between those of isolates and normal adults (Fig. 6.5b,c). Hearing **models** is therefore important in the development of the young chaffinch's song. But is that all that is important?

Chaffinch song develops through three well-defined stages. The first is known as **subsong** and is little more than a soft, featureless babble sung at the end of the bird's first summer. In the following spring, subsong matures into the second stage, **plastic**

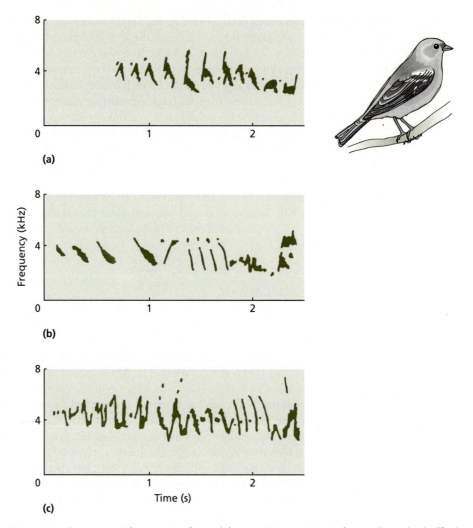

Figure 6.5 Sonograms (the pattern of sound frequencies over time) of songs by male chaffinches (*Fringilla coelebs*) when: (a) kept in auditory isolation, (b) reared with a peer male and (c) provided with an adult male tutor. After Thorpe (1961).

song, which is similar to the final **full song** in having most of the right syllables, but it lacks organisation into phrases. Once full song emerges, it remains unchanged through the rest of the bird's life. This progression suggests the bird needs to practise its song and listen to the results in order to mimic the adult male model properly. To see how important such feedback is, Nottebohm (1967) deafened birds at different stages of song development. If birds were deafened as adults, when their songs had already matured, there was little effect on the song (Fig. 6.6a), but if they were deafened during plastic song (Fig. 6.6b), the final form of the song depended on how long they had been singing plastic song before the manipulation. If they had been singing for some time, subsequent full song was almost normal; if only for a short time, the song was less complex than usual. Birds deafened during subsong, however, ended up with little more than a long screech (Fig. 6.6c). The type of song chaffinches finally sing, therefore, seems to be a combined product of the songs they hear when young, and feedback from their attempts to match them.

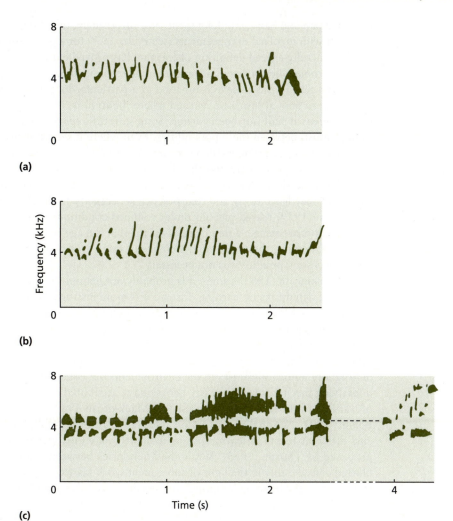

Figure 6.6 Sonograms of songs by male chaffinches deafened: (a) as an adult, (b) during the plastic song phase and (c) during the subsong phase. After Nottebohm (1967).

So experience can play a crucial role in honing behaviour. Indeed, it may be the dominant shaping force in many cases. But how exactly does experience change behaviour? Is its potential open-ended, or are there limits to what it can achieve? These questions take us into the realm of **learning**.

6.2.1 Learning

A parallel is often drawn between learning and the evolution of behaviour by natural selection. Both are means by which behaviour 'improves' in the sense of becoming more effective at solving problems of survival and reproduction. Learning shapes behaviour *within* generations, while selection acts *between* generations. Ideas about learning have developed under many different theoretical frameworks, and its pioneers include some of the greatest names in the history of the study of behaviour – Pavlov, Thorndike, Watson,

Hull, Tolman and Skinner among them. Not surprisingly, therefore, its progress has been peppered with involved disagreements (see Bolles 1979 for a good historical perspective). Even a generally agreed definition of learning has proved elusive (Mackintosh 1983). In part the problem lies in distinguishing learning from other causes of behavioural change. An animal that begins to search for food in a particular location when it did not a few hours before may at last have learned where the food is, or it might simply be hungry now whereas it had not been earlier. Some careful experiments would be necessary to distinguish between these possibilities. Even where it seems obvious that experience has led to change, care may still be needed. Young birds, for example, cannot fly at first. Instead, they appear to practise the appropriate movements as they grow, until, after a few weeks, they have developed the capacity for flight. It thus seems reasonable to conclude that they learn to fly and that practice is necessary to develop the skill. However, Grohmann (1939) reared pigeons under confined conditions that prevented them practising flight movements. When he released them at the age they would normally be able to fly successfully, there was no difference in their proficiency relative to unconfined birds, suggesting practice was not necessary.

As Mackintosh (1983) argues, it is probably not helpful to worry too much about all-embracing definitions of learning. Instead it is more instructive to focus on the different situations in which learning occurs and the conditions necessary for its expression. This may encourage the view that learning is just an umbrella term for disparate, unrelated processes, but such might indeed be the case. It is certainly not obvious that the same process underlies learning to navigate a maze, associate particular food items with feeling ill or ignore innocuous noises in the environment. But equally there could be a common underlying mechanism, and many people have attributed superficially different forms of learning to the same basic process (see Mackintosh 1983; Roper 1983; Macphail & Bolhuis 2001). We shall return to this issue later. To begin with, however, we shall discuss different forms of learning using a slightly extended version of the classification proposed by Thorpe (1963). While the boundaries between Thorpe's categories have blurred considerably in the light of subsequent work (as we shall see), their terminology still pervades the learning literature and remains a useful framework for discussion.

6.2.1.1 Habituation and sensitisation

Habituation and **sensitisation** both result from repeated exposure of the animal to a single event, and thus differ from other forms of learning where the animal learns about relationships between different events. In the case of habituation, the upshot is a *reduction* in response to the event; in sensitisation, responsiveness *increases*.

Habituation

Razran (1971) has defined habituation as 'learning what not to do'. Animals are frequently bombarded with different stimuli emanating from the environment. The cost in terms of time and energy of responding to every one would be prohibitive, particularly since only a small proportion is likely to have a significant impact on their chances of reproducing. In Chapter 3, we saw how an animal's perceptual systems filter out some of the noise in information received from the environment; habituation is a way of filtering out some more. More specifically, it eliminates responses to stimuli that are sometimes important, but that, in a particular instance, are not. Rustling leaves, for instance, may be worth reacting to, e.g. by orientating or hiding, because they sometimes indicate the approach of a predator. Repeated rustling without the appearance of a predator, however, is probably just caused by the wind and can be ignored.

Habituation is stimulus-specific, so responsiveness is reduced only with respect to the **habituating stimulus** or something else very like it. This specificity is shown very clearly by habituation in territorial male three-spined sticklebacks (*Gasterosteus aculeatus*). When they allowed captive males to establish breeding territories, Peeke & Veno (1973) found they were initially very aggressive towards neighbouring individuals but that responsiveness quickly declined. This makes sense since there is no point harrying fish that are unlikely to be intruders. However, territory owners cannot afford to generalise habituation to all other sticklebacks because some of them *will* be intruders; habituation should thus be confined to neighbours. To test this, Peeke & Veno introduced 'intruder' fish in glass tubes to various locations in a territory owner's tank. Owners were exposed to each 'intruder' for 30 minutes, given a 15 minute break, and then tested with one of four stimuli: (a) the same fish in the same location, (b) the same fish in a different location, (c) a different fish in the same location, and (d) a different fish in a different location. The results were clear cut (Fig. 6.7). Owners showed least aggression towards stimuli (a) and (b), though (b) elicited slightly more than (a), more aggression towards (c) and most towards (d). Thus decreasing familiarity with the stimulus, in terms of both individual and location, increased the tendency to override the previous habituation. However, it did not affect the rate at which habituation subsequently occurred: the lines in Fig. 6.7 all have approximately the same slope.

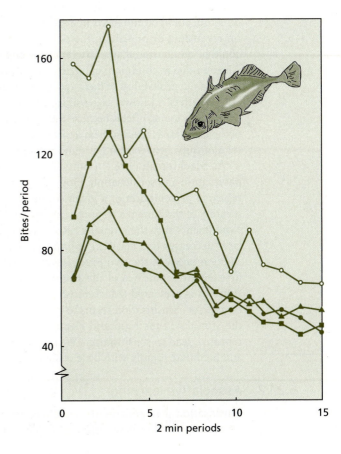

Figure 6.7 Habituation in territorial male three-spined sticklebacks (*Gasterosteus aculeatus*), measured as aggressive responses to ● the same fish encountered in the same location as previously, ▲ the same fish in a different location, ■ a different fish in the same location as a previous encounter, and ○ a different fish in a different location. See text. After Peeke & Veno (1973).

Of course, it is important to establish that habituation really is the result of learned non-responsiveness and not the upshot of something less interesting, such as fatigue or sensory adaptation. When we cease to become aware of our clothes shortly after putting them on, for example, the effect is due to sensory adaptation (sensory cells ceasing to fire) rather than habituation (learning to ignore the stimulation). Experiments with the sea hare *Aplysia* show how these other effects can be discounted (Kandel 1976). As we saw in 3.1.3.2, mechanical stimulation of the siphon in *Aplysia* elicits a gill withdrawal response. If the stimulation is applied repeatedly, however, the response rapidly wanes to give a typical habituation curve. If at this point a novel mechanical stimulus is applied somewhere else on the animal, responsiveness to stimulation of the siphon is immediately rekindled, thus ruling out fatigue or sensory adaptation as a cause of the initial decline.

Habituation seems to occur throughout the animal kingdom, and its demonstration in groups such as the Cnidaria (e.g. *Hydra*), which do not appear to be capable of associative learning (see 6.2.1.2), implies that at least some of the processes underlying habituation are not shared with conditioning. However, there is considerable debate about the distinction between habituation and more complex learning at the fundamental level of process (Mackintosh 1983).

Sensitisation

While repeated stimulation can result in a decline in some kinds of response, it can cause an increase in others. Wells (1978) found that the tendency for an octopus (*Octopus vulgaris*) to attack or withdraw from a neutral stimulus, such as a plastic disc suspended on a rod, could be enhanced by repeated feeding or shocking. The effects did not depend on rewarding or punishing the octopus for the response itself (food or electric shock could be administered at a different time and in an entirely different part of the tank), so were not due to conditioning (see below). Rather, repeated feeding or shocking seemed to increase the probability of their appropriate responses being elicited by a neutral stimulus. Such as effect is known as sensitisation. It is very important to distinguish sensitisation from conditioning effects because the two can produce overtly similar outcomes. As in the octopus example, teasing them apart requires careful control over the spatial and temporal relationships between stimulus and response (Evans 1966; Mackintosh 1983).

As Mackintosh (1983) points out, there is some conflict between the concepts of habituation and sensitisation. How can two diametrically opposite effects arise from the repeated presentation of a given stimulus? The fact is, however, they can, and careful scrutiny of the animal's response can reveal them. It is commonly the case in habituation experiments that the animal's responsiveness to a stimulus at first *increases*, and only later decreases in a classical habituation curve (Groves & Thompson 1970). One distinction between the two processes may thus be that sensitisation occurs only for the first few presentations of the stimulus, while habituation continues for as long as the stimulus is presented (Mackintosh 1983). The strength of the stimulus also appears to be a factor. Sensitisation seems to be more pronounced the stronger, or more significant, the stimulus; thus food and electric shock are typical examples of stimuli which, if repeated, lead to sensitisation. Weak stimuli appear less likely to produce sensitisation, and habituate rapidly with little evidence of an initial increase in response.

6.2.1.2 Associative learning

Sensitisation may underlie the enhanced performance of some behaviour patterns, but we usually think of new responses being acquired, or existing ones enhanced, through processes of **associative learning**. Here, some action or stimulus of hitherto little significance takes on significance as a result of suddenly having some important consequence. Accidentally treading on a mound of soft earth, for example, yields an unexpected bonanza of nutritious ants' eggs, or the silhouette of a large bird flying overhead is quickly followed by an alarming attack. If the consequences reliably flow from the action or stimulus each time it occurs, a long-term association may develop and the animal alters its behaviour accordingly. Thus it becomes more likely to trample a mound of earth on encounter, or dive for cover when a large bird flies over. In the terminology of learning theory, the animal's behaviour has become **conditioned** by the events it has experienced. Experimental approaches to conditioning typically rely on arranging temporal relationships between two events, E1 and E2, and observing changes in behaviour as a result of exposing animals to the relationship. Two categories of conditioning are generally recognised depending on whether E1 is a neutral stimulus, usually some extraneous environmental event or object, or the animal's own actions. In the first case, learned outcomes are usually referred to as **classical conditioning**, in the second, **operant** or **instrumental conditioning**. However, as we shall see, the two may not be as distinct as their labels imply.

Classical (Pavlovian) conditioning

Classical conditioning stems from the pioneering work of the Russian physiologist Ivan Pavlov in the early part of the twentieth century. A stimulus that initially does not elicit a response comes to do so by association with a stimulus that does. This new property of the stimulus is thus *conditional* on its association with an established stimulus–response relationship. The stimulus is therefore referred to as the **conditional stimulus** (**CS**) and its elicited response the **conditional response** (**CR**). By the same token, the stimulus in the established stimulus–response relationship is the **unconditional stimulus** (**UCS**) and its response the **unconditional response** (**UCR**). The relationships between these components are summarised in Fig. 6.8, using Pavlov's classic study of the salivation response in dogs as an example. Importantly, while the CR is functionally similar to the UCR, there may be subtle qualitative and quantitative differences between them that beg some questions about the process of stimulus–response association.

The temporal and/or spatial association of the CS with the UCS is known as **reinforcement** (Box 6.1), with the close proximity between the two eventually leading the animal to anticipate the UCS on perceiving the CS, and hence perform the CR. Thus pairing the bell with the presentation of meat powder in Fig. 6.8, leads the dog to expect meat powder on hearing a bell, and so salivate. The important relationship in classical conditioning is therefore that between the two stimuli; reinforcement occurs even if the animal does not perform the CR (in early exposures, the CS is almost always ineffective at eliciting the CR). This contrasts with operant conditioning where the animal experiences the reinforcing stimulus *only* if it performs the CR.

Pavlov found that almost any stimulus could become a CS, as long as it did not evoke too strong a response of its own, as might an urgent desire to escape from the experimental apparatus for example. Even pain could be used to elicit salivation in his dogs. In many cases, dogs also **generalised** from the CS to other similar stimuli. Animals conditioned to salivate in response to a pure tone of a specified frequency would also salivate when other tones were played, though the response was often weaker, a phenomenon known as **generalisation decrement**. With persistent exposure to one particular tone, however, dogs became more **discriminating**, i.e. they were less likely to generalise the CR to different stimuli. Perhaps not surprisingly, discrimination can be enhanced if, as

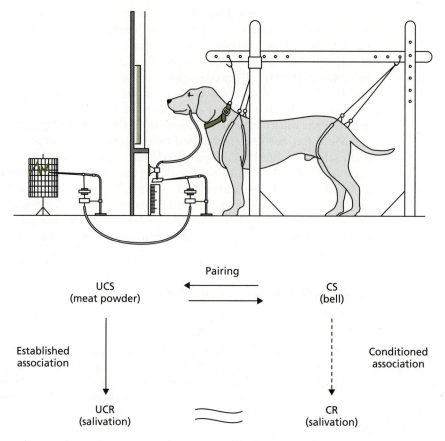

Figure 6.8 Pavlov's classical conditioning experiment with dogs, in which salivation became conditioned to the ringing of a bell. An established relationship between an unconditioned stimulus (UCS) and an unconditioned response (UCR) can be used to condition a similar (conditioned) response (CR) to a novel (conditioned) stimulus (CS) by pairing the latter with the US. See text. After Fantino & Logan (1979) and Manning & Dawkins (1998).

well as being rewarded for responding to the correct stimulus, subjects are also punished for responding to incorrect ones.

From the discussion so far, and indeed intuitively, we might expect that the CS has to precede the UCS if stimulus association is to occur. In fact, this is not so, although the relative timing of the two stimuli does affect the efficacy of the association. Figure 6.9 shows a number of possible temporal relationships between the CS and UCS. Of these, conditions (a) and (b), sometimes called **delayed conditioning**, in which the UCS at least partly precedes the CS, are the most effective. Perhaps surprisingly, **simultaneous conditioning** (d), where the CS co-occurs with the UCS but does not precede or extend beyond it, is very ineffective, more so even than **backward conditioning** (e), the effectiveness of which can be increased by making the UCS contiguous with the CS. We shall return to the issue of pairing the UCS and CS and its implications for conditioning a little later. Of some interest, particularly with respect to some rhythmic behaviours (see 3.4.1), is so-called **temporal conditioning**, where the CR relies on presentation of the UCS at predictable time intervals and the intervals themselves act as a CS (Fig. 6.9).

Underlying theory

Box 6.1   Reinforcement

Reinforcement is a central concept in associative learning. A reinforcer can be defined as *any event that increases the probability that the behaviour it follows will recur in the future*. There are two kinds of reinforcer: positive and negative. Positive reinforcers make the actions they follow more likely by their occurrence, while negative reinforcers make the actions they follow more likely by their termination or non-occurrence. An electric shock that induces a dog to jump a hurdle and thereby escape the shock, acts as a negative reinforcer because it increases the future likelihood of the response that terminates it. Logically, it can be difficult to distinguish between positive and negative reinforcement. For example, eating by a hungry animal can be viewed as either positive reinforcement (the animal gets a reward) or negative reinforcement (it escapes from deprivation or hunger). Similarly, the negatively reinforced jumping of the dog could be recast in terms of positive reinforcement as the dog escapes to safety.

A negative reinforcer, however, must not be confused with a punisher. Reinforcers *strengthen* the behaviours they follow. Punishers *weaken* them. In our jumping dog example, electric shock would be a punisher if it decreased the likelihood that the dog would remain in the cage until the floor was made live. A more contorted notion is that of a negative punisher. Here a response is made more likely by the *omission* of a positive reinforcer. Thus a dog can be trained not to jump at people by withholding a food reward each time it jumps; only by not jumping does the animal receive the reward.

Figure 6.9 Different schedules of class-ical conditioning, in which the temporal relationships between the unconditioned stimulus (UCS) and conditioned stimulus (CS) are varied. See text. After Fantino & Logan (1979).

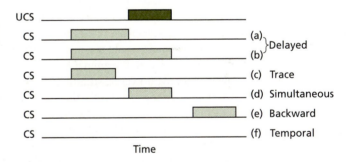

Excitatory and inhibitory conditioning   In most conditioning, the CS indicates the likely occurrence of the UCS, a situation referred to as **excitatory conditioning**. However, sometimes the CS can indicate that the UCS is likely to be *absent*, in which case the process is known as **inhibitory conditioning**.

The eye-blink response in rabbits is a good example of excitatory conditioning. The response can be conditioned to the presentation of a brief tone followed by a mild mechanical shock to the cheek. The intensity of the shock is just enough to elicit a blink, and a few pairings with the tone (which does not normally result in blinking) is enough to turn the tone into a CS for the response. Figure 6.10 shows some typical results, in which the percentage number of times a conditioned blink response occurred is plotted for successive blocks of 100 trials. The likelihood of response increases sharply at first, but then levels off in an asymptote, a stable level of responding for the particular experimental

Figure 6.10 The acquisition (left) and extinction (right) of a (tone) conditioned eye-blink response by rabbits. After Gibbs *et al.* (1978).

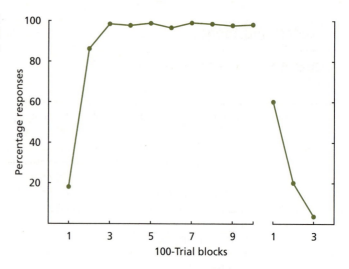

conditions. To the right of the figure, there is a rapid decline in the CR when the CS is no longer paired with shock. This is known as response **extinction**. Extinction does not simply eradicate the CR and leave the animal as if conditioning had never occurred, however. If the animal is later presented with the CS again, the CR shows **spontaneous recovery**, though not to its original level (the recovered CR also extinguishes more rapidly the next time pairing with the CS ceases). An extinguished CR can also be recovered by presenting a novel stimulus along with the CS. Thus an extinguished salivation response in dogs can be rekindled by pairing the buzzer or bell CS with a flashing light, an effect Pavlov referred to as *disinhibition*. While extinguished responses can be recovered, repetition of the UCS–CS association beyond the level that maximises performance of the CR (**overtraining**) can increase resistance of the CR to extinction in the first place.

Excitatory conditioning is a means by which one stimulus comes to indicate another, and is concerned with the animal predicting things that are likely to happen in its environment. Equally, however, stimuli may predict that certain things will *not* occur in the environment. In the real world, this may be important, for example in choosing feeding places that are safe from predators, or steering clear of migration routes that do not cross sources of water. An experiment with pigeons shows that animals are perfectly capable of this kind of inhibition conditioning.

Hearst & Franklin (1977) looked at the responses of pigeons to two illuminated keys in a Skinner box. The keys lit up one at a time, and in random order, for 20 seconds, with an average of 80 seconds between successive illuminations. Food was periodically delivered into a hopper during the interval, but never when either of the keys was lit. Hearst & Franklin observed the movements of birds around the Skinner box during each session to see whether their responses to the keys changed with time. In early sessions, pigeons seemed indifferent to the keys and disregarded them whether or not they were lit. As time went on, however, they developed a marked tendency to move away from a key when it was illuminated. One interpretation of this is that the pigeons learned that lighting a key signified the absence of food and that moving away indicated conditioned withdrawal from a negative stimulus. That the tendency to move away was stronger when food was delivered more frequently during the interval between illuminations (thus emphasising its absence during illumination) adds support to this conclusion (Fig. 6.11).

Figure 6.11 The mean approach : withdrawal ratio by pigeons with respect to an illuminated key when birds received food at (a) 0.37, (b) 0.74 or (c) 1.47 presentations per minute only when the light was turned *off*. The greater the reward rate, the more likely birds were to move away from the lit key. See text. After Hearst, E. & Franklin, S.R. (1977) Positive and negative relations between a signal and food: approach-withdrawal behaviour, *Journal of Experimental Psychology: Animal Behaviour Processes*, 3, 37–52. Copyright © 1977 by the American Psychological Association. Adapted with permission.

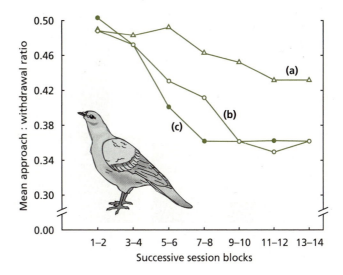

Adaptive significance Intuitively, a learning capacity that enables the animal to predict events in its environment would seem to be useful, and it is not difficult to imagine reproductive benefits accruing from it. Surprisingly, however, relatively few studies have set out to test this explicitly. While psychologists have speculated for many years about the adaptive value of conditioning, interest has generally remained theoretical rather than experimental. An exception is the work of Karen Hollis at Mount Holyoke College, Montana (see e.g. Hollis 1982, 1999). For the best part of 20 years, Hollis has been interested in the functional significance of classical conditioning and has conducted extensive experiments using territorial defence, and mating behaviour in fish as model systems. Figures 6.12 and 6.13 below show some results from her experiments with blue gouramis (*Trichogaster trichopterus*), close relatives of the familiar Siamese fighting fish (Hollis 1999).

In Hollis's experiments, male gouramis were trained to expect a rival male or a female (UCS) after the presentation of a brief light stimulus (CS). In later tests, these classically conditioned males, along with control males that had either been exposed to the light stimulus without the appearance of another fish, or had experienced other fish with no preceding light, were given a light cue and then presented with a rival or a female. The conditioned males showed pronounced differences in response compared with the two control groups. When the light came on in rival male treatments, conditioned males rapidly approached it with a full frontal threat display. When the rival itself appeared, they were therefore already responding aggressively. Consequently, conditioned males delivered more biting attacks (Fig. 6.12a), and other aggressive responses such as tail-beating, to the rival than either of the control groups, and so won more contests (Fig. 6.12b). The priming effect of the light thus appeared to give the conditioned males a marked advantage in aggressive disputes.

A similar advantage accrued in encounters with females. Males conditioned to expect a female after a light stimulus were more likely than control males to show courtship appeasement instead of aggression (Fig. 6.13a). They were also more likely to build a nest over the two hours subsequent to encountering a female (Fig. 6.13b). As a result, they coupled more often and sired more offspring (Fig. 6.13c). Conditioning therefore had a very direct effect on their reproductive success.

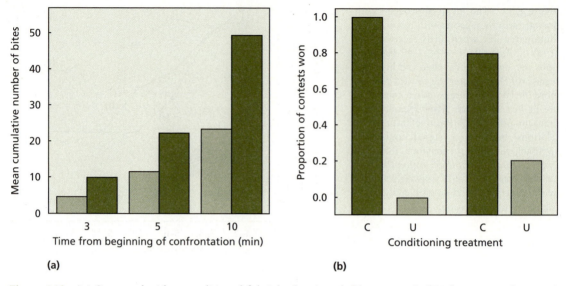

Figure 6.12 (a) Compared with unconditioned fish (plae bars), male blue gouramis (*Trichogaster trichopterus*) delivered more bites towards a rival male when they had been conditioned to expect another fish after a light stimulus (dark bars). (b) They also (dark bars) won more contests than unconditioned controls (pale bars). After Hollis (1999).

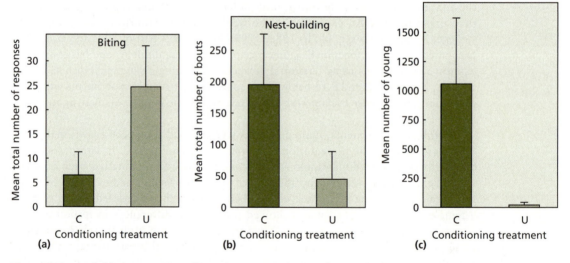

Figure 6.13 Male blue gouramis conditioned to expect a female after a light stimulus (dark bars) were less likely to be aggressive towards females (a), more likely to build a nest (b) and ended up producing more young (c) than unconditioned fish (pale bars). After Hollis (1999).

Classical conditioning underlies adaptation in many different contexts. Aposematic signalling (8.3.5.3), for example, where distasteful, or otherwise noxious, prey adopt garish colour patterns that are highly memorable, results from naïve predators having an unpleasant experience when they first encounter such prey, and rapidly learning to associate the experience with the colour pattern (e.g. Martin & Lett 1985). As a result, predators avoid anything bearing the pattern in the future, to the advantage of both themselves and the erstwhile prey. Other (noxious or non-noxious) prey species can then capitalise on this avoidance conditioning by adopting the UCS of the bright colour pattern and avoiding predation themselves, thus driving the evolution of Müllerian, Batesian and other forms of mimicry (8.3.1).

Surprise, attention and conditioning   Classical conditioning results from a developing association between a UCS and a CS, or between the CS and a response. For much of its history it was assumed that conditioning would come about whenever the UCS and CS were paired. During the 1970s, however, views began to change, and it is now believed that the important attribute of the UCS is not that it is paired with the CS, but that it is in some way surprising or unexpected, so heightening attention which can then be focused on the CS.

Animals do not attend equally to all the stimuli reaching them (see also 3.2.1). Pigeons in a Skinner box prefer to peck the lighted key of a pair, for example, and cats attend more to movement than to colour. An important effect of differential attention is that it can bias the capacity for different stimuli to form stimulus–response associations. The past history of other stimuli in the environment can be a major factor in this. If stimuli A and B are presented together, but stimulus A has been reinforced in the past, it may prevent the development of a stimulus–response association with stimulus B, a phenomenon known as **blocking**. Sometimes, the effectiveness of a stimulus is merely reduced when it is presented together with others, with the effect depending on the relative strengths of the other stimuli. Miles & Jenkins (1973), for example, found that tone stimuli were more effective than lights of different intensity when the latter were similar and difficult to tell apart. When the lights were easily distinguishable, however, lights were better than tones. In this case, one stimulus is said to **overshadow** the other (here tone first overshadowed light, then vice versa). As a rule, the more intense of any pair of stimuli tends to overshadow the less intense stimulus (see also Box 7.4).

Attention can also depend on *relationships* between stimuli. Thus pigeons presented with circles or triangles on red or green backgrounds can be trained to attend to shape when a blue light is shining, but to background colour when the light is yellow. This is known as **conditional stimulus control**. Sometimes animals can be trained to discriminate between two stimuli (see later) by using an already effective stimulus and **transferring stimulus control** to one of the new stimuli, often by 'fading in' the new stimulus over the effective one once the animal is attending to it. **Feature-value effects** (the attention-grabbing effect of some distinct feature of a stimulus – e.g. a spot on a Skinner box key) also influence the effectiveness of stimulus conditioning.

While the animal's attention can be influenced by many factors, one property that is argued to underlie attention is the element of surprise. Several theories of learning have stressed the importance of surprise in conditioning (e.g. Mackintosh 1975; Pearce & Hall 1980; Wagner 1981), but the most influential has been that of Rescorla & Wagner (1972).

Rescorla–Wagner theory: Two assumptions are fundamental to Rescorla–Wagner theory. First, repeated pairing of the CS and UCS will lead to an increase in the strength of association between them. However, this will not continue indefinitely but will be limited by the strength of the CS–UCS association relative to that of the UCS. Second, when there is an increase in the strength of the association, the increase will not be by a fixed amount. Rather, it will be determined by the *difference* between the present strength of the CS–UCS association and the *maximum* possible strength of the UCS. Thus, when the difference is large, as at the start of the conditioning process, the increase in associative strength is correspondingly large, but once a strong association has formed, there can be little further change. These assumptions can be expressed as a simple equation (Box 6.2).

Underlying theory

Box 6.2 The Rescorla-Wagner model

The Rescorla–Wagner model assumes that the degree of increase in a CS–UCS association depends on the difference in the current strength of the association and the maximum possible strength of the UCS. This can be expressed in a simple equation:

$$\Delta V = \alpha(\lambda - V) \qquad (6.2.1)$$

where V is the strength of the CS–UCS association, ΔV the change in strength of the association over a given time (or trial), λ the magnitude of the UCS (and thus the maximum strength of the CS–UCS association) and α the magnitude of the CS (α takes a value between 0 and 1 and does not vary during conditioning).

Application of the equation is very straightforward. If we assume at the outset that α is, say, 0.2, and the CS has no prior association with the UCS (i.e. is novel), then on the first CS–UCS pairing, the value of V will be zero. The value of λ can be set arbitrarily at 100. From the equation, the increase in strength of association over the first pairing will then be:

$$\Delta V = 0.2(100 - 0)$$
$$= 20$$

For the second pairing, V takes the value 20 (the increment over zero in the first pairing), so the next increment will be smaller:

$$\Delta V = 0.2(100 - 20)$$
$$= 16$$

and so on for the third ($V = 20 + 16 = 36$):

$$\Delta V = 0.2(100 - 36)$$
$$= 12.8$$

and successive pairings.

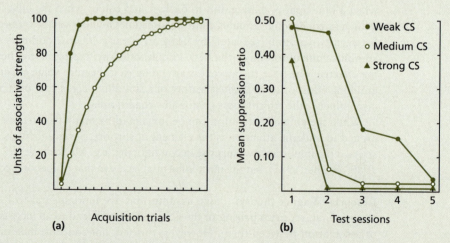

(a) Acquisition trials

(b) Test sessions

Figure (i) (a) The change in the strength of the CS–UCS association predicted by the Rescorla–Wagner equation for two intensities of CS (●: $\alpha = 0.8$; ○: $\alpha = 0.2$). After Pearce (1997). (b) The development of conditioned suppression of emotional behaviour in rats trained to the same UCS but exposed to different strengths of CS (white noise). After Kamin, L.J. and Schaub, R.E. (1963) Effects of conditional stimulus intensity on the conditional emotional response, *Journal of Comparative and Physiological Psychology*, 56, 502–7. Copyright © 1963 by the American Psychological Association. Adapted with permission.

Testing the model

Various experimental studies have tested the Rescorla–Wagner model by attempting to manipulate the value of different parameters in the equation. Figure (i)a, for example, shows the change in associative strength predicted by changing the value of α (intensity of the CS), and Fig. (i)b, the outcome of an experiment in which rats were conditioned to suppress 'emotional' behaviour (defecation) in response to different intensities of white noise (Kamin & Schaub 1963). Similar outcomes are predicted and found when the intensity of the UCS is varied (Fig. (ii)a,b)

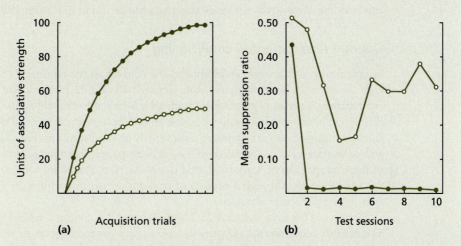

(a) Acquisition trials   (b) Test sessions

Figure (ii) (a) The change in CS–UCS strength predicted for two intensities of UCS (●: $\lambda = 100$; ○: $\lambda = 50$). (b) The development of conditioned suppression of emotional behaviour in response to a noise CS received in association with a shock UCS of 0.49 milliamps (○) or 0.85 milliamps (●). From Annau, Z. and Kamin, L.J. (1961) The conditional emotional response as a function of intensity of the US, *Journal of Comparative and Physiological Psychology*, 54, 428–32.

Based on discussion in Pearce (1997).

One interpretation of the above is that the difference in strength between the UCS and the CS–UCS association (λ–V in Box 6.2) indicates the degree to which the UCS is unexpected or surprising. The greater the difference, the greater the surprise when the UCS comes along. Various aspects of the UCS and CS are likely to influence the effect, as shown in Box 6.2. However, the situation in the real world is unlikely to be quite as straightforward as this because at any given time a range of different potential CSs is likely to be impinging on the animal when the UCS is presented. According to Rescorla & Wagner it is how well this *combination* of stimuli predict the UCS that determines how surprising it is. This becomes important more formally when animals are conditioned to a **compound stimulus**, for instance a light and buzzer presented together. As we saw above, various relationships between the components of such stimuli can influence the development of conditioned associations through blocking, overshadowing and other effects. A Rescorla–Wagner account of these differs from the conventional interpretation. Take blocking, for example. The account earlier implied that blocking is due to a previously formed association (say with a light) distracting the animal from forming an association with a new CS (say a buzzer) when the two CSs are presented at the same time.

Rescorla & Wagner would argue instead that pretraining with the light ensures the UCS is predicted by the compound light–buzzer stimulus. The UCS is therefore unsurprising, and so prevents the formation of an association between itself and the buzzer.

While the Rescorla–Wagner view has been very influential, and generally well supported by experimental evidence, it is not without problems. Not all aspects of blocking and overshadowing, for example, are consistent with it, and it does not fully account for the role of surprise in conditioning (see e.g. Dickinson *et al.* 1976). Various developments of, and alternatives to, the Rescorla–Wagner model have been proposed and have met with some success, but no single theory as yet caters satisfactorily for all aspects of conditioning. A good discussion of the issues can be found in Pearce (1997).

Operant (instrumental) conditioning

In operant or instrumental conditioning, the delivery of the reinforcer (E2) is associated with the animal's performance of some activity (E1) rather than another stimulus in the environment. Instead of starting out with a UCS-linked UCR, which becomes associated through experience with a novel CS, the animal becomes conditioned through initially chance reinforcement to respond to a previously ignored stimulus. For example, a hungry animal wandering about in search of food is likely to perform a range of behaviours. If one of these happens to procure food, and is associated with finding food sufficiently often, the animal learns, through a process of trial-and-error, to perform the behaviour regularly in that particular situation.

Pioneers of the field include E.L. Thorndike and B.F. Skinner, whose 'puzzle box' and Skinner box respectively have provided paradigms for the experimental investigation of operant conditioning. In Thorndike's puzzle box (Fig. 6.14a), an animal obtains a reward

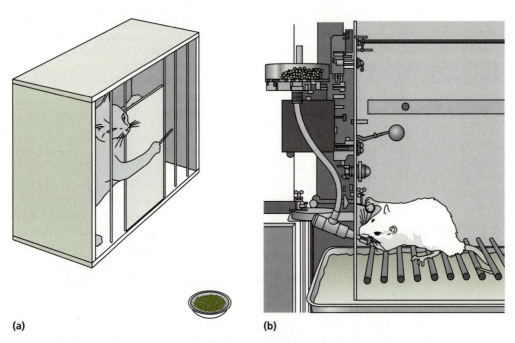

(a) (b)

Figure 6.14 Versions of Thorndike's puzzle box (a) and Skinner's technically more sophisticated derivation, the Skinner box (b), have provided the experimental environments for much of operant conditioning theory. After Fantino and Logan (1979).

by tripping a particular catch or treadle which opens a door and allows it to escape or gain access to food. In Skinner's modern equivalent of the puzzle box (Fig. 6.14b), the animal obtains food, water or access to some other resource, by activating a lever, button or panel on the wall of the box and being rewarded on some preprogrammed schedule.

Thorndike (1898) was the first to suggest that operant conditioning reflects learning about responses, and that, when a response is followed by a reinforcer, a stimulus–response (S–R) relationship is strengthened (Thorndike's **Law of Effect**). For a rat that has to press a lever for food, the stimulus may be the lever itself, and the response the action of pressing it. Each successful press strengthens the association between the sight of the lever and the act of pressing. Thus, as encounters with the lever accumulate, so the tendency to press it increases, eventually reaching some asymptote. Animals can be taught to perform novel and sometimes complicated behaviours in order to receive a reward, a process known as **shaping**. Skinner (1953) likened shaping to a sculptor moulding clay. At first any coarse approximation of the desired outcome is reinforcing, but reinforcement soon demands closer and closer matching to the required outcome to be effective.

Reinforcement schedules   In the real world, rewards rarely follow every performance of a behaviour, but are instead intermittent. Thus a bumble bee probing a flower receives nectar on most occasions, but sometimes another bee has beaten it to the reward and the flower is empty. The frequency with which a particular behaviour is rewarded is referred to as the **reinforcement schedule**. A large body of work shows that reinforcement schedules have important and predictable effects on the strength, rate of performance and resistance to extinction of conditioned responses. We shall discuss just a few examples.

Continuous reinforcement schedules, where each performance of a behaviour is rewarded, not surprisingly tend to produce the strongest and most consistent responses and are generally used in initial shaping procedures. **Fixed ratio schedules**, in which the animal has to respond a certain number of times before being rewarded, encourage high rates of response because the faster the animal responds the sooner it obtains a reward. As Goodenough *et al.* (1993) put it, fixed ratio schedules are like piecework in factories, where an employee is paid on completion of a given number of items and therefore strives for a high rate of production. **Variable ratio schedules** also generate high rates of response. In this case, the number of responses required for a reward varies randomly. The response rate is high because, once again, faster responses mean quicker rewards, but responses are more resistant to extinction than those of fixed ratio schedules because random unrewarded gaps are an inherent feature of the schedule and thus less likely to be a disincentive. Variable ratio schedules are therefore like gambling machines: there is an average level of reward, but the payout in relation to expenditure is unpredictable.

The nature of stimulus–response associations   Early views of operant conditioning saw the outcome of responses as automatically reinforcing any preceding S–R association. However, several lines of evidence suggest this is too simplistic, and contrary views abound. One problem is distinguishing empirically between operant and classical causes of conditioned responses. **Autoshaping**, where an animal is induced to respond to a Skinner box key, or other device, by highlighting the device in association with a reward, is a good example. Opinion is predominantly in favour of a classically conditioned (CS–UCS association) account of autoshaping, but the animal may instead learn that the response *causes* the reward to become available, so reinforcing an operant S–R relationship. One way to decide between these two possibilities is to look at the effects of an omission schedule (see 4.2.1.1) on the development of the response.

Figure 6.15 (a) Pigeons trained to peck a key for food, do so with the bill open and the eye closed, just as if the key *was* food. (b) When trained to peck for water, however, the bill is closed and the eye open, as if the bird was actually drinking. After Moore (1973).

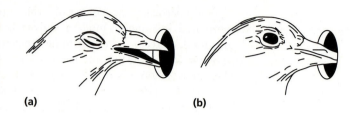

(a) (b)

Instead of rewarding a response each time it is performed, an omission schedule delivers a reward when the response *does not* occur. Thus the response will never be followed by a reward, which therefore cannot strengthen a preceding S–R relationship. If autoshaping is a result of an operant S–R association, then it should be ineffective under omission schedule conditions. Williams & Williams's (1969) experiment in 4.2.1.1 used just such a procedure to look at pigeons pecking at an illuminated key in a Skinner box. As we saw there, imposing an omission schedule on a key-light, key-peck association did little to abolish the conditioned pecking response. One conclusion, therefore, is that the remaining loose and intermittent pairing of the key-light CS with the food UCS was sufficient to generate the pecking CR, which thus came about by classical conditioning (Pearce 1997). Detailed video analysis of pigeons pecking keys during autoshaping supports this conclusion further by showing that the *kind* of peck delivered depends on what the bird expects as a reward (Moore 1973). When it is pecking a key for food, the peck is brief and forceful, the bill is opened at the instant of contact and the eyes are shut (Fig. 6.15a) – just like a peck at real food, in fact. When the same key is pecked for water, the response is longer and less forceful, the bill is opened only slightly on contact, the tongue is extended, and the eyes stay open (Fig. 6.15b) – just as when actually drinking. This suggests that the pigeon treats the key as if it *is* food or water. Thus it is not learning an association between *pecking* the key and reinforcement (operant conditioning), but between the key itself and reinforcement (classical conditioning).

Classical and operant conditioning may also act in combination, rather than as alternatives, to produce a CR. Mowrer (1960), Konorski (1967), Rescorla & Solomon (1967) and others have argued that classically conditioned motivational factors might play a role in operant conditioning. A CS might, for instance, lead to an arousing representation of the UCS which then has a preparatory effect on the CR. Experiments have supported this idea by showing it is possible to manipulate the strength of an operant response by presenting a classically conditioned CS at the same time (Lovibond 1983).

Discrimination learning   So far we have looked at the role of reinforcement and associative learning in the context of responses to one particular stimulus, albeit sometimes a compound stimulus. In many situations, of course, animals are faced with choices of stimulus. Which they choose may depend on subtle differences in the reward offered by each stimulus and the animal's ability to learn the appropriate associations. Such **discrimination learning** has been subject to extensive investigation and become fertile ground for arguments about learning theory in general. The following are just some of the issues.

Matching: Making a choice often involves experience with alternative reward schedules over a period of time, rather than an instantaneous decision. Thus the relative rate or duration of reinforcement offered by the various options becomes important. Nevertheless, preferences can sometimes be expressed quickly, after as little as one inter-reward interval in some cases (Mark & Gallistel 1994).

Figure 6.16 The relative frequency with which pigeons responded to a key A as a function of the relative frequency of reinforcement at key A (results shown for three different reinforcement schedules). The correspondence between reinforcement and response is referred to as 'matching'. The line through the origin represents a 1 : 1 relationship. See text. After Herrnstein (1961).

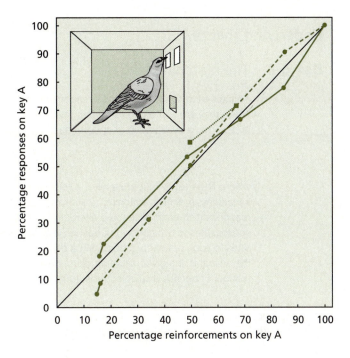

Herrnstein (1961) and Catania (1963) found that pigeons presented with two different rates of food reward (**concurrent variable-interval schedule**) in a Skinner box delivered most pecks to the key yielding the greater rate or duration of food presentation. However, they did it in a way that **matched** proportionately the distribution of rewards across the two keys (Fig. 6.16). This result has now been found so consistently in so many species and contexts that it has become referred to as the **matching law** (Box 6.3; Williams 1988; Shettleworth 1998). Animals match responses to more or less any measure of reinforcer – size, rate, duration or delay – and matching is sufficiently well established to be used as a diagnostic in assessing how animals value options (Hamm & Shettleworth 1987). The 'matching law' also has obvious functional implications, since it relates to the reward maximisation models of optimal foraging theory (e.g. Staddon 1983; Shettleworth 1998; Fantino 2001; see Chapter 8) and the distribution of predators across resource patches (Shettleworth 1998; Chapter 7). A problem with the idea that matching maximises reinforcement, however, is that rewards are sometimes maximised by allocating more (if not all) effort to the option yielding the greater reinforcement (so-called **overmatching**). At other times the animal would do better by responding more to the poorer option (**undermatching**). Which the animal should do (match, overmatch or undermatch) depends among other things on the context (e.g. social *vs* solitary foraging [Gray 1994; Thuijsman *et al.* 1995]) and predictability of the concurrent schedules. We shall return to this problem again in Chapter 8.

Choice and reinforcement schedule control: While response matching with concurrent variable interval schedules may seem intuitively sensible, there are problems beyond the departures from expectation touched on above. If very different reinforcers or reinforcement schedules are used, the subsequent measure of choice may be confounded by the nature of responses to the particular reinforcers or schedules used (so-called **reinforcer-** or **schedule-control**). For instance, a predator may spend more time hunting for food during the day than it does sleeping, but we cannot infer from this

Underlying theory

Box 6.3   The matching law

Animals tend to distribute their responses across options according to the relative rate at which they are reinforced (see text). The distribution of choices thus matches the distribution of rewards, a relationship referred to as the matching law and expressed as:

$$\frac{R_1}{R_1 + R_2} = \frac{r_1}{r_1 + r_2} \tag{6.3.1}$$

where R denotes response rate, r reinforcement rate and subscripts alternative reward schedules. A similar matching relationship applies to the amount of time an animal spends responding to each of the two alternatives. While the equation above applies to relative response rates to two concurrent options, it can be modified easily to cater for *absolute* response rates to each option by introducing a constant k, the maximum overall rate of responding by the animal in the given situation (thus, here $k = R_1 + R_2$). The constant is independent of the number of options available. Substituting k in the equation gives:

$$\frac{R_1}{k} = \frac{r_1}{r_1 + r_2} \tag{6.3.2}$$

so, the response rate for at each option is:

$$R_1 = \frac{kr_1}{r_1 + r_2} \tag{6.3.3a}$$

and

$$R_2 = \frac{kr_2}{r_2 + r_1} \tag{6.3.3b}$$

While these equations predict many aspects of choice on concurrent schedules, they do not predict all and have been modified in various ways to widen their applicability. For example, there are problems when either r_1 or r_2 are zero (i.e. there is only a single schedule). To get round this, Herrnstein (1970) argued that there is strictly no such thing as a single schedule, because there are almost always other reinforced behaviours (e.g. daydreaming, defaecating, scratching an ear) going on simultaneously with the schedule presented by the experimenter. Thus there is always a choice of some kind. Herrnstein therefore introduced a further term, r_0, into the equation to cater for 'other' reinforced responses outside the design of the experiment. The equation for R_1 thus becomes:

$$R_1 = \frac{kr_1}{r_1 + r_2 + r_0} \tag{6.3.4}$$

Herrnstein (1970) increased the generality of the equation yet further by introducing a final constant, m, which allows for interactions between multiple concurrent schedules. Thus:

$$R_1 = \frac{kr_1}{r_1 + mr_2 + r_0} \tag{6.3.5}$$

The matching law is sometimes stated in a generalised form that can be applied across experiments as:

$$\log(R_1/R_2) = \log k + b \log(r_1/r_2) \tag{6.3.6}$$

where b is an additional scaling parameter fitted to each data set.

Based on discussion in Fantino & Logan (1979) and Baum (1974).

that the predator prefers hunting to sleeping. The requirements of hunting behaviour mean that it must be performed more often and for longer than sleeping. Similarly, reinforcement schedules may produce different levels of response as a result of their intrinsic structure. We have seen this already in the relative response rates on concurrent variable ratio and **fixed interval** schedules. Inferring *choice* on this basis would clearly be dangerous.

Learned 'concepts': Experiments in which animals are required to discriminate between more or less complex stimuli have led some authors to suggest that they are able to acquire sophisticated concepts, such as the idea of shape, colour, flowers or wetness. Perhaps the most striking example is Irene Pepperberg's African grey parrot (*Psittacus erithacus*) Alex. Like many parrots, Alex has been taught to talk, but Pepperberg has used this over many years to probe his understanding of the world around him. Her experiments suggest that Alex has acquired, and can articulate, an astonishing range of concepts, from relatively basic things such as colour, shape and number, to the decidedly more abstract properties of sameness and difference (e.g. Pepperberg 1991). However, other species have also demonstrated impressive abilities in this direction, if via more conventional experimental methods. Pigeons are a noteworthy example.

Using Skinner box procedures, Herrnstein *et al.* (1976) presented pigeons with photographs of various scenes projected above a key that they had to peck to obtain a reward. The birds were given the task of distinguishing between pictures that contained a certain feature and those that did not. The feature could be anything (e.g. human beings, fish, leaves, water, abstract shapes [see Watanabe *et al.* 1993]). Pigeons turned out to be astonishingly good at the task, not just recognising the relevant feature, but recognising it in almost any form or context in which it was portrayed. Thus they could correctly identify a 'human being' whether it was a close-up of someone's face, a picture of a person sitting on the floor or a crowd scene. Pigeons are also extremely good at recognising complex abstract shapes in different orientations (Fig. 6.17; Hollard & Delius 1982), a task commonly used in human intelligence tests.

People have made various claims for these seemingly impressive examples of discrimination. Some see them as evidence that other animals are able to form concepts in much the way we can; others view them as no more than discrimination learning with unusually complex stimuli (Wasserman & Astley 1994). Yet others rule the ability of non-human species to form concepts out of court on principle, on the grounds that concepts can be defined only by language (Chater & Heyes 1994). Certainly we have to be careful how we interpret the animals' responses. We cannot, for example, conclude that Herrnstein *et al.*'s pigeons saw the slide images as pictures in any sense that we would recognise. Indeed, it is not necessary for animals to see the images as representations of objects at all in order to be able to classify them. The available experimental evidence suggests that pigeons responded to Herrnstein *et al.*'s photographs simply as patterns of coloured patches. Even monkeys tend to respond at this level. D'Amato & van Sant (1988) trained capuchin monkeys (*Cebus apella*) to discriminate between slides with and without people. Like pigeons, the monkeys performed the task very well, but careful analysis revealed they were basing their discrimination at least partly on the presence or absence of red patches. Slides containing other kinds of red patch, such as a slice of watermelon or dead flamingo, were treated as slides with people. As Shettleworth (1998) argues, what animals are doing in these kinds of experiment is best interpreted operationally rather than in terms of hypothetical mental processes, even though these processes might indeed exist. Thus **category discrimination** with broad stimulus generalisation, rather than concept learning, is probably the safer interpretation of the results.

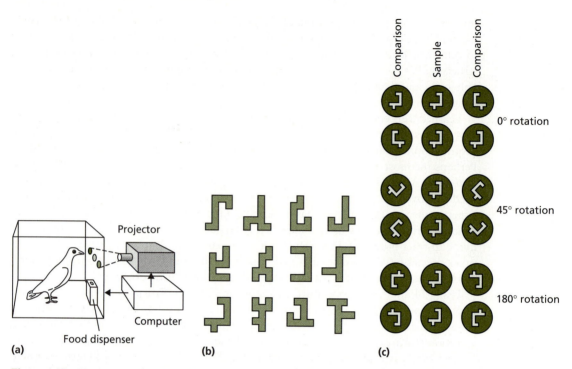

Figure 6.17 Pigeons tested with projected images (a) of complex shapes (b) were remarkably good at recognising the shapes again even when they were rotated at different angles (c). After Manning & Dawkins (1998) from Hollard & Delius (1982). Reprinted with permission from 'Rotational invariance in visual pattern recognition by pigeons and humans', *Science*, Vol. 218, pp. 804–6 (Holland, V.D. and Delius, J.D. 1982). Copyright 1982 American Association for the Advancement of Science.

We can begin to understand why pigeons might have this capacity for complex visual discrimination learning by considering their normal way of life. Pigeons have evolved for flight and rely very heavily on visual information. We might therefore expect them to be able to pick out important visual features in a fast-moving, fast-changing environment. Trees must be recognised in the distance as diversely shaped masses, but also closer to as the bird selects a roosting site and swoops into the foliage to alight. Looked at like this, the pigeons' apparent ability to categorise is arguably more to do with perception and sensory processing than it is with mental sophistication (Manning & Dawkins 1998). Nevertheless, there do appear to be some basic similarities in the way pigeons and humans classify images that suggest a degree of congruence in cognitive processing (Delius 1992).

6.2.1.3 Latent learning

The status of Thorpe's fourth category of learning, that is whether it should or should not be regarded as a form of associative learning, has been a matter of some debate. The crux of the issue is the absence of any obvious reinforcement behind the learned associations. If a rat or mouse is placed in a maze that contains no reward, the animal nevertheless investigates the apparatus, running up and down the various alleys and dead ends, and inspecting the empty goal box. When required at a later date to navigate the maze for a food reward, such individuals do better than naïve animals experiencing the maze for the first time. Figure 6.18 shows the results of a classic early experiment along these lines by

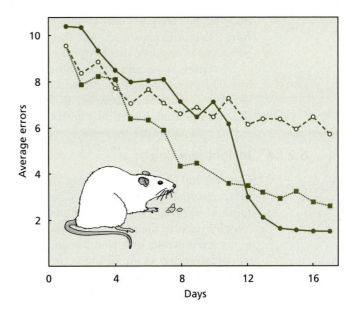

Figure 6.18 Latent learning by rats in a maze. Animals trained without food in the goal box for 11 days, then given food (●), showed a sharp drop in error rate compared with those that continued without food (○). Error rates even dropped slightly below those of rats trained with food throughout (■). The precipitate decline in errors suggests the rats must have learned something about the maze prior to being rewarded. After Tolman & Honzik (1930).

Tolman & Honzik (1930). Clearly the animals learned something on their initial visit to the maze even though they received no reward for their effort. For many years, learning theorists preferred to treat such **latent learning** as a separate category of learning, but as views about the role of reinforcement in operant conditioning changed, so did opinions as to the associative nature of latent learning. As long ago as 1970, Hinde argued that latent and operant learning differed only in the timing of their expression; both involved memorising aspects of a situation, but only in operant learning was memory translated into action straight away.

As in classical and operant conditioning, there are interactions between latent learning and other forms of learning. This is reflected in events in the CNS. Ohyama & Mauk (2001), for example, found that conditioning of the eye-blink response in rabbits (see Fig. 6.10) depended on changes in the interpositus nucleus of the cerebellum. However, these changes were preceded by latent changes in the cerebellar cortex that had important effects on the magnitude and timing of the conditioned response. Experiments in which the two regions of the cerebellum were reversibly disconnected showed that learning of the temporal characteristics of a tone CS took place in the cortex prior to the appearance of any conditioned blink response. The latter appeared only after changes in the responsiveness of the interpositus nucleus. In other cases, for example in honey bees, the neuronal basis of latent learning appears to be distributed and connected with circuits affecting other forms of learning (Menzel 2001).

Adaptive significance

The adaptive value of latent learning is not difficult to imagine. A knowledge of the home area, its food sources, hiding places and escape routes, would pay obvious dividends when needs arose. An experiment with white-footed deermice (*Peromyscus leucopus*) by Metzgar (1967) provides a rather stark illustration. Metzgar introduced pairs of mice into a room containing a screech owl (*Otus asio*). One of each pair had previously had the opportunity to explore the room for a few days, while the other was unfamiliar with it. The owl caught a mouse on 13 out of the 17 trials in the experiment; of these, 11 were mice that had not previously been in the room. An almost inescapable conclusion, therefore, is that prior experience with their surroundings gave the familiarised individuals an advantage in avoiding attacks by the predator.

Sometimes important spatial information is gleaned passively as the animal moves around its environment over time. On other occasions it seems to be acquired during specific reconnaissance activities, as, for example, in the orientation flight of the female digger wasps as they leave their burrows to hunt (Tinbergen 1951; see also 5.1.1.3).

6.2.1.4 Insight learning

Classical and operant conditioning are often referred to as 'simple' associative learning (Mackintosh 1983), implying that some animals are capable of other, more complex, forms of learning. One of these is what Thorpe and others have called **insight learning**. 'Insight' refers to the rapid apprehension of solutions to problems – too rapid to be due to normal processes of trial-and-error. It is the equivalent to our own 'Aha!' experience, where the answer to a problem occurs to us in a flash, even though we may have been wrestling with it fruitlessly for days. However, while responses may be too rapid for *physical* trial-and-error procedures to have taken place, the animal may have run through trial options quickly in its head. Once again, we are back with the problem of the animal's private mental processes and how to infer them from the indirect evidence of behaviour (see 4.2.1). Nevertheless, if we accept the possibility of mental trial-and-error, we acknowledge that the animal has at least some capacity for reason, a capacity that is doubtfully distinguishable from insight (Manning & Dawkins 1998).

Several well-known experimental paradigms have been used to assess reasoning and insight. We have seen some already in the detour experiments and cognitive maps in Chapter 4 (4.2.1.2). Perhaps the best-known, however, are Wolfgang Köhler's (1927) experiments with chimpanzees. Köhler studied a captive group of chimpanzees in Tenerife during the First World War. He set them various problems, typically requiring them to obtain food that was out of reach beyond the bars of their cage, or suspended from the roof. Various items were available to the animals in their cage, and they quickly learned to use these in novel and inventive ways to get to the food (Fig. 6.19). For example, they used bamboo poles as rakes to pull in food on the ground outside. If the poles were too short, they joined two together to make them long enough. To try to

Figure 6.19 Chimpanzees joining poles together and stacking boxes to get at food that is out of reach. After Eibl-Eibesfeldt (1975) from Köhler (1927).

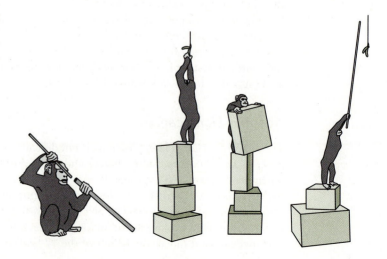

get at food suspended from the roof, the chimpanzees positioned wooden boxes underneath and climbed up on them. If one box was not good enough, they stacked another on top of it until they could jump up and grab the food. Köhler noticed that these solutions often appeared to occur suddenly to a particular animal. For instance, a male called Sultan tried unsuccessfully for over an hour to obtain food with two bamboo poles that were each too short. He appeared to give up the endeavour and sat lethargically on the ground playing with the poles. Suddenly, while he happened to be holding a pole in each hand, he apparently realised that the end of one could be inserted into that of the other, so making a longer implement. Immediately he rushed to the bars of the cage and used the connected poles to pull in the food. On a number of occasions the poles separated mid-operation, but Sultan quickly recovered the lost component and refitted it to the end of the other pole, thus suggesting he understood the functional potential of joining the two together.

Köhler interpreted the behaviour of his chimpanzees in terms of their having perceived new relationships between stimuli (the poles) and outcomes (getting at the food), relationships they had not experienced previously. In his view, they were reflecting on the problem as a whole and deducing novel solutions. However, this is not a universal view. Others have explained results such as Köhler's as the consequence of previously learned associations. For example, the chimpanzees that moved boxes into place then climbed on them might previously have learned the two responses separately, then just put them together. This idea has been tested in pigeons.

Pigeons do not usually move boxes around and climb on them, but they can be trained to do so. Epstein *et al.* (1984) rewarded pigeons for pushing a box towards a green spot on the floor, but not for pushing a box when there was no green spot. In a separate regime, pigeons were trained to sit on a box to peck at a banana suspended above. When birds that had received *both* kinds of training were put in a room without a green spot, but with both a box and a suspended banana, they responded very like Köhler's chimpanzees. While at first they attempted unsuccessfully to stretch up from the floor and get the banana, they eventually shuffled the box along underneath the banana, climbed up and pecked it. Birds trained *either* to peck a banana, but not to climb on a box to do it, *or* to push a box to a particular spot, but not to peck at a banana, were unable to put the two skills together and get the banana. They either stretched repeatedly, but ineffectually from the floor, or they pushed the box aimlessly around without climbing on it. Thus, apparently insightful behaviour can be generated from preceding learned S–R relationships.

Learning sets

Very often, we solve a problem quickly by applying our experience with similar problems in the past. Past experience allows us to perceive general rules that apply across the board, even though each individual problem we encounter may be different in detail. Thus, once we have been shown how to solve simultaneous equations with a handful of examples, we can solve any others that come our way (or some of us can). In recognising and applying a general principle like this, we are said to have formed a **learning set** (Harlow 1949). Other animals seem to be able to form learning sets too.

In his early experiments with monkeys, Harlow (1949) presented subjects with pairs of dissimilar objects, say a matchbox and an egg cup, one of which concealed food. The positions of the two objects were alternated, but food was always under the same one, and the monkeys soon learned which it was. Harlow then changed the objects for two entirely different ones, say a building block and half a tennis ball, and repeated

Figure 6.20 The formation of a learning set in a rhesus monkey. As a monkey encountered more variants of an object discrimination problem, it became quicker at solving new versions of the problem (performance improved faster in later discriminations). In this case, the monkey learned the solution was to adopt a 'win–stay, lose–shift' strategy. The first four lines represent early discriminations (1–8, 9–16, 17–24, 25–32 respectively). See text. After Harlow, H.F. (1949) The formation of learning sets, *Psychology Review*, 56, 51–65.

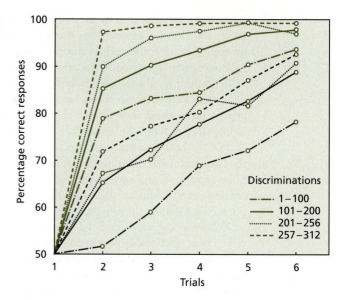

the procedure. Again the monkeys learned which object concealed food, but took about the same time to do it as previously. After several dozen such trials, however, they became much quicker at learning which was the right object (Fig. 6.20). While the objects in the problem continually changed, the monkeys had learned the general solution to it, that is: 'choose the same object each time if it yields food in the first trial, but change to the other object if it doesn't', a strategy known as *win–stay, lose–shift*. Many other species have since been shown to adopt a *win–stay, lose–shift* solution in similar circumstances. Animals can also just as easily learn the opposite strategy – *win–shift, lose–stay* – if they are subjected to so-called 'repeated reversal' training. In this case, the object concealing the reward is swapped once the animal has learned which to go for. Thus if the animal learns that A conceals a reward but B does not, the reward is hidden under B for the next few trials. Once it has learned to pick B, the reward reverts to A, and so on.

Learning sets and intelligence While learning ability as a whole features prominently in arguments about intelligence, learning sets have commanded particular attention in this respect. Harlow (1949) went so far as to say that acquiring learning sets 'transforms the organism from a creature that adapts . . . by trial and error to one that adapts by seeming hypothesis and insight. (Learning sets) are the mechanisms which, in part, transform the organism from a conditioned response robot to a reasonable rational creature.' The idea that animals can learn underlying principles rather than just immediate stimulus–response relationships has obvious appeal in the context of intelligence. Moreover, the shape of the error curve over successive trials seems to provide an objective relative measure of ability that is independent of absolute levels of response. That is, regardless of how well or poorly different species learn in the first problem, we can ask whether they improve over successive problems and eventually get it right on the second trial of each new task. The more they do, the more intelligent we consider them to be.

Early comparative studies of mammals encouraged this view. Figure 6.21 shows performance in visual learning set tasks for a range of mammal species (Warren 1965). As we might expect, Old World rhesus monkeys did better than New World squirrel

Figure 6.21 Different species of mammal differ in their ability to form demanding object discrimination learning sets. The measure of success is the percentage of correct responses on the second trial with each new set of objects (see text). Here, rhesus monkeys come out best, and rats and squirrels worst. After Warren (1965).

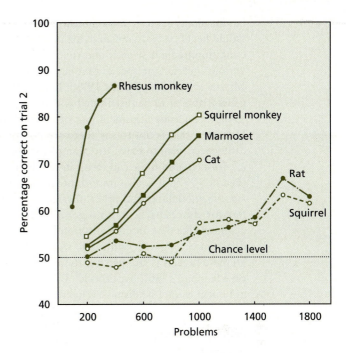

monkeys, which in turn did better than marmosets, cats, rats and squirrels. The order sits comfortably with our notion of an evolutionary ladder of intellectual prowess ascending from humbler mammals to primates. Such an interpretation, however, would be highly dangerous. For one thing, as Warren himself noted, the various species in Fig. 6.21 were tested in very different ways, thus introducing all sorts of confounding effects into the comparison (Warren 1973). Another problem is that learning sets can depend on the sensory modality of the task. The rats in Fig. 6.21, for example, did rather poorly on their visual discrimination tasks, barely rising above chance levels even after several hundred attempts. However, with spatial discrimination tasks, rats acquire learning sets within 50 trials (Zeldin & Olton 1986), and fare even better in olfactory discriminations (Eichenbaum *et al.* 1986). Further problems arise when other species are slotted into the comparison. Other mammals, for instance, do not fall where expected in the order of performance in Fig. 6.21 (Macphail 1982), while some birds seem to do as well as rhesus monkeys (Kamil 1985).

These and other difficulties led Macphail (1982, 1985, 1987) to suggest that all non-human vertebrates are in fact of equal intelligence (the so-called '**null hypothesis**' of vertebrate intelligence). Absurd as this suggestion might seem at first, diverse taxa do actually share important fundamental features of learning and memory, and the argument is not as easy to dismiss as one might imagine (Macphail & Bolhuis 2001). Take perhaps the most obvious objection to the 'null hypothesis', that species have become specialised during evolution to occupy very different niches each requiring different learning capabilities and cognitive skills (see 6.2.2). At one level this is certainly true. We have seen already that scatter-hoarding birds have a special kind of memory (3.1.3.1), and that cognitive maps tend to be possessed by animals foraging out from a central location, such as a hive (4.2.1.2). But these arguments ignore the fact that even very different niches can impose similar basic demands on mental processes. Classical conditioning is a good example. The potential for a stimulus to predict an event can exist in almost any kind of environment, and a capacity to learn the appropriate associations would benefit any organism living in them. Accordingly, the same basic rules of classical conditioning have been found to apply across vertebrates (Macphail 1982) and in the various invertebrate species that have been tested appropriately (e.g. Walters *et al.* 1981; Menzel *et al.* 1993; Dukas 1999). Indeed, the basic principles of associative learning in general appear to be distributed widely across taxonomic groups (Domjan 1983; Roper 1983). The same argument can be applied to other cognitive abilities (Pearce 1997). However, that is not to say there are no important differences between species. Subtle differences seem to exist, for example, in the precise mechanisms by which associations are formed, and these can result in different outcomes in appropriate experimental tests (Pearce 1997). It is also important to test animals with procedures appropriate to their sensory modalities and other physical attributes (tasks requiring manual dexterity are fine for chimpanzees, but no good for goldfish) in drawing fair comparisons, and to acknowledge that natural selection has shaped learning and memory to function under different ecological conditions. Unfortunately, experimental evidence that caters for all these caveats is still patchy (to say the least), so the jury will be out for some time yet on Macphail's 'null hypothesis'. But, despite its seemingly unlikely claim, Macphail's argument cannot be dismissed lightly.

Social learning

Thorpe included imitation in his definition of insight learning, by which he meant a heterogeneous range of behaviours involving some degree of copying from other individuals (Box 6.4). Social companions, or indeed any other individuals, can be a rich source of information about how to behave. By exploiting this information, an animal can save time and energy that it would otherwise have to invest in learning the task for itself. The responses of others might also save its life if they trigger appropriate action when a predator is around. Such arguments help us to understand why **social learning** might spread through a population. But social learning has also been studied for what it might tell us about the capacity for imitation and its implications for rational understanding on the part of the animal.

So what kinds of things are learned socially and how might they benefit the learner? Among the obvious contenders are skills associated with finding and consuming food, avoiding predators, choosing a mate and communicating. However, we must bear in mind that social information may be made available incidentally, such as the loud rustling made by a squirrel rooting through fallen autumn leaves, or strategically, as when a male bird of paradise displays to a female to persuade her to mate (the distinction between 'cues' or 'signs' [Seeley 1989; Hauser 1996] and signals [Dawkins 1993b]; see Galef & Giraldeau 2001 and Chapter 11). Of the contenders above, feeding behaviours have perhaps received the most detailed attention (see Galef & Giraldeau 2001 for a brief review).

Social learning and feeding behaviour Foraging efficiency appears to be one of the main selection pressures favouring social aggregation among animals (see Chapter 9), and social information about the whereabouts and quality of food is a major factor enhancing feeding efficiency in foraging groups. Norway rats (*Rattus norvegicus*) are a good case in point.

Norway rats are colonially living omnivores that eat virtually anything, but are good at avoiding things that make them ill. Naïve young rats thus have a lot to learn about what is and what is not good to eat. Not surprisingly, therefore, rats seem to be adept at

Some definitions

Box 6.4   Types of imitation

Thorpe (1963) included a range of social effects on behaviour in his definition of imitation.

Social facilitation

An animal may already possess a particular behaviour in its repertoire, but becomes more likely to perform it as a result of seeing it performed by another individual. If a bird is given water, for example, others that see it bathing may start to wash themselves even though they had water present all the time. Similarly, satiated rats may start to feed again if they see a hungry individual eating. The infectiousness of yawning is a familiar example in humans.

Local enhancement

Animals may direct their activities towards a particular part of the environment as a result of responses by others. The spread of milk bottle opening in tits is an example (see text), but the effect is widespread in foraging behaviour and can be important in locating concealed or novel feeding sites (see Chapter 9).

True imitation

Thorpe reserved the term 'imitation' for 'the copying of otherwise improbable utterances or acts'. Vocal mimicry in birds and food-washing in primates are frequently-cited apparent examples. The key distinction between true imitation and local enhancement is that, in the latter, the observer has learned something about the environment (i.e. something has happened here) but not necessarily behaviour (it happened because *A* did this). The existence of true imitation and its implications for intelligence in non-human animals are the subject of ongoing debate (see text).

picking up social information about foodstuffs. In fact, pups begin to learn about food and develop later feeding preferences while still *in utero* (Galef 1996a). They continue to learn from their mother while suckling because flavours of foods ingested by the mother find their way into her milk (Galef & Sherry 1973; see Hudson *et al.* 1999 for similar learning in rabbits [*Oryctolagus cuniculus*]). Once they have weaned, the newly independent young rats begin to leave the nest and forage near other conspecifics, or areas recently visited by conspecifics. All these routes help to ensure that young rats are educated in their choice of food. However, pioneering experiments by Jeff Galef at McMaster University have shown that there is more to the social guidance of foraging in rats than just early familiarity.

Galef surmised that the breath of an animal that had just fed could carry olfactory information about recently ingested food. Other individuals could then pick up this information and make use of it. Rats are good candidates for this because they frequently sniff the facial area of other rats they encounter. Galef & Wigmore (1983) therefore set up a simple experiment using pairs of animals in which one individual (the *demonstrator*) was fed rodent pellets that had been flavoured in a distinctive way, for instance with cocoa or cinnamon. The other rat (the *observer*) was then exposed to the

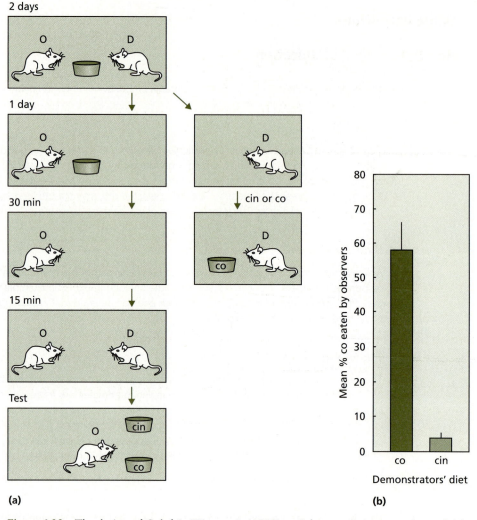

Figure 6.22 The design of Galef & Wigmore's (1983) social transmission experiment (a) in which 'observer' rats (O) acquired information about food from a 'demonstrator' (D) and used it to make a choice later on. In this case the 'demonstrator's food was flavoured with cocoa (co), and the 'observer' showed a marked preference for cocoa over cinnamon (cin) in a later choice test (b). See text. After Shettleworth (1998) from Galef & Wigmore (1983).

demonstrator before being given a choice of foodstuffs (Fig. 6.22). As Fig. 6.22 shows, when observers were exposed to cocoa-fed demonstrators, they predominantly chose cocoa in the preference test. When the demonstrator had been fed cinnamon-flavoured pellets, the observers preferred cinnamon.

Clearly, then, feeding preferences in rats can be influenced by social information. But is this any more than a brief localised effect, or can it result in 'cultural traditions' within groups of animals? To find out, Galef & Allen (1995) established rats in groups of four and trained them to prefer one of two diets (pellets flavoured with Japanese horseradish or cayenne pepper) by making then ill when they ate the other kind. They then gradually replaced animals in these 'founder' groups with naïve rats until the group comprised animals that had never experienced illness after eating either of the diets. The results were

Figure 6.23 Curio *et al.*'s (1978) apparatus in which a 'teacher' blackbird could teach a naïve 'pupil' to mob a non-predatory object. See text. After Gould & Gould (1994).

clear. Rats maintained the dietary preferences of their predecessors even though there was no longer a disincentive to reinforce it. The preference was maintained for four generations of naïve rats in one case, and was transmitted even when new group members had not fed in the presence of established animals but had merely interacted with them.

Social learning in other contexts Experiments based broadly on the demonstrator–observer principle have tested for social learning in other contexts. Eberhard Curio and coworkers, for example, looked at socially transmitted mobbing responses in birds. Blackbirds (*Turdus merula*) were set up in the 'teacher–pupil' apparatus in Fig. 6.23 (Curio *et al.* 1978). The 'teacher' bird was presented with a stuffed owl in the central chamber, and responded with the characteristic mobbing call normally elicited by predators in the wild (see Chapter 9). This stimulated mobbing behaviour in the 'pupil' out of sight on the other side of the apparatus. The 'pupil', however, was presented not with an owl, but with a non-predatory stimulus such as a harmless bird or plastic bottle. The result was that the 'pupil' became classically conditioned to mob the neutral stimulus when later encountered on its own. When conditioned 'pupils' then became 'teachers', they transmitted their novel, culturally acquired image of a predator to new 'pupils', a 'tradition' that could be perpetuated for up to six 'generations' of birds (Curio *et al.* 1978).

Mating preferences can also be influenced by social example. While partners of the opposite sex may be chosen on the basis of various individual attributes (Chapter 10), simply seeing others of the same sex attending them may be enough to influence preferences. Some of the evidence for, and debate surrounding, such **mate choice copying** is discussed in 10.1.3.2.

Figure 6.24 The spread of potato-washing in a group of Japanese macaques (*Macaca fuscata*) on Koshima Island in the 1950s. See text. From Galef (1996a).

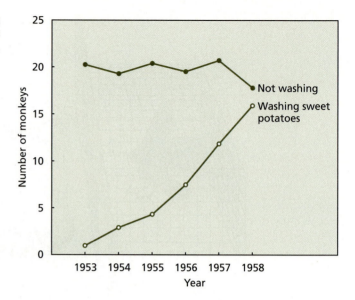

Imitation We have seen that various sources of information can lead to social influences on behaviour. In few of the above examples, however, would we be tempted to suggest that animals were directly imitating the actions of others. Imitation is at once of great interest, because it might suggest advanced appreciation of how things can be achieved, but is difficult to demonstrate unequivocally, because of the many other ways social learning can resemble it. The famous potato-washing Japanese macaques (*Macaca fuscata*) (see Itani & Nishimura 1973; Nishida 1987) are a salutory example.

A colony of macaques on Koshima Island in Japan was provided with sweet potatoes. While the monkeys happily ate them, the moist tubers frequently became covered with sand. In 1953, however, a female called Imo was noticed taking sandy pieces of potato to a stream and washing them clean. Very quickly, the behaviour spread through the colony (Fig. 6.24), initially to individuals associated with Imo, but then further afield. The spread was almost universally interpreted as an innovative behaviour by one individual being imitated by others. However, several factors militate against this conclusion (Galef 1996b; Shettleworth 1998). For one thing, transmission by imitation should start slowly, because there are few exemplars to copy, then speed up as more and more individuals demonstrate the skill. This is clearly not the case in Fig. 6.24 (though see Lefebvre 1995). The more or less constant (and rather modest) rate of spread in the Koshima macaques is more consistent with individuals learning the skill independently (Shettleworth 1998). A second point is that washing food may not be as remarkable as the macaque story implies. Indeed it may be something that individual monkeys pick up for themselves quite easily. This latter conclusion is supported by evidence from several species, in which the juxtaposition of dirty food and water, and a widespread tendency to play with objects in the water, frequently led to washing behaviour (e.g. Visalberghi 1994).

Many other claims that animals imitate have foundered on the same kinds of criticism (see Shettleworth 1998; Heyes *et al.* 2000; Galef & Giraldeau 2001), one of the best-known being the puncturing of milk bottle tops by blue tits (*Parus caeruleus*) to get at cream (Fisher & Hinde 1949; Hinde & Fisher 1951). The local spread of puncturing behaviour after it first appeared in the early–mid-twentieth century in the UK was initially attributed to birds copying one another. Later teacher–pupil experiments with

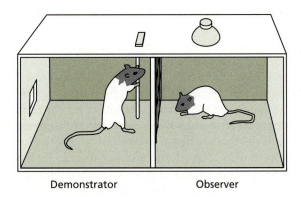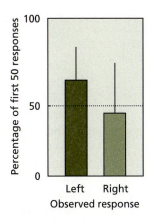

Figure 6.25 Observational learning in rats, in which an 'observer' learns from a 'demonstrator' the direction in which to swing the suspended pole. In this case, the 'observer' learned to push the pole to the left as viewed from the 'demonstrator's' perspective. See text. After Shettleworth (1998) from Heyes & Dawson (1990).

parids in the laboratory, however, suggested that birds simply learned the skill independently after having encountered already opened tops (Sherry & Galef 1984, 1990).

So *is* there any good evidence for true imitation? The question is not easy to answer, as a glance at the ongoing debate will verify (see e.g. Zentall 1996; Shettleworth 1998; Heyes 1993; Heyes *et al.* 2000 for a flavour). The most promising evidence, however, comes from a procedure known as the **two-action test** (Heyes 1996).

The principle of the two-action test goes back to Thorndike (1911), who put chicks in a puzzle box with two possible, and equally easy, escape routes. The chick was allowed to watch one of the routes being used and then given a choice itself. If it chose the route it had seen demonstrated, instead of choosing either at random, then the chick was capable of imitation. Later refinements of these kinds of test met with varying success depending on the species and task concerned (see Galef *et al.* 1986). However, an experiment with rats by Heyes & Dawson (1990) has provided strong evidence for imitation through observation. Demonstrator and observer rats were established in a partitioned cage as in Fig. 6.25. The demonstrator had been conditioned to push a hanging pole in one direction (right or left) to obtain food, and the observer was allowed to watch the demonstrator work for a sequence of 50 rewards. The observer was then moved into the demonstrator's erstwhile compartment and allowed to push the pole. Importantly, observers received a reward *whichever* way they pushed the pole, but despite this, they showed a significant tendency to push it in the direction shown by the demonstrator (Fig. 6.25). This is an important result because it suggests that the observer rat had learned not simply to push the pole for food, but to push it in a particular direction *with respect to the body of the demonstrator*. Further refinements of the experiment confirmed the outcome and showed crucially that the pole had to be moved by a demonstrator (automatically controlled movement was ineffective). The conclusion that rats were imitating the demonstrator is thus difficult to resist.

While a version of the two-action test has also provided convincing evidence of imitation by chimpanzees (Whiten *et al.* 1996), studies of other primates have not (Call & Tomasello 1995). Even in chimpanzees imitation was confined to directing attention towards appropriate targets rather than mimicking the actions needed to deal with them (**goal emulation**) (Whiten *et al.* 1996). This contrasted starkly with the performance of children in similar tests. The status of true imitation among non-human species thus remains equivocal.

Teaching   It is clear that animals of various kinds can learn from the actions of others, albeit with different implications for the mental processes involved. While it is helpful at a descriptive level to refer to demonstrators and observers, or tutors and pupils, this is a far cry from suggesting that demonstrators or tutors set out to **teach** their acolytes. They are simply doing what they do and incidentally making information available, even if the situation is deliberately rigged to that end by an experimenter. Curio *et al.*'s mobbing blackbirds above are a good example. But if social learning is advantageous, it is not difficult to imagine circumstances in which it would pay one individual actively to teach another certain skills, perhaps how to deal with a particular food, or negotiate a difficult obstacle. Parents and offspring are the most obvious candidates for this kind of relationship. So do any other animals teach?

To answer this, we must first decide what would constitute teaching. Caro & Hauser (1992) suggest that, to be regarded as teaching, an animal must change its behaviour in the presence of naïve individuals in such a way as to facilitate their learning. There should be some cost to the didactic act to the teacher, but also some reproductive benefit, hence the intuitive likelihood of it occurring between close relatives. The most likely contexts in which teaching might evolve are those where complex skills must be learned but take a while to perfect – hunting skills, for example. There is much anecdotal evidence that predators, from domestic cats to cheetahs (*Acinonyx jubatus*), meerkats (*Suricata suricata*) and ospreys (*Pandion haliaetus*) (Caro & Hauser 1992), teach their offspring about what to catch and/or how to catch it. Chimpanzees also appear to teach infants the knack of using tools to crack nuts, for instance by providing tools when infants are next to a suitable anvil, or orientating nuts properly when infants fail to crack them (e.g. Boesch 1991). Some of the examples are compelling, but evidence that the various parties are benefiting from the acts in each case is often lacking, and we must remember that many complex behaviours can mature adequately without specific experience (6.1).

Perhaps surprisingly, given their lowly reputation in the intelligence league, a promising example comes from domestic chickens. Sherry (1977) looked at food-calling behaviour in mother hens. Here, the mother bird adopts a characteristic posture and utters a distinctive call while repeatedly picking up and dropping food items and pecking the substrate with food in her bill. Chicks are attracted by the display and, through local enhancement (Box 6.4), start to peck at the food. The display appears to be costly to the mother (it takes time and energy away from her own feeding and may attract predators), but chicks do seem to learn about food from items presented by hens (Suboski & Bartashunas 1984). Food-calling thus appears to meet the important criteria for teaching.

6.2.1.5   Imprinting

The final category of learning in Thorpe's classification is **imprinting**. Imprinting was first identified by Lorenz and has long been considered to show characteristics that set it apart from other kinds of learning, including long-lasting effects, irreversibility and occurrence during some kind of **critical period** in the animal's development. However, these distinctions have been softened somewhat in the light of more recent work. It is clear that they often depend on how responses are measured, and that they vary with the life history strategy (and thus the premium on rapid early learning as opposed to ongoing development throughout life) of the animal. It is also clear that, despite the absence of obvious reward, and its somewhat unusual features, imprinting has much in common with associative learning (e.g. Hollis *et al.* 1991; Bolhuis 1999) and most people would now no longer make a hard and fast distinction between the two.

Imprinting is a narrowing of the range of stimuli that elicit a particular social response. The process takes place early on in the animal's life and seems to be directed towards a mother object (**filial imprinting**), though other social stimuli, such as siblings, can play a similar role indirectly, and sensitive learning periods also occur in other behavioural contexts (e.g. navigation [Gagliardo *et al.* 2001; see Chapter 7]). Quite what counts as a mother figure can be surprisingly open-ended. Domestic chicks and ducklings, for instance, have been imprinted on objects ranging from people and canvas hides to balloons, matchboxes and blocks of wood. However, birds often show naïve preferences for certain characteristics such as colour, shape, size and movement, and can rapidly develop preferences that steer them towards their real mother figure rather than arbitrary objects (Hampton *et al.* 1995; Bolhuis 1999).

While imprinting initially relates to a parent object, it can also influence later social preferences, particularly choice of a mate (**sexual imprinting**). Schutz (1965), for example, reared mallard (*Anas platyrhynchos*) ducklings with foster parents of a different species. When the mature birds were later released into a mixed species flock of ducks and geese on a lake, many attempted to mate with birds of their foster species. However, the tendency was more pronounced among drakes, possibly because females of many of the duck species were drab and difficult to tell apart, making it advantageous to have a reliable pre-set yardstick for the 'correct' species. Females, in contrast, can choose between drakes that are brightly coloured and differ markedly between species. In species where both sexes are drab, such as doves, both males and females acquire their sexual preferences through early imprinting. The long-lasting influence of imprinting is emphasised by the ineffectiveness of so-called 'counter-experience' experiments, where birds reared by a different species are paired for long periods with conspecifics, to overcome preferences for mates of the foster species. This behavioural evidence is now supported by neurological studies. Rollenhagen & Bischoff (2000) suggested that sexual imprinting in male zebra finches (*Taeniopygia guttata*) is a two-step process, with a period of acquisition early in life, and a process of stabilisation during the first attempts at courtship. During the stabilisation phase, they found that neurons in two areas of the forebrain (referred to as ANC and HAD) changed morphologically, increasing their spine density as the process progressed. Neurons in two other areas (MNH and LNH) showed a corresponding decrease in spine density. Both sets of changes appear to be underpinned by androgen secretion, since anti-androgen treatment abolishes them. However, further experiments by the authors suggested that, of the two, changes in the MNH and LNH played the greater role in the overall imprinting process.

Critical periods

Its occurrence during a well-defined critical (or sensitive) period was long regarded as a defining characteristic of imprinting. While imprinting undoubtedly takes place early in life, however, perceptions about the nature of critical periods have changed considerably since the early days of ethology. It has become clear, for example, that the evidence for a sharply defined period depends on how responses are measured. In ducklings, plotting the period in terms of percentage following responses after a single exposure to an object gives a much more sharply defined period than the percentage of birds following

Figure 6.26 The 'sharpness' of the critical period in imprinting depends on how it is measured. The figure contrasts measures based on following responses by mallard ducklings after a single exposure to a moving object at different ages (■), and during an initial exposure to the object (○). Drawn from data from Ramsay & Hess (1954) and Boyd & Fabricius (1965).

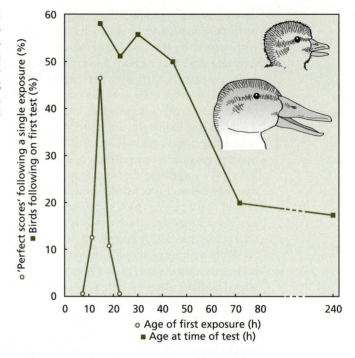

during their first exposure (Fig. 6.26). Social factors can also play a role: chicks kept singly remain responsive to moving objects much longer than those kept in groups, perhaps because the latter imprint on one another (Guiton 1959). Such imprinting between siblings may facilitate a cohesive response by the brood where only some chicks have imprinted on the mother (Boyd & Fabricius 1965). Interactions between developmental age (age since conception) and postnatal/hatching age in narrowing responsiveness can further complicate things (e.g. Gottlieb 1961; Landsberg 1976).

In the light of this kind of evidence, Bateson (1979) put forward a more open-ended model of critical periods, drawing an analogy with the carriages of a train (Fig. 6.27). The train represents a developing animal travelling one way from 'Conception' to a place where it vanishes from the tracks. Each compartment and its occupants reflect a behavioural system that is sensitive to the external world at a certain stage of development, represented in Fig. 6.27 by the sudden opening of opaque compartment windows. In the extreme situation (Fig. 6.27a), all the windows are closed for the first part of the journey, but are flung open at a particular moment to expose the occupants and then closed again. The occupants of all the compartments are thus exposed to the world at the same time. Alternatively (Fig. 6.27b), the windows of different compartments may be opened and closed at different times, so that different behavioural systems become sensitive at different stages of development. A third possibility (Fig. 6.27c) is that the windows of a compartment are opened but never closed. Here, any end to the critical period results from changes to the occupants (behavioural systems) themselves rather than a cut-off mechanism. Clearly, then, several arrangements could generate critical periods, each differing in timing and discreteness. The key to understanding them lies in the mechanisms underlying the onset and termination of sensitivity.

The onset of heightened sensitivity to environmental cues can be due to internal and/or external factors. In many cases, it begins as the relevant sensory and motor systems of the animal reach an appropriate stage of development, with sensory experience shaping

Figure 6.27 Bateson's train carriage model of critical periods in imprinting. See text. After Bateson (1979).

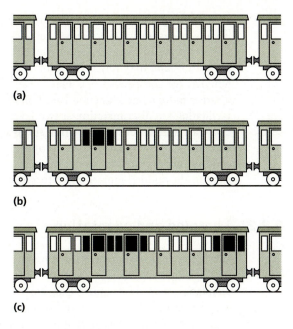

(a)

(b)

(c)

the development of neuromuscular systems (6.1.1) to impart a permanent influence on subsequent adult behaviour (Pytte & Suthers 2000). Hormonal changes can also bring about changes in sensitivity (3.3.1.2). These endogenous changes then interact with environmental cues to define a period of sensitivity in the development of particular behaviours. For instance, visual filial imprinting in chicks begins once birds are able to register and respond to visual stimuli, but experience with light is needed to seed the onset of a critical period (Bateson 1976).

Two mechanisms have been suggested to account for the decline in sensitivity at the end of a critical period. The *internal clock hypothesis* assumes a wholly endogenous mechanism (see 3.4) that switches the period off independently of external cues. The *competitive exclusion hypothesis*, on the other hand, assumes that certain external factors play a crucial role by effectively closing down the potential for others to influence responses (thus if the image of its mother influences the growth of neural connections in its brain, a chick's responsiveness becomes tailored to the maternal image and therefore less likely to respond to other images, such as that of an adult duck [Bateson 1987]. This is similar to the exclusion effects that can operate between elements of compound stimuli [Bolhuis 1999]).

Adaptive significance

So what is the function of imprinting? It seems to be a curious half-way house between hard-wired inflexibility and open-ended learning. If it is important to narrow responses to certain restricted stimuli, why not go the whole hog and evolve a hard-wired template? The answer seems to be that target stimuli for imprinting, while discrete as categories (own *vs* other species, mother *vs* not mother), vary in their characteristics within categories. Thus there is no way an individual can predict exactly what its mother, say, will look like in advance. Her unique features need to be learned. Thus there is a tradeoff between focusing quickly on the right object, and catering for unpredictable variation within classes of object.

Many explanations of imprinting suggest species recognition as the driving force, usually to ensure appropriate mating (e.g. Irwin & Price 1999). This becomes especially important in, for instance, brood parasites, where young are reared entirely by a different species and may rely on auditory imprinting to establish a 'password' recognition template for conspecifics (e.g. Hauber *et al.* 2001). Bateson (1983), however, argues for a rather more refined function based on kin discrimination (see 9.3). He sees imprinting facilitating kin discrimination at two levels: first in establishing a parent–offspring bond as a coevolution between offspring survival and discriminating parental investment, and second by allowing individuals to optimise their degree of inbreeding/outbreeding by choosing mates that differ from close kin by an appropriate degree. Several lines of evidence support the optimal inbreeding/outbreeding idea, as we shall see in Chapters 9 and 10.

6.2.2 Learning and adaptation

While the analogy between learning and natural selection as processes of adaptation (6.2.1) stands superficial scrutiny, there is an important qualification: learning is *itself* an adaptation shaped by natural selection and can be shown to have the genetic basis necessary for selection to act (Chapter 5). The ability to adjust behaviour more or less flexibly in the light of experience is adaptive in just the same sense as camouflage or defending a territory – it helps the animal survive and reproduce. To do this, however, learning must be tailored to the reproductive needs of the animal, and these, of course, vary enormously both within and between species. Thus we should expect to see evidence of adaptive *specialisation* in learning capabilities. Indeed, we have already come across some examples, for instance in food-hoarding birds, where prodigious feats of spatial memory reflect neuroanatomical specialisation of the hippocampus (3.1.3.1). But much of what we have discussed in this chapter has suggested *generalised* features of learning that apply across the board to all species and circumstances. So, is learning a general property of living organisms, or is it a heterogeneous collection of specialisations?

6.2.2.1 Ethology, psychology and laws of learning

This question has a long history that goes back to the debates between ethologists and psychologists in the middle of the last century (1.3). As with other aspects of behaviour, ethologists looked at learning in a naturalistic perspective, concerning themselves mainly with questions of function. Much information was therefore acquired about context-dependent aspects of learning, such as habituation, song learning and filial and sexual imprinting, but little attempt was made to seek common features between them. Psychologists, on the other hand, were more interested in mechanism, and adopted a highly analytical approach in contrived laboratory settings to try to uncover the basic principles that lay behind learning (Roper 1983). This **learning theory** approach focused almost exclusively on associative learning, though more recent work implies that other apparent categories of learning can be brought within its ambit (e.g. Hollis *et al.* 1991; Schmajuk & Thieme 1992).

General process theory and the principle of equipotentiality

A main plank of learning theory has been the idea that all instances of associative learning involve the same basic underlying mechanism (Roper 1983), a position known as

general process theory. This idea, while rarely stated explicitly, has provided much of the impetus for the laboratory-based approach of learning theory: if there is a common process underlying learning, it is more likely to be revealed under controlled conditions where species-specific idiosyncracies can be minimised (Dickinson 1980). It also justifies the predominance of a handful of stock laboratory species as experimental subjects, since if learning is the same in all animals, experimenters can confine themselves to the most convenient species for their work.

As Roper (1983) emphasises, it is important to be clear what 'the same basic underlying mechanism' means in this context. From the discussion in Chapter 3, one might be forgiven for thinking it meant the neural and other physiological hardware of learning. Not so. Learning theorists have traditionally eschewed 'nuts and bolts' interpretations. Instead, their 'underlying mechanisms' owe more to the 'software' explanations of Chapter 4, being concerned, as we have seen, with the relationships between input variables (stimuli and events and the way in which they become paired) and output variables (observed behaviour) as mediated in black box fashion by the internal systems of the animal.

Over time, the rules that underlay these relationships became formulated into basic **laws of learning**. These related to the effects of stimulus strength, the interval between unconditioned and conditioned stimuli, the relative intensities of the components of compound stimuli, and so on, many of which we have discussed already. Alongside general process theory, the **principle of equipotentiality** held that all stimuli and events could become associated with equal ease in all species. In other words, animals could learn to associate anything. Note that equipotentiality and general process theory are not the same thing. General process theory does not assume animals can learn anything, merely that where learning *does* take place it follows the same basic rules (Roper 1983).

Constraints on learning

The first serious challenge to general process theory and equipotentiality came from within learning theory itself during the mid 1960s (see also 1.3.4). In what turned out to be a set of landmark experiments, Garcia and coworkers (e.g. Garcia & Koelling 1966; Garcia *et al.* 1966, 1972) tested the efficacy of various punishers in causing rats to avoid certain foods. In Garcia & Koellings' experiment, two punishers (nausea and electric shock) were paired with two novel water stimuli. Water was either 'bright' and 'noisy' (there was a flash of light and a loud click when the rat contacted the spout of the bottle), or 'tasty' (the water was flavoured with saccharin). Although both punishers produced conditioned aversion, their effectiveness depended on the kind of water the rat encountered. Nausea (induced by X irradiation) was effective only when paired with 'tasty' water, while shock was effective only with 'bright–noisy' water (Fig. 6.28a). The rats thus appeared to associate only punishers and water cues that had some functional complementarity – a gustatory cue with a gastro-intestinal punisher, and a mechanical cue with a mechanical punisher. Interestingly, birds, which hunt food visually, do seem able to pair visual cues with induced sickness effects (Martin & Lett 1985). Nevertheless, birds show similar limits to association in other contexts. For example, Stevenson-Hinde (1973) trained chaffinches to perform various tasks in order to obtain food or hear a burst of recorded song. She found she could train them to land on a particular perch to hear song, and to peck a key for food, but it was extremely difficult to get them to perch for food or peck for song.

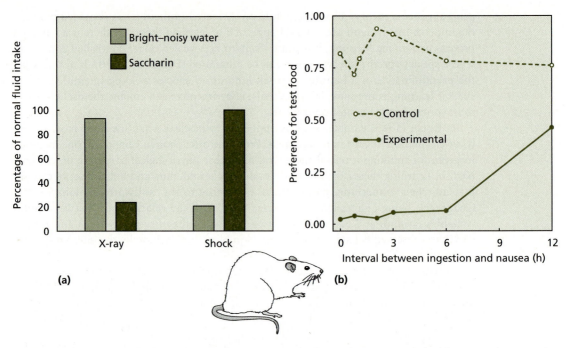

Figure 6.28 (a) Aversion learning in rats, where nausea or shock punishers were paired with sweet-tasting water or water associated with noise and flashing light. Nausea paired with sweet taste and shock paired with noise and light were the only combinations effectively suppressing water intake. After Garcia & Koelling (1966). (b) The strength of food aversion learning (suppression of eating) in relation to the interval between eating test food and nausea in rats. Control animals ate the food without being made ill. After Smith & Roll (1967).

Garcia & Koellings' results, which have since been corroborated by other studies and extend to operant conditioning procedures (see e.g. Breland & Breland 1961; LoLordo 1979), clearly challenge the principle of equipotentiality, since rats manifestly failed to associate certain pairs of stimuli. They therefore did (could?) not learn just anything. Garcia's work also challenged one of the so-called laws of associative learning under the general process theory umbrella, namely that the interval between pairs of events must be brief (no more than a few seconds long) for an association to be learned. Garcia *et al.* (1966) manipulated the delay between a rat consuming a particular food and becoming ill. Once they had recovered from the punisher, rats were offered the same food again and their response noted. The results of the experiment, and subsequently of several others (e.g. Smith & Roll 1967), showed that aversion to the food could be learned even with several hours' delay been ingestion and the animal feeling ill (Fig. 6.28b).

Garcia's work suggested that, far from being a uniform, generalised property of living organisms, learning is constrained by the particular requirements of different animals. Animals could not learn everything with equal facility; they learned what they needed to learn in a fashion that served the purpose. Biases and predispositions abounded (Box 6.5). While these effects have widely been referred to as **constraints on learning**, the phrase is perhaps not entirely apposite, since it implies that its design features serve only to *limit* learning. As Bolles (1979) points out, they can also *facilitate* it, sometimes to a pronounced degree, as in concept learning in pigeons and spatial memory in seed-hoarding birds. Perhaps the best way to think about learning is to view it as a means of working out

Underlying theory

Box 6.5   Constraints on learning

The 'constraints on learning' literature discusses a number of aspects of behaviour that bear on the outcome of learning experiments. Among the best-known accounts are those of Breland & Breland (1961), students of B.F. Skinner who used operant conditioning techniques to train animals for commercial purposes.

Using standard operant procedures, the Brelands were able to induce animals of various species to perform unlikely tricks. However, the ease with which tricks could be acquired often depended on the natural behavioural characteristics of the animal in question. In one case, a raccoon (*Procyon lotor*) was being trained to put coins in a 'piggy' bank as a shop window gimmick to attract investors. While the animal could be trained to do this, the procedure was not straightforward. Instead of simply dropping the coin in the bank, the raccoon would repeatedly dip it in, retrieve and fondle it, then dip it in again before eventually letting it go and being rewarded. The behaviour was reminiscent of the way raccoons often wash food and clean it before eating. In learning the trick, therefore, the raccoon brought some of the baggage of its normal feeding behaviour into play as well, making it an unsuitable candidate for the window job. The Brelands had similar problems with pigs, which insisted on rooting with their snouts when approaching the money box.

In other cases, it has proved difficult or impossible to condition particular responses at all. Shettleworth (e.g. 1975, 1978), for instance, tried to condition a variety of responses (rearing, scatching, digging, washing and so on) with different reinforcers (food, nest material, direct brain stimulation) in hamsters. Scrabbling, rearing and digging turned out to be easily conditioned with most reinforcers, while washing, scent-marking and scratching proved highly resistant whichever reinforcer was used. Two things could explain this. Either hamsters have a *learning deficit* and are unable to form stimulus–response associations for these behaviours, or they have a *performance deficit*, and can learn the association but not alter their response. Experiments with scratching in rats initially suggested the former (Morgan & Nicholas 1979). However, Pearce *et al.* (1978) found that scratching could be conditioned if rats were fitted with a mildly irritating Velcro collar. Thus they concluded that a performance deficit (absence of the necessary causal factor – an 'itch') was the more likely explanation.

Failure to learn may also be due to selection against learning. One of the reasons avoidance learning often depends on the kind of stimulus used (see text) is that natural hazards, such as predators and poisons, may leave little room for learning. One mistake and the animal is dead. The resulting intense selection is likely to lead to hard-wired rather than learned responses (the reaction of naïve kiskadee (Aves: Tyrannidae) chicks to the colour patterns of coral snakes is a good example [Smith 1977]). Avoidance may thus become highly stereotyped, so that animals show **species-specific defence reactions** (SSDRs). In rats, SSDRs consist predominantly of fleeing or freezing, so when expecting a threatening stimulus, these are the only responses available. This may explain why rats can associate noxious stimuli and running in a wheel, but find it difficult to pair noxious stimuli with pressing a bar: activities akin to bar-pressing do not fall within their repertoire of avoidance reactions (see Bolles 1979).

Based on discussion in Bolles (1979) and Roper (1983).

how the world works. If learning has evolved for this purpose, then the properties of learning should reflect those of the problems it is designed to overcome. Thus associative learning should be geared to real cause and effect relationships in the animal's environment, so making some associations easier to form than others. In the real world, causes usually precede effects, so conditioning is more likely when E1 occurs before E2. Certain kinds of effect (e.g. nausea) tend to result from certain kinds of cause (e.g. ingestion of contaminated food), so these are the relationships that are learned most easily. Operant conditioning seems to be easier to bring about when responses relate to technical skills (e.g. learning to deal with different types of food) needed in the natural environment.

6.3 Play

A prominent feature of development in many species (see Bekoff & Byers 1998) is the performance by young animals of seemingly purposeless activities collectively called **play** (Fig. 6.29). Kittens frequently 'attack' small, moving objects, batting them about with their paws. Young monkeys and baboons chase each other and engage in mock fights. Puppies chase their tails. Play is one of the most familiar aspects of development in animals (largely

Figure 6.29 Play in young ravens (*Corvus corax*): (a) sliding down an incline, (b) hanging by the bill and playing with a twig, (c) pulling a dog's tail. After Heinrich & Smolker (1998).

because it occurs in our most familiar companion species), even though its distribution across species is very patchy. However, it is also one of the hardest to explain.

Even defining play is difficult because it covers such a wide range of behaviours, including social interactions, manipulation of inanimate objects and individual movement patterns (so-called **social, object** and **locomotory play**; e.g. Burghardt 1998). Since it is not clear that these have the same functional or developmental significance (Burghardt 1998), Bekoff & Byers (1981) advocated a neutral definition of play as: 'motor activity performed postnatally which appears to be purposeless and in which motor patterns from other contexts may be used, often in modified form and/or altered temporal sequencing'. This highlights the central problem with play, which is its apparent purposelessness, but obvious cost. Playing consumes energy, with estimates ranging from 5% to 20% of the energy not required for growth and metabolism. In keeping with this there is evidence that play is reduced when energy is limited. Lambs, for example, play less when the ewe's milk supply is inadequate (see Fagen 1977). However, in other cases, for instance domestic cats, restricted maternal resources stimulate *more* object play by young, arguably to prepare them better for fending for themselves (Bateson *et al.* 1990). These kinds of difference between species may reflect differences in life history strategy and the likely later benefits of play (Fagen 1977). Play can also incur significant physical costs, with injuries being a common consequence of chases and mock fights, and cavorting individuals exposing themselves to predation. Harcourt (1991), for instance, found that, while young fur seals (*Arctocephalus australis*) spent only 6% of their time playing, some 86% of casualties to predatory sealions (*Otaris flavescens*) were playing when attacked. So what kinds of benefit might overcome these potential costs?

Most of the suggested benefits relate to advantages later in adult life. The fact that play often incorporates various components of adult behaviour suggests young animals may gain from practising them early on. Certainly there is evidence that early performance of motor activities enhances neuromuscular development and that play may form a developmental continuum in this respect with prenatal movements (Bekoff & Byers 1981). There is also little doubt that practice can improve specific skills (see 6.2). What is not clear, however, is whether *playful* practice enhances later performance. Some argue that playfulness is unnecessary – if you need to practise, practise seriously; playing just wastes additional time and energy. Another argument against the practice-makes-perfect hypothesis is that play sequences are often very similar to their eventual adult form, so there is not much to improve by practising. Evidence from young carnivores suggests that any later advantage from playing is small and individuals lacking experience of play soon make up any deficit (Caro 1980, 1995). Conversely, play behaviours may be so dissimilar from their later equivalents that their function cannot be one of practice. Byers & Walker (1995) review some of the relevant studies.

Another possible function of play is to generate new behaviour patterns so that the animal is better able to deal with eventualities. By allowing free combination of different activities, play might come up with novel sequences for tackling new or existing problems. Spinka *et al.* (2001) take this argument a stage further and suggest that a major function of play is actually to generate novel situations which the animal can then get used to handling – a kind of confidence-building exercise in other words. Of course, there is always the counter-argument that novel spin-offs from play are just incidental rather than an indication of underlying function (e.g. Symons 1978).

Finally, several hypotheses suggest that play enhances socialisation, the main arguments being that it helps the development of communication and reduces aggression, thus facilitating the establishment of social bonds (see Symons 1978; Biben 1998). Again, however, the evidence is patchy and equivocal.

The diversity of form and context in play defies simple pigeonholing. While it is clearly an important element in the development of many species, the bottom line is probably that there is no single function to play, but rather a mix of functions that varies from species to species. A good idea of the state of the debate can be gained from the volume edited by Bekoff & Byers (1998).

Summary

1. The development of behaviour reflects a complex interaction between organism and environment. There are two broad ways in which this interaction is expressed: through maturation of the organism's physical systems and accompanying sensory and motor capabilities, and through learning.

2. Developmental changes in motivation and behaviour may reflect development and differentiation in the nervous system and changing hormone profiles. Changes in the organism's morphology as it grows may also lead to behavioural change. However, many physical maturation effects are confounded with increased experience, and the two may be difficult to distinguish as agents of change.

3. Growth and maturation take place in the context of the organism's life history strategy. Behavioural development thus reflects the different investment priorities of different individuals. Life history strategies, and therefore behavioural development, can be influenced by maternal investment, both prenatally (through foetal programming) and postnatally.

4. Many behaviours depend on experience and learning to develop properly. There appear to be several different types of learning but there is considerable debate about the extent to which these reflect different underlying processes. Much of this debate has centred on associative learning and the nature of classical and operant conditioning, though other forms of learning may also share properties with associative learning.

5. Apparently more complex forms of learning, such as insight, concept learning and the acquisition of learning sets, have been argued to reflect intelligence. However, many seemingly sophisticated learning outcomes can be explained in terms of basic learning processes and definitive experiments are still required in many cases.

6. While psychologists long sought general laws of learning, it is clear that learning shows adaptive specialisation and that organisms differ in what they can learn and how easily they can learn it. The 'constraints on learning' literature provides numerous examples. In the context of associative learning, constraints on learning undermine the principle of equipotentiality (animals can learn associations between anything), but offer less of a challenge to general process theory (where it occurs, learning always follows the same basic rules).

7. Play is an obvious feature of behavioural development in many species, especially mammals. However, while it seems to be a costly activity, its functional significance remains unclear. Many features of play suggest some form of preparation for adult behaviour, but evidence that playfulness *per se* enhances adult performance is weak.

Further reading

Dickinson (1980) and Pearce (1997) provide excellent general introductions to learning theory. Hollis (1999) reviews her work on the adaptive value of classical conditioning. Shettleworth (1998) discusses many aspects of learning in the context of cognition and intelligence, Heyes's papers cast a critical eye over the learning–intelligence debate generally, and Heyes *et al.* (2000) and Galef & Giraldeau (2001) review social learning and its implications in the context of foraging. Roper (1983) provides a thoughtful introductory summary of the constraints on learning debate. Bateson (1979, 1983, 1987, and his many other papers) offers numerous insights into imprinting, and Fagen (1981), Martin & Caro (1985), Bekoff & Byers (1998) and Spinka *et al.* (2001) review some of the thinking about play.

Introduction

Organisms tend to be associated with particular habitats. Does this happen by chance or could it come about by organisms choosing where they live? If choice has been shaped by natural selection, organisms should choose habitats that increase their reproductive success. Is there any evidence that they do? If all individuals choose the best places, they will end up competing with each other. How will this affect their distribution? Choosing habitats implies that organisms move around. Movement from one habitat to another is often referred to as migration and contrasted with 'trivial' everyday movements such as searching for food. But is migration really a special category of behaviour, or can it be explained on the same basis as other kinds of movement? There is no doubt that some migrations are impressive, both in terms of scale and the apparent navigational skills required. How do organisms find their way? Do long-distance migrations require any special navigational mechanisms?

We have now looked at behaviour from the standpoint of each of Tinbergen's 'Four Whys'. While we have stressed the complementary nature of the 'Four Whys', it is clear that evolution provides the framework for understanding why mechanism and development work the way they do and that it does so via the functional consequences of behaviour. Differential reproductive success, even though often estimated indirectly (e.g. as the probability of survival, or some measure of performance efficiency), is the yardstick of adaptive change whatever level of explanation we are considering. However, adaptation takes place in the context of particular environments and the animal's requirements within them. So how does behaviour at each of its various different levels contribute to reproductive success in an ecological context? We shall try to answer this in the next few chapters by looking at the ecological factors underlying the choices animals have to make – where to live, what to eat, whether to associate with others, who to choose as a mate and so on. We begin in this chapter with the problem of finding a place to live.

7.1 Where to live?

Animals are generally not born with silver spoons in their mouths. With some exceptions, they have to compete for what they need, including the basic requirement of somewhere to live or spend the next period of time, whatever that may be. For all kinds of reason animals may have to search around before finding somewhere suitable. Young animals often leave their natal area to avoid competition with their parents; an area that is good for feeding may not be good for nesting or sleeping, and so on. Animals may thus move from place to place according to changing requirements and constraints. Despite this, species tend to be associated with particular kinds of habitat, and have unique ecologies that distinguish them from one another. This association between species and habitat, and the tendency for animals to maintain their movements within a species-typical range, begs a number of questions. How does the animal decide what kind of habitat is appropriate? How does it know when to move elsewhere? How does it find its way to the new location? These are some of the questions we shall examine in the present chapter.

7.1.1 Habitat choice

'Habitat' is a somewhat open-ended concept (see 7.2) that can mean different things at different times. Here, we shall use it in the broad sense of Partridge (1978) to mean simply 'the conglomerate of physical and biotic factors (e.g. shade, humidity, prey items, nesting sites) that together make up the sort of place in which an animal lives'. Clearly, the quality of the habitat in terms of these various factors, particularly resource availability and exposure to predators, is likely to affect the animal's survival and reproductive success. Association with a particular habitat might thus arise in two ways: through random dispersal and differential mortality (individuals that happen to end up in favourable habitats establish and reproduce, those that do not, die out), or through preference for places that are likely to enhance reproductive success. Given the potential advantages of the latter over pure serendipity, we might expect selection to favour mechanisms of active habitat preference. Of course, in some cases, as in many plants and zooplankton, where movement and distribution are often a function of abiotic factors such as air and water currents, there may be little opportunity to exercise a preference, though planktonic organisms can show some habitat preference – larvae of the polychaete *Spirorbis borealis*, for example, vary their choice of substrate according to exposure to wave action, and become less fussy as time spent searching increases (Knight-Jones 1953; MacKaye & Doyle 1978). Independently mobile animals, however, might generally be expected to choose where they end up. Is there any convincing evidence that they do?

7.1.1.1 Heritability and habitat choice

The simplest way to test for habitat preferences is to conduct choice experiments. Many studies have demonstrated preferences in this way and used the approach to unravel some of the habitat characteristics involved. Lindauer (1961), for example, studied the establishment of a new hive by swarms of honey bees.

In late spring, just before the new queens emerge, the old queen leaves the hive with about half her colony. The swarm flies to a nearby vantage point where it aggregates into a seething cluster. From here, scout bees fly out to investigate potential sites for a new hive, returning to the swarm and dancing on the outside to indicate the location and suitability of inspected sites, rather in the manner of the foraging recruitment dances discussed in 11.3.2.1. Other bees then fly out to inspect the location indicated, with more venturing out the better the quality of the site suggested by the dance. Eventually, the swarm 'agrees' on a location and moves in. By carrying out a series of choice experiments, Lindauer was able to show that new sites were chosen on the basis of a variety of factors, though three were particularly important: the degree of protection from weather, size (sites just big enough to accommodate the swarm were preferred) and distance from the swarm's original hive (bees preferred the farthest of otherwise equivalent sites, presumably to reduce competition for limited nectar and pollen).

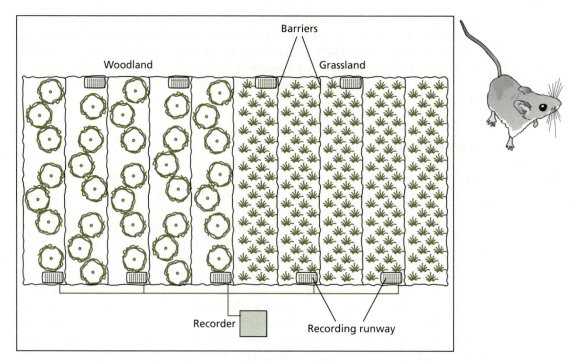

Figure 7.1 Wecker (1963) tested for habitat preferences in prairie deermice (*Peromyscus maniculatus*) by offering wild-caught and captive-bred subjects a choice between grassy and wooded areas in an enclosure. See text. From Alcock (1975), after Wecker (1964).

Choice experiments must be used with care, however. Simply allowing animals to choose between their native habitat and some alternative is unlikely to be very informative if familiarity with the former confounds the choice. While familiar habitat may, of course, provide a template for locating somewhere new, we might expect at least some evidence for heritable preferences that are independent of experience if adaptive habitat preferences have evolved by natural selection (see Chapter 2). Studies of prairie deermice (*Peromyscus maniculatus*) provide a nice example.

Prairie deermice, as their name suggests, live mainly in the open grasslands of North America. However, their grassland habitat is often mixed with low-density woodland and mice are faced with a variety of habitat types within their home range. To see whether their prevalence in grassland is the result of habitat preference, Wecker (1963) conducted a series of choice experiments in a large enclosure. The enclosure consisted of two areas of contrasting habitat: grassland and woodland, within which the movement and location of mice could be monitored with automatic data-loggers (Fig. 7.1). When Wecker introduced wild-caught *P. maniculatus* into the enclosure, he found, as might be expected, that animals spent virtually all their time in the grassland section. To control for the obvious problem of prior experience, however, he tested mice that had been born in captivity and reared in either grassland *or* woodland. Regardless of their rearing habitat, the captive-bred mice showed the same preference for grassland as their wild-caught counterparts. Deermice thus seemed to show the kind of heritable preference we might expect if habitat choice had been subject to generations of natural selection.

Interestingly, some subspecies of *P. maniculatus* live in different habitats in the wild. *Peromyscus maniculatus bairdii*, for example, lives in grassland, while *P. m. gracilis*, lives mainly in woodland. Once again, choice experiments show that both wild-caught and captive-bred mice actively prefer their native habitat when given a choice. Moreover, the two subspecies show other differences in preference, for instance for surface temperature, that accord with likely differences in their respective habitats. Habitat choice therefore seemed to be associated with a number of adaptive differences in the mice.

7.1.1.2 Habitat choice and reproductive success

If habitat preference is adaptive, we should expect it to increase the animal's reproductive success. Few studies have managed to demonstrate this directly, but several have shown that preferences are associated with enhanced survival or performance, thus providing indirect evidence for a reproductive benefit.

Feeding skills and habitat in tits

Gibb (1957) and Partridge (1974) looked at the relationship between habitat preference and foraging skills in two species of tit (Aves: Paridae) in Britain. Blue tits (*Parus caeruleus*) are largely birds of deciduous woodland, while coal tits (*P. ater*) occur mainly in coniferous woods. To see whether the two species showed a preference for their respective habitats, Gibb and Partridge performed choice experiments similar to those of Wecker above. Birds were either wild-caught or hand-reared, and chose between deciduous and coniferous 'branches' on artificial trees in an aviary (Fig. 7.2). Both wild and captive-bred birds of each species preferred the branches appropriate to their natural habitat (Fig. 7.2), suggesting a heritable genetic basis to the preference. But what did

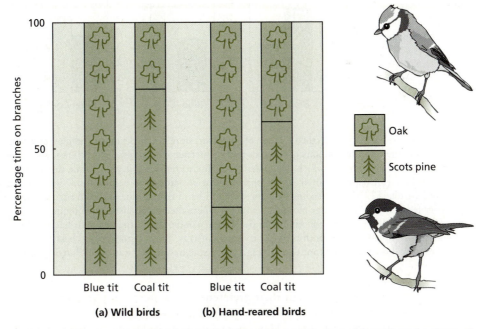

Figure 7.2 Blue tits (*Parus caeruleus*) and coal tits (*P. ater*), whether wild-caught or hand-reared, spent more time in the branches of artificial trees that came from the kind of habitat they normally occupy. Thus blue tits used mainly branches from deciduous trees, while coal tits used branches from conifers. After Partridge (1978) from Gibb (1957) and Partridge (1974).

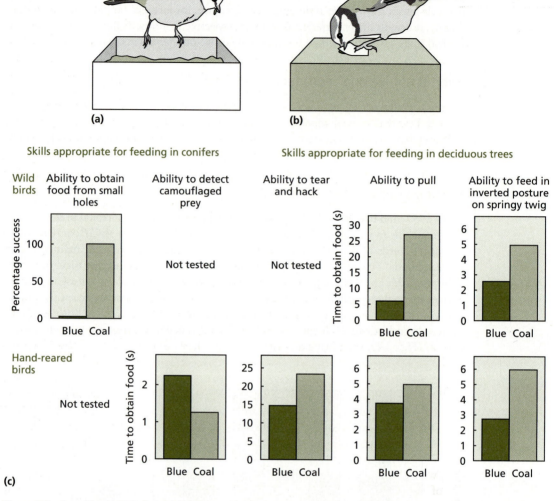

Figure 7.3 Partridge (1976) designed a range of feeders to simulate the kinds of foraging problem faced by blue tits and coal tits in their natural habitats (deciduous and coniferous woods respectively). Feeders deemed appropriate to coal tits included 'hoppies' (a), which demanded an ability to find cryptic prey, while those appropriate to blue tits included 'pullies' (b), requiring an ability to pull and tear at material to expose food. See text. Drawn from photographs in Partridge (1976). (c) Feeding skills of wild and hand-reared blue and coal tits as measured on the various laboratory feeders. Coal tits forage mainly in conifers, where their small insect prey require winkling out of crevices. Blue tits, on the other hand, forage mainly in deciduous woods, where they have to hack and tear at leaf material, and manoeuvre upside down on thin twigs, to get at insects. Birds tended to do best (had a greater percentage success, or took less time to find food) on feeders appropriate to their habitat. After Partridge (1978).

the birds get out of their preference? One possibility, reasoned Partridge, is that the two habitats demand different feeding skills and that the tit species differ in their ability to cope with them. Thus choosing the right habitat might pay dividends in terms of feeding success. To test this, Partridge (1976) designed some ingenious analogues of the foraging problems faced by tits in the two habitats and looked at the performance of captive birds on each.

Figure 7.3(a,b) shows two of the six tasks used by Partridge to simulate different kinds of foraging skill demanded by deciduous or coniferous habitats. As measures of performance, Partridge looked at the time taken for birds to obtain food and their percentage success on each type of feeder. As Fig. 7.3(c) shows, blue tits were quicker to find food and more successful overall on the feeders appropriate to deciduous woods, while coal tits fared better on those appropriate to conifers. In addition, Partridge found a significant negative correlation between the time taken to get food from each type of feeder and how long birds spent on it; tits thus seemed to prefer using the feeding skills that rewarded them the most. Feeding experience turns out to be important in the habitat preferences of other species too, including various invertebrates (e.g. Dorn *et al.* 2001).

While the tit study suggests a link between habitat preference and adaptive skills, effects of preference on reproductive success can only be assumed. The same is true for studies showing associations with life history components, such as longevity, growth rate and age of maturation (e.g. Dingle 1996; Dorn *et al.* 2001; Hendry *et al.* 2001). Is there any *direct* evidence that habitat preference can lead to increased reproduction? One study that shows it can concerns the cottonwood poplar aphid *Pemphigus betae*.

Leaf preferences in poplar aphids

Thomas Whitham (1980) studied dispersal and choice of breeding site in female *P. betae* following their emergence from overwinter eggs laid in the bark of cottonwood poplar trees in North America. After hatching, females, known as stem mothers, migrate up the trunk of their host tree and along the branches and twigs until they encounter a suitable leaf. Here they stimulate the formation of a gall on the midrib (Fig. 7.4a) and proceed to produce daughters asexually. On maturing, some of these daughters become winged migrants and disperse to new host plants. However, the extent to which stem mothers succeed in producing daughters depends heavily on the kind of leaf they choose. Perhaps not surprisingly, females prefer to establish on larger leaves, since these have a richer flow of nutrients. As leaf size increases, the risk of the gall (and thus reproduction) aborting declines and the female ends up producing a greater number of migrant daughters (Fig. 7.4b,c). But there is a problem. Large leaves are in short supply. Whitham found that, while all the largest leaves were colonised by stem mothers, they represented only 2% of the leaf crop of the host trees. Small leaves made up some 33% of the crop but were generally avoided. Thus there was stiff competition for the best sites. Since a female encountering a large leaf is very likely to find it already occupied, what should she do? She has two options: either she can settle on the leaf and try to establish a second (or even third) gall, or she can move on to search for another site. The problem with becoming an additional occupant is that the female is forced further out along the midrib where the likelihood of successful reproduction is considerably reduced. She should thus double-up only if the leaf is so good that she does better by cohabiting than by becoming the sole occupant of a smaller leaf. In fact, Whitham found that females generally doubled- or trebled-up only when leaves were *much* bigger than average. Moreover, when he calculated the productivity of cohabiting versus single galls, he found that females sharing very large leaves did just as well as those settling singly on smaller leaves, each on average producing around 80 offspring (Fig. 7.4d). Careful site selection by females thus appeared to pay dividends in terms of their seasonal reproductive success. However, while Whitham's study provides a nice example of the reproductive benefits of habitat choice, it also has something more general to say about the distribution of competitors across habitats, as we shall now see.

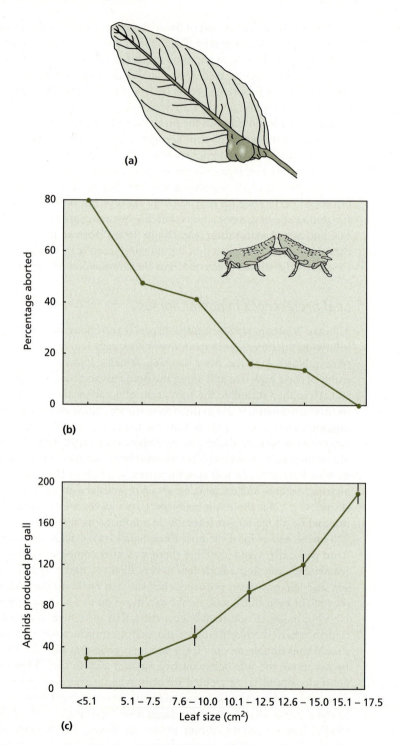

Figure 7.4 (a) Stem mother poplar aphids induce the formation of a protective gall on the midrib of the leaf, within which they then produce daughters asexually. Female poplar aphids settling on larger leaves do better in that their galls are less likely to abort (b) and they end up producing more daughters (c).

Figure 7.4 (*continued*) (d) If female aphids settle on leaves that are already occupied, they risk failure of their gall and reduced reproductive success even if the gall does not abort. However, they can overcome this by sharing only when leaves are very large, with the result that the average reproductive success for one, two and three stem mothers per leaf (horizontal line) is approximately the same. After Whitham (1978). After Whitham T.G. (1980) The theory of habitat selection examined and extended using Pemphigus aphids. *American Naturalist* **115**: 449–66, reprinted by permission of The University of Chicago Press, © 1980 by The University of Chicago. All rights reserved.

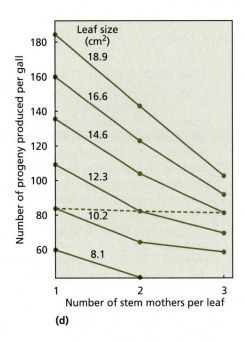

(d)

7.1.1.3   Competition, habitat choice and the 'ideal free distribution'

Whitham's poplar aphids illustrate an important point about habitat quality. While it may be possible to identify high- and low-quality habitats a priori, habitat quality changes as a function of exploitation by organisms. Thus the value of a highly prized large leaf to a female aphid is reduced once another female has settled on it. The problem is that high-quality habitats are almost certain to be exploited. Not only will this deplete resources or render them inaccessible, but it may also increase the level of interference experienced by competing individuals. So what starts out as the best place to be may soon become no better than a site of previously lower quality. At this point, therefore, it may pay an individual to go to the lower-quality site rather than struggle for access to what had originally been the best. Competition will then build up at the new site until it too declines in value and individuals would do better to go to the third best site. In theory, this process will repeat itself down the hierarchy of habitat qualities available to the population. Since intrinsically high-quality sites will be able to support more competitors than low-quality sites, a point will be reached where the distribution of competitors reflects the distribution of habitat quality. This outcome is captured by the so-called **ideal free distribution**, an evolutionarily stable strategy (ESS) theory of dispersion first proposed by Fretwell and Lucas (1970) and summarised in Box 7.1.

Several studies have borne out the predictions of ideal free theory in that competitors have distributed themselves either in proportion to the ratio of resources or so as to achieve an equal reward rate (e.g. Parker 1974; Milinski 1979; Godin & Keenleyside 1984; Fig. 7.5a,b). Where they have not been borne out, the problem can often be traced to errors of prediction rather than theory (see Earn & Johnstone 1997). However, support in terms of numerical distribution has sometimes been achieved in violation of one or more of the underlying assumptions of the theory. For example, the assumption that competing individuals will experience equal reproductive gain once they reach an equilibrium distribution

Underlying theory

Box 7.1 The 'ideal free distribution'

The 'ideal free distribution' (Fretwell & Lucas 1970) is an evolutionarily stable (see 2.4.4.2) distribution of competitors across habitats of different quality. It assumes that all individuals are equally good competitors that choose a habitat so as to maximise their net reproductive benefit. The theory is so-called because it assumes that individuals are free to move wherever they like, and are 'ideal' in having perfect knowledge of the relative value of each habitat. In practice, reproductive benefit is usually approximated as rate of food intake, number of copulations or other indirect measures of success (see text). We can use food intake as an example.

In choosing where to feed, the value (profitability) of a site depends on two things: the food intake achievable without competition, and the extent to which intake is reduced by the presence of competitors. Predators searching for prey can achieve an ideal free distribution through the interference effects of competition: any differences in food availability will be cancelled out by different levels of competition (a lot at rich sites, less at poor sites), so feeding rate will end up the same at all sites. If feeding rate becomes better at one site for any reason, more individuals will move there and the equilibrium feeding rate will be restored. The ideal free distribution of predators based on interference can be modelled as a simple equation. Assuming that feeding rate is the same at all sites, and thus a constant, the proportion of predators at the ith site (b_i) relates to the proportion of prey (a_i) for a given level of interference m as:

$$b_i = ca_i^{1/m}$$

The ideal free distribution is reminiscent of the matching law (see Box 6.3), but at the level of populations rather than individuals (Shettleworth 1998). The crucial difference, of course, is that the 'matching' distribution could arise from each individual staying at a single site the whole time (e.g. eight individuals at site 1, and four at site 2 on a 2 : 1 resource ratio) rather than distributing its time in the required ratio (8/12ths at the good site, 4/12ths at the poor one). Even so, the ideal free distribution can be cast in the generalised log ratio format of the matching law (Box 6.3, equation 6.3.6; Kennedy & Gray 1993).

Sometimes a distinction is drawn between distributions predicted under **interference competition** (as above), where a given habitat quality is reduced by interference between competing individuals, and **renewal**, where resources are continually replenished and individuals compete for their share. As Milinski (1988) and Sutherland & Parker (1992) point out, however, both ultimately rely on interference for their predicted outcomes and can be resolved into a single model.

Based on discussion in Sutherland & Parker (1985, 1992), Milinski (1988) and Shettleworth (1998).

(Box 7.1) is frequently violated (e.g. Thornhill 1980; Harper 1982; Godin & Keenleyside 1984), as are those of equal competitive ability across individuals and freedom of movement between habitats (e.g. Davies & Halliday 1979; Harper 1982). Indeed, the first violation may be a direct consequence of the second two, with dominant (high competitive ability) individuals preventing subordinates from gaining access to good sites, so reaping greater rewards (though inequality may arise for other reasons, such as differences between individuals in foraging skill, perception or learning ability).

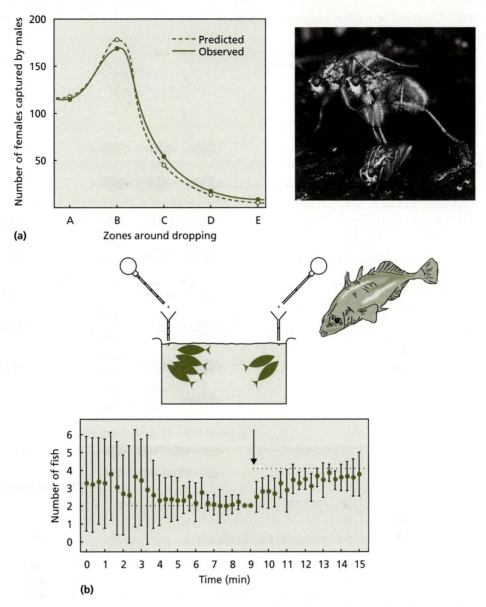

Figure 7.5 Two examples of apparently ideal free distributions. (a) Male yellow dungflies (*Scatophaga stercoraria*) search for females in different zones on and around a fresh cow pat. During the 20 minutes or so after it is deposited, females arrive at the pat to lay eggs fertilised by a previous mating and males compete to mate with them in order to fertilise the next batch. Competition soon becomes fierce on the pat itself (Zone A), so some males search in the surrounding grass (Zones B–E) where later females arrive to avoid the mêlée. Since the number of females arriving in each zone differs, so should the number of searching males. The solid curve shows the number of females captured by males in each zone (Parker 1974), and the dashed curve the number expected if the distribution of males conforms to ideal free theory (i.e. males achieve an equal capture rate). The difference between the observed and predicted curves is not significant. After Parker (1974). Photograph of male *S. stercoraria* fighting to copulate with a female courtesy of Geoff Parker. (b) Groups of six sticklebacks were provided with two feeding sites, one providing *Daphnia* at twice the rate of the other. The lower figure shows the mean number of fish at what was initially the poorer site and then at the same site when it switched to the higher reward rate (arrow). In both cases the number quickly changed to match that predicted by ideal free theory (dotted lines). After Milinski (1988).

Figure 7.6 Some consequences of unequal competitive abilities for the numerical distribution of individuals across feeding sites (a–d) compared with the prediction of ideal free theory for equal competitors (e). In the model, one feeding site provides food at twice the rate of the other, so the ideal free distribution predicts twice as many competitors at the good site when competitors are equal. In (a–d), however, large individuals can consume food at twice the rate of small individuals. As a result, there are several different ways of distributing large and small individuals across sites so that the average intake is the same on both sides (a–d). Moreover, the number of possible combinations of *individual* competitors that could give rise to the distributions differs between (a), (b), (c) and (d). Only one combination is possible in (a), but 90 are possible in (b), 225 in (c) and 20 in (d). See text. After Sutherland & Parker (1985).

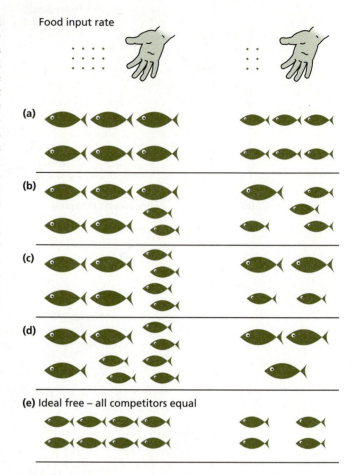

Effects of unequal competitors

Differences in competitive ability can lead to a wide range of predicted distributions of competitors under the assumption that each individual seeks to maximise its reproductive gain and all individuals gain the same at equilibrium. These are best viewed as ideal free distributions of competitive abilities, rather than individuals (Sutherland & Parker 1985; Fig. 7.6). However, as Fig. 7.6 shows, certain equilibrium combinations of competitive abilities (Fig. 7.6c) yield a numerical distribution that matches the prediction of ideal free theory for individuals. Interestingly, under the conditions in the figure, such combinations turn out to be the most likely to occur by chance – there are simply more ways of combining individuals to generate (c) than any of the other outcomes (Houston & McNamara 1988). This statistical quirk may partly account for the widespread occurrence of apparent ideal free distributions when the assumption of equal competitive ability does not hold.

Predictions become even more complicated when rewards are mobile prey, and consideration is given to intraspecific competition in both predators and prey, and an ability by each to choose where to go in relation to the other (Iwasa 1982; Hugie & Dill 1994). Under these circumstances, the predicted distribution of predators can become divorced entirely from that of their prey (Iwasa 1982; see also Lima 2002). We discuss this in more detail in the next chapter (see 8.4).

7.2 Migration

Finding a place to live implies moving around. Even if the present place is suitable now it may not be in the future; food availability may change, for example, or the animal may be challenged by a territorial intruder. When an organism moves from one place to another we often use the term **migration**. However, to many people, 'migration' conjures up images of particular kinds of movement, usually over long distances and often on a diurnal, seasonal or other return basis. Much effort has been expended over the years trying to define migration so as to distinguish it from so-called 'trivial' movement, such as a bee flying from one flower to another on a plant, or a lizard moving in and out of a patch of sunlight (see Dingle 1996). Not surprisingly, such attempts have spawned some involved classifications of animal movement (e.g. Table 7.1).

In many cases, 'migration' has been defined in terms of movement between *habitats* (e.g. Thomson 1926; Southwood 1981; Kennedy 1985; Dingle 1996). But, as Baker

Table 7.1 Levels of movement contributing to the lifetime track (see 7.2.1) of organisms. After Dingle (1996).

Movement	Characteristics	Examples
Movements home range or resource directed		
Stasis	Organism stationary	Corals, trees, hibernating mammals
Station keeping	Movements keeping organisms in home range	
Taxes and kineses	Changes in direction or rate of movement or turning	Planarian in shadow; moth in a pheromone 'plume'
Foraging	Movement in search of resources; movement stops when resource encountered	Movement in search of food or oviposition site (animals); modular growth (plants, corals)
Commuting	Movement in search of resources on a regular short-term basis, usually daily; ceases when resource encountered	Albatross foraging; diel 'migration'; vertical 'migration' in plankton
Territorial behaviour	Movement and agonistic behaviour directed toward neighbours and/or intruders; stops when intruder leaves	Many examples across taxa
Ranging	Movement over an area so as to explore it; ceases when suitable habitat/territory located	'Dispersal' of some mammals; 'natal dispersal' of birds; parasite host seeking
Movement not directly responsive to resources or home range		
Migration	Undistracted movement; cessation primed by movement itself. Responses to resources/home range suspended or suppressed	Annual journeys of birds to and from breeding grounds; flight of aphids to new hosts; transport of some seeds to germination sites
Movement not under control of organism		
Accidental displacement	Organism does not initiate movement. Movement stops when organism leaves transporting vehicle	Storm vagrancy

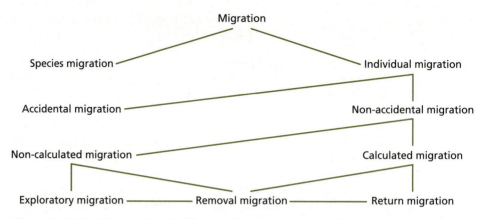

Figure 7.7 Baker's hierarchical classification of migration. See text. After Baker (1978).

(1978, 1982) points out, this simply runs into the problem of the open-ended nature of 'habitat'. What constitutes the habitat of an aphid in a wood, for example? The leaf on which it is feeding? The twig bearing the leaf? The branch bearing the twig? The tree? The entire wood? Species-specific scales and patterns of movement also make it difficult to come up with a robust and generally applicable definition of migration (Dingle 1996). In a radical departure from the traditional line, therefore, Baker (1978) abandons the quest to distinguish migration from other kinds of movement and instead defines it simply as: '. . . the act of moving from one spatial unit to another'. Thus, in Baker's view, *all* levels of movement around the environment become migration. What is interesting is not what sets migration apart from other movement, but why animals choose one level of movement (say moving from one tree to another) at one time but a different level (say moving between continents) at another.

7.2.1 'Calculated' and 'non-calculated' migration

As a framework for thinking about this, Baker envisages a hierarchy of migratory movement (Fig. 7.7). Since we are interested in adaptive decision-making, we are not concerned with the geographical shifts in species distributions sometimes referred to as **species migration**; neither are we concerned with accidental vagaries of movement caused by storms, water currents or other physical forces. We are concerned with deliberate (and thus non-accidental) decisions to move from one place to another. If such decisions are adaptive, migration should be undertaken only when there is a reproductive benefit in doing so. Baker (1978) states this more formally, predicting migration when *the potential reproductive success achieved on the way to (M) and in the spatial unit to which the animal migrates* (H_2) *exceeds that that would be achieved during the same period by remaining in the current spatial unit* (H_1) – or, more succinctly, when $H_1 < H_2M$. (Note that the cost of migration [M] is levied against the spatial unit to which the animal is moving, just as the cost of switching reduces the incentive to change behaviour in Box 4.2.) Thus formulated, the decision to migrate becomes a classic optimisation problem (2.4.4.3, and see Alerstam & Hedenström 1998). We shall come back to this shortly.

Baker recognises two forms of non-accidental migration: **calculated** and **non-calculated** (Fig. 7.7). The distinction rests on the information available about H_2 – the place where the animal is going. In calculated migration, the animal is able to assess the relative quality of H_2 before moving; in non-calculated migration it is not, and essentially takes a leap in the dark. Where possible, therefore, we should expect animals to make calculated migrations. Of course, under some circumstances, such as exploring a new area, migration must necessarily be non-calculated, but **exploratory migrations** may then provide the information on which to base a later calculated **removal migration** (Fig. 7.7). **Return migrations**, where the animal moves to and fro between the same areas (e.g. seasonal feeding and breeding grounds), are likely to involve calculated movement, except, of course, where they are being made for the first time (but see 7.2.1.1).

Baker sees all the levels of individual movement summating into what he calls the animal's **lifetime track**; that is, all the movements – local, regional, global – undertaken by the animal between birth and death (Baker 1978, 1982; Table 7.1). Since we have just argued that migrations are likely to be shaped by natural selection (see above), we should expect the form of an animal's lifetime track to reflect its ecology and life history. This seems broadly to be the case. The lifetime tracks of many species, for example, are characterised by regular return migrations between foraging and resting or breeding areas (Fig. 7.8a,b). Others reflect localised patterns of movement dictated by environmental

Figure 7.8 Predominant patterns in the life-time tracks of different organisms. (a) The tracks of savannah baboons (*Papio cynocephalus*) at Murchison Falls National Park, Uganda, are dominated by return loops of a few kilometres to regular sleeping sites (solid dots). Muttonbirds (*Puffinus tenuirostris*) in Tasmania also show return loops (b), but this time between seasonal ranges and traversing half the globe (dots show the locations of recovered individuals).

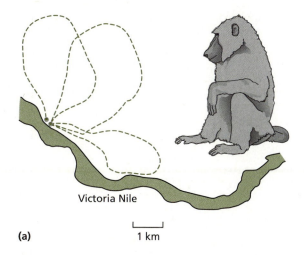

Victoria Nile

(a)

1 km

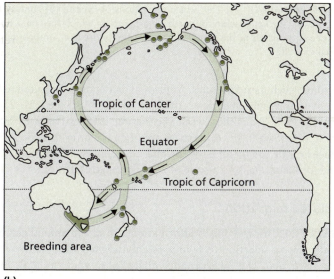

Tropic of Cancer

Equator

Tropic of Capricorn

Breeding area

(b)

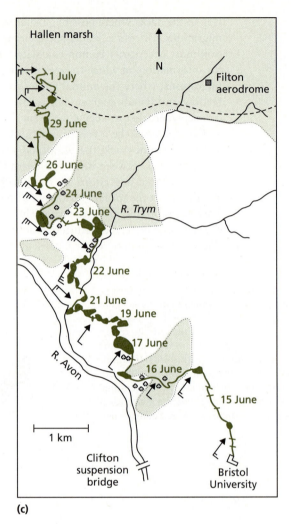

(c)

Figure 7.8 (*continued*) (c) The return migrations of (a) and (b) contrast with the more or less linear track of the small white butterfly (*Pieris rapae*) which flies on a fixed bearing relative to the perceived position of the sun. The track in the figure is a composite for several individuals. After Baker (1982).

constraints, such as tidal cycles. Yet others may comprise extended, linear movements driven by simple rules of compass orientation (Fig. 7.8c; see 7.2.1.2). An important point arising from Fig. 7.8 is that, while the lifetime tracks of various species may seem very different at first sight, the differences are often a matter of scale rather than form. Thus the tracks of the savannah baboons and muttonbirds in Fig. 7.8(a,b) are both basically return loops, but one is on a scale of a few kilometres, the other is transoceanic.

7.2.1.1 Calculated migration

Calculated migration requires information before the animal can act. While we might generally expect selection to favour calculated migration, is there any evidence that it has? In particular, can long-distance migrations, where animals remove themselves to remote destinations, be explained in terms of calculated migration? According to Baker, the answer to both questions is 'yes'. Some examples show why.

Calculated removal migration in gorillas

In social primates the composition of the group has a major influence on the foraging and anti-predator advantages of group living, and therefore on the reproductive success of individual animals (see Chapter 9). The optimal composition of the group, however, may vary with the sex, social status or other attributes of its various members. The best group structure for a dominant male may be very different from that for a juvenile female. As group membership changes from time to time, there are likely to be occasions when it would pay particular individuals to leave and seek a more favourable social environment elsewhere. One opportunity to do this might be when another social group is encountered.

In mountain gorillas (*Gorilla gorilla*), neighbouring groups often share a territory boundary. As they move round their territories, groups periodically meet and have an opportunity to compare their respective compositions. Schaller (1963) noticed that, on the occasions when adult females changed groups, exchanges tended to take place during these encounters, rather than at other times. In contrast, migration of adult males occurred independently of contact between groups. It turns out that females do better in groups that have a high ratio of mature 'silverback' males to females, because silverbacks protect females and their offspring from predators, and groups with low male : female ratios tend to have high infant mortality. If females make adaptive calculated migrations between groups, therefore, they should tend to move to a group where the silverback male : female ratio is greater than that of their own. Using data from Schaller's study, Baker (1978) compiled a table of transfers by females, allocating an 'attractiveness' score to different groups based on the number females transferring into or out of them. As the summary plot in Fig. 7.9 shows, females were indeed more likely to transfer into a group with a higher silverback ratio and to leave a group where the ratio was low.

Exploratory and calculated removal migration in prairie dogs

Inter-group calculated migrations in female gorillas appear to be based on information about alternative sites gleaned from within the current site (i.e. while still in H_1, see above). In this case, information is acquired, and the animal eventually migrates, over a relatively short distance. In other cases, while distances may still be short, information must be acquired by moving away from the present place and exploring new areas – i.e. by making non-calculated exploratory migrations. Black-tailed prairie dogs (*Cynomys ludovicianus*) provide an example (King 1955).

Prairie dogs live in 'towns', which spread over some 3 to 30 hectares. The towns are in turn divided by topographical features, such as trees and ridges, into 'wards', and the wards into territories (or coteries) of about 2000–3000 m². Each territory is occupied by a unit of dogs comprising 1–2 adult males, 2–3 adult females and anything up to 40 or so young, all of which feed on the seeds and vegetation within their territory. Emigration from territories tends to occur mainly in June and July, when the young of the year reach independence and pressure on food supplies becomes intense. Preparation for migration begins in spring, when adults of both sexes, including young of the previous year, undertake exploratory return migrations to new feeding areas outside their territory. As the spring progresses, burrows are dug and enlarged in the new area, often by animals from several different territories. By June, however, exploratory migrations

Figure 7.9 Female mountain gorillas are more likely to transfer into a new group if it has a higher ratio of silverback males to females and young than their present group. The vertical axis is an arbitrary 'attractiveness' score based on the number of females transferring into and out of a group. Plotted from data in Baker (1978) after Schaller (1963).

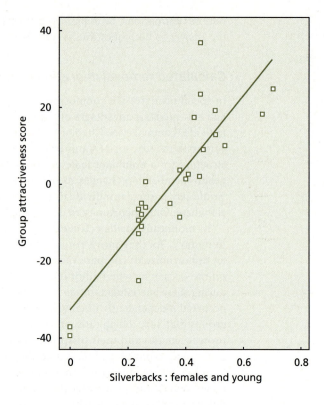

become less frequent, and adults from a particular territory begin to spend more and more time in the new area, eventually abandoning their old territory to the, now independent, young of that year, and establishing new territories. The process thus seems to be one of repeated exploratory migrations that extend the animals' familiar area, followed by a calculated removal migration based on the resulting information acquired. Whether or not removal migration takes place appears to depend on differences in food availability between the old (H_1) and new (H_2) areas, and the severity of competition between different territorial groups during the establishment of new burrows. Often it is adults from territories with very high numbers of young that emigrate (King 1955), suggesting competition for food as a factor in the decision.

Calculated migration and long-distance movement

It is easy to see how calculated migration could work in the case of gorillas and prairie dogs, where alternative sites are nearby and easy to assess. But can it account for longer-distance movements, such as the transcontinental seasonal migrations of many bird species, for example? Baker (1978, 1982) argues that it can.

Baker points out that individuals, especially the young, of long-distance migratory species frequently make deviations from their established migratory route, and that these may play a role in assessing the suitability of new sites and lead to changes in the traditional route. As a supporting example, Baker cites the results of a classic translocation experiment with starlings (*Sturnus vulgaris*) by Perdeck (1958).

Perdeck's experiment was in fact designed to test something else entirely, namely that starlings used a '**clock and compass**' mechanism (i.e. flew for a certain time on a certain

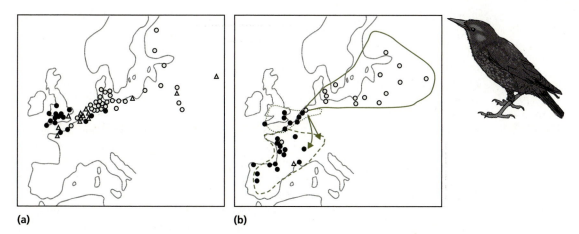

(a)   **(b)**

Figure 7.10 (a) Starlings breeding in the Netherlands (○) normally fly down to winter in Britain or France (●). Birds displaced from the Netherlands to Switzerland showed a similar southwest displacement in their wintering grounds to the Iberian peninsula. Some displaced birds found their way back to their normal breeding sites, but the next winter (b) went back to Iberia instead of returning to their previous wintering grounds. The dashed line in (b) outlines the second year wintering area and the dotted line the wintering area of control birds that were caught and ringed (like displaced birds) but not removed to Switzerland. The solid line encloses the breeding area. After Perdeck (1958) and Baker (1982).

bearing) during their autumn and spring migrations. To test this idea, he displaced birds that normally flew from the Netherlands to their western wintering grounds in France and Britain to Switzerland at the time of their autumn migration. If they used a simple clock and compass mechanism, the birds should show an equivalent displacement (i.e. south and east) in their eventual wintering grounds. In fact that is what happened: the displaced starlings wintered in the Iberian peninsula, southeast of their usual destinations (Fig. 7.10). Despite this, many of these birds found their way back to their normal breeding grounds the following spring (see Fig. 7.10). The important point, however, is what happened to them the next autumn. Instead of resuming their normal migratory route, as might have been expected, the birds went back to Spain, an outcome at odds with a clock and compass mechanism, but consistent with calculated migration to a now familiar, and evidently (from the previous year's experience) suitable, new area. While there does seem to be good evidence for clock and compass migration in some bird species (Mouritsen 1998a; Mouritsen & Larsen 1998), it is clear that migratory birds often do use experience from previous journeys to adapt their traditional migration routes (Wiltschko & Wiltschko 1999a), so Perdeck's results may reflect a more widespread phenomenon among migratory species.

Seasonal return migration in ungulates Perhaps one of the most familiar examples of seasonal return migration is that by ungulates in the Serengeti region of east Africa. Wildebeeste (*Connochaetes taurinus*), zebra (*Equus burchelli*) and Thomson's gazelle (*Gazella thomsoni*) perform an annual, long-distance return migration through the Serengeti/Masai Mara region of Tanzania and Kenya in response to seasonal changes in rainfall (Pennycuick 1975; Inglis 1976). While this is undoubtedly on an epic scale in terms of distance and numbers of migrants, close scrutiny suggests it may be based on a series of localised calculated migrations.

The three species spend the dry season (July to December) feeding in open thorn woodland, where, owing to the low rainfall, they are limited in their movements to the vicinity of waterholes. In December, when the first rains come, the herds move into the central plains of the Serengeti and migrate in a predominantly anticlockwise direction following the rains and new vegetation. The sequence in which the species move round is determined by their respective food requirements. Zebra generally move first, followed by wildebeeste and then the gazelles. Stomach analyses have shown that zebra take a high proportion of tough stem material. By removing the old stems, and generally trampling the vegetation, zebra open up the new grass leaf layer favoured by the wildebeeste. Once this has been grazed down, the gazelles can take advantage of exposed broad-leaved plants. The result is that each species moves from a depleted area to one with fresh, abundant vegetation, with individuals of the later species benefiting from the selective feeding of their predecessors. The survival value of joining the migration is shown clearly by the mortality figures. The chief cause of death among the migrating zebra, wildebeeste and gazelles is predation; few animals die of starvation. The opposite is true for non-migratory species such as impala (*Aepyceros melampus*) and wart hog (*Phacocoerus aethiopicus*), where most mortality is due to starvation. Despite the orderly and predictable progress of the migration, however, movement appears to be largely opportunistic, with animals generally moving towards areas where rain can be seen or heard falling, sometimes 100 kilometres or more away. Thus what appears to be one long seasonal return migration can instead be viewed as a series of shorter calculated removal migrations based on direct sensory information about H_2 from the present H_1.

7.2.1.2 Non-calculated migration

Of course, not all movements can be undertaken with prior knowledge of the destination. As we have already noted, exploratory movements are, by their very nature, ventures into uncharted territory and thus, in Baker's terminology, non-calculated, even though they may provide information on which a later calculated migration can be based (as in the prairie dogs above). In some cases, however, exploration may not yield appropriate information. Indeed, there may not be an opportunity to explore at all, as, for example, where populations are surrounded by large expanses of unsuitable or dangerous habitat. Here, a non-calculated removal migration may become obligatory if the quality of the present site deteriorates sufficiently.

Just such a problem faces the North Pacific sea otter (*Enhydra lutris*) (Kenyon 1969). The otters inhabit the relatively shallow coastal waters of Pacific North America. Their range extends about 16 kilometres from the shore and is restricted to water less than 60 metres deep. When a stretch of coast is colonised, the density of the population gradually increases to about 16 animals per km². A crash then usually follows, with a subsequent recovery to about half the pre-crash density, before the population finally settles at around 4–6 animals per km². Migration appears to be a response to declining food availability and mostly calculated, in that it takes place to areas explored previously during periods of low population density. Already occupied areas are generally judged the most suitable. In some cases, however, suitable alternatives may be separated by large stretches of very deep water, which otters do not normally cross. Under these circumstances, otters appear to be forced into non-calculated removal migration, sometimes crossing 100 kilometres or more of unsuitable water to reach a site of hitherto unknown quality.

Non-calculated migration on a different level occurs in the small tortoiseshell butterfly (*Aglais urticae*) (Baker 1972). Tortoiseshells, like many other species, have to exploit a range of different habitats during a day to satisfy their various requirements. Places that are suitable for resting or basking are generally not suitable for feeding or laying eggs. Each time the butterfly changes activity, therefore, it is likely to have to move somewhere else. It may also have to move to continue its present activity if the quality of the current area declines sufficiently. Such migrations occur several times a day, but because the butterfly uses a simple orientational rule of thumb – always taking up the same bearing with respect to the sun's azimuth (position relative to the observer) – movements are usually over previously unvisited areas. Thus, like the small white butterfly in Fig. 7.8(c), small tortoiseshells may travel considerable distances in a series of short non-calculated migrations.

7.2.1.3 Optimality theory, life history strategies and migration

The examples above offer qualitative support for calculated and non-calculated migration *sensu* Baker, and suggest that, where possible, animals variously acquire information in order to make adaptive decisions about when and where to migrate. As we have noted, these arguments are rooted in optimality theory, and Baker himself develops them into a series of optimisation models (Baker 1978). However, many of Baker's models remain qualitative in their predictions. Indeed, there have been surprisingly few quantitative applications of optimality theory to the problem of migration, and, where there have, models often remain to be tested empirically (see e.g. Alerstam & Hedenström 1998). In part this is because migration is rarely a unitary, all-or-nothing phenomenon, that can be analysed as a simple, discrete behavioural phenotype, even in species traditionally regarded as migrants (Dingle 1996). But it is also difficult to compare equivalent samples of migrants and non-migrants. To test functional explanations of migration, it is necessary to compare individuals that migrate with those that do not, while keeping other factors (such as size, age, sex.) constant. Unfortunately, many species are either entirely migratory or entirely non-migratory, making such comparisons impossible. Where individuals do differ in migratory tendency, migrants and non-migrants often belong to different age classes or sexes, again confounding comparison (O'Connor 1981). Nevertheless, some within and between population variation in migration lends itself to the optimality and ESS approaches introduced in Chapter 2 (e.g. Kaitala *et al.* 1993; Boriss & Gabriel 1998; Danhardt & Lindström 2001; Srygley 2001; Box 7.2). We shall now look at some examples and at the case for migration as an adaptive strategy in the sense defined in Chapter 2.

Differences between and within populations

In many species, local populations differ in their tendency to migrate, and some contain both migratory and non-migratory individuals. Closely related species can also differ in their migratory behaviour. These differences provide an opportunity for comparative studies to see whether migration is adaptive, and whether there is evidence that it has been shaped by natural selection.

Differences between populations Differences between populations or closely related species can occur in the presence or absence of migration, and/or the amount or kind of migratory behaviour shown. Often, they are associated with morphological, physiological and other behavioural differences that reflect different developmental pathways and life

Underlying theory

Box 7.2 Optimisation models and migration

Migration can be a costly business. For example, fewer than half the waterfowl migrating south from North America each autumn make it back to their breeding grounds the following spring. Many factors contribute to this, including predation, weather and disorientation. Chief among them, however, is the energy cost of the trip. A bird uses about six to eight times more energy flying than resting. A small bird flying non-stop for 100 to 150 hours across an ocean would use up as much energy as a human being running a four-minute mile for about 80 hours. Energy conservation and flying efficiency are thus likely to be key factors in the evolution of migratory behaviour and physiology, and have formed the basis of several optimisation models.

Optimal fuelling strategies

Houston (1998) considered migrants selected to minimise the time spent travelling (and thus maximise travelling speed) and asked how they should optimise fuel loading and consumption to achieve this. He used a 'continuous stopover' scenario in which the migrant could stop whenever it wished to boost its fuel reserves. The question then was how the migrant should trade off accumulating reserves prior to departure against the need to stop over and refuel. Taking on a large load at the outset would reduce the need to stop, but it would also reduce speed and increase the risk of predation. Taking on just enough to get to the first (or next) stopover, however, risks a shortfall if resource quality at stopovers varies unpredictably. Houston modelled two conditions: (a) changes in initial fuel accumulation rate (k_0) do not occur at stopover sites (there is 'local' variation in accumulation) and (b) similar changes take place at all sites ('global' variation in accumulation). Both yield a positive correlation between initial and subsequent rates of accumulation under time minimisation, but the relationship rises more steeply in the 'local' scenario (Fig. (i)). Data from migrating bluethroats (*Luscinia svecina*; Lindström & Alerstam 1992), in which food levels had been

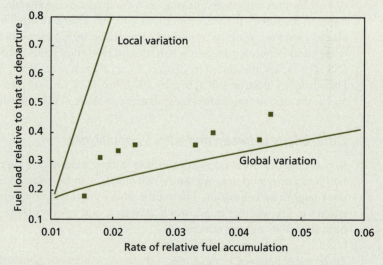

Figure (i) The relationship between fuel load relative to that at departure and the rate of fuel accumulation at stopovers in migrating bluethroats (*Luscinia svecina*) (■) compared with those predicted if there is 'local' or 'global' variation in the rate of accumulation (curves). See above. After Houston (1998).

manipulated to simulate the 'local' variation condition, showed the positive trend expected if birds were minimising migration time (Fig. (i)), but the data fell between Houston's two theoretical extremes and lay closer to the 'global' variation prediction. Houston argued that this may be reasonable if, as seems likely, the birds' knowledge of the spatial predictability of changing food availability is poor and they operate on a 'global' rule of thumb.

Optimal flight speeds

In order to maximise flight efficiency (distance travelled on a given amount of energy), aerial migrants may need to adjust their airspeed as wind conditions change. For instance, optimality theory predicts that airspeed would be reduced in a tailwind and increased in a headwind (Srygley 2001). However, the optimal strategy of airspeed regulation may differ between classes of migrant. For example, it may pay males, or dominant individuals, to get to the breeding site faster to establish the best territory, while females or subordinates may need to conserve limited resources for reproduction or simply to survive the flight. Srygley (2001) tested this by tracking cloudless sulphur butterflies (*Phoebis sennae*) migrating from Colombia towards Panama over the Caribbean Sea. Butterflies were followed in a power boat and their movement, as well as relevant environmental parameters, such as wind speed and direction, sampled as they headed west over the sea during the morning then southeasterly downwind to the shore in the afternoon. When he plotted individual airspeeds against tailwind velocity (Fig. (ii)), Srygley found the negative correlation predicted under flight efficiency maximisation in females, but no such correlation in males. By maintaining a maximum sustainable airspeed, males arriving at the breeding site earlier may gain copulations over later arrivals, while adjusting speed may minimise the consumption of lipids vital for egg production among females.

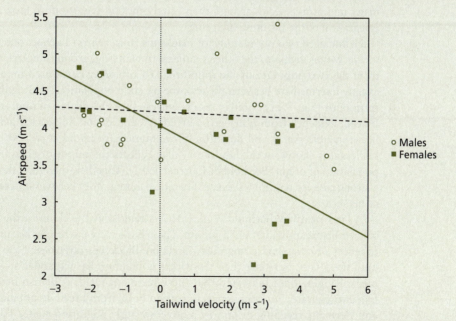

Figure (ii) Regression lines for the relationship between tailwind velocity and flight airspeed in male (O dashed line) and female (■ solid line) cloudless sulphur butterflies crossing the Caribbean. Females adjust their airspeed to a greater extent than males. See above. After Srygley (2001).

Based on discussion in Houston (1998) and Srygley (2001).

history strategies. The distinct migratory forms of many insects are good examples (Dingle 1996), as are the early- and late-developing migratory forms of different Chinook salmon (*Onchorhynchus tshawytscha*) populations (Taylor 1990). In the salmon, the different life histories appear to reflect a tradeoff between the feeding benefits of early migration to the sea and reducing the costs of the seaward journey by delaying departure until body size has sufficiently increased (see also Roff 1988; Snyder 1991).

Similar tradeoffs can be seen at the interspecific level. Birds resident throughout the year in Britain, for example, show greater sexual dimorphism in wing length than migratory species, a difference that appears to hinge on the consequences of body form for survival among the two groups (O'Connor 1981). Among residents, dimorphism may reflect niche separation between the sexes (Selander 1966), thus reducing competition for resources during winter. While migratory species probably experience similar selection pressure for divergence, the demands of long-distance flight may constrain dimorphism by favouring conservative aerodynamic qualities in body shape and size (Moreau 1972; Thomas 1997).

One of the best examples of interpopulation differences in migration concerns the blackcap (*Sylvia atricapilla*), a western Palaearctic warbler studied extensively by Peter Berthold and coworkers at the Max Planck Institute in Germany (Berthold 1999). The blackcap has a wide distribution divided into several geographical races. Not surprisingly, these races differ in many features of their migration, including timing, distance and predominant bearing. The young of each year are able to undertake the seasonal migration appropriate to their region without help from conspecifics, suggesting a genetic basis to the various regional differences. Helbig (1991) tested this experiment-ally by cross-breeding birds from different populations and exploiting their period of so-called **migratory restlessness** or **Zugunruhe** (a peak of general activity shown by many migratory species around their usual times of migration) to look at the migratory orientation of the resulting offspring.

Helbig used two populations of blackcaps from central Europe: one flying southwest to overwinter in Spain, the other southeast to wintering grounds in Africa. Crossing birds from the two populations, he found that F1 offspring orientated on a bearing almost exactly intermediate between those of the parental populations and significantly different from both (Fig. 7.11). Similar results emerged for the degree of migratory restlessness. Indeed, evidence from other cross-breeding experiments suggests a genetic correlation between restlessness and direction of orientation (Berthold *et al.* 1990). The degree of restlessness shown in the laboratory also reflects the migratory distance of the birds' population of origin (Berthold & Querner 1981). Together, therefore, these findings make a strong case for naturally selected differences in migratory behaviour between populations of blackcaps.

A later study by Berthold *et al.* (1992) strengthens the claim further. Berthold *et al.* studied blackcaps wintering in Britain. These birds were interesting because they appeared to reflect a recent shift in migratory direction. Blackcaps rarely spent the winter in Britain until the 1950s, but since then the winter population has expanded to several thousand birds. Ringing data indicated that these were not resident British blackcaps that had failed to migrate, but continental European birds from breeding populations in Belgium and Germany reaching Britain by a novel westerly migration route. To see whether the new route reflected a genetic change within the European population, Berthold *et al.* caught some wintering birds in Britain and bred them under laboratory conditions in Germany. The following autumn, the birds and their F1 offspring were tested for migratory orientation and compared with control birds from Germany. Like most of the central European population, birds from the control area normally migrated in a southwesterly

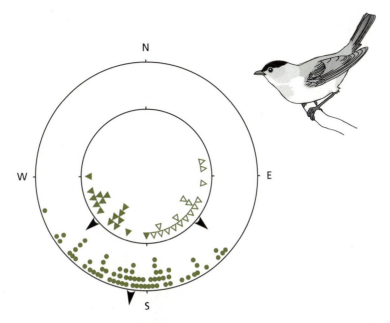

Figure 7.11 Mean bearings (arrows) taken up by hand-raised blackcaps during the period of their autumn migration. Points in the inner circle are the parental generation from Germany (▲) and eastern Austria (△). Those in the outer circle (dots) are the F1 cross generation. See text. From Helbig (1991).

direction towards the Mediterranean. Both adult British migrants and their offspring orientated just north of west, on mean bearings that were statistically indistinguishable from each other but markedly different from those of the young control birds that orientated in line with their traditional migratory route (Fig. 7.12). The heritable orientation bearing of the birds wintering in Britain thus differed from that of the bulk of the central European breeding population; it also differed from the direction taken up by birds breeding in Britain (predominantly southward) and by Scandinavian birds passing through Britain in the autumn but not overwintering (southwesterly). The obvious conclusion therefore is that, over three or four decades, some central European blackcaps evolved a new migratory route and established winter grounds some 1500 kilometres north of their previous Mediterranean sojourn. The shift has been associated with a change in gene frequency, because northwesterly migrants were not recorded before the early 1960s, but now make up some 7–11% of the breeding population in some parts of central Europe. While northwesterly migration was probably maintained at a low frequency in the past, improving winter conditions in Britain, coupled with attractions such as bird tables, a shorter migration route and less competition, may have favoured a more northerly option for overwintering (Berthold *et al.* 1992).

Interestingly, accumulating evidence suggests that the migratory behaviour of many bird species is in fact changing, quite possibly in response to global warming (Berthold 2001). The trend is towards sedentariness as climatic and resource constraints ease in previously harsh seasonal environments. Since it appears that many of the basic features of migration can be under direct genetic control (Berthold 1998 and above), these changes may well reflect rapid selection and microevolution of the kind revealed in Berthold *et al.*'s blackcaps.

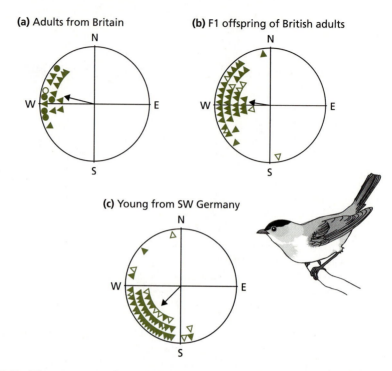

(a) Adults from Britain

(b) F1 offspring of British adults

(c) Young from SW Germany

Figure 7.12 The orientation of (a) adult blackcaps caught in Britain then tested in Emlen funnels the following autumn, (b) their F1 offspring, and (c) control birds from southwest Germany. Arrows indicate mean bearings, and solid and open symbols significant and non-significant departures from random orientation respectively. See text. After Berthold *et al.* (1992). © Nature Publishing Group (http://www.nature.com/), reprinted by permission.

Differences within populations   Populations, then, appear to differ in various aspects of life history strategy and behaviour relating to their tendency to migrate. Some of these differences can be shown to have a genetic basis suggestive of their having been shaped by natural selection. But what about differences *within* populations?

Intrapopulation variation in movement patterns is widespread, and is referred to as **partial migration** where individuals differ in the tendency to migrate, or **differential migration** where they differ in the timing or distance of movement (Alerstam & Hedenström 1998). The two, of course, are not mutually exclusive, with partial migration often involving movement at different times or over a wide range of distances (e.g. Baker 1978; Swingland 1983; Pulido *et al.* 1996).

Partial migration can be viewed as a game between alternative strategies of behaviour (migration versus non-migration), and thus as amenable to the kind of games theory analysis introduced in Chapter 2 (e.g. Lundberg 1987; Kaitala *et al.* 1993; Kokko & Lundberg 2001; Box 7.3). In some cases, migratory and non-migratory behaviours appear to reflect a genetic dimorphism (Messina 1987) or a genetic monomorphism coding for a flexible migratory threshold (Pulido *et al.* 1996; Berthold 1998); in others they reflect a conditional strategy based on, for example, age, sex or social status (Swingland 1983; Näslund 1991). Genetic dimorphisms appear to lie behind migratory differences in robins (*Erithacus rubecula*; Biebach 1983) and weevils (*Callosobruchus maculatus*; Messina 1987) and were initially suggested by Berthold's studies of blackcaps. In the latter, offspring of birds taken from partially migratory populations in southern Europe and bred in the laboratory

Underlying theory

Box 7.3   Games theory models of partial migration

Partial migration can be treated as an evolutionary game between two competing strategies: migration and non-migration. Kaitala *et al.* (1993) considered a population of birds (though the model could apply to any suitable species) that reproduce during summer. Part of the population overwinters in the breeding area, while other individuals migrate to winter grounds elsewhere and return the following year. The two strategies incur different costs. Non-migrants do not run the risks of migration, but they are likely to suffer **density-dependent** mortality (the risk of death depends on the *number* of non-migratory individuals) through competition for diminishing winter food supplies. Migrants risk dying during migration (see Baker's cost of migration in 7.2.1) but are unlikely to starve over the winter. Their mortality is thus density-independent. However, mortality is likely to be **frequency-dependent** (risk depends on the relative *proportion* of migrants and non-migrants). Using parameter values for fecundity and environmental carrying capacity derived from stable populations of birds in the field, Kaitala *et al.* produced a games theory model of the interaction between density- and frequency-dependent selection on the two strategies. Some of the output is shown in Fig. (i).

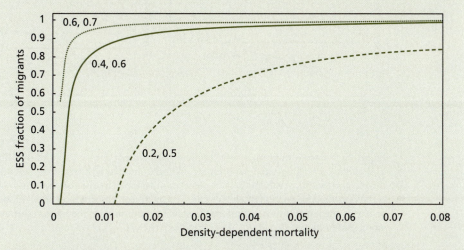

Figure (i)   The evolutionarily stable proportion of migrants in Kaitala *et al.*'s (1993) model population rises steeply then levels off as density-dependent mortality among non-migrants increases. The pair of numbers on each curve shows the survival probabilities for migrants (first number) and non-migrants (second number) in the simulation. The evolutionarily stable proportion of migrants rises more sharply as their survivorship increases. After Kaitala A. Kaitala V. and Lundberg P. (1993) A theory of partial migration. *American Naturalist* **142**: 59–81, reprinted by permission of The University of Chicago Press, © 1993 by The University of Chicago. All rights reserved.

The figure suggests that density-dependent winter mortality could be a key factor maintaining partial migration within the population. As density-dependent mortality becomes severer, the evolutionarily stable proportion of migrants increases dramatically. This proportion rises progressively more quickly as the disadvantage of migrating relative to remaining behind (i.e. the ratio of the density-independent [migrants] and density-dependent [non-migrants] probabilities of survival) decreases.

Box 7.3 *continued*

Conditional strategies

In many cases, the reproductive consequences of migrating versus remaining are likely to be conditional on the attributes of the individual. Lundberg (1987) has modelled this for birds (but, again, it could apply to other organisms) differing in social status. In Fig. (ii)(a) the expected reproductive success of both dominants and subordinates remaining behind increases with the proportion of individuals leaving the breeding area. However, success for dominants is always greater than that for migrants of either status, while that for subordinates is always lower. Thus subordinates would all do better by migrating. However, if, say, winter conditions improve, the expected reproductive success curve for subordinates is shifted upwards (Fig. (ii)b). While dominants should still stay in the breeding area regardless of the proportion of individuals leaving, the benefit of staying versus migrating to subordinates is now frequency-dependent with an evolutionarily stable proportion of migrants at p_s.

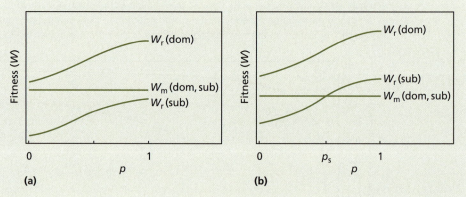

Figure (ii)   Partial migration as a frequency-dependent choice. (a) The reproductive success of dominant ($W_r(dom)$) and subordinate ($W_r(sub)$) individuals remaining behind in relation to the proportion (p) of a hypothetical population leaving the site for the winter. W_m (dom,sub) is the reproductive success of either status if it migrates. (b) If winter survival increases among residents, there may be an evolutionarily stable equilibrium (p_s) between residents and migrants among subordinates. See above. After Lundberg (1987).

Based on discussion in Kaitala *et al.* (1993) and Lundberg (1987).

divided into migratory and non-migratory fractions that correlated with those of their parental populations (Berthold *et al.* 1990). The proportion of migrants could also be increased by artificial selection (Berthold *et al.* 1990; Fig. 2.11b), providing strong suggestive evidence that frequencies of migrants in natural populations were maintained by selection. However, a later study by Pulido *et al.* (1996) suggested that both the incidence and the amount of migratory activity are aspects of the same trait, and that the phenotypic migrant/non-migrant dichotomy reflects a threshold that is adjusted genetically and in response to environmental factors. Either way, Biebach (1983) suggests that such dichotomies in migratory tendency, and the fact that they are heritable, may reflect a parental 'bet-hedging' strategy through which parents seek to maximise offspring survival in the face of varying winter conditions (but see Lundberg 1987).

While there are models in abundance, and evidence for the necessary genetic conditions, attempts to test games theory predictions on partial migrants are sparse in the extreme. One, however, is Swingland's study of giant tortoises (*Geochelone gigantea*) living on the coral atoll of Aldabra in the Indian Ocean (Swingland & Lessells 1979; Swingland 1983). At the beginning of the rainy season a proportion of the tortoise population migrates to the coast while the other remains inland. Individual tortoises are consistently migrants or non-migrants, and morphological and physiological differences suggest the dichotomy has some genetic basis. The benefit of migrating is a richer food supply and increased reproduction after the rains, but migrants pay a cost in terms of greater mortality due to the lack of shade at the coast. As a result, the lifetime reproductive success of migrants and non-migrants appears to be about the same. The tradeoff is frequency-dependent because the greater the number of migrants, the less food and shade there is and thus the greater the mortality. This frequency dependence, the consistency of individual behaviour from year to year, and the morphological and physiological differences between migrants and non-migrants is thus consistent with a mixed ESS based on a genetic dimorphism, although various factors suggest there may be a conditional element to the dimorphism.

Condition-dependent partial migration certainly occurs in other species. Migratory European blackbirds (*Turdus merula*) are an example (Schwabl 1983; Lundberg 1985). Here, whether or not an individual migrates depends on age and sex. Females are more likely to migrate than males, and juveniles more likely than adults. In aggressive encounters, males are dominant to females and adults dominant to juveniles. Because they can control resources, resident adult males are able to put on fat over the winter, something females and juveniles are not able to do. As a result females and juveniles appear to be forced into migration as the 'best of a bad job'. In keeping with this, migrants are much more likely to become residents in subsequent years (when they were older and bigger) than vice versa. In Schwabl's study for instance, about 80% of resident males and females retained resident status in succeeding years, and around half of all migrants later became residents.

7.3 Finding the way

Choosing where to settle and when and where to move somewhere else is clearly important to an animal, and much evidence suggests that natural selection has shaped these decisions adaptively. Like all adaptive decisions, however, mechanisms are required to ensure they work. In order for the animal to end up where it needs to be, it has to decide where to go and know when it has got there. In short it needs an ability to **orientate** and **navigate**.

7.3.1 Orientation and navigation

Although sometimes used interchangeably, 'orientation' and 'navigation' in fact refer to different things. Orientation in its simplest sense means taking up a particular bearing (e.g. due south) with respect to the current position, regardless of destination. Thus if an animal en route across open country is displaced laterally, it will carry on travelling parallel to its original course. Some form of directional information is required for this, but it is used only to determine the prescribed bearing. **Goal orientation,** on the other hand, involves heading towards a particular location. If our displaced traveller was using goal orientation, it would compensate for the displacement and head along a new bearing towards the same point in space as before. Two forms of goal orientation are generally recognised: (a) **pilotage**, the art of finding the way to a known destination across a *familiar* area using local sources of reference ('**landmarks**'), and (b) navigation, finding the way to a known destination across *unfamiliar* territory using any of a number of possible mechanisms (Baker 1984; 7.3.1.2). Into these broad categories falls a wide range of behaviours involving many different sensory modalities and mechanisms of processing information. We shall look at some of them shortly. Before we do, however, we must once more sound a note of caution about anthropomorphism.

Discussions of migration, particularly orientation and navigation, are often, as above, couched in terms of goals. The language of goal orientation is tempting, especially where homing (the ability to navigate home from a distant, often unfamiliar, point) is concerned, but we must reiterate the warnings about it in Chapters 1 and 4, because, once again, there is a danger of confusing functions with causes (Kennedy 1992). That homing pigeons reliably find their way back to their loft from distant release sites certainly implies they have some impressive directional capabilities, but we cannot automatically conclude that they set off from the release point with the *intention* of heading home. They may have done, but this requires independent verification. Unfortunately, discussions often lose sight of this and turn metaphorical descriptions of function into conclusions about cause (see Baker 1984; Kennedy 1992). Goal orientation can be a helpful metaphor in understanding the evolution of migration, but the metaphor by itself does not justify interpreting behaviour in terms of sentience, mental maps or other cognitive attributes implying intentionality; such interpretations demand a far greater weight of evidence. Once again, therefore, the apparent purposefulness of behaviour can lead us into unwarranted conceptual territory if we are not careful.

7.3.1.1 Kineses and taxes

Perhaps the simplest orientation response is **kinesis**, where animals reach their preferred location without reference to any directional stimulus. Instead the animal's rate (**orthokinesis**) and/or direction (**klinokinesis**) of movement varies with the strength of some non-directional stimulus that has the effect of leading it to a suitable environment (Fraenkel & Gunn 1961; Benhamou & Bovet 1989, 1992). Woodlice slowing down and turning more on moist substrate are a classic example (see 1.1.2.1). But higher organisms can show kinesis too (Bovet & Benhamou 1985). For instance, larvae of the brook lamprey (*Lampetra planeri*) bury themselves head down in the muddy bed of a lake or stream. The larvae have a light receptor near the tip of the tail and burrow into the mud until the receptor is covered, whereupon they become motionless. If the receptor becomes exposed, the larvae become agitated and squirm around until they are buried again. That the response to light is kinetic and not directional is demonstrated by a simple experiment (Jones 1955). Larvae were introduced into an aquarium with a substrate that did not allow burrowing. The aquarium was lit at one end and gradually darkened towards the other. As the larvae moved around, activity levels increased towards the lit end of the aquarium and decreased towards the dark end. They also increased with the intensity of the light, but were independent of its direction. As a result, since they could not burrow, the larvae ended up down the dark end of the tank.

Taxes are the simplest form of *directional* orientation, and generally move the animal towards or away from an external stimulus. In **klinotaxis**, the animal moves in a given direction by turning with respect to the position of the stimulus. Larvae of the house fly

(*Musca domestica*), for example, move away from a light source by swinging their heads from side to side, alternately exposing left and right photoreceptors to the light behind. When the left side is stimulated, the animal bends its body to the right and vice versa, thus moving itself away from the light. **Tropotaxis** is the result of bilaterally symmetrical receptors whose relative stimulation is compared centrally. If the receptors are stimulated unequally, the animal is not orientated towards or away from the source of stimulation. The pillbug *Armadillidium cinereum* shows positive tropotaxis to light. Intact individuals exposed to even illumination from two light sources set slightly apart will generally pass between them. If blinded unilaterally, however, their movements curve or show a spiral pattern. The **dorsal** and **ventral light reactions**, by which many aquatic and aerial animals maintain their dorsal or ventral side uppermost, also appear to be tropotactic, since they can be shown to depend on a balance of stimulation between bilateral photoreceptors (Herter 1927).

In both klinotaxis and tropotaxis the animal orientates by equalising the intensity of stimulation on the two sides of its body. **Telotaxis** does not depend on stimulus equalisation. If the animal is presented with two sources of stimulation in the same sensory modality, it orientates with respect to *one* of them, implying that the effect of the other is somehow inhibited. Hermit crabs and honey bees have been shown to respond to point stimulation telotactically, in the honey bee's case through serial adaptation of the ommatidia (units of the compound eye) and apparent motor inhibition. Tropotaxis and telotaxis generally involve the animal orientating towards or away from a stimulus. Sometimes, however, the animal orientates at an angle to it, as, for example, when using a sun compass (see below). This is sometimes referred to as **menotaxis**, but appears to be a sophistication of tropo- or telotaxis rather than a separate mechanism in itself (see Hinde 1970 for a general discussion).

7.3.1.2   Cues used in orientation and navigation

Kineses and taxes are usually relatively simple orientational responses, often on a local scale. But they can be more complex, as, for instance, in the time-compensated sun compass menotaxis of some insects. Larger-scale movements, especially among vertebrates, often depend on interactions between different orientational cues (see below) and sophisticated mechanisms for using them. This is not always the case, however, and some impressive feats of orientation have turned out to be based on relatively simple information and response rules of thumb.

One much-debated question is whether animals show 'true' navigation, i.e. the ability to establish the location of a destination regardless of current position and the availability of telltale local landmarks. True navigation requires both a **compass** to provide directional information and a **map** to tell the animal where it is in relation to where it wants to go. As we shall see, there is plenty of evidence for compass orientation, but considerably less for a map. The term 'true navigation' is perhaps slightly unfortunate since it implies that other methods of finding the way from place to place, for instance by combining sun compass and landscape information, do not constitute navigation, which manifestly they do.

Studies of orientation and navigation have focused heavily on birds, since birds provide some of the most conspicuous examples of long-distance migration and homing and are often amenable to field and laboratory experimentation. Many of the examples we shall discuss therefore come from birds, but the mechanisms involved are also used by other taxonomic groups. Studying long-distance migration presents a number of practical challenges, especially in deducing the route taken by individual migrants. While tracking technology, such as radar and satellites, can yield astonishingly detailed information about long-range movements (e.g. Emlen & Demong 1978; Gudmundsson *et al.* 1995), experimental approaches have often used local responses to infer direction of travel. **Orientation cages** are one such approach and plotting so-called **vanishing bearings** another. The former involve placing animals in a cage or chamber at a suitable time in their activity cycle (e.g. during migratory restlessness) and recording the directional bias in their movements (Fig. 7.13a). In Emlen funnels, or their aviary equivalent, a pad soaked in ink allows the preferred direction of migratory birds to be gleaned from the concentration of inky footprints (Emlen & Emlen 1966), or directionality is inferred from scratches on typing correction paper. In other cases, directional bias may be measured as the amount of activity, or the number of individuals, concentrated in different parts of the apparatus (e.g. Mather & Baker 1980, 1982; Wiltschko 1982; Fig. 7.13a). Vanishing

Figure 7.13 (a) An example of an orientation cage, in this case one used to study homeward navigation in wood mice (*Apodemus sylvaticus*). The time spent running on a wheel in each arm of the cage was used to calculate a directional preference for each individual. When the preferences are plotted (circles) and their mean bearing (arrows) calculated, mice appeared to be orientating more or less homewards. Asterisks indicate significantly non-random orientation. After Mather & Baker (1980). © Nature Publishing Group (http://www.nature.com/), reprinted by permission.

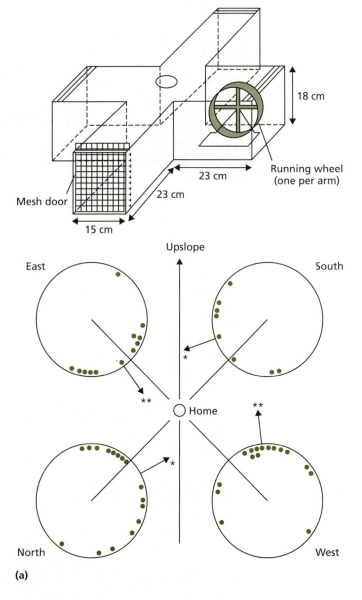

(a)

(b)

Figure 7.13 (b) A homing pigeon being released in the field so that its 'vanishing bearing' can be recorded. See text. Photograph courtesy of Wolfgang and Roswitha Wiltschko.

bearings are measures of orientation at release sites in the field and are calculated from the direction in which the departing migrant disappears from the view of the observer (e.g. Visalberghi & Alleva 1979; Wallraff & Foà 1981; Fig. 7.13b). The assumption of both techniques is that recorded orientations reflect the overall direction of migration. Happily there is evidence that this is the case (Mouritsen 1998b), and many of the results we shall discuss derive from these approaches.

Landscape topography

Landscape topography would seem to provide obvious cues by which to orientate and navigate. They certainly appear to be used on occasion, but surprisingly are often of secondary importance to other cues, such as the sun or the Earth's magnetic field. Our own landscape is dominated by topographical features that we detect visually – hills, trees, buildings and so on – but we also recognise 'landscapes' in other sensory modalities, such as sound and smell. The 'landscapes' of other species are likely to strike a very different balance between modalities, reflecting the different priorities attached to the various senses. So how important are visual landscapes here?

Observational and experimental evidence suggests that animals of many taxonomic groups, including rodents, fish, cephalopods, and many arthropods and birds respond to visual landmarks at different stages of their migratory journeys (see reviews by Cheng & Spetch, Collett & Zeil and Braithwaite in Healy 1998). Landmarks can be used in a

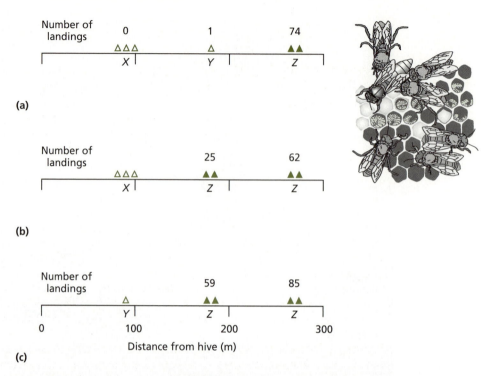

Figure 7.14 Chittka *et al.* (1995) provided honey bees with landmarks in the form of coloured tents indicating the path between the hive and feeding stations. *X* tents were yellow and initially placed 90 m from the hive, *Y* were green and placed at 180 m, and *Z* blue and placed at 270 m. (a) Once bees were trained to a feeder at 270 m, test feeders were put out at 90 m and 180 m and the number of bees landing at each of the three was recorded. (b) When blue tents were subsequently moved to the green (180 m) position, but leaving yellow at 90 m, the number of premature landings showed some increase. (c) When a green tent was substituted at the yellow position, however, premature landings increased dramatically. Thus, in (c) bees seemed to register that they had passed a green landmark and become primed to expect a blue one (and thus food) next, even though the blue tents were 90 m out of position. After Chittka *et al.* (1995).

number of ways, for instance as beacons marking a goal, or as geometrical configurations, perhaps forming a **mosaic map**, from which longer-range trajectories can be calculated (**geodetic navigation** [Baker 1984]). They are perhaps used most obviously in pilotage, for instance, in fish using features of a coral reef to patrol along regular paths (Reese 1989). Among migratory species generally, the landscape configuration appears to be important in recognising familiar sites, both once the animal has arrived, and more distantly in helping it reach a site (Collett & Zeil 1998). This may involve some complex information processing. Experiments with honey bees, for example, have shown they are sensitive to many components of landscape configuration, including size, shape, colour and, crucially, position and sequence (Chittka *et al.* 1995; Fig. 7.14). In keeping with this, many migrants appear to invest appreciable time and effort in learning landscape configurations (Cheng & Spetch 1998; Collett & Zeil 1998; Lehrer & Bianco 2000; Box 7.4; see also 6.2.1.3).

Landscape features can provide the dominant orientation cues, or they can play a supporting role. In 3.1.3.4 we saw that the desert ant *Cataglyphis* uses polarised light in the sky to navigate home after foraging. However, the polarised light compass can be only

Underlying theory

Box 7.4   Learning theory and the use of landmarks

The use of landmarks in orientation and navigation often involves learning spatial configurations. While some aspects of spatial learning appear to involve specialised mechanisms adapted for the task (see 3.1.3.1), several features of landmark learning conform to the general principles of learning discussed in Chapter 6. We briefly mention two.

The predictive value of landmarks

As we have seen (6.2.1.2) predictability plays an important part in learning, particularly in the context of compound stimuli where differences in the predictive power of stimulus components affect their role in learned associations (Rescorla & Wagner 1972). Some aspects of landmark learning are consistent with this (Cheng & Spetch 1998). The tendency for many species to rely preferentially on landmarks nearer the goal is an example. Landmarks close to the goal supply more precise information about its location, because, as the animal moves away from the goal, the compass direction and relative distance to nearer landmarks change faster than for landmarks farther away. The consequences of this have been explored in some experiments.

Cheng (1989) presented pigeons with two landmarks at different distances (20 cm and 60 cm) from a goal (food). When the landmarks were subsequently moved away in opposite directions, the pigeons headed to where the (previously) nearer mark had been placed. When marks were moved in the same direction, movement of the nearer mark had a greater effect on the birds' movement than moving the more distant mark. Similar results emerged when landmarks were presented on a touch screen monitor – movement of the mark closest to the goal had the biggest impact on the pigeons' searching accuracy (Spetch & Wilkie 1994). Rats and domestic chicks also show biases towards nearer landmarks in these kinds of test (Cheng 1986; Vallortigara *et al.* 1990).

Predictive value and overshadowing

In classical and operant conditioning, it is not just the predictive value of a stimulus that matters, but its predictive value relative to other stimuli present. Overshadowing, in which the potential for a stimulus to become associated with a response is reduced by the presence of another stimulus to which the animal pays greater attention (see 6.2.1.2) is one important influence here. To see whether control by a landmark in spatial learning could be subject to overshadowing, Spetch (1995) tested human and pigeon subjects with touch screen arrays of marks. Marks were presented in pairs where the focal mark was overshadowed (presented with another mark closer to the goal) or not overshadowed (the focal mark was the closer to goal). When the overshadowed and non-overshadowed marks were subsequently presented on their own, overshadowed marks showed less control than their non-overshadowed equivalents in both sets of subjects. Overshadowing has also been shown to affect the use of extra-maze cues in spatial learning by rats (March *et al.* 1992).

Based on discussion in Cheng & Spetch (1998).

part of the story. The ant needs to know where home is in order to use it. There are various ways this could be done, but the ant appears to use **dead reckoning** (a corruption of 'deduced reckoning' that refers to the integration of angles turned and distances moved during the foraging trip so that a course can be plotted home [e.g. Hartmann & Wehner 1995; Etienne *et al.* 1998]). But dead reckoning is prone to error and needs a backup system to compensate. In *Cataglyphis*, this seems to be provided by local landscape features. In its meanderings, *Cataglyphis* is able to memorise the configuration of landmarks around a locality, such as a nest or feeding site, and, when the need arises, orientates towards it as if matching the current retinal image with a memorized 'snapshot' (Wehner 1992, 1998). The 'snapshot' seems to be retrieved from memory only as the ant approaches the locality using its dead reckoning system. If an artificial landmark configuration around the nest site is moved to a point between the nest and the present foraging site, the ant ignores it completely. This is important for the system to work effectively and prevent the ant becoming side-tracked should it encounter similar configurations elsewhere (Wehner 1998). Just such a distraction effect has been found in birds. Dornfeldt (1982) analysed deviations from the home direction in the vanishing bearings of homing pigeons and found an association with topographical features, particularly structures such as power lines and buildings, that were similar to those near the home loft.

Wagner (1978) has shown that pigeons are also influenced by the topography at the release site itself. Birds released opposite their home loft on the far shore of Lake Constance in Switzerland took up vanishing bearings along the shore or inland rather than across the water towards home. The birds also flew along valleys rather than crossing mountain ridges, but when released in low-lying fog headed for mountain peaks visible in the distance. By fixing a small camera on a pigeon's head, Köhler (1978) was able to determine the direction of gaze when the bird held its head steady for a few seconds. He found that, on release, pigeons systematically scanned the horizon and briefly fixated topographical features, though the tendency was stronger at lower visibilities.

There is thus evidence that birds take note of landscape topography near home and at distant release sites. But does this mean they use it en route as well? For daytime migrants, the circumstantial evidence is strong, in that many migratory species are regularly observed following coastlines, mountain ranges, tracts of forest and other large-scale features (Baker 1984). But the visible landscape appears to play a role even among nocturnal migrants. Tracking night-flying birds with radar as they crossed the Swiss Alps, Bruderer (1982) found that they regularly orientated with respect to the landscape features ahead of them and, where there were consistent changes in direction (in the order of tens or hundreds of kilometres), these were often associated with topography.

Odour

For many species, olfaction can provide directional information, again either as the primary means of orientating or as an adjunct to other mechanisms. Odours may act as beacons, allowing animals to track their concentration gradients with respect to the source, or, more controversially, provide map information for true navigation. The olfactory homing capability of salmon discussed in 3.2.1.4 is a familiar example of the first, and olfactory orientation is well known in other fish as well as some mammals and numerous invertebrates (Braithwaite 1998; Ishida *et al.* 1999). One of the most surprising suggestions to emerge, however, is that odour may also play a role in birds.

While it is clear that migrating birds can detect and respond to visual landmarks at all stages of their journey, pigeons at least can still find their way when fitted with frosted lenses (Schmidt-Koenig & Keeton 1977). Orientating with respect to the sun (which is still detectable through the lenses) can explain their broadly correct direction, but it is difficult to see how it could account for birds pinpointing their loft. One possibility is that, as birds near the loft, they are able to home in on its familiar odour. But if they can use odour near the loft, perhaps they can also use it over longer distances. These suggestions seem surprising because we do not generally think of birds as having a well-developed sense of smell. However, much evidence now suggests they do (see Roper 1997) and that odours may be important in navigation.

The idea that birds might use odours to navigate was first proposed by Papi *et al.* (1972), who noticed that pigeons whose sense of smell had been removed (and so were **anosmic**) showed impaired orientation and homing ability. On the basis of this, Papi suggested that young pigeons in their home loft could learn to associate particular odours with air currents from different directions and use the information to orientate home from distant release sites. Thus odour A might arrive at the loft mainly from the north, odour B mainly from the east and so on. A bird subsequently released to the north of the loft would notice that odour A was stronger and know that it had to fly south to get home. It would duly consult one of its compass mechanisms and head off south.

In Papi's view, therefore, the pigeons inhabit an olfactory landscape, with odour landmarks providing spatial configurations. As with visual landscapes, this leads to the idea of a mosaic map (see above), a patchwork of irregularly changing odours that can be consulted wherever the bird finds itself and compared with the odour configuration at home (Papi *et al.* 1972; Wallraff 1991, 2001). Alternatively odours may be distributed in intersecting gradients that cut across those of other stimuli, such as the Earth's magnetic field (see below) to form a **gradient map** (Bingman 1998; Wallraff 2000, 2001). To say the least, however, none of these ideas has found universal acceptance (e.g. Schmid & Schlund 1993; Dingle 1996; Wiltschko 1996), and Papi's suggestion has spawned clever experimentation and controversy in almost equal measure. Nevertheless, 30 years of research has produced some convincing evidence that odour can play a role (Guilford *et al.* 1998; Papi 2001) even if the nature of any olfactory map remains elusive.

Figure 7.15 shows one of Papi's early experiments, which used wind deflector lofts to test the idea that pigeons registered the direction of odours arriving at the loft (Baldaccini *et al.* 1975). When released to the north of the loft, pigeons that had experienced deflected air currents (Fig. 7.15a,c) chose bearings to either side of control birds (Fig. 7.15b) that were consistent with the direction of wind deflection at home. Unfortunately it proved difficult to replicate these results elsewhere, perhaps because the deflector lofts incidentally altered some other factor affecting orientation, or even because birds vary geographically in their use of cues (Keeton 1981; Guilford *et al.* 1998). However, experiments changing air flow in other ways have also successfully manipulated orientation on release (Ioalé *et al.* 1978; Fig. 7.15d).

Another approach involves the use of artificial odours. Papi *et al.* (1974) raised two groups of pigeons in aviaries whose plastic and bamboo walls allowed air to enter diffusely. In addition, odour-laden air was blown into the aviaries through ducts in the roof. Birds in one group experienced the odour of olive oil blowing in from the south and turpentine blowing from the north, while the other received the same odours from the opposite directions. On release, to the east of the loft, a drop of olive oil was placed on the beak in half the birds and a drop of turpentine in the other half. Birds receiving olive oil flew south if they had experienced the odour of oil coming from the north, and north if they had experienced it coming from the south. The same was true for the birds treated with turpentine.

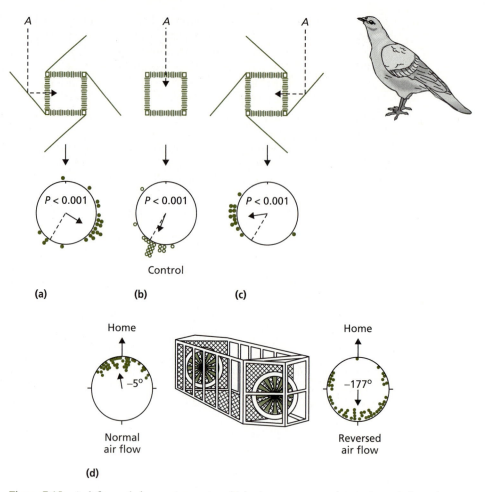

Figure 7.15 A deflector loft experiment, in which air currents entering a pigeon loft, and putatively bearing characteristic odours (in this case, odour A), are deflected in controlled directions. In (a) odour A appears to be blowing in from the west, while in (c) it appears to be coming from the east. In fact, it is coming from the north, as in the control treatment (b). When pigeons are released to the north of the loft, where odour A is presumed to be stronger, they show vanishing bearings that accord with their experimental treatment: birds in treatment (a) flew more easterly, and those in (c) more westerly, than control birds, which flew south. After Baldaccini *et al.* (1975). (d) The effect of reversing the air flow through an experimental loft on the vanishing bearings of pigeons. After Ioalé *et al.* (1978).

Since Papi *et al.*'s initial finding, a wealth of other experiments has shown predictable effects of temporary or permanent anosmia on orientation, both in pigeons (e.g. Wallraff & Foà 1981; Bingman & Benvenuti 1996; Guilford *et al.* 1998) and other birds (Wallraff *et al.* 1995). While the possible effects of induced anosmia (especially where surgery is involved) on motivation to orientate, or even brain function generally, have been suggested as alternative reasons for reduced performance (e.g. Schmid & Schlund 1993; Wiltschko 1996), some later studies have been able to rule these out. To be sure, there are still many inconsistencies, and some contradictions, but orientation and navigation involve multiple senses and high redundancy, so this may not be surprising when only a single sensory modality is considered.

The geomagnetic field

Another suggestion, sparking as much controversy as odour-based orientation, is that the Earth's magnetic field may provide directional information, certainly as a compass, and perhaps as a mosaic or gradient map.

The idea that birds, and indeed other groups, might use the Earth's magnetic field as a compass or map goes back more than a century to Middendorf (1855) and Viguier (1882). The possibility has been mired in disagreement ever since, but there is good evidence that many organisms, from bacteria and protozoans (Blakemore 1975; Kobayashi & Kirschvink 1995) to arthropods and all major groups of vertebrates, including humans (e.g Baker 1981; Lohmann & Lohmann 1996; Fischer *et al.* 2001; Papi 2001), are sensitive to natural variation in the magnetic field. As Viguier was the first to point out, components of the Earth's magnetic field could combine to provide a global reference grid. The **intensity** (strength) and **inclination** (angle of dip relative to the Earth's surface) of the field vary with global position, as does **declination** (the angle between magnetic and true north). Together these could provide information about latitude (north–south position) and longitude (east–west position), though a cruder estimate of position could be gained just from intensity and inclination. Assistance in determining true north might come from the Earth's **gravitational field**, which varies in strength over the surface of the planet owing to local variation in the density of the crust and the Earth not being a perfect sphere. Birds appear to show some sensitivity to this variation (Larkin & Keeton 1978), though the effects are weak and difficult to disentangle from other causes. Prevailing opinion is that the magnetic field provides compass information rather than a map (e.g. Baker 1984; Phillips 1996; Wallraff 1999; Papi 2001), though recent evidence from newts (*Notophthalmus viridescens*) and loggerhead turtles (*Caretta caretta*) is consistent with use of a unicoordinate (inclination only) or crude bicoordinate (intensity and inclination) map (Lohmann & Lohmann 1996; Fischer *et al.* 2001). Moreover, where it is used as a compass, it is often as a backup to celestial compasses like the sun and stars when these are unavailable due to cloud cover (Gould 1998; see below).

The first evidence that birds could register changes in a magnetic field close in strength to the Earth's very weak field came from Wolfgang Wiltschko's laboratory in Germany. Wiltschko (1968) tested European robins in an orientation cage surrounded by a Helmholtz coil (Fig. 7.16a) so that the magnetic field immediately around the cage could be manipulated experimentally. The cage was placed in an enclosed room, thus precluding the use of external visual cues by the birds. Using the coil, Wiltschko altered the apparent position of magnetic north and south during the robins' autumn and spring periods of migratory restlessness. In spring, when birds would normally migrate north, a shift in the position of magnetic south produced a corresponding rotation in the mean bearing of the robins (Fig. 7.16b). A similar rotation towards experimental magnetic south occurred when birds were tested in autumn (Fig. 7.16c). By manipulating the angle of the magnetic field, Wiltschko also showed that robins were sensitive to the inclination of the field but not its **polarity** (its north–south alignment). Thus they orientated north in spring whether the field pointed north and down or south and up, and south in autumn under either of the opposite conditions. They were unable to orientate at all in a visually cueless cage when the field was horizontal, as it is at the equator. Similar results emerged for other migratory species, including various warblers and buntings (e.g. Viehmann 1979; Bingman 1981). While robins do not cross the equator, where the horizontal field might cause a problem, some of the other species, such as the blackcap and garden warbler (*Sylvia borin*), do. So

(a)

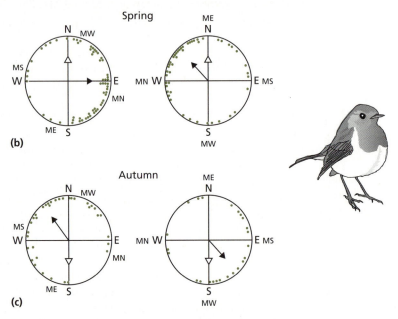

(b)

(c)

Figure 7.16 Changes in the orientation of caged European robins in response to manipulation of the magnetic field. (a) An orientation apparatus surrounded by a Helmholtz coil. (b) Birds showed deviations towards experimental magnetic north during spring when they normally migrated northwards (open arrows), and (c) towards experimental magnetic south in the autumn when they normally fly south. After Wiltschko (1968). Photograph courtesy of Wolfgang and Roswitha Wiltschko.

Figure 7.17 Pigeons with bar magnets fitted to their backs are disorientated (their vanishing bearings are dispersed) when released from an unfamiliar site in overcast conditions, but home normally if the sun is visible. Control birds fitted with non-magnetic brass bars home normally under both conditions. See text. After Keeton (1971).

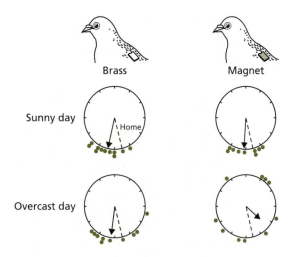

why aren't they sensitive to polarity? One suggestion is that because the polarity of geomagnetic field periodically (every few thousand years) reverses, there might be problems if birds depended on polarity for direction. However, it seems unlikely that reversals on this long-term scale could exert sufficient selection pressure on sensitivity.

Work with blind mole rats (*Spalax ehrenbergi*), solitary, subterranean rodents that spend their lives in a network of tunnels, has demonstrated a similar sensitivity of behaviour to an experimental magnetic field. Kimchi & Terkel (2001) were able to reverse the animals' north–south preference for nesting sites in a maze by reversing the surrounding magnetic field. Predicted shifts in orientation cages or other enclosed settings have been found in numerous other species, including insects (e.g. Mather & Baker 1982) and humans (Baker 1981).

Field experiments have confirmed an effect of magnetism in free-ranging migrants. If bar magnets are attached to pigeons, their ability to orientate home under overcast conditions is impaired compared with controls fitted with brass bars (Keeton 1971; Fig. 7.17). Under clear skies, however, they can use the sun and head home with no problem. Similar results have emerged when pigeons are fitted with Helmholtz coils (Walcott & Green 1974). These findings are supported by circumstantial evidence from natural (e.g. magnetic 'storms') or anthropogenic (e.g. industrial) magnetic disturbances, which have been shown to influence the bearings of migrants in several studies (Baker 1984; Phillips 1996). In humans, Baker (1981) found that magnetic disturbance (wearing electromagnetic helmets) during a long outward journey significantly reduced the ability to point in a particular direction (towards the place from which the blindfolded subjects had been driven). Interestingly, women tended to be better at navigating over short distances, but men were more adept at translating their journey-based estimates of direction into a bearing to reach a distant destination. The difference is consistent with the assumption that males have been subject to stronger selection for skills in long-distance exploration and navigation (see 3.1.3.1).

Some of the early doubts about geomagnetic cues centred first on the weakness of the field and the seemingly unlikely degree of sensitivity required to detect it, and second on the lack of an obvious mechanism of detection. The total strength of the field is around 80 000 nanotesla (nT) (100 000 nT equals 1 gauss in old units) at the poles and 29 000 nT at the equator, with natural changes being in the order of 500 nT or less – very slight, in other words (Kirschvink *et al.* 1986). However, migratory species seem able to respond to weak fields. Robins, for example, can orientate to fields of 16 000 nT (Wiltschko 1968) and pigeons in laboratory experiments are affected by even weaker pulses (Beason *et al.* 1997). Honey bees appear to be sensitive to intensities as low as 1000 nT, and possibly even 10 nT (Lindauer & Martin 1968; Martin & Lindauer 1977). As regards a mechanism of detection, mounting evidence is in favour of a magnetite (magnetic iron oxide [Fe_3O_4])-based system (Kirschvink *et al.* 2001), though experiments with a number of species indicate a critical role of light acting via the effects of magnetism on visual pigments (see Gould 1998). Behavioural and neurophysiological studies suggest magnetite and photo-pigment systems may in fact be complementary in detecting the different elements of the magnetic field (Beason *et al.* 1995; Beason & Semm 1996). Scepticism following initial suggestions that magnetite may play a role (see Baker 1984) has largely evaporated since the discovery of the material in a wide range of migratory species (Kirschvink *et al.* 2001), and evidence for concordance between magnetic sensitivity and the properties expected of a magnetite-based system (e.g. Beason *et al.* 1995). In a recent study, for example, Winklhofer *et al.* (2001) found that clusters of magnetite nanocrystals in the skin of the upper mandible of pigeons undergo characteristic changes in shape as the magnetic field changes. This and the proximity of the clusters to neural tissue encourage speculation about their role in magnetoreception.

Celestial cues

Among the earliest, and now most familiar, discoveries about animal orientation and navigation was that the sun and stars can provide important directional information. As with olfaction and magnetism, however, arguments have abounded as to the *kind* of information they provide. Both the sun and the stars could in principle allow true navigation, the sun because its apparent daily trajectory across the sky provides information about latitude and longitude, and the stars as the bicoordinate map with which human navigators have long been familiar. However, all the evidence points to their role as compasses rather than any kind of map. The moon is also used as a compass by some nocturnal migrants, such as moths (Sotthibhandu & Baker 1979) and beach hoppers (e.g. *Talitrus* and *Orchestoidea* spp.; Papi & Pardi 1963; Enright 1972), but its idiosyncratic pattern of movement across the sky, and availability for only part of each lunar month, limit its usefulness.

The sun compass   Although Matthews (1951, 1955) championed the idea that birds could use the sun's trajectory to navigate, by comparing aspects of its path and position with those at the intended destination (the **sun arc hypothesis**), there are several practical difficulties with the suggestion and no evidence that birds (or any other organisms) use the sun in this way. Instead it seems to be just the sun's **azimuth** (its position relative to the observer) that provides directional information in the form of a compass. This was first convincingly demonstrated in birds by Kramer (Kramer 1957), who used mirrors to manipulate the direction of sunlight perceived by caged starlings during migratory restlessness. The birds showed a consistent orientation with respect to the sun whatever direction it entered the cage (Fig. 7.18). Mirrors were also used in this way much earlier to demonstrate a sun compass response in the desert ant *Lasius niger* (Santschi 1911). Sun compass orientation has since been found in a wide range of other animals from amphipod and decapod crustaceans (Herrnkind 1972) to fish (Leggett 1977), amphibians and reptiles (Adler & Phillips 1985) and mammals (Lüters & Birukow 1963).

Figure 7.18 Orientation of caged starlings in relation to deflected sunlight (dashed arrows). Solid arrows indicate mean bearings. See text. After Kramer (1957).

Of course, a problem with using the sun as a compass is that its apparent position changes during the day. Animals therefore need to compensate for this using some kind of internal clock if they are to orientate effectively. One way to test this is to time-shift them so that their putative clocks are out of phase with the sun's position. Hoffmann (1954) did this with starlings by keeping them on an artificial light–dark cycle that was phase-shifted with respect to the natural cycle. He found that clock-shifted birds changed their orientation by 15° for each hour they were out of phase (15° is the apparent distance moved by the sun in an hour). A bird clock shifted by six hours thus changed its orientation by 90°. Figure 7.19 shows similar results for clock-shifted homing pigeons (Schmidt-Koenig 1960). Note, however, that the effect in the pigeons was apparent only on sunny days. When the sky was overcast, the birds orientated in the same way as non-clock-shifted controls, showing they could use a backup mechanism when the sun was not available. That magnetic disturbance appears to reduce pigeons' ability to orientate in overcast conditions (Fig. 7.17) implies that backup is provided by a magnetic compass. Magnetic orientation seems to provide a similar backup mechanism in other species deprived of a view of the sky, for instance in sockeye salmon smolt migrating under ice (Quinn & Brannon 1982). Nevertheless, despite the success of clock-shift experiments, there appears to be some variation in orientation under overcast conditions that is difficult to explain by either a sun or a magnetic compass (Benvenuti *et al.* 1996).

Despite the obvious disadvantages, not all organisms using a sun compass seem to compensate for time. As we have seen (Fig. 7.8c; 7.2.1.2), some butterflies take up a constant bearing to the sun's azimuth through the day. However, caution is needed here. For one thing, the butterflies have not been tested in clock-shift experiments. For

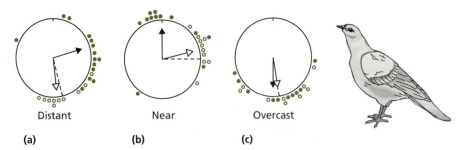

Figure 7.19 Vanishing bearings for pigeons clock-shifted by six hours (●) compared with control birds (○). Data for birds released on sunny days away (more than 1.5 km) from the loft (a), near (within 1.5 km of) the loft (b) and on overcast days (c). See text. After Schmidt-Koenig (1960).

another, experiments with monarch butterflies (*Danaus plexippus*), which have also been claimed not show time compensation (Kanz 1977), suggest directional tendencies that are independent of the sun's azimuth, and may be influenced by landscape topography and the magnetic field (Schmidt-Koenig 1985; Gibo 1986). In the monarch's case, their orientational responses serve long-distance seasonal movement; in the small tortoiseshell in 7.2.12, they serve localised movements during the butterflies' daily maintenance behaviours. Thus, sun compass orientation underlies many levels of directional movement. North American scrub jays (*Aphelocoma coerulescens*), for example, use it to find their hidden stores of piñon pine seeds (Wiltschko & Balda 1989). If jays are clock shifted and required to find previously cached seeds in an aviary, their searching behaviour deviates as predicted by their changed estimate of time, just like the orientation of homing pigeons and migrating starlings. That local movements on a daily basis can use the same orientation mechanisms as long-distance migration brings us back once again to Baker's (1978) open-ended definition of migration (7.2) and the wisdom of seeking special explanations for movements just because of their periodicity and/or the distances involved. Comparative evidence points to the movements traditionally referred to as migration having evolved from small-scale everyday movements, and the orientation mechanisms used by traditional migrants having first evolved to serve such local movements (Wiltschko & Wiltschko 1999b).

Surprisingly perhaps, even nocturnal migrants can use a sun compass. Able (e.g. 1982) has shown that night-flying birds can orientate by the glow of the sun at sunset. Using rotating polarising filters, he showed that savannah sparrows (*Passerculus sandwichensis*) can use the pattern of **polarised light** (or **e-vectors**), which produces a band of maximum polarisation at 90° to the sun's position, thus providing directional information. He also showed that the ability is learned by manipulating the orientational cues (including the sun's azimuth, the pattern of polarised light and the magnetic field) available to young sparrows prior to their first autumnal migration (Able & Able 1990).

Honey bees and the desert ant *Cataglyphis* also appear to use polarised light to determine the position of the sun (Wehner 1989). In their case, a specialised region of the retina near the upper margin of the eye is sensitive to polarised light. Receptors here work in the ultraviolet range and are arranged in a pattern that broadly matches the pattern of polarisation in the sky. As the insect sweeps the sky, the pattern of polarisation translates into changes in the rate of firing of receptor interneurons and is relayed to the CNS.

Some animals orientate to the plane of polarisation itself, rather than use it to determine the position of the sun. Tiger salamanders (*Ambyostoma tigrinum*) and various fish species appear to use polarised light in this way to direct their movement along streams

Figure 7.20 The orientation of mallard
released at night under clear and overcast
skies. After Bellrose (1958).

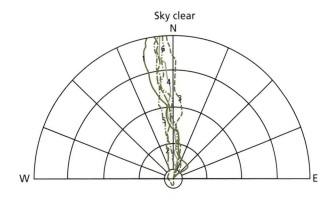

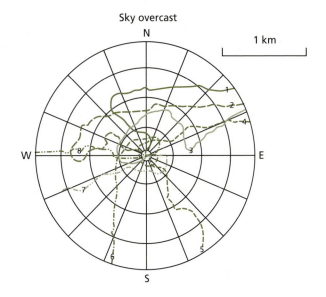

and lakes, or towards land or water (Adler 1976; Hawryshyn 1992). Some nocturnally
migrating birds, such as yellow-rumped warblers (*Dendroica coronata*), also seem to use
the plane of polarisation, only falling back on the position of the setting sun in its
absence (Moore & Phillips 1988; Phillips & Moore 1992).

The star compass   Nocturnal migrants, of course, can also orientate by the stars.
Early experiments by Bellrose (1958) attached small lights to the feet of mallard (*Anas
platyrhynchos*) and tracked the birds' orientation under clear and cloudy night skies.
On clear nights, the mallard consistently flew north, but in cloudy conditions they
orientated more or less randomly (Fig. 7.20). An ability to see the stars thus appeared to
be important in their choice of direction. Field observations such as those of Bellrose
were supported by experiments in planetaria, where captive birds could be exposed to
artificially manipulated star patterns (constellations). By rotating the planetarium sky,
Sauer (1957) and Emlen (1967) were able to induce predictable changes in the orientation
of migratory warblers (*Sylvia* spp.) and indigo buntings (*Passerina cyanea*). In a notable
series of experiments, Emlen (e.g. 1970, 1975) shifted artificial skies out of phase with

the birds' estimate of time, and selectively blocked out individual constellations and segments of sky to identify the way birds used star patterns. So, how do they use them?

Like the sun, star patterns appear to be used as a compass rather than a bicoordinate map (Baker 1984; Gould 1998; Wiltschko & Wiltschko 1999b; Mouritsen & Larsen 2001). But the nature of the compass is very different from that of the sun. Experiments with indigo buntings and garden warblers (Wiltschko *et al.* 1987) showed that young birds respond initially to the axis of apparent rotation of constellations in the night sky. The axis is aligned north–south, so can provide directional information. However, birds do not rely on the axis itself for long. Rather they learn which star patterns indicate the position of the axis and then use the patterns instead. For example, when Emlen exposed hand-reared buntings to a planetarium sky with an altered axis, the birds learned to orientate to constellations appropriate to that axis. When later tested under a normal night sky, they were unable to orientate correctly (Emlen 1972). Mouritsen & Larsen (2001) have recently demonstrated a similar use of specific stellar configurations in pied flycatchers (*Ficedula hypoleuca*) and blackcaps.

In the northern hemisphere, the axis of rotation is marked by the pole star *Polaris* (Fig. 7.21a), and several nearby constellations conveniently indicate its position (Fig. 7.21b). While the axis of rotation in the southern hemisphere is not coincident with any particular star, it lies about midway between the *Crux* (Southern Star) and *Achernar*, and can again be indicated by surrounding constellations (Fig. 7.21a). These findings imply that the axis of rotation functions as a reference cue against which the star compass is calibrated during early experience (Keeton 1981). Importantly, given the relatively stable spatial relationships between the stars, birds adjust their orientation to the same star pattern according to their required direction of migration. Thus, under the same sky, indigo buntings orientate northward in spring but southward in autumn, their preferred direction being determined by their condition, not a seasonal change in the stellar configuration. They also compensate for the apparent nightly rotation of the night sky by taking up different bearings as the night progresses (Emlen 1967).

Apart from humans, evidence for sensitivity to star patterns in other taxonomic groups is equivocal. There is some indication that night-flying moths may be able to use the stars to orientate (Sotthibandhu & Baker 1979), but experimental manipulations in planetaria have proved difficult and the ability remains uncertain. The moon appears to act as a more convincing orientational cue in this case.

Other cues

While attention has focused on landscape topography, odour, magnetism and celestial cues as the major aids to orientation, animals are capable of using other cues as well. We have discussed two of these, **electrical fields** and **sound topographies**, already (3.2.1.2 and 3.2.1.5). Yet another possibility is **infrasound** (low-frequency sounds with wavelengths below 30 Hz). Infrasound can travel many hundreds, even thousands, of kilometres, potentially allowing animals to detect distant natural sources such as waves breaking along a coastline, or wind blowing through mountains. Sound as low as this is beyond our own acoustic range, but Kreithen & Quine (1979) used conditioned heart beat responses to show that birds can detect frequencies down to around 0.05 Hz, low enough to be able to register Doppler shifts in infrasound as they fly towards or away from the kinds of source above. As with other cues, people have argued that natural infrasonic sources could provide a navigational map (Hagstrum 2001), but once again convincing evidence to support this has yet to come in.

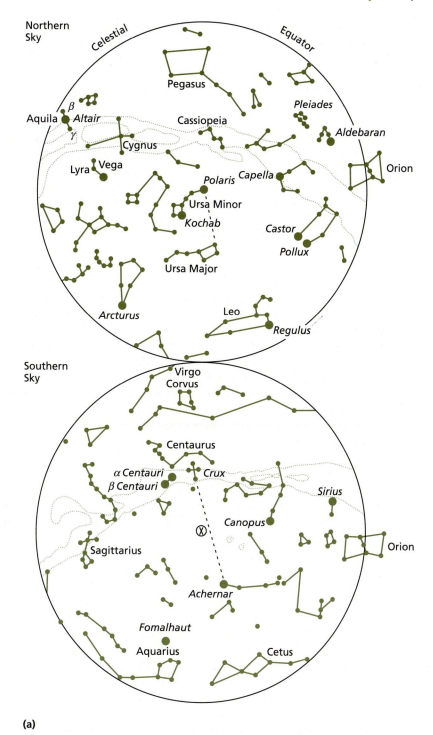

Figure 7.21 (a) Major constellations of the northern and southern hemisphere skies. The axis of rotation is indicated by *Polaris* in the north and a point (×) between the *Crux* and *Achernar* in the south. After Baker (1984).

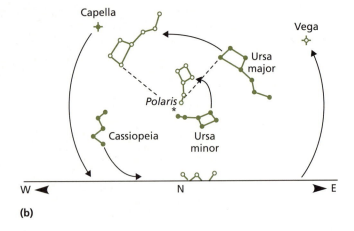

Figure 7.21 (*continued*) (b) The rotation of constellations about *Polaris* can be used to indicate north. See text. Closed circles indicate early evening positions, and open circles positions six hours later. After Goodenough *et al.* (1993).

Barometric pressure may also play a role. Again using cardiac conditioning, Kreithen & Keeton (1974) found that pigeons are sensitive to pressure changes equivalent to a 10 metre or smaller change in altitude. Such an ability would clearly be useful as an altimeter, but it would also allow detection of pressure patterns in the atmosphere, the distribution of air turbulence, wind direction and weather fronts, all helpful in long-distance orientation. There is certainly evidence that birds are good meteorologists. For example, in the eastern United States, major autumnal migrations tend to align down the east side of a high-pressure zone following a cold front, while spring movements are concentrated on the west side of high-pressure zones heralding the advance of a low-pressure cell (Richardson 1978). There are also many reports of birds showing behavioural responses to approaching weather fronts long before there is any indication of the front to human observers (Keeton 1981).

Interactions between orientational cues

It is clear from the above that animals are capable of using a variety of orientational cues and that different cues can interact in determining directional responses. Systems have high **redundancy** (several cues can do the same job) and vary considerably between times, locations, species and individuals. Thus it is not surprising that opinions have differed (to put it mildly) about the importance, or even existence, of particular mechanisms. Attention has therefore focused increasingly on the integration of different cues and the constraints that govern their use. Progress has been made in a number of areas, though, again, mainly with birds (see Dingle 1996; Gould 1998; Wiltschko *et al.* 1998, Wiltschko & Wiltschko 1999b). Interactions between celestial and magnetic cues provide some examples.

Early studies with young pigeons revealed a complex interaction between the sun compass and magnetic cues. When young birds are released for their first homing flight, they appear to require the sun compass to orientate correctly. If they are released in overcast conditions they orientate randomly (Keeton & Gobert 1970), unlike experienced birds that can switch to a (probably magnetic) backup system (see Figs 7.17, 7.19). Despite this, young birds appear to require magnetic cues to use their sun compass. If they are released wearing bar magnets, they orientate randomly even on sunny days (Keeton 1971). Thus, first flight birds seem to require both sun and magnetic cues, while experienced birds can use one or the other. However, if young birds are reared without seeing the sun at all, they are able to orientate correctly under overcast. Having never experienced the sun, therefore, the birds did not build it into their orientation system and so were not thrown by its absence on release (Keeton 1981). Young birds of other species have since been shown to rely on geomagnetic cues alone when reared without celestial cues (see Wiltschko *et al.* 1998).

Like the star compass in other species, pigeons' use of the sun compass is learned. Young birds reared permanently clock-shifted by six hours orientate normally, with no evidence of the 90° deviation shown by experienced birds when clock-shifted. The birds thus seemed to learn that the 'morning' sun was in the south, the 'midday' sun in the west and so on. When moved onto a normal photoperiod and retested, however, they showed the typical 90° rotation (Wiltschko *et al.* 1976). Taken together, therefore, these results with young pigeons suggest that the sun compass is a derivative compass that requires initial calibration using the magnetic field.

While studies of developing orientation mechanisms have provided a number of insights into interactions between cues, another approach has been to present experienced birds with conflicting information from different cue systems. Many of these cue-conflict experiments have involved testing birds under natural sky in deflected magnetic fields. The experiments have produced a range of responses, from birds choosing celestial cues over magnetic, magnetic cues over celestial and complete disorientation (Wiltschko *et al.* 1998). Referring mainly to nocturnal migrants, Able (1993) has suggested that the mixed outcomes indicate a hierarchy of cues, with magnetic cues dominating over star patterns, and sunset (polarised light) dominating both stars and magnetism. The diversity of responses above, however, is most evident during initial exposure to conflicting cues. Once birds have experienced the conflict for a number of trials, they appear to recalibrate one set of cues to harmonise its directional information with that of the other. Which cues are recalibrated, however, may depend on the phase of migration. From the available evidence, Wiltschko *et al.* (1998) conclude that, prior to migration, the magnetic compass is generally recalibrated to bring it into line with celestial information, while the reverse is true during migration itself, possibly because the geomagnetic field provides more consistent information over large distances when birds are on the move (Wiltschko & Wiltschko 1998; Sandberg *et al.* 2000).

Summary

1. Organisms are generally associated with particular kinds of habitat. While such associations might arise by chance distribution and differential mortality, it seems more likely that natural selection will have favoured mechanisms of habitat preference. This is supported by choice experiments, showing that habitat preferences can be heritable and related to the adaptive skills and reproductive success of the organism.

2. Habitat preference is likely to lead to competition for the best sites, resulting in a decline in their quality through exploitation and interference between competitors. Ultimately we should expect the distribution of competitors across sites to reflect this and reach an equilibrium based on equal reproductive gain. If individuals have perfect information about habitat quality, are of equal competitive ability and free to move

where they wish, their numbers at equilibrium should be distributed across sites as the ratio of site qualities (the ideal free distribution). However, differences in competitive ability between individuals may mean that equilibrium distributions reflect 'competitive units' rather than individuals.

3. Choosing a habitat usually means moving around. Movement from one place to another is often referred to as migration. Traditionally, 'migration' has been reserved for particular kinds of movement, often long distance and on some kind of return basis, to distinguish it from everyday 'trivial' movements such as foraging or returning to a nest. However, it is not clear that any hard and fast distinction can be made along these lines, and a more recent approach suggests the term should embrace all levels of movement and simply ask what makes different levels of movement adaptive at different times or for different individuals.

4. If migratory movements are adaptive, we should expect them, where possible, to be informed by knowledge of the relative quality of the current site and the prospective destination. Such 'calculated' migration can be contrasted with 'non-calculated' migration, in which the quality of any destination site is unknown when migration is initiated. Evidence suggests that many migratory movements, including long-distance return migrations, are indeed 'calculated'.

5. Comparative studies within and between species show that the tendency to migrate is often associated with differences in life history strategy. Migration versus non-migration can be analysed as an evolutionary game between alternative strategies in which the alternatives are maintained by frequency-dependent selection or reflect migrants making the best of a bad job. In keeping with this, several lines of evidence suggest that migratory tendency, and associated orientational responses, have a genetic basis.

6. Migration is supported by a number of orientational and navigational mechanisms. It is agreed that these provide various kinds of compass, but it is not yet clear whether any provide a true navigational map. Orientational mechanisms generally show high redundancy with complex interactions between different cues.

Further reading

Partridge (1978) still provides a good introduction to habitat choice, and Sutherland & Parker (1985) and Milinski & Parker (1991) to developments of ideal free theory. Baker's (1978) monumental treatise sets out his functional slant on migration, which is also summarised in the later pocket version of the book (Baker 1982). Dingle (1996), in contrast, provides a more traditional review of migration. The special volume of the *Journal of Avian Biology* (*Optimal Migration*) edited by Alerstam & Hedenström (1998) contains several useful discussions of optimisation approaches to migration. Healy (1998) is an excellent collection of essays on spatial representation and its implications in animals, while Gould (1998), Wiltschko & Wiltschko (1999a,b,c), Berthold (1999) and Papi (2001) provide a cross-section of current opinion about orientation and navigation in birds.

Introduction

Most animals are predators in the sense that they rely for energy on eating all or part of other organisms. As such, they have to make a broadly similar range of decisions – where to eat, what to take, when to look somewhere else and so on. Given the enormous diversity of feeding relationships, are there any general principles that predict how these decisions should be made? Do predators differ in their priorities when it comes to making choices? Predation is likely to have an impact on the survival and reproduction of prey. How does the behaviour of predators affect this? Can we predict what the impact on prey will be? The inroads of predators are likely to lead to the evolution of counterresponses by prey. What forms do these take, and how does behaviour contribute to them? Paradoxically, some anti-predator behaviours, such as alarm signals, seem to *increase* risk rather than reduce it. How could such behaviours have evolved?

Most animals are predators of one sort or another. The diversity of animal life is sustained by the transfer of energy from eating all or part of other organisms, sometimes plants, sometimes other animals, sometimes bacteria or fungi. Although we use different words, such as 'herbivore' and 'carnivore', to describe animals that eat different kinds of organism, a sparrow is as much a predator from a seed's point of view as a shark is from ours (Krebs 1978). Even detritus feeders are predators on the organic particles they sift from their environment. Whatever they feed on, predators face a broadly similar range of problems in their quest for food. Where should they look? What should they take? When should they look somewhere else, or choose something different? Since their survival and reproductive success will clearly depend on how these problems are solved, we might expect selection to have honed predators into efficient decision-makers.

Of course, predation cuts both ways. Animals may be predators, but they are frequently also prey. While enhancing their foraging skills, therefore, selection is also likely to shape their ability to thwart the foraging skills of others, and so avoid becoming food themselves. Predators and prey may thus become locked in a coevolutionary arms race of mutual adaptation and counteradaptation (Box 2.6), so that foraging decisions and anti-predator strategies run hand in hand on an evolutionary timescale (see 8.4). However, on the assumption that many races will have run long enough to reach some kind of stable outcome (see Dawkins & Krebs 1979), it is reasonable to invoke the usual optimisation and evolutionarily stable strategy arguments (see 2.4.4.2) to test hypotheses about adaptation. In this chapter, therefore, we shall use the approach to see how predators resolve the various foraging decisions confronting them, and then look at the impact of some of these decisions on anti-predator responses by their prey.

8.1 Optimal foraging theory

Foraging decisions by predators have proved to be fertile ground for testing optimisation models of behaviour, in part because the economic principles of the optimality approach seem to apply fairly directly. Predators invest time and energy seeking out, capturing and ingesting prey, and reap a reward in terms of food energy for their pains. In principle, therefore, costs and benefits can be calculated in equivalent currencies – an energy return for a time and energy investment (see Box 2.5). In practice, of course, life is not always quite as straightforward as this, as we shall see, but the assumption that predators forage to maximise their net rate of energy intake has proved to be a surprisingly robust one.

Why optimise?

Thinking about optimisation in the context of foraging also serves to make a more general point. While it is easy to see that a small endothermic vertebrate might need to be an efficient predator (for example, during winter, a small insectivorous bird such as a tit may have to find an item of food every three seconds just to stay alive [Gibb 1960]), it is not so obvious why, say, a locust sitting in a field of grass should feed efficiently. Surely, here is an animal immersed in a superabundance of food. There is no obvious competition, so as long as the locust feeds *adequately*, it will survive and reproduce. Why should it optimise? The problem with the 'adequate-will-do' argument is that it focuses on the wrong kind of competition. While individuals in a population of adequate foragers may survive to reproduce, any mutant that happens to forage more efficiently will have more time and energy to spend on other things, such as avoiding predators or finding a mate, so will reap an immediate reproductive advantage. 'Adequate-will-do' is thus unlikely to be evolutionarily stable as a foraging strategy and will be toppled by more efficient alternatives. Ultimately, selection between invading alternatives is likely to result in foragers that maximise efficiency. The important source of competition, therefore, is the evolutionary one between alternative foraging strategies rather than the physical one between individuals, and the winner is the evolutionarily stable strategy that is resistant to invasion (see 2.4.4.2).

Having said why we expect predators to forage efficiently, how can we find out whether they do?

8.1.1 'Patch' and 'prey' models

Natural food supplies are seldom distributed regularly, or even randomly, through the environment. Resource requirements, social interactions and other factors mean that prey are often clumped or **patchy** in space or time (Taylor 1961). Patches may be discrete entities, such as rotting tree stumps or swarms of insects, or local variations in density in otherwise continuously distributed items, such as earthworms in a lawn. Questions as to where predators should forage and how long they should forage there have thus been addressed largely in the context of patchy food supplies. Traditionally, studies of foraging behaviour have drawn a distinction between exploiting patches and exploiting prey (MacArthur & Pianka 1966), the former involving decisions about location and stay time, the latter about eating an encountered item or searching for another one. However, where predators exploit individual prey items over a period of time (like slowly sucking their juices out), the distinction between patches and prey begins to

Underlying theory

Box 8.1   Patch and prey models

Optimal foraging theory frequently deals with exploiting food patches and individual items of food as if they were different problems. However, both 'patch' and 'prey' models derive from Holling's 'disc equation', so-called because it was derived empirically by allowing a blind-folded secretary (predator) to search for sandpaper discs (prey) on a desktop (Holling 1959).

The disc equation assumes that time spent searching for and handling (catching and devouring) prey are mutually exclusive activities and that the number of prey encountered is a simple linear function of the time spent searching.

If T_s and T_h are the times spent searching and handling respectively, and E_f is the net amount of food energy gained in $T_s + T_h$, then the rate (R) the predator should maximise is:

$$R = \frac{E_f}{T_s + T_h}$$ (8.1.1)

However, because we are assuming encounters are linearly related to T_s, we can express both E_f and T_h as linear functions of T_s, i.e. as the gain or handling time *per encounter*. If λ is the rate of encounter with items (patches *or* prey), then λT_s is the number of items encountered. The average energy gained per encounter can then be denoted as e (where $E_f = \lambda T_s e$), and the average time spent handling as h (where $T_h = \lambda T_s h$). By substitution, equation (8.1.1) thus becomes:

$$R = \frac{\lambda T_s e}{T_s + \lambda T_s h}$$ (8.1.2)

and cancelling T_s gives:

$$R = \frac{\lambda e}{1 + \lambda h}$$ (8.1.3)

which is Holling's disc equation. The same equation is cast from the perspective of the predator's impact on prey in Box 8.5.

Based on discussion in Stephens & Krebs (1986).

blur (Cook & Cockrell 1978; Lucas 1985; Stephens & Krebs 1986). Indeed, the rate of energy gain from either can be calculated from the same simple equation (known as **Holling's disc equation**; Holling 1959; Stephens & Krebs 1986; Box 8.1). Nevertheless, it remains convenient from a functional point of view to treat 'where to forage' and 'what to eat' as different kinds of decision.

8.1.1.1   Choosing where to forage

Unless it searches randomly, the first thing a foraging predator must decide is where to start looking. As it moves around, it may encounter a range of food patches, all yielding different rates of return for time and energy spent harvesting. Clearly, if it is to maximise its feeding efficiency, the predator should select the patch with the highest net rate of return. However, differences between patches may not be readily apparent when they are first encountered. In order to choose the best, therefore, the predator may first have to **sample** different patches. This poses a tricky problem. If it spends too long sampling before making a choice, the predator's net rate of food intake for the foraging bout will be reduced because of the time wasted in poor patches. But, if it does not spend long enough, it may settle for the wrong patch, with similar consequences. Is there any evidence that animals compare the quality of different patches and opt for the best? And more to the point, do they get the balance between sampling and exploiting right?

Sampling versus exploiting

Smith & Sweatman (1974) looked at the way great tits (*Parus major*) sampled a series of foraging grids in an aviary. Six grids were set out as discrete patches containing hidden pieces of mealworm (*Tenebrio molitor*). Patches contained different densities of food and could be swapped around in different spatial arrangements by the experimenters. Tits soon learned to concentrate their feeding effort on the grid providing the highest density, but continued to spend some of their time searching the other grids (Fig. 8.1a). The value of this seemingly wasted effort became apparent when the food density on the best grid was suddenly reduced. Now, tits switched their attention primarily to the grid that had previously contained the second highest density of food (Fig. 8.1b). In a changeable environment, therefore, it is likely to be worth spending some time sampling even when the current best patch is known in order to keep track of relative patch qualities elsewhere.

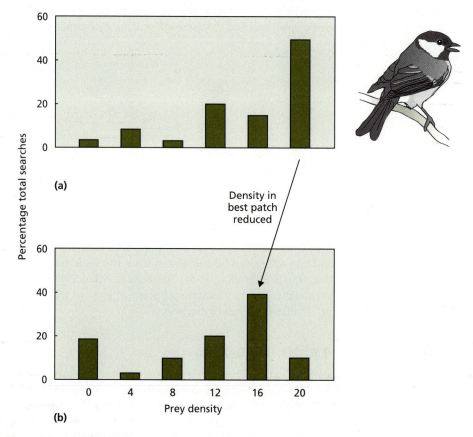

Figure 8.1 (a) Great tits learn to concentrate their foraging in the highest-density food patch (foraging grid), but (b) switch to what had previously been the second highest density patch when the quality of the best patch is reduced. See text. After Smith & Sweatman (1974).

So great tits seem to be able to compare the quality of different patches and concentrate on the best. But can they optimise the ratio of sampling to exploitation in order to maximise their returns from a foraging bout? Krebs and coworkers (1978b) tried to find out. They trained captive great tits to hop up and down on two perches at opposite ends of a large aviary. Hopping on a perch caused a computer-controlled disc to rotate and present the bird with a piece of mealworm on a predetermined schedule. The two perches were programmed to deliver mealworms at different rates, so simulating patches of different quality. Krebs *et al.* looked at how long birds sampled the patches (i.e. divided their hops between the two perches) before plumping for one or the other. A crucial element in the decision was the total number of hops the birds had to 'spend' at the perches (a function of the length of the experimental tests). The longer the test, the more hops a bird could afford to spend sampling to make sure it got the eventual choice right. A second important element was the difference in quality between the patches. The bigger the difference, the easier it would be to tell which was the best, so the fewer the hops that would have to be wasted deciding. Figure 8.2(a) shows the number of hops the birds should have spent sampling in order to maximise their feeding rate over the test as a function of the difference in quality (food reward rate) at the two perches. Curves for three different lengths of test (the number of hops available to spend) are shown. As expected, the optimal number of sampling hops goes down as the difference in patch quality increases. It also goes down as tests become shorter. Interestingly, the tits in Krebs *et al.*'s experiment conformed almost exactly to the curve predicted for bouts of 150 hops. As Fig. 8.2(b) shows, this was more or less the mean number of hops performed by the birds during the tests. The results thus suggest that great tits are able to judge investment in sampling and exploitation to get the most out of a foraging bout. The same also seems to be true of tits foraging in naturally patchy environments in the field (Naef-Danzer 2000).

The problem facing the tits in Krebs *et al.*'s experiment amounted to choice on a variable ratio schedule (see 6.2.1.2), where the number of responses needed to obtain a reward varies between options. The tits' ability to determine the better of two options and then spend all their time (Fig. 8.2c) there mirrors that of pigeons choosing on variable ratio schedules in a Skinner box (Fig. 8.2d). Note, however, that the switch from one key to the other in the pigeons' case is less clear-cut than that in the tits. Effort directed towards the better key increases continuously rather than as a step function. These differences in response remind us of some of our earlier discussion (6.2.1.2) about the rules governing choice, most notably the 'matching law' (see Box 6.3).

Matching versus maximising

'Matching' refers to the widespread tendency for animals to allocate effort to alternative sources of reward in proportion to their relative reward rates (see 6.2.1.2). One problem with matching from a functional point of view, however, is that it often does not seem to square with rate maximisation. Animals would, as we noted in 6.2.1.2, often do better by spending more time at the better (overmatching) or poorer (undermatching) site. Several factors could lead to such biases. For example, if reward schedules are changeable (an unstable environment), it may pay the animal periodically to check reward schedules at each option and so track any changes in reinforcement (matching or undermatching; see above). If reward schedules are relatively stable, however, the animal should refrain from sampling and concentrate on the option with the greatest reward (overmatching) (see Baum *et al.* 1999). A high cost of switching between options may also

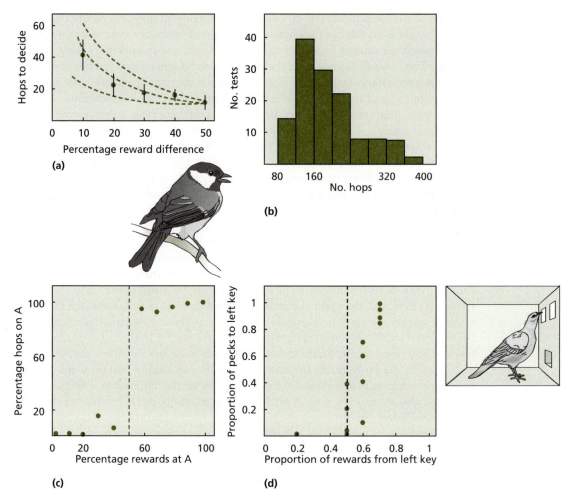

Figure 8.2 (a) The predicted number of hops spent by great tits sampling two food patches before deciding which is the best, when the total number of hops available is 250 (top curve), 150 (middle curve) and 50 (bottom curve), and the difference in reward rate at the two patches varies. Points are the mean number of hops shown by great tits in Krebs *et al.* (1974) experiment. (b) The frequency distribution of hops per test in Krebs *et al.'s* (1974) experiment. (c) Great tits switch to perch A when the relative reward rate there is greater than 50%, a response similar to that by pigeons choosing between two food keys in a Skinner box (d). See text. After Krebs *et al.* (1978b) and Herrnstein & Loveland (1975). © Nature Publishing Group (http://www.nature.com/), reprinted by permission.

result in over- or undermatching (Staddon 1983; see also Box 4.2), and could account for the differences in response between the tits (high switching cost – flying from one end of an aviary to the other) and pigeons (low switching cost – switching between keys in a Skinner box) in Fig. 8.2(c,d).

Interestingly, however, experiments that have set out to test whether animals match to or maximise reward have shown that they sometimes stick to matching in the face of opportunities to do better. For example, when pigeons were required to deliver more responses to a less reinforced option in order to change the schedule to a more rewarding one, they continued to match to the current reinforcement schedule, thus gaining less reward than if they had switched (Mazur 1981). This suggests an underlying rule of thumb that is unable to respond to the artificiality of the experimental procedure.

But, as Krebs & Kacelnik (1991) also point out, the equation for the matching law (equation (6.3.6) in Box 6.3) has a flexible output generating a spectrum of response ratios that can be counted as matching. Some of these may also *maximise* reward rate, but whether or not they do depends on the precise nature of the reward schedule. Thus matching and maximisation may not always be conflicting alternatives.

8.1.1.2 Choosing how long to stay

The patches in Krebs *et al.*'s experiment, and their Skinner box equivalents, are non-depleting. That is, food availability is unaffected by the harvesting activity of the predator. This is unlikely to be the case in many natural situations. Instead, food supplies will dwindle as the predator removes items and/or (where they are living organisms) disturbs them as a result of its activity. As the food supply declines, so does the quality of the patch relative to others around it, and there will come a point where the predator would do better to move to a new, undepleted patch. The question is, how does it know when this point has come? An elegant answer was provided by Ric Charnov in one of the landmark papers of optimal foraging theory (Charnov 1976b).

The marginal value theorem

A predator must take several factors into account in deciding how long to stay in a patch. To start with, it needs some idea of the average quality of patches in its foraging environment, so as to provide a yardstick against which to judge the wisdom of remaining in the current patch. The time and energy needed to move to another patch are also import-ant (another example of the cost of switching discussed in Chapter 4 [see Box 4.2]). Third, the predator needs some index of the declining quality of the current patch, and a rule of thumb for deciding when the appropriate time to leave has come. Charnov com-bined the principal factors in a simple model known as the **marginal value theorem**, which is summarised in Box 8.2.

The marginal value theorem is elegant and intuitive, and, like all good optimisation models, precise in its predictions. But does it work? Several studies, involving a wide range of predators, suggest that it does (e.g. Charnov 1976b; Cowie 1977; Cook & Cockrell 1978; Cuthill *et al.* 1994; Wajnberg *et al.* 2000; Nonacs 2001). We shall look at two that illustrate rather different applications of the model.

Underlying theory

Box 8.2   The marginal value theorem

The question facing the predator is how long should it stay in the present patch. If it leaves too soon, it forfeits a valuable foraging opportunity; if it stays too long it wastes time that would have been better spent elsewhere. The marginal value theorem (Charnov 1976b) predicts that the predator will stay in the current patch until its rate of food intake drops to a level equal to the average for patches in the environment as a whole. That is, the average amount of food per patch divided by the time spent in a patch plus the time taken to travel between patches. Thus the model has at its core the simple equation (8.1.1) in Box 8.1, modified slightly to cater for the time the predator has been in the patch. The

Box 8.2 *continued*

optimal stay time per patch, T_p, is that which maximises the net rate of food (here assumed to be food energy) intake, R, i.e:

$$R = \frac{E(T_p)}{T_t + T_p} \tag{8.2.1}$$

where T_t is the average time taken to travel between patches, and $E(T_p)$ is the energetic return as a function of the time spent in the patch. The optimum can be calculated from the equation (see Lendrem 1986), or found graphically. The latter is the more familiar approach.

The curve in Fig. (i) represents the predator's cumulative food intake in a patch as a function of time. The predator arrives in the patch, having just travelled for T_t seconds from the previous patch, and starts feeding. At first the intake curve rises steeply as the predator encounters a high density of undepleted (or undisturbed) prey. As items are removed, food density declines and with it the predator's rate of intake. At some point, its intake rate will drop to the average for that of the environment as a whole (in economics jargon, the *marginal value* for the environment [Charnov 1976b]) and it will pay the predator to move to a new patch and start again on the steep part of the intake curve. This point, the optimal stay time, can be found by constructing a tangent to the curve from the origin of the preceding travel time (Fig. (i)). We can see why by casting the solution in terms of equation (8.2.1) above.

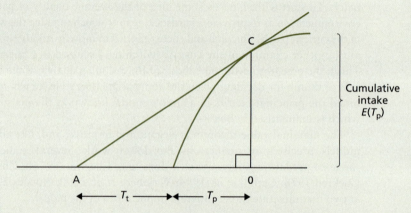

Figure (i) The marginal value theorem predicting the optimal stay time in a patch. A predator should stay in a patch until its intake rate drops to the average for all patches in its environment. The optimal stay time is found as the tangent to the intake curve from the origin of travel time. T_t, travel time to reach the patch; T_p, time spent in the patch. See above for an explanation of ACO. After Lendrem (1986).

The time axis and tangent in Fig. (i) make a right-angled triangle ACO. The gradient of the line AC is C0/A0. Note that C0 is the intake rate $E(T_p)$ in equation (8.2.1), and that A0 = $T_t + T_p$ in the same equation. Thus the gradient AC = $E(T_p)/(T_t + T_p)$. If the predator is to maximise its net rate of food intake, R, therefore, it must maximise the slope of the line AC. This turns out to be greatest when the line is the tangent to the intake curve. Dropping a perpendicular from the tangent to the time axis then gives us the optimal stay time for the patch.

Based on discussion in Charnov (1976b) and Lendrem (1986).

Staying with great tits for the moment, Cowie (1977) provided what is still one of the most cited tests of the marginal value theorem. In his experiment, captive tits were required to find pieces of mealworm hidden in sawdust-filled plastic cups (patches). The cups were arranged on the 'branches' of artificial trees set in a large aviary. Cowie tested his birds in two types of environment: one with a short travel time between patches, and one with a long travel time. In this case, travel time was manipulated by fitting tight- or loose-fitting cardboard lids to the cups (Fig. 8.3a). The effect of changing travel time on the patch stay time predicted by the model is shown in Fig. 8.3(b): patch stay time should increase the longer the travel time. How did Cowie's birds respond? Figure 8.3(c) shows his results. As expected, stay times increased with travel time, but were generally longer than predicted by the marginal value theorem when only the *time* cost of travel was taken into account. However, Cowie reasoned that the birds might also be sensitive to the *energy* spent during travel. Sure enough, when he incorporated an estimate of this into the model, the predicted curve was a much closer fit to the data (Fig. 8.3b).

The second example brings us back to the issue of individual prey as patches (see above). Waterboatmen (*Notonecta glauca*) are voracious aquatic insect predators, common on the surface of ponds and ditches. Cook & Cockrell (1978) studied them

Figure 8.3 (a) A great tit searching for food in a plastic cup. (b) Increasing or decreasing travel time in the marginal value theorem changes the slope of the tangent (AB/A′B′) and predicts longer or shorter patch stay times (t_{opt_1}/t_{opt_2}). After Krebs (1978).

(a)

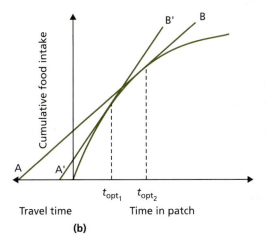

(b)

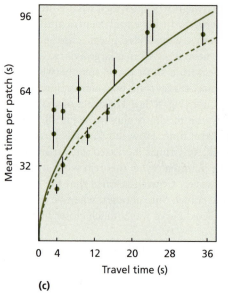

Figure 8.3 (*continued*) (c) The mean foraging time per patch (plastic cups) in great tits when travel time was varied by fitting lids to the patches. The marginal value theorem provided a better fit to the data when both the time and energy costs of travel were taken into account (solid curve) than when only time costs were considered (dashed curve). After Cowie (1977). © Nature Publishing Group (http://www.nature.com/), reprinted by permission. Photographs courtesy of Richard Cowie.

feeding on larvae of the mosquito *Culex molestus*, which involves sucking out the internal contents of the prey before discarding its hard exoskeleton. Cook & Cockrell showed that the rate at which the waterboatmen were able to extract fluids started off high but subsequently diminished as extraction became more difficult (Fig. 8.4a). The cumulative intake curve was thus a decelerating one, as envisaged for food patches by the marginal value theorem. Having calculated the intake curve for their waterboatmen, Cook & Cockrell manipulated travel time by presenting prey at different densities, then used the marginal value theorem to calculate the optimal time that should be spent on each prey item. When they compared the predicted curve with their observed values (Fig. 8.4b), they found a very close match.

In its familiar form, the marginal value theorem assumes a decelerating gain curve. But not all gain functions will necessarily be like this. Cuthill (1985) and Kacelnik & Cuthill (1987), for example, have considered linear gain functions, in which prey become available at a constant rate within patches until a maximum is reached. Under the marginal value theorem, this kind of gain function predicts no relationship between travel time and time spent searching. Instead, predators should always take the maximum number of prey in each patch. In keeping with this, starlings (*Sturnus vulgaris*) presented with such patches in an aviary took the maximum number of prey per patch on more that 95% of patch visits, irrespective of travel time (Cuthill *et al.* 1990).

While many experiments have leant support to the marginal value theorem, others have pointed to some of its apparent limitations. For example, one of the assumptions of the model is that predators have a reasonable estimate of the quality of patches in the foraging environment. This, of course, makes several further assumptions, one of

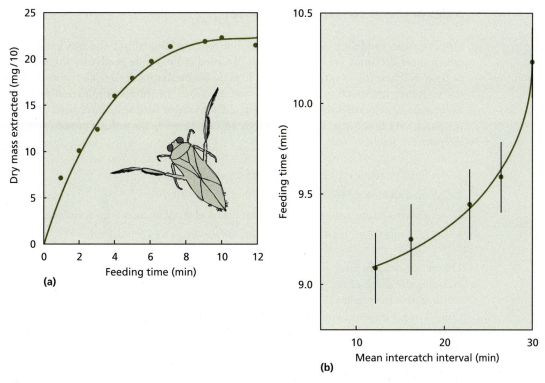

Figure 8.4 (a) The cumulative intake curve for waterboatmen feeding on mosquito larvae. (b) The optimal time spent per larva (curve) as predicted by the marginal value theorem for different intercatch intervals (travel times), and the mean times spent by waterboatmen. After Cook & Cockrell (1978).

which is that the foraging environment is more or less stable. If it is not, the predictions of the marginal value theorem begin to break down. Kamil *et al.* (1993) looked at what happened to patch stay times in blue jays (*Cyanocitta cristata*) when the predictability of patch quality was manipulated. Kamil *et al.* provided blue jays with a foraging environment in which the probability distribution of prey availability within patches was controlled experimentally. As the likelihood of finding prey in a patch decreased, birds showed a tendency to stay too long in patches containing no prey compared with the predictions of the marginal value theorem. They also adjusted stay time in patches to travel time when the latter actually had no effect on optimal stay times.

Even in a stable environment, it seems unlikely that predators will have complete knowledge of patch qualities. Indeed, in many cases, they may be entering a foraging area for the first time. This might seem fatal to the marginal value theorem, but is it? McNamara (1985) and McNamara & Houston (1985) have shown that, in principle, a predator starting off with no idea of average patch quality could arrive at an accurate estimate (and thus the optimal stay time per patch) by adjusting the time spent in a patch according to the average gain experienced up to that point. Despite complete ignorance at the outset, therefore, it seems that a predator could still end up optimising in accordance with the marginal value theorem by capitalising on its accumulating experience (Krebs & Kacelnik 1991). This leads to a more general question about the rules of thumb predators should use in deciding how to exploit food patches.

Rules of thumb for exploiting patches

Optimisation models predict decisions but say nothing about the mechanisms by which these decisions are made. Models can predict how long predators should spend in food patches, for example, but how exactly do predators determine the moment to leave? As we discussed in 2.4.4.3, animals are likely to solve optimisation problems using approximate rules of thumb (see Fig. 2.21). Several such rules have been proposed for patch stay times (e.g. Cowie & Krebs 1979). These fall into four broad categories (Stephens & Krebs 1986):

1. A number rule – e.g. 'leave after catching n prey'.
2. A time rule – e.g. 'leave after t time units in the patch'.
3. A giving-up time rule – e.g.'leave after t time units of unsuccessful searching'.
4. A rate rule – e.g. 'leave when the capture rate drops to a critical value r'.

Theoretical studies have shown that the rule that does best depends on several things, including the shape of the gain curve, variation in the quality of patches and the extent to which it pays the predator to sample (e.g. McNair 1982; Green 1984; Olsson *et al.* 2001). Conducting a rigorous comparison can thus be a tricky task. Waage (1979a), however, working with the parasitoid wasp *Nemeritis canescens* hunting patches of moth larvae (*Plodia interpunctella*), managed to show that stay times on *Plodia* obeyed a sophisticated giving-up time rule rather than any of the alternatives (1, 2 and 4) suggested above. Krebs *et al.* (1974) also found evidence for a giving-up time rule rather than one based on the number of items found in captive black-capped chickadees (*Parus atricapillus*). As predicted for an optimising predator, chickadees showed longer giving-up times in poor quality environments (where intercatch intervals were long) than in rich environments (where they were short). Similar relationships have been found in other tit species (Winkler & Kothbauer-Hellmann 2001). Unfortunately, because giving-up times were short relative to the intercatch intervals in Krebs *et al.*'s experiment, the differences in their case could have arisen simply by chance (e.g. through random disturbances such as a door slamming; Cowie & Krebs 1979). The conclusion that birds varied their giving-up times adaptively is thus not entirely watertight. In a field study of spotted flycatchers (*Muscicapa striata*), however, Davies (1977) found that birds moved between foraging perches (from which they flew out to attack swarms of insects) with giving-up times of around 1.5 times the intercatch interval at the current perch. This turned out to be just long enough to maximise the birds' net rate of food intake, given the prey handling and interperch travel times involved.

Giving-up time rules of thumb beg a question as to the past foraging history on which they are based. One early suggestion was that predators might use some form of sliding '**memory window**' (Krebs 1978), thus updating their giving-up times on the basis of a given period of previous foraging experience. Convincing evidence for memory windows is thin on the ground. However, there is some indication that parasitoid wasps may regulate patch stay times according to a sliding window of past encounters (Roitberg & Mackauer 1993), and somewhat more evidence that other predators may use them in decisions about diet (see 8.3) (Shettleworth & Plowright 1992; Mackney & Hughes 1995). Nevertheless, while appealing as an idea, memory window models are difficult to test (Cowie 1977; Giraldeau 1997), and attention has focused more on other kinds of experiential rules, such as **linear operator** learning models (Kacelnik *et al.* 1987; Giraldeau 1997).

Linear operator models work by devaluing outdated information and using a moving weighted average of past outcomes to predict decisions in changeable environments (Houston & Sumida 1987). Several studies have provided evidence consistent with the predictions of linear operator models. Kacelnik & Todd (1992) and Cuthill *et al.* (1994), for example, found that starlings subjected to different experimental travel times varied the number of prey taken per patch in accordance with intake maximisation, but did so in relation to different numbers of preceding travel episodes depending on the conditions of the experiment. In some cases decisions appeared to be based on only the immediately preceding visit (Kacelnik & Todd 1992).

Time estimation and patch stay times The marginal value theorem, and the various rules of thumb predators might use to exploit patches, assume that predators are able to measure time, remember intervals and adjust behaviour according to the time elapsed (Brunner *et al.* 1992). But what if animals do not assess time accurately? We have seen that uncertainty in the foraging *environment* has an important bearing on predicted responses, but what happens if there is uncertainty in the predator's judgement of time? A theoretical framework for looking at this comes from the motivation literature and the role of **expectancy** (a combination of the motivational value of an event and its estimated likelihood of occurring at a given time) in predicting responses on different reinforcement schedules. **Scalar expectancy theory** (Gibbon 1977), as it is called, assumes that animals measure and remember the time elapsing to a significant event (such as a food reward), but that their memory of the time period is imperfect. For each period experienced, the animal remembers a distribution of values centred around the actual value. When required to recall the time period, the animal draws randomly from the distribution so that the variance of the distribution (the spread of values within it) determines the scope for memory error (Giraldeau 1997; see Box 8.3).

Alex Kacelnik and coworkers have tested some of the predictions of scalar expectancy theory in experiments with starlings (e.g. Kacelnik *et al.* 1990; Brunner *et al.* 1992, 1996). Several of their results accord with the idea. For example, when starlings were offered patches of prey with a range of intercatch intervals, the variability of the birds' so-called 'giving-in' time (the time at which they ceased trying to obtain food at one patch before travelling to another) increased with the length of the intercatch interval (Fig. 8.5a), as predicted by scalar expectancy theory (see Box 8.3; Kacelnik *et al.* 1990). Since the patches provided prey at an unchanging rate (within treatments) that suddenly depleted at some unpredictable point, a bird with a perfect estimate of time would simply have given up once the treatment intercatch interval had been exceeded. Moreover, when the giving-in times were scalar transformed (i.e. standardised) with respect to intercatch interval, the frequency distributions turned out to be very similar (Fig. 8.5b), again as predicted by the theory (Box 8.3). Kacelnik and coworkers' studies thus show that underlying psychological constraints can have an important influence in decisions about timing. In doing so, they emphasise once again the need to understand mechanism when testing predictions about function.

8.1.1.3 Choosing what to eat

So far we have discussed how predators respond to the spatial properties of their food supply in deciding where to feed and when to look somewhere else. While foraging at a given site, however, predators are also likely to be faced with decisions about what to eat. Prey do not normally come in neat, standardised packages. They vary in size, agility, nutritional value, toxicity and a host of other characteristics that affect their attractiveness

Underlying theory

Box 8.3   Scalar expectancy theory

Scalar expectancy theory assumes that animals remember time intervals as distributions of estimates rather than accurate values. It further assumes that the coefficient of variation (the ratio of the variance to the mean) is the same for all memorised distributions, so that variance increases with S (the time interval to a food reward; Fig. (i)). Since the coefficient of variation is constant, however, the memorised distributions can be made identical by simple scalar transformation. When applied to foraging decisions based on a constant S (e.g. a particular travel time or intercatch interval), scalar expectancy theory predicts a distribution of errors (positive and negative) about S, and a positive correlation between variation in error and the magnitude of S (see text).

Figure (i)   Hypothetical memory distributions arising from experience with three different time intervals (S) to a food reward, as assumed by scalar expectancy theory. After Giraldeau (1997).

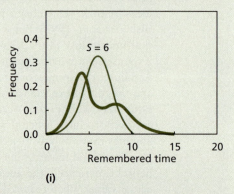

(i)

Of course, in many cases S does not remain constant. Intervals between successive prey items, for example, are likely to vary. How will this affect the animal's memory of them? Scalar expectancy theory assumes that the memorised distribution of time intervals will be an aggregate of the distributions associated with each experienced S. Thus, a foraging bout that is likely to yield rewards after, say, 4 or 8 seconds, will result in a memorised distribution combining those around $S = 4$ and 8 (Fig. (ii)). However, one consequence of the scalar property that draws out the variance as S increases, is that the combined distribution will be negatively skewed relative to the distribution for a fixed equivalent value of S (Fig. (ii)). Thus, when predators are presented with a choice between a fixed and a variable option with the same mean interval, they will be more likely to recall a short interval from the combined distribution than from the fixed distribution. This has implications for the functional significance of preferences for fixed versus variable foraging options (see 8.4).

Figure (ii)   An aggregate of the distributions for $S = 4$ and $S = 8$ in Fig. (i) (bold line) compared with the distribution for a fixed value of $S = 6$ (thin line). See above. After Giraldeau (1997).

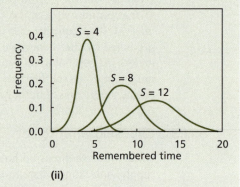

(ii)

Based on discussion in Giraldeau (1997).

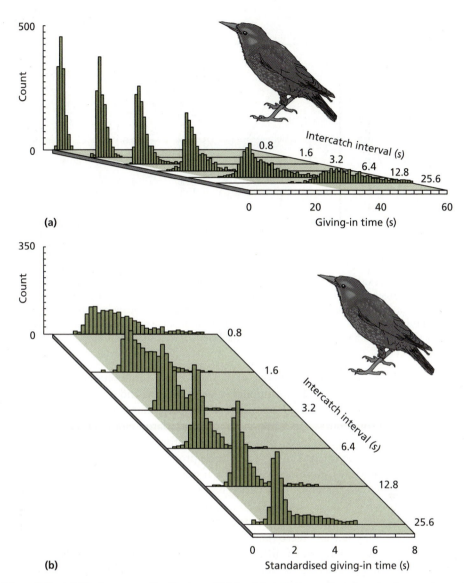

Figure 8.5 (a) The frequency distribution of 'giving-in' times by starlings foraging in experimental patches with different intercatch intervals. Note the increase in variance (spread) as intercatch intervals increase. (b) The same distributions standardised with respect to intercatch interval. Note that most of the transformed distributions are very similar. See text. After Kacelnik *et al.* (1990).

and value to the predator. Which should the predator take? Once again, we might expect predators to choose so as to maximise their net rate of food intake, so the answer will depend on the costs and benefits of taking the different types of prey.

Benefit may come in various forms. Usually it is assumed to be food energy, but it might be a particular nutrient, such as a vitamin or vital trace element. Whatever it is, items with more of the important ingredient(s) – the biggest, perhaps – would seem to be the obvious ones to go for. Unfortunately, there is a catch. While they may contain lots of resource, the biggest items may also take a long time to process, perhaps because they fight back or have tougher skins. A predator might be able to eat two or three smaller

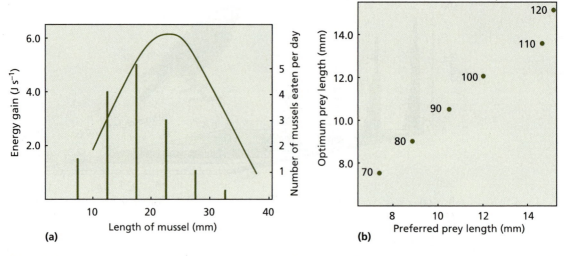

Figure 8.6 Some examples of predators preferring profitable prey. (a) Shore crabs (*Carcinus maenas*) take mainly the most profitable sizes of mussel (*Mytilus edulis*). Curve, rate of energy gain from mussels of different size; bars, the number of mussels eaten per day. After Elner & Hughes (1978). (b) The sizes of shrimp (*Neomysis integer*) preferred by 15-spined sticklebacks (*Spinachia spinachia*) as a function of the optimal size in terms of dry mass of prey per unit handling time. The number next to each point is the length of the fish in mm. After Kislaliogu & Gibson (1976).

items in the time taken to tackle a single large one. This means it should take into account both the nutritional value of an item and its **handling time** in deciding whether or not to take it. The nutritional return per unit handling time of an item is known as its **profitability**, and a rate-maximising predator should clearly go for the most profitable prey. By and large, that is what predators seem to do (Fig. 8.6 and 2.20). But there is another catch. The most profitable prey may be thin on the ground. If the predator concentrated on these exclusively, it may do worse than if it just took anything it came across. **Searching time** between prey items is thus another cost that is likely to influence what is taken. In Figs 2.20 and 8.6(a), for example, it is clear that lapwings and crabs were not taking just the most profitable prey, but were taking some smaller and larger ones as well. Were these just mistakes, or did they reflect adaptive choices based on the availability of different classes of prey? In other words, were predators choosing an optimal *breadth of diet* rather than an optimal kind of prey? To find out, we need to calculate the net rate of return from including different classes of prey in the diet. Once again it was Charnov who showed the way forward (Charnov 1976a).

Charnov's optimal diet model

Charnov's optimal diet model is summarised in Box 8.4. Like the marginal value theorem (Box 8.2), it is based on Holling's disc equation (Box 8.1). The model makes three key predictions:

1. A predator will *always* take the most profitable prey.
2. It will take or ignore less profitable prey according to the inequality in Box 8.4 equation (8.4.1).
3. Whether or not the predator includes less profitable prey will be independent of its encounter rate with such prey.

Underlying theory

Box 8.4   Charnov's optimal diet model

Charnov's (1976a) optimal diet model derives from Holling's disc equation as in Box 8.1. To see how it arrives at its predictions, we shall consider the simplified case of a predator foraging for just two kinds of prey. Following the terminology in Box 8.1, the prey yield e_1 and e_2 units of net food reward respectively, are encountered at the rate of λ_1 and λ_2 items per second and take h_1 and h_2 seconds to handle. Their profitabilities are therefore calculated as e_1/h_1 and e_2/h_2.

Let us suppose that prey type 1 is more profitable than prey type 2. On the basis of equation (8.1.2), the predator would maximise its net rate of food reward by taking exclusively prey type 1 when:

$$\frac{\lambda_1 e_1}{1 + \lambda_1 h_1} > \frac{\lambda_1 e_1 + \lambda_2 e_2}{1 + \lambda_1 h_1 + \lambda_2 h_2} \tag{8.4.1}$$

that is, when the rate of gain from prey type 1 alone is greater than that from both types together. The equation can be rearranged to give the threshold encounter rate at which it pays to specialise on prey type 1:

$$\frac{1}{\lambda_1} = (e_1/e_2)(h_2 - h_1) \tag{8.4.2}$$

where $1/\lambda_1$ is the time it would take to search for the next item.

Based on discussion in Charnov (1976a) and Krebs & McCleery (1984).

The third prediction may seem surprising at first, but it is easy to see why it holds. If the most profitable prey are sufficiently abundant for the predator to benefit by specialising, any time spent handling less profitable items will be time lost from searching for the next profitable one. No matter how common they are, therefore, the predator should ignore less profitable items and continue to specialise. The decision to include less profitable prey should depend *only* on the encounter rate with profitable prey. This is explicit in Box 8.4 equation (8.4.2) in which encounter rate with less profitable prey is no longer part of the inequality. Is there any evidence that these predictions are borne out?

Many studies have provided partial support for Charnov's model, for instance in showing that predators prefer profitable prey or become more selective when encounter rates with profitable prey increase (see the summary table in Krebs *et al.* 1983), but still relatively few have tested all the predictions rigorously. One that has is Krebs *et al.*'s (1977) study of great tits.

Krebs *et al.* looked at tits taking food from a conveyor belt moving past their cage. Birds were presented with large (eight-segment) and small (four-segment) pieces of mealworm moving past them one at a time in sequence. The small pieces had little strips of card attached to them to increase their handling time (h in Charnov's model [Box 8.4]) and so reduce their profitability (E_2/h_2) relative to the large pieces (E_1/h_1). Encounter rate with each type of prey (λ_1 and λ_2) was controlled via the conveyor belt. The predictions and results of the experiment are summarised in Fig. 8.7. When their encounter rate

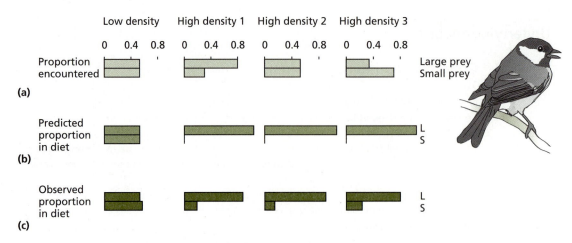

Figure 8.7 Testing the predictions of Charnov's optimal diet model in great tits (*Parus major*). (a) Relative encounter rates with large (profitable) and small (unprofitable) prey on a conveyor belt. (b) The proportions of large and small prey predicted to be taken by the model. (c) The proportions taken by birds in the experiment. See text. After Krebs (1978).

with both types of prey was low ('low density' treatment), birds were unselective, as predicted by the model. When their encounter rate with large (profitable) prey increased to the point where they would do best by taking large prey only, birds switched to taking predominantly large prey, again as predicted. Even when encounter rate with large prey was kept constant and that with small prey increased to twice the rate for large prey ('high density 3'), birds remained selective. However, agreement with the predictions of the optimal diet model was not perfect. Birds still took some small prey when they should have taken none, and the number taken was not entirely independent of their encounter rate (proportionally more were taken in the highest encounter rate ['high density 3'] treatment). We shall come back to this shortly.

Krebs *et al.*'s great tits were tested in a highly controlled laboratory environment where it is relatively easy to manipulate the various components of the model. But what about in the field? Can we test the model in the uncontrolled complexities of a natural foraging environment? A study by Thompson & Barnard (1984) suggests we can.

Thompson & Barnard looked at lapwings taking earthworms from agricultural pasture in the Midlands region of the UK. We have seen already (Fig. 2.20) that lapwings take mainly the most profitable size class of worms, but also take worms from other size classes. By calculating encounter rates with different sizes of worm from soil samples and observations of foraging birds, and using bomb calorimetry to work out the energy content of the worms, Thompson & Barnard applied Charnov's model to see which combination of size classes would yield the highest net rate of energy intake. Figure 8.8(a) shows that birds would do best by taking the *three* most profitable sizes (the expected cumulative intake curve rises to a peak at size class 3). When the birds' actual cumulative intake was plotted (Fig. 8.8b), the first three size classes turned out to constitute about 90% of their diet. Moreover, their intake rate for the three profitable classes correlated with the density of these worms in the soil, whereas intake rate for less profitable classes showed no correlation with their own density but instead correlated with the intake rate of profitable worms (Thompson & Barnard 1984). The main predictions of the optimal diet model thus seemed to be borne out. Importantly,

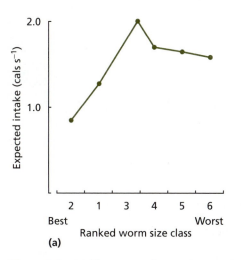
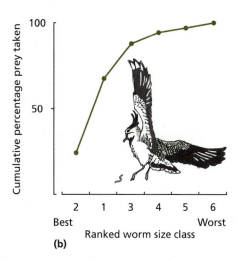

Figure 8.8 (a) The expected rate of energy intake for lapwings when progressively more size classes of earthworm are included in the diet. (b) The cumulative contribution of each size class to the birds' diet in the field. See text. After Thompson & Barnard (1984).

lapwings did significantly better by taking this mixture than they would have done by taking worms randomly.

While both great tits and lapwings thus accorded closely with the predictions of Charnov's model, they both continued to take some unprofitable prey when they should have specialised on profitable items. Thus, in the tits' case, instead of a stepwise switch from unselective to selective feeding, birds showed a more graded shift (Fig. 8.9); in the jargon, they are said to have shown **partial preferences**.

Partial preferences   Are partial preferences fatal to the optimal diet model? Not necessarily. They could happen for several reasons. One possibility is that perceptual constraints lead to confusion as to what are profitable and unprofitable prey, so that predators simply make mistakes. There is good evidence that such confusion occurs in great tits under experimental conditions similar to those of Krebs *et al.* (1977) and can account for the partial preferences in their case (Rechten *et al.* 1983). Difficulties in recognising profitable prey imposes an additional time and energy cost on specialising, so that it may pay a predator to broaden its diet. Erichsen *et al.* (1980) and Houston *et al.* (1980) demonstrated this experimentally in tits when they hid prey in pieces of drinking straw which birds then had to distinguish from similar straws containing string or other types of prey. When different types of prey are encountered randomly, a 'run of bad luck' may cause the predator to reappraise relative prey availabilities and change its diet, again leading to partial preferences over the foraging bout as a whole. The shore crabs in Fig. 8.6(a) seem to do this; they are more likely to take prey of low profitability if they have just had a run of encounters yielding few profitable prey (Elner & Hughes 1978). Finally, if prey communities are likely to change, then, as in the case of patches, it may also pay predators to sample different prey from time to time, to check that relative profitabilities have not changed. These, and other factors (see e.g. Krebs & McCleery 1984), may thus cause departures from the predictions of simple optimal diet models but not undermine the principle of optimal foraging *per se*.

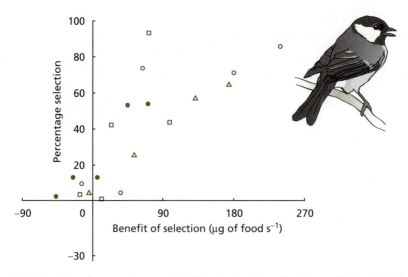

Figure 8.9 Partial preferences in great tits. The graph shows the degree of selectivity for large (profitable) prey by great tits in Krebs *et al.*'s (1977) experiment (see Fig. 8.7) as a function of the benefit of selective foraging. Birds showed a gradual increase in selectivity instead of the stepwise switch to large prey predicted by Charnov's optimal diet model. The symbols represent four different individuals. See text. After Krebs & McCleery (1984).

Rules of thumb and optimal diet choice

As with choosing patches, predators are likely to use approximate rules of thumb in deciding what kind of prey to take. We have seen an example already in Fig. 2.21, where common shrews (*Sorex araneus*) used prey size as a rule of thumb for profitability, and stuck to the rule even when it led to the wrong choice (Barnard & Brown 1981). Shrews appear to use approximate rules in other contexts too, for example when choosing prey in the presence of competitors. One effect of competition is to reduce the predictability of the food supply. Both the absolute and relative abundances of different kinds of food are likely to be affected, so that the predator's past experience no longer provides a reliable guide to the future. Under these conditions, we might expect predators to become less selective. Barnard & Brown (1981) tested this prediction in their shrews.

Shrews were presented with large and small pieces of mealworm on a grid as in Fig. 2.21. Encounter rates were such that, on the basis of their size rule of thumb, the shrews should have specialised on large prey (see above). In some trials an apparent competitor was introduced onto the grid, 'apparent' because competitor individuals had not been trained to forage on the grid and therefore did not deplete the food supply. In this way, Barnard & Brown could measure the effect of the subject animal's *expectation* of competition rather than simply its response to dwindling food availability. The results were clear-cut. Shrews took significantly fewer large prey than expected from their encounter rate when an apparent competitor was present (Fig. 8.10).

The lapwings in Figs 2.20 and 8.8 also appear to use a simple rule of thumb to select prey. The more profitable size classes of earthworm tend (within the birds' pecking range) to be deeper down in the soil. To locate them, lapwings have to spend time in a characteristic crouching posture before pecking (Barnard & Thompson 1985). The longer the crouch, the more profitable the prey eventually retrieved. However, crouching

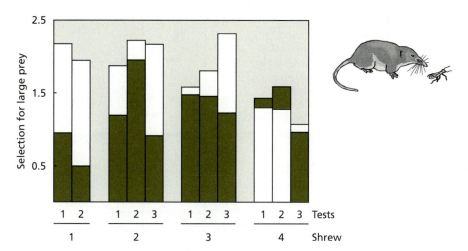

Figure 8.10 A rule of thumb response to apparent competition by foraging shrews (*Sorex araneus*). Shrews generally took fewer large prey (shaded bars) than expected from their encounter rate (open bars) when an apparent (non-foraging) competitor was present. From Barnard (1983).

seems merely to facilitate targeting rather than assessment of worm size and profitability *per se*. As an amusing illustration of this, lapwings occasionally discard worms they have spent several seconds, and much effort, hauling out of the ground. Why? Because flocks are frequently attended by kleptoparasitic (food-stealing) gulls (see 9.1.1.2). The gulls readily detect feeding postures in lapwings and attack the birds as they attempt to handle prey. The larger the prey, the more vulnerable the lapwing to attack. When very large worms are inadvertently pulled up, therefore, the birds jettison them before they attract unwelcome attention (Barnard & Thompson 1985).

8.1.1.4 Maximising intake rate or minimising risk?

The examples we have discussed so far assume that predators are designed to maximise their net rate of food intake. This seems a sensible enough starting assumption if we are going to apply optimality theory to foraging decisions, and we have seen that it accords well with the behaviour of several different predators. However, it is not difficult to envisage situations where rate maximisation may not be the best option.

Imagine a small bird nearing the end of a long, cold winter day. It has the opportunity to visit one last foraging site before dark to get the remaining food energy it needs to survive the night. Suppose it requires four more units of food and has a choice of two possible sites. One yields three units every time it is visited, and is thus of *constant* quality; the other yields five items on half the occasions it is visited, but nothing at all on the other half, so is therefore of *variable* quality. For a rate-maximising predator, the constant site would seem to be best, because its average yield per visit is 3 units compared with only 2.5 at the variable site. Unfortunately, were our bird to be tempted by this it would be dead by morning. Clearly the only way it is going to see another dawn is to go for the variable site and gamble on getting 5 units instead of none. Such a decision would be based on minimising the risk of starvation rather than maximising net rate of intake, and our bird would be said to forage in a **risk-sensitive** fashion (Caraco 1981).

Figure 8.11 Risk-sensitive foraging in common shrews. When their expected intake was less than requirement (vertical line), shrews tended to opt for the variable of two feeding stations (were risk-prone). When expected intake exceeded requirement, they went for the constant station (were risk-averse). After Barnard & Brown (1985a).

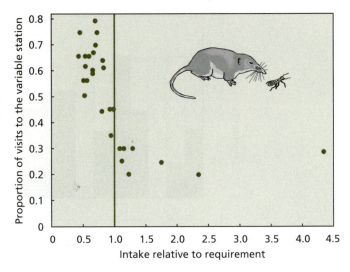

Risk-sensitivity, taking account of the variability of reward as well as its mean, is a well-known feature of economic decision-making, and evidence suggests it plays a role in foraging decisions by other animals too. In an almost direct test of the scenario above, Caraco *et al.* (1980a) provided yellow-eyed juncos (*Junco phaeonotus*) (small North American buntings) with two feeding stations yielding the same mean reward rate. One (the constant station) provided two seeds every time, the other (the variable station) four seeds half the time and none the other half. When birds were meeting their energy requirement, Caraco *et al.* found they always preferred the constant station. That is, they were **risk-averse**. When their requirement was greater than the mean reward at the stations, however, they preferred the variable station, i.e. became **risk-prone**. As our hypothetical bird should have done, therefore, the juncos took account of the variability in reward as appropriate to their food requirement at the time. Barnard & Brown (1985a) found a similar switch in common shrews when their intake changed from meeting to failing to meet their food requirement (Fig. 8.11). Interestingly, the shrews also showed a rule of thumb switch to risk-proneness in the presence of apparent competitors. Even when meeting their requirement, shrews preferred the variable station if they could detect another (non-foraging) shrew behind a perforated Perspex partition (Barnard & Brown 1985b). Thus, as with choosing prey (see above), shrews seemed to adjust their risk-sensitive foraging decisions on the basis of *anticipated* unpredictability (and potential shortfall), rather than a change in requirement or food availability.

Psychology and risk sensitivity

While risk-sensitive foraging, especially in relation to energy budgets, can be accounted for in functional terms like this, tendencies towards risk-aversion and risk-proneness can also be explained in terms of underlying psychological mechanism. As always, the functional and mechanistic accounts are complementary, not rivals for the truth, and neither on its own can explain fully the responses of predators to variance in reward (Kacelnik & Bateson 1996). However, a general tendency for animals to be risk averse when confronted with variance in the *amount* of reward, but risk prone when there is variance in *delay* to reward (two factors that have frequently been confounded in experimental tests of risk-sensitivity) is reminiscent of well-established preferences on concurrent reinforcement schedules (Krebs & Kacelnik 1991; see 6.2.1.2). Several studies have shown that animals prefer variable delays to fixed delays for any given mean reward value (e.g. Herrnstein 1964; Mazur 1984; Reboreda & Kacelnik 1992). Careful experimentation suggests this may be associated with an increased probability of experiencing a short delay. By 'titrating' fixed delays to a reward against a variable delay schedule until subjects became indifferent, Mazur found that animals devalued a given reward disproportionately the longer they had to wait for it (see also Fig. 4.8b). Apparent risk-proneness under these conditions thus has an underlying explanation (randomly recalling short delays on a variable delay schedule) that is very different from, but nevertheless compatible with, the functional interpretation of risk-proneness (minimising the risk of starvation) above (Krebs & Kacelnik 1991).

8.1.1.5 Constraints on foraging decisions

Many things may prevent a predator conforming to the predictions of simple optimisation models. Some of these may be intrinsic to the predator itself, perhaps a limitation of its perceptual system or memory; others may be complications thrown up by the environment, such as the presence of competitors. As we pointed out in Chapter 2 (see Box 2.5), taking such **constraints** into account is an important part of developing and testing appropriate models. We shall look at two examples to illustrate the point.

Nutritional constraints

Although it is convenient, and probably in most cases reasonable, to assume food energy as the currency in optimal foraging models, there are undoubtedly times when other factors play a role in foraging decisions. A requirement for particular nutrients might be one. Belovsky (1978) provides a nice example in moose (*Alces alces*) feeding on the shores of Lake Superior in Michigan.

Nutrient constraints are likely to be particularly important in herbivores because individual plant species often lack essential dietary requirements. Herbivores may therefore need to select a range of species to be sure of getting what they need. For the moose along Lake Superior, it is sodium that constitutes a limiting requirement. The moose have two principal sources of food: forests, where they browse the leaves of deciduous trees, and shallow lakes, where they crop aquatic plants. The leaves are rich in food energy, but deficient in sodium, while the reverse is true for aquatic plants. In order to meet both their energy and sodium requirements, therefore, moose must take a mixture of the two kinds of material. But what mixture? To find out, Belovsky used an optimisation approach known as **linear programming**, in which constraints can be represented graphically as boundaries to the available options (see Belovsky 1990) (Fig. 8.12).

The axes of the plot are the intake of terrestrial and aquatic plant material. Both kinds of plant provide food energy, even though terrestrial material provides more. It is thus possible to identify the combinations of terrestrial and aquatic plants that provide the moose's minimum energy requirement. This is shown as the 'energy constraint' line in Fig. 8.12. The moose's diet therefore has to be on or above this line to satisfy its energy demand. It also has to include enough aquatic plant material to meet the animal's sodium

Figure 8.12 Moose along the shores of Lake Superior must choose combinations of aquatic and terrestrial plant food that meet their energy and sodium requirements within the limitations of their rumen capacity. Their diet should therefore lie somewhere in the shaded triangle bounded by the three lines. In fact it lies at the far right-hand point of the triangle (asterisk), suggesting moose maximise their rate of energy intake subject to the sodium and rumen constraints. After Belovsky (1978).

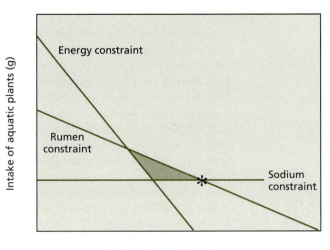

Intake of terrestrial plants (g)

requirement, so must be on or above the 'sodium constraint' line. However, moose are subject to a third constraint: rumen capacity. Plant material is bulky, aquatic plants particularly so, and the rumen, which ferments material prior to digestion, can only take so much at a time. Thus the combined intake must lie *below* the 'rumen constraint' line in the figure. When all three constraints are taken into account, only the diets within the small shaded triangle meet all the conditions. But where exactly in this triangle should the diet fall? We can use the constraint boundaries to find out. For example, if moose had been selected to maximise sodium intake, the optimal diet would be up in the top left corner of the triangle (the maximum possible intake of aquatic plants). If, on the other hand, they had been selected to maximise energy intake, the diet should be down in the far right-hand corner. When Belovsky (1978) looked at what moose actually took, he found their diet was in the right-hand corner (Fig. 8.12). Moose thus appeared to select their diet to maximise energy intake subject to the constraints of sodium requirement and rumen capacity.

Predation risk constraints

Decisions about foraging, like those about any other activity, need to bear in mind that many predators can themselves become prey. Time spent in vigilance, or a reluctance to visit sites where risk is high, may impose constraints on what and how foragers choose. Milinski & Heller (1978) investigated just such an effect in sticklebacks.

Three-spined sticklebacks readily attack swarms of water fleas (*Daphnia pulex*) in an aquarium. When hungry fish are given a choice between swarms of different density, they prefer the highest density (Fig. 8.13) where capture rates are expected to be greatest. But when they are less hungry they tend to go for lower densities (Milinski & Heller 1978). One interpretation of this is that foraging on high-density swarms demands a lot of attention because individual prey are hard to target (see 9.1.1). Attention thus cannot be devoted to other things such as approaching predators. If a fish is very hungry, and risks starvation if it does not take in food quickly, the price may be worth paying. As the risk of starving recedes, however, it may pay to be more vigilant for predators, in which case it is better to go for low-density swarms that demand less attention.

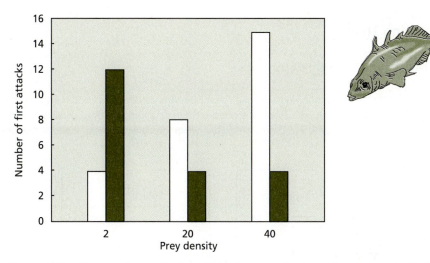

Figure 8.13 Hungry three-spined sticklebacks normally (open bars) went for high-density swarms of *Daphnia*, which are likely to yield food more quickly. If a predator (a model kingfisher) threatened, however, they switched to lower density swarms (shaded bars), where catching prey required less attention. After Milinski & Heller (1978). © Nature Publishing Group (http://www.nature.com/), reprinted by permission.

Such a strategy is fine as long as predation risk remains reasonably low. If it increases, however, hungry fish may be forced to spend more time being vigilant too and thus not be able to pay sufficient attention to foraging to benefit from high-density swarms. If so, they should also prefer lower density swarms. Milinski & Heller (1978) tested this prediction by 'flying' a model kingfisher (*Alcedo atthis*) over a tank of hungry fish. As Figure 8.13 shows, the fish exposed to the kingfisher switched their preferences towards low-density prey. Apparent predation risk therefore constrained the fishes' ability to exploit what should have been the most profitable site.

8.2 Predators and prey

So far, we have discussed foraging decisions as they affect the predator's own foraging efficiency. But choices about where to forage and what to take have an impact on prey, and may lead to adaptive responses on the part of the prey to reduce it. So what sort of effect might foraging behaviour have on prey, and what sort of counterresponses might be elicited?

8.2.1 Effects of foraging behaviour on prey

By concentrating foraging effort where returns are greatest, predators are likely to inflict differential mortality on prey as a function of their profitability, whether this is in terms of return per unit foraging time in patches or per unit handling time for individual prey. Intuitively, we might expect the impact on prey to increase with their density, but in a manner that is limited by searching and handling times. Such relationships have been dubbed **functional responses** (Solomon 1969) and do indeed show characteristic forms depending on the time costs of foraging.

Underlying theory

Box 8.5   Functional responses

The term 'functional response' describes the relationship between prey density and the number of prey attacked by a predator during a foraging bout. Holling (1959) divided functional responses into three categories, which he called Types I, II and III (Fig. (i)).

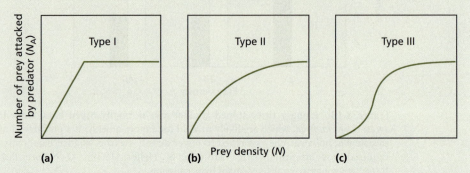

Figure (i)   Generalised Types I, II and III functional response curves. See below. From Hassell (1976).

Type I responses are characterised by a sharp cut-off following an initial increase in the number of prey taken (Fig. (i)a). They assume a constant rate of encountering prey (the attack coefficient a') and a threshold density N_x above which the predator ceases to feed. Thus:

$$N_A = a'T_sN_x \tag{8.5.1}$$

where N_A is the number of prey attacked, T_s the time spent searching and N the number of prey available. Type I responses apply to the behaviour of some filter feeders, where food particles are taken in relation to their density until the predator is satiated, whereupon it abruptly stops feeding (Hassell 1976).

Type II responses (Fig. (i)b) show a negatively accelerating rise to a plateau and are described by Holling's disc equation (see also Box 8.1, where the same equation is cast in terms of average gain per encounter to the predator) as:

$$N_A = \frac{a'NT}{1 + a'T_hN} \tag{8.5.2}$$

where T is the total time available and T_h the time taken to handle a prey item. The deceleration of the response is thus due to the increasing proportion of available time taken up handling prey as more of them are encountered.

The sigmoid Type III response (Fig. (i)c) is generally thought to be a consequence of learning by the predator in the face of changing densities of different types of prey, or when searching for cryptic prey. In the first case, low densities of one type of prey may cause the predator to concentrate on an alternative type, so depressing intake of the rare prey and creating a lag in the intake curve. As the density of the initially rare prey increases, however, it becomes more likely that the predator will switch from the alternative, so the intake curve suddenly rises more steeply. As in Type II responses, the increase then slows as handling time starts to constrain intake rate. A similar curve may be generated by cryptic prey as the predator first has to learn to spot them, something that becomes easier as encounter rates increase with prey density. This underrepresentation of rare prey in the diet is sometimes referred to as **apostatic prey selection** (Clarke 1962). However, sometimes predators seem to go for prey that are odd in some way and stand out from the crowd (Mueller 1971), thus creating a selection pressure *against* rare prey, exactly the opposite of apostatic selection. This apparent contradiction may be a function of prey density, with apostatic selection working where prey are dispersed and encountered sequentially, and 'oddity' selection where they are clumped and comparisons can be made simultaneously (Allen 1972).

Based on discussion in Hassell (1976).

8.2.1.1 Functional responses

A functional response can be defined as a change in the numbers of prey taken in a given period of time by a single predator as the density of prey is changed (Solomon 1969). Holling (1959) formalised the limiting effects of searching and handling time in a simple model, and suggested there were three categories of functional response, known generally as Types I, II and III, each typified by a particular shape of intake curve (Box 8.5).

Type I responses are limited to a rather specialised range of predators (see Box 8.5). Type II and III responses, however, appear to be fairly widespread. Figure 8.14 shows an example of each. Type II responses reflect the rate-limiting effects of the predator's handling time (Box 8.5), but the more complex Type III responses can arise in a number of ways. One long-standing explanation hinges on the idea of **search images**, i.e. the predator 'getting its eye in' for prey of a particular type as it encounters more of them in the environment. How good is the evidence for this?

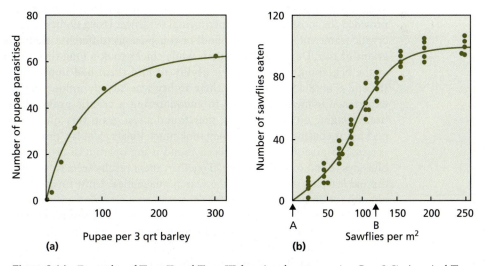

Figure 8.14 Examples of Type II and Type III functional responses (see Box 8.5). A typical Type II responses is shown by the parasitoid wasp *Nasonia vitripennis* attacking housefly (*Musca domestica*) pupae (a), while Type III responses is shown by deermice (*Peromyscus*) taking sawfly (*Zaraea*) pupae buried in sand (b). After DeBach & Smith (1941) and Holling (1965).

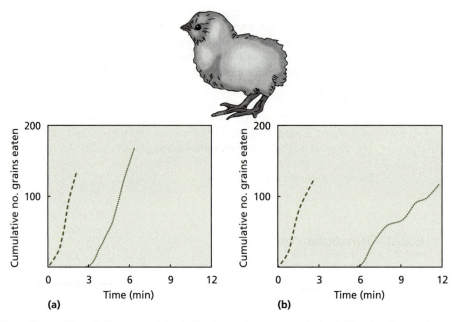

Figure 8.15 The relative ease with which chicks found cryptic (dotted line) and conspicuous (dashed line) grains of rice when (a) searching for green (conspicuous) or orange (cryptic) grains on an orange background, and (b) searching for green (cryptic) and orange (conspicuous) grains on a green background. In both cases chicks took longer to start finding cryptic grains, but then found them quickly, implying they had formed a search image. After Dawkins (1971).

Search images

A number of early studies provided evidence consistent with the formation of search images (e.g. Tinbergen 1960; Croze 1970; Dawkins 1971). Of these, Dawkins (1971) provided the most extensive experimental evidence. Using domestic chicks searching for cryptic (same colour as background) or conspicuous (different colour from background) grains of rice, Dawkins found that, while chicks took a long time to find cryptic grain at the beginning of a test, their ability to find them had improved dramatically by the end (Fig. 8.15). However, their ability was heavily influenced by their searching experience immediately prior to encountering a cryptic grain. If they had already been foraging on cryptic grain, they found a test grain very quickly. If they had been pecking at conspicuous grain, they took much longer. Similar 'priming' effects of cryptic versus conspicuous prey have been found using operant discrimination procedures in blue jays (Bond & Kamil 1999). Together, these results seem to suggest the kind of perceptual learning envisaged by the search image idea. Unfortunately, though, things are not quite so straightforward.

As Dawkins (1971) herself points out, there could be several explanations for intake rate improving with prey density or foraging experience that do not rely on predators formating a search image. Predators might learn where or how to search, for example, or their efficiency in capturing and handling prey might improve (Giraldeau 1997). One explanation relating to cryptic prey that gained considerable experimental support for a time suggests that crypticity forces predators to slow down and search more carefully, thus becoming more likely to spot prey (Gendron & Staddon 1983; Guilford & Dawkins 1987).

The 'search image' and 'search rate' hypotheses lead to a number of common predictions and can be difficult to separate empirically. However, they differ when it comes to predicting how predators with experience of one kind of cryptic prey should respond to different, but equally cryptic, prey. If they rely on a search image, predators should find only the prey with which they are familiar. But, if they are simply searching more carefully, they should be able to find novel cryptic prey too. Various studies have used carefully controlled operant discrimination experiments to try to tease these two possibilities apart. The upshot seems to be more in favour of search images than a reduction in search rate (Reid & Shettleworth 1992; Plaisted & Mackintosh 1995). However, the results imply that search images arise from changes in *attention* rather than perceptual learning, thus departing from the initial assumptions behind the search image hypothesis (Reid & Shettleworth 1992).

8.2.1.2 Area-restricted searching

When food supplies are patchy, predators can focus their foraging effort in another way: by altering their pattern of movement after a find. By slowing down and turning more often, the predator is likely to spend more time in areas of locally high food density, and so increase its chances of finding more food. Such **area-restricted searching** appears to be fairly widespread, and has been demonstrated in both vertebrate and invertebrate predators (e.g. Murdie & Hassell 1973; Andersen 1996; Hill *et al.* 2000; Leising 2001). However, since the adaptiveness of area-restricted searching depends on the patchiness of the food supply, we might expect predators to adjust their tendency to show it accordingly.

Smith (1974) tested this by watching thrushes (*Turdus* spp.) foraging for earthworms in a meadow. Thrushes show typical area-restricted searching after finding a worm when foraging on natural prey distributions, so Smith conducted an experiment in which birds were presented with 'populations' of cryptically coloured artificial prey set out in regular, random or clumped (patchy) distributions. As expected if area-restricted searching is adaptive, thrushes showed a greater tendency to turn after a find on random and clumped distributions. Interestingly, the difference was apparent only when overall food densities were low. At high densities, the birds tended to turn more all the time, not just after a find. Barnard (1978) found a similar effect of distribution in captive house sparrows (*Passer domesticus*), with area-restricted searching being least evident on regular and most evident on clumped distributions of mealworms on a grid (Fig. 8.16). Note, however, that sparrows showed *some* increase in turning after a find on all distributions, perhaps suggesting a movement rule of thumb based on an assumption that prey will generally be clumped.

8.2.1.3 Aggregative responses

It is clear from the preceding discussion that various aspects of foraging behaviour, such as preference for more profitable patches and area-restricted searching, conspire to focus the attention of predators on rich sources of prey. This means that predators tend to spend more time where prey are abundant. As a result, their numbers are likely to build up, so that high densities of prey eventually attract large aggregations of predators (see also 7.1.1.2 and Box 7.1), an effect known as an **aggregative response** (Hassell & May 1974). This will become important again later when we discuss some of the costs and benefits of living in groups (see 9.1.1.2), but for now it highlights the fact that the responses of *individual* predators are only part of the picture when it comes to their impact on prey.

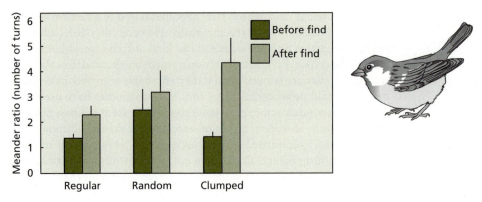

Figure 8.16 Mean meander ratios (the straight line distance between two points divided by the distance actually travelled between the points) for house sparrows before and after finding a mealworm on regular, random and clumped distributions of prey. Birds showed greater meander ratios (more area-restricted searching) after finding food as clumping increased. Plotted from data in Barnard (1978).

8.3 Anti-predator behaviour

Given what we have discussed so far, it is not surprising that many prey species have evolved elaborate morphological, physiological and behavioural adaptations to reduce their risk of predation. However, it is important to stress that predation need not be a limiting factor in prey populations for anti-predator responses to evolve (Harvey & Greenwood 1978). Like the herbivorous insect at the beginning of the chapter, any mutant individual that is just a little better at avoiding predators is likely to have more time or resources for other things and so enjoy a reproductive advantage. In addition alleles enhancing anti-predator responses may reap indirect (inclusive) fitness benefits (see 2.4.4.5 and below) by safeguarding the reproductive success of their other bearers.

Here, we are concerned with anti-predator adaptations that rely on behaviour. However, many adaptations, such as camouflage, distastefulness and mimicry, that are primarily morphological or physiological, often depend for their efficacy on the animal behaving in the right way. There is no point looking like a bird dropping, for example, if you blow your cover by running up and down tree trunks all day. We shall return to some of the examples below, such as warning displays and group defence, in other chapters, because they illustrate wider aspects of the evolution of behaviour. For the moment, however, we shall focus simply on their role as defence mechanisms.

8.3.1 Avoiding detection

The best way of thwarting predators is to avoid encounters with them in the first place, for instance through **crypsis** or **mimicry** of noxious species that predators do not regard as food. Here, behaviour is often an important adjunct to morphological adaptation. The Caribbean fish *Lobotes surinamensis*, for example, lives among mangroves (*Rhizophora* and *Avicennia* spp.) when juvenile. Periodically, leaves from the mangrove trees fall into the water and float for a while before sinking. Juvenile *L. surinamensis* are roughly the same shape and size as mangrove leaves, and also share their speckled yellowish-brown coloration. When a leaf falls into the water close to a fish, the latter often swims towards it and floats on its side with its head tilted slightly down, closely resembling the gently curved profile of the leaf. Similar crypsis-enhancing postures are found in many other species, such as the various grass-, stick- and leaf-mimicking phasmids, mantids, grasshoppers and butterflies (see Edmunds 1974; Endler 1984). In some bark-mimicking moths (*Melanolophia* and *Catocala* spp.), correct alignment of wing markings to the pattern of the bark is achieved using tactile cues or gravity (Sargent 1969).

Crypsis, of course, depends on being viewed against an appropriate background. Sometimes this occurs passively, simply because of where the animal happens to live. In other cases, it seems to be a result of choice. The California yellow-legged frog (*Rana muscosa*), for example, which is conspicuous when sitting on dry rocks above water, jumps onto submerged rocks covered in yellow-brown algae when alarmed. The frog's coloration matches the algal bloom almost perfectly, and the animal appears to vanish into its background as it submerges (Norris & Lowe 1964). In some insects, matching is achieved visually by comparing certain parts of the body with the substrate, the insect coming to rest on backgrounds that match best (Sargent 1968). Choice of substrate may vary with the risk of predation. Stonefly (*Paragnetina media*) nymphs, provided with alternative (light and dark) backgrounds in an aquarium, chose the dark background, against which they were cryptic, during the light phase of the diurnal cycle, but wandered freely over either background when it was dark (Feltmate & Williams 1989). A subsequent experiment, in which nymphs were exposed to rainbow trout (*Salmo gairdneri*), their natural predator, showed that those on a dark background were far less likely to be taken (3 out of 24) than their counterparts on a light background (19 out of 24). In some cases, cryptic coloration may coordinate with the background more dynamically. Some octopus species, for instance, fine-tune their coloration almost continually (changing on average nearly three times a minute!) as they move over different substrates while foraging (Hanlon *et al.* 1999). In other cases, such as caddis fly larvae (Insecta: Plecoptera) and spider crabs of the family Majidae, animals may physically attach debris from the environment to their bodies, so ensuring appropriate camouflage (e.g. Wicksten 1980). Crypsis-enhancing behaviours can also be very crude, however. Moth species living in oak-hickory forest in New Jersey, for example, rely for crypsis on flying between certain dates when suitable backgrounds are available (Endler 1984). Here, cryptic behaviour is largely just a matter of timing general activity.

Mimicry and behaviour

Not surprisingly, mimicry of other species can also involve behaviour. **Locomotor mimicry** (the convergence of general activity patterns between distantly related prey species) occurs in several Müllerian mimetic relationships (mimicry between noxious prey species), where it may reduce selection against odd individuals (Srygley 1999; see Box 8.5). It also occurs in Batesian mimics (harmless prey that mimic a noxious model). Several species of hoverfly (Diptera: Syrphidae) mimic bees or wasps. The dronefly *Eristalis tenax* mimics honey bees and is highly convergent with its model in colour and body form. However, recent studies suggest that its behaviour while foraging is also mimetic. Observations of *E. tenax* feeding on a range of flowers suggested that the time spent on flowers, or travelling between them, was more similar to that of honeybees than of other flies (Golding & Edmunds 2000). Video footage of foraging *E. tenax*, honey bees and other, non-mimetic, flies also revealed convergence in flight velocity and trajectory, and the amount of time spent hovering (Golding *et al.* 2001). Locomotor mimicry may sometimes be very subtle and require special technology to detect it. Using

Figure 8.17 Locomotor mimicry in *Heliconius* butterflies. Wingbeat frequency and asymmetry are more similar between mimetic species of different lineages than between species within the same lineage. Open symbols, species in lineage 1 (top), closed symbols, species in lineage 2. After Srygley & Ellington (1999).

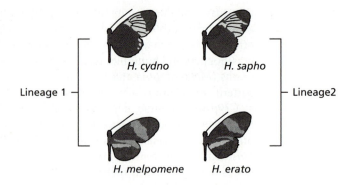

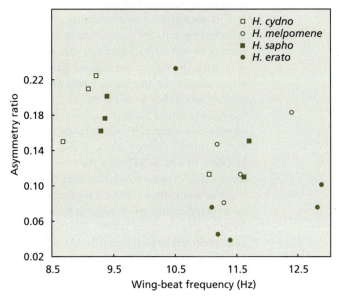

high-resolution kinematic techniques, for example, Srygley & Ellington (1999) were able to show that mimetic passion vine butterflies (*Heliconius* spp.) were more similar in left–right wingbeat asymmetry than even sister (extremely closely related) species (Fig. 8.17). Wingbeat asymmetry is not detectable to the human eye, but may nevertheless be an important component of locomotor mimicry in the butterflies.

The preference of predators for profitable prey (8.1.1.2) may lead to selection for mimicry of their unprofitable counterparts. **Escape mimicry** is one form of this that depends on prey being difficult to capture (Srygley 1999). Like distasteful, or otherwise noxious, species, there is evidence that some prey that are difficult to catch have evolved bright warning (**aposematic**) coloration (see 8.3.5.3) to advertise their unprofitability (Pinheiro 1996). Frustrated attempts to catch them leads predators to associate the aposematic phenotype with unprofitability, thus generating selection for Batesian mimicry of unprofitable models (Srygley 1999).

Avoiding detection is sometimes referred to as a **primary defence mechanism**, to be distinguished from **secondary mechanisms,** such as escape or distastefulness, that come into play once potential prey has been discovered (Edmunds 1974). Of course, few prey can avoid detection completely. Even highly cryptic individuals can be discovered by chance or systematic searching, and search image formation (see 8.2.1.1) can then focus the predator's attention on the newly recognised prey, rapidly stepping up mortality. Some species counter this by diversifying their appearance, so that successively encountered individuals look very different (**polymorphism**). While avoiding predation is by no means the only explanation for polymorphism, there is certainly evidence that it reduces mortality by predators (Croze 1970). Many other species rely on backup secondary defence mechanisms to get them out of trouble. Young northern water snakes (*Natrix sipedon*), for example, have a banded appearance that is cryptic when lying still in vegetation. But, when startled, and the snake moves off quickly, the bands blur together, increasing the impression of speed. When the snake comes to a halt and lies still again, human observers (and presumably visually hunting predators), suddenly lose sight of it and are fooled into thinking it has travelled further than it really has. As a result, they resume their search some way ahead of the snake's real position, so giving it a chance to escape (Pough 1976). While water snakes combine crypsis and visual illusion to thwart predators, many other animals rely almost entirely on secondary defences. So what kinds of behaviour are involved here?

8.3.2 Withdrawing to cover

Many animals live in burrows, crevices or other sheltered places into which they can retreat in times of danger. Such animals are sometimes known as **anachoretes** (from the Greek for 'retire'). Sometimes retreats are constructed by the animals themselves, but often they are natural havens such as hollow trees, rock crevices or even just dense vegetation. Wherever they secrete themselves, most animals must emerge from seclusion at some point to feed or carry out other important activities. Emergence may be only partial, as in tube-dwelling polychaete worms such as *Sabella*, but even where complete, it is often confined to a limited area around the retreat. This is also true of other species that, while not anachoretes, rely on cover for protection, and can impose quite a constraint on the animal's activity. A study of foraging grey squirrels (*Sciurus carolinensis*) illustrates the point. Squirrels frequently forage on open ground, but when they find an item, they often take it up a nearby tree rather than eat it on the spot. This seems a remarkably inefficient way to feed, so why do it? An obvious possibility is that the squirrel is taking account of the risk of predation should it handle food out in the open. But if it always runs back to cover, its rate of food intake is going to be seriously reduced. Lima *et al.* (1985) therefore predicted that squirrels would be more likely to dash for cover when they were feeding nearby, so that the impact on feeding rate was reduced. They also predicted that they would tend to run back with larger items, which would otherwise take a dangerously long time to handle in the open. Lima *et al.* tested these predictions by putting out pieces of biscuit on bird tables situated at different distances from cover. As expected, squirrels ran back to cover more often from nearby tables and when they had taken a large piece of biscuit (Fig. 8.18a).

Like Lima *et al.*'s squirrels, some bird species also punctuate foraging with flights back to cover. House sparrows on farmland, for example, often feed close to hedgerows. Characteristically, flocks fly out from a hedge, feed on the ground nearby for a few seconds, then fly back to the hedge again, repeating their to-ing and fro-ing as they move up and down the hedge. The further out birds feed, the more risk they run from aerial predators such as sparrowhawks (*Accipter nisus*). Rather like Milinski & Heller's sticklebacks in Figure 8.13, therefore, we might expect them to become more vigilant as the risk increases. In line with prediction, Barnard (1980a) found that scanning rates (head

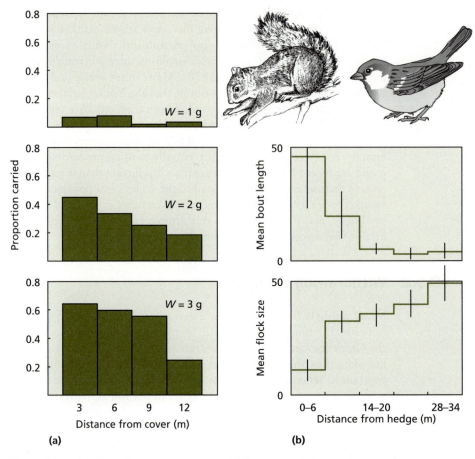

Figure 8.18 (a) When food was put out on bird tables at different distances from cover, grey squirrels (*Sciurus carolinensis*) were more likely to run back with items if they were large (increasing value of *W*) and found close to trees. After Lima *et al.* (1985). (b) House sparrows (*Passer domesticus*) fed in shorter bouts and bigger flocks the further they were from a hedgerow. After Barnard (1983).

cocks per second) by individual sparrows increased with distance from a hedge once other factors affecting the time allocated to different behaviours (such as food density and flock size [see Fig. 9.9a]) were taken into account. Birds also foraged for a much shorter time further out and then only when they were in large flocks (Fig. 8.18b). Proximity to cover therefore had a profound effect on foraging behaviour as a result of sparrows having to invest in anti-predator defences, in this case vigilance, when away from it.

One way around such constraints, of course, is to take your cover with you. Hermit crabs (e.g. *Clibariarius* and *Pagurus* spp.) do just this by acquiring gastropod (snail) shells into which they insert their vulnerable bodies. Shells are assessed carefully before being adopted (see Elwood & Briffa 2001 and 11.1.1.5) and crabs move into a succession of larger ones as they grow. Like many other kinds of retreat, however, a seemingly safe shell can turn into a death trap if a predator manages to gain entry. Some hermit crabs (e.g. *Diogenes* spp.) have therefore evolved an enlarged chela (claw) which occludes the opening of the shell when the crab withdraws inside. Edmunds (1974) provides examples of such protective 'blockading' in several other species.

8.3.3 Flight behaviour

The response of many animals when faced with a predator is to flee, and selection has acted in a variety of ways to enhance the efficacy of flight behaviour. Speed and agility are perhaps the most obvious adaptations, and the coevolved relationship between long-legged herbivores such as zebras and gazelles, and running carnivores such as lions and cheetahs, is a familiar example.

Sustained speed, however, is very costly, and involves sacrifices in other attributes, such as body mass. Another way of enhancing flight is to move in an erratic and unpredictable way. Many species, such as ptarmigan (*Lagopus mutus*), snipe (*Gallinago gallinago*), rabbits, antelopes and gazelles flee from predators in a characteristic zig-zag fashion, with rapid, unexpected changes of direction making it difficult for a predator to keep track of its target (Fig. 8.19; see also 9.1.1.1). Driver & Humphries (1988) refer to these as **protean behaviours** (after *Proteus* of Greek mythology, who frustrated would-be captors by constantly changing shape; see Chance & Russell 1959), a heterogeneous range of behaviours that seems to imbue actions with a degree of uncertainty and unpredictability from the viewpoint of an observer.

A different way of enhancing escape by flight is to use so-called **flash** behaviour. Here, an alarmed animal flees for a short distance then freezes, like the water snakes in 8.3.1. Many predators are unexcited by immobile prey, and a sudden burst of activity appearing from nowhere then vanishing again can create considerable confusion. The

Figure 8.19 Examples of protean behaviour: (a) the erratic flight of an alarmed snipe (*Gallinago gallinago*), and (b) the scattering behaviour of a herd of impala (*Aepyceros melampus*). From *Protean Behaviour: The Biology of Unpredictability* by P.M. Driver and D.A. Humphries (1988). Reprinted by permission of Oxford University Press.

(a)

(b)

Figure 8.20 The butterfly *Thecla togarna* at rest, showing a 'false head' ('antennae' and 'eyes') at the tips of the hindwings. See text. From Edmunds (1974) after Wickler (1968).

sudden, conspicuous jumps of many frog and grasshopper species are other familiar examples. Flash responses are often embellished with bright markings which, like the coloured hindwings of some grasshoppers and mantids, are exposed during flight but disappear once the animal resumes its resting position (e.g. Edmunds 1972). The silvery flashing of fish when a shoal scatters, and the bright patches of plumage exposed on the wings, back or tail of many flock-feeding birds as they take flight (Brooke 1998), may also serve to confuse predators in this way (see 9.1.1.1).

8.3.4 Diversion

Another way of dealing with predators is to divert their attention, either away from the prey or its young, or away from vital or vulnerable areas such as the head. Many fish and insects have evolved 'eye' spots or false heads at the posterior end of their body, which, it has usually been assumed, deflect attack away from their true head (Robbins 1981; DeVries 1997). In lycaenid butterflies, such as *Thecla togarna*, the diverting effect of false 'eyes' and 'antennae' on the rear of the wings (Fig. 8.20) is enhanced by the butterfly orientating with its real head down (instead of up like most butterflies) when resting on vertical surfaces. Recently, however, Cordero (2001) has suggested that false 'eyes' and other 'head' features may function in entirely the opposite way: that is to direct attention *towards* the real head. Reviewing some of the experimental evidence, Cordero argues that predators in fact *avoid* going for the head because that is where the prey's sense organs are concentrated, and any attack is likely to be detected. It is therefore to the prey's advantage to fool the predator into going for its head by creating a false (and usually more prominent [Cordero 2001]) 'head' at the opposite end.

Sometimes diversion is aided by attention-grabbing movements, such as the tail-twitching displays of many lizards and snakes. Some lizards are able to break off the tail (**autotomy**) should an attack be carried through. The detached tail then squirms with a life of its own, keeping the predator occupied while the rest of the lizard makes its escape (Arnold 1988). Autotomy of body parts is also known in other organisms, including annelids, molluscs and arthropods, while tail-stripping (shedding the skin and flesh of the tail) may serve a similar function in some small rodents.

A different kind of diversion display is designed to protect nests or offspring rather than the performer itself. Various ground-nesting birds, such as killdeer (*Charadrius vociferus*) and meadow pipits (*Anthus pratensis*) perform convincing 'broken-wing' displays when danger threatens (e.g. Brunton 1990). The adult bird meanders away from its nest trailing an 'injured' wing on the ground. The predator follows this seemingly easy meal only to have it take off when it reaches a safe distance from the nest. Suitably diverted, the predator continues on its way without further danger to the bird's offspring.

8.3.4.1 Death-feigning

Rather than diverting the attention of a predator elsewhere, an animal may try to make it lose interest altogether by playing dead (**death-feigning** or **thanatosis**). Many predators will not take carrion or respond to immobile prey (see also 8.3.3), so appearing to be dead may be a good ploy. Death-feigning certainly occurs in a wide range of species, including several insects and spiders as well as reptiles, birds and mammals. The best known example is probably the Virginia opossum (*Didelphys virginiana*), which has given rise to the phrase 'playing possum', meaning playing dead. Hognosed snakes (*Heterodon platyrhinus*) feign death as one of many anti-predator responses in their repertoire. If hissing, writhing and making false strikes at the predator fail to discourage it, the snake rolls over on its back and lays motionless with its mouth open and tongue lolling out. Once the predator loses interest, the snake slowly rights itself and moves away. But how long should the snake wait before giving up its sham? The answer seems to depend on the level of threat posed by the predator. Burghardt & Greene (1988) tested newly hatched hognosed snakes with a range of predatory stimuli, which included a stuffed screech owl (*Otus asio*) and human observers either staring at the snake or averting their gaze. Figure 8.21 shows the time taken by the snakes to recover from death feigning in each of the experimental conditions. Snakes took longer to start moving again in the presence of a stuffed owl or a human observer staring at them than in respective control (stimulus absent) conditions. They showed intermediate recovery times if a human observer was present but looking away. Thus the young snakes seemed to adapt the time they remained 'dead' to the prevailing apparent risk.

8.3.5 Deterrence

Diversion and distraction, then, can be effective ways of shaking off the attentions of a predator. Another ploy, however, might be to threaten (or appear to threaten) the predator, or deter it from going any further by some other means. Displays designed to intimidate a predator are sometimes referred to as **deimatic** or **dymantic** (from the Greek for 'frighten'), and include a wide range of morphological, behavioural and physiological adaptations. But predators can also be deterred in other ways, for instance by being made to think an attack will fail and is therefore not worthwhile. We discuss a number of deterrence effects below.

8.3.5.1 'Startle' responses

We have seen that flight and diversionary responses can involve an element of surprise – 'flash' coloration and autotomy being familiar examples. In some cases, however, surprise is the entire *raison d'être* of the response. For example, many cryptically coloured animals such as rabbits and partridges (*Perdix perdix*) remain motionless until an approaching predator is almost upon them, whereupon they erupt noisily from concealment and speed away to safety. Anyone who has experienced such a **startle response** will be in no doubt as to its unnerving effect on the hitherto oblivious approacher.

In rabbits and partridges, the startle response is little more than an embellishment of flight behaviour. But in other cases animals have evolved highly specialised and sometimes elaborate displays that appear to be designed to startle or intimidate predators and buy time for escape. Some nice examples are shown by saturnoid and sphingid moths.

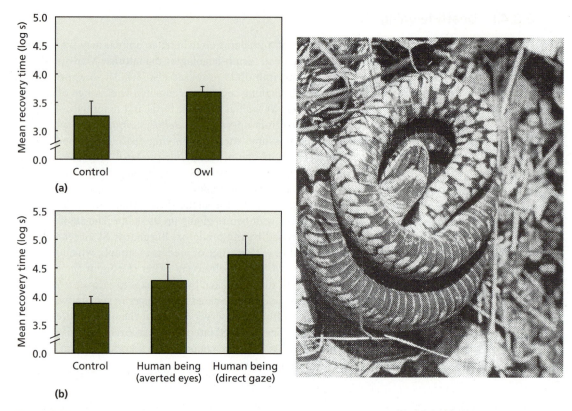

Figure 8.21 Young hognosed snakes (*Heterodon platyrhinus*) took longer to stop feigning death when a stuffed owl was present (a) or a human observer was gazing at them (b). See text. After Burghardt & Greene (1988). Photograph of a dice snake (*Natrix tessellata*) feigning death courtesy of Peter Davies.

When disturbed by an interfering prod or peck, resting moths perform a vigorous wing-flapping and body-rocking display. This causes a predator to hesitate for a few seconds during which time the moth has used the display activity to warm its flight muscles for takeoff. Other moths possess brightly coloured spots on their underwings, which resemble vertebrate eyes. These 'eye' spots are much larger and more conspicuous than those contributing to the false head effects in 8.3.4, and generally remain hidden until the animal is alarmed, whereupon they are suddenly revealed. One suggestion is that alarm effect of the markings lies in their passing resemblance to the eyes of owls, which are predators of the insectivorous birds attacking the moths. However, it may be that the sudden appearance of bright colours *per se* is all that is required. Red and yellow underwing moths (*Catocala* and *Triphaena* spp.) are a case in point. These moths are a cryptic dull brown when at rest, but when they fly, their brightly coloured underwings make them highly conspicuous. While they might simply act as flash markings, the dramatic appearance of the underwings could startle a predator. Schlenoff (1985) tested this possibility by presenting captive blue jays with model moths made of piñon nuts, in which variously painted 'hindwings' sprang into view as the bird picked the nut off a presentation board. During training, jays learned to catch and eat model moths with plain grey hindwings. They were then tested with a row of eight models, seven of which had grey wings, and one (randomly placed in the sequence) wings painted in bold

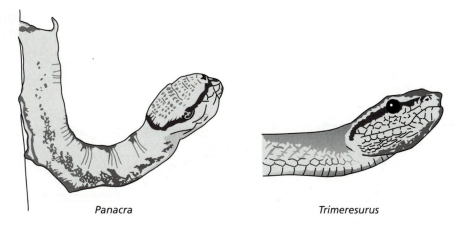

Panacra   *Trimeresurus*

Figure 8.22 The caterpillar of the hawkmoth *Panacra mydon* closely resembles the head of a young Wagler's pit viper (*Trimeresurus wagleri*), a local predator. After Edmunds (1974).

colours like the underwing moths. When jays picked up a model with bold wings, they raised their crests and gave alarm calls. Sometimes they even dropped the model and flew away from it. The sudden appearance of bright markings thus had a very clear startling effect on the birds.

While 'eye' spots may or may not mimic predators, there are some adaptations for which this interpretation is inescapable. The caterpillar of the neotropical hawkmoth *Leucorampha*, for example, swings its anterior end free of its supporting twig and expands it into a highly convincing snake's head when disturbed. In the caterpillar of another hawkmoth, *Panacra mydon*, a similar snake head display closely mimics the head pattern of a young Wagler's pit viper (*Trimeresurus wagleri*), a common local predator of birds (Fig. 8.22).

8.3.5.2   Intimidation

The snake head displays of the caterpillars above may startle a predator, or they may intimidate it by appearing to pose a threat. **Intimidation** is used widely by many species to deter predators. Several vertebrate groups have evolved displays that exaggerate body size, for instance by inflating the body, as in some fish, frogs and toads, erecting frills, spines or crests, or spreading wings, as in various reptiles and birds, or raising hair or quills, as in some mammals. Visual effects may be accompanied by sounds, such as loud clicks, hisses, growls or calls, or by odour emissions (see 11.2.1.1). In some cases, sounds or odours may be the principal means of deterrence (8.3.5.3). Predators can also be intimidated by displays of weaponry. Teeth, horns, antlers and spines can all be presented to good effect and are prominent components of threat displays in many mammals. The spines of various fish species, such as sticklebacks and trigger fish (*Balistes* spp.) deter predators by making ingestion difficult. Three-spined sticklebacks are able to lock their spines, causing them to jam in the mouths of predators such as perch (*Perca fluviatilis*) and pike (*Esox lucius*). When presented with a mixture of minnows (*Phoxinus phoxinus*), ten-spined sticklebacks (*Pygosteus pungitius*) and three-spined sticklebacks in an aquarium, pike ate the spineless minnows first, followed by the small-spined *P. pungitius* before taking the three-spined sticklebacks with their large

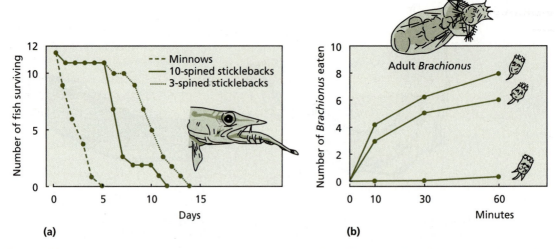

Figure 8.23 (a) The survival of minnows, three-spined sticklebacks and ten-spined sticklebacks when hunted by a pike in a large aquarium. Three-spined sticklebacks, with their large, lockable spines, survived longest. After Hoogland *et al.* (1957). (b) The number of each of three different morphs of *Brachionus calyciflorus* eaten by the predatory rotifer *Asplanchna*. Again, the spiniest morph survived best. After Edmunds (1974) from data in Halbach (1971).

unpalatable spines (Hoogland *et al.* 1957; Fig. 8.23a). Similar relative survivorship effects have been shown in long-spined versus short- or non-spined morphs of the rotifer *Brachionus calyciflorus*, which is preyed upon by other rotifers of the genus *Asplanchna* (Halbach 1971; Fig. 8.23b).

8.3.5.3 Chemical deterrence

Various species have evolved defence mechanisms based on noxious chemicals. Sometimes these are associated with special glands which secrete or evacuate chemicals to dissuade predators from attacking. In other cases, they may have an effect only after the animal has been bitten or otherwise injured by the predator.

Defensive secretory glands are found in a wide range of animals. Gastropod molluscs secrete a variety of noxious substances, including proteinaceous compounds, mucopolysaccharides and sulphuric acid. Many predatory fish are deterred by the secretions and mollusc species with secretory glands are usually rejected as food (Thompson 1960). Sometimes the delivery of a defensive substance can be quite involved. The harvestman *Vonones sayi*, for example, repels ants with a carefully mixed concoction of body fluid exuded from the mouth, and a quinonoid secretion from the carapace glands. The mixture is then picked up on the legs and brushed onto attacking ants, repelling them (Eisner *et al.* 1971). In other cases, chemicals are actively sprayed onto the predator. Whip-scorpions (*Mastigoproctus* spp.) and bombardier beetles (*Brachinus ballistarius*) have specialised anal glands that can direct acidic or hot spray with great accuracy at interfering predators (Eisner & Meinwald 1966; Dean *et al.* 1990). Perhaps the best-known emission chemical defence mechanism is that of the skunks *Spilogale putorius* and *Mephitis mephitis*. When first alarmed, skunks perform a deimatic display, stamping the forefeet on the ground and, in *S. putorius*, sometimes rearing up in a handstand to display the full length of its striking black and white back to the predator. If this fails to dissuade, the animal squirts a nauseating pungent fluid from its anal glands.

Aposematism

An entirely different kind of chemical deterrent relies on the toxic, distasteful or otherwise noxious qualities of the prey's own tissues. While not in itself a behavioural attribute, noxiousness is often associated with characteristic warning (aposematic) displays, like those of the skunks above (we shall discuss aposematic displays again in Chapter 11 when we consider the evolution of signals generally) or depends on other aspects of the prey's behaviour. Monarch butterflies (*Danaus plexippus*) are a good example of the latter. While some distasteful species synthesise their own noxious chemicals, others, such as monarchs, sequester them from their food, the highly toxic milkweed plant (*Asclepias curassavica*). Caterpillars feeding on milkweed store the toxins in their tissues and retain them when they metamorphose into adults (Brower & Brower 1964). Naïve predators (scrub jays, *Aphelocoma coerulescens*) feeding on milkweed-reared monarchs vomit soon after ingestion and reject further presentations, while those fed monarchs reared on cabbage show no ill effects and continue to accept the butterflies (Brower & Brower 1964).

Sequestering toxins, however, may come at a cost, resulting in reduced investment in other components of life history, for example growth (e.g. Camara 1997, but see Fordyce 2001). There may therefore be a frequency-dependent temptation for individuals to exploit the investment of their sequestering companions by feeding on toxin-free material. As long as there are sufficient noxious individuals to maintain predator aversion, non-sequesterers can freeload their way to a potential reproductive advantage, protected by their resemblance to their noxious counterparts. Such **automimicry**, as Brower called it, is a form of producer–scrounger relationship (see 9.1.1.2) and seems to be a feature of populations with toxin-based chemical defences (e.g. Guilford 1994; Ritland 1994; Moranz & Brower 1998).

Defences such as distastefulness and associated aposematism also pose another problem from an evolutionary point of view. Since in the first instance their defence mechanism is perceived by a predator only *after* the prey has been sampled, any rare noxious mutant in a population of naïve predators would seem to be at an immediate disadvantage. So how could the trait establish? The most likely way appears to be through kin selection (2.4.4.5; e.g. Fisher 1930; Turner 1971; Alatalo & Mappes 1996). Many distasteful butterfly species, for example, are gregarious both as larvae and as adults. Their low rates of dispersal mean that individuals are likely to be living alongside close relatives. Distasteful *Heliconius* butterflies in particular have restricted home ranges, roost communally and are long-lived, ideal conditions for the formation of kinship aggregations with their potential for indirect fitness (see 2.4.4.5) benefits. Experimental evidence with novel predator–prey systems also supports the idea that aggregation was an important precondition for noxious defences and aposematism to take off (Alatalo & Mappes 1996; Mappes & Alatalo 1997). However, other studies suggest that noxiousness could spread without the assistance of kin selection under certain circumstances, for instance where the predator inflicts only limited damage on sampled prey (Ohara *et al.* 1993). Furthermore, phylogenetic analysis of some warningly coloured groups indicates that warning coloration preceded gregariousness, implying that noxiousness is not driven by close association and kin selection in these cases (Sillén-Tullberg 1988).

Once predators have experienced noxiousness, stimulus generalisation confers protection on other, similar, prey, a process facilitated by aposematism and leading to mimicry. If the cost of sampling is sufficiently high, then hard-wired avoidance responses are likely to replace bitter experience as a driving force for aposematism and mimicry, an example being the first-time response of avian predators to brightly coloured ring patterns reminiscent of coral snakes (e.g. Brodie & Janzen 1995).

8.3.5.4 Aggregation and deterrence

Finally, there are various ways in which social aggregation can deter predators. Intimidation, for one, can be a matter of numbers, rather than size or weaponry. In some cases, such as the defensive rings formed by adult musk oxen (*Ovibus moschatus*) around their calves, all three elements combine. While it may also serve other functions (see e.g. Curio 1978 and below), '**mobbing**' is a widespread form of group intimidation found in many birds and mammals and some invertebrates. It involves harassing or directly attacking the predator, and frequently appears to place the perpetrator(s) in considerable danger. However, several studies suggest that the boldness and vigour of mobbing is carefully geared to the level of risk involved (Curio 1978; Swaisgood *et al.* 1999); indeed, in some cases, such as California ground squirrels (*Spermophilus beecheyi beecheyi*) attacking northern Pacific rattlesnakes (*Crotalus viridis oreganus*), mobbers may have a degree of immunity to the predator's main weapon – poison (Swaisgood *et al.* 1999). Mobbing is often associated with characteristic calls, which, unlike many alarm calls (see below), are often easy to locate (Marler 1955; Jones & Hill 2001) and serve to recruit others to the fray. *Like* alarm calls, however, mobbing calls can be effective across the species boundary. Curio *et al.* (1978), for example, found that caged blackbirds could be induced to mob arbitrary 'predator' stimuli (see Fig. 6.23) if these were presented while a multi-species mobbing chorus was being played. Interestingly, mobbing responses are very resistant to habituation (6.2.1.1). Even minor changes to a predator stimulus, such as moving it to a slightly different location, is sufficient to overcome the small degree of habituation sometimes encountered in laboratory environments (but, interestingly, rarely in the field; Shalter 1978a).

But deterrence need not necessarily require threat or alarm. Simply indicating that an attack is unlikely to be successful might be enough. Another explanation for mobbing is that it does just this by robustly informing the predator it has been spotted so may as well give up. However, aggregation on its own can be sufficient to persuade a predator that its chances are not good, perhaps because predators find a lot of potential targets confusing, or more individuals means more sets of sense organs scanning the environment and a greater probability of being spotted (see 9.1.1). That grouping has a deterrent effect was shown nicely by Neill & Cullen (1974), who found that fish and cephalopod predators preying on small fish did better when hunting small schools or singletons than when hunting large shoals. When they looked at the predators' tendency to attack, they found it was much lower against large schools.

Alarm signals

While early detection may account for reduced success against groups in itself, the effect is often enhanced by special **alarm signals**. When an approaching predator is spotted, one or more individuals in the group may give a signal that alerts the others. Alarm signals may be visual, auditory, chemical or even mechanical, but their effect is to broadcast the presence of the predator and either trigger escape or garner support to confront the attacker. In the case of chemical alarm signals, such as the alarm substance released by injured tadpoles of the western toad (*Bufo boreas*), they may both elicit escape responses in other conspecifics and act as a direct deterrent by being noxious. In the case of *B. boreas* tadpoles, there is clear evidence that the substance reduces the likelihood of attack by giant water bugs (*Lethocercus americanus*) and dragonfly (*Aeshna umbrosa*) nymphs (Hews 1988). Somewhat more controversially (see Magurran *et al.* 1996; Smith

Figure 8.24 Stotting by a Thomson's gazelle (*Gazella thomsoni*) after having spotted a pack of hunting dogs (*Lycaon pictus*). After Manning & Dawkins (1998).

Table 8.1 Hypotheses to explain the function of stotting. From Caro (1986a).

Benefits to the individual
  Signalling to the predator
    1. Pursuit invitation
    2. Predator detection
    3. Pursuit deterrence
    4. Prey is healthy
    5. Startle
    6. Confusion effect
  Signalling to conspecifics
    7. Social cohesion
    8. Attract mother's attention
  Signalling not involved
    9. Anti-ambush behavior
   10. Play
Benefits to other individuals
   11. Warn conspecifics

1997), a similar substance, so-called '**Schreckstoff**' (von Frisch 1941), released by injured fish may also trigger escape responses in companions (see Chivers & Smith 1998), but there is no convincing evidence that it has a deterrent effect on predators. Indeed, it may even attract them by suggesting an easy meal (Wisenden & Thiel 2002).

Stotting **Stotting** or **spronking** among ungulates, such as deer, pronghorn and antelope, has also been argued to both warn conspecifics and deter predators. Stotting is a curious stiff-legged jumping display in which all four feet leave the ground simultaneously (Fig. 8.24). It is performed in a variety of situations in which predators threaten, so perhaps not surprisingly, several anti-predator functions have been proposed to explain it. In a *tour de force* pair of papers, Caro (1986a,b) reviewed 11 of these (Table 8.1), and assessed the evidence for each by recording the responses of Thomson's gazelles (*Gazella thomsoni*) to cheetahs (*Acinonyx jubatus*) and other predators. By looking carefully at the behaviour of both predator and prey, and comparing hunts in which gazelles stotted with those where they did not, Caro whittled the field down to a firm front runner – that stotting informed the predator it had been spotted. When a gazelle stotted, hunts were much more likely to be abandoned, a wise decision given that they were very likely to fail if the predator persisted (Caro 1986b). However, two other explanations also received support: mothers sometimes stotted to divert cheetahs away from their fawns, while fawns appeared to use it to alert their mother when they had been disturbed from hiding (Caro 1986b). In addition, a later study by Fitzgibbon & Fanshawe (1988) suggested that, when faced with predators that run down their prey over long distances, stotting conveyed information about the prey's ability to outrun the predator. In keeping with this, hunting dogs (*Lycaon pictus*) appeared to base their decision to chase gazelles at least partly on the rate of stotting by potential targets. Gazelles that were chased stotted on average 1.64 times per second, while those that were not had an average stotting rate of 1.86 per second (Fitzgibbon & Fanshawe 1988).

Vocal alarm signals   Vocal alarm signals too can have a deterrent effect. Zuberbuhler *et al.* (1999), for example, studied six species of monkey in The Tai Forest of the Ivory Coast. The monkeys gave characteristic alarm calls when predators appeared, but gave markedly more when the predator was a leopard (*Panthera pardus*). By following a radio-collared leopard, Zuberbuhler *et al.* found that the high rate of calling by the monkeys caused the animal to break cover and leave the group sooner than would be expected by chance. The strategy thus seemed to be geared to deterring predators that depended on surprise. We shall discuss alarm calls tailored to particular predators again in Chapter 11.

Costs and benefits of alarm signals   One problem in explaining the evolution of alarm calls and other conspicuous alarm signals is that they would seem to draw a predator's attention to the signaller. Even if calling effectively thwarts an attack, why should the caller run such a risk in the first place? While rarified forms of group selection (see 2.4.4.1) have been suggested as an explanation (Wilson 1980), these have not met with widespread acceptance. Instead, as with noxiousness and aposematism, suspicion initially falls on kin selection. The caller's apparent altruism might make sense if the beneficiaries were its close relatives. However, the transient membership of many social aggregations where alarm calling is widespread, such as many flocks of birds, rules this out as a general explanation. Nevertheless, kin selection does appear to have shaped alarm calling in some mammals. Hoogland (1983), for example, looked at the tendency for black-tailed prairie dogs (*Cynomys ludovicianus*) to call in response to a stuffed natural predator, the American badger *Taxidea taxus*. As we saw in 7.2.1.1, prairie dogs live in socially structured populations in which group territories, or coteries, form the basic unit. Young females tend to remain in their natal territory along with yearling males. Thus females and young males within territories are usually closely related. Hoogland found that alarm calling was significantly more likely if either offpsring or non-descendant relatives of the caller were present in the territory than if no relatives were present (Fig. 8.25). However, while Hoogland's results suggest an influence of kin selection, this cannot be the whole story because prairie dogs clearly give alarm calls when they have no relatives to benefit. This suggests there must be some *direct* benefits to calling which outweigh its immediate cost. What could these be?

As we have seen, one possibility is that alarm calling deters the predator by announcing it has been detected. As long as the predator desists without further ado, the caller may benefit along with everyone else, though the caller has incurred some cost in calling while the others benefit for nothing. However, the caller has one piece of crucial information that is not immediately available to its companions: it knows where the predator is. If calling triggers mass escape, individuals not in the know are as likely to head into danger as out of it. The caller, however, is in a unique position to choose the safest route away, perhaps by placing itself on the opposite side of a group to the predator. Alarm calling may thus

Figure 8.25 When presented with a stuffed predator, male (pale bars) and female (dark bars) black-tailed prairie dogs (*Cynomys ludovicianus*) were more likely to give alarm calls if relatives were present in their territory (B and C). There were significant differences between A and both B and C, but no differences between B and C. Bars present means with standard errors. After Hoogland (1983).

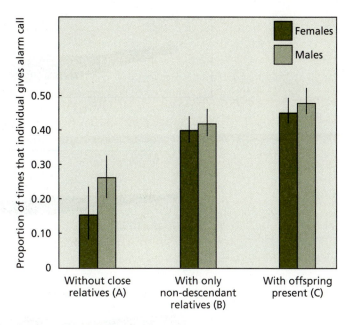

represent selfish manipulation of companions to provide a smoke screen for the caller's escape (Charnov & Krebs 1975). Another possibility is that it pays to warn everyone else, even if they are not relatives, because if the predator was successful once then it might be tempted to come back later and constitute a risk in the future (Krebs & Davies 1993).

While much effort has gone into explaining how alarm callers might overcome the cost of their apparent altruism, some features of alarm calls suggest the risk of calling might be less than it appears. A striking feature of alarm calls among birds, especially small passerines, is their similarity. Most are high-pitched 'seeet' sounds, which Marler (1957) explains in terms of the functional requirements of the calls. Since a problem with giving an alarm call is the possibility of directing the predator's attention to the caller, an optimally designed call would be loud enough to act as a warning, but difficult for a predator to locate. The physical qualities of 'seeet' calls appear to be ideally suited to solving this conflict, as least as far as location by aerial predators such as hawks is concerned. As Catchpole (1979) points out, birds locate sounds binaurally by comparing phase, intensity and time differences in each ear (see also 3.2.1.2). Phase differences are useful only at low frequencies because their information becomes ambiguous once wavelength is less than twice the distance between the receiver's ears. Intensity differences, however, are more effective at high frequencies because the receiver's head forms a sound 'shadow' when its size exceeds the wavelength. The further apart the ears of the receiver, the more apparent time differences become. Time differences can be used across the frequency range, but their effectiveness is improved by repeating, interrupting and modulating the call. From calculations, it appears that calls with a frequency of around 7 kHz would be too high for using phase differences and too low for using the sound 'shadow' effect. As Fig. 8.26 shows, this is precisely the frequency range of many small passerine calls when real or dummy hawks fly over. Perrins (1968) has suggested that some calls may not just be difficult to locate, but may actually direct the predator's attention elsewhere; that is, they may be **ventriloquial**. While human listeners certainly have trouble locating passerine alarm calls, and find some ventriloquial, evidence from the listeners that matter – birds of prey – is equivocal. Some, such as red-tailed hawks

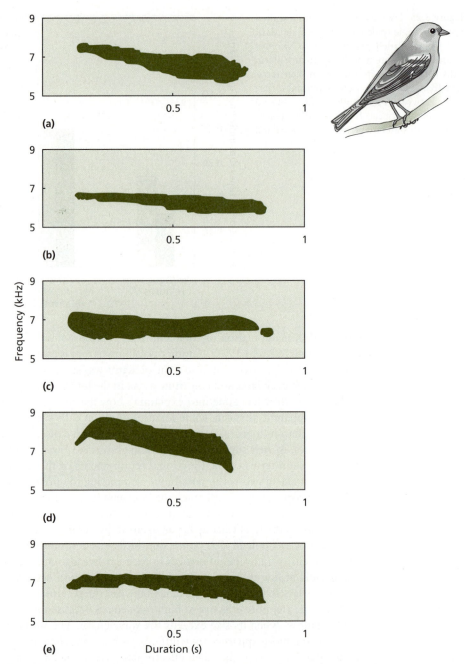

Figure 8.26 Sonograms of the alarm calls of five species of passerine bird: (a) reed bunting (*Emberiza schoeniclus*), (b) blackbird (*Turdus merula*), (c) great tit (*Parus major*), (d) blue tit (*Parus caeruleus*) and (e) chaffinch (*Fringilla coelebs*). See text. After Marler (1957).

(*Buteo jamaicensis*) and great-horned owls (*Bubo virginianus*) concur with humans and find locating callers troublesome (Brown 1982). Others, such as goshawks (*Accipiter gentilis*), barn owls (*Tyto alba*) and pygmy owls (*Glaucidium passerinum*) find them no more difficult to locate than other kinds of call (Shalter 1978b).

8.4   Mutual responses by predators and prey

Something may have struck you as a little odd about this chapter. We began by reminding ourselves that predators and prey enjoyed an evolutionarily dynamic relationship with one another. Predators exerted a selection pressure on prey, whose response exerted a counter-pressure on predators, often resulting in an escalating arms race. When it came to discussing foraging behaviour and anti-predator strategies, however, the dynamism largely disappeared. The prey stalked by our optimally foraging predators occurred in this or that density, and had this or that profitability, but remained unresponsive to the activities of their predators. Similarly, when prey *did* exhibit anti-predator responses, the predator was implicitly assumed to be a 'fixed risk' facing the prey (see Lima 2002).

Holding one side constant while looking at the other is, of course, a convenience that makes problems tractable for models and experiments. Things become complicated by 'moving goalposts' syndrome if the behaviour of two sides is allowed to coevolve. As a result, very few studies have attempted to cater for coevolving responses in their approach to the behaviour of predators and their prey (but see Sih 1984; Abrams 2000). Not surprisingly, this is likely to be an important omission. As Lima (2002) points out, some of the cherished predictions of classical foraging theory become imperilled when even relatively minor behavioural latitude is given to the protagonists. For example, theory predicts that predators will choose patches where the density of (immobile and unresponsive) prey is greatest (8.1.1.1). If there is competition for patches, such predators may reach an ideal free distribution determined by the numerical distribution of prey across patches (7.1.1.3). Another body of theory, however, looks at patch choice by mobile prey under the fixed constraint of an immobile and unresponsive (to the evasion tactics of the prey) predator (e.g. Houston & McNamara 1999). Here, prey should prefer the patch with the lowest risk. If prey are allowed to compete, the resulting distribution across patches is considerably more complex than anything predicted by classical ideal free theory, and may result in multiple stable outcomes (Moody *et al.* 1996). Combining the two – choosing and competing predators and choosing and competing prey – predicts that predators will go where prey are most abundant, while prey will go where they are most likely to avoid predators. Both therefore respond to thwart the other. This can lead to some highly counter-intuitive outcomes.

Iwasa (1982), for example, used a games theory approach (2.4.4.2) to explore the interaction between fish, their zooplankton prey and the patches of phytoplankton on which the latter in turn depended. In the model, both predator and prey were free to move between patches that, from the zooplankton's perspective, varied from rich but dangerous to poor but safe. The outcome suggested that zooplankton should distribute themselves according to the risk imposed by fish, rather than the density of their own phytoplankton prey, but that fish should respond to phytoplankton density even though they do not consume phytoplankton (see also van Baalen & Sabelis 1993, 1999). Further studies where the distribution of predators depends on the resources of their prey suggest that the extent to which predators and prey match the distribution of resources depends strongly on the level of intraspecific competition at the two trophic levels (Sih 1998). Weak competition among predators, for instance, leads to undermatching of resources by prey (prey avoid rich patches), while strong competition favours overmatching (prey prefer rich patches).

Predators might also 'manage' the anti-predator behaviour of their prey so as to maintain their vulnerability to attack (Charnov *et al.* 1976; Lima 2002). Essentially, this means spreading their hunting effort, and thus the risk to prey, over their various feeding sites, for example, by adopting relatively simple 'win–shift' or 'lose–shift' (see 6.2.1.4) rules of thumb. As we saw above, anti-predator responses such as alarm calling might have an influence on a predator's decision to return to a site. Lima (2002) considers several other ways in which the behaviour of predators might influence the evolution and ecology of predator–prey relationships, many of which remain in need of rigorous testing.

Summary

1. Most animals are predators of some sort, and face a broadly similar range of problems in their search for food. Since their survival and reproduction depend on how these problems are solved, we might expect selection to have honed predators into efficient decision-makers. Perhaps not surprisingly, therefore, foraging behaviour has turned out to be very productive in terms of testing optimisation models of behaviour.

2. Most natural food supplies are patchily distributed, so decisions about where to forage and how long to stay there have been addressed largely in the context of food patches. While, intuitively, a distinction is usually made between decisions about where to forage as opposed to what to take, the distinction begins to blur when prey items are handled over a period of time. In fact, both 'patch' and 'prey' decisions can be collapsed into a single, simple model.

3. Optimisation models based on predators maximising their net rate of food (usually energy) intake successfully predict behaviour in food patches and prey choice in a wide range of species. In some cases, however, predators may be selected to minimise risk of starvation rather than maximise intake rate. Decisions by such risk-sensitive foragers are influenced by the variance in reward as well as its mean.

4. While many aspects of foraging conform to the predictions of optimisation models, complementary explanations emerge from understanding some of the psychological mechanisms underlying foraging decisions, such as the means of estimating time. Foraging decisions also require operational rules of thumb, which can sometimes act as constraints on optimisation.

5. By concentrating their foraging effort where returns are greatest, predators are likely to inflict differential mortality on prey as a function of their profitability. Functional responses describe relationships between the impact of the predator and prey density, and vary in shape according to the constraints of searching and handling time. The formation of search images, area-restricted searching and aggregative responses are some of the ways the behaviour of predators can increase their impact on prey.

6. Given its impact, it is not surprising that predation has resulted in some elaborate anti-predator adaptations among prey. Crypsis, mimicry, startle responses and alarm signals are a few examples that involve behaviour. Some, such as warning displays and alarm signals, may have evolved through indirect fitness benefits, but in many cases direct advantages accruing to the performer are sufficient to account for their existence.

7. Considerations of foraging and anti-predator behaviour often implicitly assume that either predators or prey are unresponsive, evolutionarily speaking, to the behaviour of the other. Studies allowing mutual responsiveness have led to some counter-intuitive predictions about the behaviour of both parties that have important implications for foraging theory generally.

Further reading

Curio (1976), Harvey & Greenwood (1978) and Endler (1984, 1991) are good general discussions of predator–prey relationships from behavioural and evolutionary viewpoints. Stephens & Krebs (1986) is a thorough but readable introduction to optimal foraging theory, while Lendrem (1986) provides a gentler tutorial on some of the basic models. The extensive collection of papers by Krebs, Kacelnik, Cuthill and coworkers from the mid 1970s to the 1990s are well worth sampling as exemplars of theory-driven experimentation. Edmunds (1974) still provides a valuable review of anti-predator strategies in animals, while Caro (1986a,b) shows what is involved in critically testing hypotheses about anti-predator behaviour.

9 Social behaviour

Introduction

Spacing within populations is unlikely to be a matter of chance. An animal's reproductive success can be affected in many different ways by the proximity of other individuals, whether of their own or other species. Selection should thus favour strategies for regulating spacing adaptively. Social aggregation and territoriality are two apparent consequences. But how does being in a group or defending a territory affect reproductive success? Can we predict when animals should aggregate or defend territories, and how big groups or territories should be, from a knowledge of how spacing affects individual reproductive success? Are social aggregation and territoriality mutually exclusive alternatives, or can animals join groups and be territorial at the same time? How do differences between individuals affect the costs and benefits of being in a group or defending a territory? Several aspects of social behaviour involve apparent cooperation. How can cooperation evolve when selection is expected to favour self-interest and exploitation?

Animals do not act in isolation in their environment. What they do is often affected by the activities, or even simply the presence, of other individuals around them. Foraging, mating, migrating, avoiding predators, practically any aspect of behaviour, is likely to be influenced in some way by who else is around. In some cases, the presence of others may facilitate behaviour, by providing opportunities for mating, say, or indicating where food might be. In other cases, it may constrain it, perhaps by increasing competition or the risk of infection. The spatial relationship between neighbours is thus unlikely to be a matter of chance. Instead we should expect selection to favour different degrees of aggregation or dispersion according to their reproductive costs and benefits to the individual. On some occasions it will pay to be in a group (social aggregation), on some it will be better to exclude others from the vicinity altogether (territoriality). Social aggregation and territorial behaviour are thus two sides of the same coin. Here, we shall look at the various costs and benefits that determine which side of the coin prevails. However, we should bear in mind that a particular social structure may be the result of constraint rather than choice, as, for example, when individuals are unable to disperse from a natal area (e.g. Emlen 1991; see below). We should also remember that aggregation and territoriality are not really mutually exclusive; as we shall see, both can occur at the same time, as for instance, when individuals or coalitions defend territories within a group. Nevertheless, they provide a convenient contrast for considering the selection pressures shaping social behaviour.

9.1 Living in groups

Social aggregation takes a wide variety of forms, from the loose, milling herds of savannah herbivores to the complex, structured societies of eusocial insects. Even cultures of

Table 9.1 Some suggested costs and benefits of social aggregation. See text.

Benefits	Costs
Aggregative response to prey density means that groups indicate rich sources of food (Krebs 1974)	Greater interference from competitors (Forrester 1991)
Individual risk of predation diluted by joining a group (Treherne & Foster 1981)	Greater risk of contracting disease (Hoogland 1979)
Groups can tackle larger prey than single individuals (Kruuk 1972)	Greater chance of being cuckolded, or mistakenly feeding someone else's offspring (Bray et al. 1975; McCracken 1984)
Grouping confuses predators, making it harder for them to target prey (Milinski 1977)	
Huddling in groups helps thermoregulation (Trune & Slobodchikoff 1976)	Investment in foraging, courtship, or other activities exploited by other group members (Barnard & Sibly 1981)
	Young may be cannibalised by neighbours (Sherman 1981)
Energetic advantages to swimming or flying in a group through 'slipstreaming' (Weihs 1973, but see Partridge & Pitcher 1979)	Greater risk of inbreeding (Faulkes & Bennett 2001)

microorganisms can show a degree of sociality (Oleskin 1994; Crespi 2001). The costs and benefits of aggregating are thus likely to be diverse (Krause & Ruxton 2002). Table 9.1 lists just some of the suggestions, along with studies that have addressed them, and we have already met others in Chapter 8. The list in Table 9.1 could go on, but rather than simply catalogue the various ways in which grouping may or may not confer a reproductive advantage, we shall focus on how different costs and benefits might integrate to determine social behaviour.

9.1.1 Predators and prey

While grouping may serve a variety of functions, many appear to hinge on a reduction in the risk of predation or increased feeding efficiency (e.g. Bertram 1978; Krebs & Davies 1993). We have mentioned one of these, predator deterrence, in 8.3.5.4, but it is important to look at some of the others because they show that similar advantages can arise in very different ways, and that these can interact in determining whether or not animals form groups (see also Krause & Ruxton 2002).

9.1.1.1 Effects of grouping on predation risk

The deterrent effects of grouping in 8.3.5.4 arise from a number of consequences of grouping by prey for the likely success of a predator, increased risk of injury during attack and loss of the element of surprise among them. But deterrence is just one of the ways in which the risk to individual prey can be reduced by forming groups. Bertram (1978) recognises a number of others.

Avoiding detection

Paradoxically, joining a group may reduce the risk of being encountered by a predator at all. Intuitively, we might expect a group to be more conspicuous than a single individual, and thus more likely to be detected. Indeed, in many cases, detectability does seem to increase with group size (Andersson & Wiklund 1978; Wrona & Dixon 1991; Mooring & Hart 1992). The characteristically strong smell of large bird roosts, for example, appears to make them more attractive to predators (Andersson & Wicklund 1978). Some aspects of group structure, however, may lessen conspicuousness to predators that hunt visually, as seems to be the case in clusters of tropical spiders preyed upon by wasps (Uetz & Hieber 1994). Because groups are likely to be scarcer than scattered individuals (more of the population is concentrated at fewer points in space), they may be missed more often by predators wandering across their home range. Models of randomly searching predators that are likely to capture only one individual per group at best, suggest that success rate may be considerably reduced by this 'encounter dilution' effect (Vine 1971; Treisman 1975; Mooring & Hart 1992).

Diluting individual risk

Even if a group is discovered, the fact that an individual is associating with others may still reduce its risk of capture. Because the predator has a number of possible victims from which to choose, the probability of any given individual drawing the short straw is the reciprocal of the group size (assuming all individuals are equally vulnerable). Moreover, those individuals not caught during an attack can escape while the predator handles its victim. In many cases anti-predator responses, such as mobbing or alarm calling (see 8.3.5.4), confer additional protection, but some prey do appear to benefit from a simple dilution effect (Calvert *et al.* 1979; Duncan & Vigne 1979). A nice example comes from a marine insect, the skater *Halobates robustus* (Foster & Treherne 1981).

Like their more familiar freshwater relatives, *H. robustus* skim about on the surface of the water, often forming large groups or 'flotillas'. Their predators are small fish (*Sardinops sagax*) that snap them up from below, thus leaving little scope for any early warning advantage of being in a group (see below). Foster & Treherne found that the attack rate by fish did not vary with group size, so the rate of attack per individual *H. robustus* changed as a simple function of the number in the flotilla, i.e. by dilution (Fig. 9.1).

The dilution effect assumes all individuals are equally vulnerable to attack, but this is often not the case. Young or sick individuals, for example, usually present easier targets than healthy adults, and are disproportionately more likely to be attacked (Kruuk 1972; Estes 1976). However, just being in the wrong place in a group may be enough to render an individual more vulnerable. Prey on the edge, for instance, are likely to be more exposed than those in the middle, as suggested by increased attacks or mortality on the edges of groups (e.g. Kruuk 1964; Patterson 1965; Mooring & Hart 1992), and the fact that individuals on the periphery are often more vigilant (Jennings & Evans 1980; Colagross & Cockburn 1993; Burger & Gochfeld 1994; Fig. 9.2). In general, therefore, we might expect a certain amount of competition for a place in the middle of a group, a prediction supported in some species by the tendency for socially dominant individuals to be at the centre while subordinates are squeezed out to the periphery (Coulson 1968;

Figure 9.1 The dilution effect in flotillas of the marine skater *Halobates robustus*. When fish attack from below, the risk to any given individual skater is a simple decreasing function of the number in the flotilla. Thus the attack rate per individual (solid line) does not differ from that expected if risk is entirely determined by dilution (dashed line). After Krebs & Davies (1993), Foster & Treherne (1981). © Nature Publishing Group (http://www.nature.com/), reprinted by permission.

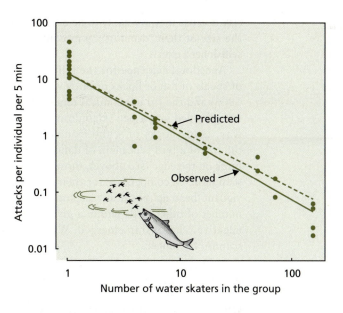

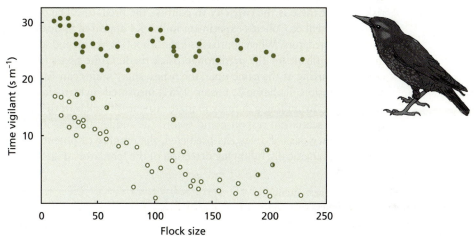

Figure 9.2 Vulnerability to predators is likely to vary with position in a group, and is often reflected in vigilance behaviours. Starlings on the edge of a flock (●) scan at a higher rate for any given flock size than those in the centre (○). Birds midway between the edge and the centre scan at an intermediate level (◑ circles). From Jennings & Evans (1980).

Hall & Fedigan 1997). Hungrier individuals may also tolerate the heightened risk at the periphery in order to avoid competing for food at the centre (Romey 1995).

The **edge effect** arises because vulnerability can depend on the spatial relationship between group members. Hamilton (1971) modelled this more generally in his classic paper 'Geometry for the selfish herd'. Hamilton assumed that the vulnerability of any given individual to a predator appearing at a random point in the environment would be a function of the unoccupied space between itself and its nearest neighbour, its **'domain of danger'** as he called it. The 'domain' encloses all points, and only those points, that are closer to the individual in question than to any other. When a predator suddenly appears in their midst, individuals with the largest 'domains' have the greatest risk because they are likely to be nearest to the predator. Prey should therefore minimise the size of their 'domain' by moving towards their nearest neighbour, the so-called **selfish herd** effect.

Anecdotal evidence for the selfish herd effect includes the characteristic bunching of shoals of fish or flocks of birds when predators threaten (e.g. Tinbergen 1951; Breder 1976) and the (to us) comical tendency for sheep to jump on top of each other when being herded into a pen (Hamilton 1971). Lazarus (1978) tested the idea in a different way. Using flocks of white-fronted geese (*Anser albifrons*), Lazarus predicted that the size of a bird's 'domain of danger' would be reflected in its degree of vigilance. Like birds on the edge (see above), those that were relatively isolated within the flock should scan more to compensate for their increased vulnerability. When the time spent in the vigilant 'head up' position was plotted against the size of their 'domain' (measured as the number of other birds within a given distance of the observed individual), a significant negative relationship emerged. When they stopped to scan, therefore, isolated birds scanned for longer.

Confusing the predator

Other properties of social aggregation can enhance safety beyond a simple dilution of risk. One of these is the capacity to confuse the predator. The deterrent effect of large schools in Neill & Cullen's experiment in 8.3.5.4 appeared to be due to confusion on the part of the predators. When approached, the schools broke up and scattered in all directions, making it difficult for the predator to track them. Quite apart from lots of individuals suddenly moving at the same time, the fishes' shiny scales can flash as they turn, adding to the confusion. Treherne & Foster (1981) measured such an effect in the marine skater *H. robustus*. By pulling a dummy predator towards flotillas of the insects, they were able to trigger a scattering response and record the sudden flashes of light from the insects' bodies as they dashed away. As predicted, the number of potentially distracting flashes increased dramatically following detection of the 'predator' (Fig. 9.3).

Figure 9.3 The change in the mean number of light reflections from the bodies of marine skaters as they scatter before a dummy predator (indicated by the arrow). From Treherne & Foster (1981).

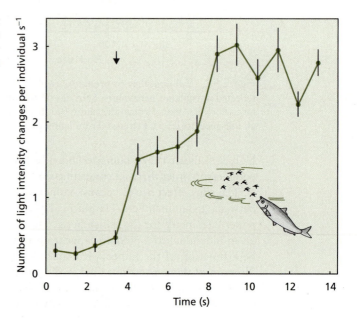

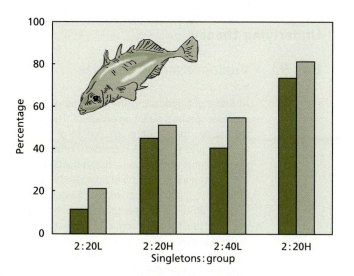

Figure 9.4 Three-spined sticklebacks showed a greater preference for attacking pairs of *Daphnia* ('singletons') the denser the swarm next to them. *Daphnia* were presented in 'singleton' : swarm ratios of 2 : 20 and 2 : 40, where swarms were of either high (H) or low (L) density. High densities were created by reducing the amount of water in the swarm's tube. Dark bars, approaches to the 'singleton' tube; pale bars, bites to the 'singleton' tube. After Milinski (1977).

While circumstantial evidence such as this is suggestive, Milinski (1977) tested the confusion effect more directly in a carefully designed experiment with three-spined sticklebacks. He provided fish with a choice between a 'swarm' of *Daphnia* in a glass tube and a 'singleton' (actually two individuals to encourage movement) in another tube, and measured their preference by counting the number of bites or visits to each tube. Milinski found that sticklebacks attacked 'singletons' more frequently the denser the swarm next to them (Fig. 9.4) (see also Jakobsen *et al.* 1994 for related field evidence in zooplankton). Interestingly, however, it was the swarms that the fish tended to approach first, presumably because they were easier to detect. But when they arrived at a swarm, the fish seemed unable to decide which individual to attack, and, after a few seconds, abandoned the tube and vigorously attacked the 'singleton'. Similar hesitant behaviour, coupled with tell-tale changes in head orientation, has been shown in geckos (*Eublepharis macularius*) confronted with groups of prey (Schradin 2000).

The problem of tracking a target, assumed to underlie the confusion effect, has been explored more formally using neural network models (Krakauer 1995; see Chapter 3), which show how a predator's ability to perceive the position of individual prey degrades as group size increases. When prey are relatively uniform in appearance, the confusion effect (reduction in success) describes a smooth decreasing and decelerating curve with increasing group size (Krakauer 1995). If there are 'odd' individuals in the group, however, the rate of decline is reduced and predators are able to target prey more easily. An obvious prediction, therefore, is that prey should prefer to associate with others of similar appearance, a prediction for which there is some supporting evidence with respect to size in fish (e.g. Peuhkuri *et al.* 1997). We have already discussed selection against odd individuals in the context of apostatic selection in Box 8.5.

Early detection

As we saw in Chapter 8, many predators depend on surprise for a successful attack. If it is spotted too early, a predator's chances of success are likely to drop sharply. This is exactly what can happen when prey form groups because, with more individuals scanning the environment, the chances that one of them will be looking in the right direction at the right time increase (see 8.3.5.4). The effect has been modelled formally by Pulliam

Underlying theory

Box 9.1   Vigilance and group size

At first sight, modelling the chances of an individual in a group spotting an approaching predator seems a daunting prospect. They are likely to depend on so many things: visibility in the habitat, the direction and nature of the predator's approach, the concealment afforded by cover and so on. In his classic model, Ron Pulliam (1973) boiled all this 'noise' down to a single, simple assumption – that detecting a predator depended on whether any individual in a group scanned during a critical period lasting T seconds, where T is the time taken for the predator to make its final uncovered dash at the group. Before the predator breaks cover, it is assumed to be undetectable; once it begins its approach it is always spotted provided a prey individual scans in time. The question can then become 'Given a scanning rate λ, what is the probability of an individual scanning during an interval lasting T s?' (Lendrem 1986). Assuming that the timing of both a predator's attack and decisions by prey individuals to scan are random, the answer to this for a single prey individual becomes:

$$P = 1 - e^{-\lambda T} \tag{9.1.1}$$

where P is the probability of detecting the predator. As group size increases, the number (N) of individuals scanning randomly and independently of each other also increases, so the probability of a group of size N detecting the predator becomes:

$$P = 1 - e^{-N\lambda T} \tag{9.1.2}$$

Thus, the more individuals in the group, the more likely it becomes that the group will spot a predator for any given approach time. The corollary of this is that as group size increases individuals can afford to reduce their scanning rates without also reducing the overall vigilance of the group (see Fig. 9.7). Rearranging equation 9.1.2 to calculate scanning rate by group size, gives:

$$\lambda = -\ln(1 - P)/TN \tag{9.1.3}$$

Based on discussion in Lendrem (1986), Pulliam (1973) and Elgar & Catterall (1981).

(1973; Box 9.1). While simple in its assumptions, Pulliam's model leads to two important predictions: first that the probability of spotting a predator will increase with group size, but, second, that individuals will be able to reduce their investment in vigilance without compromising the likelihood that the predator will be spotted. This is the so-called **double benefit of grouping**: by aggregating, individuals are both safer and have more time to do other things, such as find food (Roberts 1996).

Group size and early detection   Several studies have borne out the predictions of Pulliam's model, though, as we shall see, considerable care is needed in interpreting some of their results (see particularly Elgar 1989). That increasing group size can lead to earlier detection of an approaching predator was shown clearly by Kenward (1978), who flew a trained goshawk (*Accipiter gentilis*) at flocks of wood pigeons (*Columba palumbus*). As flocks became bigger, so did the distance at which the pigeons took flight from the approaching hawk (Fig. 9.5a), thus reducing its chances of success (Fig. 9.5b). One problem with this kind of analysis, however, is that it uses the prey's response to

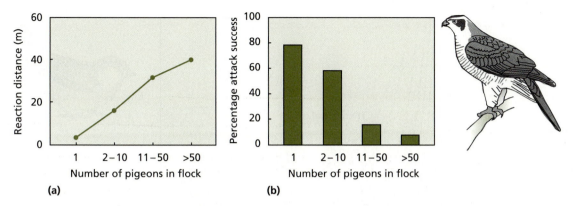

Figure 9.5 (a) Woodpigeons (*Columba palumbus*) in large flocks reacted sooner to an approaching trained goshawk (*Accipiter gentilis*) which was therefore less successful when attacking larger flocks (b). After Kenward (1978).

infer that the predator has been detected. But if prey feel safer in larger groups, they may delay their response to check whether the predator is really worth bothering about, thus confounding the expected relationship between group size and apparent detection distance. Lazarus (1979) provides the best experimental evidence for this.

Using captive weaver birds (*Quelea quelea*), Lazarus found that the number of birds responding to an alarm stimulus (a small light coming on in a cage) in flocks of different size did not differ from that expected if birds were detecting the light independently (as opposed to reacting to others detecting the light [see below]). However, analysis of the *type* of response shown by the birds revealed a pronounced effect of flock size. As flocks became bigger, birds were more likely simply to orientate towards the light rather than show flight intention movements or take off. Greig-Smith (1981) found a similar effect of flock size on the responses of barred ground doves (*Geopelia striata*) to a human observer.

While Lazarus's weaver birds appeared to respond to his light independently, the heightened responsiveness of large groups more generally often seems to be due to individuals reacting to one another rather than to the threatening stimulus itself. Responses to alarm signals (see 8.3.5.4) are one example of this, but the triggers are often much subtler. In the skater *H. robustus*, for example, the movement of individuals closest to a predator approaching on the surface of the water creates ripples that fan out and alert neighbours. Neighbours move in turn, sending out further ripples, so that, very quickly, the entire flotilla reacts to the danger and moves away. Treherne & Foster (1981) call this the *Trafalgar effect*, after the ship-to-ship signalling system that informed Nelson the French and Spanish fleet was leaving Cadiz before the famous battle. Video analysis of bird flocks taking off in alarm also shows that individuals cue in to the movements of their neighbours, and that increased interneighbour distances can delay the flight response (Hilton *et al.* 1999).

Group size and vigilance There is plenty of support for the second prediction of Pulliam's model, that individuals in groups can afford to reduce their commitment to vigilance. In the main, this has relied on relationships between group size and the frequency or duration of identifiable 'vigilance' behaviours, like the head-up scanning posture of birds. There is good evidence that these decline as groups get larger, and that individuals are responding to the number of others in the group and not some other factor correlating with group size (Lima 1995). However, while the amount of time each

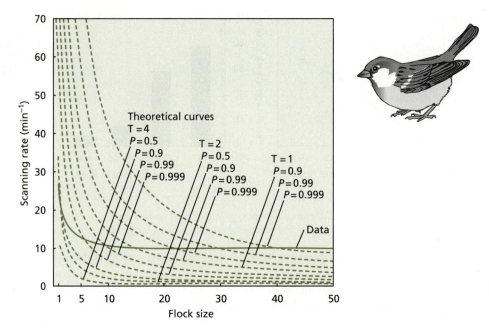

Figure 9.6 The observed relationship between vigilance (scanning rate) and flock size in house sparrows (bold curve), and relationships predicted for various parameter values by Pulliam's (1983) model (dashed curves; see Box 9.1, equation 9.1.3). See text. From Elgar & Catterall (1981).

individual spends vigilant often decreases with group size (Elgar 1989; Fig. 9.6; but see Catterall *et al.* 1992; Beauchamp 2001), this may be achieved without reducing the group's overall chance of spotting an approaching predator – its so-called **corporate vigilance** (Barnard 1978; Bertram 1980; see Fig. 9.7).

Trends in vigilance behaviours thus conform qualitatively to Pulliam's prediction, but attempts to test the model more quantitatively point to some problems. One of these has been highlighted by Elgar & Catterall (1981) in their study of house sparrows in Australia (Fig. 9.6). Elgar & Catterall found the expected negative relationship between scanning and flock size in their sparrows, but when they compared the curve with that predicted by Pulliam's model, the fit was not a good one, as Fig. 9.6 shows: birds in small flocks scanned less often, and those in large flocks more often, than they should over a wide range of parameter values in Pulliam's model. The reason for the discrepancy lay partly in a violation of one of Pulliam's basic assumptions, namely that scanning is an event of negligible duration. Raising the head and looking around obviously takes time (just over half a second in Elgar & Catterall's birds), which limits how often it can be done within any given period. Such a constraint is not catered for in the model.

Two other assumptions of the model are also undermined by the evidence. First, it is clear that, while the timing of scanning events is usually *irregular* (Scannell *et al.* 2001), it is often not *random* (e.g. Hart & Lendrem 1984, Sullivan 1985; cf. Box 9.1). Indeed, random scanning could be both wasteful (scans after very short intervals would be relatively uninformative) and dangerous (predators could exploit very long intervals). Only if predators monitor the variation of interscan intervals and attack accordingly, does something approaching a random scanning strategy appear to pay off (Scannell *et al.* 2001). Second, individuals in a group frequently do not scan independently; if one looks up, its neighbours follow suit to see what is going on (Lendrem 1986), an effect that tends to reduce overall levels of vigilance (Lazarus 1979).

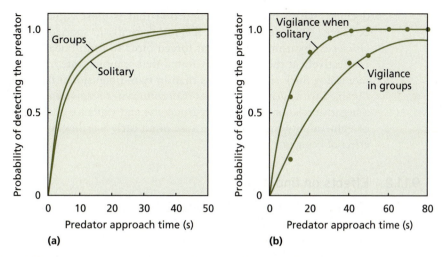

Figure 9.7 Relationships between the probability of ostriches detecting an approaching predator and the time taken by the predator to make its final uncovered dash (predator approach time), as predicted by later versions of Pulliam's model (see text and Box 9.1). When a predator attacks at random with respect to prey scanning behaviour, groups of three or four ostriches are more likely to spot a predator (i.e. have greater 'corporate vigilance') than solitary individuals (a). This is despite *individuals* in groups of three or four being less vigilant than solitary birds (b). From Lendrem (1986) after Hart & Lendrem (1984).

Later developments of Pulliam's model have tried to cater for non-instantaneous, non-random and non-independent scanning by focusing on relationships between inter-scan intervals (rather than scanning rate) and predator approach time, and regarding the group as the unit of vigilance rather than the individual (Hart & Lendrem 1984; Lendrem 1986). Using this approach with data from ostriches (*Struthio camelus*; Bertram 1980), and assuming that predators attacked groups at random with respect to prey scanning behaviour, Hart & Lendrem (1984) showed that, over a wide range of predator approach times, the probability of detection by groups (i.e. their corporate vigilance) was greater than that for solitary individuals (Fig. 9.7a), even though, *individually*, birds in groups were less vigilant than solitary birds (Fig. 9.7b).

Some shortcomings of simple vigilance models can thus be overcome by suitable modifications to the model. Others may be less tractable. One assumption underlying all the studies above, for example, is that prey are vigilant when they adopt a characteristic scanning posture but are unable to detect an approaching predator when feeding (or doing anything else) with their heads down. Recently, however, Lima & Bednekoff (1999) have shown that dark-eyed juncos (*Junco hyemalis*) are still able to detect an approaching model hawk when it could be seen only in the head-down feeding posture (head-up scans were obstructed). Indeed, under some conditions, detection distances were only marginally shorter than those of birds able to make unobstructed scans. As a result, Lima & Bednekoff suggest that, instead of occurring only when prey are actively scanning, vigilance in fact consists of bouts of low-quality detection (head-down) interspersed with bouts of high-quality detection (scanning). If so, this clearly complicates traditional interpretations of the relationship between group size and vigilance.

Finally, associations between group size and vigilance may arise for reasons other than those implicit in Pulliam's model. The edge effect (see above) is one example. We have seen that individuals at the periphery of a group are often more vigilant (Fig. 9.2). As groups become bigger, however, the proportion of individuals at the periphery declines, which means the proportion that is most vigilant also declines. Sadedin & Elgar (1998) demonstrated this rather neatly in spotted turtle doves (*Streptopelia chinensis*) by using feeders that forced birds to form linear flocks (where everybody is effectively at the edge). They found that scanning rate in such flocks declined with increasing flock size much less than in two-dimensional flocks. Similar results were obtained with evening grosbeaks (*Coccothraustes vespertinus*) by Bekoff (1995a). Thus a negative relationship between group size and vigilance could arise from the geometry of vulnerability rather than because of an early warning, or other safety in numbers, effect of being in a group.

9.1.1.2 Effects on finding food

Quite another problem with vigilance arises from confounding relationships between group size and feeding behaviour. While joining a group may help an animal avoid becoming food, it might also help it *find* food. But because anti-predator behaviour and foraging are likely to compete for the animal's time, the two benefits can be inter-dependent, making it difficult to decide cause and effect in their respective relationships with group size. Problems can arise in a number of ways.

Finding better feeding areas

A hungry animal may range over a number of potential feeding sites before deciding which to exploit. We saw in Chapter 8 that time spent exploring the environment may be an essential downpayment if the animal is to forage efficiently. Recognising good feeding sites from food availability alone can be difficult because many food sources are cryptic or concealed. However, the accumulation of other foragers at rich sources of food (the *aggregative response*; see 8.2.1.3) might provide a simple cue.

The use of aggregations in this way has been shown clearly by Krebs (1974) in a study of great blue herons (*Ardea herodias*) on tidal flats in Canada. Herons fly over large areas of mudflat searching for suitable tidal pools in which to feed. The birds typically feed in flocks and those in larger flocks experience a greater rate of food intake, so landing in pools where there are already lots of other birds should be a good way to rich pickings. To test this, Krebs put out life-sized models of herons in different 'flock' sizes to see whether birds flying over would prefer to land where 'flocks' were bigger. The results were striking. Not only was there a strong positive correlation between 'flock' size and the likelihood of passing birds dropping down to the pool but birds tended to land close to one of the models and immediately start foraging. Since there was no relationship between model 'flock' size and food availability, this implies that herons were working on the expectation of such an association.

If group size is to provide an accurate indication of food availability, however, it should track the changes in availability that are likely to result from depletion and prey mobility. Barnard (1980b) looked at this in house sparrows by baiting three hedgerows on a farm with different densities of millet seed. The number of birds foraging from each hedge were counted prior to baiting and then again after a high-, medium or low-density strip of seed had been laid along the hedge. The procedure was repeated at weekly intervals until all hedgerows had been baited with each seed density. As Fig. 9.8 shows, the change from pre- to post-baiting counts was dramatic, and post-baiting counts closely matched the relative density of seeds at the three hedgerows regardless of which density was at any particular hedge. The reason flock size tracked seed densities

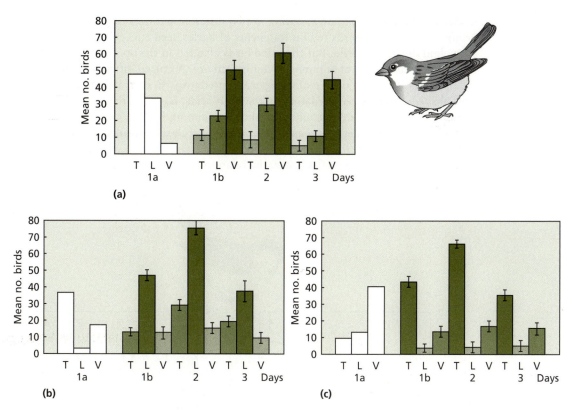

Figure 9.8 The number of house sparrows feeding from each of three hedgerows (T, L and V) increased with the density of millet seed laid along it. Day 1a, the number of birds foraging from each hedge just prior to baiting; Day 1b, the number of birds over the two hours following baiting; Days 2 and 3, the number over two hour samples on the subsequent two days. The densities were rotated round the hedges at weekly intervals (a–c). Light shading, low density (300 seeds m^{-2}) medium shading, medium density (600 seeds m^{-2}), heavy shading, high density (1200 seeds m^{-2}). From Barnard (1980b).

round the hedges in Barnard's study was because individual birds fed for longer as seed density increased, so providing more opportunity for flocks to build up.

The association between group size and food availability has obvious implications for how individuals within groups spend their time, since it is likely to affect, among other things, the opportunity (and thus incentive [see 4.1.1.2]) to feed and the level of competition experienced. This becomes important when we try to infer the relative importance of anti-predator and feeding benefits in driving the formation of groups, as we shall now see.

More time to feed

The cause and effect arrow in the 'double benefit' predicted by Pulliam's model of group size and vigilance implies that feeding benefits arise because individuals in larger groups can spend less time scanning. Various studies have shown that time spent feeding increases with group size as time spent in vigilance declines (e.g. Powell 1974; Barnard 1980a; Bertram 1980). As we have just seen, however, this relationship is likely to be confounded by correlations between group size and food availability. Do individuals in large groups feed more because they scan less, or scan less because they can feed more? Surprisingly few studies have addressed this problem. One that has is Barnard's (1980a) study of house sparrows.

Barnard watched sparrows feeding in two different habitats on a farm. In one, a group of cattlesheds, birds fed on fragments of barley seed concealed in the bedding straw laid down for cattle. But the same birds also fed in the open fields around the farm, taking the debris of autumn harvesting activity. A crucial difference between the two habitats lay in the degree of exposure to predators. In the cattlesheds, the birds were sheltered from aerial predators. In the open fields, however, they were completely exposed except when taking refuge in the surrounding hedges, and sparrowhawks (*Accipiter nisus*) and kestrels (*Falco tinnunculus*) regularly patrolled the fields where the birds fed. When the frequency with which birds scanned and pecked at seeds on the ground was plotted against flock size, the relationships predicted by Pulliam's model

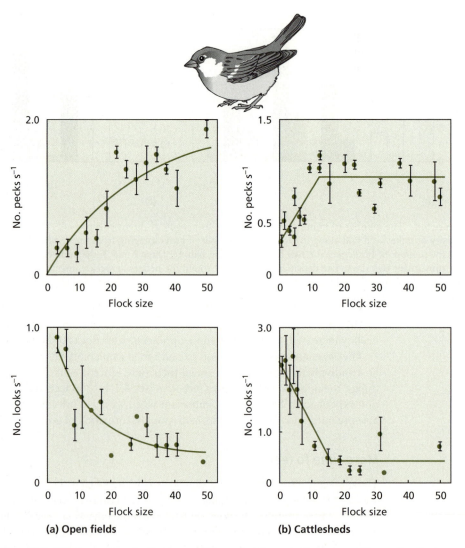

Figure 9.9 House sparrows feeding in two different habitats on a farm scanned less and pecked more in both as flock size increased. However, relationships described smooth curves in open fields (a), but levelled off more sharply in cattlesheds (b). This turned out to reflect differences in the relative importance of food density and flock size in determining behaviour. See text. From Barnard (1980a).

emerged in both habitats (Fig. 9.9a,b): birds scanned less and pecked more in larger flocks wherever they were. However, the shapes of the curves in the two habitats were suspiciously different, those in sheds being better approximated by two intersecting straight lines rather than a smooth curve. When the density of food on which observed birds had been feeding was taken into account statistically, the relationships with flock size vanished in sheds but remained in the open fields. Thus in the safe environment of the sheds, the Pulliam-like associations were an artefact of bigger flocks feeding on higher densities of seed, and birds therefore spending less time looking around. In the riskier environment of the fields, the relationships with flock size existed *independently* of food density. While Barnard controlled for food availability statistically, others have controlled for food availability experimentally by holding it constant while flock size varied (e.g. Studd *et al.* 1983; Elgar *et al.* 1984).

The different cause and effect relationships in Barnard's (1980a) study reflected adaptive shifts in the sparrows' time budgeting priorities between habitats. Birds could afford to let feeding constrain vigilance in the sheds because they were relatively safe from predators. However, feeding may constrain vigilance for entirely different reasons. As we have seen, one consequence of groups getting bigger is that competition is likely to increase. As a result, individuals may be forced to struggle harder to get their share of the food and devote more time to feeding (Clark & Mangel 1986). Clark & Mangel use the example of two people sharing the same milk-shake, where a 'tragedy of the commons' effect tempts each to suck on their straw more vigorously than if they both had their own milk-shake. In keeping with this, studies of various species have shown that rates of interaction increase with group size (e.g. Caraco 1979; Barnard 1980a; Monaghan & Metcalfe 1985; Watts 1985), and that competition can elevate rates of food intake (Barnard 1978; Grand & Dill 1999).

Local information about food

A key assumption behind relationships between group size and vigilance, is that vigilance reflects looking out for predators. But this may not be so. We have already seen that animals in groups may respond to their neighbours rather than to events outside the group, so vigilance may in fact be directed towards companions and have little or nothing to do with predators. Two obvious (and interrelated) reasons why it might be worth keeping an eye on companions are the risk of competition and locating resources.

When animals feed in sensory contact with one another, information about their foraging success is incidentally (and, in some cases, deliberately) transmitted between them. Foraging thus generates what Valone (1989) calls **public information** about the food supply. Depending on the type of food, individuals may be able to enhance their foraging efficiency by exploiting this information. Exactly how can vary. In some cases, individuals may copy the behaviour of a successful forager (social facilitation). In others they may direct their attention to the place where it was successful, or search in places that are similar (local enhancement; see Box 6.4). Once again, the best studies have been with flocks of birds.

In a laboratory study of captive great tits, Krebs and coworkers (1972) found that birds searching for patchily distributed (see 8.1.1) food hidden on artificial trees did better in flocks of four than in pairs or singly. The effect appeared to be due to local enhancement in that, immediately after a find, both the successful individual and other flock members searched on the perch where the food had been found. Birds in flocks were thus able to direct their attention to profitable feeding sites more quickly. Of course, this will only work if food is clumped. Little would be gained from copying companions on a dispersed food supply. When Krebs *et al.* tested their birds on dispersed food, they indeed found that copying was less likely to occur.

In mixed-species groups, where even closely related species can differ in some aspect of foraging behaviour or food requirement, such local information effects can lead to significant changes in the foraging strategy and diet ('**foraging niche**') of associating species (e.g. Morse 1970; Krebs 1973; Barnard & Thompson 1985; Hodge & Uetz 1996; Hino 1998).

Producers and scroungers   The rich opportunities for cashing in on public information can lead to some individuals specialising in exploiting others. Barnard & Sibly (1981) referred to these as **scroungers**, while their diligently foraging providers were termed **producers**. Producers and scroungers can be thought of as alternative strategies or tactics (see 2.4.4.2), and their maintenance in the population can be modelled in the same way as the *Digging* and *Entering* wasps in 2.4.4.2 (Box 9.2).

Several studies have tested the predictions of the model in Box 9.2. In captive flocks of house sparrows studied by Barnard & Sibly (1981), for instance, scrounger individuals (called *Copiers*) obtained most of their food by searching in the vicinity of successfully foraging companions (*Searchers*). *Searchers* and *Copiers* maintained their characteristic differences (active searching versus copying) regardless of the *Searcher* : *Copier* composition of the flock. By looking at how long different combinations of searchers and copiers persisted, Barnard & Sibly found that the most stable flocks were those with ratios of *Searchers* : *Copiers* in which the two types had roughly equal capture rates, suggesting that, like the behaviours of Brockmann *et al.*'s digger wasps, they could be maintained by frequency-dependent selection (but see Barnard & Sibly 1981 for some qualifications).

While the sparrows in Barnard & Sibly's study appeared to adopt fixed producer or scrounger roles, it is unlikely that these would be maintained over all foraging conditions. As we saw earlier in Krebs *et al.*'s great tits, copying pays only when food is clumped, so we should generally expect scrounging to be flexible and influenced by factors such as food distribution (e.g. Rohwer & Ewald 1981) and how much of a find is available only to producers (the so-called *finder's advantage*; e.g. Ranta *et al.* 1996). Several studies have now demonstrated such flexible scrounging in a number of bird species (Marchetti & Drent 2000; Rohwer & Ewald 1981; Giraldeau & Lefebvre 1986; Giraldeau *et al.* 1994) and some mammals (Held *et al.* 2000). It is also clear that producer–scrounger roles can be adopted on the basis of individual foraging experience (Flynn & Giraldeau 2001).

Scrounging and spacing within groups: Apart from its effects on individual food intake, scrounging is likely to have an impact on several other aspects of group foraging. Spacing within the group is one example. To maximise returns on their efforts, producers might be expected to distance themselves from potential competitors, while scroungers should actively seek the vicinity of other foragers. A good example comes from a mixed-species producer–scrounger relationships in flocks of lapwings (*Vanellus vanellus*) and golden plovers (*Pluvialis apricaria*) (Barnard & Thompson 1985). Such flocks are a common sight on agricultural pasture in northern Europe and are frequently exploited by klepto-parasitic (food-stealing) black-headed gulls (*Larus ridibundus*). The gulls pick up on the characteristic crouching posture adopted by plovers just before they pull an earthworm out of the turf, and time their attacks to reach the bird and steal its prey before the latter can be ingested. These attacks can be injurious, or, if the plover tries to escape with the worm, involve costly aerial chases. Not surprisingly, therefore, plovers take a number

Underlying theory

Box 9.2   Producers and scroungers

Many interactions between organisms can be viewed as **producer–scrounger** relationships. One species or individual (scrounger) in some way exploits the investment of another (producer) to obtain a limited resource. Examples are legion: kleptoparasitic skuas (*Stercorarius skua*) steal hard-won fish from puffins (*Fratercula arctica*) as they return to their burrows, brood-parasitic cuckoos (*Cuculus canorus*) exploit the parental investment of their reed warbler (*Acrocephalus scirpaceus*) hosts, young red deer (*Cervus elephus*) stags sneak matings from hinds while their harem owner is busy defending them from would-be usurpers and so on (see Barnard 1984a for many more examples). Scroungers appear to dodge some of the costs of exploiting a resource by allowing producers to put in all the effort then usurping the fruits of their labours. However, the success of scroungers within any group or population is likely to depend on their frequency relative to producers, since, if there are too many, the rate at which producers make resources available will be insufficient to support them. The frequency-dependent payoffs to producers and scroungers can be illustrated in a simple model (Fig. (i)).

Figure (i)   Payoff to individual producers and scroungers as a function of the *ratio* of the two in a group (here of six). The intersection of the two curves is the point of equal payoff and thus indicates the evolutionarily stable ratio of producers to scroungers. From Barnard & Sibly (1981).

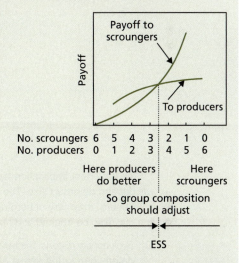

The model represents reproductive payoff in terms of the amount of resource (food, nests, mates, etc.) acquired by individual producers or scroungers as a function of the number of each in the group. The payoff to scroungers increases with the number of producers and thus the opportunity to usurp resource, while the payoff to producers declines with increasing numbers of scroungers as more of their resource is lost. One difference between the curves is that the payoff to scroungers starts at zero, because if there are no producers, scroungers get nothing, but the payoff for producers may reach zero only if there are very large numbers of scroungers. Where the curves intersect, the two approaches do equally well: to one side producers do better, to the other, scroungers do better. The intersection thus represents the evolutionarily stable combination to which groups will converge.

However, the situation is unlikely to be as simple as this because *density*-dependent (group size) effects are likely to be at work as well. Payoffs to producers and scroungers are thus represented more accurately as *surfaces* instead of curves (Fig. (ii)). The same principles as in Fig. (i) apply, except that the intersection between surfaces produces a *line* of equilibrium rather than a point.

Box 9.2 *continued*

Figure (ii) Payoffs as a function of the *number* of producers and scroungers in a group yields two surfaces. The intersection of the surfaces is a line indicating the evolutionarily stable combination at each group size. From Barnard & Sibly (1981).

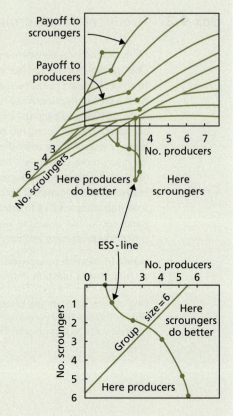

The important implication here is that the ratio of producers to scroungers at equilibrium will depend on group size. The exact shape of the payoff surfaces, and therefore the line of equilibrium, may vary according to the nature of the producer–scrounger relationship. The line may also be tracked at a number of different levels:

☐ *Evolutionary history.* The tendency to be either a producer or a scrounger may be genetically determined in the sense of being a species-typical strategy in mixed-species associations, or an individual strategy (pure or mixed [see 2.4.4.2]) within single-species groups.

☐ *Developmental history.* Individuals may be producers or scroungers for a limited part of their life cycle. Male red deer, for example, are usually scroungers on harem-guarding stags only when they are young bachelors, and switch to being harem-guarding producers later on (Clutton-Brock *et al.* 1979).

☐ *Individual capability.* Individuals may choose to scrounge on the basis of relative competitive ability, and thus the risk of challenge. In flocks of Harris sparrows (*Zonotrichia querula*), for instance, it is mainly dominant birds that become scroungers (Rohwer & Ewald 1981).

☐ *Opportunity.* The decision to be a producer or a scrounger may depend on the opportunity for scrounging within the group.

☐ *Internal state.* Individuals that are, for example, hungrier may be prepared to risk aggression and injury to usurp food rather than use more energy searching for it themselves.

Based on discussion in Barnard & Sibly (1981).

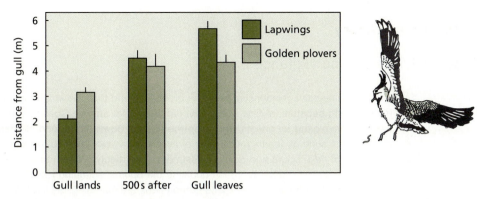

Figure 9.10 When a kleptoparasitic gull lands in a flock of lapwings and golden plovers, birds gradually move away from it to reduce the risk of prey being stolen. From Barnard & Thompson (1985).

of precautions when gulls arrive in the flock. One of these is to move away from a gull when it lands nearby. As Fig. 9.10 shows, the average distance between plovers of either species and a newly arriving gull gradually increases until the gull takes off and moves to a new vantage point in the flock where the process begins all over again. Flynn & Giraldeau (2001) report similar avoidance of scroungers in flocks of spice finches (*Lonchura punctulata*). This suggests that foraging strategies/tactics have an important effect on spatial decisions within groups.

Scrounging and vigilance: The spacing effects driven by scroungers may influence predation risk by increasing the domain of danger of pro-ducers (Barta *et al.* 1997), and perhaps even contribute to the edge effect discussed earlier (Flynn & Giraldeau 2001). However, the investment by scroungers in scanning for feeding opportunities may offset these negative effects by increasing the overall vigilance of the group (Ranta *et al.* 1998; Beauchamp 2001). Just such an effect has been reported by Thompson & Barnard (1983) in their flocks of gulls and plovers. Gulls may be a thieving nuisance from the plovers' viewpoint, but they are also highly vigilant and conspicuous, two characteristics that make them an effective early warning system. By walking slowly towards feeding flocks, Thompson & Barnard showed that the distance from the observer at which birds took off in alarm was much greater if one or more gulls was present. Tellingly, gulls tended to take off first, implying they had spotted the observer. Several different cost–benefit currencies may thus need to be considered in predicting the stability of producer–scrounger relationships (Ranta *et al.* 1998).

The heightened vigilance of scroungers is part and parcel of their scrounging lifestyle. However, vigilance by others may itself become an exploitable commodity, since there would seem to be an obvious temptation for individuals to let companions scan while they themselves feed, or do whatever else they need. Box 9.3 looks at the potential for parasitising investment in vigilance.

Scrounging and social learning: The interference effect of scrounging may also have some more subtle social consequences. As we saw in Chapter 6, being in a group provides an opportunity to learn from individuals, and there is plenty of evidence that skills are acquired via social learning. But if social information leads to scrounging, the spread of learned traits through the group may be curtailed because scroungers acquire a pro-ducer's resource by usurping it rather than learning how the producer acquired it in

Underlying theory

Box 9.3 Exploiting vigilance

The classic negative relationship between group size and the time individuals spend vigilant is interesting because, at first sight, it would seem that animals scanning in large groups risk being exploited by cheats who keep their heads down and spend all their time doing other things, such as feeding. One way of avoiding this would be to monitor the scanning behaviour of neighbours and adjust investment accordingly. Pulliam *et al.* (1982) modelled this using a games theory approach because the optimal scanning rate for any given individual depends on how frequently its companions are scanning. Several different optimal strategies are possible. One solution might be for individuals in the group to cooperate and scan at the rate that maximises the survival of all cooperators. However, this is vulnerable to cheats who scan at less than the cooperative rate. An alternative possibility might therefore be to scan at a selfish rate which, if adopted by all group members, would result in any individual deviating from it having a lower probability of survival. Pulliam *et al.* calculated the expected payoffs (probability of surviving the day) accruing to *Cooperator* and *Selfish* scanners and compared the outcomes with scanning rates in flocks of yellow-eyed juncos (*Iunco phaeonotus*). To their surprise, the relationship between scanning rate and flock size in juncos was close to the *Cooperator* optimum.

One explanation for this unexpected finding is that juncos were in fact adjusting their scanning to that of their neighbours by playing a conditional strategy (see 2.4.4.2) along the lines of 'cooperate along as others do, but defect if they don't', dubbed the *Judge* strategy by Pulliam *et al.* A *Judge* thus scans at the cooperative rate if it is associating with *Cooperator*s or other *Judge*s, but at the selfish rate if it detects a cheat. In simulations, *Judge* did as well as *Cooperator* in flocks of *Cooperator*s and/or *Judge*s, and better than *Cooperator* in flocks containing *Selfish* strategists. *Judge* also did as well as *Selfish* scanners when in company with them, and better against another *Judge* than *Selfish* did against itself. Of the three strategies, therefore, only *Judge* was evolutionarily stable. It may be, therefore, that Pulliam *et al.*'s juncos scanned at the cooperative rate because *Judge* had gone to fixation in the population. But there is another possibility. It could simply be that there is an advantage to being the first bird to spot a predator (Charnov & Krebs 1975; see 8.3.5.4), in which case birds would benefit by scanning at a rate higher than the selfish equilibrium (Pulliam *et al.* 1982).

Based on discussion in Pulliam *et al.* (1982).

the first place. Giraldeau & Lefebvre (1987) provided a clear demonstration of this in captive pigeons, where scrounging was simulated by allowing observer birds to receive food passively from demonstrator individuals. Normally social learning is a potent force in skill acquisition among pigeons in observer–demonstrator tests, but when scrounging was allowed, demonstrators failed to enhance learning by observers (Giraldeau & Lefebvre 1987). Comparable negative effects of scrounging have been found in capuchin monkeys (*Cebus apella*; Fragaszy & Visalberghi 1989); jackdaws (*Corvus monedula*; Partridge & Green 1987) and zebra finches (Beauchamp & Kacelnik 1991). However, they were not evident in similar experiments with chickens (Nicol & Pope 1994).

Groups as local information centres   Individuals may be able to capitalise on information about resources while in a foraging group, but useful information may also be available when groups are not actively foraging. For species relying on patchy resources that vary in their spatial and temporal availability, day-to-day foraging may be a chancy business: one day a bonanza, the next, nothing at all. If there was some way of finding out how well other individuals had done, however, it might be possible to exploit the information and improve success. Ward & Zahavi (1973) have proposed just such a function for roosting aggregations in birds.

The nub of Ward & Zahavi's idea is that, by returning to a communal roost at night, unsuccessful birds get the opportunity to follow more successful companions the next morning, perhaps by responding to the decisiveness with which different individuals appear to set out. While successful individuals are essentially parasitised in this scenario, they may benefit from being exploited if, for example, their risk of predation at the feeding site is diluted as a result. Although not in the context of information centres, there is good evidence that individuals may actively recruit others to a feeding site to offset the risk of predation and gain time to feed, though the tendency depends on the divisibility of the food supply and the likelihood of competition from recruits (Elgar 1986). Is there any evidence for Ward & Zahavi's information centre idea?

The best experimental evidence comes from a study of captive weaver birds (*Quelea quelea*) by Peter de Groot (1980). De Groot allowed two groups of weaver birds to roost together in one compartment of an aviary, but trained each group to find either food or water in one of four other compartments. When the two groups were deprived of water, thirsty birds from the food-trained group tended to follow those from the water-trained group to the water compartment. When they were deprived of food, however, the water-trained birds followed those trained to food. In a second experiment, two groups were trained to find food supplies of different quality. After a period of food deprivation, the birds trained to the poorer supply followed those trained to the better one. The conclusion from these experiments is that, when the need arises, naïve birds can somehow recognise 'knowledgeable' birds and follow them to a resource they themselves have never visited.

De Groot's results thus strongly suggest that birds in communal roosts can glean information about foraging success from companions, though how they did it in de Groot's case remains unclear. The mechanisms may not be so elusive in other species, however. As we saw in Galef & Wigmore's (1983) experiment with rats in Chapter 6 (Fig. 6.28), food odours can provide the necessary cues in some mammals. A number of field studies also support the information centre idea (e.g. Krebs 1974, Sigg & Stolba 1981; Ydenberg *et al.* 1983). Using radio-tagged individuals, Sonerud *et al.* (2001) showed that free-ranging hooded crows (*Corvus corone cornix*) were likely to return on successive days to a superabundant food patch set out by the experimenters. Furthermore, crows that had not visited the patch on one day visited it the next if they had roosted overnight with a 'knowledgeable' crow that had been there the previous day. Evans (1982) considers some other mutual benefits that might arise from information transfer and exploitation in communal gatherings.

Other consequences of grouping for finding food

Associating in groups can confer a number of other feeding advantages. Larger groups can sometimes tackle prey that are beyond the capability of a single individual, as in various carnivores (e.g. Estes & Goddard 1967; Mech 1970), birds of prey (e.g. Bednarz

1988) and insects (e.g. Traniello & Beshers 1991). Like mixed-species associations, therefore, single-species grouping can extend a predator's foraging niche. Groups may be able to make multiple kills, or kill more quickly, so increasing overall foraging efficiency (Pulliam & Caraco 1984). However, foraging efficiency may be improved in more subtle ways. Where depleted food patches replenish over time, it is important that a predator does not return to a patch too soon, or leave it too late, since either will reduce its expected rate of intake. Both exclusive ownership of resources (territoriality; see below) and foraging in a cohesive group might enable return times to be optimised by minimising competitive interference between independently foraging individuals (Charnov 1976a; Cody 1971). Grouping may also reduce variance in intake rate, for instance by reducing the risk of outright failure (e.g. Wyman 1967) or the variance in searching time between, and capture rate within, food patches (Pulliam & Millikan 1982). If reduced variation in foraging success reduces the predator's risk of starving, grouping may reflect a risk-sensitive foraging strategy (see 8.1.1.3).

9.1.1.3   How big should a group be?

While there are many ways in which grouping can confer an advantage on associating individuals, it is important to bear in mind that advantages are unlikely to increase with group size indefinitely. As is clear in Fig. 9.9, benefit curves usually start to decelerate beyond a certain number of individuals, local limitations in resource availability and increasing competition being two likely causes. In some cases, the advantage to be gained from joining a group disappears at extremely small group sizes. Major (1978), for instance, found that the average food intake of predatory fish did not increase beyond a group size of three (see Caraco & Wolf 1975 and Nudds 1978 for similar limits in carnivores). However, the size of group that maximises individual intake may vary with circumstances. Hunting group size in spotted hyaenas (*Crocuta crocuta*) varies with the kind of prey being stalked, something to which the prey themselves are sensitive; zebra (*Equus burchelli*) generally show alarm responses only to large packs, and rarely to pairs or singletons, which are more likely to be hunting gazelle (Kruuk 1972). In lions, groups may be bigger than expected from hunting success because larger groups are better able to stop hyaenas usurping the kill (Packer *et al.* 1990). Sometimes group size reflects the need for an important division of labour. Packs of African hunting dogs (*Lycaon pictus*), for example, leave an adult behind to guard pups when they go off to hunt. This reduces the hunting efficiency of the pack because it has fewer dogs to tackle prey and carry food back to the den. If pack size drops too low, it may mean that the dogs have to make many more hunting trips to bring back enough food. Or, worse, pups may have to be left unguarded. Recent work in Zimbabwe by Courchamp *et al.* (2002) suggests that five adults is the minimum pack size for sustaining the division of labour between hunting and babysitting. Below this, reduced hunting efficiency and increased pup mortality is likely to drive the pack to extinction. Is it possible, then, to make general predictions about how large groups should be?

Optimal group sizes

The kinds of relationship between group size and behaviour highlighted above suggest there may a group size that maximises the reproductive payoff to individuals. One problem with testing this, however, is that the many different factors correlating with group size can pull the optimum in different directions, leading to a lack of overall relationship between size and reproductive benefit. Colonies of cliff swallows (*Hirundo pyrrhonota*) are a good example. Swallows become infested with blood-sucking swallow bugs (*Oeciacus vicarius*) that live in the birds' nests. Larger colonies have more bugs per nest, so increased group size carries a cost that is reflected in reduced growth by nestlings. Brown & Brown (1986) were able to demonstrate the cost to growth by fumigating nests and killing the parasites. Chicks in cleansed nests grew faster and ended up bigger than those in infested nests. Moreover, clean birds in large colonies grew faster than their counterparts in smaller ones, giving rise to a positive correlation between group size and growth. By manipulating colony size experimentally, Brown (1988) was able to control for potential confounding effects, such as larger colonies being near richer food supplies, and show that the effect was due to group size *per se*, probably because bigger colonies brought in food at a greater rate. When he looked at unfumigated colonies, however, no such correlation emerged. The susceptibility of large colonies to high parasite burdens thus appeared to offset their growth rate advantage.

As well as showing that conflicting selection pressures might operate on group size, Brown's study highlights the very different currencies (see Box 2.5) in which the costs and benefits are measured. Integrating these into a single predictive model can be difficult. One way round the problem is to cast them all in terms of a single currency and ask how the organism should allocate this across its various activities. Caraco *et al.* (1980b) used **time budgeting** (the allocation of time to different activities) as a means of doing this and explored the effects of group size using juncos as a model system (Box 9.4).

While flock size in juncos accords qualitatively with the predictions of Caraco *et al.*'s model (Box 9.4), there are good reasons why optimal group sizes might not prevail as a general rule. In a simple model, Sibly (1983) argued that, while a certain group size might be optimal in terms of expected individual reproductive success, individuals might still do better by joining the group than remaining alone, thus causing the group to grow beyond the optimal size. To use Sibly's example, suppose a number of birds leave their overnight roost alone to feed, and, seeing all the flocks that have formed up to that time, join the one that maximises their rate of food intake. The first 20 birds to leave will form an optimal flock of 20 (say), but the 21st bird will still do better by joining the flock than by feeding alone, so flock size moves away from the optimum to 21. This process continues until the flock reaches, say, 55 birds, when the benefit of joining becomes less than that of feeding alone. The 56th bird therefore feeds alone, the 57th joins it, and so on. A similar argument applies if groups split and reform as smaller or larger units (Kramer 1985; see also Ranta 1993): optimal group sizes are unstable and groups will tend to increase beyond the optimal size. However, it *is* possible to predict stable optimal group sizes. Sibly's model is an example of a simple ideal free distribution (see Box 7.1), and, as we saw in Chapter 7, sophistications of the basic ideal free model lead to more complex outcomes. Giraldeau & Gillis (1985), Kirkwood (in Thompson & Barnard 1985) and others have varied the shape of group size-related cost–benefit curves and other factors affecting the incentive to join or remain in a group, and generated a number of relationships between optimal and stable group sizes, including cases where the two coincide. An important influence here is the distribution of resources, with optimal/stable group sizes being larger or smaller according to the level of clumping and fragmentation in the environment (Fig. 9.11).

Underlying theory

Box 9.4 A time budgeting approach to optimal group size

Looking at how individuals allocate time to different activities can give some insight into how they deal with the various selection pressures acting on them. Tom Caraco and coworkers have used the approach to model optimal group sizes.

Basing their model on small overwintering birds, Caraco *et al.* assumed that individuals faced two main risks: starvation and predation, and therefore divided their time between behaviours that reduce these risks, namely scanning (to avoid predators) and feeding and fighting (to gain food). From their previous studies of small buntings called juncos, they assumed that time spent scanning would decline with increasing flock size and that spent fighting would increase (because of increased competition). As a result, the time available for feeding would peak at some intermediate flock size (Fig. (i)a).

If the risks change, however, the allocation of time to the different behaviours should change, and with it the flock size at which time spent feeding is maximised. Thus, if a

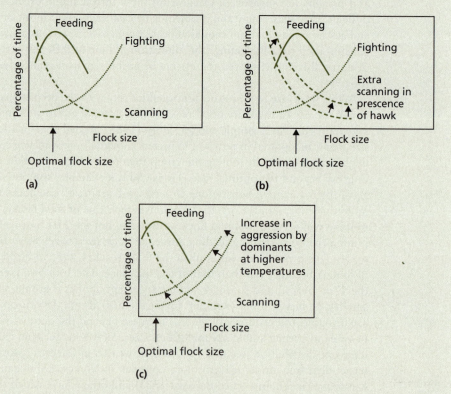

Figure (i)   A model of optimal flock size in yellow-eyed juncos based on time budgeting. (a) As flock size increases birds spend less time scanning for predators but more time fighting over food. The amount of time for feeding is thus greatest in medium-sized flocks. (b) If predation risk increases (a hawk is present), birds scan more, so shifting the optimal feeding flocks size upward. (c) Conversely, if food demand is reduced by higher temperatures, birds can afford to compete more so feeding flock size goes down. From Caraco *et al.* (1980b) and Krebs & Davies (1993).

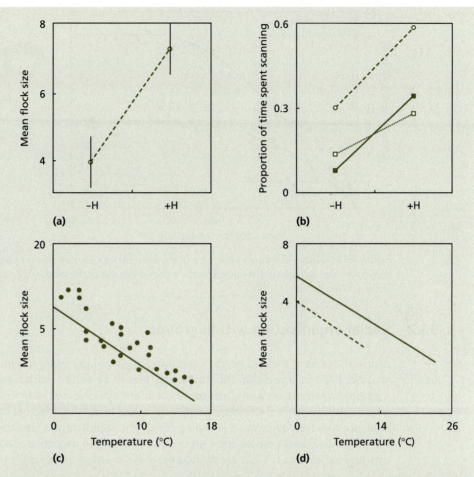

Figure (ii) Both flock size (a) and scanning rate (b) in yellow-eyed juncos increase after a hawk has flown over (+ H). ○ sollitary birds, □ flocks of 3–4 birds, ■ flocks of 6–7 birds. Flock size decreases with increasing temperature (c) and when supplementary food is made available (dashed line in (d)). See text. After Caraco (1979) and Caraco *et al*. (1980b).

hawk suddenly appears in the vicinity, birds should increase their commitment to scanning. The scanning curve thus shifts upward (Fig. (i)b) and the optimal flock size for feeding increases. Conversely, if the temperature increases and food demand becomes less urgent, birds can afford to compete more aggressively (Fig. (i)c), so the competition curve is raised and time spent feeding is greatest in smaller flocks. Does flock size respond as predicted by the model? Results from field experiments with yellow-eyed juncos suggest it does (Fig. (ii)a–d). When apparent predation risk was increased, by flying a hawk over flocks, scanning rate and flock size both increased as predicted (Fig. (ii)a,b). Flock size also decreased with increasing temperature (Fig. (ii)c) and was lower at any given temperature when supplementary food was put out (Fig. (ii)d), again as predicted under conditions of relaxed food requirement.

Based on discussion in Krebs & Davies (1993), Caraco (1979) and Caraco *et al.* (1980b).

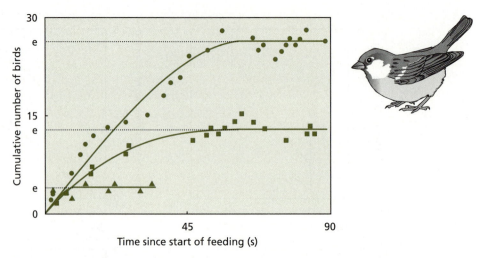

Figure 9.11 Flocks of house sparrows reach a different equilibrium size (e – where arrival and departure rates are equal) determined by the size and density of the food patch on which they are feeding. From Barnard (1980c).

9.1.2 Social organisation within groups

Although we have focused on the costs and benefits of grouping to individuals, most analyses have assumed that all individuals benefit or suffer similarly from being in a group, something we know already is often not the case. Sex, age, spatial position, social status, producer–scrounger relationships and many other factors can profoundly influence the consequences of grouping for different individuals (see also 7.2.1.1 and Fig. 7.9). The reasons for joining a group are thus likely to depend as much on its social structure as on its size. What are the important factors likely to be here?

9.1.2.1 Social dominance

Many social groups are characterised by sustained aggressive/submissive competitive relationships between individuals. Where encounters are frequent enough, individuals can often be ranked according to their tendency to win or lose against each of the others in the group. Such **rank orders** are referred to as **social** or **dominance hierarchies** (Schjelderup-Ebbe 1935). Within a hierarchy, an 'alpha' individual is dominant to all the others in the group, a 'beta' individual is dominant to all but the 'alpha', and so on down to the lowest ranker who is dominant to no one, though there are often triangular relationships (e.g. A beats B, B beats C, but C beats A) that disrupt the linearity (**transitivity**) of the hierarchy. Sometimes, dominance relationships are **despotic**, with a single individual, or an **alliance** between two or more individuals (e.g. de Waal 1996; see below), being dominant to all the others, but no rank order among the rest.

It is important to stress that dominance hierarchies are constructs of the observer, sometimes under conditions far removed from the animals' normal social context (e.g. Goodall 1968), and may have no functional significance in themselves. Indeed, *post hoc* classification of individuals into discrete rank categories may disguise underlying continuity in behavioural and physiological characteristics (Creel 2001). Furthermore, hierarchies may not necessarily reflect adaptive decision-making by individuals. They can be generated simply because individuals differ in size and bigger individuals beat smaller ones in random encounters (Barnard & Burk 1979). In many cases, however, hierarchies *do* reflect assessment of the chances of winning or losing in an encounter, and decisions about whether to persist or withdraw accordingly. Assessment may be based on cues relating to competitive ability, such as size or weaponry (see 11.1.1.4), or simply on accumulated 'confidence' from past wins and losses (Barnard & Burk 1979; Simmons 1986; Bekoff & Dugatkin 2000; Dugatkin 2002).

Dominance is not equivalent to aggressiveness. Dominants may not be the most aggressive individuals; indeed, aggression may be more characteristic of middle-ranking individuals where relative competitive abilities are often less clear-cut (e.g. Enquist & Leimar 1990; Hurst *et al.* 1996; 11.1.1.4). What matters is that dominants tend to 'win' encounters, even though this may be due to passive, sometimes barely discernible, deference by lower rankers. Furthermore, dominance status is not just a function of behaviour, but is often associated with differences in physiology, particularly levels of steroid hormones such as androgens and glucocorticoids (Christian & Davis 1964; Sapolsky 1992). Early studies found higher levels of glucocorticoid (so-called stress) hormones in low-ranking individuals, giving rise to a general assumption that being of low social rank was stressful (Creel 2001). However, many studies reporting associations between social rank and glucocorticoids have been conducted on confined, captive animals where levels of aggression can be abnormally high. A widespread view is that these relationships are an artefact of low rankers not being able to escape, and that social dominance in fact evolved as a means of reducing the cost of encounters when the outcome of aggression is predictable. In keeping with this, levels of aggression in the field, or in more spacious captive conditions, are generally much lower (e.g. Creel 2001). A further complication is that glucocorticoid levels are sometimes higher among dominants, not subordinates, suggesting dominants may tradeoff high costs against high reproductive benefits (Frank *et al.* 1995; Barnard & Behnke 2001; see also 4.3.2). Rather than reflecting a dichotomy in stress between high and low social ranks, glucocorticoid levels may in fact be more indicative of turbulence among middle rankers. The stability of dominance relationships varies enormously within and between species, but all are subject to some degree of disturbance as individuals die, leave, grow bigger, become sick, or change in competitiveness for other reasons. Interestingly, in free-living baboons (*Papio anubis*), glucocorticoid levels tend to rise when middle-ranking individuals start losing to those below them in the hierarchy, but show little change when challenges are directed to individuals above (Sapolsky 1992). Thus, battling your way up in the world may be less stressful than being forced down.

Not surprisingly, dominance status can have an important effect on access to resources. Dominants often obtain more food or mates, or have access to resources at crucial times, such as the period of ovulation in females (Le Boeuf 1974; Hausfater 1975; Dittus 1977). Dominant females may mature earlier, produce more offspring or nurture young more effectively (Sade *et al.* 1977; Wilson *et al.* 1978; McCann 1982). The relationship between dominance and resource acquisition is not always clear-cut, however, and rank order in relation to one resource may not be the same as that in relation to another (e.g. Rowell 1974). Moreover ownership of resources can itself *confer* dominance (see 11.1.1.4), so that, like the *Bourgeois* strategy in 2.4.4.2, dominance can be conditional on ownership and change as resources change hands (see Packer & Pusey 1985 for a nice example in lions). Whatever the cause-and-effect relationship between resource ownership and dominance, commanding resources can have a pronounced impact on the reproductive success of dominant individuals, biasing reproduction in their favour

Underlying theory

Box 9.5   Dominance and reproductive skew

Reproductive skew (unevenness in the distribution of reproduction among breeding individuals) in social species is often a consequence of dominance relationships within groups. Competitive social relationships and socially mediated physiological (e.g. hormonal) changes can suppress reproduction in subordinate individuals in a number of ways, for instance by reducing the viability of sperm or preventing ovulation or implantation of the embryo. Suppression via external social influences is often referred to as the **dominant control model** of reproductive skew (Clarke *et al.* 2001). In contrast, subordinates of species such as Damaraland mole rats (*Cryptomys damarensis*) appear to exercise 'voluntary' reproductive suppression – the so-called **self-restraint model** – possibly as a means of avoiding inbreeding (Clarke *et al.* 2001). Interestingly, the physiological mechanisms underlying dominant control and self-restraint suppression seem to be the same, though exactly how they are triggered under difference circumstances remains to be clarified (Faulkes & Bennett 2001).

Reproductive suppression is by no means a universal feature of dominance relationships. Some social species, such as the banded mongoose (*Mungos mungo*), have a relatively even spread of reproductive opportunity across individuals (de Luca & Ginsberg 2001), and thus tend towards an **egalitarian** society. Ecological (mainly breeding resource) constraints are assumed to drive the different degrees of reproductive skew seen across species, with optimisation models considering a continuum of social systems (the *eusociality continuum* [Sherman *et al.* 1995]) in which the distribution of reproductive opportunities is skewed to a greater or lesser extent according to circumstance (Faulkes & Bennett 2001). The question then becomes one of how the appropriate distribution is established. **Concession models** argue that it sometimes pays dominants to allow some subordinates to breed as an inducement to stay in the group and help defend its resources (see 9.2.2.1), or to prevent them fighting for reproductive control (Sherman *et al.* 1995; Reeve *et al.* 1998). **Incomplete control models** suggest that subordinates simply get away with it, and breed because dominants do not exert effective control (Clutton-Brock 1998). A third category, **threat of eviction models**, reflects the self-restraint referred to earlier, and argues that subordinates regulate their reproductive activity to avoid expulsion from the group. Clutton-Brock (1998) and Faulkes & Bennett (2001) provide incisive reviews of these and other issues relating to reproductive skew.

Based on discussion in Clutton-Brock (1998) and Faulkes & Bennett (2001).

and leading to **reproductive skew** within the group (Vehrencamp 1983; Emlen 1997a; Clutton-Brock 1998; Cant & Johnstone 2000; Box 9.5).

If so many advantages flow from being dominant, why do some individuals put up with being subordinate? One view, of course, is that subordinates simply have to put up with their lot, and make the best of the bad job of being poor competitors (see 2.4.5.1). Along with this may come a number of life history adjustments, such as the tendency to conserve immune responsiveness (see Fig. 3.37) or adopt so-called passive coping and 'sneaky' breeding strategies (Clutton-Brock *et al.* 1979; Koolhaas *et al.* 1999). However, if low status is a function of age or size, subordinates may have to sit it out only until they are old or large enough to compete successfully. A completely different view is that dominants and subordinates are alternative strategies or tactics (2.4.4.2) that reap the same net benefit. Instead of being put upon weaklings, subordinates in fact do just as well as dominants but in a different way (e.g. Rohwer & Ewald 1981; Mendl & Deag 1995). One problem with testing the adaptive alternatives hypothesis, however, is that dominants and subordinates may accrue benefits in different currencies, making equivalences hard to judge. Despite this, the idea has important implications for the evolution of social relationships, and also, arguably, for coping and welfare in captive groups (Mendl & Deag 1995; Koolhaas *et al.* 1999; see 4.3).

9.1.2.2 Mating systems and family groups

Social dominance may translate into reproductive benefits, but reproductive relation-ships can impose structure on groups in their own right in the form of **mating systems**.

'Mating system' generally refers to the monopolisation of mates, i.e. whether one male mates with several females, one female with several males, the sexes form stable pair bonds and so on, though it sometimes encompasses a wider spectrum of sexual and parental behaviour (e.g. Vehrencamp & Bradbury 1984; see Chapter 10). Sometimes, the distribution of mates means that several can be defended against rivals, as in the harem system of many ungulates (Jarman 1974; Emlen & Oring 1977; 10.1.2.2). In some other cases, the demands of parental care or the scarcity of breeding sites results in non-breeding individuals delaying their own reproduction to act as helpers in their natal group (e.g. Emlen 1984, 1997a; see 9.3). These different outcomes all have con-sequences for social structure within populations, but delayed reproduction has special significance in that it can lead to the formation of **family groups**.

Family groups

As Emlen (1997a) points out, many species, indeed a majority among birds and mammals, live in multigenerational family groups, that is groups where offspring continue to interact into adulthood with their parents. Family groups arise because young do not disperse from their natal area, often because of a shortage of breeding sites or a high-quality natal area (Fig. 9.12). The formation of family groups can have several consequences for social evolution. One is that the degree of relatedness among group members increases, leading to opportunities for kin-selected altruism (see 2.4.4.5 and 9.3.1), but also problems for inbreeding avoidance. Another is the potential for offspring to inherit the breeding position of their parents, or acquire part of a natal area expanded by the activities of the group. Territorial 'budding' in Florida scrub jays (*Aphelocoma coerulescens*) is good example of the latter (Woolfenden & Fitzpatrick 1978; Fig. 9.13), but large families can have a competitive advantage even within non-territorial groups by dominating disputes with smaller families or single individuals (Lazarus & Inglis 1978). Inheritance of family resources can lead to the long-term occupancy of an area by a given genetic lineage, forming, in effect, a family dynasty (Emlen 1997a). Such dynasties are likely to be widespread among social species, but their ecological and evolutionary implications have only recently begun to be recognised.

Inheriting the family fortune may be one benefit of staying with your nearest and dearest, but benefits may flow from family relationships in other ways. For example, female Old World monkeys (Primates: Cercopithecidae) and spotted hyaenas form stable hierarchies in which daughters rank directly below their mother in reverse order of age

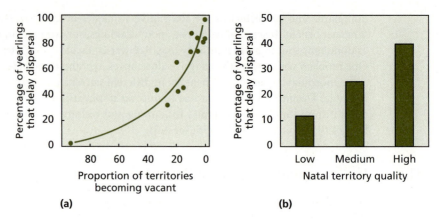

Figure 9.12 The tendency towards the formation of family groups among acorn woodpeckers (*Melanerpes formicivorus*) depends on ecological opportunities and constraints. The proportion of yearling birds that remain on their natal territory goes up as (a) the chances of finding a vacant territory decrease, and (b) the quality of the natal territory increases. See text. After Clutton-Brook (1997) from data in Stacey & Ligon (1987). Stacey P. and Ligon J.D. (1987) Territory quality and dispersal options in the acorn woodpecker, and a challenge to the habitat saturation model of cooperative bleeding. *American Naturalist* **130**: 654–76, reprinted by permission of The University of Chicago Press, © 1987 by The University of Chicago. All rights reserved.

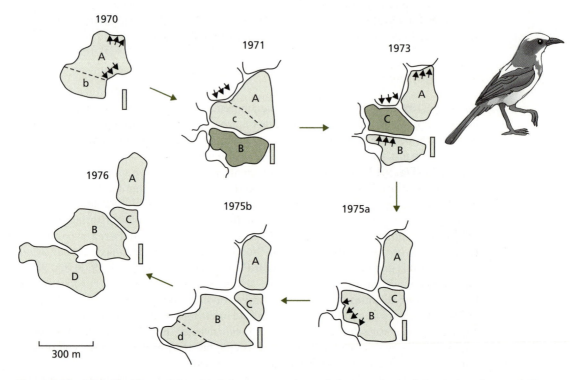

Figure 9.13 Male Florida scrub jays (*Aphelocoma coerulescens*) that remain on their natal territory to help may later inherit part of the territory for themselves. The diagrams show how three helpers, B, C and D, acquired territories over successive years by 'budding off' (as b, c and d) part of the expanded territories or established owners (A). From Krebs & Davies (1993) after Woolfenden & Fitzpatrick (1978).

(see also 2.4.5). Ranking is then between matrilines rather than individual competitive abilities. Daughters inherit their mother's rank through the latter supporting them in disputes with their older sisters and females from lower-ranking matrilines (e.g. Cheney 1977; Datta 1983). Matrilineal hierarchies may have evolved as a result of females protecting their daughters at a vulnerable stage in their social development when rank relationships were determined between individuals, and mothers, daughters and other descendant kin banded together to repel challenges from high-ranking females (Chapais 1992).

Alliances such as those in matrilines play a crucial role in the social organisation of primates, but can be very fluid and both stabilise and destabilise social relationships. For instance, among male chimpanzees, male A, say, may rely on support from male C to beat male B, and lose to B if C is absent. Individuals can then exploit such alliances for their own ends, as in subordinate males, where C sometimes supports A against B, but at other times supports B against A, thus playing the two off to gain greater access to resources itself (de Waal 1982).

9.1.2.3 Roles and castes

Dominance, mating systems and family units impose a certain structure on social groups, but structuring can sometimes extend to a true **division of labour**, in which different individuals perform specific **roles**. A 'role' is a pattern of behaviour that is characteristic of particular individuals within a group and occurs repeatedly in different groups of the same species. The behaviour(s) characterising a role affect those of other individuals in the group in predictable ways, either through specific signals (Chapter 11) or the physical effects of the performer's activity. For instance, in some primate and mongoose species certain individuals may spend a lot of time on the periphery of the group, or on exposed vantage points, scanning for predators. These can be said to perform a 'scout' or 'lookout' role. Other individuals may function in 'leadership', 'baby-sitting' or 'guarding' roles. 'Role' therefore describes the predominant behaviour of an individual as it contributes to the social dynamics of the group, though given individuals may switch from one role to another at different times (see below and Fig. 9.14). However, the designation of role is to some extent subjective, and there is a danger of creating as many roles as there are behaviours in an animal's repertoire, thus forfeiting any explanatory value. An exception to this is the well-defined **caste system** of eusocial species.

Eusociality

Division of labour, social roles and reproductive skew are exemplified most clearly in the highly structured societies of eusocial species. **Eusociality** is characterised by division of labour based on a caste system, in which some members of the group lose their ability to reproduce altogether and become a sterile 'worker' caste toiling on behalf of other, reproductive, individuals (often referred to as 'kings' and 'queens'). Other key features of eusocial systems are cooperation in the care of young, and the maintenance of at least two overlapping generations capable of sharing the group labour (Wilson 1971), though these features are also shown by some non-eusocial species, such as birds and mammals with helper-at-the-nest systems (see above).

The best known eusocial systems are those of the social insects, particularly the Hymenoptera (ants, bees and wasps) and termites, but it also occurs in some aphids (Aoki 1977; Benton & Foster 1992), beetles (Kent & Simpson 1992) and thrips (Crespi 1992). Differentiation into castes can be extreme (Fig. 5.2). In some species of ant, for

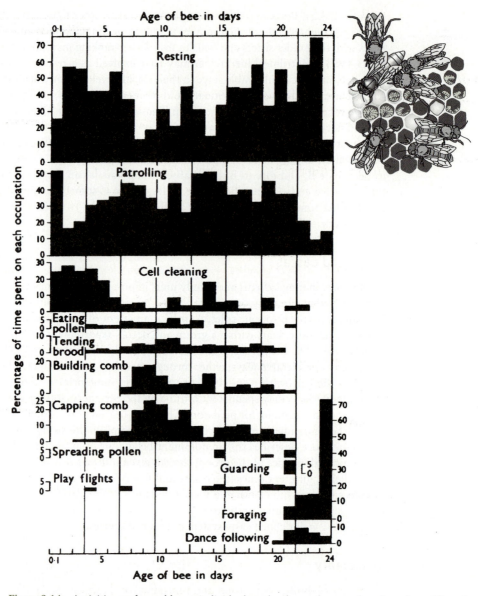

Figure 9.14 Activities performed by an individual worker honey bee as a function of age. There is a recognisable age-related sequence from cell cleaning to foraging, but throughout life a great deal of time is spent patrolling the hive or resting, perhaps allowing the bee to acquire information about colony activity and adjust her own behaviour appropriately. From Lindauer (1961). Reprinted by permission of the publisher from *Communication Among Social Bees* by Martin Lindauer, p. 20, Cambridge, Mass: Harvard University Press, Copyright © 1961 by the President and Fellows of Harvard College.

example, certain workers, called 'repletes', are little more than food casks, hanging statically from the roof with their abdomens distended with sugar solution (Hölldobler & Wilson 1990). Until relatively recently, eusociality was thought to be exclusive to insects, but it has now been discovered in two species of mammal, the naked mole rat (*Heterocephalus glaber*) and Damaraland mole rat (*Cryptomys damarensis*) (Jarvis

1981; Jarvis & Bennett 1993), and a crustacean, the snapping shrimp *Synalphaus regalis* (Duffy 1996). Some microorganisms also show a division of labour reminiscent of eusociality, with individuals specialising in different tasks, such as biosynthesis, selective cell death or nutrient capture (Crespi 2001).

The roles played by worker castes vary both within and between species, but broadly fulfil the tasks of brood care, foraging, nest construction and guarding the colony. Sometimes, a task is the sole preserve of a specialised caste, as in ants and termites, where a soldier caste with specially enlarged jaws guards the colony. In other cases, individuals move through a succession of roles as they age. In their six week lifespan, for example, worker honey bees start off cleaning and feeding the brood, before moving on to brood cell construction, then guard duty and, finally, towards the end of their lives, foraging (Fig. 9.14). Sterility is not the only reproductive sacrifice made by worker honey bees. If they are called upon to attack during their time as guards, they are likely to die in the attempt, since their stings (modified ovipositors) are barbed and remain in the flesh of the victim, tearing out the contents of the guard's abdomen as it flies off. We shall return to the problem of altruism in eusocial species in 9.3.

Eusocial species generally, but insects in particular, can display staggering complexity in their behaviour and coordinated activity within the colony, inviting the time-honoured, but wholly inaccurate, *superorganism* analogy that compares individuals within colonies to the cells of a single organism (Seeley 1989; Wilson & Sober 1989). A large ant or termite colony, for example, can contain tens of millions of workers and construct an abode that is ten thousand million times bigger than any of the individuals that built it (Seger 1991; Fig. 9.15a). Foraging army ants (*Eciton burchelli* and *E. hamatum*) can rapidly adapt the formation of their swarms to the distribution of resources (Fig. 9.15b), a single *E. hamatum* worker returning from a find to the main foraging swarm recruiting 50–100 new individuals, and establishing a continuous stream of ants, in a matter of seconds (Chadab & Rettenmeyer 1975). The honey bee dance language (see 11.3.2.1), used by successful foragers to tell recruits about the direction and distance of food sources, is one of the few known systems of communication in non-human animals to refer to distant

Figure 9.15 (a) A nest of the South American leafcutter ant *Atta vollenweideri*, showing the fungus garden and spoil dump chambers. Person drawn to scale. After Hölldobler & Wilson (1990).

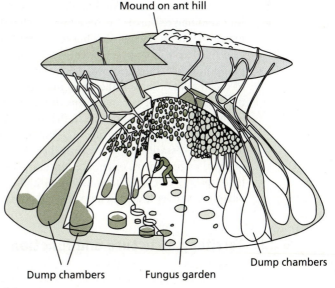

Mound on ant hill

Dump chambers   Fungus garden   Dump chambers

(a)

Figure 9.15 (*continued*) (b) Patterns of raiding by colonies of army ants: column raiding by *Eciton hamatum* (left) and swarm raiding by *E. burchelli* (right). From Heinrich (1978) after Rettenmeyer (1963).

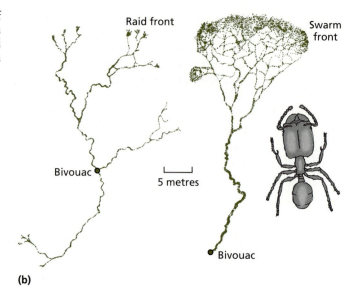

(b)

objects using an abstract code. Despite their outward sophistication, however, these feats are based on relatively simple rule of thumb responses to the physical environment and other workers. The towering edifices that are termite mounds are not the product of visionary architects, but the upshot of millions of blind decisions to place grains of sand in a particular relation to other grains of sand. There is no suggestion these actions, or any other accomplishments by social insects, are based on intelligent insight.

9.2 Territoriality

So far we have concentrated on the costs and benefits of social aggregation as a strategy for exploiting resources. As we emphasised at the beginning, however, social relationships may lead, not to aggregation, but to spacing out and **territorial behaviour**. A territory can be defined as 'a more or less exclusive area defended by an individual or group' (Davies & Houston 1984), thus acknowledging aggregation and territoriality as part of a continuum of social relationships rather than mutually exclusive alternatives. 'Defence' can involve overt aggression or various forms of 'keep out' signal, such as song or odour cues (e.g. Krebs *et al.* 1978a; Gosling 1982), that result in little more than an intruder acknowledging present occupancy and moving on (Davies 1978a,b; 11.1.1.4). Of course, spacing by itself does not necessarily imply territorial behaviour; sometimes it is due to the structure of the physical environment, or chance mortality, as in some sessile organisms. However, even here, spacing can be suspiciously non-random, as in barnacle (*Balanus crenatus*) larvae, which seem to settle at least a body length away from their nearest neighbour (Crisp 1961).

9.2.1 Territory structure and function

As Wilson (1975a) pointed out, territories are more than just defended areas. They are dynamic, often changing in size and shape with season, population density, age or other factors. Huxley (1934) compared them with elastic discs, having the resident animal as their

Figure 9.16 Elastic disc territories in North American dunlin (*Calidris alpina*). Territories are small when population density is high (Kolomak), but large with 'buffer zones' in low-density areas (Barrow). After Holmes (1970).

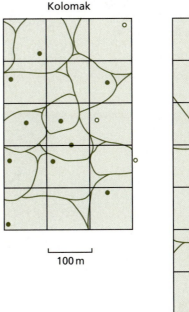

Kolomak

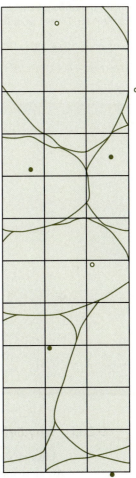

Barrow

100 m

centre and responding to changing population density by expanding and contracting. Huxley recognised limits to this elasticity. Contraction would ultimately be resisted by aggression among residents, or lead to a breakdown in the territorial system, while expansion would reach a point where neighbouring boundaries were no longer contiguous, and either became diffuse, or separated by 'buffer zones' or 'no man's land'. The territories of north American dunlin (*Calidris alpina*) appear to fluctuate in this way (Fig. 9.16). Sometimes the pattern of space use within a territory changes as it expands or contracts. Tree sparrows (*Spizella arborea*), for example, use the whole of their territory with equal intensity when it is fully contracted, but use only the central area intensively when it is expanded.

If several elastic discs are allowed to expand in close proximity so that they meet, their originally circular boundaries distort to form hexagons. An important point about perfectly matching, contiguous hexagons is that they leave no space between them. By allowing their boundaries to conform to hexagons, therefore, territory owners would retain the maximum possible space within them. A clear example of hexagonal territory boundaries is found in the mouthbrooding cichlid *Tilapia mossambica*, where the borders of neighbouring male territories are sometimes visible as sandy ridges surrounding each owner (Fig. 9.17).

Figure 9.17 Hexagonal territory boundaries in male mouthbrooders, *Tilapia mossambica*. See text. Drawn from Barlow (1974).

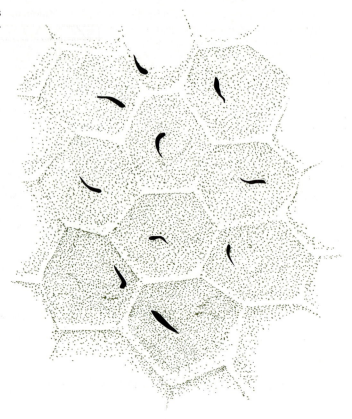

9.2.1.1   Why defend a territory?

Huxley's elastic disc analogy is, of course, merely descriptive and says nothing about underlying function or mechanism, though it continues to provide a basis for modelling patterns of spacing and territory-packing within populations (e.g. Byers 1992). But what are territories for, and when should animals defend them? As usual, the answer comes down to reproductive costs and benefits.

Whether territories are defended through violent border disputes, or by more subtle signalling systems, defence is a cost on the owner. The time, energy and injury costs of fighting are obvious, but ownership signals can be costly too. A bird sitting conspicuously on an exposed twig and singing for hours to advertise its presence suffers a serious drain on its time and energy, and may risk death from predators. Odour signals can be expensive metabolically; urine marking by territorial male house mice, for example, quickly leading to significant loss of weight (Gosling *et al.* 2000). What reproductive benefits might offset such costs? The two most obvious relate to food and mates.

Much evidence suggests that territorial behaviour is linked with the spatial and/or temporal distribution and availability of food (Horn 1968; Johnson *et al.* 2002). Correlations between territory size and the biomass of individuals or groups in many species imply that defended areas are decided on the basis of food availability, giving rise to the so-called **resource dispersion hypothesis** of social structure (see Johnson *et al.* 2002). In social primates, for example, species that include a lot of foliage in their diet tend to have smaller territories than those taking mainly fruit, presumably because fruit is more sparsely distributed through the environment (Clutton-Brock & Harvey 1977). Comparative studies of ant species have also revealed associations between spatial defence strategies by colonies and food distribution (Hölldobler & Lumsden 1980). A nice experimental demonstration of a link between territoriality and food comes from Zahavi's (1971) study of white wagtails (*Motacilla alba alba*) in Israel. When food was provided in small clumps, wagtails defended them aggressively, but when the same food was spread out more thinly, the birds fed in peaceable flocks. Similar effects of experimentally manipulated food distribution have been found by Davies & Hartley (1996) in dunnocks (*Prunella modularis*) and Rubenstein (1981) in captive groups of pygmy sunfish (*Elassoma evergladei*).

Territoriality also plays an important role in mating success, often because of its relationship with food availability. Males with good territories can acquire several females while their less well-provisioned neighbours remain unmated (Davies & Hartley 1996). Sometimes mating preferences are based not on the feeding quality of a territory, but on the availability of suitable breeding sites, or the protection afforded from predators. In yet other cases, territories do not contain resources at all, and preference depends on the quality of the male. We shall return to some of these issues in Chapter 10. Various other benefits may accrue from defending a territory, such as reduced cannibalism of offspring by neighbours, access to safe roosting sites, and, in reptiles, access to good sunning areas for thermoregulation. Davies (1978a) and Davies & Houston (1984) provide good general reviews.

While all this is good evidence that territoriality provides benefits, it is largely qualitative and does not allow us to make precise predictions about when defence is worthwhile or how big a defended area should be. But, if animals are weighing up costs and benefits, and we in turn can measure them, we should be able to use optimality theory to specify the conditions under which territoriality will emerge. Brown (1964) did this in a general way in his principle of **economic defendability** (Box 9.6).

Underlying theory

Box 9.6 Economic defendability and territoriality

Brown (1964) recognised that defending a territory will incur both costs and benefits in terms of reproduction for the owner. To be favoured by natural selection, the benefits must exceed the costs (the principle of economic defendability), but the conditions under which this happens will depend on how costs and benefits vary with territory ownership. The simple graphical model in Fig. (i) makes this clear. Here hypothetical reproductive costs (C) and benefits (B) are plotted as a function of territory size. As the territory becomes bigger, so does the cost of defending it, because there is a larger area to cover and a bigger boundary provides more opportunities for intrusion. Indeed, the cost may escalate disproportionately when territories become very large (the slope of the curve increases). Benefit may also increase with territory size, but eventually it is likely to level off as resources become superabundant relative to the owner's needs. As a result, the cost/benefit curves intersect at A and B, which means that benefit exceeds cost only between the two points. While there is therefore some advantage to defending a territory anywhere between p and q, the maximum advantage is to be gained at point X, which thus indicates the *optimal* territory size. All other things being equal, therefore, selection should favour owners that defend territories of size X.

Box 9.6 continued

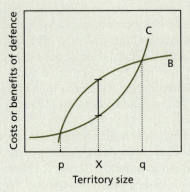

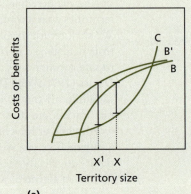

Figure (i) Hypothetical cost/benefit curves for defending territories of different size. See above. After Davies & Houston (1984).

Figure (ii) Shifts in either (a) the benefit (B > B′) or (b) the cost (C > C′) curve alter the range of territory sizes over which there is a net benefit in Fig. (i), causing a change in the optimal size (X > X′). (c) Changing the shape of the curves, however, can cause the optimal territory size to shift in the opposite direction. See above. After Davies & Houston (1984) and Krebs & Davies (1993).

(a)

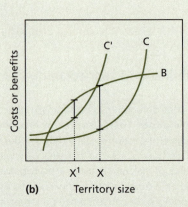

(b) Territory size

(c) Territory size

However, various factors are likely to alter the cost/benefit curves, and thus the optimal territory size, in Fig. (i). An increase in resource quality, for example, will shift the benefit curve up so that the optimal territory size decreases (Fig. (ii)a). Increased levels of competition, on the other hand, will accelerate cost, also reducing the optimal territory size (Fig. (ii)b). But, if the shape of the cost/benefit curves changes, as in Fig. (ii)c, increased resource quality could lead to an *increase* in the optimal size. Optimal territory size is thus likely to be fluid, increasing and decreasing as cost/benefit functions vary.

Based on discussion in Brown (1964), Davies & Houston (1984) and Krebs & Davies (1993).

9.2.2 Economic defendability

Brown's model is just a graphical restatement of the verbal optimisation argument that animals should do whatever maximises benefit relative to cost. The trick in applying it is knowing the currency in which costs and benefits should be measured (see Box 2.5). This can be complicated. For example, male great tits sometimes defend territories in late winter that are neither critical for feeding, nor function in attracting females (Perrins 1979). Instead, the payoff comes the following summer, when territories allow nests to be spaced out, thus reducing the risk of nestling predation (Krebs 1971). The male therefore spends time and energy in territorial defence during winter in order to enhance the survival prospects of its offspring later in the year (Davies & Houston 1984). As in virtually all other aspects of behaviour, reproductive costs and benefits tend to be measured partially and indirectly, for instance as feeding rate or the number of visiting females. Testing for optimisation under the principle of economic defendability is therefore unlikely to be straightforward. Perhaps not surprisingly, given its potential for accessible currencies (see 8.1), food intake has provided the most fertile ground for testing the idea.

9.2.2.1 Optimal territory size

Carpenter & MacMillen (1976) studied changes in territorial behaviour in Hawaiian honeycreepers (*Vestiaria coccinea*), small nectar-feeding birds in which territorial defence seems to be based on the need to maintain access to sufficient flowers to ensure survival. By carefully quantifying the energetic costs and benefits of defence, Carpenter & MacMillen were able to predict exactly when honeycreepers should and should not be territorial. Defence should begin when the amount of energy gained without defending a patch of flowers drops below that needed to meet the bird's basic cost of living. At this point, birds have to exclude competitors to prevent their additional depletion of the nectar supply. However, defence should cease if nectar productivity drops so low that, even with exclusive access to flowers, it is insufficient to meet the bird's needs. From their experimental measures, Carpenter & MacMillen calculated the upper threshold (defence should begin) to be around 207 flowers, and the lower threshold (defence should cease) around 60 flowers. Of 10 birds in their study, 9 conformed to prediction by defending territories only between the two thresholds (Fig. 9.18).

While Carpenter & MacMillen's study looks convincing, it has some difficulties. For example, the lower threshold, as characterised in their model, is actually a limit to existence, not just territoriality. Birds would be expected to abandon a territory long before this. The model also fails to allow for likely variation in pressure from competitors as the number of flowers in a territory changes (see Box 9.6). Later studies by Carpenter, however, provided more compelling evidence for optimal territory sizes in nectar-foragers, this time in rufous hummingbirds (*Selasphorus rufus*). Carpenter *et al.* (1983) studied hummingbirds defending patches of flowers during a migratory stopover in California to replenish fat reserves. By putting out perches attached to sensitive balances, Carpenter *et al.* were able to monitor the mass change in birds defending territories of different size. Since the object of the stopover was to put on fat to make it through the rest of the migration, birds might have been expected to defend territories that maximised mass gain. Figure 9.19(a) shows the pattern of mass change in relation to territory size for one of Carpenter *et al.*'s birds. As can be seen, the bird began with a small territory that afforded little in the way of mass gain, then enlarged it without

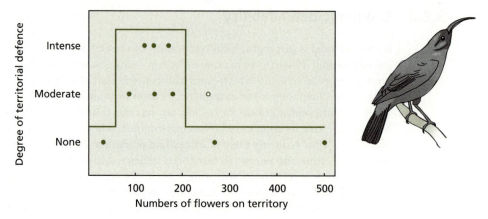

Figure 9.18 Thresholds of territorial defence in Hawaiian honeycreepers. Estimation of the energetic costs and benefits of defence predicted that birds should defend territories of between 60 and 200 flowers. Most birds in an experimental test conformed to the prediction (●); one did not (○). See text. From Davies (1978a) reprinted with permission after 'Threshold model of feeding territoriality and test with a Hawaiian honeycreeper', *Science*, Vol. 194, pp. 639–42 (Carpenter, F.L. and MacMillen, R.E. 1976 American Association for the Advancement of Science.

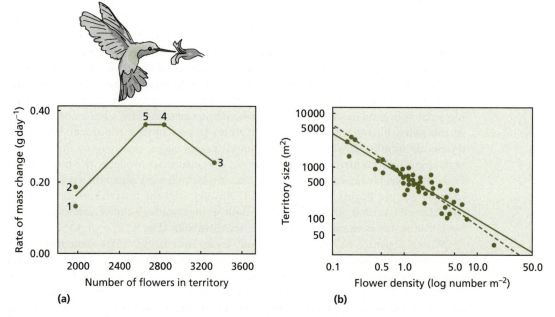

Figure 9.19 Optimal territory size in rufous hummingbirds. (a) The daily mass gain of a bird when it defended different numbers of flowers over five days. The bird started with a small territory (Days 1–2), then expanded it on day 3 before settling on an intermediate size (Days 4–5) where daily weight gain was maximised. From Krebs & Davies (1993) after Carpenter *et al.* (1983). (b) The size of territory defended varies inversely with the density of flowers. The slope of the data (solid line) is not significantly different from that expected if birds were simply defending the same number of flowers (dashed line). See text. After Kodric-Brown & Brown (1978).

much improvement, before finally settling on an intermediate sized territory that yielded the maximum gain. That territory size related to resource availability is also supported by the tendency for birds to vary the area defended in line with daily changes in the density of flowers (Kodric-Brown & Brown 1978; Fig. 9.19b).

One reason intermediate territory sizes might yield the maximum payoff is because greater intruder pressure in larger territories increases the cost of defence. Praw & Grant (1999) tested this in convict cichlids (*Archocentrus nigrofaciatus*) by manipulating the area of defendable food patches and exposing owners to intruders over a 10-day period. They found that the growth rate of owners increased with patch area up to a point, then began to decrease. The initial increase was due to increased food intake in larger patches, but the subsequent decline resulted from escalating investment in driving away intruders. However, studies of a variety of species suggest that optimal territory sizes based on tradeoffs between resource quality and intruder pressure by individual owners may be something of an exception. Instead, simultaneous interactions between neighbouring owners, and resulting pressure on contiguous territory boundaries may have more important roles to play (Adams 2001). Models incorporating simultaneous effects of multiple competitors have recently begun to emerge but have yet to be rigorously tested (Adams 2001).

Sharing a territory

Territories need not be defended exclusively. Sometimes it may pay to share a territory with another individual that at other times would be regarded as an intruder. Davies and coworkers studied such an arrangement in pied wagtails (*Motacilla alba yarrellii*) feeding along a river bank in southern England. Some wagtails defended winter territories along the bank, exploiting insects washed up in the mud by the river. The insects provided a renewing food supply, meaning that feeding sites were replenished by the river after having been depleted. A major benefit of territoriality appeared to be that owners could regulate their return times to each site so as to ensure it had had time to replenish before being visited again. This meant that owners foraged by conducting a systematic circuit of their territory (Fig. 9.20a). Defence was worthwhile because, if intruders got onto the territory, they would deplete the food supply and interfere with the anticipated cycle of replenishment (Davies 1981).

Unlike Carpenter & MacMillen's honeycreepers, wagtails maintained their territories even when they could do better feeding elsewhere; in this case in nearby flocks of non-territorial birds. When food availability on territories was low, owners would join the flocks, but periodically return to their territories to resume defence. The reason for this seemed to be that the food supplies exploited by flocks were ephemeral; birds in flocks sometimes did very well, but at other times they did very badly. Territorial birds, in contrast, always had access to at least some food, so benefited from a sure, if not always the best, food supply throughout the winter.

Strangely, however, given this important advantage, owners did not always drive intruders away. When food availability was high, they often allowed another bird, usually a juvenile or female from a neighbouring flock, to share the territory as a **satellite**. Owners and satellites usually coexisted by walking round half a circuit behind each other (Fig. 9.20b), so that the effective return time to any given feeding site for the owner was halved. However, this foraging cost was offset by the additional defence carried out by the satellite when the rich pickings on the territory attracted more intruders. The owner thus accepted help with defence as '**payment**' from the satellite for the privilege of foraging on its territory. By monitoring food availability and the effects on intake rate of foraging alone or sharing, Davies & Houston (1981) were able to predict the conditions under which wagtails should exclude or share with another bird. Test observations showed a convincing split across the predicted threshold (Fig. 9.20c).

Figure 9.20 Pied wagtails defending territories along a river bank (a) sometimes shared their territory with a 'satellite' (b), which followed the owner around about half a circuit behind. (c) Owners were predicted to share their territory with a 'satellite' when the abundance and rate of renewal of food were above a threshold curve. In a field test most territory owners conformed to prediction (● shared territories; ○ unshared territories). From Krebs & Davies (1993) after Davies & Houston (1981).

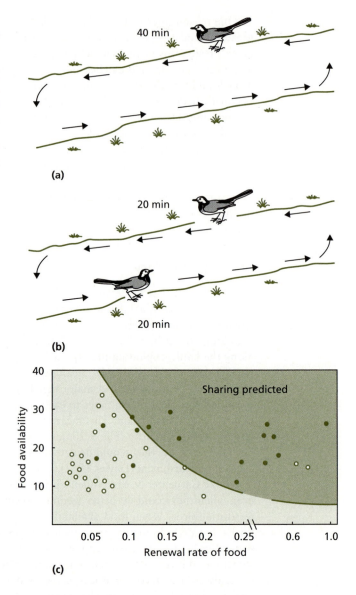

Allowing a satellite to remain on a territory as a tradeoff between feeding and defence might, under the right conditions, be expected to lead to the evolution of group territoriality. Many species defend territories as groups, often with the reproductive cost of additional residents being borne by dominant breeding individuals in return for various payments, such as help with defence or vigilance (e.g. Barnard *et al.* 1991). Indeed, games theory models suggest that allowing supernumerary individuals onto territories may be an evolutionarily stable strategy of territorial competition, through increasing the pressure on neighbouring boundaries (Craig 1984). Whether or not there is a general advantage in having more individuals on a territory, however, reproductive conflicts of interest are likely to arise and can have an important influence on the structure and dynamics of territorial

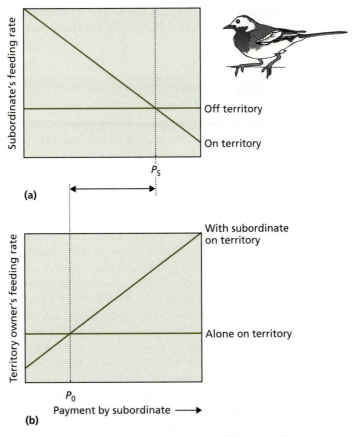

Figure 9.21 Hypothetical relationships between a 'satellite' wagtail's payment to a territory owner and the feeding rate of the 'satellite' (a) and the owner (b) on the territory. The horizontal lines represent expected feeding rate of the 'satellite' if it fed elsewhere (a) and the owner if it did not allow the 'satellite' on the territory (b). See text. After Davies & Houston (1984).

groups, as we have seen already in mountain gorillas (7.2.1.1; Fig. 7.9). Davies & Houston (1983) have modelled this for the owner/satellite association in pied wagtails.

Figure 9.21 represents hypothetical feeding rates for owners and satellites as a function of the satellite's defence payment. Satellites do well on territories as long as they do not pay too much, but the more they pay, the less they get, until they would do better by leaving and feeding away from the territory (Fig. 9.21a). Conversely, the owner's feeding rate is depressed if satellites take food but do little to repel other intruders, and may drop below the point at which the owner would be better off alone (Fig. 9.21b). The key point is that the threshold payment at which it would be better for the satellite to stay (the intersection of the lines) differs between the satellite and the owner. We should expect the owner to try to squeeze the maximum acceptable payment out of the satellite (i.e. go for a point just below P_s in Fig. 9.21a), but the satellite to hold out for the minimum acceptable payment (just above P_0 in Fig. 9.21b). Various factors are likely to complicate things, however. For instance, a satellite may not be able to gauge the value of P_0, or competition among potential satellites might drive up the going rate for payment (Davies & Houston 1984). The resolution of these conflicts will depend on local circumstances and are unlikely to be easy to predict.

9.3 Cooperation

We now come to one of the major evolutionary issues surrounding social behaviour, indeed a seminal issue in the science of behavioural ecology itself: **cooperation**. Both grouping behaviour and territoriality can lead to, or, indeed, depend on, a degree of cooperation between individuals. In an evolutionary context, cooperation can embody many things, from the coordinated hunting of a pride of lionesses, to the sterile worker castes of hymenopteran societies. It may thus involve mutualistic interactions with obvious shared benefits, or seemingly unilateral self-sacrifice (altruism; see 2.4.4.5). Like sex (see 10.1), cooperation, but particularly phenotypic altruism, is something of an evolutionary paradox. How can natural selection, which, by its very nature, should foster selfishness, produce individuals that sacrifice their own reproductive potential while increasing that of others? The question was not lost on Darwin, who was acutely aware of the problem it created for his argument. Such traits posed '. . . a special difficulty, which at first appeared to me insuperable, and actually fatal to my whole theory' (Darwin 1859). In his efforts to encompass them, Darwin suggested that selection might sometimes act at the level of the family or colony, thus foreshadowing the group/kin selection debate a century later. Like many philosophers and natural historians before him, Darwin also mused on the implications of cooperation in other organisms for cooperation and conflict in our own society, especially traits such as bravery and self-sacrifice in battle, which forced him into a more explicitly group selectionist stance (Darwin 1871; see also Dugatkin 1997).

Of course, group selection provides a tempting framework for understanding cooperation, but, for the reasons we discussed in 2.4.4.1 and 2.4.4.2, it is unlikely to provide a satisfying one. Instead, the major front-runner is kin selection, which could drive the evolution of cooperation among close relatives through the indirect fitness benefits to cooperative alleles from helping copies of themselves into the next generation (see 2.4.4.5). While kin selection has received the lion's share of the attention, however, there are other, not necessarily mutually exclusive, routes by which cooperation and phenotypic altruism might have evolved. We shall now look at kin selection and its alternatives, and at some of the evidence that they have had important roles to play.

9.3.1 Cooperation and kin selection

The principle of kin selection has already been introduced in 2.4.4.5. We have also mentioned various forms of cooperation – sterile worker castes (2.4.4.5), alarm signals (8.3.5.4) and reproductive helpers on territories (9.1.2.2) – in which it might have been a shaping force. The most widespread example of kin-selected altruism, of course, is parental care (10.2), where help and self-sacrifice by the parent pay obvious genetic dividends through the eventual reproductive success of closely related offspring. Just because cooperation takes place between relatives, however, does not necessarily mean it has evolved by kin selection (e.g. Griffin & West 2002). The haplodiploidy story in Box 2.8 is a good example.

9.3.1.1 Kin selection and eusociality

Hamilton's demonstration that haplodiploid sex determination could account for worker altruism in hymenopteran insects (Box 2.8), by increasing the degree of relatedness between sisters, was one of the formative triumphs of modern evolutionary theory. For kin selection to provide a general explanation for the evolution of helper altruism in eusocial animals, helping must have evolved according to Hamilton's rule (Box 2.7). In haplodiploid workers, the benefit multiplier in the rule is hiked upwards by an increased coefficient of relatedness (r) (see 2.4.4.5), so making it more likely that the cost of helping will be outweighed. But haplodiploidy is clearly neither essential, nor an inevitable trigger, for the evolution of eusociality and worker altruism. Both occur in diploid species, such as termites and mole rats (see 9.1.2.3), and several haplodiploid species are not eusocial. Furthermore, the coefficients of relatedness in Box 2.8 hold only if a colony of haplodiploid insects is founded by a single queen that has mated once, as happens in ants. Even if she has mated with only two unrelated males, the relatedness between workers that do not share a father drops to 0.25, a far cry from the 0.75 of Hamilton's scenario. In honey bees, queens may mate up to 20 times, although the effect of multiple matings may be offset by the ability of workers to discriminate between sisters of different degrees of relatedness to themselves (Page *et al.* 1989; see below).

The question mark over the role of haplodiploidy in the evolution of eusociality leaves all options open as far as kin selection is concerned. As Bourke (1997) points out, the relatedness bias created by haplodiploid sex determination is not necessary for the conditions of Hamilton's rule to be met, as studies of many diploid species have shown. Furthermore, haplodiploidy may have genetic consequences that predispose towards eusociality independently of its effects on relatedness, for instance by reducing the loss of alleles coding for altruism through genetic drift (Reeve 1993). Hymenopteran insects also share features other than haplodiploidy that might facilitate the emergence of eusocial organisation, such as the tendency for solitary wasps and bees to build nests, which may constitute a resource worth cooperating over and possibly inheriting, or the possession by females of a sting, which might facilitate the emergence of a defensive caste (Evans 1977; Andersson 1984a; Crespi 1994; Bourke 1997). Thus haplodiploidy may be bound up with several factors, genetic and non-genetic, that have resulted in its present association with eusocial behaviour.

9.3.1.2 Constraints and kin selection

Kin selection makes clear predictions as to when and where cooperative behaviour should emerge. For instance, it should be more prevalent in family groups than in less closely related aggregations, and, within families, it should happen more often between close relatives (Emlen 1997a). Favourable conditions for kin-selected cooperation are widespread, and seem frequently to have been exploited. For example, around 96% of birds and 90% mammals that live in family groups show cooperative breeding (helping others rear offspring; Emlen 1995; see 9.1.2.2). Indeed cooperative breeding in birds and mammals seems to be largely *limited* to species with a family structure (Emlen 1997a). There is also evidence that helping is directed towards more closely related recipients within family groups (e.g. Emlen & Wrege 1988; Fig. 9.22; but see 9.3.1.3). But kin-selected cooperation is not just a matter of an appropriate kinship environment. The reproductive costs and benefits that tip the balance can reflect important constraints. An obvious one is ecological opportunity.

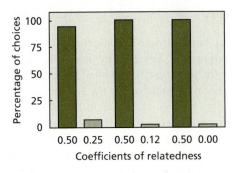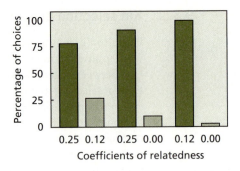

Figure 9.22 White-fronted bee-eaters (*Merops bullockoides*) are cooperative breeders that live in family groups. Within the family, however, birds are more likely to help at the nests of more closely related breeders (dark bars). The figure compares percentage preferences for pairs of available breeders with their coefficients of relatedness to the helper on the *x*-axis. After Emlen (1997a).

Table 9.2 Florida scrub jays often stay on their natal territory to help relatives rear more offspring. Calculations, however, suggest they would do much better by going off to breed on their own – if they were able to find a suitable territory. See text. After Woolfenden & Fitzpatrick (1984).

Option		Result
1 Stay on natal territory and help	Young produced by experienced parents with no help	1.80
	Young produced by pair with helpers	2.38
	Extra young due to presence of helpers	0.58
	Average number of helpers	1.78
	Genetic equivalents to a helper (*r* to nestlings = 0.43)	0.14
2 Go off and rear own young	Young reared by first time breeders	1.24
	Genetic equivalents (*r* to own offspring = 0.5)	0.62

Ecological constraints

In 9.1.2.2, we mentioned the example of Florida scrub jays, in which young birds remain on their natal territory to help the breeding pair until they can inherit a portion of the (by then expanded) territory later on (Fig. 9.13). But what of the intervening time? Are the birds doing better by remaining or would they be better off leaving to breed on their own territory? The answer depends on the consequences of the two options for the spread of an allele for helping compared with one for breeding independently.

Woolfenden & Fitzpatrick (1984) calculated the future 'genetic equivalents' likely to be generated by birds under the two conditions (Table 9.2). In the vast majority of cases, helpers assist their own parents, and thus contribute to rearing full siblings, though a few end up helping less closely related, or even unrelated, pairs. From a genetic point of view, rearing a full sibling is just as good as (or thereabouts) rearing an offspring: $r = 0.5$ either way. On the basis of painstaking fieldwork, Woolfenden & Fitzpatrick were able to compare the number of young produced by helping or setting up as a novice breeder, and, knowing *r* in each case, the estimated payoff in genetic terms. As Table 9.2 shows, jays would do much better by leaving home and setting up on their own. These calculations are somewhat over-simplified, and the 'genetic equivalents' cannot be taken as absolute, but even allowing for sizeable error, the conclusion that birds would be better off producing offspring themselves is unlikely to change. So why don't they? The answer seems to be that there is nowhere to go. The availability of suitable breeding territories is limited and habitats frequently saturated. This is supported by the fact that, as soon as vacancies do arise through mortality, birds leave their home territories to occupy them. Staying to help birds may therefore be a way of making the best of a bad job until there is an opportunity to inherit coveted breeding space within the home territory at a later date.

A similar picture emerges, though for different reasons, in naked mole rats. Helping by members of the worker caste seems to be facilitated by the high degree of relatedness between colony members – a result of inbreeding within more or less closed groups. However, there is also a high cost to dispersal, since the underground tubers on which the animals depend are sparsely distributed and unpredictable in location. The time and energy cost of tunnelling to seek them out, especially for a small group, may thus be extremely high (Lovegrove 1991). The extreme ecology of the species may therefore be a crucial factor maintaining their equally extreme social organisation.

In scrub jays and mole rats, then, staying to help may be partly a response to a constraint on dispersal. The opposite problem affects helping in black-backed jackals (*Canis mesomelas*). Patricia Moehlman (1979) studied jackals in the Serengeti region of Tanzania, where monogamously breeding pairs can have up to three young from previous litters acting as helpers (regurgitating food for pups and lactating mothers, and grooming and guarding litters). Breeding success on territories increases sharply with the number of helpers (Fig. 9.23), so, theoretically, the more helpers the better. This time, however, it is resources within the territory itself that constrain opportunity, the young from previous litters often being forced to leave rather than remain behind to help because of limited food availability.

Life history constraints

We mentioned that full siblings were worth the same as offspring in terms of r in diploid, sexually reproducing species. This is true, but a little misleading when it comes to predictions about kin selection. Why? Because equivalence in relatedness does not necessarily mean equivalence in value from an inclusive fitness point of view. Our suspicions should be raised by the fact that apparent altruism towards siblings is much less common than that towards offspring. To see why, we need to reflect again on the components of Hamilton's rule: r, b and c. While siblings and offspring may share r, the reproductive costs and benefits (c and b) of helping them may be very different. A young offspring is likely to be more or less defenceless and in no position to reproduce until nurtured successfully to adulthood. An older sibling, however, may be perfectly capable of competing for mates and reproducing without any help at all. Thus the net benefit in inclusive fitness from helping will be very different in the two cases. Only if helping really helps, as in the extra offspring production by scrub jays and jackals, will the effort be worth it.

The sibling versus offspring argument hinges on differences in the reproductive value of potential beneficiaries. As we saw in 2.4.5, reproductive value varies with many life history parameters that are likely to affect the inclusive fitness benefits from helping. A grandparent past its reproductive prime, for example, is likely to be a less valuable candidate than an equivalently related grandoffspring, with its reproductive life before it. On the other hand, even a distantly related individual with modest reproductive

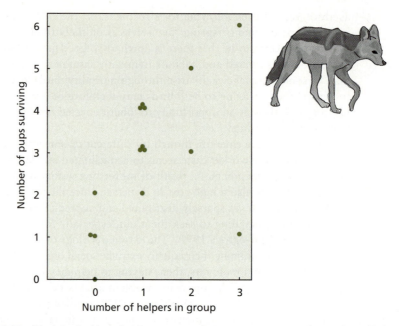

Figure 9.23 Black-backed jackals (*Canis mesomelas*) rear more pups as the number of helpers in the group increases. After Moehlman (1979). © Nature Publishing Group (http://www.nature.com/), reprinted by permission.

prospects might be worth helping if an animal has no chance of reproducing itself. Thus, whether or not a trait is favoured by kin selection will depend on many factors: genetic, developmental and environmental.

One further point, first emphasised by Altmann (1979), is worth making, which is that the coefficient of relatedness is not a prescription for dividing help proportionally across available relatives; i.e. giving a first cousin half as much help as a grandchild, or a grandchild half as much as a sibling. This would lead to the *reductio ad absurdum* of ever-diminishing acts of assistance being dispensed to ever more distantly related recipients, including, ultimately, those of other species (Dawkins 1989). As Altmann points out, what the helper should do, other things being equal (see above), is dispense *all* its help to the most closely related beneficiary; any other strategy simply dilutes the potential indirect benefit to the helper. The fallacy of proportional dispensation arises from an error of thinking that confuses proportion with probability. What *r* should determine is the *probability* that help will be dispensed to a particular individual, not the proportion of available help it will receive (Dawkins 1989). Similar confusion over *r* as the proportion of alleles shared between individuals, rather than the probability that they share a particular allele, has resulted in other misleading arguments about the allocation of help (Fagen 1976; Partridge & Nunney 1977).

9.3.1.3 Kin selection and kin recognition

Kin-selected cooperation leads naturally to the possibility of kin recognition (Hamilton 1964b), since in order for kin selection to drive the allocation of help, it is necessary for kin to be the beneficiaries (see 10.1.3.2 for a discussion of kin recognition in relation to mate choice). While a wealth of studies now purport to demonstrate kin recognition (see

Fletcher & Michener 1987; Hepper 1990; Sherman *et al.* 1997; Figs 8.25 and 9.24), such claims are often fraught with difficulty (e.g. Waldman 1987; Grafen 1990a; Barnard *et al.* 1991). One problem is that, while kinship is undoubtedly the commonest cause of genetic similarity between individuals, there are several ways that genetic similarity could influence their behaviour that have nothing to do with kin recognition, or, indeed, recognition of any sort (Box 9.7).

From the point of view of kin selection, biasing helpful responses towards relatives means that help is likely to be given to those sharing the allele for helping. But how is this to be achieved? Is relatedness *per se* a recognisable feature of individuals? Probably not. However, relatedness is likely be associated with other things that *are* recognisable, such as having been raised in the same nest, or suckled by the same mother, or perhaps looking or smelling similar (Fig. 9.24; Box 9.7). Thus many different levels of recognition could be used to discriminate in favour of kin, so a special ability to recognise kin – 'kin

Figure 9.24 (a) The frequency of social investigation (a prelude to aggression) decreased and that of passive body contact increased the greater the degree of relatedness between unfamiliar male mice interacting in dyads. FS, full siblings; HS, paternal half siblings; UR, unrelated individuals. Pale bars, juveniles; dark bars, adults. Full siblings were unfamiliar only postnatally. From Barnard *et al.* (1991). (b) If mice are familiarised (reared together), however, the effects of relatedness disappear (dark bars, social investigation; pale bars, passive contact). From data in Kareem & Barnard (1982). (c) The likelihood of unfamiliar female sweat bees (*Lasioglossum zephyrum*) getting past guard bees and entering a colony as a function of their relatedness to the guards. Closely related bees were more likely to be allowed access: □ unrelated individuals; △ cousins; ○ aunts/nieces; × sisters. Numbers are sample sizes for each point. Reprinted with permission after 'Genetic companent of bee odor in kin recognition', *Science* Vol. 206, pp. 1095–7 (Greenberg, L. 1979). Copyright 1979 American Association for the Advancement of Science.

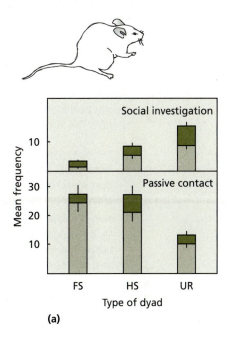

(a)

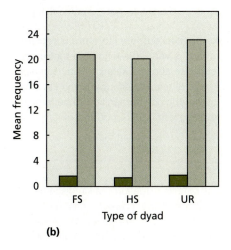

(b)

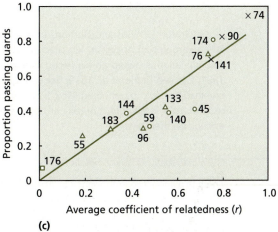

(c)

Difficulties and debates

Box 9.7   The problem of kin recognition

An ability to recognise kin and discriminate in favour of them seems to be a logical prediction of both Hamilton's theory of inclusive fitness (kin selection) and the idea that mating preferences reflect an optimal degree of outbreeding (see 10.1.3.2). However, we immediately encounter two problems when we come to test this. The first concerns the meaning of 'kin recognition', the second, the potentially confounding effects of genetic similarity on behaviour.

What is 'kin recognition'?

Deciding whether an animal recognises something almost always depends on recognition leading to an act of discrimination. If the animal does not change its behaviour as a result of recognition, the latter remains a private, externally inaccessible, state. A failure to respond therefore does not necessarily imply a failure to recognise (we recognise apples when we see them, but we always do not eat one as a result). By the same token, discrimination that seems to be based on some property of a stimulus need not reflect a special ability to recognise the property – sweet apples may be discriminated from sharp apples, but sweetness may not be the property recognised when selecting them. Terms therefore need to be defined carefully. The following are important in the case of kin recognition (see Barnard 1991):

☐ *Kin bias*: a neutral term reflecting the tendency for related individuals to show some response to each other, or to be involved in a particular context together, more or less than would be expected by chance. It implies nothing about the functional significance of any bias, or the mechanism that causes the bias.

☐ *Kin recognition*: the externally unobservable neural process of classing individuals as kin by recognising genetic relatedness *per se* (e.g. Byers & Bekoff 1986; Waldman 1987; but see particularly Grafen 1990a, 1991). Kin recognition may or may not result in some kind of differential response towards kin.

☐ *Kin discrimination*: kin bias based on kin recognition or recognition that individuals belong to some other category (such as a local group) correlating with kinship. Whether kin discrimination results from a special ability to recognise kin or from another level of recognition that leads to the same end, it is the resulting kin bias that drives its evolution. However, kin recognition and kin discrimination are manifestly not the same thing.

As Grafen (1990a) and Barnard *et al.* (1991) have pointed out, the literature is replete with examples of apparent kin discrimination, but there is very little evidence that such discrimination is based on kin recognition in the strict sense above (Grafen 1990a; Barnard 1991; but see Petrie *et al.* 1999). Indeed, given the range of mechanisms by which kin discrimination could be effected (Box 9.8), we might expect strict kin recognition to be largely redundant – familiarity, context or other cues correlating with kinship often being good enough on their own.

Is kin bias really kin discrimination?

There is a widespread tendency to assume that kin bias reflects some form of kin discrimination. As Grafen (1990a) and Barnard *et al.* (1991) have emphasised, however, this may well not be the case. Genetic similarity could influence the responses of individuals with respect to one another in a variety of ways. All of these are likely to produce kin bias (since kinship is the major cause of genetic similarity), but only some involve discrimination (of any kind), and even fewer kin discrimination. The problem is summarised in Fig. (i).

As the figure shows, some kin bias, say in social aggregation, could be due to relatives sharing resource requirements (Pfennig 1990) or behavioural attributes (e.g. a tendency to move around less at low temperature), and not involve recognition or discrimination of genetic similarity at any level (*non-discriminatory effects*). Even when it comes to discrimination, the significance of kin bias may vary. In some cases, bias may arise *incidentally*, as a result of, e.g., species recognition, where natal groups have low dispersal and close relatives provide the standard against which novel individuals are later identified as conspecifics. Kin bias caused by this cannot be called kin discrimination, even though it is based indirectly on recognising genetic similarity.

Even where genetic similarity is recognised directly (*direct discrimination*), it may not constitute kin recognition. A discrimination allele coding for an ability to recognise copies of itself in other individuals and respond accordingly (*direct cobearer discrimination*), or to recognise that individuals are similar at other loci (*direct similarity discrimination*) may cause kin bias if the matching alleles are rare, but this cannot be regarded as kin recognition if its purpose is simply to match genotypes at those loci, since non-relatives that happen to carry the same allele(s) will be treated in the same way. If such matching is used as an indirect means of estimating (otherwise undetectable) similarity at other loci (*indirect discrimination*), however, then it can reasonably be viewed as kin recognition, since it will work only if individuals are indeed close kin. Where genetic similarity is estimated by recognition at other levels, such as location or familiarity (Box 9.8), animals are showing kin discrimination, but not based on kin recognition.

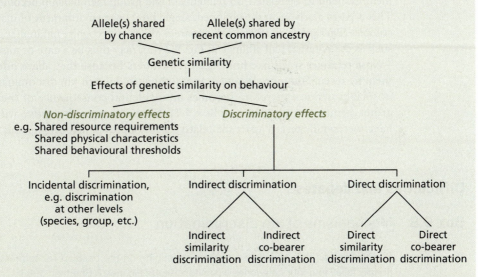

Figure (i) Ways in which genetic similarity between individuals could lead to kin bias in behaviour. From Barnard *et al.* (1991).

Based on discussion in Grafen (1990a) and Barnard *et al.* (1991).

vision', as Grafen (1991) characterises it – is likely to be redundant. This hard-line view continues to cause debate, but opposition to it is largely the result of failing to distinguish between what is *discriminated* and what is *recognised* (see Box 9.7 and Barnard *et al.* 1991). What sort of difficulties does this failure create?

Familiarity and kin discrimination

Many studies have shown that early familiarity can play a role in apparent kin recognition (Box 9.8). In Fig. 9.24(b), familiarising unrelated mice resulted in levels of social tolerance (little aggressive investigation and frequent passive contact) characteristic of siblings (Kareem & Barnard 1982). Similar results were obtained by Holmes & Sherman (1982) (Fig. 9.4(a)) with aggression in pairs of unrelated young ground squirrels. The effect is akin to imprinting between offspring and parent in 6.2.1.5. However, in both mice and ground squirrels, relatedness affects social tolerance between *unfamiliar* individuals (Fig. 9.24a; Holmes & Sherman 1982), suggesting either **phenotype matching** (comparing cues between self and unfamiliar others) or a hard-wired 'genetic' recognition mechanism may be operating (Box 9.8). So does this mean animals are using two different mechanisms of kin discrimination under different conditions? Not necessarily. One interpretation of the responses in mice is that kin bias among unfamiliar animals is caused by the narrower range of genotypes experienced in single sibship litters. Thus the seemingly contradictory outcomes with respect to relatedness in unfamiliar and familiar animals may just be due to differences in the template against which familiarity is judged (Barnard 1990). But, if so, are unfamiliar individuals really discriminating relatedness at all? A plausible alternative is that mice are simply learning who belongs to their home group.

In the wild, commensal house mice defend territories as family groups (Hurst 1989). Within groups, the social network is underpinned by a complex system of chemical communication based largely on inherited urinary odour cues (Hurst 1990a,b,c; Hurst *et al.* 2001). Since the dominant male usually sires most offspring (Hurst 1987), group members tend to be related, so relatedness and group membership become confounded. This is particularly so in the confined, single litter cage environment of many laboratory colonies (Barnard *et al.* 1991). The apparent ability to recognise unfamiliar close relatives, such as the paternal half siblings in Fig. 9.24(a), may thus be a case of mistaken identity – close relatives smell like home group members because they share odour cues with them by recent common descent – and nothing to do with kin discrimination. Kin bias in social tolerance therefore becomes an incidental consequence of (here, mistaken) group member recognition (see Box 9.7). Experiments with captive wild house mice have successfully teased apart the relative roles of group membership and relatedness in

Difficulties and debates

Box 9.8 Mechanisms of kin discrimination

Suggestions as to how animals might discriminate on the basis of relatedness have traditionally fallen into four categories (Hamilton 1964b; Holmes & Sherman 1983):

Location/timing

If relatives are reliably associated with a particular location, such as a nest or exclusive home range, then treating any individual encountered there as kin may be a reasonable rule of thumb for kin discrimination (see Hepper 1986). However, such a rule can be easy to exploit, as when cuckoo chicks usurp the parental care of their host 'parents'. Location-based discrimination may depend on context or timing. Male house mice, for example, often kill unrelated pups, but are less likely to do so if they are encountered in the nests of females with which the male has previously copulated (Huck *et al.* 1982).

Familiarity/association

Prior familiarity, perhaps during a critical period of development (see 6.2.1.5), may also provide a reliable guide to relatedness if previous encounters are likely to have been with relatives. Littermates or members of a territorial family group provide obvious models, and several studies have revealed an important role of previous experience in kin-biased behaviour (see text).

Phenotype matching

Animals sometimes appear to discriminate kinship without any prior experience of the individual concerned, or when all individuals are equally familiar (Waldman 1987). One possibility is that they somehow compare appearance, sound or odour cues with their own, or with a template derived from other relatives. This has variously been referred to as 'comparing phenotypes', 'signature matching', 'phenotype matching' or 'the armpit effect' (e.g. Beecher 1982; Holmes & Sherman 1982; Dawkins 1999).

Recognition alleles

Phenotype matching generally assumes that kinship templates are derived experientially, e.g. by self-inspection or learning the cues of relatives. But animals could in principle inherit genetically encoded templates and discriminate kinship without any prior experience. A 'genetic recognition' mechanism in this form would simply represent a hard-wired version of phenotype matching. However, a second, entirely different, mechanism of 'genetic recognition' has also been suggested. Here, 'recognition alleles' code pleiotropically for (a) a phenotypic marker to betray their presence, (b) an ability to recognise the marker and (c) a helpful response towards cobearers of the marker. Dawkins (1989) envisages a typically memorable scenario in which such an allele caused bearers to sprout a green beard and behave nicely towards other individuals with green beards. The point about a 'green beard' system, however, is that it is a means by which a discrimination allele can favour copies of itself *directly*, without recourse to relatedness between cobearers. While it clearly benefits from kin *selection*, therefore, it has nothing to do with kin *discrimination*.

Are these distinctions helpful?

Although these categories are often presented as alternatives, they are really nothing of the sort. As Waldman (1987) pointed out long ago, the distinction between discrimination based on familiarity, phenotype matching and recognition alleles is not matching of phenotypes. In all cases, relatives are identified by matching their phenotypic traits to those expected of kin. What differs is the extent to which the matching template is hard-wired or modifiable by experience, the specificity of the template, and the degree of stimulus generalisation in the matching process (Barnard 1990). Unfortunately the traditional distinctions continue to be used, often with obfuscating effects on the debate about the significance of kin-biased behaviour (Barnard 1999).

Based on discussion in Waldman (1987) and Barnard (1990).

social tolerance, and demonstrated convincingly that group member recognition offers the more parsimonious explanation (Hurst & Barnard 1992, 1995).

'Recognition alleles'

Apparent recognition between unfamiliar close relatives has sometimes been used to suggest a genetically encoded template for kin recognition, one of the two forms of **'recognition allele'** mechanism in the traditional classification (Box 9.8). As well as their dubious claim to be a separate mechanism (Box 9.8), genetically encoded templates suffer from the same levels of recognition problem as familiarity, phenotype matching or any other potential cause of kin bias. Greenberg's (1979) study of sweat bees in Fig. 9.24(c) is a good example. The striking correlation between the chances of admission to the nest and the relatedness of unfamiliar intruders to guard bees has been widely interpreted in terms of recognition alleles. However, like honey bees, colony members are closely related, so just such a correlation might be expected if relatedness determined the degree of overlap with colony recognition cues. Just like the mice, therefore, guards may be mistaking unfamiliar relatives for colony members (see also Carlin 1989).

'Green beard' recognition   The second kind of 'recognition allele' is the hypothetical, and unlikely, **'green beard'** allele (Dawkins 1976), so-called because it codes pleiotropically for all three components of the discrimination system: a phenotypic marker (e.g. a green beard) to show an individual carries it, an ability to recognise the marker, and an appropriately helpful response towards the bearer. A 'green beard' allele thus codes for direct cobearer discrimination (Box 9.7), achieving directly what kin discrimination achieves indirectly – enhanced transmission of the discrimination allele. 'Green beard' recognition thus has nothing to do with either kin recognition or kin discrimination because genetic similarity at other loci (i.e. relatedness) between cobearers is irrelevant. Nevertheless, is still often, and incorrectly, mentioned in the same breath.

While the idea of 'green beard' alleles provides an instructive thought experiment, is there any evidence that they actually exist? Various candidates have been proposed, such as the selective association of tunicate (*Botryllus schlosseri*) larvae by matching to a single histocompatibility locus (Grosberg & Quinn 1986; Hamilton 1987), but, until recently, none had met all the criteria for a 'green beard' system (although see Haig 1996 on 'green beard'-like linkage effects across loci in mother–foetus relationships). In a study of selective queen-killing in ant colonies, however, Keller & Ross (1998) have provided compelling evidence for one. In multiqueen colonies of the red fire ant (*Solenopsis invicta*), all egg-laying queens are *Bb* heterozygotes at a particular locus known as *Gp-9*. Earlier studies had shown that *bb* homozygotes die prematurely from congenital causes, thus preventing the *b* allele going to fixation, but what about *BB* individuals? Keller & Ross discovered that, if any *BB* queens try to initiate reproduction, they are killed by workers, but specifically those with the *Bb* genotype. Discrimination is mediated by odour cues associated with the *b* allele, which appear to induce workers to kill all reproductives except those bearing the '*b*' odours. Careful experimentation revealed that either *Gp-9* itself, or another gene in complete linkage disequilibrium (i.e. always segregating in association) with it, was responsible for both the salient odour and the aggressive response of the workers. Thus the gene seemed to have all the attributes of a 'green beard' system, coding for a detectable phenotypic feature (an odour), the ability to recognise the feature, and a differential response towards individuals possessing or not possessing it (Keller & Ross 1998).

Do any animals really recognise kin?

We have dwelt on some of the problems surrounding kin recognition because there is a danger of tacitly accepting it happens without ruling out other potential causes of kin bias. Grafen's (1990a) important paper on the issue is often cited, but its cautionary message seldom acted upon. Grafen himself distilled the burgeoning evidence for 'kin recognition' down to a single example that met his stringent criterion for the term: the tunicate *B. schlosseri* again. While Hamilton (1987) viewed matching at the histocompatibility locus by *B. schlosseri* larvae as a possible 'green beard' system (see above), Grafen saw the locus as a marker for kinship. The locus is highly polymorphic and individual alleles rare, so each is likely to be shared only by close relatives. The crux of Grafen's argument is that fusing with other individuals to form a colony risks exploitation, unless both partners invest equally. By choosing a close relative, a 'cooperative' larva increases the chances that its partner will also possess the 'cooperative' allele, thus safeguarding its investment. The histocompatibility marker allows true kin recognition, and thereby kin discrimination in favour of cooperative partners (a case of indirect similarity discrimination [Box 9.7]). Attractive though it seems, this scenario unfortunately remains speculative; it could just as easily be that there is some fitness advantage at the matching locus itself, in which case relatedness between larvae is irrelevant (Barnard 1991). Thus a question mark has to remain over the *B. schlosseri* example.

More recently, however, another promising candidate has emerged. Marion Petrie and coworkers studied mating aggregations (leks, see 10.1.2.2) among peacocks (*Pavo cristatus*) at Whipsnade Park in England. From DNA fingerprinting evidence, Petrie *et al.* (1999) discovered that birds within leks were more closely related to each other than to birds in other leks. This might not be surprising if, say, birds tended to remain with their familiar brood mates into adulthood, or remained in a familiar part of the habitat. A carefully controlled experiment, in which eggs from related and unrelated birds were hatched together in incubators and the chicks subsequently reared in mixed groups, however, suggested there was more to it than this. Despite not having had the opportunity to become familiar with either their relatives or the environment, captive-reared birds released into the park eventually established display sites close to their (full or half) brothers (Fig. 9.25). This extraordinary outcome strongly suggests kin discrimination based on kin recognition, though exactly how birds achieve it remains unclear. Shared habitat preferences seem unlikely on this scale (Petrie *et al.* 1999); perhaps some form of visual self-matching or auditory cue is involved. But why do the birds do it in the first place? Again the answer is not yet clear, though it is possible that females are attracted to larger aggregations, and associating with relatives is a way for males to offset a reduced individual chance of mating by contributing to the success of others in the related group (see Shorey *et al.* 2000).

9.3.2 Cooperation and reciprocity

A second route by which cooperation might evolve relies on the likelihood that helpful acts will be repaid by the recipient, perhaps at a later date (Trivers 1971). As long as the benefit of the act to the recipient is greater than the cost to the actor, both individuals

(a)

(b)

Figure 9.25 Peacocks tend to form leks with close relatives. The tendency remains even when birds have never encountered each other before or had experience of the local environment. See text. Photograph of displaying peacock reproduced with permission of James Osmond (www.jamesosmond.co.uk); photograph of four brothers on their lekking ground courtesy of Marion Petrie.

stand to gain from such **reciprocity**. Reciprocity is a familiar feature of human social behaviour (see Chapter 12), but accounting for its evolution (in humans or anything else) seems to run into an awkward problem: the temptation to cheat. What is to stop a recipient accepting help now but neglecting to return the compliment when called on later? Reciprocal cooperation seems to be inherently unstable from an evolutionary point of view. To find out whether it is, Axelrod & Hamilton (1981) adopted a games theory approach (2.4.4.2), using a model well known to economists and political scientists: the **Prisoner's Dilemma**.

The Prisoner's Dilemma, and its *Tit for tat* solution (Box 9.9), show that repeated interactions between given individuals can lead to reciprocal cooperation, but that the temptation to defect is an ever-present problem in getting cooperation started. The possible early dependence of *Tit for tat* on kinship clustering (Box 9.9) is interesting in the light of the discussion in Box 9.7, since it shows that relatedness may facilitate selective interaction between cobearers of reciprocal cooperation alleles, but not itself have any fitness consequences – clustering of otherwise unrelated cobearers would have exactly the same effect.

9.3.2.1 Evidence for reciprocity

So is there any evidence for reciprocal cooperation, and any that it works along the lines of *Tit for tat*? Studies in both the laboratory and the field suggest there is.

Tit for tat *and predator inspection*

One context in which *Tit for tat*-like reciprocity has been demonstrated experiment-ally is predator-inspection behaviour in fish. Several species of fish show characteristic slow, jerky movements away from companions and towards a potential predator

---

Underlying theory

Box 9.9   Reciprocity and the Prisoner's Dilemma

The Prisoner's Dilemma, first popularised by von Neumann & Morgenstern (1953), is based on a conflict of interests between two hypothetical prisoners being interrogated in separate rooms. Cooperation and defection are defined in terms of loyalty between the two. If both keep quiet and refuse to blame the other for the putative offence (i.e. both cooperate), then each receives a sentence of 1 year in prison, since the police have enough evidence to convict without a confession. If both give the game away and blame the other (both defect), they each go down for 3 years. However, if one blames the other, but his friend remains loyal, the 'squealer' gets away scot free, while his duped friend goes to gaol for 5 years. Thus each prisoner would be better off if both cooperated (1 year in prison) than if both defected (3 years in prison), but there is always a temptation to defect because of the chance of going free. Hence the dilemma.

We can apply the same logic to cooperation and defection in other species, except the rewards and punishments are not freedom or years in prison, but relative reproductive

Box 9.9 *continued*

success. The situation above can be formalised as a fitness payoff matrix, like the *Hawk–Dove* game in Chapter 2:

		Player B	
		Cooperate	Defect
Player A	Cooperate	$R=3$ Reward for mutual cooperation	$S=0$ Sucker's pay-off
	Defect	$T=5$ Temptation to defect	$P=1$ Punishment for mutual defection

The payoffs (with respect to Player A) in the matrix are, of course, arbitrary, but reflect the likelihood that an allele coding for the response in question will be transmitted to the next generation. As in our prisoners scenario, both players do better if they both cooperate than if they both defect, but if one cooperates, it always pays the other to defect. At face value, therefore, defection is the ESS. This conclusion holds under the conditions:

$$T > R > P > S, \text{ and } R > (S + T)/2$$

So, is there any way animals could escape the dilemma and find a way to cooperate and avoid the punishment for mutual defection? Not if they only meet once, or meet a fixed number of times that is known at the outset (because defection will always be the best response on the last encounter and therefore on the one before that and so on back to the first encounter). However, if there is simply some given probability that they will meet again, the picture becomes more interesting, with various combinations of cooperate and defect looking likely. But will any of them be evolutionarily stable against the 'always defect' option?

To find out, Axelrod (1984) took the novel approach of inviting scientists around the world to submit candidate strategies that could be run against each other, and against a randomly cooperating strategy, in a round-robin tournament. Sixty-two strategies entered the game all told, and the probability of players meeting again was set at just over 0.99. Despite a number of very sophisticated strategies being included, the eventual winner was the simplest one of all, a strategy called '*Tit for tat*' that cooperated on its first move, then copied whatever the other player did in response. *Tit for tat* was successful because it reciprocated, 'retaliating' against defection and discouraging it, but 'forgiving' defection if a player subsequently cooperated, thus allowing cooperation a way back in. However, *Tit for tat* was not strictly an ESS, because, while it was stable once established, an alternative strategy, *All defect*, did just as well and imposed the 'sucker's' payoff on cooperators at their first move. So the problem was how to get *Tit for tat* started. Two possibilities are that it could emerge first between relatives, where relatedness was recognised on the basis of reciprocal cooperation (which could then be generalised to non-relatives playing *Tit for tat*), or where *Tit for tat* individuals occurred in clusters, and could interact with one another. Of course, clustering could be associated with kinship, so the two possibilities are not mutually exclusive.

Based on discussion in Dugatkin (1997).

when the latter appears in their vicinity (Dugatkin & Godin 1992; Pitcher 1993). Such behaviour seems bizarre at first sight because fish appear to be putting themselves at risk unnecessarily. If it looks as if there is a predator, surely it is better to keep out of the way. Experimental evidence certainly suggests that predator inspection carries a greater risk of predation (Dugatkin 1992b) and it incurs other costs such as reduced food intake (Magurran & Girling 1986). However, inspection behaviour may bene-fit the fish by enabling it to decide whether a predator is actively hunting. If it is not, other important activities, such as feeding and courting, can be resumed (Magurran & Pitcher 1987). Inspection approaches may also deter the predator from attack (Magurran 1990; see 8.3.5.4).

In three-spined sticklebacks, fish often carry out predator inspection in pairs, possibly because individual risk is diluted and/or the predator is more likely to become confused should it attack (9.1.1). Milinski (1987) has suggested that this may lock the pair in a Prisoner's Dilemma, since circumstantial evidence indicates the crucial inequalities in payoff ($T > R > P > S$ and $R > (S + T)/2$, see Box 9.9) are probably met. Mutual co-operation (both inspect [R]) is likely to yield a greater reward than mutual defection (both retreat [P]) because inspectors are known to transmit useful information about predators and failure to inspect increases the risk of predation for all individuals in a school (Magurran & Pitcher 1987; Godin & Smith 1989). However, there is always a temptation (T) for one fish to hang back and let the other take the lead, so that it can gauge the level of danger without putting itself at risk. Thus T will be greater than R, and R, P and T will all be greater than S because any deserted fish left as the sole inspector will be exposed to the full risk of predation. The condition $R > (S + T)/2$ is also likely to be met, argues Milinski, because a group of two fish should spot an impending attack sooner as well as benefit from any confusion effect. So, do predator inspectors solve the problem using *Tit for tat*? Milinski (1987) designed an ingenious experiment to find out.

Milinski used strategically placed mirrors to simulate 'cooperation' and 'defection' by a fish's own image (Fig. 9.26a). In the 'cooperation' treatment, a mirror ran the length of the aquarium from a refuge at one end to a compartment containing a predator (behind a clear partition) at the other. Thus as the fish moved towards the predator, its image dutifully followed alongside. In the 'defection' treatment, a shorter mirror was set at an angle, so that the reflected image gradually dropped behind as the fish moved forward, then finally disappeared altogether. If inspecting fish played *Tit for tat*, they should have approached closer to the predator in the 'cooperation' treatment than when their apparent companion 'defected'. This turned out to be the case (Fig. 9.26b). But Milinski went further. He divided fish into 'bold' and 'cautious' individuals based on how closely they approached the predator in the 'cooperation' treatment. Since cautious individuals approached more slowly, their image in the 'defection' treatment would appear to stay with them for longer. Thus, if inspecting fish were really playing *Tit for tat*, 'cautious' individuals should be affected much less by the 'defection' treatment than 'bold' individuals. Again, the prediction was borne out. 'Bold' fish 'retaliated' by defecting when their image did so, but resuming inspection once they caught up with their image again (i.e. 'forgiving' it when it 'cooperated' once more). 'Cautious' fish, on the other hand, showed little change. Inspecting pairs thus showed many of the key features of *Tit for tat* reciprocity. While Milinski's interpretation has since been the subject of some debate, the prevailing view seems to be that pre-dator inspection in fish is a good approximation of *Tit for tat* (see Dugatkin 1997 for a good resumé).

Figure 9.26 (a) The reflected image of a stickleback was made to 'cooperate' (top) or 'defect' (bottom) during predator inspection by altering the position of a mirror in the aquarium. (b) Fish with a 'cooperating' image (O) approached closer to a predator than fish whose image 'defected' (●). See text. After Dugatkin (1997) and Milinski (1987). © Nature Publishing Group (http://www.nature.com/), reprinted by permission.

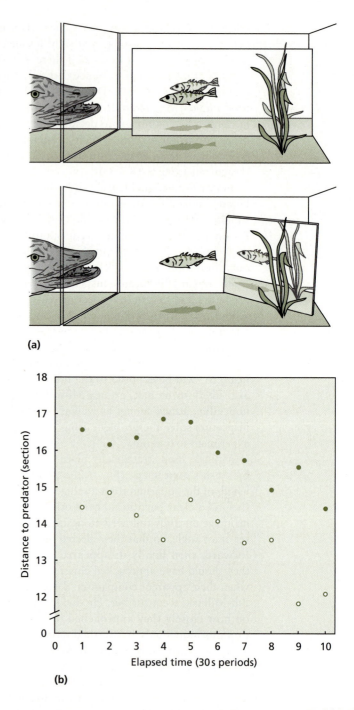

(a)

(b)

Reciprocal feeding relationships

Field studies have also provided evidence for *Tit for tat*-like reciprocity. Vampire bats (*Desmodus rotundus*) (Fig. 9.27a) fly out from their daytime roosts (Fig. 9.27b) at night to gorge on a meal of blood. Sometimes individuals fail in their quest and return to the roost hungry. Remarkably, however, these unlucky bats may not have to remain

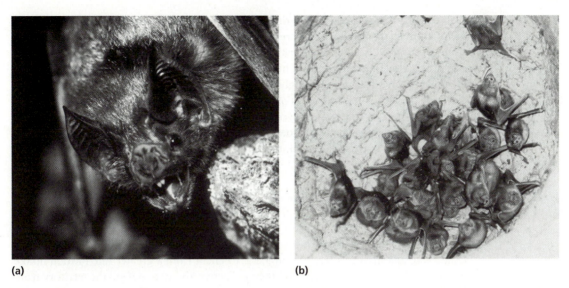

Figure 9.27 (a) Vampire bats (*Desmodus rotundus*) spend the day in communal roosts (b) where they may share blood meals with other, unrelated, individuals. See text. Photographs courtesy of Gerald Wilkinson.

hungry. By studying a marked population of vampire bats in Costa Rica, Gerry Wilkinson (1984) discovered that animals failing to find blood during the night were often able to beg it from their companions during the day. Interestingly, regurgitation occurred only between close relatives or unrelated bats that tended to roost together regularly in the colony, conditions likely to favour reciprocal cooperation. To test this more rigorously, Wilkinson carried some carefully controlled experiments in the laboratory using a mixed colony of bats from two different roosts. By removing individuals from the colony, depriving them of food for a short time, then reintroducing them, Wilkinson was able to monitor their success in soliciting regurgitations. He found that 92% of regurgitations occurred between bats that had come from the same roost in the field and were thus likely to be familiar. Moreover, deprived bats that received food tended to reciprocate later on. Thus all the evidence points to reciprocal cooperation between repeatedly interacting individuals. Does this matter to the bats? Data on starvation times suggests it does. A relatively small degree of food deprivation (resulting in only a 5–10% loss of body weight) is enough to endanger a bat's life (Wilkinson 1984). A night without food is therefore serious, so blood donors back at the roost provide a vital safety net.

Reciprocity and 'raising the stakes'

One limitation of the Prisoner's Dilemma and *Tit for tat* is that they allow only discrete options of cooperation or defection. As Roberts & Sherratt (1998) point out, however, cooperation is rarely all-or-nothing; animals may cooperate or cheat to varying degrees. Under these conditions, a more subtle form of cheating could arise whereby, instead of going for all out defection, individuals simply invest slightly less than their partner. As a result, reciprocal cooperation could be undermined by 'short-changing'. In an important theoretical paper, however, Roberts & Sherratt have shown that cooperation could still thrive under conditions of short-changing if individuals limited their initial investment in cooperating, then increased it as partners reciprocated. Such a strategy of *Raising the stakes* turns out to be capable of invading a population of non-cooperators and resisting subsequent exploitation. Thus, as Roberts & Sherratt put it, 'testing the water' and adjusting investment accordingly greatly enhances the effectiveness and stability of reciprocity over the quantum 'leaps of faith' inherent in *Tit for tat*. Reciprocal grooming between partners in impala (*Aepyceros melampus*) appears to accord with the predictions of *Raising the stakes*, as does the formation of friendships in humans (Roberts & Sherratt 1998, but see 12.2.1.1), though the idea has yet to be tested extensively in other contexts.

9.3.3 Mutualism

Sometimes individuals cooperate simply because each gains something as a result. Thus the pied wagtails sharing a territory in 9.2.2.1 benefit from a greater rate of food intake than they could achieve alone because their collective ability to defend the territory from other competitors outweighs the cost of sharing the food supply. Since there is mutual gain, such relationships may sometimes seem reciprocal, but the resemblance to true reciprocity ends there, because there is no temptation for participants to defect, and any reciprocal benefit is merely an incidental by-product of individual gain. Thus, Brown (1983) refers to *by-product* **mutualism**, and Connor (1986) to *pseudo-reciprocity*. Indeed, various people have argued that mutualism is not really cooperation at all, since cheats would do worse than cooperators. But, since it requires some degree of coordinated action between participants, and high reproductive payoffs cannot be obtained without other cooperators, Dugatkin (1997) concludes that mutualism falls legitimately within the definition of cooperation.

As well as resembling reciprocity, mutualism can also be difficult to distinguish from kin-selected cooperation. For example, male lions form coalitions to take over and defend prides of females. Larger coalitions fare better on both scores, so it would seem to pay males to team up in large cooperative groups. However, as coalition size increases, the disparity in reproductive success across individuals also increases. When a pair of males is resident in a pride, each gains from the association, so coalitions appear to be mutualistic. But in larger coalitions subordinate males gain little or nothing, so why do they join them? Genetic studies have shown that males in coalitions of four or more are generally closely related, so the reduced reproductive success of subordinate males may be compensated by indirect benefits arising from their successful dominant kin (Packer *et al.* 1991). The driving force for cooperation among males therefore appears to shift from mutualism to kin selection as coalitions increase in size.

Mutualistic interactions can be thought of in terms of **biological markets** (e.g. Noë & Hammerstein 1995) in which 'traders' of 'commodities' (benefits that one individual can offer another) compete to obtain the highest return for their investment (reproductive payoff). Just as in human markets, there are conflicts over the exchange value of commodities, as in the threshold payment for feeding on an owner's territory in wagtails (Fig. 9.21), so payoffs are likely to depend on careful choice of partners and the risk of exploitative cheating. The idea of biological markets and the mutual exchange of commodities provides a much more flexible framework for understanding cooperation between unrelated individuals (both within and between species) than classical reciprocity.

9.3.4 Manipulation

While mutualism can comfortably be interpreted in terms of cooperation, there *are* behaviours that look cooperative but, on closer inspection, turn out not to be. A nice example comes from Vehrencamp's (1978) study of groove-billed anis (*Crotophaga sulcirostris*), neotropical members of the cuckoo family.

For many years, anis were regarded as paragons of cooperative behaviour. They live in small groups in which individuals collectively defend a year-round territory. Each group comprises one to four monogamous breeding pairs and the odd unpaired bird that acts as a helper. All individuals contribute to building a communal nest within which all females lay their eggs, and incubation and parental care are shared between members of the group. However, things are not quite as they appear on the surface.

Problems arise because there is a limit to the number of eggs that can be incubated efficiently in the nest. If there are a lot of eggs, some become buried and not turned and warmed sufficiently, and so fail to hatch. Where there are large communal nests, therefore, there is likely to be pressure on females to ensure that it is not their eggs that suffer. One suspicious indication that females may compete in this way is the number of eggs that end up on the ground under active nests. By individually marking birds, Vehrencamp found that females had a linear dominance hierarchy, and that dominant females tended to lay their eggs later than subordinates. Moreover, dominants visited the nest several times prior to laying and often tossed an egg that had already been laid out onto the ground. Once dominants had begun to lay themselves, however, they no longer removed eggs. As a result, late-laying dominant females were less likely to suffer egg removal than earlier laying subordinates, and therefore enjoyed greater hatching success (Fig. 9.28).

As might be expected, subordinate females have begun to adopt some counter-strategies to this **manipulation** by dominants. As group size increased in Vehrencamp's study, early-laying females laid disproportionately more eggs than their companions

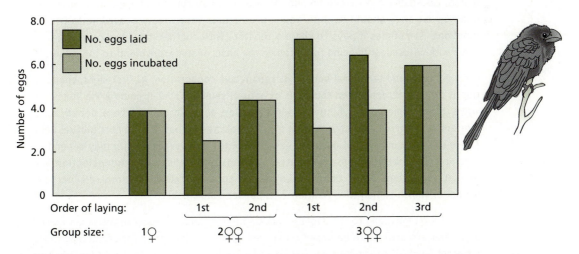

Figure 9.28 As group size in groove-billed anis (*Crotophaga sulcirostris*) increases, females laying eggs earlier tend to lay more eggs than those laying later. However, females that lay later have greater hatching success. These trends can be interpreted in terms of dominant females manipulating reproductive success within communal nests in their own favour, and subordinate females counter-responding. See text. From Emlen (1978) after Vehrencamp (1977).

(Fig. 9.28) and left a longer period between successive eggs. They even sometimes produced a late egg, well after their main clutch, and at a time when dominant females were likely to have started to lay their own eggs. The veneer of cooperation within ani groups thus disguises competitive relationships based on manipulation, and an arms race between exploiters and victims to rig things in their own favour.

Summary

1. Reproductive success is likely to be influenced by the proximity of other individuals. If the effect is positive, selection is likely to favour some degree of social aggregation. If it is negative it is likely to favour territoriality. However, aggregation and territoriality are not mutually exclusive, and both can occur in the same social system.

2. Many factors favour the aggregation of animals into social groups, but the most common seem to be feeding or anti-predator benefits. These can arise in several ways, and their covariation with group size can make it difficult to infer cause and effect. In principle it is possible to predict optimal group sizes from relationships between fitness functions and group size, though this may depend on finding a common currency, and optimal group sizes may be inherently unstable.

3. Reproductive success within groups can depend as much on their social structure as their size. Dominance relationships, mating systems and kinship structure are some of the factors that are likely to be important. Individuals may perform specific roles within groups, particularly in the highly structured societies of eusocial species.

4. Territoriality is often based on food resources and/or the acquisition and defence of mates. Optimality theory can be used to predict when territories should be defended and how big they should be, but, again, predictions vary widely according to fitness functions. Sometimes it pays to share a territory with another individual or as a group, often with co-occupants being tolerated on the basis of some 'payment' to the owner.

5. Both grouping and territorial behaviour can involve cooperation. At first sight, cooperation is difficult to explain on the basis of natural selection, but it could arise in several ways. Most explanations hinge on kin selection or reciprocity, where benefits are indirect or deferred, but cooperation may sometimes be based on direct mutual gain. These explanations are not mutually exclusive. For example, reciprocity may depend on kinship to get started, and cooperative systems may reflect both kin selection and mutualism. In some cases, cooperation may be apparent rather than real, a superficial upshot of selfish exploitative relationships between individuals.

6. Kin selection may lead to the evolution of kin recognition, but many forms of kin-biased behaviour are unlikely to be based on kin recognition in the strict sense. Instead many different levels of recognition may be exploited to discriminate in favour of kin. Kin bias might also arise incidentally, as a result of relatedness correlating with such things as home group membership or imprinted templates for species recognition.

Further reading

Pulliam & Millikan (1982), Pulliam & Caraco (1984), Davies & Houston (1984) and Krause & Ruxton (2002) provide good reviews of the costs and benefits of social aggregation and territoriality, and Hamilton's (1996) collected papers are an excellent starting point for the theoretical background to the evolution of social behaviour. Wilson (1971) is still unsurpassed as an introduction to the social insects and insect sociality, while Andersson (1984a) takes a critical look at the evolutionary significance of haplodiploidy. Dugatkin (1997) provides a readable review of the cooperation literature and the debates therein, and Grafen (1990a) and the ensuing exchanges in *Animal Behaviour* (Volume 41, 1991) are a must-read for a grasp of the arguments surrounding kin recognition. The papers by Noë & Hammerstein (1995) and Roberts & Sherratt (1998) discuss some emerging ideas about mutualism and reciprocity.

10 Mating and parental care

Introduction

From an evolutionary point of view, reproducing and passing on copies of its genes is the most important act of an animal's life. Yet the mode of reproduction adopted by most animals – sex – appears to be genetically very costly. Is sex really costly, and if so, what is the nature of the cost and why is sex so widespread? In most cases, there are two sexes: males and females. Why is this, and what is the fundamental difference between them? Males and females often differ markedly in their appearance and behaviour. Can we understand these differences from the respective reproductive roles of the two sexes? In particular, can we understand why many morphological and behavioural characteristics associated with mating are extreme and apparently costly? Sometimes males and females form stable pair bonds; on other occasions males may mate with several different females or *vice versa*. What determines the number of partners each sex acquires? Finding a mate is only one step on the way to reproducing successfully. In many species, mating is followed by a more or less protracted period of parental care. How do the different reproductive interests of parents and offspring affect the relationship? If there are conflicts of interest between the two parties, who wins?

We have discussed various reasons why animals might benefit from associating with, or distancing themselves from, others. Aggregation and territoriality are options that allow animals to optimise their use of resources, and trade off conflicting demands, such as finding food and avoiding predators. For many organisms, however, there is one compelling reason why they should associate with another individual, even if only fleetingly, and that is to reproduce. Most organisms reproduce sexually, and, even where hermaphrodite (individuals are both male and female), this usually means the behaviour of two otherwise independent individuals must be orchestrated in order to mate. Social contact may continue after mating and extend to resulting offspring if there is a need to protect and provision young until they are capable of fending for themselves. From an evolutionary point of view, reproducing and so passing on copies of its genes is the most important act of an animal's life. Surviving to reproduce, and ensuring viable, reproductive offspring enter the next generation, is the ultimate driving force behind the organism's morphological, physiological and behavioural design. The fact that most choose to do it sexually is thus extremely surprising. Not only does finding and persuading another individual to mate introduce uncertainty into the process, but, at first sight, sex seems actually to *reduce* an individual's reproductive capacity relative to an asexual competitor (Box 10.1).

This **'fertility' cost** of sex (Maynard Smith 1978; Box 10.1) arises because a sexual female divides her offspring into (usually) two types: males and females, neither of which is capable of reproducing independently of the other. All other things being equal, therefore, an asexual female will have twice the fecundity of a sexual rival. But all other things are

Difficulties and debates

Box 10.1 The costs and benefits of sex

The widespread existence of sexual reproduction is often regarded as one of the great paradoxes of evolution because, at first sight, it seems to impose a two-fold genetic disadvantage on its practitioners. Instead of reproducing independently, and faithfully (bar chance mutations) transmitting her evidently successful genotype to the next generation, a sexual female depends on mating with a male, and, worse, halves her genetic contribution via meiosis and shackles her offspring with the unknown quantity of the male's genes. In contrast, an asexual female avoids the vicissitudes of mating and effectively photocopies herself into the future. Yet, somehow, sex seems to prevail. Why? Before answering this, we first need to be clear what sex and its evolutionary costs really are.

What is sex?

The somewhat surprising answer is that it is not reproduction. Sex is a device for mixing genes. It does this in two ways: through **recombination** (the exchange of genetic material within individuals, usually between homologous chromosomes during meiosis) and **outcrossing** (the combination of genetic material from different individuals). In most organisms, genetic mixing occurs as part of reproduction, but the two processes can be separate, as in some bacteria.

What is the cost of sex?

The two-fold cost alluded to above comes in two forms, one generally accepted, the other a still frequently promulgated fallacy.

The 'cost of meiosis' fallacy

One version of the cost of sex has it that a sexual female throws away half her genotype each time she reproduces, thus imposing a 50% cost of meiosis which has to be overcome by whatever advantages sex confers. As Treisman & Dawkins (1976) long ago pointed out, this is one of several fallacies that arise from thinking in terms of individual selection (see 2.4.4.5). The preservation of a female's individual genotype matters not a jot from an evolutionary perspective; what matters is the probability that an allele for a given trait will find its way into future generations. As Treisman & Dawkins show, whether considering an allele for sex (versus asexual reproduction), or an allele for any other phenotypic character, meiosis *per se* does not impose a 50% cost on the probability of transmission (see also Maynard Smith 1984).

The 'fertility cost' of sex

The true cost of sex lies in reducing the capacity of females to produce offspring (Maynard Smith 1978). By dividing her offspring into males and females, which cannot reproduce independently, a sexual female effectively halves her fertility. The problem is particularly acute in anisogamous (males and females produce different-sized gametes) species, where males appear to invest vastly less in the offspring, both at the gamete stage (sperm are much smaller than eggs, but see 10.2) and, often, during any period of parental care. Isogamous species, where males and females produce gametes of the same size, avoid the cost of differential investment in gametes, but still pay a time and metabolic (but not

Box 10.1 *continued*

genetic [see above]) cost of undergoing meoisis (Maynard Smith 1984). The task is thus to find advantages to sex that outweigh the apparently serious handicap of anisogamy.

The advantages of sex

The proposed advantages of sex are many, but they divide into two broad camps: those to do with overcoming deleterious mutations, and those conferring an advantage in terms of evolutionary rate.

Mutational load

The genes in an organism's DNA are not inviolate. Many factors, inherent and environmental, can cause changes, most of them deleterious. By providing new copies of genes from other individuals, sex may facilitate the correction of any damage or copying errors that arise (the **DNA repair** hypothesis). However, while this may indeed be of some advantage, it is more an argument for polyploidy (see 2.3.2.3) than sex (Maynard Smith & Szathmáry 1999).

An entirely different argument hinges on the potential for asexual lineages to accumulate deleterious mutations, a process known as **Muller's ratchet** (Muller 1964). Should a damaging mutation arise, the only ways an asexual line could lose it are by (highly unlikely) back mutation or by a mutation arising elsewhere that masks its effects. Deleterious mutations would thus steadily build up while mutation-free individuals dwindled to extinction. Sex, on the other hand, could facilitate new combinations of genes through recombination and outcrossing, and increase the chance of overcoming unwelcome mutations through epistasis (e.g. dominance effects; see 2.3.1.1).

Evolutionary rate advantages

New combinations of genes mean more opportunity to respond to selection pressures, especially those imposed by the coevolving biotic environment, such as predators, competitors and parasites. The ever-changing nature of the biotic environment, and the coevolutionary arms races (see Box 2.6) it induces, put a premium on genetic diversity among progeny, since this increases the likelihood of an appropriate response to novel pressures in each generation (see the Red Queen effect in 2.4.4.3). In selfish gene terms, an allele for sex can thus be thought of as 'hitch-hiking' on the back of the beneficial consequences of recombination and outcrossing at other loci. Dunbrack *et al.* (1995) provide an experimental demonstration of the effect in flour beetles (*Tribolium castaneum*), and Rice (2002) reviews the experimental evidence for it more widely. There is also circumstantial evidence for sex as a response to environmental unpredictability, for instance in the timing of sexual reproduction in the life cycles of some aphids (Williams 1975).

By focusing mating preferences on fitness characteristics of potential partners (see 10.1.3.2), sexual selection is likely to add greatly to the efficacy of the process (e.g. Trivers 1976), and much attention has centred on parasites, with their relatively short generation times and rapid rates of evolution, as key agents in the evolution of both sex (Hamilton 1980; Lively 1992; Camacho *et al.* 2002) and mating preferences among their hosts (Hamilton & Zuk 1982; 10.1.3.2). Maynard Smith (1984) and Maynard Smith & Szathmáry (1999) provide good discussions and review some of the evidence relating to the evolutionary rate advantage of sex.

Based partly on discussion in Maynard Smith (1984) and Maynard Smith & Szathmáry (1999).

often not equal. For instance, the potential two-fold cost can be offset if a male helps a sexual female to raise more offspring. 'Help' may come in many forms, from paternal care to the female mating selectively with males that pass on an ability to survive and attract lots of partners. If this results in the female producing twice as many successfully reproducing offspring as she could unaided, the fertility cost of sex (or the **cost of 'male laziness'**, as it is sometimes called) would effectively disappear. The question is: *does* the male's contribution overcome the cost? Numerous suggestions have been made as to how it might, and much of this chapter will be concerned with them. However, our focus here is not so much on the evolutionary advantages of sex *per se*, but on the behavioural mechanisms underlying sex and the selection pressures that have shaped them.

10.1 The evolution of sexual behaviour

To understand why and how sexual behaviour has evolved, we need to understand how the sexes differ. This is not as straightforward as it may seem. Certainly we have little difficulty identifying males and females among most mammals. Males have a penis, are often larger and may possess distinctive secondary sexual characters such as deep voices, antlers or manes. Go further down the phylogenetic scale, however, or look at any number of plant species, and the distinction is often much less obvious. Short of dissection or molecular genetics, there may be no way of telling the sexes apart until they mate. Moreover, among hermaphrodites, the sexes are combined in the same individual, so the question may not arise in the first place.

Clearly, then, external appearance is an unreliable guide. So what *is* the basic distinction between males and females? From the point of view of divergence in morphology and behaviour, much of the answer lies in their gametes. Generally, females produce relatively few, large, well-provisioned gametes called **eggs**, whereas males produce tiny, more or less unprovisioned, gametes called **sperm**, though they may produce them in prodigious numbers. This has frequently prompted the misleading metaphor of the male as a 'parasite' on the female's investment. The metaphor is misleading because, while individual sperm may be cheap, their number, and the paraphernalia frequently involved in transferring them to the female may not be (e.g. Gwynne 1981). Furthermore, males are often involved in time-consuming and damaging contests with each other. Thus males pay considerable reproductive costs too, and do not simply freeload on investment by females (Parker 1984).

The difference in gamete size between males and females is known as **anisogamy** and is the fundamental driving force behind the evolution of many, though not all (see below) of the differences between the sexes. But why are there *two* sexes and why are they generally anisogamous?

10.1.1 Why two sexes and why anisogamy?

10.1.1.1 The evolution of two sexes

For most people, sex means copulation between individuals of two different sexes. But why only two sexes? Why not 3, or 26? The short answer is that there are not always just two sexes – some protists, such as *Paramecium* and *Chlamydomonas*, have multiple mating types ($\equiv$ sexes) – but the two-sex condition is vastly predominant. Various reasons have been suggested for this, but the front-runner hinges on the fact that, where present, important cytoplasmic factors such as mitochondria and chloroplasts (cell organelles that power cellular metabolism) tend to be inherited from only one mating type (Maynard Smith & Szathmáry 1999). In the green alga *Chlamydomonas*, for example, mitochondria are inherited only from the – mating type and chloroplasts only from the + mating type. In animals, mitochondria are inherited from the mother.

An acid test of this idea comes from the ciliate Protozoa, single-celled organisms, such as *Paramecium* and *Stentor*, that bear hair-like processes called cilia (Hurst & Hamilton 1992). In most ciliates there is no fusion between gametes. Instead two cells establish contact and '**conjugate**', a process involving the mutual exchange of haploid nuclei but no mixing of cytoplasm. The cells then separate as diploid individuals, possessing chromosomes from both conjugating partners but only their own mitochondria. In keeping with the uniparental inheritance argument above, these organisms show multiple mating types instead of the more usual two. However, one group of ciliates, known as hypotrichs, show both conjugation and a second form of sexual reproduction that does involve gamete fusion. Tellingly, multiple mating types exist where hypotrichs conjugate, but where gamete fusion occurs, there are only two (Hurst & Hamilton 1992).

10.1.1.2 The evolution of anisogamy

So the widespread existence of two sexes looks like being a consequence of inheriting cytoplasmic organelles. But what about anisogamy? Why do sexually reproducing organisms produce different-sized gametes instead of going for the apparently less costly (Box 10.1) option of isogamy? Again, some organisms *do* reproduce isogamously, but whether or not this happens depends, at least partly, on size at adulthood and the need to provision developing offspring (Maynard Smith & Szathmáry 1999; Bulmer & Parker 2002). As so often with questions about evolution, evidence for this comes from comparative studies of closely related species. *Volvox*, a genus of green algae related to *Chlamydomonas*, forms multicellular colonies in the form of a hollow ball of cells. Species that form small colonies produce motile gametes that are all the same size. Species forming colonies of intermediate size also produce motile gametes, but this time of varying size. Species forming large colonies, however, produce two kinds of gamete: one small and motile, the other large and non-motile – effectively sperm and eggs (Knowlton 1974). Similar, but less clear-cut associations with size have been found in other groups of algae (Bell 1982).

Geoff Parker and coworkers have examined how the need to provision the developing offspring might lead to anisogamy from a somewhat different perspective. Their model suggests that anisogamy is an inevitable consequence of natural size variation in primitive isogametes shed into a fluid medium (Parker *et al.* 1972; Bulmer & Parker 2002). Like any biological entity, ancestral isogametic cells are likely to have varied slightly in size, some being a little larger and some a little smaller than average. Parker *et al.* suggest this variation is unlikely to have been selectively neutral. An immediate advantage would accrue to slightly larger gametes if their greater size translated into better provisioning for the developing zygote (offspring). If it did, we might expect selection to favour larger and larger gametes. But there is a catch. The additional provisioning of larger-than-average gametes can be exploited by smaller-than-average gametes. While large gametes would do better by fusing with other large gametes, they may have little choice in the matter. Their greater motility means that small gametes can encounter their larger

Figure 10.1 Size variation in ancestral gametes is likely to lead to disruptive selection because virtually all fusions will involve small gametes, owing to their greater number and motility. If resulting zygotes need substantial provisioning, only those fusions involving larger gametes will be viable. Thus small and large gamete producers will persist, while intermediate-sized gamete producers will be selected out, thus leading to anisogamy. See text. After Parker (1984).

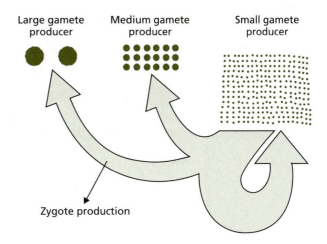

Large gamete producer

Medium gamete producer

Small gamete producer

Zygote production

counterparts faster than the latter can encounter each other. This exploitative relationship triggers disruptive selection for smaller and smaller (i.e. more motile) gametes on the one hand, and large gametes whose size approximates the optimum for the developing offspring (to compensate for the lack of provisioning by the small gametes) on the other (Fig. 10.1). Intermediate sizes are penalised because they have neither the motility of the small extreme nor the provisioning of the large. The net result is the evolution of two, highly dimorphic, kinds of gamete, and thus anisogamous sex.

10.1.2 Anisogamy and parental investment

Whether or not anisogamy is the evolutionarily stable outcome of size variation among primitive isogametes, it has had profound implications for the subsequent evolution of the two sexes. At their root are the consequences of gamete dimorphism for egg and sperm production and thus mating opportunity for males and females. Because eggs are bigger, females produce relatively few of them during their lifetime. They also frequently invest more than males in other aspects of parental care (see below). As far as females are concerned, each fertilisation therefore represents a major commitment, not to be undertaken lightly. Males, on the other hand, generally produce small gametes in much larger numbers and are potentially capable of fertilising eggs at a greater rate than they become available (the emphasis on relative reproductive rate, rather than gametic size *per se*, is important because it is relative reproductive rate that determines levels of mating competition; see Sutherland 1985; Clutton-Brock & Vincent 1991). Thus males can increase their reproductive success by fertilising more females, while females can do so only by investing more in eggs or offspring (Fig. 10.2). This can be rephrased as males putting most of their reproductive investment into **mating effort**, while females put most of theirs into **parental effort**. Parental effort amounts to the total **parental investment** made by an individual over its lifetime, where parental investment is the cost to a parent of rearing a given offspring in terms of its reduced ability to rear offspring in the future (Trivers 1972). The upshot of this disparity between the sexes is that females are effectively a limiting resource for which males have to compete (Trivers 1972), and there is a fundamental conflict of interest between males and females in optimal mating strategy (see also 5.3.1). Care is needed here, however, because, as we shall see in 10.1.3, it is

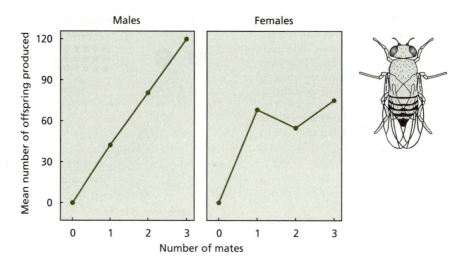

Figure 10.2 Bateman's (1948) classic experiment with fruit flies (*Drosophila melanogaster*) showing that males, but not females, can increase their reproductive output by mating with more individuals. Equal numbers of the two sexes were put in bottles and the offspring produced by each individual were scored using genetic markers. From Trivers (1985) after Bateman (1948). © Nature Publishing Group (http://www.nature.com/), reprinted by permission.

only *some* males (those that compete successfully) that can capitalise on multiple mating opportunities (Wade & Shuster 2002). Generalising the argument to the potential reproductive success of *all* males violates the constraint that males and females overall must necessarily have the same reproductive output (every offspring sired requires one male and one female) (Wade & Shuster 2002; Kokko & Jennions 2003). Bearing this caveat in mind, however, much follows from the relative investment argument, as we shall see shortly.

10.1.2.1 Sex ratios

One aspect of parental investment that has attracted a good deal of attention is the **sex ratio** of offspring. Since males can potentially fertilise so many females, an obvious question is why do there tend to be so many of them? The ratio of males to females in most populations is around 1 : 1. Surely it would be more efficient to have, say, one male for every 20 or 30 females. The argument has a certain logic to it, but only if we consider the reproductive success of the population as a whole. As we saw in Chapter 2, however, selection is unlikely to act at the level of the population. Instead it acts at the level of alternative alleles, in this case coding for females that produce different sex ratios among their offspring. Once again, the question is one of evolutionary stability. It was R.A. Fisher who (while not using ESS terminology) first pointed out that the evolutionarily stable sex ratio was likely to be 1 : 1 (Fisher 1930). The argument is straightforward and depends on the familiar principle of frequency-dependent selection (Box 10.2).

Adaptive adjustment of investment in the two sexes

While Box 10.2 shows how selection will tend to maintain equal parental investment in the sexes, there are circumstances in which it would seem to make sense to adjust investment in favour of one sex or the other. Environments vary, but Fisher's equal investment theory assumes that males and females will benefit or suffer similarly in

Underlying theory

Box 10.2 Evolutionarily stable sex ratios

We can understand why many populations have a sex ratio of unity by invoking a simple ESS argument first put forward by Fisher (1930).

Suppose a female exploited the reproductive capacity of males by producing only daughters. Since her daughters would have no trouble mating, an allele for daughters only would spread through the population, and the sex ratio would become biased towards females. However, when the ratio of females to males exceeded equality, there would be an incentive for females to produce sons, since a son could capitalise on the glut of females by fertilising many of them. Thus the sex ratio would begin to drift back again. While the number of males and females might oscillate backwards and forwards like this for a time, the eventual result would be an evolutionarily stable sex ratio of unity.

Fisher's argument holds as long as the two sexes cost about the same to produce. If one sex is more costly than the other, however, the evolutionarily stable sex ratio in terms of individual males and females may shift to maintain a 1 : 1 ratio of *investment*. Say a male costs twice as much to produce as a female, because he grows bigger and needs twice as much food to attain adult size. If a female produced an equal number of sons and daughters, both would produce on average the same number of offspring if the sex ratio was 1 : 1. But the son's offspring would have been produced at twice the cost of those of the daughters – a bad investment for the mother. Parents would thus do better to bias their progeny towards females so that the population sex ratio will swing towards females and increase the reproductive potential of males. By the same argument as earlier, therefore, the ratio will oscillate, but eventually settle on a stable ratio of two females per male – sons cost twice as much to produce, but now they generate twice the reproductive output in return. Good supporting evidence for this comes from social wasps of the genus *Polistes*, in which species differ in the degree of sexual dimorphism. Where females are smaller than males, the sex ratio is biased in their favour. Where the sexes are the same size, there is no bias. In both cases, though, the ratio of investment in the two sexes is 1 : 1 (Metcalfe 1980).

consequence (Clutton-Brock & Godfray 1991). However, it is easy to imagine they might not, and that selection would favour investing more in the sex that benefits most (so-called **environmental** [or **conditional**] **sex determination** [Charnov 1979; Bull 1981]). Trivers & Willard (1973) were the first to voice this idea formally, using the example of maternal condition as a factor biasing investment towards high resource-demanding males or less demanding females (see Box 10.3 and Fig. 10.3). Any number of environmental factors could exert such a pressure, and some examples are summarised in Box 10.3. However, there has been considerable debate as to the likely generality of Trivers & Willard's predictions (Charnov 1982; West & Sheldon 2002), in part because, in many groups, including vertebrates, sex determination is controlled by independently segregating chromosomes (such as the XX/XY system in mammals), which would seem to fix the sex ratio at 1 : 1 from the outset (Williams 1979). Manipulation of the sex ratio under these conditions would have to rely on some form of intervention, such as selective abortion, which is likely to be costly (Clutton-Brock & Godfray 1991). In contrast, sex ratios would seem to be relatively easy to manipulate in groups such as the haplodiploid Hymenoptera, where sex is determined by whether a female chooses to lay a fertilised or unfertilised egg.

Underlying theory

Box 10.3 Adaptive sex ratio manipulation

Various environmental selection pressures are likely to favour adjustment of parental investment in the two sexes away from the 1 : 1 ratio initially predicted by Fisher (1930) (Box 10.2). The following are some examples.

Maternal condition

In many species, competition between males for mating opportunities involves physical contest between rivals in which it is an advantage to be big. Size is likely to depend partly on how well a male was nourished during its development, so parental investment will play a crucial role. This means a male's competitive ability may well depend on that of his parents, since this will determine their access to resources and capacity to invest. We might therefore expect more competitive parents to bias investment towards males and reap the reproductive benefits of competitive offspring that will fertilise many females. This is exactly what seems to happen in red deer. Over a 20-year study, Clutton-Brock *et al.* (1986) found that high-ranking hinds consistently biased their calves towards males, while subordinate hinds produced an excess of daughters (see Fig. 10.3), perhaps because males are more likely to die *in utero* under poor conditions. Interestingly, Cameron *et al.* (1999) have demonstrated a similar relationship between maternal condition and sex ratios in wild horses (*Equus equus*) in which there appears to be no sexual dimorphism that might lead to a greater likelihood of one sex dying *in utero*.

Mate quality

In sexually dimorphic species, one sex, usually males, often bears traits (such as bright plumage) that the other finds attractive (see 10.1.3). As long as attractiveness is heritable, a female mating with an attractive male might do well to bias her offspring towards males. Evidence to support this comes from collared flycatchers (*Ficedula albicollis*) (Ellegren *et al.* 1996) and blue tits (Sheldon *et al.* 1999) in which the attractive traits are a white forehead patch and the ultraviolet reflectance of the male's plumage respectively. In both species, females mating with more attractive males produced more male offspring. However, such relationships have not emerged in some other species (Grindstaff *et al.* 2001). In a notable study, Burley (1986) found that the attractiveness of captive zebra finches could be manipulated using coloured leg rings. Males with red rings, and females with black, were especially attractive. Astonishingly, but in keeping with adaptive sex ratio theory, females produced male-biased broods if they had mated with a male bearing attractive combinations of rings. They also increase the amount of testosterone in their eggs (Gil *et al.* 1999). This latter point is important because it suggests females may influence the quality of offspring sired by different males, so potentially confounding qualities inherited from the males themselves (10.1.3.2).

Local mate competition

Where competition for mates takes place between relatives, Fisher's theory predicts a shift in ratio away from the competing sex. Thus, if two brothers compete for the same female and there is only one chance to mate, only one of them can be successful. From their mother's point of view, the other is a wasted investment. When sons compete for matings, therefore, their value to the mother is reduced and she should produce more daughters. An extreme example of where such **local mate competition** has produced a highly biased sex ratio comes from the extraordinary viviparous mite *Acarophenox*, in which males mate with their sisters while still inside the mother and die before they are ever born. In keeping with the local mate competition hypothesis, female *Acarophenox* produce extremely female-biased broods of 1 son and up to 20 daughters (Hamilton 1967).

Local resource competition

In species where one sex remains in the natal area (is **philopatric**) and the other disperses, competition for limited resources (**local resource competition**) may favour investment in the dispersing sex (Clutton-Brock & Iason 1986; Emlen 1997b). Clark (1978) found just such an effect in the prosimian *Galago crassicaudatus*. Here, as in many other mammals, males are the dispersing sex, while females remain behind and compete with their mothers, and each other, for limited sources of food. Field and museum studies revealed a widespread male-biased sex ratio among *G. crassicaudatus* subspecies (Clark 1978).

Reproductive helpers

In contrast to local resource competition, individuals staying behind in some species may enhance, rather than reduce, parental reproductive success by helping breeders rear more offspring (see 9.3.1.2). Under these conditions, it may pay females to rear proportionally more of the helper sex. A good example comes from the Seychelles warbler (*Acrocephalus sechellensis*), in which it is females that sometimes remain to help in their parents' territory. Komdeur *et al.* (1997) found that whether the warblers invested more in helper females or dispersing males depended on environmental conditions. On high-quality territories (those with more insect food), one or two helpers increased reproductive success, but on low-quality territories they reduced it. More than two helpers on good territories also reduced success. Komdeur *et al.* manipulated territory quality by moving some birds to better quality habitat on a different island. They found that pairs that previously had either no helpers, or just one, switched from producing almost only males to producing almost only females. Pairs that previously had two helpers, however, switched to producing mainly males. Sex ratios could also be manipulated in the expected direction by removing helpers from territories.

Population sex ratios

If the sex ratio in the population as a whole departs from 1 : 1, selection may favour a bias towards the rarer sex. In the wasp *Polistes metricus*, for example, Metcalfe (1980) found that some nests produced only male offspring as a result of workers laying unfertilised eggs following the death of the queen. The remaining nests in the population, however, turned out to have highly female-biased sex ratios.

Despite appearing to impose a constraint, recent analyses suggest that chromosomal sex determination has not been quite the problem for sex ratio manipulation initially envisaged (West & Sheldon 2002). However, there are difficulties in testing the theory rigorously, at least in vertebrates. These stem from the need for detailed life history data, relating in particular to the reproductive costs of sex ratio manipulation (e.g. selective

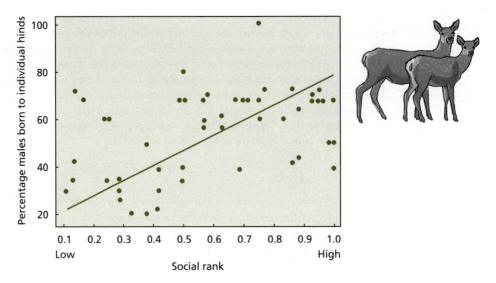

Figure 10.3 The proportion of male calves born to red deer hinds of different social status over their lifetime. Status was determined from observed threats and displacements between individuals. High-ranking (dominant) hinds were more likely to produce sons. See text and Box 10.3. From Clutton-Brock *et al.* (1986).

abortion) and differences in the cost of rearing sons and daughters (Clutton-Brock & Godfray 1991; Lummaa & Clutton-Brock 2002). For example, in red deer, hinds that produce a male calf in one year, tend to calve later the following year and are almost twice as likely not to calve at all (Clutton-Brock *et al.* 1981). The small litter sizes and relative longevity of many vertebrates make such costs tricky to measure. For these reasons, studies of vertebrates have tended to focus on effects of mate quality and reproductive helper systems (Box 10.3).

Conflict over the sex ratio　A further problem is that individuals may disagree as to the optimal sex ratio. A classic example comes from the haplodiploid social insects. As we have seen already (Box 2.7), there is an asymmetry here in the relatedness of (diploid) daughters to their diploid mother ($r = 0.5$), haploid father ($r = 1$) and sisters ($r = 0.75$), assuming there is one colony queen that mates only once. This has interesting consequences for the expected sex ratio in the colony. For a queen, the relatedness ratio of daughters : sons is 0.5 : 0.5, i.e. 1 : 1, so, all other things being equal, her stable investment ratio under Fisher is the same as for any other diploid organism. For worker females, however, the relatedness ratio of sisters : brothers is 0.75 : 0.25, or 3 : 1. Under these conditions, therefore, queens and workers will disagree as to the optimal sex ratio in the colony, queens favouring 1 : 1, workers a female bias of 3 : 1 (Trivers & Hare 1976). So, how should the conflict be resolved?

Trivers & Hare (1976) have argued that the outcome should usually favour the workers. They have the whip hand in the conflict because they outnumber the queen and rear the larvae, so are in a position to manipulate broods in their own interest. Where there is one queen per colony, therefore, the sex ratio should be around 3 : 1. In multi-queen colonies, however, additional queens dilute the workers' relatedness asymmetry because fewer worker females are full sisters. Here, the sex ratio should be closer to 1 : 1. The ratio should also be closer to 1 : 1 in single queen colonies that

Figure 10.4 The ratio of investment in male and female offspring in 21 species of ants. Investment is measured as mass (to take account of varying size differences between the sexes) plotted against number. The lower line represents equal investment in the two sexes, while the upper line represents a 3 : 1 bias in favour of females. Note the inverted scale on the *y*-axis. Reprinted with permission after 'Haplodiploidy and the evolution of the social insects', *Science*, Vol. 191, pp. 249–63 (Trivers, R.L. and Hare, H. 1976). Copyright 1976 American Association for the Advancement of Science.

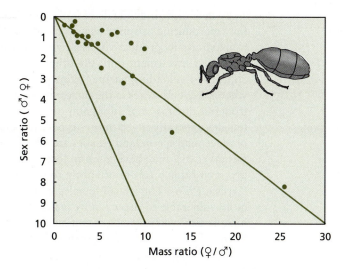

are slave-makers, as in some species of ant. Slave-maker workers kidnap pupae from the nests of other species and tend them until they mature, whereupon the slaves take over responsibility for rearing the slave-makers' brood. Thus, the workers of the home colony have effectively given up their potential to control sex allocation within the colony, and the ratio should default to that of the queen.

An analysis of 21 species of ant that appear to meet the criteria of one queen and one mating, and in which workers should achieve control, showed that the ratio of investment in males and females was indeed closer to the expected 3 : 1 figure (Trivers & Hare 1976; Fig. 10.4). Later studies have both confirmed these results for single-queen colonies and shown that investment ratios in multi-queen and slave-maker colonies approximate 1 : 1 (e.g. Pamilo 1990; Bourke & Franks 1995). Nevertheless, ratios show considerable scatter about the prediction, most probably because the assumptions of Trivers & Hare are not always fully met. In normally single-queen species, for instance, there may be a degree of multiple mating by queens, some colonies with more than one queen and occasions when workers themselves reproduce (Bourke 1997; Mehdiabadi *et al.* 2003). There have also been counter-arguments to Trivers & Hare. Alexander & Sherman (1977), for example, suggested that female-biased ratios among single-queen species could be a result of local mate competition (see Box 10.3) rather than worker control, since it would appear that related males compete for matings. However, in most cases, ants mate in large swarms comprising individuals from several colonies, so the conditions for local mate competition seem unlikely to arise. Furthermore, without knowing its exact extent it is also difficult to predict how local mate competition should affect the sex ratio. For the moment, therefore, Trivers & Hare's hypothesis is tentatively accepted as the best explanation for sex ratio variation in social insect colonies.

10.1.2.2 Mating systems

Sex ratios are one important reflection of parental decision-making, and, as we have just seen, there can be scope for disagreement as to the relative investment in the two sexes. In fact, disagreement between interested parties is a fundamental feature of parental investment, as Trivers (1972) was the first to recognise.

By definition, individuals of both sexes will be selected to produce as many successfully reproducing offspring as possible. At the same time, however, each will be selected to maximise its productivity relative to investment. One way to do this is to shift the burden of investment onto the other parent. As long as the duped parent is able to cope, the freeloading partner gains progeny at a bargain rate. But who should dupe whom? Are both sexes in the same position to exploit one another? As we might suspect from our discussion of anisogamy above, the answer in many cases will be no. Because females generally invest more in parental effort overall (eggs are more expensive to produce than sperm, and females are often committed to gestational and other developmental costs), they have more to lose if an offspring dies, or, more correctly, more to invest to bring a future offspring to the same stage of development (see Dawkins & Carlisle 1976). This places females in a 'cruel bind' (Trivers 1972), and renders them more vulnerable to desertion by males than *vice versa*. Exceptions that prove the rule include fish, where gametes can be shed into the water, thus obviating the need for internal fertilisation with its potential to burden the female by default. Here, the partner that sheds gametes first – usually the female – has the greater opportunity to desert, thus, perhaps, accounting for the widespread incidence of paternal care among fish (Dawkins & Carlisle 1976).

While it is possible to make general predictions about sex differences in vulnerability to desertion, however, it is not difficult to imagine that the decision to desert might depend on the circumstances of each sex at the time. In penduline tits (*Remiz pendulinus*) and Florida snail kites (*Rostrhamus sociabilis*), for example, which sex deserts seems to depend on whether the opportunity for further mating at the time is greater for males or for females (Beissinger & Snyder 1987; Persson & Öhström 1989). Maynard Smith (1977) has explored the problem more generally using games theory, and suggests there may be a number of evolutionarily stable combinations of caring and desertion among males and females (Box 10.4).

Asymmetries in parental investment between the sexes form the basis for understanding the variety of mating systems (see also 9.1.2.2) in different species because, in conjunction with ecological factors, they determine the pattern of **dispersion** in males and females. Since a few matings, indeed sometimes only one, are all a female may need to fertilise all her eggs, reproductive success among females is limited not so much by access to males as access to resources (Davies 1991). Dispersion in females is thus influenced largely by the distribution of food, nesting sites and other commodities necessary for reproduction, while that in males is influenced mainly by the distribution of mating opportunities, and so females. This is illustrated nicely in some sequentially hermaphroditic (changing from one sex to the other) fish, where individuals switch from female to male only when they become large enough to defend a number of females (Hoffman 1983). Where males provide paternal care, as in many fish and birds, however, they may themselves become an important resource for females, with females competing for access to males in a reversal of the more usual sex roles. Depending on the pattern of dispersion of the sexes, and the requirement for paternal care, mating systems can thus change between **monogamy** (pair bonds involving one individual of each sex) and various forms of **polygamy**, i.e. **polygyny** (pair bonds involving more than one female per male), **polyandry** (pair bonds involving more than one male per female), **polygynandry** (pair bonds involving more than one individual of each sex) and **promiscuity** (multiple mating without the formation of pair bonds) (Emlen & Oring 1977; Clutton-Brock 1989; Davies 1991). The relative roles of dispersion and paternal care in this respect are broadly exemplified by mammals and birds (Fig. 10.5a,b).

Underlying theory

Box 10.4 An ESS model of desertion

While ecological factors and physiological constraints (e.g. internal fertilisation, lactation) play an important role in determining relative parental investment by the two sexes, the optimal decision for one sex at any particular time is likely to depend on what the other chooses to do. For instance, if the female looks as if she will stay to look after the young, it may pay the male to desert and copulate with another female. If it looks unlikely that she will stay, the male may do better by caring for the young himself. Maynard Smith (1977) developed a games theory model to explore the consequences of each sex's strategy depending on that of the other. The model works as follows.

The probabilities that offspring will survive if they are (a) given no parental care, (b) cared for by one parent, and (c) cared for by two parents are respectively P_0, P_1 and P_2, where $P_0 < P_1 < P_2$. A male that deserts has a probability p of mating again, while a female that deserts lays W eggs compared with w (where $W > w$) if she invests care. The payoff matrix for the game is shown in the table below, and yields four possible ESSs:

1. *Both sexes desert.* For this to hold, WP_0 must be greater than wP_1, or the female should care, and $P_0(1 + p)$ must be greater than P_1, or the male should care.

2. *Female deserts, male cares.* The conditions for this are $WP_1 > wP_2$, or the female should care, and $P_1 > P_0(1 + p)$, or the male should desert.

3. *Female cares, male deserts.* Here, $wP_1 > WP_0$, or the female should desert, and $P_1(1 + p) > P_2$, or the male should care.

4. *Female cares, male cares.* Here, $wP_2 > WP_1$, or the female should desert, and $P_2 > P_1(1 + p)$, or the male should desert.

1 and 4, and 2 and 3 can be viewed as two pairs of alternatives, with different conditions determining which strategy emerges over the other. For instance, 2 is favoured where females can lay many more eggs if she does not care (i.e. $W \gg w$), one parent is much better than none ($P_1 \gg P_0$), but two parents are not much better than one ($P_2 \approx P_1$), while 3 holds if a deserting male has a much greater chance of mating again.

Maynard Smith's (1977) ESS model of parental investment. Each sex can choose to invest or desert, and the matrix shows their predicted reproductive success (payoff) from each strategy.

Male	Payoff to	Female	
		Cares	Deserts
Cares	Female	wP_2	WP_1
	Male	wP_2	WP_1
Deserts	Female	wP_1	WP_0
	Male	$wP_1(1 + p)$	$WP_0(1 + p)$

Based on discussion in Krebs & Davies (1993) and Maynard Smith (1977).

Female dispersion	Male dispersion	Mating system		Examples
1	**Male assistance required for successful rearing of young**			
Solitary in territory defended against other females	One male defends one female's territory	♀	♂ — Obligate monogamy	Gibbons, saki, titi, night monkey, silver-backed jackal, prairie vole, klipspringer
	Several males defend one female's territory	♀	♂ ♂ — Polyandry	Saddle-backed tamarin, African wild dog
2	**Male assistance not required for successful rearing of young**			
(a) Female range defendable				
(i) Solitary with small ranges sometimes defended	One male defends one or more female territories	♀ / ♀ ♀	♂ — Monogamy or Polygyny	Galago, several voles
(ii) Small social groups within a small range, sometimes defended	One male defends the females' range (sometimes subordinate males in group but only one breeding male)	♀ ♀ ♀	♂ — Unimale polygyny	Columbian ground-squirrel, yellow-bellied marmot, black-tailed prairie dog, grey langur
(iii) Larger social groups within larger ranges, sometimes defended	Several males defend the females' range	♀ ♀ ♀ / ♀ ♀ ♀	♂ ♂ ♂ — Multimale polygyny	African lions, chimpanzees, red colobus
(b) Female range not defendable: stable female groups				
(i) Small, stable groups in large undefended ranges	One male defends female group as a harem (sometimes subordinate males present). No long-term defence of resources	♂ ♀ ♀ ♀	Harem polygyny	Red deer, gelada baboon, hamadryas baboon
(ii) Large, stable groups in large undefended range	Several males associate with female group. Individual males defend receptive females. No long-term defence of resources	♂ ♂ ♂ ♀ ♀ ♀ ♀ ♀ ♀	Multimale polygyny	Yellow and olive baboons, Cape buffalo
(c) Female range not defendable: unstable female groups				
(i) Large, unstable groups in large undefended range	Males defend large resource-based territories within female range, and mate with females as they pass through	♀ ♀ ♀ ♀ ♀ ♂ ♂ ♂	Polygyny	White rhino, Grevy's zebra, waterbuck, puku
	Or, when local female density high, males defend small, clustered mating territories (no resources), which females visit solely for mating	♀ ♀ ♀ ♀ ♀ ♂ ♂ ♂ ♂ ♂	Lek	Fallow deer, Uganda kob
(ii) Solitary or small unstable groups, widely and unpredictably distributed	Solitary males search widely for receptive females which male may guard temporarily	♀ ♂ ♀ ♀	Scramble competition polygyny	Thirteen-lined ground squirrel, moose, polar bear
(iii) Large, unstable migratory herds; mate on migration	Very variable; may defend individual females, harems or temporary territories when migration halts		Polygyny	Wildebeest, reindeer

(a)

Figure 10.5 Factors influencing dispersal and mating systems in (a) mammals and (b) birds. Dashed lines represent male territories. From Davies (1991), partly after Clutton-Brock (1989).

Mating system	Male and female dispersion		Parental care
1 Monogamy (one male forms pair bond with one female)	*Resource defence* Many passerines, shorebirds and seabirds	Multipurpose territory defended by male or both sexes, or nest site only defended (seabirds)	Male and female
	Mate defence Some passerines, many ducks and geese	Male guards female and may defend nest site	Male and female (e.g. finches), or female only (many ducks)
2 Polygyny (one male forms pair bond with several females simultaneously)	*Resource defence* Some passerines, and shorebirds	Male defends large multipurpose territory within which several females defend exclusive nest sites or smaller territories	Male and female, or mainly female
	Some passerines (Icteridae, Ploceidae)	Male defends clumped nest sites	Male and female, or mainly female
	Polyterritoriality (Pied flycatcher)	Male defends separate nest sites or territories, often several hundred metres apart, with one female at each site	Male and female, or mainly female
	Harem defence Ring-necked pheasant	Male defends group of females who nest solitarily	Female
3 Polyandry (one female forms pair bond with several males simultaneously)	*Cooperative* Galapagos hawk, Tasmanian native hen	Several males defend one female's territory	All the males may help the female
	Resource defence Jacanas Spotted sandpiper	Female defends large multipurpose territory within which several males defend smaller, exclusive territories	Mainly or exclusively male
4 Polygynandry (several males form pair bonds with several females simultaneously)	Dunnock	Several males defend a territory, within which several females may each defend smaller, exclusive territories, or share the whole territory and nest communally	All the males may help the female
	Acorn woodpecker		
5 Promiscuity (no pair bond – female and male meet briefly to copulate)	*Resource defence* Orange-rumped honeyguide, some hummingbirds	Males defend territory containing food and copulate with females who visit to feed	Female
	Display site defence Bowerbirds, manakins, some birds of paradise, grouse and shorebirds, kakapo	Males defend small display site, containing no resources, which females visit solely to copulate. Males may be dispersed, loosely aggregated ('exploded leks') or densely aggregated ('leks')	Female
6 Sequencial polygamy (male or female form pair bond with one mate and then desert to find one or more mates in succession)	*Sequential polygyny* Woodcock	Males guards one female and then, after clutch completion, deserts to find another female. No territorial defence	Female
	Sequential polyandry Phalaropes Dotterel	Female guards one male and then, after clutch completion, deserts to find another male. No territorial defence	Male
	Sequential polygyny and sequential polyandry Temminck's stint	Female lays clutch in one male's territory, which male cares for, then deserts to lay for one or more other males. Males may gain a second female, who incubates the clutch herself if the male is already incubating a first female's clutch	Male or female
	Simultaneous polygyny and sequential polyandry Rhea Tinamou	Group of females lay in a communal nest, cared for by a male, and then desert to lay in the nests of other males	Male

(b)

The ecology of mating systems

Since the dispersion of the sexes is ultimately driven by environmental resources, it should be possible to relate resource distribution to the different kinds of mating system above. The key, as Fig. 10.5 suggests, is the extent to which mates can be defended and thus monopolised: the **environmental potential for polygamy** (**EPP**), as Emlen & Oring (1977) called it in their classic paper. Emlen & Oring recognised two main preconditions for the evolution of polygamous (polygynous, polyandrous, polygynandrous) or promiscuous mating systems: (a) multiple mates or resources sufficient to attract multiple mates must be energetically defendable, and (b) animals must be able to capitalise on their defendability.

Polygyny   When the EPP is high, and animals are able to take advantage of it, polygamous mating can arise in several ways (Emlen & Oring 1977). **Resource defence polygyny** is likely to occur where males are able to defend areas of resource vital to females. Here, the quality of a territory, rather than of the male himself, can play an important role in a female's decision to mate (see 10.1.3.2 and 9.2.1). An extreme example occurs in orange-rumped honeyguides (*Indicator xanthonotus*), in which beeswax forms an essential part of the diet and males maintain year-round territories at the location of hives (see 5 in Fig. 10.5b). Since hives are sparsely distributed on exposed cliffs, a small number of males can monopolise the entire supply and mate promiscuously. Courtship centres on the hives, and mating success is high for territory owners, with one male copulating 46 times with 18 different females.

In **female defence polygyny**, males are able to gain possession of a number of females directly and defend them against rivals. Female defence polygyny is more likely if females have a tendency to be gregarious, as they do in many ungulate species. Red deer are a good example (see 2bi in Fig. 10.5a). Here, stags compete for ownership of harems of females during the breeding season, with defence being most aggressive at times when hinds are most likely to conceive (Clutton-Brock *et al.* 1979; Fig. 10.6). In species such as Burchell's zebra (*Equus burchelli*) and gelada baboons (*Theropithecus gelada*), where breeding seasons are much longer, harems may be defended all year (Dunbar 1984; Rubenstein 1986).

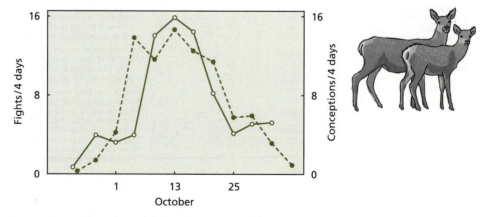

Figure 10.6   Red deer stags fight more intensely (solid line) at times when hinds are most likely to conceive (dashed line). After Clutton-Brock *et al.* (1979).

Sometimes males defend neither resources nor females, but instead aggregate and establish dominance relationships among themselves. Females may then choose between them on the basis of their competitive ability and a few males can once again end up with most of the matings. The form this **male dominance polygyny** takes depends largely on the extent to which females synchronise their sexual activity. So-called **explosive breeding** occurs when all the females come into breeding condition simultaneously and arrive at the male aggregation together. Explosive breeding is common among frogs and toads, where, as in the European frog (*Rana temporaria*), mating can sometimes occur on only one night in the year. If females are not synchronised in their periods of sexual activity, only a small proportion of them visit the male aggregation at any one time, and competition among males is intense. Such aggregations are referred to as **leks** (Box 10.5). Since males at a lek control neither resources useful to females nor the females themselves, females are free to choose males solely on their physical characteristics. Selection in lekking species thus tends to favour extreme epigamic (mate attraction) characters, such as bright plumage and attention-grabbing behaviour, to which we shall return later (10.1.3).

Monogamy In Emlen & Oring's scheme, monogamy arises either when the EPP is low, or animals are unable to capitalise on it. The prevalence of monogamy among birds, for example, is probably due to the latter as a result of the high demand for paternal care in many species (e.g. Bart & Tornes 1989; Hatchwell & Davies 1990). However, stiff competition for mates may also limit the opportunity for polygyny among males (Davies 1991). Importantly, molecular genetic information over the past two decades has shown that many apparently monogamous relationships are only superficially so, with social monogamy often concealing extra-pair matings by one or both sexes in the partnership (Birkhead & Møller 1998; Griffith *et al.* 2002). Extra-pair matings appear to be commoner in colonial species, where the opportunity to solicit copulations is greater. In the ostensibly monogamous swallow (*Hirundo rustica*), for instance, extra-pair paternity within colonies has been estimated to be as high as 24% of chicks (Møller 1987a). In some cases, extra-pair copulations reflect the high cost to females of resisting the advances of other males. In others, females actively seek copulations from other males, sometimes to gain additional paternal care for their offspring (Davies 1992), or to mate with a male of higher quality (e.g. higher ranking or more attractive) than their present partner (Smith 1988; Møller 1988a).

Polyandry and polygynandry If parental assistance by males is extensive, there may be scope for polyandry or polygynandry. Emancipation from care gives females the chance to increase their reproductive output through multiple matings, but only, of course, to the extent that males are receptive and able to care for resulting offspring. Thus males become a limiting resource and females end up competing for access to them (**sex role reversal**). The conditions under which polyandry emerges are therefore likely to follow principles similar to those for polygyny.

Resource defence polyandry is found in American jacanas (*Jacana spinosa*) (Jenni & Collier 1972; see 3 in Fig. 10.5b). Here, females parcel up suitable ponds and lagoons into 'superterritories', within which a number of males set up their small, individual, breeding territories. Females successfully defending superterritories often have several males incubating clutches at the same time, and readily replace any clutches that are lost to predators. Female jacanas have even been recorded killing other females' chicks to gain access to more males. The trigger for polyandry in these birds may be a limit on the number of eggs that can be laid in a single clutch, so that females can increase their reproductive success only by laying additional clutches, and thus mating with more males. Like resource defence polygynous males, females in resource defence polyandrous

Difficulties and debates

Box 10.5 Why lek?

Although relatively rare, lekking has attracted a lot of attention because males form aggregations solely for the purpose of attracting a mate, and females seem to choose between males when the latter have nothing to offer but their own intrinsic individual qualities. So why do males aggregate and where do they choose to do it? Four (not mutually exclusive) main hypotheses have been suggested (Bradbury & Gibson 1983):

1. *Aggregation at 'hotspots'*. Sometimes males aggregate at sites used regularly by females for other reasons, such as to feed or drink (e.g. Leuthold 1966). In topi (*Damiliscus korrigum*), for example, males cluster in areas of short grass where females prefer to rest with a good chance of spotting an approaching predator (Gosling 1986). In Lawes's parotia (*Parotia lawesii*), a bird of paradise, females occupy overlapping home ranges, and males aggregate where the zone of overlap involves the greatest number of female ranges (Pruett-Jones & Pruett-Jones 1990).

2. *Aggregation around 'hotshots'*. Another possibility is that males benefit by clustering around particularly attractive individuals and exploiting their superior 'pulling power' (Beehler & Foster 1988). Some evidence for this comes from Arak's (1988) study of natterjack toads (*Bufo calamita*) where large males have louder calls and can attract females over a greater distance. Small males, with weak calls, settle next to large ones and try to intercept females as they home in on the loud signal (see also 2.4.5.1).

3. *Reduced risk of predation*. As we have seen in Chapter 9, joining a group can reduce an individual's risk of predation. Mating aggregations in some chorusing frogs may serve this purpose, with individuals benefiting from a dilution effect (9.1.1.1) when subject to attack by aerial predators (Ryan *et al.* 1981). However, predation risk cannot be a general explanation for lekking because predation seems to be rare at many bird leks (Davies 1991).

4. *The attractiveness of aggregations of males, or particular sites*. It may be that aggregations of males *per se* attract females, either because aggregations increase signal strength and grab the attention of females over greater distances, or because there may be a better chance of finding a high-quality mate (e.g. aggregated males may fight and reveal their competitive ability) or mating more quickly. Females may also benefit from a reduced risk of predation in a cluster of males. Many insect leks occur at distinctive landmarks, such as a prominent tree or bush, a clearing in a forest, or the top of a hill. Such behaviour may reduce the searching time required to find a mate when populations are sparsely distributed (Thornhill & Alcock 1983).

There is evidence for and against all the above suggestions in different groups of animals, and it seems likely that characteristics of both location and males have an important influence on females at leks. Teasing these apart in the field, however, is a considerable challenge. Davies (1991) provides a good discussion.

Based on discussion in Bradbury & Gibson (1983) and Davies (1991).

species are the larger, more aggressive sex, by some 50–70% in American jacanas, but a more modest 20–30% in other polyandrous species, such as the spotted sandpiper (*Tringa macularia*) (Oring & Maxson 1978).

Females may be mated to several males at the same time (**simultaneous polyandry**), such as jacanas, or mate with different males at different times (**sequential polyandry**). Sequential polyandry is found in some insects, such as belastomatid water bugs (Smith 1980), and in phalaropes (*Phalaropus* spp.), wading birds in which females compete for access to males directly in a manner similar to male dominance polygyny (Hilden & Vuolanto 1972; Schamel & Tracy 1997). Male and female phalaropes congregate at bodies of water to feed, display and copulate, and aggressive competition among females determines which of them succeed in mating with males. Males may then incubate the eggs alone and care for the young after hatching. Emlen & Oring (1977) refer to such systems as **female access polyandry**, and, once again, females tend to be the larger, more adorned sex.

In **cooperative polyandry**, females have several mates simultaneously, but, instead of farming out offspring to be cared for by different males, care for a single brood jointly with them. Examples include the Tasmanian native hen (*Tribonyx mortieri*) and the Galapagos hawk (*Buteo galapagoensis*) (Ridpath 1972; Faaborg *et al.* 1980; see 3 in Fig. 10.5b). In yet other cases, females mate with several males, but neither sex cares for the offspring. Such **non-parental polyandry** is probably common in species such as pelagic fish which do not form long-term associations with conspecifics or particular locations (Bradbury & Vehrencamp 1984).

Finally, there may be simultaneous polyandry, but with males caring for the broods of several females. Thus both sexes mate polygamously, leading to a polygynandrous mating system. Polygynandry is particularly prevalent in substrate-nesting fish that guard their eggs (Perrone & Zaret 1979), but it also occurs in various birds, such as ratites, accentors and woodpeckers (Bruning 1974; Davies 1991, 1992; Fig. 10.5b).

Mating systems and conflict between the sexes

While mating systems can be characterised in these various ways, and explained in terms of species ecology, it is not difficult to see that there is scope for conflict between the two sexes. The mating system that is best for females may not be the best from a male point of view. Given the disparity in parental investment discussed above, for example, we might expect females to show a greater preference for monogamy than males. If a female has a male all to herself, and he contributes parental care, she may be able to rear more offspring than if she has to share him with another female. Surely, therefore, the female should always prefer single males and avoid those that are already mated. In some species, such as the marsh wren (*Cistothorus palustris*) this indeed seems to be the case; females settle with mated males only after all single males have paired (Leonard & Picman 1987). But a moment's thought suggests this is unlikely to be a universal rule.

Certainly females are likely to suffer a cost if they mate polygynously because they will have to share the male and his resources with other females. However, if males, for example, defend territories, and these vary in quality, a female mating polygynously with the owner of a good territory may do better than one mating singly with the owner of a poorer territory. Thus there may be a threshold territory quality above which it pays females to mate polygynously (the **polygyny threshold model**; Verner & Willson 1966; Orians 1969). The polygyny threshold model (Fig. 10.7) is a form of ideal free distribution (see Box 7.1), in that females are assumed to be free to choose between males, to have

Figure 10.7 The Orians–Verner–Willson polygyny threshold model. A female has a choice of settling with an unmated male on territory B, or with a male that is already mated, but on a territory of higher quality (A). The polygyny threshold model assumes that female reproductive success increases with territory quality, but there is a cost (*c*) of sharing a male with another female (the curve for a polygynous female lies below that for a monogamous female). Only if the difference in quality between territories A and B exceeds the polygyny threshold (PT) will it pay a female to choose an already mated male. After Davies (1991) from Orians (1969). Orians G.S. (1969) On the evolution of mating systems in birds and mammals. *American Naturalist* **103**: 589–603, reprinted by permission of The University of Chicago Press, © 1969 by The University of Chicago. All rights reserved.

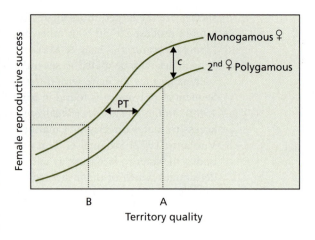

perfect information about the relative reproductive value of settling with any one of them, and to select a partner that maximises their reproductive success. There is some evidence that females behave as predicted by the model. Female great reed warblers (*Acrocephalus arundinaceus*), for example, sample several territorial males before settling, and often choose a male that is obviously already mated after having visited a number of single males (Bensch & Hasselquist 1992). Such polygynous females seem to do no worse than those winding up with single males on poorer territories (Ezaki 1990), suggesting that the difference in territory quality between mated and unmated males takes females over the polygyny threshold (Fig. 10.7).

As with the ideal free distribution itself, however, the conditions for the polygyny threshold model to work as suggested in Fig. 10.7 are unlikely to be met in many cases. For instance, competition between females for the best males may lead to some individuals being excluded from territories they would otherwise have chosen. Competition between males may also affect things. A particularly nice example comes from Davies's (1985) study of dunnocks (*Prunella modularis*), a superficially unremarkable little brown bird, sometimes (and inaccurately) referred to as the hedge sparrow.

Dunnocks have a highly variable mating system that includes monogamy, polygyny, polyandry and polygynandry. As might be expected, females do least well under polygyny and best under polyandry, while the opposite is true for males. Males and females therefore conflict in their relative preference for polygynous, monogamous and polyandrous matings (Fig. 10.8), and this is very apparent from their behaviour. Dominant females in polygynous relationships often attempt to drive the other cohabiting female away, while the male does his best to keep his females apart to ensure they both remain on his territory. Equally, monogamous females solicit copulations from other males on the territory, so that they might help to raise her brood. Not surprisingly, dominant males invest much effort in trying to keep other males away and ensure their own paternity. While they are doing this, however, females attempt to evade their mate guarding activities and sneak matings with the rival male (Davies 1985). Which mating system eventually prevails depends on ecological factors and the relative competitive ability of the protagonists. Sometimes neither sex gets the upper hand; the dominant female is unable to evict the second female, and the dominant male cannot drive off his rival. The result is a polygynandrous foursome, with two males sharing two females (Fig. 10.8).

Figure 10.8 Conflict over mating systems in dunnocks. Female territories do not overlap and may be defended by one or two males. However, males and females are likely to disagree over which mating system is best. The arrows show the directions in which these conflicting interests tend to push things. See text. From Davies (1991).

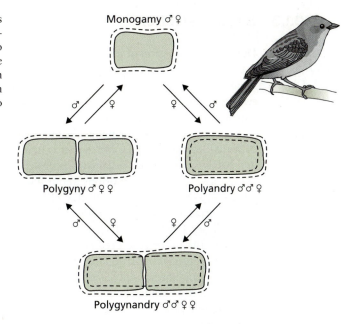

Conflict and deception In some cases, one sex seems to be conned into a mating system that is against its reproductive interests. A good example is the pied flycatcher (*Ficedula hypoleuca*), an insectivorous bird studied by Alatalo *et al.* (1981) in Sweden (see also Catchpole *et al.* 1985).

Flycatchers nest in holes in trees, and are easily persuaded to take up residence in nest boxes, making them amenable to experimental manipulations in the field. Males defend suitable cavity sites and sing to attract a female. Once he has mated and the female is incubating her eggs, however, the male goes off to another site, some way away, and attempts to attract another female. If he is successful, he leaves the second female to rear her brood alone and returns to the first to help feed her chicks. Thus the male's polygyny impacts relatively little on the first female, but is very costly to the second. So why does the second female accept a polygynous mating, when she could do much better by mating monogamously? One possibility is that there just are not that many unmated males around, so the female is having to make the best of a bad job. Another is that, although her present brood is penalised, any sons the female produces may inherit their father's tendency to polygyny and pay dividends later. Yet a third is that the second female is an unwitting dupe. Since the male establishes his second territory some distance from the first, the female may have no way of knowing he is already paired. By strategically positioning nest boxes in suitable habitat, Alatalo *et al.* (1981) were able to provide females with the opportunity to sample both paired and unpaired males in areas of similar quality and show that, despite suffering a significant reduction in young raised, they were just as likely to mate with paired as unpaired males. Thus, the deception hypothesis seems to win over the lack of unpaired males hypothesis. It also seems to win over the polygynous sons possibility, since male flycatchers turn out to vary in their tendency to polygyny from year to year, implying the future benefits would be meagre. Thus female flycatchers just seem to be hoodwinked into polygyny.

10.1.3 Sexual selection

During our discussion of mating systems, we referred several times to individuals of one sex competing for access to those of the other, or choosing their partner from among them. Either way, by fighting for mating opportunities or attracting attention, there is strong selection for characteristics, such as weaponry and striking appearance, that enhance an individual's chances of mating. Following Darwin (1859, 1871), selection acting via competition for mates (as opposed to competition for survival) is referred to as sexual selection (2.4.1). Much that is fundamental about the morphology, physiology and behaviour of living organisms is driven by it. However, great care is necessary to avoid unwarranted assumptions about what is and what is not sexually selected, and which apparently sexually selected characters are used in competition for access to mates (intrasexual selection) rather than choice of mate (intersexual selection) (Harvey & Bradbury 1991; see 2.4.1). Darwin himself was acutely aware of the problem. For example, the prehensile organs used by some male insects and crustaceans to grasp the female during copulation can be viewed as products of natural selection in the same way as organs specialised for grasping food. But if the organs used in mating become elaborated to prevent competitors dislodging a paired male, or the female leaving to copulate with a rival, they are viewed as products of sexual selection. Identifying the relative contributions of natural and sexual selection can thus be tricky.

Similar difficulties arise over the distinction between intra- and intersexual selection. Take the male pheasants in Fig. 1.13, for example. One might put good money on the characteristically ornate tails of the group having evolved through female choice. It also seems reasonable to suppose that the sharp tarsal spurs also possessed by males have been selected as weapons in intermale competition. In many cases these assumptions would be proved right. In male golden pheasants (*Chrysolophus pictus*), however, the long, elaborate tail turns out to be used for support during fights between males, while the spurs of male ring-necked pheasants (*C. colchicus*), not their plumage, are used by females to choose a mate (von Schantz *et al.* 1989). Similar counterintuitive findings have emerged from studies of the plumage displays of male birds of paradise (LeCroy 1981). Hard and fast distinctions between the kinds of character favoured by intra- versus intersexual selection may thus be difficult to draw. Nevertheless, the two processes lead to some general predictions, as we shall now see.

10.1.3.1 Intrasexual selection

For the reasons with which we are now familiar, intrasexual selection is usually more intense among males. Its most obvious manifestations are the stylised displays or out-and-out fighting of rival males. A general expectation, therefore, is that intrasexual selection will lead to increased body size, weapons, armour, vocal capacity and other accoutrements of an aggressive lifestyle, and so to greater sexual dimorphism (see Fig. 2.10). It can also lead to gaudy 'badges of status' (see 11.1.1.4) like the plumage in LeCroy's birds of paradise above, and thus to dichromatism (the sexes differ in colour; see Kimball & Ligon 1999). However, dimorphism and/or dichromatism can also arise for other reasons, such as sex differences in feeding ecology, allometric (relatively disproportionate) patterns of growth, or intersexual selection, so making broad generalisations difficult (see Andersson 1994). Games theory models suggest that males should invest in such traits more when the ratio of monopolisable females to competing males (and thus the likely reproductive payoff) is high, and the distribution of trait sizes in the population is biased towards low values (e.g. when trait size increases with age, and large, old males are rare, as in many vertebrates), thus making it easier to gain an advantage (Parker 1983a).

Sperm competition

While intrasexual competition is most apparent prior to mating, because fighting or displaying individuals are usually conspicuous, it can also occur afterwards. Indeed, postcopulatory competition may be even more severe than its precopulatory equivalent. This is because females frequently mate with more than one male, even in socially monogamous pairings (10.1.2.2), so that the sperm of different males compete to fertilise her eggs (Parker 1970). We shall come back to why females might do this later. However, the **sperm competition** that results from multiple mating has led to various strategies by males to gain the upper hand.

One approach is to prevent rivals from inseminating the female in the first place, for instance by staying close to the female, or even attached to her, before and after copulation until her eggs have been fertilised. Precopulatory **mate guarding** is shown by males of the freshwater amphipod *Gammarus pulex* (Birkhead & Clarkson 1980), while males of various damselfly and bird species guard their females after copulating and through the period of egg-laying (Birkhead & Møller 1998; Simmons 2001). Mate guarding is also shown during the oestrus period in mammals, where it is often referred to as **consortship** (Packer & Pusey 1983; Sherman 1989). However, males can deal with the risk of sperm competition in many other ways as well.

In various species, including humans, where sperm lingers or is stored in the female reproductive tract (as in the spermathecae [sperm storage sacs] of various female insects) males often use specially adapted penises to rake out the sperm of previous males, or squeeze it out of the way, before inseminating the female with their own (Waage 1979b; Eberhard 1985; Gallup *et al.* 2003). Spines on the penis may even be designed to damage the female's reproductive tract to cause discomfort and discourage her from further mating (Crudgington & Siva-Jothy 2000). The chances of further mating can also be reduced by plugging up the female after copulation. In some species, both vertebrate and invertebrate, the male seals the female tract with a plug formed from the ejaculate. The plugs in rodents and bats appear to contain malformed sperm, for instance with two heads or tails, that have become tangled together to form a mesh (Fenton 1984; Baker & Bellis 1988). The usual explanation for the sometimes high proportion (up to 40% in humans) of malformed sperm in the ejaculates of males of different species has been developmental error, a limitation on quality control in a system of extreme mass production. Baker & Bellis (1988), however, suggest these malformations may not be errors at all, but adaptive diversity (polymorphism) in sperm design to meet the needs of sperm competition (see also Silberglied *et al.* 1984). In their view, apparently deformed sperm have sacrificed their chances of fertilising an egg to perform other roles, such as plug formation, or entangling the sperm of rival males (the **kamikaze sperm hypothesis**). While plausible, however, the idea remains controversial and in need of rigorous testing. Simmons (2001) reviews some other suggested functions of sperm polymorphism.

Males of some species have become more inventive with their plug-forming capabilities. In the parasitic acanthocephalan worm *Moniliformis dubius*, males sometimes copulate with other males and seal up their genitals, thus reducing the competition at source (Abele & Gilchrist 1977). Males of the bug *Xylocoris maculipennis* also copulate homosexually, inseminating rivals, as they do females, simply by piercing their body wall and injecting sperm into the body cavity. In females the sperm then swim around until they encounter her eggs, but in other males they make their way to the testes where they may

Figure 10.9 Males of various species adjust their production of sperm to take account of sperm competition. (a) Males of the crickets *Acheta domesticus* increase the number of sperm in their spermatophores if apparent competitor males are present. Points are medians with interquartile ranges. After Gage & Barnard (1996). (b) Men ejaculate more sperm when copulating with their regular partner as the amount of time they have spent with them since the previous copulation decreases (i). However, the correlation is not present when samples are produced by masturbation (ii), suggesting something about the copulatory relationship influencing sperm transmission. After Baker & Bellis (1989).

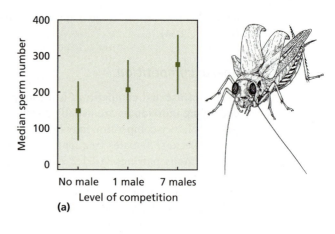

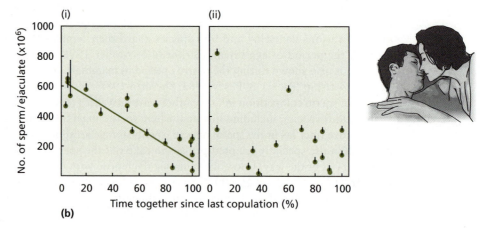

be passed on the next time the recipient mates (Carayon 1974). Homosexual inseminations thus exploit the potential for sperm competition to produce offspring vicariously through the mating effort of would-be rivals.

More prosaically, males can deal with the risk of competition simply by increasing the amount of sperm they transfer and swamping the ejaculates of rivals. Evidence from several species, including insects, rodents and humans, suggests that males respond to the threat of sperm competition by increasing the number of sperm in their ejaculates (e.g. Baker & Bellis 1989; Gage 1991; Stockley *et al.* 1997). Gage & Barnard (1996), for example, showed that the spermatophores of male crickets (*Acheta domesticus* and *Gryllodes sigillatus*) contained more sperm if males were exposed to competitors when they mated (Fig. 10.9a). In humans, men appear to produce more sperm per ejaculate when they copulate with their regular partner as the time the pair have spent together since the last copulation decreases (i.e. when there is a greater chance that their partner will have indulged in an extra-pair copulation) (Baker & Bellis 1989; Fig. 10.9b). In keeping with this, comparative studies have shown that species in which females mate polyandrously, or partners often spend long periods apart, have bigger testes and/or copulate more frequently than their monogamous relatives or those where mated pairs remain close together most of the time (Harcourt *et al.* 1980). Birkhead & Møller (1998) and Simmons (2001) discuss many other examples of sperm competition and adaptations to promote or reduce it.

Infanticide

In some cases, competition between males may continue even beyond fertilisation. Lions are a good example.

Lions are unique in being social cats. Their social structure is based on a group (pride) of related females (mothers, daughters, sisters, aunts, etc.) and their cubs, and usually two or three adult males that are related to each other but not to the females (Bertram 1978). This situation arises because females tend to remain with the pride into which they were born, while males, on maturing, move out and search together for a new pride. The search is not easy, because resident males defend their prides against marauding usurpers. If the resident males beat off a takeover bid, the questing pair or group must continue on its way and search further afield. If the residents lose, however, they are evicted by the new males, which then have exclusive access to the females until they themselves are eventually displaced, usually within two to three years (Pusey & Packer 1983).

Because their likely residence time is short, male lions must make the most of their brief reproductive opportunity and sire cubs as soon as possible. One problem, however, is that the females may still be suckling cubs by the previous males and thus not be ready to mate. If they waited for the females to return to breeding condition normally, the new males could lose a substantial portion of their potential reproductive output, and, worse, benefit the offspring of rival males. They therefore take the logical, if drastic, step of killing all the unweaned cubs. This not only removes the competition, but it also brings females back into oestrus more quickly, and, importantly, synchronously (Packer & Pusey 1983). Synchrony matters because it ensures male cubs will mature together and fare better in coalitions when they eventually leave the pride. Males clearly benefit from this, but what about the females? Certainly the loss of their cubs is costly to lionesses, which sometimes attempt to defend them from infanticidal males. However, even if females are successful in their defence, the cubs rarely survive more than a few weeks after a takeover. Thus it pays females to cut their losses by abandoning existing cubs and transferring their investment to those of the new males. Infanticide by males therefore seems to be a strategy for maximising paternity by ensuring that competition for female resources is removed and resources are channelled solely into the resident males' offspring. The so-called Bruce effect, in which the presence of a strange male causes females to abort, or implantation of the embryo to fail, leads to a similar outcome in rodents.

10.1.3.2 Intersexual selection

At various points in our discussion we have referred to members of one sex choosing among members of the other. Usually it has been females choosing males, but, as we have seen, it can sometimes be the other way round. This still involves intrasexual competition for fertilisations, but through preference by the choosing sex rather than fighting or outcompeting the sperm of rivals (though success in fights may well be a basis for choice).

Intersexual selection via mating preferences is widely assumed to have resulted in the elaborate adornments and behaviours that are the textbook examples of secondary sexual characters, the peacock's extraordinary tail being the most familiar. While intuitively satisfying as an explanation for such characters, however, there is still much debate as to how mating preferences might have evolved. In some cases, mate choice appears to influence the chooser's reproductive success, and is thus easy to understand. In others, the reproductive benefits remain obscure, and the basis of the preference a matter of (sometimes contentious) speculation. Advancing theoretical and genetic understanding of sexual conflict is also beginning to suggest more complex consequences for the evolution of intersexual interactions than those envisaged by traditional sexual selection theory (e.g. Parker 1979; Rice & Holland 1997; Chapman *et al.* 2003; see also 5.3.1). Instead of viewing coyness on the part of females as a device for screening male quality, the emphasis is more on differences between the sexes in optimal outcomes (e.g. relative parental investment) from mating (see 5.3.1) and the costs of mating (e.g. Parker 1979; Arnqvist & Nilsson 2000; Crudgington & Siva-Jothy 2000). The fact that females sometimes incur heavy costs, even dying, resisting mating attempts suggests there may be more to female mating resistance than screening for male quality (the advantages of which would have to be improbably large in some cases). As yet, however, the relative importance of sexual selection and this more fundamental kind of sexual conflict in driving mating decisions is far from clear, and, indeed, there may not be a clear boundary between them (Chapman *et al.* 2003). Nevertheless, it is important to bear the issue in mind in evaluating the evidence for apparently sexually selected mating preferences.

In the discussion that follows, we shall mostly assume that it is females that choose males. While Darwin (1859, 1871) attributed a significant role to female choice in the evolution of secondary sexual characters, he was not able to provide a convincing explanation as to *why* it should happen. What did females get out of it? Although the principle of female choice was generally accepted following Darwin, it was widely assumed that it did little more than ensure females mated with males of the right species (Trivers 1985). Male courtship displays and exaggerated secondary sexual characters were regarded simply as attempts to stimulate generally reluctant females into some enthusiasm for mating. Once aroused, the female would mate with any male in sight.

Sexy sons and runaway female choice

Ronald Fisher (1930) was the first to provide a convincing mechanism through which female choice could drive the evolution of extreme characters in males, and so do justice to Darwin's assumption as to its evolutionary importance. Fisher's model is based on the idea that females, for reasons that do not matter for the moment, find the features of certain males, perhaps a longer than average tail, attractive. As long as these features are heritable, females mating with the males stand to pass the attractive qualities on to their sons, which will then be attractive to females in their turn (the **sexy sons** effect). Males with longer tails, and the females preferring them, thus enjoy greater reproductive success, with the result that both long tails and preference for them spread through the population. If female preference is based on longer than average tails (i.e. *relative* trait size; see O'Donald 1983), the covariance between trait and preference will lead to the evolution of longer and longer tails over successive generations (the **runaway** effect). Ultimately, however, the process will be halted by opposing natural selection, as trait size reached the point where its reproductive advantage is outweighed by its survival, or other utilitarian, costs. Fisher's idea is explained in more detail in Box 10.6, while Andersson's (1982) study of long-tailed widowbirds (*Euplectes progne*) provides one of the classic experimental demonstrations that females can choose on the basis of relative (though whether strictly Fisherian has still to be established) character elaboration in males (Fig. 10.10).

Figure 10.10 Experimental manipulations of tail length in male long-tailed widowbirds affected the males' attractiveness to females. Two groups of males with initially similar levels of attractiveness, measured as the number of active nests in their territories (a), had their tails lengthened or shortened by cutting and glueing the feathers. Two control groups had their tails cut and glued back together again, or were unmanipulated. (b) After the manipulations, most females preferred the males whose tails had been lengthened and avoided the males with shortened tails. After Andersson (1982).

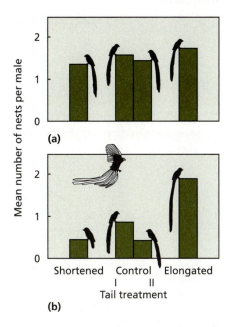

Underlying theory

Box 10.6 Fisher's runaway effect

Fisher's runaway effect starts with the assumption that males in an ancestral population varied in some trait, such as tail length, crest size or the number of yellow spots on the rump, and that females varied in their preference for the trait. We will use tail length as an illustration, so that some males in our population have longer than average tails, some shorter. At the same time, some females prefer longer-tailed males and some prefer those with shorter tails. The important point is that when females mate with their preferred kind of male, their offspring inherit both the male's tail-length *and* the female's preference. Each sex carries both characters, but the characters are expressed in a **sex-limited** fashion; that is, only male offspring express the trait, and only females the preference.

If equal numbers of females prefer long and short tails, there will be no net change in male tail size in the population. But if, for some reason, females preferring longer tails have a slight majority, the covariation between male trait and female preference will propel the population towards longer and longer tails (Fig. (i)). This is because at each generation, longer than average tails are preferred against a baseline of increased size. Precisely the same would happen in the opposite direction if females preferring short tails were in the majority. In Fisher's initial conception, the direction of female preference initially had some utilitarian benefit – perhaps a slightly longer than average tail allowed greater aerial manoeuverability, so that better-endowed males were more able to escape predators. Later models, however, suggest a utilitarian advantage is not necessary; a chance bias in preference is all that is needed (Lande 1981).

The runaway effect is thus driven by a positive feedback effect arising from the covariation between male trait and female preference. Whatever the reasons for the bias in preference in the first place, the attractiveness *per se* of the male trait quickly becomes the driving force behind the process, to the extent that it eventually begins to compromise the survival of males (as, for instance, in túngara frogs (*Physalaemus pustulosus*) where the loud calls of males

Box 10.6 continued

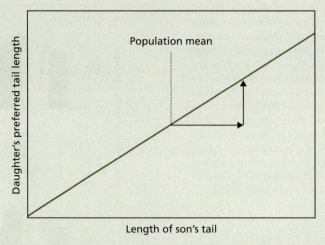

Figure (i)   Genes for a long tail in males and preference for a long tail in females are inherited together by offspring, with a degree of association (covariance) reflected by the slope of the line. If equal numbers of females express preferences above and below the mean tail length in the population, there will be no net change. But if a slight majority prefer longer (or shorter) than average tails, a runaway effect is triggered in which both tail length itself and preferred tail length increase (or decrease), as indicated by the arrows. After Krebs & Davies (1987).

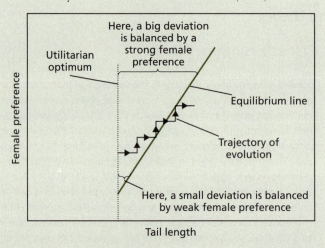

Figure (ii)   Tail length may reach an equilibrium where the strength of runaway sexual selection is balanced by opposing utilitarian natural selection. How far sexual selection can push tail length away from its purely utilitarian optimum depends of the relative strengths of the opposing selection pressures. After Krebs & Davies (1987).

increases their vulnerability to frog-eating bats [e.g. Ryan 1985]). At some point (or range of points, depending on the relative strengths of the runaway effect and opposing utilitarian selection [Kirkpatrick & Ryan 1991]), the detrimental effects of directional selection on the trait are likely to outweigh its attractiveness advantage and the runaway process will cease (Fig. (ii)). The strength and final outcome of the runaway effect are likely to depend on several factors, but an important one is the cost of choosing to females, for instance in predation risk or delayed reproduction (Andersson 1994; Houle & Kondrashev 2002). Pomiankowski *et al.* (1991) discuss some of the implications of preference costs for Fisher's model.

Based on discussion in Andersson (1994), Ryan (1997) and Krebs & Davies (1993).

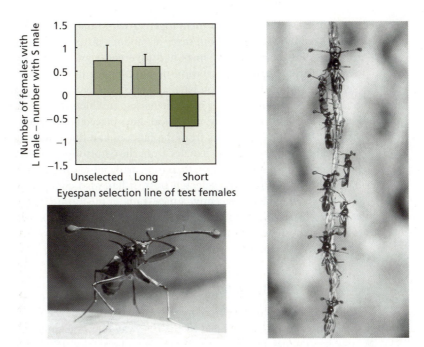

Figure 10.11 When male stalk-eyed flies were artificially selected for longer (L) or shorter (S) eyestalks, female preference for eyestalk length changed too. Given a choice, females in unselected control and 'long' lines preferred long-stalked males, while those from 'short' lines preferred short-stalked males. Preference was decided as the male with which the female roosted overnight. See text. After Wilkinson & Reillo (1994). Photographs courtesy of Gerald Wilkinson: photograph of single male fly by Phil Savoie (BBC Natural History Unit), photograph of roosting flies by Gerald Wilkinson.

The key to the runaway effect is the genetic covariance between male trait and female preference. While this underpins other models of female choice as well (see below), so is not unique to Fisher's process, it is nevertheless a pivotal assumption of the runaway effect and deserving of experimental test. Several studies have attempted to provide such a test. One of the most convincing is Wilkinson & Reillo's (1994) selection experiment with stalk-eyed flies (*Cyrtodiopsis dalmanni*).

Stalk-eyed flies, as their name suggests, have eyes mounted on long appendages (eyestalks) extending from the head (Fig. 10.11). While the length of the eyestalks increases with body size in both sexes, males generally have longer stalks than females. The flies live along streams in Malaysia, where they rest overnight in small groups clinging to vegetation hanging under the banks. At dawn and dusk, flies within the aggregations often mate, and males dispute with one another in an attempt to drive away competition and gain exclusive access to the cluster of females. In their head-to-head confrontations, males with the most widely separated eyes, and thus the longer eyestalks, tend to win, so stalk length seems to be driven partly by intrasexual selection (Burkhardt & de la Motte 1987; Panuis & Wilkinson 1999). However, female choice also has an influence, with females preferring males with longer eyestalks (Burkhardt & de la Motte 1987). So is there any evidence for genetic covariance between stalk length and female preference as Fisher's model would predict?

To find out, Wilkinson & Reillo collected flies from the field and established three breeding lines in the laboratory. In the first two, they picked males with the longest or shortest eyestalks and put them together with randomly chosen females. In the third, they picked individuals of both sexes at random. After 13 generations of artificial selection, the 'long' and 'short' lines had diverged significantly in eyestalk length, thus demonstrating heritable variation in stalk length in the original natural population (see 2.4.2.1). But did selection have any effect on female preference? To see, Wilkinson & Reillo carried out a series of choice tests, in which females were allowed to choose between males from the 'long' and 'short' lines placed either side of a clear plastic barrier. As Fig. 10.11 shows, females from the random and 'long' lines showed a preference for long-stalked males, whereas females from the 'short' line preferred short-stalked males. In keeping with Fisher's theory, therefore, artificial selection for eyestalk length in males had not only changed the selected trait, it had also changed preference for the trait on the part of females. Recently, a nice study of the moth *Utetheisa ornatrix* by Iyengar *et al.* (2002) has shown that female preference for larger males is inherited via the paternal Z sex chromosome, thus promoting the evolution of exaggerated male traits by ensuring preference for them is transmitted to all sons.

'Indicator mechanisms' and female choice

Wilkinson & Reillo's experiment provides strong evidence for the covariance between trait and preference demanded by the runaway effect, as, indeed, have studies of several other species (e.g. Houde & Endler 1990; Gilburn *et al.* 1993). However, demonstrating covariance does not necessarily mean the male trait has evolved because it is arbitrarily attractive to females. It is also consistent with other hypotheses about female choice that hinge on preferences having utilitarian benefits (e.g. enhancing survival or fecundity). Indeed, there is evidence that Wilkinson & Reillo's female stalk-eyed flies might benefit from choosing males with long eyestalks for two utilitarian reasons: first because long-stalked males are generally in better condition (Wilkinson & Taper 1999) and second, because they are more likely to carry an allele on the Y-chromosome that counteracts a tendency to produce too many female offspring (Wilkinson *et al.* 1998, see below). Thus long eyestalks may not be attractive for purely Fisherian reasons.

Utilitarian benefits fall into two camps: **genetic benefits**, heritable qualities of the male that improve the survival prospects, competitive ability and reproductive success of any offspsring, and **non-genetic benefits**, such as food resources, protection from predators or good nesting sites that might arise from, say, a male defending a high-quality territory. These are not mutually exclusive, and females may use both in assessing the attractiveness of a male. This seems to happen in female bitterlings (*Rhodeus sericus*), where males are chosen on both their coloration and the quality of oviposition site they are defending (Candolin & Reynolds 2002). Explanations based on genetic benefits are frequently referred to as '**good genes**' explanations, and contrasted with Fisher's runaway effect that invokes no utilitarian benefit (at least once underway). As several authors (e.g. Andersson 1994; Dawkins 1995) have pointed out, however, Fisher's model is really a 'good genes' explanation too, but the genes confer an advantage through male attractiveness *per se* rather than some other quality of the male affecting reproductive success (e.g. survivorship) that happens to be associated with the preferred trait. Thus the 'good genes' epithet is not a useful one in comparing the two kinds of explanation. Instead, Andersson (1994) suggests the term '**indicator mechanism**' to convey the idea that a trait provides information about the quality – genetic or non-genetic – of the male as a mate, and is not simply attractive in its own right as assumed in Fisher's model.

Sometimes the distinction is made between **direct selection** on female mating preferences, where mate choice has a direct effect on female fecundity, for instance through copulation efficiency or investment by males in parental care or nutrients in nuptial gifts, and **indirect selection**, where secondary sexual characters are favoured through their association with other traits that *are* under direct selection (such as competitive ability or immune function) (see Ryan 1997). The distinction corresponds broadly, but not entirely, with that between non-genetic (≡ direct) and genetic (≡ indirect) utilitarian benefits. As mentioned in Box 10.3, it is also important to remember that females may themselves influence the quality of offspring sired by a male, for instance by varying the amount of resource in the eggs he fertilises (Gil *et al.* 1999).

Genetic benefits   Indicator mechanisms relating to genetic benefits are based on correlations between a sexually selected trait and the condition of its possessor. A well-developed trait should indicate good condition, and thus viability (Andersson 1994). The implication here is that the trait is costly to the male, so only robust individuals in good condition can afford to elaborate it. For this reason, Zahavi (1975, 1977; Zahavi & Zahavi 1997) referred to costly traits as **handicaps**, introducing a paradoxical twist into the argument by suggesting that females prefer 'handicapped' males because they have demonstrated their ability to survive and compete *despite* their encumbrance. The **handicap principle** has been the subject of much controversy (see Box 10.7), and debate still flourishes as to when costly traits should be regarded as handicaps *sensu* Zahavi (e.g. Getty 1998a,b). However, there is now broad acceptance of the term as a metaphor for robust males being able to afford more costly traits (see Collins 1993).

Part of the early case against the handicap principle stemmed from **Fisher's fundamental theorem of natural selection** and the problem of dwindling heritable variation in strongly selected traits (see also Box 10.7). As happens time and again under artificial selection, consistent directional selection for a particular trait eventually exhausts the genetic variation in the trait because favoured alleles go to fixation. Any variation remaining is therefore purely environmental. In the context of mating preferences for viability indicators, this would mean that preferred traits eventually had nothing to say about heritable condition or viability and should be abandoned as a basis for mate choice. The fact that females seem to persist in their preferences under these conditions demands explanation. This has generally become known as the **lek paradox** (e.g. Kirkpatrick & Ryan 1991), since lekking species, where males provide no parental care or resources (see 10.1.2.2), and females stand to gain only genetic benefits from their mating preferences, are the extreme case.

While there have been many attempts to explain the paradox – for instance, suggesting that the costs of choice on leks are very low, or that females somehow gain direct fecundity benefits from mate choice (e.g. Reynolds & Gross 1990; Kirkpatrick & Ryan 1991) – a novel analysis by Pomiankowski & Møller (1995) suggests that the assumption of depleting variation under strong sexual selection from mating preferences may in fact be incorrect. Their comparative study concludes that additive genetic variation is actually *greater* among sexually selected characters than their non-sexually selected equivalents. Far from becoming less variable, therefore, sexually selected traits appear to be more variable than non-sexually selected traits. Pomiankowski & Møller suggest this is because strong directional selection for extreme characters favours modifier alleles (see 2.4.3.2) that increase the phenotypic variation about a mean trait size. The arms race that drives the elaboration of secondary sexual traits thus generates the genetic fuel to sustain itself

Difficulties and debates

Box 10.7 The handicap principle

Darwin invoked the idea of sexual selection because many secondary sexual characters in males – bright colours, elaborate displays, cumbersome weapons and so on – are likely to be an impediment to their survival, and thus seem to fly in the face of natural selection. While this special aspect of sexually selected characters has long been recognised, Amotz Zahavi (1975, 1977; Zahavi & Zahavi 1997) went a step further and suggested it was precisely *because* they were detrimental that such characters had evolved.

In Zahavi's **handicap principle**, females prefer males with costly traits because they reliably demonstrate that the male can survive despite them, and so must be of exceptional quality in other ways. As long as these exceptional qualities are heritable, females choosing on the basis of a handicap will pass them on to their offspring. Paradoxically, therefore, a character that is detrimental to survival can be used as an index of 'good genes'. It is essential that an index of quality is costly, and disproportionately so to low-quality males, otherwise it could easily be mimicked by low-quality cheats and would lose all predictive value.

Unfortunately, Zahavi's original formulation of the handicap principle was verbal and lacked a rigorous quantitative framework. Early attempts by other authors to model it formally all suggested that handicaps would not be evolutionarily stable. This was partly because they would be inherited by, and thus penalise, offspring that did not also inherit their father's other, superior, qualities, but partly because Fisher's fundamental theorem of natural selection predicts that heritable variation in strongly selected traits will eventually be exhausted (see text), thereby (in this case) removing the diagnostic value of any handicap. However, this early negative view has now changed dramatically. Why?

The conversion began when people considered some qualifications to Zahavi's idea. Zahavi's original concept can be thought of as a **qualifying** handicap (Grafen 1990b), where males survive or die as a result of developing the trait and females know the survivors must be good quality. Alternatively handicaps may be **conditional** (only high-quality males that can afford them develop the handicap at all) or **revealing** (males develop the handicap to demonstrate some otherwise hidden quality). The difference between conditional and revealing handicaps lies partly in the nature of the cost paid by their bearers (Dawkins 1995). The cost of conditional handicaps is paid in the currency that is being signalled. Thus roaring for a long time indicates high energy reserves by demonstrably burning up energy. The cost of revealing handicaps, in contrast, is not paid in the currency being signalled. For instance, bright plumage might indicate a good state of health, but not itself detract from it.

Both conditional and revealing handicaps remove the problem of saddling poor-quality offspring with a burden they are unable to sustain. However, disagreement remained as to whether these special cases could account for exaggerated traits independently of Fisher's runaway effect (e.g. Pomiankowski 1987; but see Collins 1993). The real breakthrough came when Grafen (1990b) used both games theory and a formal genetic approach to model selection for costly traits where males were free to choose levels of advertisement according to their assessment of their own condition, and females were free to believe or disbelieve them. The outcome could have gone in any direction, but in fact the stable solution turned out to be costly traits that accurately reflected individual male quality (and were disproportionately costly to poor-quality males), and female choice that trusted male signals. Grafen (1990b) referred to these honest, costly traits as **strategic choice** handicaps, the crucial point being that their honesty resided in their costliness, and their costliness provided their selective advantage, exactly as envisaged originally by Zahavi. However, as Collins (1993) points out, conditional, revealing and strategic choice handicaps are really all forms of conditional handicap, since their expression depends directly or indirectly on some aspect of the bearer's condition.

Collins (1993) and Getty (1998a,b) provide further discussion of the different types of handicap model and the assumptions behind them.

as well. A recent study by Kotiaho *et al.* (2001) bears out their argument. Kotiaho *et al.* looked at courtship activity in male dung beetles (*Onthophagus taurus*) and found that courtship rate was a condition-dependent trait for which females showed a preference. Importantly, male condition turned out to have a high genetic variance and to be genetically correlated with courtship. Thus female preference for high courtship rates conferred heritable genetic benefits on their offspring. What kinds of viability qualities in general might be indicated by secondary sexual traits?

(i) **Nutritional status and survivorship**. There is considerable evidence that sexually selected characters develop in relation to the nutritional status of the male. Antlers are a familiar example (e.g. Clutton-Brock *et al.* 1982; Goss 1983). Nutrition has a strong effect on antler growth, and male deer experiencing low food availability often have antlers that are small relative to their body size. Similarly the size of the nuptial crest of male newts declines under conditions of food shortage, while it increases with improved body condition (Green 1991; Baker 1992). The same is true for some secondary sexual characters in invertebrates, such as horn length in the beetle *Podischnus agenor* (Eberhard & Gutiérres 1991). The dependence of some colourful displays on hormones, or dietary elements such as carotenoids (see Olson & Owens 1998), also link sexually selected characters with nutrition (Andersson 1994). Testosterone levels in birds, for example, which are responsible for some secondary sexual adornments (but see Owens & Short 1995 and Kimball & Ligon 1999), are sensitive to nutritional constraint and fall sharply during periods of food shortage.

Trait development can also correlate with age, usually with older males bearing larger or more elaborate adornments or weapons (Smith 1965; Manning 1989; 12.2.1.1; but see Potti & Montalvo 1991). By choosing on the basis of trait development, females effectively choose older males that have demonstrated their ability to survive, or are likely to have experience in finding and defending resources and/or parental care. However, there is little evidence so far that females mating with older males pass on the longevity of their partners to offspring.

(ii) **Health**. An idea that has generated considerable interest, and a research industry to match, is that secondary sexual characters might indicate the state of health of the bearer (Hamilton & Zuk 1982). As Hamilton & Zuk note, general poor health and infection with parasites are often conspicuously apparent in the animal's external appearance. Bare patches of skin, matted or lacklustre fur or feathers, reduced vigour of displays and so on can all reveal something about current health and condition. By their very nature, bright and elaborately patterned secondary sexual characters seem especially likely to betray such evidence, and Hamilton & Zuk argue that that is exactly why they have evolved, as a diagnostic aid to mate choice (a 'revealing handicap' in the language of the handicap principle [Box 10.7]).

However, the hypothesis has a second rationale, in that it overcomes the apparent (see above) obstacle of Fisher's fundamental theorem. The arms race between parasites and their hosts is likely to generate genetic cycles of parasite infectivity and host resistance (Box 10.8), so that heritable fitness variation associated with indicator traits will be maintained in the host population despite persistent female choice. This, and the fact that challenges from parasites and pathogens are both widespread and costly, makes the Hamilton–Zuk hypothesis attractive as an explanation for the evolution of elaborate secondary sexual characters. So how good is the evidence for it?

Supporting evidence for Hamilton–Zuk has come from both comparative studies of different species and studies of relationships between sexual adornments, parasites and

Underlying theory

Box 10.8 The Hamilton–Zuk hypothesis

Hamilton & Zuk (1982) have argued that elaborate secondary sexual characters may reflect an individual's resistance to disease. The evolutionary arms race between parasites and their hosts is likely to maintain heritable variation in resistance, so females that choose males on the basis of traits indicating resistance stand to pass the attribute on to their offspring. The underlying principle is as follows.

Imagine a disease-causing parasite occurs in two genotypes, *P* and *p*. Its host also occurs in two genotypes, *H* and *h*. Parasites of genotype *P* can infect hosts of genotype *H*, but not hosts of genotype *h*, while parasites of genotype *p* can infect hosts of genotype *h* but not *H* (see table below). Imagine further that *P* and *H* individuals are initially the more abundant in their respective populations. *P* parasites attack *H* hosts, which means that hosts with the rarer *h* genotype will be at an advantage because they are resistant to *P*. The *h* genotype will therefore begin to spread. As *h* hosts increase in frequency, however, parasites of genotype *p* begin to gain an advantage over those of *P*, which swings the pendulum of advantage back to hosts of genotype H. Thus, there will be a frequency-dependent cycle of change in the predominant infection, which will tend to maintain heritable variation in resistance across hosts.

	Host genotype	
Parasite genotype	**H**	**h**
P	Can infect	Cannot infect
p	Cannot infect	Can infect

Based on discussion in Krebs & Davies (1987).

mating preferences within species. However, there are many conflicting results (e.g. Møller 1989; Milinski & Bakker 1991; Birkhead *et al.* 1998) and difficulties in testing the hypothesis (see Read 1990, Getty 2002 and Rolff 2002 for good discussions). One problem is that, as with many other evolutionary hypotheses, tests of Hamilton–Zuk have tended to be partial, in that they do not address all the components of the hypothesis. At least the following need to be confirmed for a robust test (Andersson 1994): (a) the reproductive success of the host decreases with increased parasite infection, (b) the condition of the indicator secondary sexual trait decreases with increased parasite burden, (c) there is heritable variation in resistance to parasites, (d) female choice favours males with the best developed indicator trait, and (e) females choose males that are the least parasitised. Most studies have tested only one or two of these, but some have managed to test all five (e.g. Kennedy *et al.* 1987; Møller 1988b, 1989, 1990a). Møller's study of the barn swallow is probably the best known.

Although socially monogamous, there is an advantage to male barn swallows in mating early in the breeding season, so males compete for early access to females who choose between them partly on the basis of the length of their characteristic outer tail feathers (Møller 1988b). By manipulating the length of the tail feathers with scissors and glue (following Andersson 1982 [Fig. 10.10]), Møller showed both that females

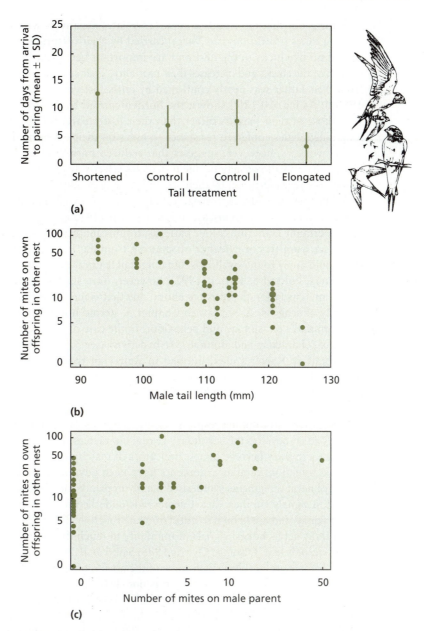

Figure 10.12 Long tail forks in male swallows are preferred by females and indicate a male's likely ectoparasite load. (a) Males with artificially elongated tails paired earlier than those with shortened or unaltered tails (control I, feathers cut and reglued; control II, feathers unmanipulated). © Nature Publishing Group (http://www.nature.com/), reprinted by permission. (b) Tail length in males correlated negatively with the number of blood-sucking mites carried by their offspring, here fostered onto another pair to avoid direct infection from their parents. (c) Cross-fostering showed that mite load on chicks correlated with that of their real father rather than their foster parent, so susceptibility to mites is heritable. From Møller (1988b, 1990a).

preferred (mated sooner with) males whose tails had been elongated (Fig. 10.12a), and that runaway elongation was likely to be halted by reduced foraging efficiency. But, was choice for long tails a Fisherian preference, or did tail length indicate some other quality of the male that benefitted offspring?

It turns out that tail length in males correlates negatively with the number of blood-sucking mites (*Ornithonyssus bursa*) carried by their offspring (Fig. 10.12b). Mites are picked up by chicks in the nest and are important because a high burden restricts the growth of the chicks and increases their mortality. That susceptibility to mites was inherited from their father was neatly confirmed by cross-fostering chicks between nests (Møller 1990a). As Fig. 10.12(c) shows, the burden carried by chicks correlated significantly with those of their parents rather than their foster parents. Thus, females mating with long-tailed males conferred resistance to mites on their offspring. Recent work by Saino and coworkers, however, suggests that reduced parasitism in the offspring of high-quality male swallows may be due to females selectively enhancing the level of immunoglobulins in eggs fertilised by their sperm (Saino *et al.* 2001).

While there is evidence to suggest that secondary sexual characters can reflect parasite burdens and resistance to disease, it is not always obvious how such an association might arise. Certainly elaborate colour patterns, trailing plumes and so on might make it easier to see ectoparasites or evidence of scouring from gut infections, but it is not clear that this would apply to the swallow's thin outer tail feathers or to, say, vocal secondary sexual characters. Folstad & Karter (1992), however, have suggested a general mechanism by which an association might come about, one that is mediated by steroid hormones, particularly androgens. As we saw in Chapter 3, steroid hormones can have several effects on an animal's physiology and behaviour. In the case of androgens, it is their dual effect on sexual characters and immune system activity (see 3.5.2.1) that is of interest here.

Folstad & Karter's idea, dubbed in Zahavian tradition the **immunocompetence handicap hypothesis,** is that elaborate secondary sexual characters may be sustainable only at some cost to immune function (Fig. 10.13a). Since androgens, such as testosterone, can have a depressive effect on immune responsiveness, only robust individuals that are able to mount an immune response despite high levels of immunodepressive hormone can afford to develop the secondary sexual characters that depend on them. Elaborate characters thereby become honest indicators of mate quality because they denote ability to resist infection. Similar arguments have been advanced about **carotenoids**, a large family of natural pigments synthesised mainly in plants and algae and acquired by animals for use in signals via their diet. Carotenoids underlie the coloration of many secondary sexual signals and are known to affect immune function. There is evidence that carotenoid-based traits act as honest signals of immunity in much the same way as those influenced by sex steroids (e.g. Olson & Owens 1998; Saino *et al.* 1999).

While there are problems with some of the assumptions behind the immunocompetence handicap hypothesis – for example, it is not clear that androgens are, in fact, generally immunodepressive (Box 10.9; e.g. Alexander & Stimson 1988; Hasselquist *et al.* 1999), and testosterone itself influences only a limited range of secondary sexual characters (Owens & Short 1995; Kimball & Ligon 1999) – there has been some support for it.

Marlene Zuk and coworkers looked at relationships between secondary sexual characters, testosterone concentration and immune function in red jungle fowl (*Gallus gallus*), the wild ancestor of the domestic fowl. Female jungle fowl prefer to mate with males possessing longer, redder combs, but pay little attention to plumage characteristics (Zuk *et al.* 1992). Zuk *et al.* (1995) found that comb size correlated positively with testosterone concentration (Fig. 10.13b) and negatively with lymphocyte (cells of the immune system) count (Fig. 10.13c). Moreover, lymphocyte count and testosterone concentration also showed a negative correlation (Fig. 10.13d). Thus, a testosterone-dependent secondary sexual trait used in mate choice appeared to be associated with immune depression, as predicted by the immunocompetence handicap hypothesis. These

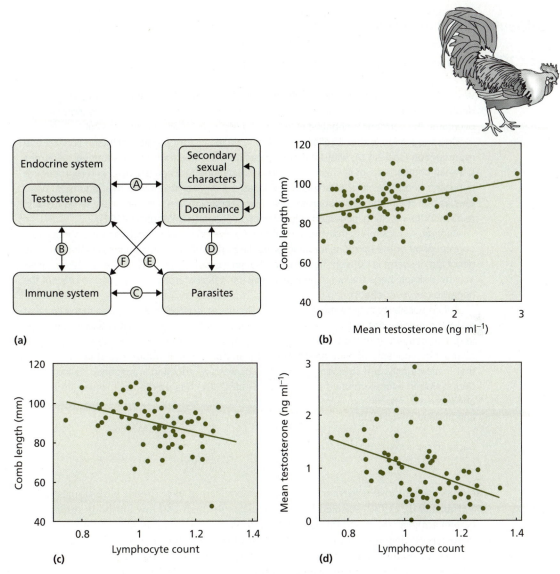

Figure 10.13 (a) Schematic representation of Folstad & Karter's (1992) immunocompetence handicap hypothesis. Testosterone, the primary androgen in vertebrates, has a positive effect on the development of secondary sexual characters and dominance (arrow A), while depressing immune response pathways (arrow B). Parasites interact with the immune system (arrow C), have a negative effect on secondary sexual characters (arrow D) and cause a reduction in immunodepressive testosterone secretion (arrow E). The development of testosterone-dependent secondary sexual characters is also compromised by reduced immunocompetence (arrow F). There is thus a feedback system linking the development of secondary sexual characters to individual resistance to parasites. After Folstad I. and Karter A.J. (1992) Parasites, bright males and the immunocompetence handicap. *American Naturalist* **139**: 603–22, reprinted by permission of The University of Chicago Press, © 1992 by The University of Chicago. All rights reserved. (b) Comb length in red jungle fowl correlates positively with testosterone concentration and (c) negatively with lymphocyte count. (d) Lymphocyte count in turn is negatively correlated with testosterone concentration. See text. After Zuk *et al.* (1995).

Difficulties and debates

Box 10.9   Why is testosterone immunodepressive?

While there is good evidence that testosterone secretion is modulated in relation to its depressing effects on immune function, the question remains as to why selection has not overcome the hormone's negative effects in the first place (Penn & Potts 1998a).

One suggestion is that it reflects the inevitable life history tradeoff between survival and reproduction (see 2.4.5), with testosterone acting as a mechanism for shunting resources between components of life history, in this case immune function (survival) and reproductive behaviour and physiology (Wedekind & Folstad 1994). A problem with this argument, however, is that the metabolic resources saved by depressing immunity are likely to be trivial in relation to the resulting increased risk of infection (Hillgarth & Wingfield 1997).

Another suggestion is that the immunodepressive effect of testosterone is an incidental consequence of the hormone's role in protecting sperm from autoimmune attack (since the genotype of sperm is different from that of the male producing it) (Hillgarth *et al.* 1997). However, this simply begs the question again: why can't testosterone be modified to prevent the negative carryover effect?

A third possibility may be more promising. Braude *et al.* (1999) have pointed to the fact that testosterone may not in fact be an immunodepressant, but, like glucocorticoid hormones (e.g. corticosterone and cortisol) may be responsible for redistributing resources across different components of immune response (the **immunoredistribution hypothesis**). Studies that measure only one component may thus misread a reduction in response as indicating depression in overall immune capacity. As Rolff (2002) has pointed out, other life history characteristics affecting investment in immunity may also be confounded with hormone status and mislead interpretation.

Based on discussion in Hillgarth *et al.* (1997).

---

results are supported further by an artificial selection experiment with domestic fowl. Verhulst *et al.* (1999) selected lines of chickens for high and low antibody response to a benign antigen (sheep red blood cells) and looked at the development of the comb (again testosterone-dependent) in each. They found that comb size was larger in males from the low antibody line than those from the high line and was intermediate in a randomly selected control line. Testosterone levels differed similarly between lines. The outcome thus strongly supported the idea that a hormone-mediated tradeoff with immune function constrains the development of a secondary sexual character in chickens.

Many tests of the Hamilton–Zuk and immunocompetence handicap hypotheses have focused on visual displays, particularly in birds. However, the ideas apply equally to other sensory modalities and taxonomic groups. Many species use odours as mate attractants, and sexual dimorphism in costly odour production can result in an olfactory equivalent of the peacock's tail (Penn & Potts 1998a). The urinary scent marks of male house mice, for example, are influenced by testosterone and expensive to produce, judging by the mass lost by heavily marking dominants (Gosling *et al.* 2000). They also carry information about the male's infection status, with females showing less interest in the urine of infected males (Kavaliers & Colwell 1995). In addition, urinary odour is influenced by the highly polymorphic genes of the major histocompatibility complex (MHC), a crucial component of the immune system (Hurst *et al.* 2001; see below). Females are able to respond to the MHC component of male urinary odour, and since

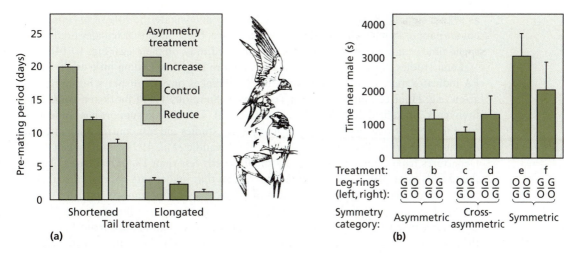

Figure 10.14 (a) The time taken for male barn swallows to pair with a female varied in relation to both the length and the asymmetry of their tail feathers. Males with longer or more symmetrical tails paired more quickly. The length and asymmetry of feathers were manipulated experimentally. After Møller (1992). (b) Female zebra finches prefer (spend more time near) males with symmetrical colour rings on their legs. After Swaddle & Cuthill (1994). Bars are means with standard errors. © Nature Publishing Group (http://www.nature.com/), reprinted by permission.

the expression of MHC genes increases when the immune system is activated, they can glean information about the male's current ability to mount an immune response (Penn & Potts 1998b).

Acoustic displays can also convey information about immunocompetence (e.g. Buchanan *et al.* 1999; Ryder & Siva-Jothy 2000; Duffy & Ball 2002). Ryder & Siva-Jothy (2000) found that a sexually selected component of song in male house crickets (*Acheta domesticus*) correlated positively with a measure of the ability to encapsulate pathogens. The degree of encapsulation was also associated with body size, and both traits were heritable. Females preferring song characteristics associated with a strong encapsulation response thus passed both this and large body size on to their offspring (Ryder & Siva-Jothy 2001).

(iii) Developmental stability. Another influential suggestion has been that females choose males on the basis of developmental stability; that is, evidence of an ability to resist environmental insults, or of possessing an otherwise competent genotype. As we saw in 5.3.3, a putative measure of developmental stability is provided by the random departures about planes of morphological symmetry known as fluctuating asymmetry (FA). FA has been shown to increase under various conditions of genetic and environmental stress, of which strong directional selection is one (Møller & Swaddle 1997; 5.3.3; but see Bjorksten *et al.* 2000). Since elaborate secondary sexual characters have been subject to strong directional selection, they should be especially vulnerable to developmental instability and thus FA. In principle, therefore, the degree of asymmetry in such traits might provide a ready indicator of the underlying quality of a male and his ability to resist developmental stress.

Support for females basing their choice of mate on character asymmetry has had a chequered history. Several early studies were positive. Møller (1992), for instance, found that female barn swallows chose males on the basis not only of tail length, but also the degree of asymmetry in the outer tail feathers (Fig. 10.14a). Importantly for the 'good genes' argument, asymmetry in tail feathers *decreased* with increasing length, suggesting only high-quality males developed long feathers. In a cleverly designed experiment, Swaddle & Cuthill (1994) used the fact that female zebra finches treat coloured leg rings on males as sexual signals (Burley 1986; see Box 10.3) to investigate the role of asymmetry in colour pattern in mate choice. When given a choice, they found that females preferred to spend time near males with symmetrically coloured rings on their legs (Fig. 10.14b).

Later studies, however, have tended to be more negative, finding little evidence that FA indicates mate quality, affects mating preferences or even displays heritable variation (e.g. Hunt & Simmons 1998; Tomkins & Simmons 1999; Bjorksten *et al.* 2000; Breuker & Brakefield 2002; Goncalves *et al.* 2002). This shift in support has come about for a number of reasons. One is that many early studies did not control adequately for measurement error (Palmer & Strobeck 1986; Swaddle *et al.* 1994; van Dongen 2001). Departures from symmetry are usually small, mostly less than 1% of trait size. They are thus highly susceptible to measurement error, so that inadequate controls are likely to inflate the number of reported instances of FA. It is also important to control for other forms of asymmetry, such as directional or antisymmetry (see Box 5.4; van Dongen 2001), because these have a quite different functional significance from FA. Again, these analyses were often lacking. Less than rigorous controls, coupled with selective reporting of positive outcomes early on, may therefore have contributed to a substantial publication bias in favour of the predicted effects of FA (Houle 1997; Palmer 1999; Simmons *et al.* 1999; but see Thornhill *et al.* 1999), a bias that is now being counterbalanced by more critical analyses (Simmons *et al.* 1999). As Simmons *et al.* (1999) point out, the reverse in fortune of FA as a vehicle for sexual selection reflects a common pattern in scientific bandwagons.

Increasing scepticism about FA as an indicator mechanism, however, does not necessarily rule out an association between FA and mate choice. The attractiveness of symmetry may have a significance that is more deeply rooted than quality indicator mechanisms. Symmetry is one of the cornerstones of Gestalt perception (the perception of entities as wholes rather the sums of their parts) in humans, and symmetrical stimuli have repeatedly been shown to have greater efficacy than asymmetrical contrasts in discrimination tasks in other species (e.g. Giurfa *et al.* 1996; Horridge 1996). The tendency to home in on symmetry is even evident in neural network models (3.1.3.2, 3.5.1.1) trained for arbitrary pattern recognition (Enquist & Arak 1994; Johnstone 1994). A possibility that arises from this is that preference for symmetry reflects a historical sensory bias (see 3.5.1.1 and below), perhaps evolving originally to distinguish the typical symmetries of living forms from non-living backgrounds that generally lack symmetry.

It may also be that mating preferences are based on developmental stability, but through its effects on an animal's performance (see 5.3.3) rather than FA in secondary sexual characters. In the decorated field cricket (*Gryllodes sigillatus*), for example, FA in a non-sexually selected character (the hind femur) correlates negatively with sperm production and positively with the time taken to copulate with a female (Farmer & Barnard 2000). However, the magnitude of FA is vastly too small to be resolved visually by the crickets and act as an indicator, so the impact on mating performance presumably arises from other cues relating to individual quality.

(iv) **Genetic complementarity/compatibility**. Females may choose on the basis of genetic quality, but in a relative rather than absolute sense. Instead of choosing males with 'good genes' *per se*, a female may choose a male whose genes appear to complement her own. Here, it is the particular combination of male and female genes that determine offspring viability and reproductive success, so a male that is best for one female may not be best for another (Tregenza & Wedell 2000). Choosing on the basis of **genetic complementarity** has the additional bonus of not falling foul of Fisher's fundamental theorem and the lek paradox because there is no consistent preference for a particular genotype.

Table 10.1 Possible costs of inbreeding and outbreeding. From Bateson (1983).

Costs of inbreeding
1 Deleterious genes more likely to be fully expressed.
2 Beneficial interactions (heterozygous advantage or over-dominance) between different alleles at same genetic locus lost.
3 Offspring insufficiently variable for one or other to cope with a varying environment.
4 Offspring more like each other and so compete more intensely.

Costs of outbreeding
1 Genes required for adaptation to particular environment lost or suppressed.
2 Coadapted gene complexes broken up by recombination.
3 In polygynous species advantage of having extra-closely related offspring lost and parental genes less well represented in next generation.
4 Infection from pathogens carried by mate more likely.
5 Travelling into another population costly and dangerous.
6 Acquired skills useful in one environment not appropriate in another.
7 Mismatch of habits acquired by mates in different environments disrupts parenting.

Apart from choosing a mate of the right species, perhaps the most obvious examples of choice for complementarity are mating preferences for different degrees of relatedness. Extremes of both inbreeding and outbreeding may incur genetic penalties by increasing the likelihood of offspring being homozygous for deleterious recessive alleles (inbreeding), or breaking up coadapted gene complexes (outbreeding) (Table 10.1; see Bateson 1983). Thus it may pay individuals to compromise between the extremes and go for a mate of intermediate relatedness (the **optimal inbreeding/outbreeding hypothesis** [Shields 1982; Bateson 1983]). As well as serving phenotypic altruism (2.4.4.5, 9.3), therefore, kin recognition/discrimination (see 9.3.1.3) may also play a role in mate choice. Various experimental studies have supported the idea. Bateson (1982), for instance, offered male or female Japanese quail (*Coturnix japonica*) the opportunity to window shop (Fig. 10.15a) for mates of different degrees of relatedness, ranging from full siblings to unrelated individuals. He found that both males and females preferred (spent most time viewing) first cousins (Fig. 10.15b). Similar results have emerged with outbred laboratory mice, in which males preferred females that were first or second cousins, depending on prior familiarity, and produced larger litters from such matings, so suggesting a reproductive advantage to the preference (Barnard & Fitzsimons 1988, 1989). The odour of cousins and more distant relatives also accelerated puberty in females of the same strain of mice (Lendrem 1985). That Darwin himself showed a preference for (indeed, married) a first cousin (Emma Wedgwood) might be taken as the clincher for the hypothesis.

While relatedness may provide an indirect means of optimising the genotype of offspring, there are circumstances in which loci linked directly to immune function also influence traits used in mate choice. MHC loci are a good example.

The vertebrate MHC is a cluster of genes principally producing proteins that present foreign peptides to cells of the immune system. It is inherited as a unit (a **haplotype**) and is highly polymorphic and heterozygous. Variation in the MHC between individuals is associated with differences in resistance to parasites and autoimmune disease (Apanius *et al.* 1997). But it also contributes to differences in individual odour by producing

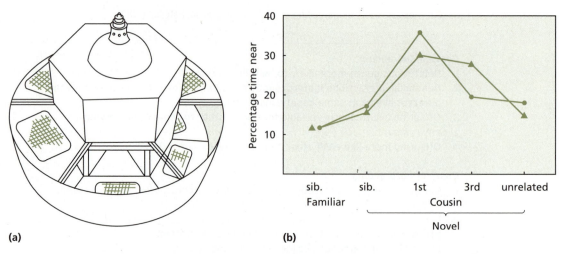

(a) **(b)**

Figure 10.15 Relatedness can play an important role in mating preferences. (a) Bateson's 'Amsterdam' apparatus in which female Japanese quail (*Coturnix coturnix*) 'window-shopped' for potential mates. The apparatus was so-called because of the somewhat similar arrangement for choosing prostitutes in Amsterdam's famous red-light district. (b) Both male (triangles) and female (circles) quail showed a preference for first cousins. See text. After Bateson (1983).

proteins that are soluble or bind to volatile molecules, or affecting gut bacteria (Tregenza & Wedell 2000). Because of its high variability, this second property of the MHC allows subtle levels of olfactory social discrimination, including individual (Hurst *et al*. 2001), and possibly kin (but see 9.3.1.3), recognition. There is substantial evidence that the MHC plays a role in mate choice between unrelated individuals, with MHC-associated mating preferences emerging in mice (e.g. Yamazaki *et al*. 1976, 1988; Potts *et al*. 1991), rats (e.g. Brown *et al*. 1987) and humans (e.g. Wedekind *et al*. 1995; Ober *et al*. 1997; see 12.2.1.1). Studies of laboratory mice have often used so-called **congenic** strains that are genetically identical except for their MHC haplotype, and have repeatedly demonstrated disassortative mating with respect to the MHC in both males and females, homozygotes showing the strongest preferences (Yamazaki *et al*. 1976; Penn & Potts 1999). Such effects may account for the marked deficiency of MHC homozygotes in some natural populations (Tregenza & Wedell 2000), including those of humans (e.g. Degos *et al*. 1974; Markow & Martin 1993) where it appears that sharing certain MHC alleles, or entire haplotypes, can lead to increased foetal mortality (Ober *et al*. 1998). Interestingly, while individuals can use odour cues to mate disassortatively by MHC haplotype, the MHC may also influence postcopulatory mate choice. There is evidence in mice that MHC-derived proteins expressed on sperm influence the likelihood that a given spermatozoan will fertilise the egg, thus making choice of MHC genotype possible at the gametic level (Wedekind *et al*. 1996; Rulicke *et al*. 1998).

More recently, attention has turned to the potential role of selfish genetic elements (see 2.4.3.2) in mate choice. As we saw in Chapter 2, selfish genetic elements promote their own transmission at the expense of other genes in the genome, selecting for counteracting modifier genes in the process. Several studies suggest that females may choose between males on the basis of whether they carry such elements or their modifiers. In Wilkinson's stalk-eyed flies (see above), females appear to prefer males carrying modifiers that suppress selfish elements linked to the X chromosome (Wilkinson *et al*. 1998), while female house mice avoid males that are heterozygous for so-called *t* alleles, a multi-locus segregation distorter (2.4.3.2) that is lethal in the homozygous recessive state (e.g. Lenington 1991). In the case of *t* alleles, the gene influencing female mating preference seems to lie within the *t* complex itself, which also contains genes associated with specific odours (Drickamer & Lenington 1987). This allows both homozygous and heterozygous *t* females to avoid the penalty of mating with a heterozygous *t* male, though heterozygous females, which run the greater risk of producing homozygous offspring, show stronger discrimination.

Non-genetic benefits   In many cases, females may gain benefits in kind, rather than 'good genes', from mating preferences. Males may defend territories containing crucial resources, for instance, or they may differ in their ability to fertilise the female's eggs. Her choice of mate may therefore have a direct impact on a female's fecundity.

In North American bullfrogs (*Rana catesbeiana*), males defend territories in ponds and lakes where females come to spawn (see 2.4.5.1). Their eggs, however, are susceptible to predation by leeches (*Macrobdella decora*), and females appear to discriminate in favour of territories offering greater hatching success. How do they do it? Two factors that seem to be important are water temperature and vegetation density. The warmer the water, the faster eggs develop and the less time they are exposed to leeches, while a relatively low density of vegetation allows eggs to coalesce into a ball and become more difficult for leeches to attack. Territories meeting these criteria are preferred by females and thus contested vigorously by males (Howard 1978).

Males can also boost female fecundity more directly. So-called **nuptial gifts** are a good example. Like those of several other insect species, male hanging flies (*Bittacus apicalis*) in the woodlands of North America catch prey that they then offer to females. Having secured a 'gift', the male hangs from a twig and emits a pheromone to attract a female. When a female approaches, the male offers his 'gift' and, if the female accepts it and begins feeding, the pair copulates. The larger the 'gift', the longer it takes the female to eat it (Fig. 10.16a), and the longer she takes, the more sperm she acquires from the male (Figure 10.16b) (Thornhill 1976). By choosing a male offering a large 'gift', the female benefits in two ways: first she gains nutrients that allow her to lay more eggs, and second she has less need to hunt for food herself, a dangerous activity that frequently leads to death in a spider web.

Of course, nuptial gifts are likely to cost the male something, so we should expect males to regulate them so as to maximise their net reproductive success. The males of some cricket species produce a nuptial gift in the form of an accessory gland secretion, known as a **spermatophylax**, attached to the ampulla of the spermatophore (package of sperm). This can be substantial, male Mormon crickets (*Anabrus simplex*), for example, losing 27% of their body mass in its production, an investment that makes them a valuable resource for females that then compete to mate with them (Gwynne 1981). When the spermatophore is transferred to the female reproductive tract, the female bends round and begins to eat it (Fig. 10.17a), but first she has to eat her way through the spermatophylax. The principal role of the spermatophylax from the male's point of view appears to be to protect the ampulla and its cargo of sperm from ingestion while sperm transfer takes place. Analyses of different species have shown that the size of the spermatophylax is just large enough to allow the complete transfer of sperm (e.g. Vahed & Gilbert 1996; Fig. 10.17b,c), suggesting that the male's investment is finely tuned to the job in hand. While their nutritional value to the female appears to vary across species (Gwynne 1981; Simmons 2001), spermatophylaxes are often rich in amino acids that stimulate feeding, implying exploitation of the female's feeding response as part of the male's management of sperm transfer (Simmons 2001).

Figure 10.16 Nuptial gifts and mating in hanging flies (*Bittacus apicalis*). (a) Larger nuptial prey result in pairs copulating for longer, though copulation generally ceases after about 20 minutes. (b) Longer copulations lead to more sperm being transferred to the female. Note that a male must offer a 'gift' that takes at least five minutes to consume, otherwise no sperm transfers to the female. After Thornhill R. (1976). Sexual selection and nuptial feeding behavior in *Bittacus apicalis* (Insecta: Mecoptera). *American Naturalist* **110**: 529–48, reprinted by permission of The University of Chicago Press, © 1976 by The University of Chicago. All rights reserved.

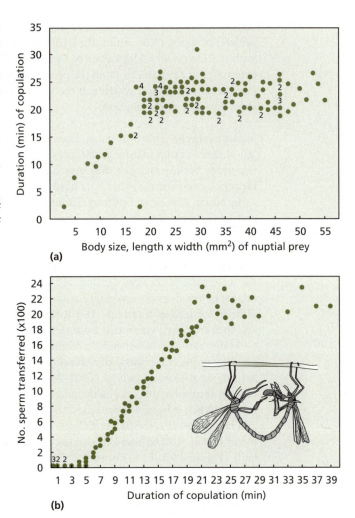

Direct fecundity benefits may underlie some mating preferences on leks, thus going some way to overcoming the apparent lek paradox (see above). A number of studies suggest that females may choose males according to their ability to fertilise the female's complete complement of eggs. This seems to be the case in the frogs *Uperolia laevigata* and *Ololygon rubra*, for instance, where females prefer males of a size that maximises the fertilisation of their eggs (Robertson 1990; Bourne 1993), probably due to an effect of relative size on the mechanical efficiency of external fertilisation (Ryan 1997). Fertilisation efficiency may also be affected by parasites, thus providing a direct incentive for females to avoid parasitised males. In the bush cricket *Requena verticalis*, the male's ability to donate nutrients via the spermatophore is affected by gut parasites, so that females mating with more heavily infected males had reduced fecundity (Simmons *et al.* 1994). Effects of parasite infection on mate choice may thus sometimes reflect direct fertility costs of infection rather than Hamilton–Zuk style indicator mechanisms.

Direct benefits can arise in various other ways. For example, in moths of the genus *Utetheisa*, adults and eggs are protected from predators by distasteful alkaloids that have to be sequestered from plants as the larva feeds (Eisner & Meinwald 1995; see 8.3.5.3). When they mate, males transfer some of their alkaloids to the female in their spermatophore from whence they find their way into the eggs. Males attract females by

(a)

Figure 10.17 Males of the tettigoniid cricket *Requena verticalis* produce a spermatophylax that protects the spermatophore, and thus the sperm inside, from being eaten by the female (a) before sperm transfer is complete. (b) By removing spermatophores from females at different times after attachment (open circles), Vahed & Gilbert (1996) found that the spermatophylax was just big enough to allow all the sperm to be transferred before it was eaten (the large closed circle shows the mean time to consume the spermatophylax). Small, closed circles are females that removed spermatophores themselves after an uninterrupted mating. (c) The relationship between sperm number and spermatophylax size across tettigoniid species. The spermatophylax is bigger in those species producing more sperm, as expected if it functions to protect sperm during transfer to the female. See text. After Vahed & Gilbert (1996). Photograph of female *R. verticalis* consuming a spermatophore courtesy of Leigh Simmons.

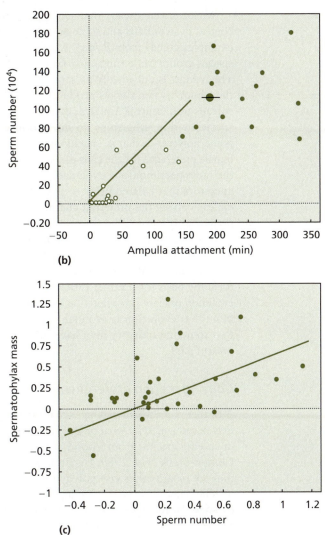

emitting a pheromone, which itself is derived from the sequestered alkaloids. Thus, males effectively advertise their alkaloid status to females that are then in a position to choose on the basis of protective benefits for the offspring.

Sensory exploitation

The runaway effect and indicator mechanisms are both ways in which traits can be selected as the basis for mating preferences. However, such traits might become involved in mate choice as a consequence of selection acting on other attributes that have nothing to do with mating. Sensory systems have evolved under many different selection pressures: to find food, avoid predators, identify dominants, recognise offspring and so on (Ryan 1997). Once tuned to particular traits for whatever purpose, a female's sensory systems may be predisposed to exploit them during mate choice (sensory exploitation; see 3.5.1.1). Thus preference for a trait may reflect historical function rather than *de novo* selection for choosing a mate.

We have seen already that such predispositions can influence preferences in a mating context in swordtails and some other organisms (3.5.1.1), but one of the best documented examples comes from Ryan's work on túngara frogs of the *Physalaemus pustulosus* species group. Male túngara frogs produce a call composed of a 'whine' and a 'chuck'. The 'whine' is universally present and is essential for mate recognition by females, but the 'chuck' is not always produced. The *P. pustulosus* group straddles the Andes and spreads into Central America, dividing broadly into two phylogenetic subgroups. Some species within subgroups do not produce 'chucks', while others do. The important point, however, is that if 'chucks' are added digitally to the calls of non-'chuck' producers, female preference for the calls is enhanced. From the phylogenetic evidence for the group, the most parsimonious explanation is that the preferences are derived from common ancestors that existed prior to the divergence of the present call types, implying that males evolving the preferred characteristics did so in response to a pre-existing female preference (Ryan 1997).

It is clear that sensory exploitation has not driven the evolution of all secondary sexual traits used in mate choice, but the accumulating evidence suggests that it has had an important role in many cases. Moreover, as we have seen in 5.3.1, it could form the basis for a rather different interpretation of the role of sexual selection in the evolution of elaborate traits. As we pointed out in 2.2, therefore, it is essential to bear in mind that evolution is as much about history as it is about current adaptation, and that sometimes history may have more to say about how things have arrived at their present pass.

Copying mating preferences

A further point worth making is that mating preferences may not be independent. Females may be influenced by what other females choose. Mate-choice copying appears to take place in several species, from isopod crustaceans to deer (e.g. Marconato & Bisazza 1986; Clutton-Brock *et al.* 1989; Dugatkin 1992a; Schlupp & Ryan 1997), but especially those that lek or are otherwise polygynous (Andersson 1994). Obvious benefits of copying include reduced costs of sampling and choice, particularly where male quality is difficult to assess, or there is a high risk of predation (e.g. Boyd & Richerson 1985; Gibson *et al.* 1991). While copying may be difficult to distinguish from social attraction for other reasons (Kraak 1996; Lafleur *et al.* 1997; Westneat *et al.* 2000; Table 10.2), ingenious recent field studies of sailfin mollies (*Poecilia latipinna*) have managed to control for such potential confounding effects and strongly suggest that both sexes copy mating preferences (Witte & Ryan 2002). Work with the closely related guppy, *P. reticulata*, has also shown that an initial act of copying can influence the choice of several subsequent females, so that socially acquired preferences can spread through the population (Dugatkin *et al.* 2002). Copying is not universal, however, and, even in some other poeciliid fish, appears not to play any significant role (Applebaum & Cruz 2000).

Apart from generating non-independence in mate choice within populations, copying may influence the strength of sexual selection for male traits (Agrawal 2001). If copying boosts the reproductive success of particular males, then sexual selection for their traits may increase (Andersson 1994). However, there is an obvious counter-argument: with fewer females assessing males independently, there is more scope for chance to influence acquired preferences, so correlations between particular trait characteristics and reproductive success may in fact be weakened (Gibson *et al.* 1991). Agrawal (2001) models some of the evolutionary consequences of copying more formally.

Table 10.2 Some potential causes of non-independent mating preferences. From Westneat *et al.* (2000).

Mechanism	Change in probability that focal female makes a particular mate choice	
	Increase	**Decrease**
Non-learned		
Stimulus enhancement/ reduction	Presence of female near male makes him more likely to be detected (and therefore assessed) than lone males	Pairs go into hiding to copulate; copulating males are therefore less likely to be noticed than males not copulating
Contagion/inhibition	Male postcopulatory display increases female receptivity to all males	In a dominance hierarchy, mating by the alpha female induces reproductive suppression in subordinates
Stimulus response	Females prefer males with eggs to those without (fathead minnow, *Pimephales promelas*: Unger & Sargent 1988)	Paired males advertise less (burying beetle, *Necrophorus defodiens*, in which female attacks her mate if he advertises; Eggert & Sakaluk 1995)
Learned		
Association-to-location	Females copulate at sites where they earlier observed high display rates, which are correlated with copulation success (some lekking birds: Gibson *et al.* 1991)	Female avoids copulating at sites where she previously saw a male-female pair (pied flycatcher, *Ficedula hypoleuca*: Alatalo *et al.* 1981)
Association-to-male	Observing female prefers male previously seen near a female over a male that appeared to be alone (Japanese medaka, *Oryzias latipes*: Grant & Green 1996)	Cue that model female resisted a copulation leads focal female to avoid that male in a later encounter
Association-to-trait	After observing the mate choices of older females, inexperienced females prefer similar males	Female prefers males that do not resemble her father (the male chosen by her mother) and consequently avoids inbreeding depression (Japanese quail, *Coturnix japonica*; Bateson 1982)
Cognition	Observing female adopts a form of idiosyncratic courtship that was used successfully by the model female	After observing a male compromise his offsprings' survival, females avoid mating with that male in order to avoid a similar outcome

Sex differences in mate choice revisited

Throughout our discussion we have assumed females as the choosing sex because they generally have the greater parental investment and lower reproductive rate. Indeed, cases where males have done the choosing (sex role reversals; 10.1.2.2) have been exceptions that prove the rule, in that males in these instances have made an unusually substantial investment in paternal contribution (as in Gwynne's [1981] Mormon crickets), or there has been a shift in the relative payoff to females from having additional broods rather than investing more in any one of them (as in some shorebirds [see 10.1.2.2]). However, as Johnstone *et al.* (1996) point out, while the relative investment argument can account for sex differences in the benefits of multiple mating and thus competition for access to mates, it is less clear that it accounts for differences in choosiness. In fact, careful consideration of the benefits of choice highlights an apparent paradox. The payoff from choosing depends on variation in quality between members of the opposite sex. If individuals are all of more or less the same quality, there is little point in being choosy. Now, if females are the higher-investing sex they are likely to exhibit greater variation in parental quality (Johnstone *et al.* 1996), so, arguably, the benefits of choice should be greater for the lower-investing sex, i.e. males. This is likely to be especially so in lekking species, where females stand to differ in both genetic quality and ability to produce and provision offspring, but males differ only in genetic quality (since males on leks provide no parental care).

However, this argument ignores potential sex differences in the cost of choosing (Pomiankowski *et al.* 1991 and Box 10.6). Since their lower investment in parental care allows a greater reproductive rate, any time spent searching for another mate after rejecting a previous one will be more costly to males than to females. Moreover, the lack of paternal care will tend to produce a male-biased operational sex ratio (the ratio of reproductively active males and females; Emlen & Oring 1977), so that the time needed to find another mate will be greater for males than females. Overall, therefore, the costs of choosiness are likely to be greater for males. The opposing considerations of variation in individual quality and costs of choice thus make general predictions about sex differences in choosiness more complicated than usually assumed (Parker 1983b). The issue is trickier still when both sexes are predicted to be choosy (e.g. Crowley *et al.* 1991) or the freedom to choose is curtailed by harrassment (Clutton-Brock & Parker 1995) or **forced copulation** (almost always inflicted by males; e.g. Thornhill 1980; McKinney *et al.* 1983; LeBoeuf & Mesnick 1990; see also 12.2.1.1). Games theory models incorporating these kinds of factor suggest that costs of choosing a mate are likely to have a much stronger bearing on which sex is choosy than sex differences in variation in mate quality. Moreover, when males and females show similar levels of parental care, differences in mate quality can drive up both choosiness and competition for mates in the *same* sex (Johnstone *et al.* 1996).

Multiple mating by females

The fact that females, as well as males, often mate with several different partners (polyandry) seems to be testimony to some of the points above. On the basis of the traditional parental investment argument, multiple mating by females should be relatively unusual. So why do some females do it? Various explanations have been proffered (Tregenza & Wedell 1998). Mating with more than one male might increase the chances of gaining a high-quality mate, for example, or it might promote sperm competition and benefit the female's sons by passing on a capacity for winning out in the scramble to fertilise eggs (Kempenaers *et al.* 1992; Keller & Reeve 1995). There may even be a correlation between the competitiveness of a male's sperm and the viability of his offspring (Madsen *et al.* 1992). Yet other possibilities are that multiple mating is a means of mitigating the potentially negative effects of genetic incompatibility between mates (see above), or of generating genetic diversity among siblings (Ridley 1993a). Unfortunately, while there are plenty of plausible hypotheses, there have, as yet, been few rigorous attempts to distinguish between them. However, Tregenza & Wedell (1998) have provided one such test using the field cricket *Gryllus bimaculatus* in which females regularly mate with more than one male (Simmons 1986).

By allowing females to mate the same number of times, but with different numbers of males, Tregenza & Wedell found that the hatching success of eggs increased with the number of mates. This was not due simply to polyandry increasing the quantity or fertilising ability of sperm, because control females receiving half the number of copulations showed no reduction in hatching success (Tregenza & Wedell 1998). Moreover, when given males were mated with different females, they had no consistent effect on hatching success across partners. The outcome thus suggests that improved hatching success among polyandrous females was not the result of females being more likely to mate with high-quality males that enhanced egg viability or fertilisation efficiency, but was due to an increased likelihood of encountering a genetically *compatible* male (Tregenza & Wedell 1998). Thus polyandrous mating in female *G. bimaculatus* may be driven by the benefits of matching individual genotypes in order to secure complementary, and so fruitful, partnerships.

10.2 The evolution of parental care

Finding a mate is, of course, only one step in the process of reproducing successfully, but it has commanded a great deal of attention over the last three decades, largely because of the difficulty in distinguishing between rival hypotheses empirically. In many species, however, successful mating is followed by a more or less protracted period of parental care. Despite this, the evolution of parental care itself has not attracted the level of interest devoted to other aspects of reproductive decision-making (Clutton-Brock 1991), a curious imbalance that is partly attributable to the rise of parental investment theory and a shift in attention to its role in sexual selection.

10.2.1 Parental care and parental investment

The distinction between parental care and parental investment is frequently blurred, but it is important to make sure it is clear. Clutton-Brock (1991) defines parental care as any form of parental behaviour that seems likely to increase reproductive potential of an individual's offspring. This is purely a descriptive definition and makes no assumptions about the reproductive cost of care to the parent. Trivers's (1972) notion of parental investment (10.1.2), on the other hand, which drives much of the established thinking about sex differences in behaviour, is entirely concerned with the impact of current care on the parent's future reproductive potential. In its broadest sense, parental care embraces everything from building nests and digging burrows to producing yolked eggs and caring for eggs or young inside or outside the parental body, even, in some cases, beyond the stage of nutritional independence (Clutton-Brock 1991; Box 10.10). In its narrower sense, it is confined to the care of eggs or young once out of the parental body. Either way, the time and resources lavished on parental care is sometimes referred to as 'parental effort', which, as we have seen (10.1.2), is a term relating to the reproductive costs of care and thus parental investment. Clutton-Brock (1991) therefore uses the term 'parental expenditure' to avoid the confusion. One reason why it can be misleading to

Background information

Box 10.10 Types of parental care

Animals show many different kinds of parental care, with taxonomic and ecological factors playing a major role in determining parental strategies. Within these, however, a number of broad forms of care can be recognised (Clutton-Brock 1991).

Establishing and maintaining nests, burrows or territories

Males and/or females of many species, both invertebrate and vertebrate, dig, build or otherwise prepare nest constructions of some kind, whether they are burrows in the ground or woven edifices in trees. In some cases, males call from (West & Alexander 1963), or display near (Tinbergen 1951), the constructions to attract females, which then lay their fertilised eggs in them. In other cases, females construct burrows or nests themselves (as in digger wasps; see 5.1.1.3). Males may also defend territories around resources needed by females to rear offspring (10.1.2.2). However, it can be difficult to know whether these behaviours represent *parental effort* or *mating effort* (10.1.2), since females often refuse to mate with males that do not possess appropriate constructions or territories.

Production and provisioning of gametes

The production of well-provisioned gametes by females constitutes the main form of parental expenditure in many species. Gamete size is often directly related to offspring growth and survival (Fig. (i)a,b), though there are exceptions, and larger eggs can sometimes take longer to hatch, so incurring a greater risk of mortality (e.g. Karlsson & Wiklund 1984). Males can contribute to the production of female gametes in various ways, including defending resources needed for eggs, offering nutritious nuptial gifts, passing on nutrients in ejaculates (Rutowski *et al.* 1983) or even allowing themselves to be eaten by the female (e.g. Buskirk *et al.* 1984).

Care of fertilised eggs

Care of fertilised eggs by one or both parents is found across a wide range of taxonomic groups. Care may be necessary to guard against predation or brood parasitism, or to maintain appropriate environmental conditions (temperature, salinity, oxygen levels, etc.) for pre-hatching development. Various forms of care have evolved, including laying eggs on carefully chosen substrate or in nests or burrows, one or both parents carrying eggs (attached or unattached) around with them, the retention of fertilised eggs within the female reproductive tract, and protection within some other part of the parent's body, such as the mouth (in mouth-brooding cichlids and some amphibians), specialised brood chambers (as in seahorses and pipefish) or the stomach (as in some frogs).

Care of offspring without provisioning

Some species care for young without further provisioning. This often takes the form of guarding aggregations of offspring, or allowing offspring to shelter beneath, or cling to, the body of one or other parent. Removal experiments in a range of species have shown that such

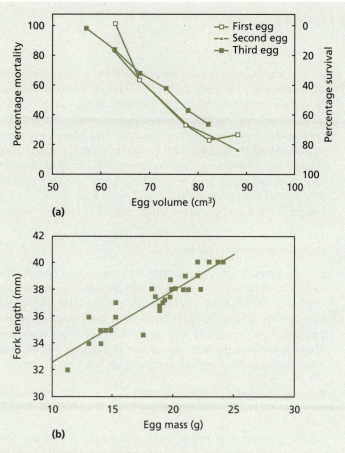

(a)

(b)

Figure (i) Increasing gamete size is often associated with improved survival and growth of off-spring. (a) The relationship between egg volume and percentage post-hatching survival for the first, second and third egg laid in herring gull (*Larus argentatus*) clutches. After Parsons (1970). © Nature Publishing Group (http://www.nature.com/), reprinted by permission. (b) The correlation between egg size and growth (measured as fork length) in Coho salmon (*Oncorynchus kisutch*). After Holtby & Healey (1986).

parental guarding can be vital in reducing offspring mortality (Wilson 1971; Dominey 1981). In many cases, parents are protecting offspring from predators or parasites, but sometimes they associate with offspring to maintain suitable environmental conditions, for example by ventilation or cleaning (Bro Larsen 1952; West & Alexander 1963).

Provisioning offspring before hatching or birth

Various mechanisms have evolved to provision young prior to their emergence into the world. In some insects, this takes the form of provisioned burrows where eggs are laid on stored food supplies (see 5.1.1.3; Fig. 1.1). In other cases, offspring are nourished from the mother's own body tissues, by eating her other eggs (oophagy) or embryos (adelphagy), by specialised secretions from the female reproductive tract or from brood chambers, or, as in some fish, amphibia and all therian mammals, from the female's blood supply via a placenta or pseudoplacenta.

Box 10.10 continued

Provisioning after hatching or birth

In many species, offspring are fed after hatching or birth. Sometimes this is with the same diet as the parents, as in some insects, crustaceans and birds, and sometimes it is with a more or less specialised diet tailored to the needs of the young. Some birds that feed on grain as adults, for example, feed their chicks with small insects. In a wide range of species, food is often predigested or preprepared in some other way. Many carnivorous mammals, for instance, regurgitate partially digested meat for their offspring. In naked molerats, young are even fed on the faeces of other colony members (Jarvis 1981). In other cases, offspring are nourished with specialised secretions, such as milk in mammals, so-called 'crop milk' (a creamy regurgitation of sloughed cells from the crop) in pigeons and doves, and various liquid secretions in insects. In some fish, young are fed on cutaneous blood or mucus from a parent.

Care after nutritional independence

In some long-lived species, parental care may extend beyond the time offspring are able to feed for themselves. In some sloths and primates, for example, parents may help offspring acquire or defend a territory until they can compete for themselves and acquire a mate (Montgomery & Sunquist 1978; Tilson 1981). Care may even extend to suppressing potential competition by threat and aggression. In some social mammals, mature daughters may establish home ranges that overlap with that of their mother, or stay in the mother's social group, thus allowing them to benefit from access to resources.

Based on discussion in Clutton-Brock (1991).

---

equate parental care and parental investment is that the absolute investment made by the parent at any given time (parental care) is likely to vary in its reproductive cost with age, condition, environmental contingencies and many other factors. Also, parental investment need not necessarily relate closely to the amount of resource received by the offspring, since the impact of any given amount of resource on the subsequent reproductive potential of offspring may vary with, for instance, the abundance of the resource (Clutton-Brock 1991).

10.2.1.1 Models of parental care

The parent–offspring relationship has been viewed in a number of ways in recent decades, partly reflecting changing perspectives on the evolution of social relationships generally. We shall look at two of these here. A third, the so-called **symbiosis model** (Rosenblum & Moltz 1983), is a misnomer arising from the fact that parents sometimes recoup some of the costs of parental care, as for example when female rats drink the urine of their pups to recover fluid and metabolites lost in milk production (Alberts & Gubernick 1983). A summary account can be found in Goodenough *et al.* (1993).

The altruistic provider model

The traditional view of parental care is of a one-way flow of resources and/or protection from the parent to its dependent young (Goodenough *et al.* 1993). The parent is a

Figure 10.18 A cost of parental care. Post-breeding survival in willow tits (*Parus atricapillus kleinschmidti*) declines with increasing numbers of young fledged during the breeding season. After Ekman & Askenmo (1986).

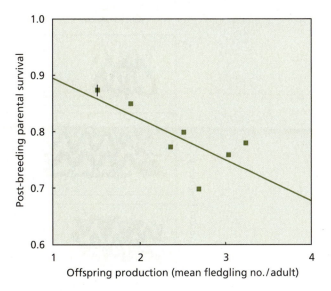

provider, and the relationship one of phenotypic altruism (now interpreted in terms of kin selection [see 2.4.4.5 and 9.3]). As Box 10.10 indicates, the nature of parental care differs enormously from species to species, in some cases providing nutrition from bodily tissues or round-the-clock foraging, in others a protective shelter or bodyguard. Whatever form it takes, however, there is no doubt that parental care can be costly.

The energetic costs of provisioning eggs are usually substantial. In reptiles, egg production can account for up to 20% of the annual energy budget (Congdon *et al.* 1982), while in birds the daily energetic cost of egg laying is around 29–35% of basal metabolic rate (BMR), with a hike of up to 230% in daily protein requirement (Robbins 1983). Caring for eggs once laid, or offspring once hatched or born, can incur severe lost opportunity costs, especially in terms of foraging. In the neotropical frog *Eleutherodactylus coqui*, for example, males that guard eggs lose around 20% of their body weight during the period of care (Townsend 1986). In birds and mammals, the energetic costs of feeding offspring usually exceed those of egg production or gestation by a considerable margin, peaking at around four times BMR in some birds (Drent & Daan 1980), a cost equivalent to that of heavy labour in humans (Clutton-Brock 1991). Female mammals generally lose weight during lactation, sometimes dramatically, and may undergo substantial physiological changes, such as hypertrophy of the liver, kidneys and digestive tract. Some of these costs can be offset by timing breeding seasons appropriately and laying down fat reserves in advance, but there is nevertheless good evidence that investment in care reduces parental survival (Fig. 10.18), either through loss of condition or from increased vulnerability to predators (e.g. Shine 1980) or parasites (Festa-Bianchet 1989).

Offspring recognition The costs involved in parental care might be expected to select for strong parental discrimination to ensure care is being lavished on the parent's own offspring. While this may seem straightforward in most circumstances, various factors can inject a degree of uncertainty into the proceedings. The risk of sperm competition from rivals (and thus uncertainty of paternity) is one problem from a male point of view (10.1.3.1), but crowded social conditions can be another, this time for either sex. In the Mexican free-tailed bat (*Phyllostomus hastatus*), for example, mothers leave their single

Cliff swallow   Barn swallow

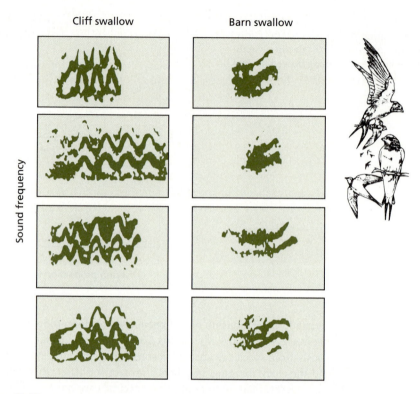

Figure 10.19 Recognising offspring by their vocalisations. These sonograms show that the chicks of colonial cliff swallows produce complex, structured calls that enable parents to distinguish them as individuals. In contrast, chicks of the less-colonial barn swallow produce much simpler calls, which may be more difficult to tell apart. After Medvin *et al.* (1993).

dependent offspring in crèches of anything up to 4000 bats per square metre while they go off to forage. When they return to the spot where they last nursed their pup, they are beset by a seething mass of pups all attempting to beg food. Could mothers possibly single out their own pup from such a throng? The seemingly self-evident answer is no, and it was long assumed that pups were fed indiscriminately. Careful studies based on analysis of blood enzymes, however, showed that feeding is in fact far from indiscriminate. Using variants of the enzyme superoxide dismutase, McCracken (1984) found that females and the pups they were feeding were highly likely to share the same variants, suggesting that mothers located their own offspring at least 80% of the time. Subsequent work has revealed that they may do even better than that, homing in on vocal and olfactory cues to pinpoint their own pups (McCracken & Gustin 1991).

Vocal signatures also turn out to be important in parent–offspring recognition in birds, as demonstrated in various species of swallow (e.g. Beecher 1982). Swallow species differ in their tendency to nest in colonies. Given the kind of problem facing free-tailed bat mothers, we might expect colonially nesting species to have more discriminating offspring recognition systems than non-colonial species. In keeping with this, comparative studies of chick calls have shown that colonial birds produce calls that are much more structured and individually distinctive than those of non-colonial birds (Medvin *et al.* 1993; Fig. 10.19). They also appear to be much better at distinguishing between different chick calls (Loesche *et al.* 1991). Furthermore, cross-fostering experiments with colonial bank

(*Riparia riparia*) and non-colonial rough-winged (*Stelgidopteryx serripennis*) swallows showed that rough-winged swallows accepted foster chicks of both their own species and the colonial barn swallow (Beecher 1982).

The apparently refined offspring recognition systems of colonial swallows are a response to the increased risk of mistaken identity in a crowded environment. But uncertainty can creep in for other reasons too. **Brood parasitism**, both within and between species, is one obvious route, since a parent is then at risk of caring for completely unrelated offspring.

Offspring recognition and brood parasitism: Interspecific brood parasitism in birds has provided one of the abiding mysteries of parental behaviour. Why do hosts feed chicks of brood parasites which are so conspicuously different from their own offspring? Their acceptance is all the more puzzling because many of these same hosts appear to have discriminated against parasite eggs in the past, so driving the evolution of present-day egg mimicry (e.g. Rothstein 1982). As we saw in 2.4.4.3, much of the evolution of egg discrimination can be understood in terms of the severity and evolutionary timescale of brood parasitism in different host species (Davies 1999), but, since the only 'discrimination' option for small birds is usually to abandon the current clutch, it can also depend on the availability of other nesting sites (Petit 1991).

Several explanations have been proposed as to why discrimination seems to break down at the chick stage. It may be, for instance, that the cost to the parent of mistakenly discriminating against one of its own offspring at this relatively late stage in the rearing process is simply too high, since the chances of getting back to the same position with another attempt in the current season may be slim. But discrimination may also incur other costs. European cuckoos (*Cuculus canorus*) sometimes 'punish' hosts that harm their offspring by returning to nest and destroying their clutch or brood (the '**Mafia effect**'; Soler *et al.* 1995). Another suggestion is that parasitic chicks exploit the host's response to begging signals by presenting a 'supernormal' (see 3.5.1.1) gape, and even mimicking the begging calls of an entire host brood (Davies *et al.* 1998; Fig. 10.20). Such tactics are likely to be facilitated in some brood parasites by turfing the eggs and/or chicks of the host out of the nest, thus removing both competition and the possibility of unfavourable comparison. Even where they share a nest with the home brood, however, the earlier hatching and larger size of most parasitic chicks may allow them to take advantage of the host's tendency to favour the most prominent, actively begging chick in the brood (Lichtenstein & Sealy 1998; see below).

Discrimination against *intra*specific brood parasites also varies (Andersson & Eriksson 1982; Andersson 1984b; Bertram 1992). Males of several bird species, such as western bluebirds (*Sialis mexicana*) and dunnocks, decide whether or not to feed a partner's chicks on crude rules of thumb relating to the phase of egg-laying, or having copulated with the female in question (Burke *et al.* 1989; Dickinson & Weathers 1999). By experimentally removing male western bluebirds from pairs, and allowing new males to move in, Dickinson & Weathers (1999) showed that replacement males would feed the female's chicks if she was still laying, and thus potentially sexually receptive and fertile, but not if she had finished laying. The fact that either way any chicks would almost certainly belong to the previous male shows once again that rules of thumb can easily be fooled by experimental manipulations (2.4.4.3).

Adoption: Successful manipulation by an inter- or intraspecific brood parasite is one way animals can end up as the adoptive parents of genetic strangers. But sometimes adoption (or **alloparental care**) can take place much later in the parental cycle and as a result of young seeking alternative carers. In some colonial gulls, chicks whose parents

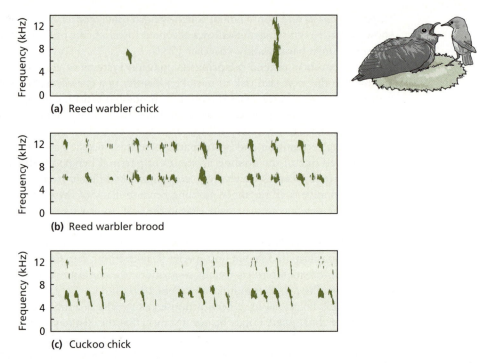

(a) Reed warbler chick

(b) Reed warbler brood

(c) Cuckoo chick

Figure 10.20 The calls produced by a European cuckoo (*Cuculus canorus*) chick parasitising the parental care of a reed warbler host (c) are more like those of a brood of reed warbler chicks (b) than a single chick (a). This appears to be part of the cuckoo's strategy for manipulating the parental responses of the warbler. After Davies *et al.* (1998).

are failing to feed them adequately sometimes voluntarily leave their natal nest and solicit food from neighbouring pairs. If they are tentative and behave in a frightened fashion, the hopeful adoptees are likely to be attacked and driven off, but if they crouch submissively and beg they are sometimes taken in and are more likely to survive than if they had stayed with their genetic parents (Brown 1998). What adoptive parents get out of the arrangement is less clear, but suggestions include diluting the risk of predation to their own offspring (see 9.1.1.1) and using inflated brood sizes to attract future mates (Rohwer *et al.* 1999). In fathead minnows (*Pimephales pimelas*), for instance, males sometimes evict a territory-holding rival from his territory and acquire the batch of eggs he had been guarding. Oddly, the usurper male retains the eggs and continues to guard them, even though they cannot be his progeny. Why? Because it turns out that his chances of acquiring a female after a take-over are much increased if he has eggs on show (see 'copying mating preferences' above). However, looking after eggs is a costly business, so males appear to optimise their investment by eating some of the eggs and leaving just enough to remain attractive (Sargent 1989). Usurper males thus stand to gain both nutrients and a mate from their adoptive behaviour.

Parent–offspring conflict

Most models of parental care assume that parents have the whip hand and are free to adjust expenditure on offspring to suit their own interests (Clutton-Brock 1991). But this is unlikely to be the case because what the parent does is inevitably constrained by the behaviour of its offspring, a problem exacerbated by the fact that the genetic interests of parent and offspring will usually be different. As we saw in Chapter 5, conflicts of interest between individuals of the same species are fundamental to understanding the evolution of decision-making at a variety of levels, and relationships between parents and offspring are no exception. Wherever parents are not genetically identical to their offspring (i.e. in all cases except clones), conflicts of interest are likely to influence both parental expenditure and investment and the demand for investment by offspring.

The scope for disagreement between parents and offspring about levels of investment was first recognised in a landmark paper by Trivers (1974). Assuming a sexually reproducing, diploid species, Trivers pointed out that the coefficient of relatedness (2.4.4.5) between a given offspring and its future siblings will be 0.5, while its relatedness to itself is 1.0. Thus the offspring should favour continuation of parental care until the cost to the parent is more than twice the benefit to itself. We can represent this schematically in Fig. 10.21(a). The figure shows the change in the ratio of maternal cost to offspring benefit from continued care as the offspring develops. Early on, the cost to the mother is relatively small, but the benefit to the offspring is high, so the relationship does not lead to conflict. As the offspring grows, however, it becomes more and more expensive to nurture, until the cost to the parent exceeds the benefit to the offspring (the cost : benefit ratio exceeds 1). A conflict of interest arises at this point because, while the parent should cease caring and reserve its investment for future offspring, the current offspring still stands to gain from further care. The conflict continues until the cost : benefit ratio reaches 2, at which point the offspring also benefits more by allowing the parent to invest in future siblings. The shape of the cost : benefit curve will vary between species, and even between individual parent–offspring relationships within species. As it does so, the timing and duration of the period of conflict will also vary (Trivers 1974).

One obvious prediction of the parent–offspring conflict model is that disagreements over continuing care will show up as a shift in the relative proportion of care-related interactions initiated by the two parties. Although long preceding Trivers's model, studies of weaning conflict in domestic cats by Schneirla and coworkers (Schneirla & Rosenblatt 1961; Schneirla et al. 1963) bear out the prediction. As Fig. 10.21(b) shows, female cats initiate almost all bouts of suckling until their kittens are about 20 days of age. After this, there is a sharp decline in initiations by the mother and a corresponding increase in those by kittens, until suckling is solicited almost entirely by kittens and the mother begins to avoid or aggressively discourage them (Fig. 10.21c).

The conflict of interest between parent and offspring will lead to an arms race of manipulation and resistance (11.1.1.4), but which party if any will gain the upper hand? Trivers argued that parents should evolve attentiveness to signals from their offspring, say calling to be fed, because these will reflect the offspring's own (and thus, presumably, accurate) estimate of its needs. This, of course, opens the way for cheats to exaggerate their apparent needs and secure an unfair share of attention. The dilemma for the parent is then whether to treat these more extreme signals as honest and respond accordingly, thus running the risk of being duped, or disregard them as manipulation and risk ignoring a genuinely needy individual. While studies of offspring begging behaviour have suggested that signals may indeed be honest reflections of need (and thus consistent with parents controlling the allocation of resources [see Kilner & Johnstone 1997]), the key predictions of honest begging are also consistent with scramble competition between siblings, and thus manipulation of parents by offspring (Royle et al. 2002). Other authors, notably Alexander (1974), have also suggested that the parent will generally be in control, but the basis of Alexander's argument – that a gene for extorting more care

Figure 10.21 (a) Trivers's (1974) schematic model of parent–offspring conflict (see text). From Wilson (1975a) after Trivers (1974). (b) Weaning conflict in cats, in which the relative frequencies of nursing bouts initiated by mother and kittens changes with the age of the kittens. Eventually (c) the mother seeks refuge from the kittens' attention on a shelf. After Schneirla & Rosenblatt (1961) and Schneirla *et al.* (1963).

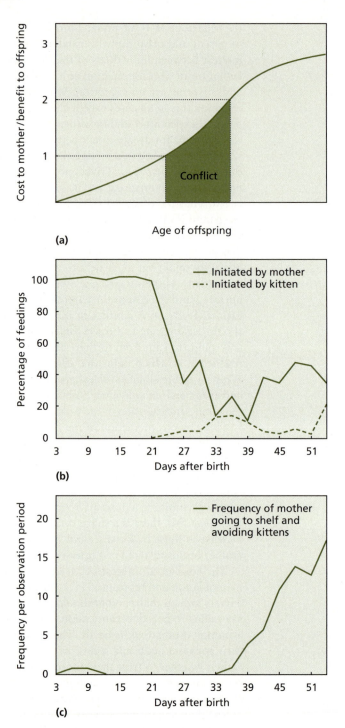

than a parent wishes to give will backfire on a manipulative offspring when it later becomes a parent itself – has been shown to be incorrect (see e.g. Dawkins 1989). Later games theory models by Parker and Macnair (e.g. Parker & Macnair 1978, 1979; Parker 1985) have provided more formal proof that there is no inherent parental advantage. In fact, the evolutionarily stable level of investment in their models appears to lie somewhere between that which is best for the parent and that which is best for the offspring, though the 'balance of power' may shift towards one party or the other according to prevailing circumstances, such as food availability and competitive asymmetries within broods (Royle *et al.* 2002).

Parental favouritism and sibling rivalry The conflict model is based on a competitive arms race between parents and offspring in which each party seeks to maximise its own reproductive interests. Weaning conflict is one manifestation of this, but the fact that an offspring is more closely related to itself than to either its parent or its siblings also leads to conflict within broods, since, all other things being equal, each offspring should vie for favoured attention. At first sight, we might expect parents to be indifferent to this as, genetically speaking, all offspring in the brood will be equivalent from their perspective (assuming there are no foreign intruders, of course). But, as we pointed out in 9.3.1.2, genetic relatedness is not the only arbiter of caring relationships. Reproductive value also counts. If some individuals in the brood appear more likely to make it to reproductive independence, it may be worth the parent focusing resources on them rather than their higher-risk siblings. Begging intensity is one obvious cue parents could use, but there is also evidence that plumage markers may convey information about offspring quality, as in the curious orange plumes of American coot (*Fulica americana*) chicks (Lyon *et al.* 1994), which may indicate something about their state of health. Thus both **sibling rivalry** and **parental favouritism** are predictable from parent–offspring conflict theory. Indeed, one may work through the other if parents allow sibling rivalry to weed out or identify offspring with poor prospects in the first place.

Sibling rivalry can be taken to extremes. In great egrets (*Casmerodius albus*), for example, chicks fight vigorously for the food their parents bring to the nest. In the process, bigger, more robust chicks frequently peck their weaker siblings to death, or cause them to fall out of the nest. Despite the obvious reproductive loss, adult egrets do nothing to police the competition or prevent deaths. Perhaps they are unable to exert any effective influence and the losses are a cost they simply have to bear. On the other hand, if food is short, it may be in their interests to allow sibling rivalry to whittle the brood down and ensure the tougher individuals get the limited resources that *are* available. Two lines of evidence suggest that sibling rivalry is indeed of some benefit to the parent. First, females appear to endow the initial couple of eggs in a clutch with higher levels of androgen (Schwabl *et al.* 1997; see also Box 10.3). The more androgen, the more aggressive the resulting chick is likely to be. Second, female egrets begin to incubate eggs as soon as they are laid, so not only do early eggs get a boost in androgen levels, they also hatch sooner so that their chicks have begun to grow by the time later siblings emerge. That this hatching asynchrony is adaptive from the parent's point of view has been neatly demonstrated experimentally. By manipulating the asynchrony of egret broods, Mock & Ploger (1987) showed that 'parental efficiency' (the number of chicks eventually surviving divided by the amount of food brought to the nest each day) was greatest for broods with natural asynchrony (Fig. 10.22). When asynchrony was artificially reduced, chicks squabbled more, demanded more food and were more likely to die. Aggressive sibling rivalry is not tolerated in all cases, however. In blue-footed boobies (*Sula nebouxi*), southern hemisphere relatives of gannets, early aggression by chicks is actively suppressed by parents, even when chicks of the closely related, but normally siblicidal, masked booby (*S. dactylata*) are introduced into the nest (Lougheed & Anderson 1999).

While the genetic asymmetry between self and siblings leads us to expect sibling rivalry, the nevertheless high degree of relatedness between siblings may exert some

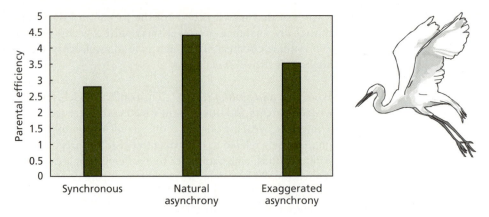

Figure 10.22 'Parental efficiency' (the mean number of surviving young per nest as a function of the daily amount of food brought to them) in cattle egrets following an experimental increase or decrease in hatching asynchrony. Broods with 'natural asynchrony' were created by putting chicks together that differed in age by the 1.5-day interval typical of the species. Plotted from data in Mock & Ploger (1987).

counterpressure. Various experiments manipulating levels of hunger in chicks (Leonard & Horn 1998) or the intensity of begging calls received by the parent (Ottosson *et al.* 1997) have shown that parents respond to differences in apparent need. It also turns out that, in some birds at least, the energetic cost of begging is small relative to the potential gain from parental favouritism (Bachmann & Chappell 1998). So why don't chicks always exaggerate their begging signals and reap the parental bonanza? One answer may be the cost of disadvantaging their closely related siblings. If this is the case, exaggerated begging should be more likely when relatedness within broods is low, as when there is mixed paternity. A comparative study of begging loudness in relation to the prevalence of mixed paternity or interspecific brood parasitism in different bird species showed exactly the kind of positive correlation expected on this basis (Briskie *et al.* 1994; Fig. 10.23).

Figure 10.23 The amount of noisy begging, assumed to reflect selfish attempts to outdo nestmates, increases as the degree of relatedness between brood members declines. The 100% unrelated point is a brown-headed cowbird chick, an interspecific brood parasite in the nest. After Briskie *et al.* (1994).

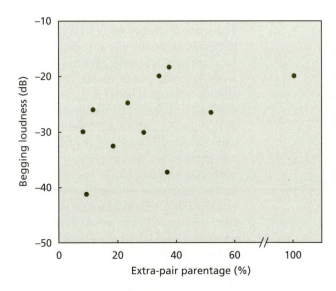

Summary

1. Reproducing successfully is the ultimate driving force shaping decision-making by animals. Yet most species reproduce sexually, a mode of reproduction that appears superficially to be genetically costly. While several suggestions have been made as to the exact nature of the cost, the generally accepted view is that sexual females suffer reduced fertility relative to an asexual competitor as a result of investing in (usually) two separate sexes that cannot reproduce independently. However, the cost may be offset if males help the female to rear more offspring than she could on her own. Since males often invest little in offspring care, even at the gamete stage, the fertility cost of sex is often referred to as the cost of anisogamy (the production of different-sized gametes by the two sexes) or the cost of male 'laziness'.

2. The fact that there are usually two sexes, rather than any other number, may be a consequence of uniparental inheritance of cytoplasmic factors vital for cell function, such as mitochondria and chloroplasts. The ratio of the two within natural populations can then be understood in terms of evolutionarily stable strategies of parental investment by females.

3. Anisogamy leads to differences in the potential reproductive rates of males and females, such that females become a limiting resource for which males compete, and exercise choice over which male is successful. However, shifts in relative parental investment by the two sexes may sometimes result in sex role reversal with females competing to mate with resource-rich males.

4. Mating competition driven by sex differences in potential reproductive rate fuels sexual selection and the evolution of secondary sexual characters. Intrasexual selection within the competing sex (usually males) generally leads to the evolution of traits helpful in physical competition, such as large body size, weaponry and aggressiveness. Intersexual selection, on the other hand, generally favours attributes that are attractive to the choosing sex (usually females), such as striking displays and plumage characteristics. However, decorative adornments can also function in intrasexual competition, while weaponry and other seemingly intrasexually selected traits can be used in mate choice.

5. Although mate choice has long been regarded as an important driving force in sexual selection, the criteria on which choice is based have been the subject of vigorous debate. Most attention has focused on potential genetic benefits to the chooser through offspring inheriting the desirable qualities of their chosen mate. Longevity, immunity to disease, developmental competence and ability to acquire resources are some of the heritable benefits suggested. Non-genetic benefits may also accrue, such as better provisioning for offspring, or increase fertilisation efficiency. However, choice may also be based on arbitrarily attractive features, or traits that reflect ancestral perceptual predispositions and do not indicate any particular adaptive quality in a potential partner.

6. In many species, successful mating is followed by a more or less protracted period of parental care. Traditionally, parental care has been seen as phenotypic altruism on the part of the parent, with both parent and offspring having the same stake in the offspring's survival and later reproduction. However, asymmetries in relatedness and reproductive value between parent and offspring, and between siblings, are likely to produce conflicts of interest that manifest themselves in parental bias and sibling rivalry, both of which can have various outcomes in terms of offspring survival and parental reproductive success.

Further reading

Hamilton's (2001) second volume of collected papers contains his seminal discussions of sex, sex ratios and sexual selection; Maynard Smith (1984) and Maynard Smith & Szathmáry (1999) provide excellent introductory discussions of the evolution of sex; Rice (2002) reviews some experimental approaches to the problem; and Ridley (1993b) is a good popular scientific treatment of the subject. Andersson (1994) remains a key general work on sexual selection, and the volume edited by Clayton & Moore (1997) looks at the role of parasites in sexual selection and host evolution generally. Read (1990) discusses the Hamilton–Zuk hypothesis in particular, and offers various insights into the difficulties of testing it. Emlen & Oring (1977) is a good starting point for mating systems, and Davies's (1992) book on the dunnock a masterly field study of mating systems in a model species. Birkhead & Møller (1998) and Simmons (2001) provide good recent reviews of sperm competition. Grafen (1990b,c) provides a searching analysis of the handicap principle, and Zahavi & Zahavi (1997) a characteristically iconoclastic view of the same. Swaddle *et al.* (1994), Møller & Swaddle (1997), Palmer (1999), Simmons *et al.* (1999) and Thornhill *et al.* (1999) collectively summarise many of the issues in the fluctuating asymmetry debate, while Clutton-Brock (1991) and Mock & Parker (1998) review the evolution of parental care and sibling rivalry respectively.

11 Communication

Introduction

Interactions between organisms are often facilitated by specialised behaviours, markings, chemicals and other attributes that we think of as signals or displays. Together with the response of recipients, these constitute a system of communication. But what exactly is communication and how does it differ from other ways organisms affect one another? Do signallers and recipients both benefit from communication, or can the benefit be one-sided? Many signals and displays are highly stylised and exaggerated. How have they evolved? What factors have shaped their design? Communication is often involved in debates about intelligence. Why is this? Do other animals possess language in the sense that we think we do? Can they be taught language skills? If they can, does it reveal anything about their underlying cognitive abilities?

In the last two chapters we have been concerned with various kinds of social responses between animals. Coming together to form social aggregations, defending territories or choosing a mate all require some kind of directed response towards other individuals. Sometimes this happens incidentally as individuals encounter one another by chance, but often it relies on specialised behaviours, chemicals, markings or morphological attributes that appear specifically designed to facilitate it. These specialisations are usually referred to as **signals** or, where behaviour patterns are involved, **displays,** and the collective act of deploying and responding to them as **communication**. While most people have an intuitive idea of what is meant by communication, however, it is in fact extremely difficult to define rigorously and disagreement abounds as to what does and does not qualify as communication. Some idea of the problem can be gleaned from Box 11.1, and we shall return to the issue a little later in the chapter. But to start with, we shall look at the elements that make up an act of communication and some views on how they might have evolved.

11.1 Signals and information

Formally, communication systems have a number of discrete components that define the participants, what is being communicated, how and in what context, and the background against which the act is taking place (e.g. Sebeok 1962; Table 11.1). Variation in any of these components is likely to affect the way the **receiver** of a signal (sometimes referred to as the **reactor**) responds to its **sender** (sometimes referred to as the **actor**). This, of course, fuels the diversity and flexibility of many signalling systems, but are there any general principles to be gleaned within all this variety?

Difficulties and debates

Box 11.1   What is communication?

The most obvious manifestation of what we think of as communication is a stylised signal or display by one individual modifying the response of another (see Box 11.2). A rival backing off from the tooth-baring display of a dominant male wolf is a ready example. Signals in these cases are distinctive, often exaggerated, actions or features that have evolved to be noticeable and attention-grabbing (Tinbergen 1952; Huxley 1966). While this may be the classical view, however, things are not always so straightforward.

For a start, signals are not always stylised, or even, necessarily, obvious to a third party observer like us; in many cases, the behaviour of receivers changes in response to subtle movements or slight changes of posture in another individual, especially in social species (see text). The signaller need not even be present for its signal to have an effect. Odour signals deposited on the substrate, for example, can influence the behaviour of recipients hours or days after the departure of the signaller. The effects of receiving a signal can also be delayed, the ovaries of some female birds developing in response to male song over a period of several days. Indeed the effects of a signal may be entirely unobservable externally. Even defining a signal as being designed to influence the response of a recipient is problematic. Few would dissent from the bright and threatening deimatic displays in 8.3.5 being classed as communication. But what about crypsis? Crypsis arguably influences the behaviour of would-be predators just as much as startle displays, and is clearly specialised to do so. Are cryptic features therefore signals? Most would argue not (but see text). The 'specially designed' criterion makes a big difference to what is and is not included within the definition of a signal. Altmann (1962) and Hinde & Rowell (1962), for instance, both tried to identify the repertoire of social signals used by rhesus monkeys. Altmann listed around 50 signals, while Hinde & Rowell managed only 22. Why the difference? Because Hinde & Rowell included only those actions that appeared specifically designed to influence the behaviour of other monkeys, while Altmann included any action that had an effect on the behaviour of others. This reflects the more recent distinction between signals and **cues**, the latter comprising features that influence behaviour without being specially designed to do so.

Another key element in distinguishing communication is economy of effort. Metabolically at least, signalling seems likely to be a less costly way of manipulating the behaviour of another individual than using physical force (e.g. Cullen 1972). A male cricket calling from its burrow to attract a female, rather than sallying forth to hunt for one, for example, shifts the burden of effort from its own leg muscles to those of its potential mate (Dawkins & Krebs 1978). However, it is far from clear that signals are in fact economical. The roaring contests of rutting red deer stags, which often lead to the exhaustion of one or other combatant (Clutton-Brock *et al.* 1979), can hardly be described as a low-cost option. Moreover, there is a problem in defining signals as being economical as well as stylised and obvious, since one property is likely to conflict with the other (Dawkins 1995).

Definitions of communication have also been caught up in the debate over intentionality (4.2, 11.3). Several workers have argued that signallers must *intend* to influence the response of a recipient in order for their actions to qualify as communication (e.g. Batteau 1968). As we have seen in Chapter 4, however, intentionality is not necessary to account for apparently purposeful behaviour, and its inclusion as a criterion would rule out some strong candidates for signalling systems such as the pheromonal and morphological insect attractants of some plant species (e.g. Lunau 1996; Peisl 1997).

Table 11.1 The components of a communication system. After Sebeok (1962).

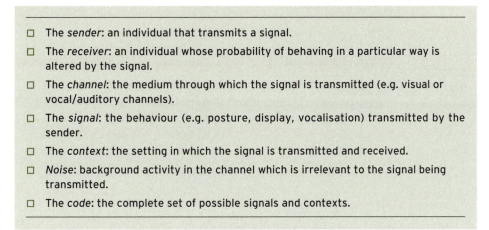

- ☐ The *sender*: an individual that transmits a signal.
- ☐ The *receiver*: an individual whose probability of behaving in a particular way is altered by the signal.
- ☐ The *channel*: the medium through which the signal is transmitted (e.g. visual or vocal/auditory channels).
- ☐ The *signal*: the behaviour (e.g. posture, display, vocalisation) transmitted by the sender.
- ☐ The *context*: the setting in which the signal is transmitted and received.
- ☐ *Noise*: background activity in the channel which is irrelevant to the signal being transmitted.
- ☐ The *code*: the complete set of possible signals and contexts.

11.1.1 The nature of signals

While an act of communication is defined in terms of two classes of participant: sender and receiver, it can represent very different things to each (Slater 1983). Smith (1965) summed this up in his distinction between **message** and **meaning**. The message is what the signal encodes about the sender: what it is, what it is up to and what it might do next. The meaning is what the receiver construes from the signal, something that we as external observers can infer only from the receiver's response, if it shows any. Meaning can vary greatly between recipients, as we shall see later. To begin with, however, we shall consider the nature of signals and their potential for conveying meaning.

11.1.1.1 The arbitrariness of signals

In most cases, the form of a signal bears no obvious relation to the meaning it conveys. The string of notes produced by a songbird to proclaim ownership of its territory does not literally depict a territory with a bird sitting in it any more than the word 'cat' resembles the animal to which it refers. To that extent, signals are arbitrary. In principle, we could choose any one of limitless combinations of letters to mean the same as 'cat'. But at another level, signals are not arbitrary at all. To begin with, they must be effective in the environment in which the sender and receiver live. This inevitably imposes constraints on the form of communication used. Visual signals are unlikely to be much use in dense forest, for example, but may be more effective than sounds and odours in face-to-face confrontations over food (11.2.1.1; Table 11.2). Also, as we saw in Chapter 10, signals *can* be related to the attribute they convey. Thus females seeking disease-free males impose a selection pressure on males to display their state of health or likely immunity status. But to what extent signals generally convey information about an animal's state, and how honestly they convey it, is something we shall return to later. First, we shall look at some of the characteristics of signals and how they convey 'messages' from the sender.

11.1.1.2 Discrete versus graded signals

Sebeok (1962) distinguished two fundamental types of signal: **discrete** (or **digital**) and **graded** (or **analogue**). Discrete signals are those that operate in an on/off manner, such

Table 11.2 Advantages of different sensory channels of communication. After Alcock (1984).

Feature of channel	Type of signal			
	Chemical	Auditory	Visual	Tactile
Range	Long	Long	Medium	Short
Rate of change of signal	Slow	Fast	Fast	Fast
Ability to go past obstacles	Good	Good	Poor	Poor
Locatability	Variable	Medium	High	High
Energetic cost	Low	High	Low	Low

as the flashing sequences of fireflies (*Photinus* spp.). By their nature, discrete signals can convey only a single, simple message. Members of a species with just one, unvarying (*stereotyped*) threat display, for example, can indicate only that they are aggressive, not *how* aggressive they are (Halliday 1983). The stereotyped nature of many apparently discrete signals gave rise to the concept of '**typical intensity**' (Morris 1957), the idea that signals and displays were always performed with the same vigour and complexity no matter how strong the stimulus evoking them. A classic example is the courtship display of male cut-throat finches (*Amadina fasciata*), which involves more or less the same posture and degree of feather-ruffling every time it is performed, regardless of the 'quality' of the female stimulus (Morris 1957). Such signals, indeed signals in general, often also show high **redundancy**, in that the same elements are performed again and again, or there is a high degree of predictability between different elements. However, assumptions about the discreteness of signals and their typical intensity and redundancy must be made with care. Several signals that appear superficially to be all-or-nothing turn out to show appreciable variation when analysed in detail. The aggressive calls of male common toads (*Bufo bufo*) competing for females are a good example. These calls are simple and sound highly stereotyped, but in fact they contain variations in pitch that relate to body size so that males are able to adjust their aggressiveness according to the size of their opponent (Davies & Halliday 1979; see 11.1.1.5).

Signals that can be transmitted with varying intensity and complexity are said to be 'graded'. Variation in graded signals can often be related to the strength of the evoking stimulus, or the putative motivational state of the animal. Ants, for instance, release quantities of alarm substance that are roughly proportional to the strength of provocation. Graded signalling is shown nicely by aggressive displays in rhesus monkeys (*Macaca mulatta*). Low arousal is manifested in a fixed stare, but as arousal increases, new components are added to the display. These include opening the mouth, bobbing the head up and down, vocalising, slapping the hand on the ground and lungeing forward (Altmann 1962). If all these elements appear, the monkey is very likely to launch an attack. An obvious reason why signals might be graded is that performing them at full intensity may be expensive, not just in time and energy, but in terms of risk as well. Starting off a confrontation with a highly aggressive signal, for example, might invite injurious retaliation from an opponent. It would be better to size up the opponent and test its likely resolve first with some low-grade signals (see 11.1.1.4).

11.1.1.3   Complex signals

While signals appear to be associated with particular messages, a given signal may convey several different messages at different times, in different contexts, or to different individuals. Male and female western grebes (*Aechmophorus occidentalis*), for example, use subtle variation in the timing and duration of advertisement calls to assess the sex, pairing status and stage of the breeding cycle in other individuals. Complexity here is thus a matter of differences in the delivery of a particular signal. But signals can also be complex in the sense of being multicomponent. Many sexual displays in birds, for instance, involve combinations of components, often simultaneously exploiting different channels (Table 11.2) by using both visual and auditory elements (Rowe & Guilford 1999; Fig. 11.1). One argument is that the multiple components serve to enhance the

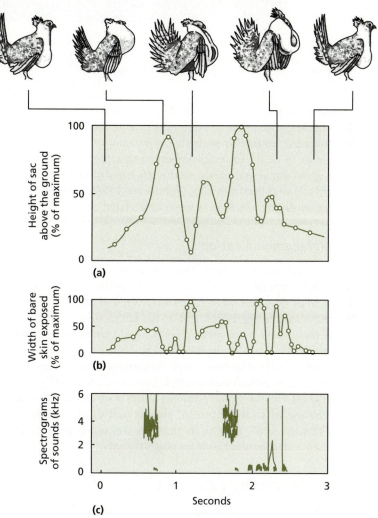

Figure 11.1 The multicomponent strut display of the male sage grouse (*Centrocercus urophasianus*) lasts about three seconds and consists of both visual and acoustic signals. The male inflates his oesophageal sac, heaving it upwards and letting it fall twice (a), so exposing bare patches of skin on his breast (b). He then compresses the sac releasing the air explosively to produce a ringing pop and low coos (c). After Wiley (1983).

Figure 11.2 Metacommunication in the form of a play bow by a dog. The bow informs a companion that the actions that follow are playful, despite being similar to, for instance, aggression or sexual behaviour. Photograph courtesy of Marc Bekoff.

accuracy with which receivers are able to assess signals (the **'backup' hypothesis**), while another is that they provide information about different qualities of the signaller (the **'multiple message' hypothesis**) (Johnstone 1996). Games theory models suggest that the choice of single- or multicomponent signals may depend partly on the cost of escalating a given signal and partly on the efficiency of employing different components by individuals of different quality (Johnstone 1996). However, the nature of the receiver also has much to do with it, as we shall see later.

Metacommunication

Displays may have more than one component for another reason: one or more of the components may act as a 'scene-setter' for those that follow. An example is the 'play bow' used by dogs and other carnivores. Many of the behaviours used in play are the same as those used in aggression and sexual interactions, so there is scope for misinterpretation if the context is not made clear. Several carnivores therefore use the characteristic bow in Fig. 11.2 as a prelude to play to ensure their subsequent behaviour is interpreted appropriately. The bow, often accompanied by a wagging tail, is not used in any other context, so provides an unambiguous indication of the motivation behind what follows (Bekoff 1995b). Some primates adopt a so-called 'play face' to achieve the same end. The use of such contextual signals and displays is known as **metacommunication** (Altmann 1967) to indicate that it is concerned with signals about other signals rather than with communicating messages directly.

11.1.1.4   Signal specificity and stereotypy

While signals may be complex, they can still convey very specific messages and be delivered in a stereotyped fashion. In many species, particularly insects and lower vertebrates, each signal elicits only one or a very small number of responses, and each response can be elicited by only one or a very few signals. A truly astonishing example of signal specificity can be found in the mate attraction pheromones of moths. The pheromones of two species of *Bryotopha* moth differ in the configuration of a single carbon atom, making them geometric isomers. Observations in the field have shown that *Bryotopha* males respond solely to the isomer of their own species. Not only that, the male's response is actually *inhibited* if the isomer of the other species is present.

Many vertebrate species are able to recognise individuals on the basis of small variations in visual, vocal, chemical or electrical signals. Playback experiments with territorial North American bullfrogs (*Rana catesbeiana*), for example, have shown that territory-owning males are able to discriminate between the calls of neighbours and strangers in the absence of the normal contextual cues relating to location (Bee & Gerhardt 2002). Similarly, male banded wrens (*Thryothorus pleurostictus*) can still distinguish the calls of territorial neighbours when these are displaced to the wrong location or presented alongside a synthetic call that closely mimics that of the neighbour (Molles & Vehrencamp 2001). Interestingly, banner-tailed kangaroo rats (*Dipodomys spectabilis*), which use foot drumming signatures in neighbour recognition, appear to modify their signals to maintain distinctiveness if they change territories and move into a new social environment (Randall 1995).

Stereotypy and ritualisation

Signal specificity is helped by the fact that animals perform many of their signals and displays in what seems to be a highly stereotyped fashion. But while there is agreement that selection has widely favoured stereotypy, there are different views as why. Huxley (1923) referred to an evolutionary process by which signals become stereotyped as **ritualisation**. The early ethologists, such as Lorenz, Tinbergen and Huxley (see 1.3.1), argued that many signals have evolved by ritualisation from pre-existing behaviours that served other functions entirely, but incidentally gave something away about how an animal was likely to respond next (the **principle of derived activities** [Tinbergen 1952]). For example, if a dog bares its teeth just before biting, a potential recipient can use this to anticipate the attack and move away before it happens. Selection will then favour dogs that make a show of baring their teeth because rivals will move away without the need for a costly fight. Teeth-baring thus becomes ritualised into an exaggerated threat display. The derivation of courtship displays from food-pecking in pheasants (Fig. 1.13) is another example of an incidentally informative behaviour becoming exaggerated through its effects on observers (in this case females). Conflict behaviours (Box 4.3), performed when an animal is apparently torn between mutually exclusive motivational states, appear to have been a particularly rich source of signals and displays. For instance, the mammalian habit of marking a territory boundary with urine may have evolved from urination as an autonomic fear response when rivals were encountered, while the oblique threat posture of black-headed gulls (*Larus ridibundus*) seems to be a stylised ambivalent behaviour containing elements of both attack and retreat (Fig. 11.3).

In the classical ethological view, ritualisation was seen as a process of reducing the ambiguity of signals (Cullen 1966). Signallers benefited from ritualisation because it distinguished one signal from another and reduced the risk of confusing the receiver. A stereotyped threat signal clearly stated 'I am likely to attack', and the receiver was unlikely to misread it as 'I am scared stiff of you', or 'I'm the guy who should fertilise your eggs'. Indeed, as Darwin (1872) noted, signals conveying opposite messages frequently take opposite forms (the **principle of antithesis**). Darwin's classic example compared a dog in a typical threat posture, standing erect and facing a rival, with a submissive dog, which crouches low to the ground and flags its head. However, as we saw above, there is a price to pay for reduced ambiguity, and that is a concomitant reduction in the **information**

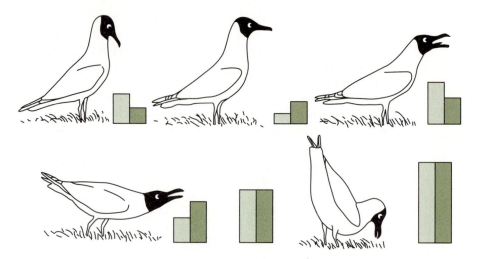

Figure 11.3 Many signals and displays are believed to have evolved from other responses that gave away information about an animal's motivational state and likely future behaviour. The ambivalent attack/escape postures in the aggressive display of black-headed gulls is assumed to reflect the different relative strengths of attack (dark bars) and escape (pale bars) tendencies. After Eibl-Eibesfeldt (1975).

content of signals (Box 11.2). While incidental movements and conflict behaviours might say quite a bit about an animal's internal state, a highly stereotyped signal performed with 'typical intensity' is likely to convey much less. Ethologists such as Morris (1957) saw the reduction in information as an unfortunate, but unavoidable, cost of ritualisation. As Krebs & Davies (1993) point out, however, the cost is unfortunate only if there is some advantage to the sender in communicating information about its state and likely course of action. There are good reasons for doubting that this is so, as we shall now see.

Ritualisation and manipulation   The classical ethological view of communication largely saw senders and receivers as partners in a mutually cooperative system of information transfer. Receivers were designed to read information from senders, and signals were designed to inform receivers as clearly as possible about the state of the sender and its likely actions. Generally (though by no means always), both parties benefited from the exchange. However, in a characteristically stimulating pair of papers, Richard Dawkins & John Krebs departed from this view and argued that, far from being a mutually cooperative venture, communication was more likely to be characterised by selfish manipulation and countermanipulation (Dawkins & Krebs 1978; Krebs & Dawkins 1984). This conclusion arises from the familiar reproductive cost–benefit considerations of earlier chapters.

We have seen in Chapter 10 that signals can be costly to the sender. Indeed, their cost may be an intrinsic part of their effectiveness as signals (Box 10.7). The downside, however, is that it is also a constraint on their form and diversity. Thus we should expect animals to signal only when the reproductive benefits of doing so outweigh the costs. Since the benefits accrue from the response of the receiver, we should expect signals to stimulate responses that are in the best interests of the sender. In other words, signals become the sender's way of **manipulating** receivers, whose responses are thereby part of the sender's extended phenotype (2.4.3.2). But receivers are not passive dupes in all this. Manipulation can work only because receivers are themselves attempting to exploit the social environment

Difficulties and debates

Box 11.2 Communication and information

The idea that signals and displays transmit information is central to much of the debate about animal communication. One reason for this is that there has been considerable confusion about exactly what kind of information is involved, with the result that very different senses of the term 'information' are frequently conflated. We briefly look at two here.

Shannon–Weaver information

Mathematicians and engineers define the information content of a message in terms of the reduction in uncertainty caused by sending it. Imagine, for instance, that a travelling salesman did business in four cities, Birmingham, London, Cardiff and Swansea, and was going to visit one tomorrow, but you did not know which. Suppose, however, that the salesman mentions he will be going to Wales. You still do not know which city he will visit, but you now know it must be either Cardiff or Swansea because the other two are not in Wales. Your uncertainty as to where the salesman will be going has just gone down from 1 in 4 cities to 1 in 2, a reduction of 50%. In engineering terminology, the announcement that the salesman is going to Wales has provided you with one *bit* of information. If you then ask whether the city is Cardiff, a yes or no answer provides the final bit of information since it identifies unequivocally the city to be visited. Thus your uncertainty is reduced from an initial two bits to one and then zero (Dawkins 1995).

In the same way, a dog baring its teeth at a rival reduces our uncertainty about what the rival will do next: it is more likely to move away than it was prior to seeing the other dog's teeth. We also suspect that when a foraging bird suddenly crouches down it is very likely to take off in the next few seconds. Thus an animal's behaviour at a given time reduces our uncertainty about what others will do as a result, or what the animal itself will do next – the difference between so-called **transmitted** and **broadcast** information (Wiley 1983). Following Shannon & Weaver (1949), this uncertainty, *H*, can be calculated as:

$$H = -\sum p_i \log_2 p_i \qquad (11.2.1)$$

where there are *i* possible behaviours and p_i is the probability that the *i*th behaviour will occur. The reason for using logarithms to the base 2 is that, as in our travelling salesman example, the reduction of uncertainty conventionally involves questions with a binary outcome – 'yes' or 'no'. If the behaviour of a receiver is truly influenced by that of a signaller, the reduction in uncertainty about its behaviour is given by:

$$H_T = H_R - H_{R/S} \qquad (11.2.2)$$

where H_R is our uncertainty about the receiver's behaviour when the signaller's behaviour is not known, and $H_{R/S}$ is our uncertainty when its behaviour is known. Various studies have calculated H_T from sequences of behaviour by signallers and receivers (e.g. Hazlett & Bossert 1965; Dingle 1972). However, empirical values of H_T must be interpreted with caution because they rely on, among other things, our categorisations of behaviour corresponding exactly to those perceived by the animals, which may not be the case.

Semantic information

Shannon–Weaver information is all about reducing uncertainty from the viewpoint of an external observer, i.e. what we can deduce about behaviour in the future from watching

Box 11.2 continued

behaviour in the present. It is not confined to specialised signals and displays, and may have little or nothing to do with what we regard as communication. Animals may also use information in this sense, but their signalling systems are often concerned with information of a quite different sort.

Information in this second sense is what passes between animals rather than to an external observer. It is information in our everyday sense – 'semantic' information – not the technical one of reduction in observer uncertainty. Krebs & Dawkins (1984) recognise three kinds of semantic information in animal communication systems:

1. *Intention*. A signaller may communicate information about what it will do next, but in a strategic way that differs from the purely statistical predictive information of Shannon & Weaver. For one thing, animals may 'lie' about their intentions, mounting a bluff that suggests they are, say, more likely to attack than is really the case (see text). They may also withhold information about their intentions (for example, to flee) in order to maintain a 'bargaining' position. Thus, while signals or displays may predict behaviour in a Shannon–Weaver sense, they may do so with varying reliability in different contexts (Caryl 1979; see text). Shannon–Weaver predictability is therefore not the same thing as communicating intention, though debates about the information content of signals have frequently confused the two (see Dawkins 1995).

2. *Quality assessment*. Communication is often about persuading a receiver that the signaller is worth paying attention to, perhaps because it is powerful or healthy. Thus signals convey information about fighting ability, freedom from parasites or other qualities in order to win disputes or gain a mate. Signals may also 'sell' an animal as a member of the right species or social group (Halliday 1983).

3. *Information about the environment*. Signalling systems may convey information about the environment, as when the dances of worker honey bees direct recruits to sources of food (von Frisch 1953), or the alarm calls of vervet monkeys warn of the approach of different kinds of predator (Seyfarth *et al.* 1980).

Based on discussion in Halliday (1983), Krebs & Dawkins (1984) and Dawkins (1995).

around them. Just as our ancestral dog warily backed off when it saw the incidentally exposed teeth of a snarling rival, animals might be expected to tune in to the myriad postures, odours and other cues around them that predict what other animals are likely to do. To use Krebs & Dawkins's (1984) metaphor, receivers are effectively 'mind readers', not in the anthropomorphic sense of guessing what others are thinking, but in being sensitive to the probability that one action by an animal will lead to another. It is this sensitivity that manipulative senders can exploit to their own advantage. There will thus be an arms race (Box 2.6) between mind readers and manipulators to bias the outcome of any information flow between them in their own, respective, interests. Of course, any given individual will be both a mind reader and a manipulator at the same time, so the two attributes will be honed side by side as they are deployed in their different contexts.

Manipulator/mind reader arms races: So how might manipulator/mind reader arms races affect signalling? Two general tendencies might be expected. On the one hand, to avoid being second-guessed as to its actions, an individual might attempt to obscure its true intentions or internal state, and reduce the amount of useful information transmitted to receivers. This, of course, is entirely opposite to the predictions of classical ethology based on enhancing the clarity of information transmission. On the other hand, it might attempt to exaggerate its case in order to persuade an otherwise reluctant receiver to respond, for instance by puffing itself up to look bigger than it really was. Exaggerated begging displays in broods of chicks (10.2.1.1) are an example. In this view, ritualisation and 'typical intensity' reflect selection, not for reduced signal ambiguity, but for disguising real intentions or cranking up a sales pitch.

The extent to which these subterfuges are pursued, however, is likely to depend on the cost to the sender of being mind-read, and the cost of manipulation to the receiver. In some cases, it may actually be in an animal's interests to be mind-read or manipulated, so there will be little selection for resistance. A dog baring its teeth benefits from the retreat of its mind-reading rival, while a bird manipulated into taking off by the alarm call of a companion (Charnov & Krebs 1975; 8.3.5.4) may enjoy a reduced risk of predation. However, while manipulation may sometimes have 'willing' victims, this does not diminish the manipulative nature of the act *per se* (Krebs & Dawkins 1984). While these general principles apply across all forms of communication, they have been particularly well studied in the context of conflict over resources.

11.1.1.5 Signals during conflict

The signalling conventions that animals use to settle disputes have long intrigued ethologists and behavioural ecologists because they are often highly stylised and restrained despite many species possessing potentially lethal weaponry. While the stock explanation for this apparent restraint used to rely on group selection (2.4.4.1), we saw in 2.4.4.2 that it can be accounted for in terms of orthodox neoDarwinism using a games theory approach. But this still leaves the question as to what exactly passes between contestants during these stylised interactions to enable them to settle the dispute. Why should a display by one individual persuade another to give up and sacrifice the coveted reward?

Disguising intentions in waiting games

Where animals lack weaponry with which to escalate an attack, contests are usually decided by persistence (e.g. Kemp & Wiklund 2001). The winner of such a 'war of attrition' (Maynard Smith 1974; see 2.4.4.2) is the contestant that hangs on the longest. The problem for each individual is then to persuade its rival that it will keep going longer than they can so that the rival will give up. Of course, it may not really be able to keep going longer, but it would be foolish to give any indication to that effect because the rival would know it had only to extend its intended display time a little longer to win. As we saw in 2.4.4.2, a simple solution to this kind of 'war of attrition' waiting game might be to display with a random distribution of persistence times, since this would make the choice of display time in any given dispute unpredictable. There is some evidence that animals do this. The duration of contests between female land iguanas (*Iguana iguana*), conform roughly to a random (negative exponential) distribution (Rand & Rand 1976). However, like most general models, the classical 'war of attrition' model is based on some highly simplifying assumptions, not least that contests are symmetrical (see 2.4.4.2) and choice of persistence time is the only material difference between contestants. This is highly unlikely to be the case, and most contests between animals are probably settled on the basis of one or more key differences (**asymmetries**) between contestants (see below). Even ignoring this problem, there are other reasons for being cautious about the predictions of the simple symmetrical 'war of attrition'. The cost of displaying is one.

Figure 11.4 The way the cost of display increases over time (cost function) affects the frequency distribution of display times expected during an aggressive contest. The random distribution predicted by the classical symmetric war of attrition model holds only if the cost function is linear (a). Other cost functions (b, c) predict different patterns of modality in the distribution. Hazard functions reflect the probability that a display will end at any given time, and thus relate to the cost of display. See text. After Norman *et al.* (1977).

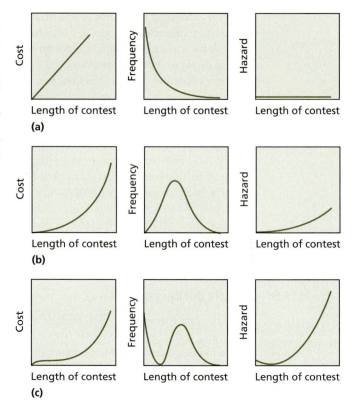

In a stimulating reanalysis of data from the literature, Caryl (1979) found that the distribution of contest durations in various species departed from random. One reason for this could be that the *cost function* for displaying – i.e. the change in cost over time (see 4.1.1.3 and Fig. 4.10) – varies between individuals and circumstances and affects the predicted distribution of persistence times. The random distribution predicted by the 'war of attrition' model is based on the cost function being linear (Fig. 11.4a). When the shape of the cost function departs from linearity, however, the predicted distribution of display times changes radically (Fig. 11.4b,c; Norman *et al.* 1977; Payne & Pagel 1996). As a result, the **'hazard' function** (the change in the probability that a display will end at any given time) also changes (Fig. 11.4a–c). Caryl's comparisons of empirically derived hazard functions and display distributions in Siamese fighting fish (*Betta splendens*) and cowbirds (*Molothrus ater*) were in broad concordance with the relationships predicted by Norman *et al.*'s (1977) model (Fig. 11.5), thus bearing out Caryl's suspicions about the effect of varying display costs. Parker (1979) and Haigh & Rose (1980) discuss some other factors complicating the predictions of classical 'war of attrition' models.

Predictability of displays While the relationship between signal unpredictability and symmetrical 'war of attrition' models may be debatable, Caryl's analysis provided other evidence that animals might conceal their intentions during disputes. Data from displays during contests in several species showed correlations between particular components of a display and the future behaviour of both sender and receiver. However, the extent to which any given component predicted subsequent behaviour varied with motivational context. In blue tits (*Parus caeruleus*), grosbeaks (*Pheucticus ludovicianus*) and skuas

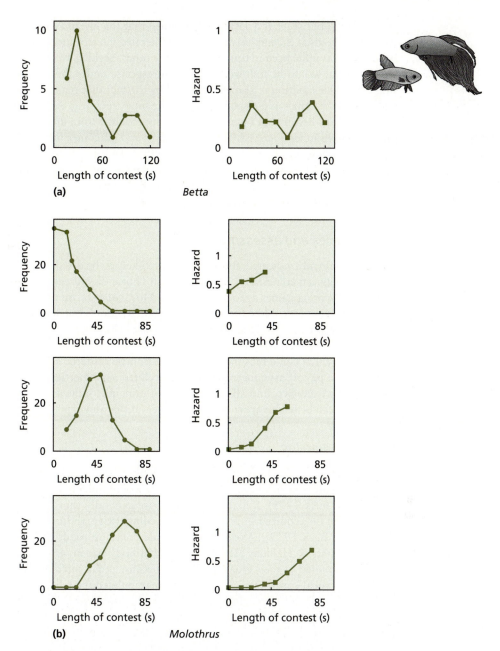

Figure 11.5 Relationships between hazard functions and distributions of display time in (a) Siamese fighting fish (*Betta splendens*) and (b) cowbirds (*Molothrus ater*) broadly conform to those predicted in Fig. 11.4. See text. After Caryl (1979).

(*Stercorarius skua*), for example, displays predicting attack did so with a much lower probability than those predicting other behaviours (Caryl 1979). Displays early on in contests in Siamese fighting fish and cichlids (*Nannacara anomala*) also turn out to be poor predictors of ensuing behaviour (Simpson 1968; Jakobsson *et al.* 1979). The idea that this indicates strategic concealment, however, has had its opponents, particularly

Hinde (1981; but see Caryl 1982), who argued that the interplay between signaller and receiver in agonistic encounters was akin to an iterated interrogation, in which each party's behaviour was influenced in turn by that of the other. Seeking an overall probability that a given signal would be followed by a given response under these conditions was thus simplistic. The situation is also complicated by the fact that many species seem to use a wide variety of signals and displays to mean the same thing (e.g. Andersson 1980; Enquist 1985), perhaps because an opponent can be approached from different directions, or in different environments, so that actions, postures and outcomes inevitably vary from one encounter to another (e.g. Jacques & Dill 1980). Despite these objections, however, comparative studies of different threat displays have found appreciable consistency in the extent to which they predict both the signaller's future behaviour and the response of the receiver (Paton 1986; Paton & Caryl 1986).

Asymmetries and assessment

As we have already indicated, most contests are likely to be decided on the basis of some difference between contestants other than their choice of strategy or tactic. This applies as much to waiting games as to all-out fights. Most contests are therefore asymmetrical, rather than symmetrical like the classical 'war of attrition'. But what *are* these asymmetries and how are they likely to influence signalling behaviour?

In a seminal paper, Maynard Smith & Parker (1976) recognised three broad classes of asymmetry that could bias the outcome of disputes: **resource holding potential (RHP)** asymmetries, **payoff** asymmetries and **uncorrelated** asymmetries (Box 11.3). They are not mutually exclusive, and all three could, in principle, influence a contest. However, their effects can be teased apart by careful observations and experiments.

RHP asymmetries

Where there are differences in RHP, it seems likely that assessment will be based on reliable indicators of prowess, since any others would quickly be devalued by cheating (Barnard & Burk 1979). Reliable indicators are generally those that are too costly to fake and can be acquired only via superior competitive ability (**'status-limited' cues** [Barnard & Burk 1979]). Vocal pitch in male common toads (*Bufo bufo*) is a good example. Davies & Halliday (1979) showed that body size in male toads is reliably predicted by the pitch of their croaking. This is because croak pitch depends on the size of the animal's vocal cords. Davies & Halliday manipulated the relationship between body size and croak pitch by silencing males amplexed (paired) with a female using an elastic band passed through their mouths and playing tape-recorded croaks of different pitch. Rival males were much less likely to attack a defending paired male when deep croaks were played, but the size of the defender still played some role because attacks were more likely when defenders were small (Fig. 11.6). The size effect was probably due to differences in the strength of kicking by small and large defenders.

Two common features of contests suggest opponents assess one another. First, as we mentioned in 11.1.1.1, contests often develop in graded stages. Red deer stags contesting harems start off by roaring at each other, slowly at first, but escalating the rate as the dispute develops (Clutton-Brock *et al.* 1979). Roaring is a reliable index of RHP because stags have to be in good condition to do it. If roaring does not deter a rival, however, the harem owner switches to the second level of display, which involves approaching the rival and engaging it in a parallel walk. By walking up and down together, each animal

Underlying theory

Box 11.3   Asymmetries affecting contests between animals

In most disputes, contestants are likely to differ in ways that bias the outcome in favour of one or other of them. Since these **asymmetries** are often unknown to the contestants at the outset, much of their behaviour during contests is thought to reflect **assessment** of opponents and thus their likelihood of winning. Maynard Smith & Parker (1976) suggested three kinds of asymmetry were likely to be important.

Resource holding potential (RHP) asymmetry

Individuals are likely to differ in their fighting ability. Parker (1974) coined the term **resource holding potential**, or **RHP**, to incorporate the spectrum of factors – size, condition, experience, weaponry, etc. – influencing the ability to contest resources. If two individuals differ in RHP, the weaker one should withdraw as soon as it assesses its chances of winning as low. Persisting beyond this point, or escalating (becoming more aggressive) before a decision has been reached, wastes time and risks injury to no good effect. Various assessable features of an opponent may be related to its RHP, such as body size, vocal pitch and general vigour, so signals designed to advertise RHP are likely to emphasise these characteristics.

Payoff asymmetry

Even when contestants are evenly matched in terms of RHP, one may be prepared to persist for longer or escalate further because it has more to gain from winning. A disputed food item, for example, might be more valuable to the hungrier of a pair of contestants, or a resident and intruder might estimate the value of the resident's territory differently. In many cases, payoff asymmetries will simply result in the same signals and displays being performed with different intensities or for different times, but sometimes, as in resident–intruder asymmetries, they may lead to special signals of ownership (see text).

Uncorrelated asymmetries

Contestants need not differ in either RHP or expected payoff for a dispute to be settled. In principle, contests could be decided on the basis of some purely arbitrary convention, much as we might toss a coin and call 'heads' or 'tails'. If the resource being contested is not in particularly short supply, such a convention would save time and reduce risk. The *Bourgeois* strategy in Table 2.4 shows how an arbitrary asymmetry, in this case resource ownership, can be evolutionarily stable. *Bourgeois* plays either *Hawk* or *Dove* during contests depending on whether it 'owns' the disputed resource or is an 'intruder'. Under the payoff conditions of the model, playing *Hawk* and *Dove* conditionally does better than playing either of the strategies on its own.

is able to gauge the other more closely, and many challenges end at this point. But, if matters are still not settled, the encounter escalates into a fight, where contestants lock antlers and push against each other. Weight and footwork are what count here, but fights carry a considerable risk of injury for both winner and loser. That contests usually end at one of the earlier stages of display, and only occasionally escalate into a full-blown fight, leads us to the second line of evidence for assessment: escalated aggression is most likely when contestants are evenly matched and any differences in RHP difficult

Figure 11.6 Male toads were more likely to attack those already paired if the high-pitched calls of small males were broadcast from a speaker than if the low-pitched calls of large males were broadcast. Paired males were silenced with an elastic band through their mouths. However, attacks were more likely overall when the paired male was small. See text. After Davies & Halliday (1978). © Nature Publishing Group (http://www.nature.com/), reprinted by permission.

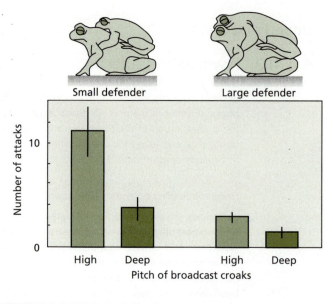

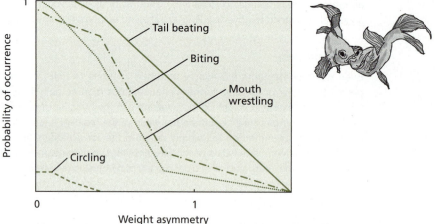

Figure 11.7 Aggressive contests appear to be more difficult to settle when individuals are closely matched in RHP. Contests between male cichlids (*Nannacara anomala*) go on for longer and become more escalated (there is more tail-beating, biting and wrestling) when opponents are closely matched in weight. After Enquist *et al.* (1990).

to discern. For example, several studies have shown that fights are more likely when opponents are of similar size or weight (Fig. 11.7). However, it is important to point out that polarized outcomes typical of RHP asymmetries *could* arise simply through contestants differing in the cost they are prepared to pay (which is likely to be related to their RHP), and thus purely through self-assessment rather than assessment of the opponent (see Taylor & Elwood 2003 for an analysis of this problem).

Clearly, there is mileage in persuading an opponent as early as possible in a contest that is it unlikely to win. Selection will therefore favour signals and displays that emphasise or exaggerate characteristics suggesting high RHP. This accounts for secondary sexual characters such as manes and large horns or antlers that have evolved in males of various species through intrasexual selection (10.1.3.1). Many threat displays also involve puffing up feathers or fur, or adopting postures (e.g. an arched back) that increase apparent size

Figure 11.8 Social status in male house sparrows is reflected in the size of their black bib. See text. After Møller (1987c).

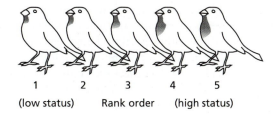

1   2   3   4   5

(low status)   Rank order   (high status)

still further. But bluffs like this may eventually be called, so they are likely to be backed up with underlying competitive ability (see below). At first sight, however, there is one category of assessment signals that seems particularly puzzling from this point of view.

Badges of status In some social systems, assessment appears to be based on signals that do not relate in any obvious way to RHP. For example, the males of various flock-feeding passerine birds, such as house sparrows, Harris sparrows (*Zonotrichia querula*) and white-crowned sparrows (*Z. leucophrys gambelii*) assess one another on the basis of plumage variation (Rohwer & Rohwer 1978; Fugle & Rothstein 1987; Møller 1987b). In male house sparrows, dominant birds have a bigger black 'bib' which acts as a **status badge** to which subordinate males (with smaller black bibs) usually defer (Møller 1987b; Fig. 11.8). Since the difference between dominant and subordinate birds appears to be just a matter of a bit of pigment in the feathers, it is not clear why subordinates do not simply develop bigger bibs and cheat. In Harris sparrows, which also use black breast and head plumage as a status badge, all males develop extensive black plumage during the breeding season, so there is no obvious physiological constraint on subordinates having bigger badges during the winter. So what else might stop them? One possibility is social 'policing'.

Rohwer & Rohwer (1978) tested this by creating 'cheats' in flocks of Harris sparrows. To start with, they simply extended the area of black on subordinate birds with dye and watched what happened to them. Would they be more competitive as a result of their artificially inflated status? Far from it. The manipulated subordinates failed to live up to appearances and were attacked vigorously. In their next experiment, the Rohwers left the subordinates' status badges alone, but injected the birds with testosterone. This made them behave aggressively when released, but their opponents took little notice of them, so again they failed to rise in status. When subordinates were both dyed *and* injected, however, the manipulation worked: the birds were treated as higher-status individuals and won more fights. Thus, simply manipulating plumage badges by themselves was not sufficient to earn the respect of flockmates; declaration of high status had to be supported by appropriate behaviour. These effects are not confined to birds. Status badges are also found in insects, for example, where experimental manipulations of badge size similarly lead to social discrimination (Beani & Turillazzi 1999). However, other studies have failed to find any evidence for social policing (Fugle & Rothstein 1987; Gonzalez *et al.* 2002) leaving the question as to what *does* maintain the stability of the status badge system open in these cases.

While social policing seems to explain why subordinate Harris sparrows do not cheat by making themselves blacker, it does not explain why they cannot both add pigmentation *and* elevate testosterone levels. Perhaps the effects of increased hormone activity would use resources a subordinate could not afford, or maybe other effects of testosterone, such as immune depression, would be too detrimental (see 10.1.3.2). Experimental manipulations of testosterone levels in male house sparrows have shown effects on bib size, but also on overall metabolic rate and antibody production, suggesting the bib status badge is an honest reflection of underlying male quality (Evans *et al.* 2000; Buchanan *et al.*

2001). Indeed, high-status individuals may bear a number of costs that are independent of aggressive contests and cannot be borne by individuals of lesser quality (Gonzalez *et al.* 2002). Although status badges may look cheap and easily faked, therefore, they may in fact be costly to maintain, or reflect other costs that have to be sustained. Even if status badges do not reflect such costs, the high-status-mimicking cheats that will then inevitably arise are likely to impose an interference cost on genuinely high-status individuals as their status is tested (**probed**) for reliability (see below). An assessor–cheat arms race will ensue, favouring recognition of any feature of genuine dominants that distinguishes them from cheats. If these features are just inexpensive chance associations, discrimination will lead to a new wave of cheats and the cycle will repeat itself for a different feature (Barnard & Burk 1979).

Payoff asymmetries

Differences in the reproductive value of winning are a second factor influencing aggressiveness and signalling between contestants. In female land iguanas, for instance, the intensity of disputes over burrows depends on each animal's estimate of the depth of the burrow. Deeper burrows are worth more because there is less work to do to prepare them for egg-laying. Thus both owners and would-be usurpers escalate more when the burrow is deep (Rand & Rand 1976). Contests over deep burrows tend to involve bites, lunges and chases, while those over shallow ones are resolved by head-swinging and gaping. However, the correlation between degree of escalation and burrow depth is better for owners than intruders, presumably because owners have better information about the value of the burrow. This may be one reason why owners tend to win more often in owner/intruder contests generally. That resource quality *per se* can influence the outcome of disputes has been shown in a number of studies. For example, when two common shrews were established separately as owners of the same experimental foraging territory and then allowed to meet, the winner was the animal that had experienced the greater reward rate during its period of occupancy (Barnard & Brown 1984).

A particularly intriguing payoff asymmetry seems to operate in female golden digger wasps. As we saw earlier (1.1.2.1; 2.4.4.2), female wasps, like Rand & Rands' iguanas, acquire burrows in which to lay their eggs by digging them or taking over those of other females. On about 5–15% of occasions, takeovers result in two females co-occupying a burrow, though they rarely meet because most of their time is spent hunting. When they do meet, however, a fight ensues. So, who wins? Both females lay an egg, and both provision the burrow with paralysed katydids to sustain the larvae once they hatch. On the face of it, therefore, the burrow should be worth the same to both females. Dawkins & Brockmann (1980), however, found that wasps did not base their decision to fight on the total number of katydids in the burrow (its actual value to the winner), but on the number they had stored themselves, the winner generally being the female that had stored more (Fig. 11.9). Why should females behave in this way?

One possibility is that female wasps commit the so-called '**Concorde fallacy**' (Dawkins & Carlisle 1976), the principle that says 'so much has been invested in this venture that we should invest yet more to ensure it isn't wasted'. By analogy with the costly Concorde aircraft project, therefore, female wasps might fight harder when they have provided more prey to make all their past hard work worth it. This, of course, risks 'pouring good money after bad' and is fallacious economics. What a good business person, indeed any rational decision-maker, should do is ask what the payoff from currently available options will be in the *future* (see Dawkins & Carlisle 1976; Dawkins 1989). So how might we account for the 'Concorde'-like behaviour of the wasps? Another explanation

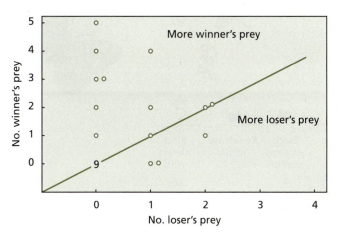

Figure 11.9 When female digger wasps (*Sphex ichneumoneus*) contest a jointly owned burrow, the winner tends to be the female that has provisioned it with the most katydids. The line represents equal provisioning by the two females. See text. After Dawkins & Brockmann (1980). Photograph of two female *S. ichneumoneus* fighting courtesy of Jane Brockmann.

is that the wasps simply cannot count all the katydids in a burrow, perhaps because of the cost of developing the appropriate neural machinery. Instead they are working on a crude rule of thumb. If they themselves have provided a lot of katydids, then the burrow probably has a lot overall and is worth fighting for. Thus females fight harder in relation to their past investment, but only because it is an approximate guide to the future value of the burrow.

As with the assessment of RHP, contests involving payoff asymmetries may escalate in a graded sequence of signalling intensity. In fulmars (*Fulmaris glacialis*) and little blue penguins (*Eudyptula minor*), disputes over pieces of fish tend to begin with low-intensity and low-cost (in terms of the risk of injurious retaliation) displays, then progressively escalate, sometimes to the point of fighting (Enquist *et al.* 1985; Waas 1991). The level to which individuals are prepared to go appears to depend on how hungry they are and thus how much they value the disputed food. In many cases, the value of a resource may not be readily apparent to contestants at the outset of a dispute. As they interact, however, each gains information about the resource, as well, of course, as the relative RHP of their opponent(s). In a series of detailed studies of disputes over shell ownership in hermit crabs (*Pagurus bernhardus*), Robert Elwood and coworkers have shown that the nature of 'shell rapping' signals (where contestants tap on each other's shells to gauge their quality) changes as opponents gain information about the value of the other individual's shell and the likelihood of taking it over (see Elwood & Briffa 2001). However, gaining information takes time and energy, and there is evidence that efforts may be curtailed when the metabolic costs, as measured by accumulating lactate in the haemolymph, become too high (Briffa & Elwood 2001). Where payoff asymmetries arise from hunger, or other temporary differences in motivational state, they are likely to show greater temporal variation than those arising from, say, territory ownership, where the odds can remain stacked in the owner's favour for considerable periods. This can lead to a wide range of displays being used in the same context, once again making it difficult to discern overall pattern in the rules of assessment.

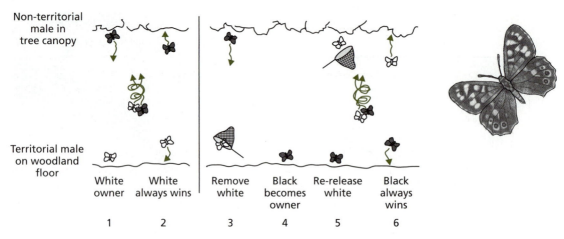

Non-territorial male in tree canopy

Territorial male on woodland floor

White owner	White always wins	Remove white	Black becomes owner	Re-release white	Black always wins
1	2	3	4	5	6

Figure 11.10   When male speckled wood butterflies (*Pararge aegeria*) are given temporary ownership of a sunspot in an experiment, the temporary owner always wins disputes. Ownership may thus constitute an uncorrelated asymmetry that can be used to settle contests quickly. See text. After Davies (1978b).

Uncorrelated asymmetries

Even if there is no difference in RHP or expected payoff between contestants, it may pay animals to settle a dispute using some arbitrary convention and avoid costly displays or fights. Resource ownership is a good example. We have seen that an owner advantage can arise through both RHP and payoff asymmetries, but it could come about because ownership itself is used as an arbitrary rule of thumb to decide outcomes (Box 11.3). While the use of such uncorrelated (with RHP or payoff) asymmetries is likely to be rare, since contests will usually involve some degree of asymmetry in RHP or payoff, there are some suggestive examples (e.g. Davies 1978b; Barnard & Brown 1982; Ranta & Lindstrom 1993).

Probably the best known is Davies's (1978b) study of territorial disputes in speckled wood butterflies (*Pararge aegeria*). Davies found that male speckled woods defended sunspots on the floor of a wood. Here, they courted females that were attracted to the spots. When other males came along, however, Davies noticed that they always retreated after a brief aerial encounter with the owner. Most importantly, the outcome could be reversed simply by allowing the intruder a few seconds' residency; if the owner was removed and the intruder allowed to take over the spot, the intruder won when the pair next met (Fig. 11.10). If two males were trained separately to regard the same sunspot as their territory, the aerial disputes when they eventually met lasted much longer than usual, suggesting the normal convention for settling contests had been overridden.

Davies concluded that the contests in speckled woods were settled according to an arbitrary ownership convention, because RHP and payoff asymmetries seemed to be ruled out by the brief period of residence necessary to establish owner superiority. Later work, however, has suggested that owners may in fact gain a thermal advantage from sitting in sunspots. Stutt & Willmer (1998) found that flight endurance during aerial contests was an important factor in winning, and that butterflies that were warmer could keep going longer. Sunspots may therefore sometimes have a subtle effect on the relative RHP of owners and intruders.

Of course, if conventions are truly arbitrary, there is no reason why an ownership convention should not work in exactly the opposite direction, so that 'intruder wins' becomes the rule. Interestingly, just such a rule seems to have been adopted by the ground-dwelling New World spider *Oecobius civitas*. Here, intruders into occupied crevices trigger desertion by the incumbents, which then intrude on the crevices of others with the same result. The effect is a chain reaction of spiders all switching crevices (Burgess 1976).

Contests with more than one asymmetry

We have discussed each of Maynard Smith & Parker's suggested asymmetries in turn, but, of course, all could have an influence on the outcome of a dispute. This seems to be the case in foraging disputes between common shrews, for example (Barnard & Brown 1982). Where this is so, predictions about signalling and aggression necessarily become more complicated. Disputes involving both RHP and payoff asymmetries in spiders illustrate the point.

Male bowl and doily spiders (*Frontinella pyramitela*) spend most of the brief (about three days) adult life visiting the webs of females for mating opportunities. Not surprisingly, they frequently come across rivals and end up competing for females. Experiments by Steven Austad (1982, 1983) have shown that the outcome of such encounters is likely to depend on both the relative size of the contestants and the changing value of the female in relation to any ongoing copulation.

Mating begins with a pre-insemination phase during which the male assesses the female to decide whether to attempt copulation (Fig. 11.11a). Once copulation starts, the male transfers sperm from his pedipalps to the female. The proportion of eggs fertilised increases sharply to begin with, but then levels off (Fig. 11.11a). The reproductive value of the female from the male's perspective thus changes as mating proceeds. If the female is a virgin, the male stands to fertilise, on average, 40 eggs, and her value to the male thus increases to this point during the pre-insemination phase as the male discovers his good fortune (Fig. 11.11b). Once copulation starts, however, the value of the female declines as more and more eggs are fertilised, reaching less than one egg by the time the pair has copulated for 21 minutes (Austad 1983; Fig. 11.11b).

While virgin females are worth 40 eggs, not all the females encountered are virgins. If virgins and non-virgins are considered together, their average value to a given male drops to 10 eggs, because, of course, non-virgins will already have had a proportion of their eggs fertilised. Imagine another male comes along while the pair above is copulating. Since he is not able to assess the value of the female himself, he must assume she has an average value of 10. Depending on the stage of copulation at which the intruder arrives, therefore, there will be a greater or lesser disparity in the value of the female to the intruder and the paired male (Fig. 11.11b). This allowed Austad to manipulate the relative value of a female to males by introducing an intruder at different stages of mating by a paired male. So what should happen? From Fig. 11.11(b), the prediction would be that, as long as intruder and paired males have similar RHP, paired males will fight harder to keep the female from the pre-insemination phase until about seven minutes later (Fig. 11.11c). Austad established that RHP was directly related to body size, so could control for this by staging interactions between males of similar size. When he compared the predicted success of paired males in such contests with that observed at different times during mating, he found a very close match (Fig. 11.11c).

Of course, not all contests were between similar sized males. Outcomes then depended on which male (paired or intruder) was bigger, and when the two encountered each other. If males of different size were introduced simultaneously at the start of the pre-insemination phase, the larger male won over 80% of encounters. Where paired

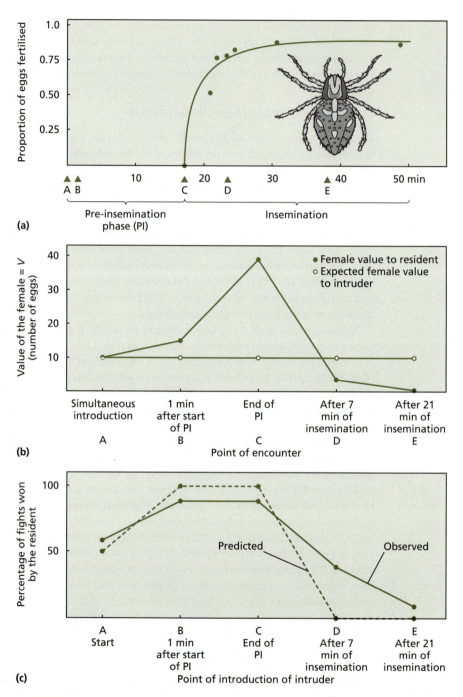

Figure 11.11 Effects of RHP and payoff asymmetries in contests between male bowl and doily spiders (*Frontinella pyramitela*). (a) The proportion of eggs fertilised by a male as a function of time spent with a female. A–E represent different stages of the courtship and insemination process. (b) The value of the female (estimated fertilisations to be gained) to already paired (●) and intruder (○) males when the intruder is introduced at different stages. (c) The percentage of fights won by the paired male at different stages of interaction with the female, and the percentage predicted from differences in the value of the female to paired and intruder males at each stage. See text. From Krebs & Davies (1993) after Austad (1983).

males were the smaller of the two, they fought harder (to the point of serious injury in some cases) when the value of the female was greatest, just after the pre-insemination phase. The most persistent fights, however, occurred when differences in body size (RHP asymmetry) and the value of the female (payoff asymmetry) cancelled each other out, as, say, when a small resident at the end of the pre-insemination phase encountered a large intruder (Austad 1983). Under these circumstances, males stood to gain equally from persisting in the dispute.

Signals and honesty

Disputes, then, can be settled with signals or cues denoting RHP, resource ownership and motivational state. We have also seen that contestants may be tempted to lie about their true state, but that dishonesty can be punished. So just how honest are signals likely to be? In Chapter 10, we argued that sexually selected signals used in mate choice can be honest indicators of individual quality by reflecting, for example, health or developmental stability. Moreover, Zahavi's handicap principle suggested that honesty is ensured by a selective advantage to costliness in signals of quality (Box 10.7). But would that mean all honest signals are handicaps? If we adhere to the strict definition of a handicap, which assumes that its efficacy lies in its cost, the answer is clearly no. As Getty (1998b) has argued, the effect of a differential handicap is to reduce the difference in reproductive potential between high- and low-quality individuals, though still maintain the superiority of high-quality individuals. But high-quality individuals could carry a costly signal that reduces their reproductive potential *less* than a cheaper signal reduces the potential of lower-quality counterparts. The signal would still be honest (high-quality individuals have a costlier one), but not a handicap (the difference in reproductive potential between high- and low-quality individuals increases rather than decreases.)

These arguments apart, however, it is clear that honesty is not in any case a universal property of signalling systems. The examples of mimicry (2.4.5.1; 8.3.1), distraction displays (8.3.4) and tactical deception (4.2.1.4) discussed in earlier chapters are obvious cases in point, but even signals relating to RHP can sometimes be flagrantly dishonest. Mantis shrimps (*Gonodactylus bredini*) provide an example.

Mantis shrimps are stomatopod crustaceans that aggressively contest ownership of burrows and cavities in which to live. To aid this, they possess large, clubbed forelimbs, called raptorial appendages, that can shoot forward with enormous force in the manner of a praying mantis (hence their common name). A well-aimed blow can literally shatter an opponent. In common with many other species possessing lethal weaponry, mantis shrimps often settle contests with a non-injurious threat display, here called a meral spread. The raptorial appendages are spread out to expose a depression, known as the meral spot, and the less offensive maxilliped appendages are extended in a circular arrangement. Normally, the meral spread reliably predicts attack (Caldwell & Dingle 1976) and acts as an honest signal of size and RHP. Sometimes, however, it appears to be used dishonestly. This is because mantis shrimps regularly (every two months or so) shed their hard exoskeleton and have to wait a few days for the new one to toughen up. During this time they are both vulnerable to attack themselves and in no position to inflict damage on anyone else. Despite this, the shrimps continue to use the meral spread display to threaten opponents. Since it does not reflect the individual's current RHP (Adams & Caldwell 1990), the display is dishonest. Interestingly, shrimps in this vulnerable phase appear to hedge their bets a little, since the bluff is more likely to be used against a small opponent, where the cost of having it called is reduced.

Honesty and the cost of assessment   But why should individuals ever accept a dishonest signal? A brief challenge, surely, would quickly reveal a con-trick. True, it might, but challenges come with a cost. If the recipient is *not* lying, there is a risk of serious retaliation from a superior competitor. The cost of challenging, or 'probing', a signal will depend on how often it is likely to be true; that is on the relative frequency with which honest and dishonest signallers are encountered. This can lead to a frequency-dependent advantage to cheating and thus the maintenance of cheats at some level or other in population (Barnard & Burk 1979; Számadó 2000). The argument is as follows. If most or all individuals in the population signalling high status are honest, the frequency of probing will be low because probers risk retaliation. The fact that probing is rare, however, creates an incentive to cheat, because the likelihood of detection is low. Cheats will therefore start to increase in frequency. But as cheating spreads, probers will discover that some high-status signals are really bluffs, so the frequency of probing will also start to increase. As more and more signals are challenged, the cost of discovery will limit the spread of cheating, which will eventually be maintained at some frequency-dependent equilibrium level, or overtaken by the emergence of a new high-status signal (Barnard & Burk 1979). This may explain why some seemingly easily fakeable signals such as status badges, for example, are not always policed to a state of honesty (e.g. Fugle & Rothstein 1987; Gonzalez *et al.* 2002).

The cost of probing signals, however, is just one example of the costs to receivers of assessing whether signals are honest. As Dawkins & Guilford (1991) point out, receivers may pay a variety of time, energy and lost opportunity costs if they put a lot into assessing signals. It may therefore pay them to be slightly less than thorough, and accept a margin of risk that a signaller may not be entirely truthful. This, of course, allows for a degree of cheating, and Dawkins & Guilford's point is that simple cost–benefit considerations on the part of receivers lead us to expect strict honesty to be an exception rather than the rule. This brings us neatly to the other half of Smith's 'message and meaning' distinction at the beginning: the detection and interpretation of signals by receivers and their influence on the evolution of signals.

11.2   Signals and receivers

Consider a male bird singing from a vantage point in its territory (Slater 1983; Fig. 11.12). In many species, males sing only during the breeding season when their testosterone levels are high, and they sing much more when they are unmated. The message of the song signal could thus be 'I am an unmated male of species x in breeding condition sitting in my territory'. Its meaning, however, will differ across the various receivers likely to be listening to it. A passing unmated female of the same species might treat it as an invitation to approach and mate, while a conspecific male is more likely to regard it as a threatening exhortation to keep away. A passing male of a different species might simply ignore the singing and carry on regardless, unless, of course, it was a predator, in which case it might home in on the sender as a source of food (Fig. 11.12). This raises an important point. By their very transmission, signals become available to any receiver able to detect them (see 11.2.1.1). Some of these, such as the conspecific female above, will respond in ways that are in the reproductive interests of the sender. Others, such as the predator, will not. As McGregor (1993) has emphasised, signallers and receivers thus become part of broader **communication networks,** in which unwelcome, or otherwise unintended, receivers can be thought of as **eavesdropping** on the sender's signal. A signal can therefore carry risks as well as secure benefits for the sender, and these have to be weighed with the

Figure 11.12 Broadcast signals are potentially available to a range of receivers, each of which may exploit the signal in a different way. In this case the territorial song of a male great tit is detected by passing females, rival males and predators. See text. After Slater (1999).

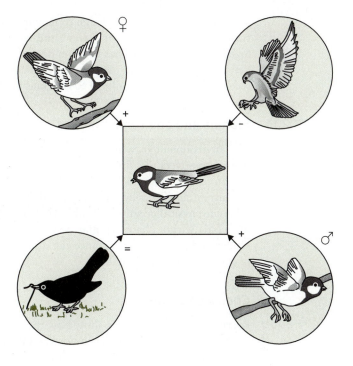

time and energy costs of transmission in determining its net reproductive benefit. The diversity of receivers will thus be reflected in the selection pressures shaping signals. But this is not just a matter of whether a given receiver benefits or penalises the sender, it is also affected by the way the receiver detects and interprets the signal, as we shall now see.

11.2.1 Receivers and signal design

As Guilford & Dawkins (1991) put it in their influential paper, receivers have generally been regarded as little more than signal detectors, 'most analyses of signal design (seeming to stop) at the level of the eye or ear, as if that were the final destination of the signal' (Guilford & Dawkins 1991). Detection by the sense organs, of course, is only part of the story, ignoring as it does the associated, and often specialised, processing, integration and storage of signal information in the CNS. We know already from Chapter 3 that neither sense organs nor central nervous systems are passive transcribers of information from the environment. Together, they filter, select and interpret the information they receive, creating models of the world that allow the animal to function adaptively (3.2.1.1). The upshot is that some features are rendered more detectable, or easier to discriminate, remember or even invoke (see 3.2.1.1), than others. The perceivers' response to them in turn selects for further enhancement or reduction (depending on the consequences for the sender) of their detectability and retention. Thus signals and cues are honed both by the needs of the sender and the selective filter of the receiver.

Guilford & Dawkins recognise two components to the design of signals: **strategic** design and **tactical** design. Strategic design reflects what the signal has evolved to communicate and how the receiver responds to it. One male may seek to impress another with his apparent stamina, for instance, and the recipient to judge what to do as a result. The roaring contests of red deer stags might reflect strategic design in such cases. But signals are also subject to tactical design to maximise their *efficacy*, that is how effectively they transmit through the environment and are received and interpreted by the receiver. Signal efficacy thus also has two components: first how well suited the signal is to being detected in the receiver's environment, and, second, how effectively it is exploited by the receiver's CNS, or, as Guilford & Dawkins put it, by **receiver psychology**.

11.2.1.1 Signals and environment

The environment in which signals are used is clearly likely to influence their form because receivers will find it easier or more difficult to detect particular kinds of signal in different surroundings. This affects fundamental features of signals, such as choice of sensory channel, since the transmission properties of different channels suit each to different physical conditions (Table 11.2; 11.1.1). The choice of channel, and variation in signals within channels, may also be influenced by the biotic environment of the sender and receiver, for instance the distribution of food resources, or the likelihood of eavesdropping by predators.

Choice of channel

Recruitment signals in ants show how the channel can be tailored to different environments (Hölldobler 1977). When worker ants discover food, they often recruit others to help exploit it. How this is done, however, differs between species in ways that reflect the kind of foraging problem facing the ants.

Where food comes in packages that are too big for one ant to drag back to the nest, but can be managed by two, recruitment communication may be tactile. In harvester ants (*Leptothorax* spp.), for instance, which feed on immobile prey such as dead beetles, successful foragers return to the nest, regurgitate some food and secrete a chemical attractant from their abdomen to recruit other workers. One of the recruits establishes antennal contact with the forager's abdomen and is led in this way (called 'tandem-running') to the food. In contrast, fire ants (*Solenopsis* spp.) take large active prey that require several workers to deal with them. Here, the ants use chemical signals in the form of odour trails. After locating prey, a worker returns to the nest depositing a trail of volatile scent from a specialised abdominal gland as it goes. The trail attracts other workers that follow and reinforce it with their own odours if they find the prey. As long as the food supply lasts, therefore, the trail accumulates in strength. However, once the food is finished, recruitment ceases and the trail quickly decays. The chemical trails thus provide *Solenopsis* with a rapid-response recruitment system that responds flexibly to the changing location of food.

Other species of ant exploit more stable sources of food. Species that feed on plant material, such as foliage (e.g the leaf-cutter ants [*Atta* spp.]), or seeds (e.g. *Pogonomyrmex* spp.), may use the same trail for days or even years on end. Here, workers use long-lasting odours to mark out trails, or physically cut paths through the vegetation, both techniques maintaining long-term routes to the food supply.

These interspecific differences in ants are reflected intraspecifically in many other species where communication has to be tailored to changing demands on signal transmission. African elephants (*Loxodonta africana*), for example, use a variety of sensory channels, both complementary and redundant, to cater for this. The majority of long-distance communication (extending for 10 km or more) involves infrasonic vocalisations and chemical signals, whereas higher-frequency vocalisations, visual and tactile signals are used in shorter-range communication (Langbauer 2000).

Acoustic communication

Acoustic signals face two fundamental problems in their transmission through the environment: **attentuation** and **degradation** or **distortion**, both of which diminish or corrupt the signal as it travels between sender and receiver.

Attenuation In an ideal acoustic environment, with a homogeneous, frictionless medium and no obstacles or boundaries, sound waves diverge in a perfect sphere from a point source. As the sound spreads out, however, its intensity drops at a rate of 6 dB for each doubling of distance owing to the inverse square law. However, animals are not ideal point sources, which means that propagation tends to be distorted from the spherical. This can either reduce or increase the level of attenuation of sound as it spreads (Michelsen 1978). Natural environments are also far from homogeneous, and compound attenuation effects through sound waves being absorbed by the air and colliding with vegetation and other obstacles. The further the receiver from the sender, therefore, the fainter (more attentuated) an acoustic signal usually becomes. Higher-frequency sounds tend to show greater attenuation because they are more vulnerable to absorption by the atmosphere, especially in warm, humid conditions. They are also scattered more easily by obstacles.

Scattering itself is a complex function of the frequency (and thus wavelength) of the sound signal, the size, shape and rigidity of objects in its path and heterogeneities in the air (Gerhardt 1983). If objects are large and hard, total reflection of the signal may occur. Small objects, unless smaller than the wavelength of the sound, scatter or redirect the signal in different ways that can either increase or decrease its level depending on the distance from the source and the phase relationship of the direct and scattered or deflected sound waves (Gerhardt 1983). In many habitats, vegetation accounts for much of the scattering, but temperature gradients in the air (especially likely in open habitats) can also have important effects. Warmer air close to the ground, for example, allows sound to travel faster, causing sound waves to bend upwards and create a sound 'shadow' (Wiley & Richards 1978), while temperature variation can cause turbulent eddies of hot air to rise up, scattering and absorbing sound in an unpredictable fashion (Gerhardt 1983).

Many animals signal close to some kind of boundary, such as the ground or the surface of water. The nature of the boundary, and the position of the sender and receiver relative to it have profound implications for the attenuation of acoustic signals. If both sender and receiver are close to a reflecting boundary such as water or a smooth rock surface, most if not all sound frequencies in the signal are reflected and there is little attenuation relative to that expected from spherical spread (Embleton *et al.* 1976). However, if signals are sent at an angle to a porous substrate such as grass or soil, there is a pronounced reduction in high-frequency relative to low-frequency elements. These can be restored to some extent by raising the receiver (or the sender) to different heights above the substrate (Fig. 11.13). Two obvious strategies for overcoming these problems might be to produce signals of low frequency or signal from an elevated position.

Studies of various species that are unable to produce low-frequency acoustic signals suggest they frequently select high vantage points instead. The advantage of this is neatly demonstrated by the cricket *Anurogryllus arboreus* (Paul & Walker 1979). Despite its name, *A. arboreus* is a burrowing cricket, but one in which males sometimes ascend trees or shrubs to call. Paul & Walker estimated the broadcasting efficiency of these individuals compared with those calling at ground level by measuring the area

Figure 11.13 Effect of elevating the signaller on attentuation of high-frequency sound where a signal is emitted above a grassy substrate. If the signaller is close to the substrate, there is pronounced attenuation of the higher frequencies (curve 3 [signaller 3 m above the substrate]). Raising the signaller correspondingly reduces the attenuation effect (curves 2 [60 m above] and 1 [120 m above]). See text. After Embleton *et al.* (1976).

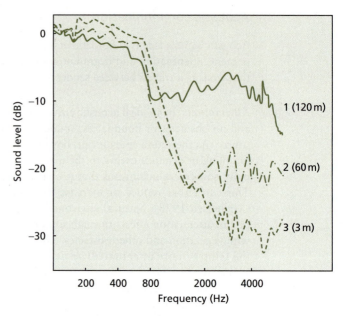

around the cricket in which sound pressure levels were above the behavioural threshold of the receiver. They found that crickets calling from trees or shrubs had broadcast areas at least 14 times greater than their counterparts on the ground. Moreover, males calling from a half to two metres off the ground attracted the most females, suggesting that receivers were indeed able to detect elevated calls more effectively.

Various studies suggest low-frequency sounds are used to overcome problems of signal attentuation. The infrasonic long-distance vocalisations of elephants are one example, while bird song may be another. Comparative studies of birds singing in different habitats suggested that species singing in high-attenuation habitats, such as dense forest, produce songs of lower average frequency than those singing in open grassland (Morton 1975; Slabbekoorn & Smith 2002). However, while vegetation and other obstacles in forests undoubtedly cause attenuation at high frequencies, some habitat effects have turned out to reflect the attenuation of lower frequencies by the ground rather than overall differences in the sound transmission qualities of the habitats as a whole. The range of physical environmental factors affecting sound transmission in different habitats makes the story more complicated than this (Wiley & Richards 1978).

Degradation   As we indicated at the beginning of the section, attenuation is not the only problem facing acoustic signallers and receivers. Degradation (distortion) of sound by the environment is at least as important. If signals become degraded they may be confused with other signals or sounds, or simply ignored altogether. Wiley & Richards (1978) suggest that the kind of habitats studied by Morton differ more strongly in sound degradation than attenuation, since air turbulence in open habitats attenuates sound just as strongly as the vegetation in forests. The major causes of degradation in forests are echoes and reverberations from branches and leaves, while in open habitats the major cause is unpredictable variation in amplitude caused by gusts of wind (the effect we experience when someone tries to shout a message to us from far away on a windy day).

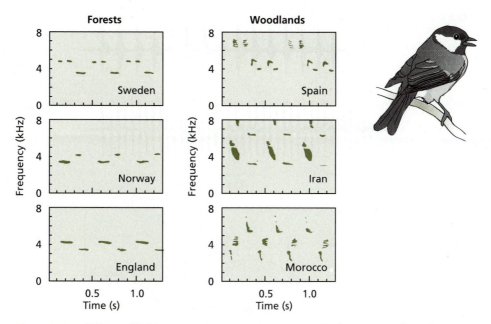

Figure 11.14 Effects of habitat structure on song in great tits. Birds singing in dense forest produce songs with a narrower range of frequencies, a lower maximum frequency and fewer notes than those singing in more open woodland. The differences are consistent despite samples for each habitat type coming from a range of geographical locations. After Hunter & Krebs (1979).

Acoustic signals can be adapted to cater for these problems. Reverberation effects are worse for high-frequency sounds because these can be deflected by smaller objects, such as leaves. They are also a problem for rapidly repeated elements, such as the 'trills' in some bird songs, because the reflected sound can become confused with newly produced notes. Thus acoustic signals in forests should be of low frequency and avoid trills unless their notes are widely spaced. The erratic fluctuations in signal amplitude in open habitats, such as grassland, on the other hand, are likely to favour rapidly repeated structures such as trills (and redundant signals generally) because these can be detected during brief periods of good transmission. By and large, these were the characteristics found in different habitats by Morton in his birds (but there interpreted in terms of attenuation), and since confirmed by other comparative studies, such as Hunter & Krebs's (1979) comparisons of song in great tits (Fig. 11.14). Degradation effects need not be all bad, however. Recent work suggests that reverberation may in fact have an enhancing effect on some acoustic bandwidths, leading to louder and longer signals after transmission (Slabbekoorn *et al.* 2002).

Acoustic communication and temperature   While temperature can influence the quality of sound transmission through the environment (see above), it can also affect the production of sound by the signaller. In ectothermic ('cold-blooded') animals, variation in temperature affects metabolic rate and neuromuscular mechanisms. It is thus very likely to affect mechanically produced signals such as sounds. There is a folk-saying that one can tell the temperature by listening to cricket song (Gerhardt 1983), and evidence from crickets and various taxonomic groups suggests it is well-founded. Figure 11.15, for example, shows the effect of temperature on trill (pulse-repetition) rates in two species

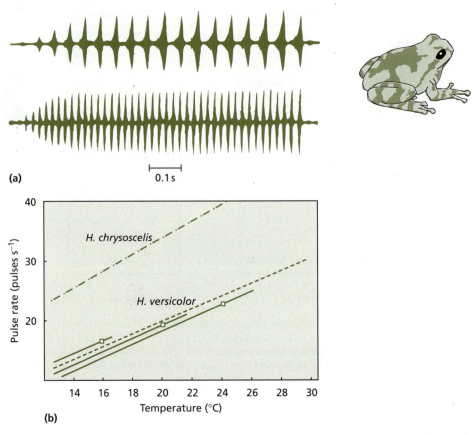

(a)

0.1 s

(b)

Figure 11.15 (a) Oscillograms showing the difference in trill structure between the treefrogs *Hyla versicolor* (top) and *H. chrysoscelis* (bottom). (b) As temperature varies the pulse-repetition rate of the two species changes and creates the potential for overlap. See text. After Gerhardt (1983).

of tree frog, *Hyla versicolor* and *H. chrysoscelis*. In both cases, there is a linear increase in rate as conditions get hotter. Not surprisingly, this has important implications for the function of calling.

Both species of frog are widely distributed through eastern North America, and breed at the same time and in the same places. The sound frequency of their calls is similar, but, at any given temperature, the trill rates of the two species differ significantly (Fig. 11.15a). This is important for females, which use the difference to home in on males of the right species. As temperatures vary, however, there is scope for the call character-istics of the two species to overlap: the trill rate of a warm *H. versicolor* becomes similar to that of a cool *H. chrysoscelis* (Fig. 11.15b). This poses a problem for females. At low temperature, a female *H. versicolor* responding simply to trill rates within her normal species range would run the risk of mating with slowed down male *H. chrysoscelis*. Similarly, a female *H. chrysoscelis* prospecting on a warm night might mistakenly choose a warmed up male *H. versicolor*. In either case, the result of cross-species mating would be sterile offspring – a severe penalty. How might females avoid the problem? One way would be to compensate for variation in temperature when responding to calls. Is there any evidence that females do this? Experimental studies in which recorded calls were played back to females suggest there is. Gerhardt (1978, 1982) found that females of both species altered their preference for different trill rates as temperature changed. Moreover, the effect was strongest when the species identification problem was most acute. Thus, female *H. versicolor* discriminated strongly against higher trill rates when tested at low temperatures, while female *H. chrysoscelis* particularly discriminated against low trill rates when tested at high temperatures. Conversely, increasing the temperature of female *H. versicolor* made higher trill rates more attractive, and decreasing that of female *H. chrysoscelis* made lower rates more attractive. Similar temperature-dependent shifts in receiver response have been found in other groups, including insects and fish, and have been referred to by Gerhardt (1978) as **temperature coupling**.

Effects of the biotic environment

Of course, the physical elements of the animal's world are not the only environmental factors affecting signals. The biotic environment is at least as important, and often imposes strongly conflicting selection pressures because of the range of potential receivers for any given signal. Noise and eavesdropping are two factors that play an important role.

Biological noise   Major sources of noise in the physical environment include such things as running water and wind blowing through vegetation. These sounds contain a broad band of frequencies that are difficult to counteract from a signalling point of view. The only solution may be to avoid signalling when and where it is a problem, or to combine or switch between different channels (Gerhardt 1983). Depending on the degree of overlap between signal and noise, however, it may be possible to adjust signal characteristics to compensate for the latter (Aubin 1994). To compound things, the biological world can also be a significant source of noise. Mixed species choruses of birds, frogs or insects, for example, can involve large numbers of species and vast numbers of individuals. The resulting **signal interference** poses a problem for any animal trying to make itself heard, and there is good evidence that it limits the ability of receivers to discriminate appropriately (Wollerman 1999; Schwarz *et al.* 2001). Interference can thus reduce the attractiveness of otherwise suitable times and places for signalling in much the same way that feeding competition reduces the value of rich food patches (7.1). What might animals do to overcome it?

   In the context of choruses, perhaps the most obvious response is to shift the sound frequency or timing of a signal to reduce overlap. There is some evidence from frog and insect choruses that associating species adopt different broadcast frequencies or modulate their calls to distinguish themselves from each other (Drewry & Rand 1983; Cooley & Marshall 2001; Sueur 2002). However, the potential for this is likely to be subject to mechanical constraint from the vocal apparatus. The dominant frequencies among frogs, for example, tend to concentrate between 1500 and 3000 Hz, with different species often calling at almost the same frequency (Loftus-Hills & Johnstone 1970). Varying the location and/or timing of calls may be a more feasible way of reducing interference. In multispecies choruses of cicadas (Hemiptera: Cicadidae) in the Mexican rainforest, each species adopts a different calling height, with males following a 'call–fly' or 'call–stay' strategy to help partition the acoustic environment (Sueur 2002). This variation in location is sometimes combined with species-typical time and frequency patterns in the call. Indeed, relative timing appears to be one of the most widespread and versatile mechanisms for dealing with interference across a range of species, as we shall see shortly. Strategies of signal variation are not deployed only between species, however. In dense and fluid penguin colonies, for example, individuals show enormous variation in vocalisation, which proves vital when birds are trying to locate their mates or chicks in the clamorous throng (Aubin & Jouventin 2002). The use of individually characteristic calls in this situation has been dubbed the **cocktail party effect** (Aubin & Jouventin 1998), after our own ability to home in on a familiar voice or mention of our name in a room buzzing with conversation.

Avoidance and jamming: In some chorusing frog species, females discriminate against males whose calls overlap (Schwartz 1993). There is thus strong selection on males to avoid overlapping. In the neotropical treefrog *Hyla microcephala*, males that are interrupted by the call of another frog increase the spacing between the elements of their own call. They respond most strongly to the loudest neighbour, and some males respond only to this individual (Schwartz 1993). This seems to be because the volume of interfering calls is a significant cause of degradation in a male's signal to a female. If calls are overlapped, a difference in volume of some 6 dB is needed to overcome the degradation effect on either call to the female. Experiments with synthetic calls showed that males can respond to variations in sound intensity as brief as 20 ms, enabling them to alternate their calls within pairs on a very fine timescale. As a result, the amount of acoustic overlap between pairs of males in natural choruses is usually less than 10% of an individual's calling time during bouts. Indeed, males can spend anything from 30% to over 90% of their calling time free of acoustic interference (Schwartz 1993).

In some cases, however, females may actively select for a degree of overlap in male calls. In the running frog *Kassina fusca*, for example, the average degree of overlap within pairs of males is around 21%, a level that corresponds almost exactly with the preference of females in playback experiments (Grafe 1999). When overlap was manipulated, females appeared to switch their preference for 'leader' (the first to call) and 'follower' (the overlapping caller) males according to the extent of the overlap, preferring 'follower' males when it was low (10–25%), but 'leader' males when it was high (75–90%). As in *H. microcephala*, males are capable of adjusting the timing of calls very quickly, so may be able to control the degree of overlap with neighbours to their advantage (Grafe 1999). Such receiver-driven overlap suggests that chorusing synchrony might act to enhance a male's chances of gaining attention from females. This contrasts with the alternative view that it is an epiphenomenon of attempts by males to jam each other's signals (e.g. Greenfield *et al.* 1997).

Greenfield *et al.* (1997) note that females of many insect and anuran chorusing species selectively orientate towards the first call of a sequence, and thus favour 'leader' males. Consequently, they argue, males have been selected to jam their neighbours' calls by overlapping them, so causing the signal rhythm in the neighbourhood to be reset and creating an opportunity for the jammer to take the lead in a new sequence. In this way, patterns of synchronised calling become an emergent property of acoustic competition between callers. However, metabolic resources may set a limit on the ability to respond to acoustic inference. For example, female *H. microcephala* tend to pair with males that have higher overall rates of calling, but females are themselves available for pairing for many hours (Schwartz *et al.* 1995). To compete continuously for the entire period of female availability would be enormously costly for males. One compromise, therefore, might be to break off periodically and conserve resources so that effort can be strung out to cover the time females are around. Analysis of calling in *H. microcephala* in relation to muscle glycogen reserves suggests that males do indeed pace themselves in this way (Schwartz *et al.* 1995). Thus temporal patterning in choruses may sometimes partly reflect energetic constraints.

Not surprisingly, given the potential consequences of jamming, signallers appear to have evolved various strategies for avoiding it. Several species of weakly electric fish (see 3.2.1.5) produce electric displays during aggressive and intersexual interactions (e.g. Hopkins 1977). Such displays, as well as the electrical discharges used in orientating and navigating round the environment, are vulnerable to jamming by emissions from other fish. To overcome this, some species alter the discharge frequency of their electric organs in a reflex response to detecting signals of similar frequency from conspecifics; that is, they possess a jamming avoidance mechanism (Rose & Canfield 1991). The gymnotid *Eigenmannia*, for example, increases its discharge frequency if jammed by a signal of lower frequency, and decreases it if jammed by one of higher frequency. Interestingly, some other species, such as *Sternopygus*, share many of the sophisticated signal discrimination and processing capabilities of *Eigenmannia*'s anti-jamming system but do not themselves show the jamming avoidance response (Rose & Canfield 1991).

Eavesdropping   Multiple signallers pose one problem for sending and extracting information in a complex biotic environment. Multiple receivers pose another. As we mentioned earlier, a broadcast signal is potentially detectable by any set of sense organs that happens to be out there in the environment. Only some of these are likely to belong to the intended receiver(s). Other receivers can eavesdrop on signals and acquire information about both senders and intended recipients, information that can be exploited, sometimes to the benefit of eavesdropped individuals, sometimes to their detriment (see above). The composition of the network of receivers is therefore likely to impose diverse selection pressures on the design and performance of signals.

Eavesdropping and signals: For many signallers, predators or, in some cases, parasites form a broad class of unwelcome eavesdroppers. In the field cricket *Gryllus integer*, for instance, stridulation ('calling') by males to attract females can also attract both other males and parasitic tachinid flies (*Eupasiopteryx ochracea*) (Cade 1979). Both kinds of eavesdropper impose a reproductive cost on the caller, in the flies' case a terminal one (the larvae develop inside the cricket and slowly devour it). That the flies were homing in on the calling signal was shown neatly by the fact that they happily laid eggs on loudspeakers when pre-recorded calls were played. Broadcasting a signal can thus be a risky business. What can signallers do to reduce the risk?

One thing they can do is make signals difficult for unwelcome eavesdroppers to detect or locate. We have already seen examples of this in alarm calls (8.3.5.4), where, in some cases, sound frequencies appear to have evolved to be difficult to locate, or even ventriloquial, from the viewpoint of predators. Better still, signals can be selectively unavailable to predators. This seems to be the case with the use of red pigmentation for intraspecific communication in poeciliid fish, where red signals are difficult or impossible for the fishes' invertebrate predators to pick up because they are effectively red-blind (Endler 1983). The structure of colour patterns may also be important in conferring crypticity. If the grain size or scale of colour patterns in a signal is smaller than a predator can resolve at its normal detection distance, then colours that are bright close up may blend together to render the pattern inconspicuous against the background (Endler 1991; Fig. 11.16). These effects are more pronounced at low light intensities, so bright courtship or territorial displays over short distances in low light may accomplish their function while at the same time affording protection from predators further away. There is good evidence that this happens in guppies (Endler 1987) and three-spined sticklebacks (Moodie 1972; see 2.4.5).

Figure 11.16 Changes in the number and length of spots in the marking patterns of male guppies in Trinidad as a function of predation intensity. The *x*-axis represents a transect from the north coast of Trinidad up the northern range of mountains (E–MR in the figure), then down the southern slope (R–Cr). The intensity of predation is indicated by the degree of hatching and shading, so is generally greater at lower elevations and greatest for E and Cr. The main predators of guppies in each case are: *Eliotris pisonis* (E), *Agonosotmus monticola* (Ag), *Macrobrachium crenulatum* and *Rivulus hartii* (MR), *Aequidens pulchur* (Aq), the characins *Astyanax bimaculatus* and/or *Hemibrycon dentatum* (Ch), *A. pulchur* and characins (AC), and *Crenicichla alta* and other species (Cr). After Endler (1991).

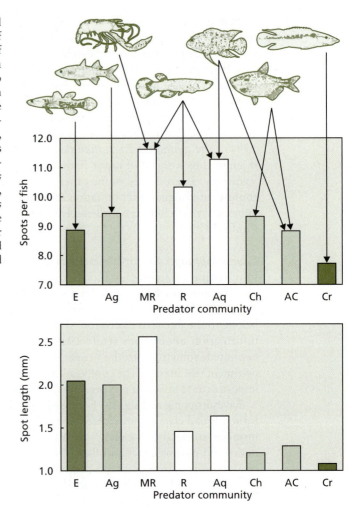

Another approach is to direct signals to specific individuals, and thus control their availability to unintended receivers. One example is the ability of squid to produce different visual displays on the two sides of their body when flanked by different individuals (Moynihan & Rodaniche 1977). However, most signalling systems do not afford this kind of flexibility, and selective direction is probably limited to producing signals when appropriate individuals happen to be around.

Eavesdropping and receivers: The idea that signals should be modulated to take account of eavesdropping assumes that eavesdroppers will act on the information they pick up. The tachinid flies afflicting Cade's crickets above show clearly that they can. The use of information by eavesdropping receivers has also been studied in some detail in various intraspecific interactions. Song overlapping in birds is a good example.

We saw above that overlapping the signal of a rival may be a means of competitive interference. In the males of some song birds, overlapping the song of a conspecific appears to indicate the readiness of the overlapper to escalate the contest, and thus indicates his level of arousal (e.g. Naguib & Todt 1997). In nightingales (*Luscinia megarhynchos*), playback experiments have shown that overlapping causes birds to sing at a greater rate and interrupt more songs themselves (Naguib 1999). These changes in behaviour can

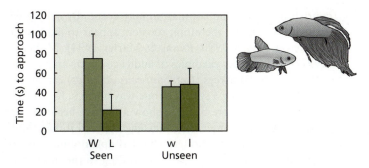

Figure 11.17 The effect of having previously seen interactions between stimulus males as shown by the latency for subject male Siamese fighting fish to display to seen (W, L) and unseen (w, l) winners and losers in dyadic encounters. Subjects were significantly slower to display to fish they had previously seen winning, but there was no effect of having won if stimulus fish had not been seen. After Oliveira *et al.* (1998).

be heard by other birds in the vicinity, which thus stand to gain information about the competitive ability and motivational state of the singers. This may put them at an advantage if they encounter the singers subsequently. Do nightingales use eavesdropped information in this way? To find out, Naguib & Todt (1997) exposed birds in the field to two 'interacting' speakers that overlapped or otherwise the song broadcast by the other. 'Eavesdropping' males responded more strongly towards the overlapping speaker during tests rather than the speaker where song was overlapped. Importantly, however, the bias in response persisted in later tests when songs from only *one* of the previously interacting speakers was played. Thus, 'eavesdropping' nightingales were able to remember attributes of 'singers' from overhearing their previous 'encounters', and use the experience later in their own interactions with them. Recent field studies of male great tits (*Parus major*) have confirmed the effect and shown that birds are also able to combine eavesdropped information with their own assessment of individual rivals during encounters (Peake *et al.* 2002).

Broadcast information is, of course, available to eavesdropping conspecifics of both sexes. Laboratory experiments with Siamese fighting fish (*Betta splendens*) have shown that males allowed to eavesdrop on encounters between other males are more cautious in subsequently approaching winners they had observed compared with winners they had not (Oliveira *et al.* 1998; Fig. 11.17), and that females are influenced by previously observed encounters in their tendency to approach males as potential mates (Doutrelant & McGregor 2000). This suggests it might pay males to be mindful of who's watching when deciding how to behave with rivals. Such changes in behaviour according to who is looking on are known as **audience effects**.

Audience effects: The presence of particular onlookers can make behaviours either more or less likely to be performed according to who the onlooker(s) perceiving them are and what they are likely to do. For example, alarm calling when a predator appears may make sense only if other individuals are around to benefit from it, since calling is likely to put the caller at risk (see 8.3.5.4). Experiments with domestic roosters, in which the birds were presented with a predator stimulus, showed clearly that alarm calls were more likely when an audience was visible (Evans & Marler 1995), especially if the audience was a hen (Karakashian *et al.* 1988). An audience effect is apparent in food-calling too. Here the rooster's call imitates that of a mother hen bringing food to its young, so when a rooster calls on finding an item of food, nearby hens are attracted to it and courtship activity and copulation may ensue. By calling when an appropriate female audience is at hand, therefore, roosters appear to manipulate mating opportunities for themselves (Gyger 1990; Evans & Marler 1994). Møller (1990b) and Baltz & Clark (1997) provide other examples of audience effects and manipulation in bird calls.

Male vervet monkeys (*Cercopithecus aethiops*) can also engineer benefits by playing to an audience. Using a system of one-way mirrors, Keddy-Hector *et al.* (1989) found that adult males were more likely to show affiliative behaviours (passive contact, grooming, play, etc.) towards any young they encountered if they thought the youngster's mother was watching. If they thought she was not looking, however, they were likely to be aggressive. Adjusting behaviour according to the mother's presence turns out to be a good idea on two counts. First, females that see a male being affiliative towards their offspring are more likely to mate with him later on, perhaps because they take such responses to indicate paternal qualities in the male. Second, any aggression towards their offspring is punished severely if a mother happens to see it. In one set of experiments, mothers were concealed behind a one-way screen, so they could see a male interacting with their offspring, but could not themselves be seen by the male. If a male had been rough with the offspring while being observed, he was vigorously attacked when the mother was eventually released from the observation chamber (Keddy-Hector *et al.* 1989).

11.2.1.2 Receiver psychology

It is clear, then, that the environment in which signals are transmitted imposes a variety of selection pressures on their form and timing. These derive from the ability of receivers (intended or unintended) to detect, interpret and respond to signals in the prevailing conditions. As Guilford & Dawkins (1991) have emphasised, mediating between detection and response are the receivers' central nervous mechanisms of signal discrimination, information processing and decision-making, which together make up the psychological filter (or 'landscape', as Guilford & Dawkins put it) through which signals must pass to trigger a response (Box 11.4). Receiver psychologies can be both specialised and flexible,

Underlying theory

Box 11.4   Receiver psychology

'Receiver psychology' refers to the information processing mechanisms of receivers that lie beyond stimulus perception in the CNS (Guilford & Dawkins 1991). Guilford & Dawkins argue that central processing (the passage of information through the receiver's 'psychological landscape') affects each of the key requirements for signal reception: *detectability*, *discriminability* and *memorability*.

Detectability

Much about the detectability of a signal has to do with its physical parameters, such as intensity, duration and repetitiveness, which are shaped by the environment and the sense organs of its various receivers (11.2.1.1). However, central processing by receivers is important too. For example, visual perception is often the result of compiling the responses of specialised feature detectors into recognisable objects, or even conjuring objects out of partial information (see 3.2.1). Fitting information into an adaptive model of the world may predispose receivers to respond to certain kinds of stimuli. Many signals and displays have different components, often exploiting different sensory modalities such as vision and hearing. Some components, such as sudden movements or piercing sounds, appear to act as attention-grabbers, focusing the receiver on the signaller and thus the rest of the display. Mimicry may also be used to enhance detectability via receiver psychology. The eye-spots on the peacock's tail, for example, may draw the female's attention because she is predisposed to notice things that look like staring eyes and might indicate a predator. Many signals used in aggressive contests and sexual display trade on attention-grabbing features such as size, brightness and mimetic patterns.

Discriminability

Discrimination involves recognising that the stimuli making up a signal belong to some discrete category, such as those associated with distastefulness or predation risk. Signals should thus be designed to conform as unambiguously as possible to the relevant discrimination categories of the receiver. For example, most palatable prey present patterns that are difficult to detect against their natural backgrounds, while warning signals comprise bright colours, often black, white, reds or yellows, that make discrimination from palatable prey especially easy. Bright patterns may also enhance discriminability by increasing consistency of appearance, making it easier to distinguish unpalatable objects from different angles or partial glimpses (Guilford 1990). Signals can aid discrimination of other features of organisms. For instance, the dark lines and spots used in many visual displays, especially among fish, may accentuate body size parameters, such as length or surface area, that are important in discrimination by rivals and mates (Zahavi 1987). Such **amplifiers** (Taylor *et al.* 1999) may then, of course, be put to good effect in deceiving receivers.

Memorability

Many signalling systems rely on learning by receivers for their effectiveness. Assessment, mate choice and defensive communication about predators are three contexts in which receivers may depend on learning the attributes of individuals sending out different kinds of signal. Signals that are striking and memorable may thus be at a premium. Several aspects of signalling systems are consistent with having been selected for memorability. Strongly coloured and demarcated patterns, for example, may be easier to learn because they gain attention and offer simple visual stimuli. Signals that provide a sudden contrast with the background, and thus a degree of novelty, may also be easier to learn. There is a wealth of evidence that novelty facilitates learning (6.2.1.2), and many displays, such as the large song repertoires of some bird species and highly variable secondary sexual plumage of many others, may partly rely on novelty to be effective. Another aspect of signalling that suggests it exploits learning potential in the receiver is the use of **potentiating displays**. The ease with which two stimuli can become associated may depend on the presence of a third stimulus, which can overshadow (prevent) or facilitate the association (see 6.2.1.2). The apparent attention-grabbing elements of some signals (see above) may enhance memorability, for example, and multimodality generally could reinforce learning, as in the emission of distinctive odours or harsh sounds during defensive displays by some warningly coloured insects (Rowe & Guilford 1999) (see text).

Based on discussion in Guilford & Dawkins (1991).

in that they are tailored to the reproductive requirements of the individual, but capable of modification through learning. They are also frequently based on approximate rules of thumb (see 2.4.4.3 and 3.2.1). Thus the full panoply of factors governing central stimulus filtering, memory and learning in different species (see 3.2 and 6.2) are likely to be brought to bear on signal processing, helping to explain the enormous diversity of signalling behaviour within and between species (Guilford & Dawkins 1991).

So far, much of the evidence for receiver psychology shaping signals has been circumstantial. However, a number of theoretical and experimental approaches have addressed the possibility more directly. Studies of multimodal warning displays relating to aposematism by Candy Rowe (Rowe 1999; Rowe & Guilford 1999) are a nice example. Rowe used domestic chicks as subjects, and fed them artificially coloured chick starter crumbs that were rendered palatable or unpalatable by the omission or addition of bitter-tasting quinine. The crumbs could then be paired with other cues to see whether these enhanced discrimination learning. Rowe was particularly interested in the role of pyrazines, odorous chemicals emitted by many aposematic insects as part of their warning display (e.g. Rothschild & Moore 1987). Earlier experiments by Guilford *et al.* (1987) had shown that chicks could use the odour of pyrazine to discriminate between quinine-tainted and untainted water from a distance once they had learned the association between odour and tainting. Subsequent tests with pyrazine-tainted water, however, showed that chicks were not aversive to pyrazine itself; rather it was its combination with other cues that led to the conditioned effect. Importantly from the point of view of aposematism, pyrazine turned out to enhance performance in colour discrimination tasks: chicks were quicker to avoid distasteful yellow crumbs when paired with palatable green ones in the presence of pyrazine (Rowe & Guilford 1999). But why should this be?

The initial interpretation was that pyrazine acted literally as a learning enhancer, perhaps by increasing the amount of attention directed towards the food stimulus. But later experiments, using palatable crumbs only, suggested pyrazine in fact facilitated an underlying inherent colour bias against yellow. Yellow, of course, is a common element of aposematic displays, so the underlying bias makes intuitive sense. But did the same thing happen with other colours common in aposematic displays? Experiments with red crumbs showed that pyrazine also enhanced discrimination against these (Rowe & Guilford 1996). So what about colours that are not generally aposematic, such as green or purple? The answer here turned out to be no; the presence or absence of pyrazine had no effect on discrimination against these colours (Rowe & Guilford 1996). Thus the effect of pyrazine seems to be to enhance unlearned biases against specific colours. But that is not all. Further experiments revealed that pyrazine enhanced aversive responses to novel colours (neophobia) and to colours that were conspicuous, effects that swamped inherent colour biases *per se*. It seems, therefore, that the emission of pyrazine may have evolved to enhance the learning potentiating effects of novelty and conspicuousness, so increasing the efficacy of dramatic aposematic warning patterns. That this is an effect of adding another signal modality, rather than a special property of pyrazine itself, has also been demonstrated, since other chemicals, as well as sounds, are effective in the same way (Rowe & Guilford 1999; Rowe 2002). More recently, Speed (2001) has modelled the effects of novelty and conspicuousness among prey on learning and memory retention in predators, and shown how predator receiver psychology may have played a pivotal role in the long-recognised association between aposematism and aggregation (see 8.3.5.3).

While receiver psychology is undoubtedly an important shaping force in signals, the shaping process is likely to be two-way. It is clear from our discussion in Chapter 3 that central information processing has been influenced by the need for selective discrimination and response on the part of stimulus perceivers. It is clear too from Chapter 6 that learning capabilities have been shaped by the reproductive requirements of the organism. Signals and cues are components of the environment that demand special attention, and receivers often require experience in order to respond to them appropriately (Guilford & Dawkins 1991). Thus we might expect significant features of central nervous system evolution to reflect a history of selection pressure from signals. One aspect of central nervous function that has attracted both attention and controversy in this respect is intelligence. It is not difficult to imagine that communication, with its scope for flexibility and manipulation, might have been an important influence in the evolution of cognitive skills. Indeed, we have already touched on the possibility in 4.2.1.5. But what are the arguments and why the controversy?

11.3   Communication and intelligence

Communication has become embroiled in the debate about cognition and consciousness for a variety of reasons, but often with a heavy dose of anthropomorphism in underlying assumptions. At one level, the idea of manipulator/mind-reader arms races suggests a driving force for the evolution of mental complexity as ever more subtle and devious methods of manipulation come along to be countered. 'Machiavellian intelligence' in socially advanced species is one upshot of this process (4.2.1.5). However, other lines of argument have their roots in more literal comparisons with our own system of communication, in particular the properties of intentionality and language.

11.3.1   Communication and intentionality

Just as with the term 'motivation' (4.1.2), labelling behaviour 'communication' risks confusing functional outcome with proximate causation (Kennedy 1992). Signalling behaviour may well elicit a predictable response from another individual, but whether the signaller set out with the intention of eliciting the change is a matter for careful further investigation. We cannot assume it did just from the behavioural outcome, any more than we can assume that an animal stopping to eat a piece of food had set out hungrily to find it (4.1.2). We have rehearsed this problem in 1.1.2.1, 4.2.1 and 7.3.1, so will not go over it again here, but it is clear that much of what makes signalling seem as if it needs special explanation is its apparent purposefulness. Signallers behave as if they are *trying* to influence the behaviour of another organism, hence the 'specially designed' criterion to distinguish signals from other response-influencing features of organisms, such as crypsis (Box 11.1). But these other 'response-influencing features' have evolved for precisely the same reason as signals: they manipulate the behaviour of another organism in the interests of the possessor and reflect the historical discriminations of their receivers (Barnard 1984b). Crypsis, for example, manipulates predators into regarding potential prey as background. It is driven just as much by receiver psychology, sensory biases and other stimulus-processing attributes of receivers as any signal. The only difference is that crypsis is designed to deflect attention, while signals are designed to attract it. The distinction between signals and cues (Box 11.1) is also tenuous. The two can be distinguished relatively easily at their extremes – for example, chemical traces of prey in the faeces of predators (to which some aquatic organisms respond with predator avoidance behaviour), are probably not 'specially designed' signals, while the complex vocal repertories of many song birds clearly are – but many cues that provide useful information are likely to be candidates for receiver-driven elaboration into a signal. The problem is deciding where response to passive information ends and that to 'special design' takes over.

The 'special design' criterion for signals thus arguably reflects an implicit anthropomorphism based on the apparent intentionality of signals. As such, it tempts cognitive interpretations of communication which then bear on discussions about intelligence and consciousness more generally (e.g. Griffin 1981, 1984, 1991). But this is not the only influence of anthropomorphism linking communication and intelligence. In a more explicit form, it plays a pivotal role in the debate by focusing attention on **language**.

11.3.2 Language and intelligence

Donald Griffin, in his much-cited books (Griffin 1981, 1984), has suggested that the communication systems of animals provide a window on their thoughts and feelings. After all, our limited insights into the thoughts and feelings of our fellow humans come about only if people tell us about them, either verbally or non-verbally. If we could only understand and interrogate the 'languages' of other species we might gain a similar appreciation of theirs. But language bears on mind and intelligence in other ways too. As we saw in 1.1.2.2, it is one of the cultural tools that can aid the internalised model-building and hypothesis-testing of Dennett's 'Popperian creatures' and turn them into the 'Gregorian' pinnacle of his decision-making hierarchy. In other words, language might potentiate levels of cognitive activity that would simply not exist without it. An extreme, and enduringly debated, position on this is that language is necessary for thought itself, thus bolstering the view that cognition and awareness begin and end with humans. It certainly seems to be the case that coaching other species in language skills can enhance their performance in unrelated tasks (so-called **language training effects**; e.g. Premack 1983; Savage-Rumbaugh & Brakke 1996), but this does not by itself imply the discontinuity in the previous sentence. Indeed, from an evolutionary perspective it is difficult to imagine how such a radical break might arise. Consequently, there is a vigorous counterargument that it in fact has not, and that the origins of human language, like those of other aspects of human behaviour, are reflected in the relevant systems of other species (see 12.1 and Dennett 1995, Hauser 1996 and Blackmore 1999 for introductory summaries). So, to what extent *do* other species share language-like capabilities, and which capabilities are they?

11.3.2.1 Features of language in the communication in other species

Cross-species comparisons with human language centre on the so-called design features of language proposed originally by Hockett (1960). Numerous attempts have been made at across-the-board comparisons with Hockett's design features on behalf of different species, and Table 11.3 shows one based on an analysis by Thorpe (1972). Many of Hockett's features can be found in different species, and some species show several (Table 11.3). But comparisons with human language have largely focused on the capacities for 'openness' and 'displacement'; that is the extent to which the communication systems of other species show unlimited scope for coining new messages, and an ability to refer to things remote in time and space from the signaller.

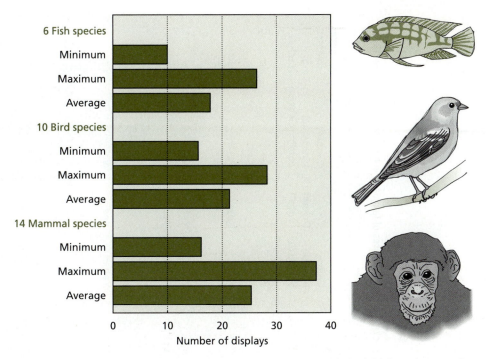

Figure 11.18 Although there is considerable variation in the number of displays shown within classes of verteb-rate, the *average* differences between the classes is small. Thus the number of displays appears to have increased very little during the evolution of the vertebrates. See text. After Wilson (1975b).

Openness

Most non-human species communicate about a very limited set of things, revolving mainly around sex, predators and food (Shettleworth 1998). Moreover, despite the apparent diversity of displays in the animal kingdom, their ways of doing so seem limited to a very small repertoire of signals, even in advanced mammals. Moynihan (1970) surveyed the displays of various species of vertebrate and discovered that the number used increased only very slightly up the phylogenetic scale, from an average of 16.3 in fish, to 21.2 in birds and 24.8 in mammals (Fig. 11.18). This is in sharp contrast to the limitless scope of human language, which is illustrated quite simply by the fact that we can coin a word for any number between zero and infinity. 'Googol', for example, indicates a 1 followed by 100 zeros. Humans, therefore, have an **unbounded** signal set. There are underlying structural limitations on this unboundedness, however. As Wilson (1975b) points out, the sound structure of language is based on between 20 and 60 **phonemes** (distinct kinds of sound), which are strung together into words and sentences with sufficient redundancy to make then easily distinguishable. It may be that 60 fundamental types of sound is the maximum number the human auditory system can reliably distinguish, just as 40 or so displays might be all the sensory-perceptual systems of other species can cope with (Wilson 1975b).

While an unlimited capacity for coining words is one measure of the openness of human language, another, equally important, one is the propensity for stringing words together using rules of **syntax**. Familiarity with the grammar of our own language allows

Table 11.3 A comparison of the communication systems of non-human species with those of Hockett's design features of language in humans. After Thorpe (1972).

Design features (All of which are found in verbal human language)	Crickets, grass hoppers	Honey bee dancing	Doves	Buntings, finches, thrushes, crows, etc.	Mynah	Colony nesting sea birds	Primates (vocal)	Canidae non-vocal communication	Primates – chimps, e.g. Washoe
1. Vocal-auditory channel	Auditory but non-vocal	No	Yes	Yes	Yes	Yes	Yes	No	No
2. Broadcast transmission and directional reception	Yes	Yes	Yes	Yes	Yes	Yes	Yes	Partly Yes	Partly Yes
3. Rapid fading	Yes	?	Yes	Yes	Yes	Yes	Yes	No	No
4. Interchangeability (adults can be both transmitters and receivers)	Partial	Partial	Yes	Partial (Yes if same sex)	Yes	Partial	Yes	Yes	Yes
5. Complete feedback ('speaker' able to perceive everything relevant to his signal production)	Yes	No?	Yes	Yes	Yes	Yes	Yes	No	Yes
6. Specialisation (energy unimportant, trigger effect important)	Yes?	?	Yes	Yes	Yes	Yes	Yes	Yes	Yes
7. Semanticity (association ties between signals and features in the world)	No?	Yes	Yes (in part)	Yes	Yes	Yes	Yes	Yes	Yes

Property									
8. Arbitrariness (symbols abstract)	Yes	No	Yes	Yes	Yes	Yes	No	No	?
9. Discreteness (repertoire discrete not continuous)	Partial	Partial	Partial	Yes	Yes	Yes	Yes	No	Yes
10. Displacement (can refer to things remote in time and space)	Yes	No	Yes	No	Time No Space Yes	Time No Space Yes	No	Yes	–
11. Openness (new messages easily coined)	Yes?	No?	Partial	No?	yes	Yes	No	Yes	No
12. Tradition (conventions passed on by teaching and learning)	Yes	?	No?	In part?	Yes	Yes	No	No?	Yes?
13. Duality of patterning (signal elements meaningless, pattern combinations meaningful)	Yes	Yes	Yes	No?	Yes	Yes	No	No	?
14. Prevarication (ability to lie or talk nonsense)	Yes	Yes	No	No	No(?)	No	No	No	No
15. Reflectiveness (ability to communicate about the system itself)	No	No	No	No	No	No	No	No	No
16. Learnability (speaker of one language learns another)	Yes	No	No?	No	Yes	Yes (in part)	No	No(?)	No(?)

us to use the same words to convey very different messages. Thus 'Brian sends a parcel to Monty' and 'Monty sends a parcel to Brian' convey different things, but use exactly the same words to do it. In contrast, 'Brian is sent a parcel by Monty' and 'Monty sends a parcel to Brian' mean the same thing even though the sequence of auditory stimuli in the former is very like 'Brian sends a parcel to Monty' (Shettleworth 1998).

Syntactical structure is also important in some other species, especially song birds. In Marler & Peters' (1989) classic study of song (*Melospiza melodia*) and swamp (*M. georgiana*) sparrows, birds were exposed to both conspecific and heterospecific song during development. The experiment showed that the species differed both in their tendency to incorporate heterospecific song into their eventual repertoire (swamp sparrows consistently produced only swamp sparrow song as adults, while song sparrows produced some swamp sparrow song as well as their own), and in the importance of the syntactic structure of the song. Thus the key unit of learning in swamp sparrows turned out to be the individual syllable. When exposed to composite songs containing both swamp and song sparrow syllables, they learned only the syllables of their own species. Song sparrows, on the other hand, depended on both syllabic structure and the syntactical order of syllables; the capacity of song sparrows to reproduce the correct syntax declines with age of exposure and its timing relative to the crucial stage of song learning. Interestingly, if song sparrows are exposed *only* to swamp sparrow song, they will learn this and reproduce it in the correct swamp sparrow syntax. But syllabic structure and syntax are not inviolable, and songs within species generally show considerable variation, sometimes through simple copying errors in the learning process (Slater & Ince 1979). Indeed, in several songbirds there are distinct local dialects which clearly characterise local populations and may even act as a reproductive barrier between them (e.g. Marler 1970; Petrinovich *et al.* 1981; McGregor & Thompson 1988; see Fig. 1.8). The point, however, is that despite the demonstrable importance of syntax in some other species, changes in syntactical structure do not appear to alter the *meaning* of the signal. Thus a territorial 'keep out' signal means 'keep out' whether it is delivered in syntactical order *A* or syntactical order *B*. Unlike the situation in human language, therefore, syntax does not appear to be used as an open-ended vehicle for generating new meaning (Shettleworth 1998).

Displacement

The animal signalling systems we have dealt with so far generally relate to the here and now. Individuals display their freedom from parasites, likelihood of attacking, ownership of a territory, distastefulness and so on as things stand at the present time and in the present place. This does not necessarily mean that a receiver can perceive directly what it is that is being signalled about, but the signal itself is given reliably on the basis of the information available to the signaller. Thus, an individual detecting a predator alarm call flees even though it may not be able to detect the predator that triggered the call in its companion. The alarm call in this case is said to have **functional reference**; that is it refers to something elsewhere to which the receiver is geared to respond (e.g. Marler *et al.* 1992). But humans can communicate about objects and events remote in time and space when neither signaller nor receiver can directly perceive what is being signalled about. Functional reference in human language thus has the additional quality of **displacement** or **situational freedom** (Shettleworth 1998). As well as allowing communication about events and objects in the past or future, displacement creates the opportunity for lying (i.e. intentional deceit). However, while there is anecdotal evidence that animals may deliberately deceive (e.g. Byrne & Whiten 1988), we have seen in 4.2.1.4 that alternative explanations are usually possible, and that evidence for lying among non-human animals remains inconclusive. Although the jury is still out on the issue of lying, however, there are aspects of signalling systems in other species that strongly suggest functional reference and other important attributes of language. We shall look briefly at two examples, and then at the potential for language tuition to reveal some of these capacities in non-human species.

The dance 'language' of honey bees

One communication system that has provoked particularly heated debate about its language-like qualities is the celebrated foraging dance of the honey bee (see Dyer 2002). The fact that worker honey bees somehow 'recruit' other workers to sources of food has been known for centuries and was remarked upon by Aristotle, but how this came about was not unravelled until the mid-twentieth century and the work of Karl von Frisch (see von Frisch 1976). By training bees to artificial feeders, von Frisch showed that when a bee returns to the hive from a food source, she performs characteristic patterns of movement known as a foraging dance. The exact form of the dance depends on the distance of the food away from the hive. If it is within 50 m, the forager performs a so-called **round dance** (Fig. 11.19a), a repeat loop roughly describing a circle. Other foraging

Figure 11.19 The foraging dance of the honey bee (*Apis mellifera*). (a) The 'round dance' generally performed on the vertical face of the comb when food is within 50 m of the hive. (b) The 'waggle dance' performed on a horizontal surface when food is further away acts as a simple pointer, but when performed on a vertical surface (c) direction with respect to gravity inside the hive is translated into direction relative to the sun outside (see text). After Curtis (1968).

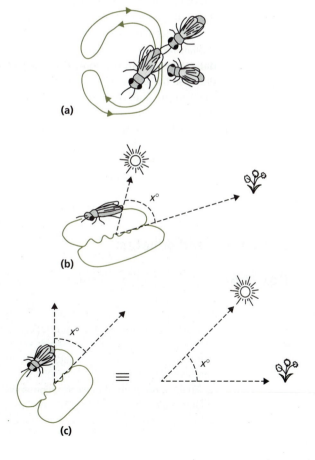

workers attracted to the dance appear to be stimulated by it to fly out and search within a 50-m radius of the hive. If the food source is further away than this, the forager performs a different set of movements known as a **waggle dance** (Fig. 11.19b,c). The waggle dance describes a somewhat flattened figure-of-eight, with a straight-line element at the cross-over during which the bee waggles her abdomen from side to side and vibrates her wings to make a buzzing sound as she progresses. At the end of each straight run she turns alternately right or left and runs back to the beginning in a semicircle to make another. Through painstaking observation, von Frisch established that the waggle dance contained detailed information about the location of the food from which the forager had just come. The *direction* of the source was indicated by the orientation of the waggle run. If the dance was performed on a horizontal surface, the forward component of the run pointed directly towards the source (Fig. 11.19b). If it was performed on a vertical surface, its angle with the vertical inside the hive translated into the same angle relative to the sun's azimuth outside to give a bearing for the food source (Fig. 11.19c). The *distance* of the source from the hive could be deduced from the length and tempo of the waggle run and the amount of buzzing accompanying it. The further away the food, the longer and slower the waggle phase. Potentially, therefore, other workers attending the dancer in the hive could acquire directions for finding the food themselves and join in the foraging effort. If this was the case, the dance would appear to have at least some of the properties of a sophisticated language, including symbolism, openness and displacement (Box 11.5). Two questions arise from this, however. First, do bees actually use the dance information in this way (a celebrated debate summarised in Box 11.5), and, second, if they do, does this necessarily mean they have a sophisticated language?

The interest value of the honey bee dance has lain in its apparent parallels with human language, particularly the capacity for displacement and the possibility that the dance operates in the context of a cognitive spatial map (Shettleworth 1998). That bees may use the latter was given impetus by the observation that dancers seemed to report the straight line distance from hive to food source rather than the distance actually flown (e.g. von Frisch 1953). We have seen earlier (4.2.1.2), however, that the evidence for cognitive maps in honey bees is still questioned, and the 'straight line distance' finding may indicate nothing more than a capacity for path integration (Dyer & Seeley 1989; see Chapter 7). The fact that bees flying against the wind, or when weighed down with a small amount of ballast, translate this into longer distances flown according to

Difficulties and debates

Box 11.5   The honey bee dance language controversy

The honey bee (*Apis mellifera*) dance language has been the subject of debate at two levels. First whether bees use the information apparently transmitted by the dance, and second, whether they use a cognitive map or some kind of path integration (dead reckoning) system to locate indicated food. The latter is discussed in 4.2.1.2. Here we briefly look at the dance information issue.

There is no doubt that information about direction and distance *can* be gleaned from the waggle dance of honey bees. The debate that followed von Frisch's work was over whether other bees actually did so and then acted on it. Early experiments by von Frisch, setting out feeders in various spatial arrangements around hives, certainly suggested they did, but there were several important caveats.

For one thing, von Frisch himself showed that bees could find food sources simply by means of odour cues. Dancing workers periodically pause to regurgitate nectar, which contains olfactory and gustatory cues from the food source. They also carry odours from food plants stuck to the waxy hairs on their body. Recruits could easily pick these up and use them to orientate when they leave the hive. Experiments in the 1960s, using similar approaches to von Frisch, quickly established odour orientation as a serious alternative to the dance language hypothesis (see Wenner & Wells 1990 for a review). That recruits were generally much slower to arrive at sites than experienced foragers, frequently made errors, and tended to favour the centre of odour gradients within food arrays all stacked up against the language hypothesis and in favour of odour orientation.

It was not until a seemingly definitive experiment by Gould (1976) that the dispute was settled in favour of the dance language in most people's eyes (but see e.g. Wenner et al. 1991). In a brilliantly conceived design, Gould used the fact that bees will respond to a small point of light inside the hive as if it is the sun. A dancing forager on the vertical face of the comb thus uses the light as it would the sun were it dancing on a horizontal surface outside the hive. In other words, it treats the light as a simple pointer: a waggle run at 20° to the left of the light corresponds to food at 20° to the left of the sun. Gould also exploited the finding that blacking out some of the simple eyes (ocelli) on the dorsal surface of the head reduces the bee's sensitivity to light. His experiment then hinged on having ocelli-blacked foragers dance in the hive where they could not see the light source. Their dances were therefore orientated with respect to gravity, and so, by extrapolation, the sun. Recruits in the hive, however, *could* see the light. If they were interpreting the dance as suggested by von Frisch, they would therefore do so with respect to the light rather than gravity. By changing the position of the light, Gould was able to manipulate the direction in which the fooled recruits should go if they were interpreting the forager's dance as a set of directions. Figure (i) shows that most bees went to the food source indicated by a misreading of the forager's waggle dance. Gould's experiment thus provided convincing evidence that the waggle dance contains information about direction and distance and that other bees are able to use it.

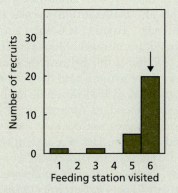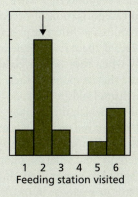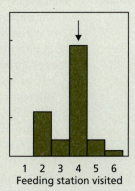

Figure (i)   Recruit bees tricked into orientating with respect to a small light source in the hive fly to feeding stations (arrowed) that could only have been indicated by the waggle dance of returning ocelli-blacked foragers. See above. After Gould J.L. (1976). The dance-language controversy. *Quarterly Review of Biology* **51**: 211–44, reprinted by permission of The University of Chicago Press, © 1976 by The University of Chicago. All rights reserved.

their dance detracts somewhat from the cognitive map idea. But what about the apparent property of displacement itself? Obviously, there is displacement in the sense of the dancer being some distance, perhaps several kilometres, away from the source of food at the time of signalling, but, as Shettleworth (1998) points out, if the dance simply reflects a journey that has just been completed, it is no more displaced than an alarm call triggered by a predator that has just been glimpsed in the grass. Seen this way, equating the displacement in the honey bee dance with the temporal and spatial displacement capacity of human language seems to be overstating things.

Shettleworth (1998) also points to other factors that make the dance 'language' less extraordinary than it might appear at first sight. For one thing, while the system may seem uniquely sophisticated among invertebrates, its components have something in common with behaviours in other insects that are not involved in communication. For instance, various species can translate between direction relative to gravity and direction relative to the sun (Gould & Towne 1987), and many show so-called 'wind down' wing vibration after flight that varies with the duration and effort of the flight. These and other observations suggest an evolutionary origin of the dance in non-communicatory movements, with the key innovation being an ability to use information in the movements to direct foraging (Dyer & Seeley 1989; Shettleworth 1998). Interestingly, different races of bee translate the distance information in the waggle dance according to different scales of flight distance. Thus, races foraging over longer distances scale up the number of metres represented by each unit of the waggle run (Dyer & Seeley 1991).

Functional reference in vervet monkeys

Following earlier observations by Struhsaker, a landmark study of alarm calls in vervet monkeys by Seyfarth *et al.* (1980) found that monkeys gave different calls to different kinds of predator (Fig. 11.20). Calls on seeing eagles or other raptors were different from those given when a leopard was seen, which were different again from those given when spotting a snake. The discriminations by vervets are quite refined. Adult monkeys are capable of distinguishing dangerous raptors passing overhead from other large birds such as storks, for example, in contrast to some other alarm-calling species such as chickens, which respond to a wide range of innocuous stimuli as well as predators. The apparent categorisation of predators also makes intuitive sense, since each kind of predator requires a different precautionary response. It is not a good idea to head up into the trees if there is an eagle around, for instance. Monkeys certainly respond appropriately when they hear a given call, but is this because they 'understand' the call, or because they respond to their own information about the predator? By playing different calls from loudspeakers, Seyfarth *et al.* (1980) were able to show that it was indeed the calls to which the monkeys were responding. Animals on the ground when a 'leopard' call was played generally leapt up into the trees, but they scuttled for bushes or long grass when a 'raptor' call was played and so on. This is consistent with calls having functional reference, but what meaning do monkeys actually extract from them? Is there just a simple stimulus–response specificity, or do calls conjure up a representation of the relevant predator or response in the receiver? Cheney & Seyfarth (1990a) used response habituation (6.2.1.1) tests in an attempt to find out.

The question in Cheney & Seyfarth's tests was whether monkeys that had been habituated to a repeatedly presented call transferred the habituation to a call that was different but meant the same thing. If they did, it would be strong evidence for some degree of representational meaning on the part of the monkeys. To do this, Cheney & Seyfarth made

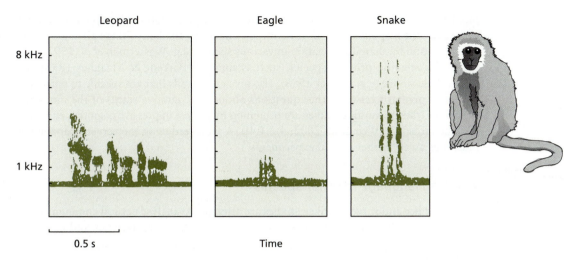

Figure 11.20 Sonagrams of the 'leopard', 'raptor' and 'snake' alarm calls by an individual vervet monkey (*Cercopithecus aethiops*). After Seyfarth *et al.* (1980).

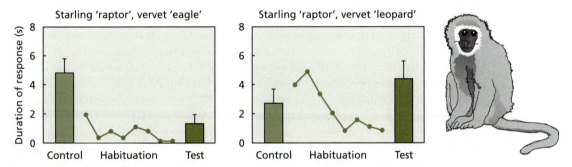

Figure 11.21 Habituation tests with different alarm calls in vervet monkeys. (a) Monkeys habituated to 'raptor' alarm calls by starlings (left bar and solid circles) generalised the habituation to the 'raptor' call of vervets. However, habituation to starling calls did not generalise to 'leopard' calls by vervets (b). After Shettleworth (1998) from Seyfarth & Cheney (1990).

use of the fact that the vervets also responded to alarm calls by a savannah bird species, the superb starling (*Spreo superbus*). Like the monkeys, starlings give a distinctive 'raptor' alarm to threatening stimuli from the air. Would habituating a monkey to a starling's 'raptor' alarm generalise to the 'raptor' call of its own species, so implying a concept of 'aerial predator' on the part of the monkey? Figure 11.21(a) shows it did. Moreover, the generalisation was specific to the appropriate kind of vervet call, since there was no habituation effect on monkey 'leopard' calls (Fig. 11.21b). Similar habituation tests showed that monkeys could distinguish between different individuals making calls, effectively learning that particular individuals were reliable or unreliable at particular times.

All this suggests that vervet monkeys have some concept of what is represented by the different calls. But do they? As Shettleworth (1998) points out, generalisations such as that from starling to vervet 'raptor' calls could come about simply because the calls share certain properties and have become associated with the same stimulus (so-called *mediated* or *secondary generalisation*), not because vervets have a concept of raptor *per se*. Indeed, Cheney & Seyfarth (1990b) acknowledge that several aspects of the monkeys' behaviour are consistent with such an associative learning interpretation. If this is the case, the responses would be no more remarkable than any others arising through generalisation by shared association in various species (e.g. Wasserman *et al.* 1992). The fact that some non-primate species, such as squirrels (Greene & Meagher 1998) and the starlings above (e.g. Hauser 1988), also give alarm calls that are specific to particular classes of predator raises further questions about the cognitive status of the monkeys' behaviour (though various studies are beginning to suggest impressive cognitive attributes in some birds [e.g. Weir *et al.* 2002; 4.2.1.4]). A final verdict on the vervets' capabilities must thus await further critical experiments.

Teaching language to other species

The honey bee and vervet monkey studies are attempts to analyse natural communication systems for what they might reveal about language-like characteristics and cognitive skills in other species. A very different approach is to see whether it is possible to *teach* other animals aspects of language and use it to explore cognitive potential via the skills so acquired. Most of the effort has gone into trying to teach language skills to apes, on the questionable basis that, as our closest living relatives, they are most likely to show some propensity for human-like skills (e.g. Rumbaugh & Savage-Rumbaugh 1994). A problem with this is that living apes are not part of the hominid evolutionary lineage, and are thus likely to be specialised in other ways. Objectively, therefore, 'closest relative' does not in itself justify any special expectations about shared features (Pinker 1994). If modern non-human primates did not exist, some other mammal would be our closest living relative, and expectations of human-like qualities would correspondingly diminish (Shettleworth 1998). There is also another problem. Views about the essential features of human language have been changing over the years, especially in relation to how language is learned by children (with which apes have most frequently been compared), so the acid tests for animal language training programmes have changed alongside them, presenting something of a moving target for their advocates (e.g. Kako 1999; Shanker *et al.* 1999). These problems notwithstanding, however, the various training programmes with apes have generated much information about the animals' language learning capabilities and an equal measure of debate as to their significance. So what exactly has emerged and what can we conclude from it?

Language learning in chimpanzees The earliest training programmes, by the Kelloggs and Hayeses in the 1930s and 1940s, attempted to rear two young chimpanzees (Gua and Viki) like children and teach them to use spoken language (e.g. Hayes & Hayes 1951). While both ended up being able to communicate and solve problems, neither learned to talk. Six years of protracted and painstaking work resulted in Viki uttering only four sounds that vaguely resembled English words, in all probability because chimpanzees lack the neural and anatomical wherewithall for speech.

A very different approach was taken by Gardner & Gardner (1969). They took an 8–14-month-old female chimpanzee, Washoe, and set about teaching her American sign language (ASL), a gesture language for deaf people that uses manually produced visual symbols as analogues of words in spoken language. ASL is a sophisticated natural language that is acquired and used just like its spoken equivalent (Pinker 1994). Just as spoken language can be analysed in terms of phonemes (see above), signs can be analysed in 'cheremes'. ASL has some 55 cheremes: 19 relating to the configuration of the hand(s) making the sign, 12 to the place where the sign is made, and 24 to the action of the hand(s). Thus a pointing hand means one thing near the forehead but another by the chin. At any given place, its meaning varies according to whether it is moved towards or away from the signer, vertically, horizontally and so on.

Washoe was kept in a room containing a range of articles normally found in a human dwelling, and during her waking hours was constantly in the company of people. Although her human companions took part in her everyday activities – feeding, bathing, dressing, etc. – the only form of 'verbal' communication was by ASL. Shaping and reinforcement procedures (see 6.2.1.2) were used, however, with Washoe being prompted by trainers and sometimes having her hands manipulated into the required configurations. By two years into the project, Washoe appeared to know some 30 signs, and it was eventually claimed she could use over 100. The use of signs was noted during spontaneous everyday behaviour and during structured tests, with special interest focusing on Washoe's attempts to combine signs and use them in novel ways, as, for instance, in combining signs for 'water' and 'bird' into 'water bird' to indicate a swan. Washoe was also able to use pronouns. When she transferred a sign from herself to another person she frequently accompanied it with gestures indicating 'yours' (usually pointing at possessions of the referent). Three or four years after starting her training, Washoe was signing simple phrases, like 'you me out' or 'you me go out hurry' when passing through a doorway.

Other non-vocal language projects involving chimpanzees also began in the 1960s, but using invented visual symbols rather than sign language, to avoid some of the methodological problems of assessing conformation to signs. The chimpanzee Sarah was trained to associate plastic tokens of different colours and shape (Fig. 11.22a) with particular objects, such as an apple or stick (Premack 1971). Once acquired, the token language was used to test Sarah's grasp of concept. Could she distinguish between things that were the 'same' or 'different', of one colour or another, or possessing different names? Rather than testing linguistic skills *per se*, therefore, Sarah's programme focused more on whether language training helped her abstract conceptual and problem-solving abilities, which Premack argued it did (see Premack 1983).

Another programme, initially with the chimpanzee Lana, used geometric designs on keys connected to a computer as a symbolic language ('Yerkish', or 'Language Analogue

(a)

What is A the same as? A/B

(b)

Figure 11.22 Visual symbols used in language training programmes with chimpanzees. (a) Questions about 'same versus different' objects presented to Sarah. (b) Lexigram keyboard symbols used by Lana. After Shettleworth (1998) from Premack & Premack (1983) and Rumbaugh (1977).

System' – for which the name Lana is part acronym; Rumbaugh 1977; Fig. 11.22b). Each design, or **lexigram**, represented a 'word' and the computer was programmed to dispense rewards in response to the production of grammatical strings of 'words', such as 'please machine give drink'. Lana was followed as a student of Yerkish by Austin and Sherman, in which the focus shifted from straightforward lexicographical association to using the language socially and to express concepts and intentions (Savage-Rumbaugh *et al.* 1983; Savage-Rumbaugh 1986). For example, the chimpanzees were tested in situations where they had to request particular tools to get into a sealed food box, or could share information with each other and choose how to respond to it, in other words communicate symbolically (Savage-Rumbaugh *et al.* 1978). This second capacity appeared to be taken to new heights by a later Savage-Rumbaugh project with a male pygmy chimpanzee or bonobo (*Pan paniscus*) called Kanzi (e.g. Savage-Rumbaugh *et al.* 1986).

Kanzi was also taught Yerkish, but distinguished himself in two ways from his common chimpanzee predecessors. First, he learned his lexicography by observing his foster mother being taught it; second, he understood human *speech*. He thus appeared to pick up his Yerkish and English language skills simply through observation, developing understanding to a point well beyond his capacity to produce, much in the manner of early language comprehension in children. In a series of carefully controlled experiments, therefore, Savage-Rumbaugh and coworkers compared Kanzi's performance in responding to specified spoken instructions with that of a 2-year-old child (Savage-Rumbaugh *et al.* 1993). Both bonobo and child were asked to comply with various more or less complicated directives, such as 'get the telephone that is outdoors'. In these, Kanzi performed as well as the child, or to a level comparable with a child six months younger. Clearly, one interpretation of these outcomes is that Kanzi used and understood words and sequences of words as representing aspects of the world at large. Despite the impressiveness of the results, however, there are still cautionary voices that cast doubt on the grander claims for Kanzi's performance (e.g. Kako 1999). These are part of an established tranche of criticisms that has been levelled at the ape language-training programme as a whole, indeed attempts to teach any of the various species that have now been exposed to language training schemes, including Pepperberg's famous parrots (see 6.2.1.2). Shettleworth (1998) provides a good summary, on which the following is based.

Problems with language-training programmes

Two broad classes of criticism have been levelled at the conclusions drawn from language-training programmes. The first reared its head in the 1970s, in connection with the early chimpanzee experiments. Terrace *et al.* (1979) trained an infant chimpanzee, mischievously named Nim Chimpsky (after the celebrated linguist Noam Chomsky), to use sign language in a similar way to the Gardners' Washoe. Like Washoe, Nim picked up a large repertoire of signs and deployed them in various combinations. Some of these combinations were short, just two or three signs, but others were longer. However, only the shorter combinations showed the non-random structure expected if Nim was stringing together comprehended meanings. While this is comparable with the very early stages of language acquisition by young children, Nim's performance quickly departed from the latter's when it came to longer combinations of signs. Vocabulary and the capacity for forming syntactical multiword sentences increases dramatically as a child matures, but Nim stubbornly carried on producing only short 'phrases' as he got older. If he did combine more than two or three signs, this generally involved repeating signs already included in the combination, as if emphasising, rather than adding to, the message (Shettleworth 1998).

Nim's limited syntactical ability raised questions as to the significance of his sign-learning capabilities, but more telling by far was the outcome of Terrace *et al.*'s careful analysis of filmed interaction between Nim and his trainers. The analysis revealed that, rather than showing spontaneous use of signs to put messages across, Nim was often just repeating signs that had been made by the trainer. The same turned out to be true of Washoe when some of the Gardners' film was examined. The learned signs thus seemed to be simple operantly conditioned responses (see 6.2.1.2). This undermined confidence in the apparently syntactical combinations of signs the chimpanzees *had* managed to put together, but also in the comparisons with language learning in children. Children engage in conversation, taking it in turns to exchange and develop information about the world generally. In contrast, the chimpanzees communicated 'out of turn' and confined their exchanges to the limited signed responses to which they had been conditioned; there was no general 'conversation'.

While the later studies were mindful of Terrace *et al.*'s penetrating criticism and controlled for inadvertent trainer effects, they have been less successful at getting over the 'it's all just conditioning' problem. Sarah, Lana, Austin and Sherman were all taught using standard operant techniques (see 6.2.1.2). In the cases of Sarah and Lana, the chimpanzees learned a classic symbolic matching-to-sample task, while Austin and Sherman were trained in both matching to sample and category learning. That the chimpanzees' behaviour could be accounted for in terms of operant discrimination learning was emphasised by a tongue-in-cheek study of pigeons that mimicked the tasks asked of Austin and Sherman (Epstein *et al.* 1980). Using standard shaping and reinforcement procedures, one pigeon (Jack) was trained to press a 'what colour?' key to 'ask' another pigeon (Jill) the colour of a light hidden behind a curtain. Jill was trained to respond by pressing an appropriate colour key, which Jack then 'acknowledged' by pressing a 'thank you' key (Fig. 11.23).

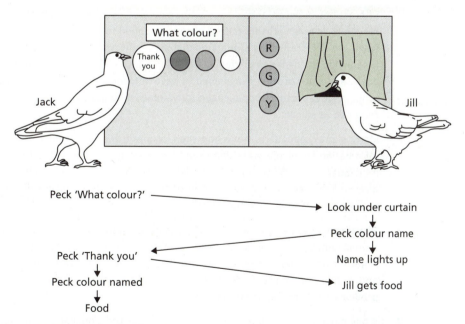

Figure 11.23 The experimental design and procedure for Epstein *et al.*'s demonstration of 'conversation' between two pigeons. See text. After Shettleworth (1998) reprinted with permission after 'Symbolic communication between two pigeons (*Columba livia domestica*)', *Science*, Vol. 207, pp. 543–5 (Epstein, R. *et al.* 1980). Copyright 1980 American Association for the Advancement of Science.

On performing correctly, each bird received a food reward. Parodying Savage-Rumbaugh *et al.* (1978), Epstein *et al.* concluded that the birds' successful performance showed pigeons '. . . can learn to engage in a sustained and natural conversation without human intervention, and that one pigeon can transmit information to another entirely through the use of symbols'.

Epstein *et al.*'s results imply that Austin and Sherman's superficially remarkable behaviour amounts to nothing more than the training effects of reinforcement, and does not require any special linguistic or conceptual abilities. That is not to say there *are* no differences in the cognitive underpinnings of the chimpanzees' and pigeons' performances, simply that the principle of parsimony does not yet demand them. While Kanzi's abilities might allow more pause for thought, he still acquired his understanding of spoken English only after extraordinarily intensive interaction with, and attention from, human companions. Like Nim, he also failed to show the rapid development of language use that characterises young children. Of course, as Shettleworth (1998) points out, there is a slight duplicity in the 'it's all just conditioning' argument, since, on the one hand, conditioning is seen as allowing animals to generate sophisticated models of the world (see Chapters 4 and 6), while, on the other, it is accused of diminishing cognitive capacity by reducing it to a set of simple principles. Nevertheless, regardless of the alternative spin, serious questions remain as to the real significance of language training programs for ape mentality.

Summary

1. Interactions between animals are often mediated by specialised behaviours, chemicals, markings or other attributes that appear designed to facilitate them. These specialisations are referred to as signals or displays, and the collective act of deploying and responding to them as communication. Communication is difficult to define unambiguously, and much of what is assumed to have influenced the evolution of signals has also influenced aspects of appearance, physiology and behaviour that are not generally regarded as signals.

2. Many signals appear to have had their origin in features that incidentally provided information about the state and likely actions of the bearer. The resulting response of individuals to such incidentally available information is assumed to have resulted in directional selection for elaboration of the features into a stylised signal, a process dubbed 'ritualisation' by ethologists.

3. Ritualisation was originally seen as reducing the ambiguity of signals in the context of mutual information sharing between signaller and receiver. More recent views, however, regard it as the upshot of an exploitative arms race between manipulative signallers and 'mind-reading' receivers. The evolution of signals used in aggressive interactions provides a rich source of examples.

4. While much attention has focused on the role of the signaller in the evolution of communication systems, it is clear that receivers have also exerted important selection pressures on signals. The fact that broadcast signals can be detected by receivers other

than those intended by the signaller means that signals are shaped in the context of more or less complex communication networks, not simple dyadic signaller–receiver relationships. 'Eavesdropping' by unintended receivers may advantage competitors and predators of the signaller, but may also lead to advantages for the signaller through attracting more welcome attention, such as that from prospective mates.

5. The effect of receivers on signals is mediated by peripheral and central mechanisms of signal detection and processing. The expression 'receiver psychology' has been coined to express the collective effects of centrally mediated detection, discrimination and memory of signals on signal evolution. The cost to receivers of assessing signals is one of a number of factors that might limit the evolution of honesty in signalling systems.

6. Communication systems are frequently invoked in debates about cognitive ability. In part this reflects anthropomorphic comparisons with human language and purpose-fulness, but signaller–receiver arms races provide an a priori reason for expecting communication to drive the evolution of mental complexity. Studies of various species have revealed language-like elements in their signalling systems, but the significance of this for cognitive status and the evolution of language *per se* remains equivocal, even in closely related primates.

Further reading

Hauser (1996) is a lengthy but readable summary of many of the issues in animal communication generally. Pinker (1994), Dennett (1995) and Shettleworth (1998) discuss some of the cognitive and evolutionary arguments surrounding language, while Maynard Smith & Parker (1976), Dawkins & Krebs (1978), Krebs & Dawkins (1984), Guilford & Dawkins (1991) and Dawkins & Guilford (1991) are good starting points for current debates about the evolution of signalling. The third volume in the series edited by Halliday & Slater (1983) provides a broad introductory perspective on some of the mechanisms underlying signals, as well as discussing their functional significance. Seeley (1995) and Dyer (2002) summarise much of the work on the honey bee dance language, while the papers of Seyfarth, Cheney and coworkers are well worth plumbing for their field experimental approaches to referential calling in vervet monkeys.

12 Human behaviour

Introduction

Human behaviour has been shaped by both genetic and cultural evolution, but unravelling the relative contributions of the two is not straightforward. How do they interact, and how potent a force is each in the evolution of behaviour? Human beings are characterised by marked sex differences in behaviour, and a widespread tendency to be altruistic. Can we explain these using the same evolutionary theories we apply to other species, or has cultural evolution imposed different selection pressures on them? Why have humans evolved traits such as adoption, celibacy and homosexuality that appear to impose severe reproductive costs? Why do they sometimes abuse and kill their children? While cooperation is a ubiquitous feature of human societies, how can we rationalise it with the fact that we frequently wage war on each other?

'Let us now consider man in the free spirit of natural history . . . In this . . . view the humanities and social sciences shrink to specialized branches of biology; history, biography, and fiction are the research protocols of human ethology, and anthropology and sociology together constitute the sociobiology of a single primate species'

Edward O. Wilson (1975a)

This book has been concerned with the behaviour of animals, and, more particularly, the general principles behind how and why they behave as they do. It has been written, of course, by another animal, which means that the arguments and interpretations presented are themselves the distant upshot of the processes they attempt to illuminate. As just another traveller in life's evolutionary parade, my views are potentiated and constrained by the mechanisms of my central nervous system just as surely as the toad's interpretation of the objects in its world in Figs 3.21–3.23. However, one difference between myself and a toad (hopefully) is that the information handling capacity of the CNS has moved on a bit from the situation loosely represented by modern amphibians. But there is another important difference. The views in this book have not sprung spontaneously and fully formed from my head. They have been shaped by reading, listening to and reflecting on views emanating from many other heads. This is because, unlike toads, we humans have a rich cultural life that fosters our thoughts and behaviour and propels them in many different directions. We are, as we put it in Chapter 1, the ultimate Gregorian creatures of Dennett's hierarchical schema (Fig. 1.3), assisted or hampered in what we do by the cultural medium in which we find ourselves. And therein lies a fundamental difficulty. To what extent can we explain our behaviour using the evolutionary principles we happily apply to all other species, and to what extent has it been taken over by cultural forces that have little if anything to do with biological (genetic) evolution? To say the least, there are different opinions on the issue! As we saw in 2.5, culture generates its own evolutionary momentum, fast and seemingly unpredictable in the features it seizes upon. Yet it must have its origins in our biological heritage somewhere, so, in principle, cultural evolution should be susceptible to the same lines of enquiry as its genetic counterpart. Is there any evidence that it is?

12.1 Culture and human behaviour

Parallels have frequently been drawn between genetic and cultural evolution (Box 12.1), and it is not difficult to find similarities in the evolution of cultural traits and biological adaptations. Darwin himself likened the evolution of languages, those ultimate cultural traits, to that of species (Darwin 1859; see Blackmore 1999). Like species, argued Darwin, languages show homologies as a result of common descent. Like species, they can go extinct, never to reappear. He even conceived of words competing for survival within their lexicon. Modern comparative linguists can trace words back through their evolutionary history and the many levels of change that have brought them to their present form (e.g. Fig. 12.1). Family trees of languages can be drawn up that look remarkably like the phylogenetic trees of taxonomic groups (see Figs 2.1–2.3). Arguably, cultures even have their own replicators – memes – that drive their rich panoply of traditions, much as genes drive the diversity of life (Dawkins 1976; Brodie 1996; Lynch 1996; Blackmore 1999; see 2.5 and 12.1.1.2). But what exactly *is* culture, and how does it fit into an overall evolutionary framework?

Figure 12.1 Mutational drift in the words of Micronesian languages as a function of the distance between islands on which they are spoken. The graph shows that the proportion of identical words on any two islands declines exponentially with distance, implying that words mutate at a more or less constant rate as distance (and probably time) of separation increases. The effect is broadly analogous to genetic drift in biological characters. After Cavalli-Sforza & Feldman (1981).

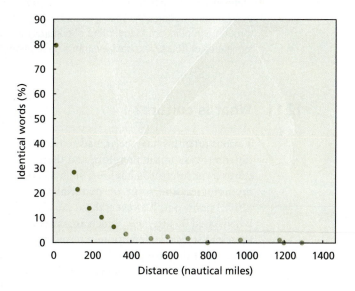

Underlying theory

Box 12.1 Cultural versus genetic evolution

Cultural evolution shares some basic features with its genetic counterpart, indeed it is an epiphenomenon of it, but it also has some distinctive properties of its own which set it apart (Table (i)).

Table (i) Some of the main characteristics of genetic and cultural evolution. Modified after Barrett *et al.* (2002).

	Genetic evolution	Cultural evolution
Unit of selection	Gene (DNA)	Meme (ideas, beliefs)
Rate of mutation	Slow, constant	Fast, variable
Mechanism of inheritance	Biological reproduction	Imitation, teaching
Mode of transmission	Vertical (between generations)	Vertical, horizontal, oblique (between and within generations)
Heritability	Low	Low-moderate
Generation length	Moderate	Variable (instantaneous to millennia)
Rate of evolution	Slow	Variable (instantaneous to very slow)

As Bonner (1980) and others have pointed out, cultural and biological evolution differ in the mode of transmission of information. The molecular process of gene transmission in genetic evolution necessarily restricts the scope of transmission so that genetic information can pass from the donor to only one other individual. For any given organism, the number of times it can transmit its genetic information is limited to the number of offspring it produces. This restriction does not apply to cultural transmission. Here, any individual can teach or learn from many others, either simultaneously or over a period of time. They can also acquire or disseminate information over large distances. Information can be recorded and made available to distant generations and other cultures, thus extending its influence beyond immediate recipients. One result of this is that cultural evolution can be very rapid in comparison with genetic evolution. A favourable mutation arising in an individual will usually, even under moderately strong selection pressure, take several generations to spread through a population. Many forms of culturally transmitted information, however, can spread in minutes or hours, and easily within a generation.

12.1.1 What is culture?

Various attempts have been made to define culture, but the term has been used in many different ways in the literature and there is no simple consensus. Some definitions have centred on rules of behaviour, such as social customs and rituals, some on ideas and mythologies, and some on material artefacts or artistic traditions. As Barrett *et al.* (2002) point out, however, these mostly come down to ideas in people's heads – even when based on artefacts, which are, after all, simply the material products of ideas. They therefore suggest a definition in terms of knowledge, beliefs and ideas (and thus rules of behaviour) that are passed on from one individual to another by some form of social learning (see 2.5; 6.2.1.4 and Box 6.4).

(a)   (b)

Figure 12.2 Cultural 'icons': certain individuals are more likely to be imitated because they are particularly skilled, or admirable in other ways. (a) England football captain David Beckham at a press conference before leaving Awaji Island after England's 2–1 defeat by Brazil in the World Cup quarter finals. (b) A 20-year-old 'Beckham fan' shows his mohican hairstyle in downtown Osaka, 19 June 2002. See text.

Social learning, and in particular imitation, provides the rapid horizontal and vertical transfer of information that gives cultural evolution its enormous rate advantage over its genetic counterpart (Box 12.1). Humans, especially when they are young, appear to be particularly geared to acquiring information and instruction by imitation. This circumvents the need for potentially laborious and error-prone independent learning, but, of course, ultimately depends on innovation by someone somewhere to generate new skills (just as natural selection ultimately depends on new mutations). Independent learners and their imitators can thus be viewed as a form of producer–scrounger relationship (Barnard & Sibly 1981; see 9.1.1.2), with the availability of learners (producers) imposing a frequency-dependent limit on the benefit of imitation (scrounging) (imitators will eventually dilute the number of learners and end up copying other imitators, thus risking acquiring suboptimal skills; Rogers 1989; see also Giraldeau & Lefebvre 1987). This frequency-dependent limit could be overcome, however, if imitators copy selectively, choosing as role models those individuals that show particularly adept behaviours (e.g. Boyd & Richerson 1994; see 12.1.1.1). Our own cultures abound with examples of high achievers, innovators and skilled performers (cultural 'icons' in current media-speak) that are emulated by others (Fig. 12.2).

The distinction between learners and imitators has led to differences of opinion as to how cultural transmission should be recognised in practice, with psychologists focusing on imitation of novel, ecologically irrelevant traits, to minimise the possibility that these reflect adaptive independent learning, but ethologists and behavioural ecologists doing just the opposite (Barrett *et al.* 2002).

For psychologists, the only form of social learning that counts unequivocally as cultural transmission is true imitation (Box 6.4), because it does not involve the imitator learning anything for itself (Tomasello *et al.* 1993). This is an exacting criterion, and many apparent acts of copying fall short of the mark. The spread of milk bottle opening by tits in 6.2.1.4 is a good example (see also Tomasello & Call 1997). Ethologists and behavioural ecologists, on the other hand, focus on naturally selected adaptive responses, so ecological relevance is inescapable in their case. Their approach thus includes responses that fall outside the psychologists' stricter definition of cultural transmission (Barrett *et al.* 2002). To muddy the waters still further, anthropologists focus more on the role of *teaching* (see 6.2.1.4), on the grounds that children are actively imbued ('**enculturated**') with cultural traditions through example and instruction. For them, cultural transmission is therefore more than simply a matter of imitation. While teaching clearly has an active component to it, however, it is questionable whether its effects are fundamentally different from those of 'passive' imitation since the upshot of both processes is simply copying what someone else does (Barrett *et al.* 2002).

Transmission by imitation is at the heart of comparisons with genetic evolution, since it provides an obvious analogy with the transmission of genetic information (Box 12.1). But how close is the parallel? Are there cultural equivalents of genes, and, if there are, does recognising them enhance our ability to explain cultural phenomena? Moreover, do genetic and cultural evolution act independently, or do they mutually facilitate or constrain one another?

12.1.1.1 Genes and culture: models of cultural evolution

The relationship between genetic and cultural evolution has long been debated and, perhaps not surprisingly, people have arrived at very different conclusions. One of the crucial questions is whether cultural evolution is based on its own replicator system (see 2.4 and 2.5) that evolves independently of the genetic replicator system of genetic ('biological') evolution, or whether the two mutually influence each other. Blackmore (1999) and Barrett *et al.* (2002) provide good discussions of the issues. Here we shall look briefly at some of the main positions.

Gene-culture coevolution 1: culturgens

In 1981, the renowned Harvard evolutionary biologist Edward O. Wilson teamed up with a physicist, Charles Lumsden, to develop a theory of coevolution between genes and culture (Lumsden & Wilson 1981). At its heart their theory assumes that cultural traits can show some degree of independent evolution but are inextricably tied to the bodies, and thus the genes, that express them, a position Wilson (1978) characterised as 'culture on a leash'.

The units of cultural evolution in Lumsden & Wilson's scheme are dubbed '**culturgens**'. Culturgens influence the fitness of underlying genes by coevolving with the neuro-psychological mechanisms they encode. The relationship is modelled on classical host–parasite coevolution, in which host and parasite gradually become adapted to each other, for example through increased resistance on the part of the host, and reduced virulence in the parasite. By analogy in Lumsden & Wilson's model, neuropsychological mechanisms become honed to the cues appropriate to ongoing cultural behaviour, while culturgens adapt to the mechanisms available to them. An entertaining example of a cultural trait that has evolved in relation to underlying perceptual mechanisms is the teddy bear. Since its

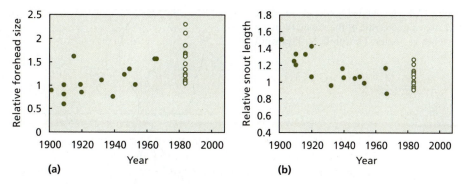

Figure 12.3 Cultural evolution of head shape in teddy bears. The heads of teddy bears have become progressively more baby-like over time as a result of an increase in the relative size of the forehead (a) and a reduction in the size of the snout (b). Solid circles, data from museum specimens; open circles, data from toy shop specimens. After Hinde & Barden (1985). See text.

first appearance in the early 1900s, the shape of teddy bears' heads has gradually changed from a pointy-snouted affair with a low forehead to the present baby-like (paedomorphic) object with a high forehead and short, button snout (Hinde & Barden 1985; Fig. 12.3a,b). Interestingly, choice tests with young children showed that they did not prefer baby-like bears over the more 'primitive' shape (Morris *et al.* 1995). Instead, the preference increased with age, suggesting it was *adult* preferences that had driven the evolution of baby features. A similar paedomorphic trend can be seen in the evolution of Mickey Mouse and other famous cartoon characters (Gould 1980). The obvious possibility here is that the producers of teddy bears and cartoon characters are cashing in on our predisposition to nurture and feel affection for things that remind us of dependent offspring. The young of many vertebrates broadly share the round headed, short-faced features of human babies (Fig. 12.4), which is why we also find them attractive. Baby cues and the associated response of adults thus have a long-established pedigree that now seems to extend its influence to some of our cultural artefacts.

Although intuitively attractive, Lumsden & Wilson's argument is unfortunately couched in some difficult mathematics and has been questioned at a number of levels. Nevertheless, it offers an interesting contrast with other models of gene–culture evolution (see below), and continues to figure in the general debate.

Gene–culture coevolution 2: semi-independence

A family of models pioneered by Cavalli-Sforza & Feldman (1981), and developed in a long series of later papers (see Feldman & Laland 1996 for a brief summary), sees cultural traits evolving rather more independently of genes, but still ultimately being selected through their effect on genetic fitness. In Wilson's analogy, the cultural 'leash' is still on, but it has been let out further and allows greater freedom of movement. These models therefore share some features with Lumsden & Wilson's model, but differ in considering the fitness of a genotype/cultural phenotype amalgam (the so-called **phenogenotype**) rather than a mutually coevolved relationship between cultural trait and genotype.

Phenogenotype models have been applied to a wide range of problems from language and altruism to mating preferences and sex ratios. As an example from the field, Feldman & Laland (1996) cite the effect of yam cultivation on sickle cell anaemia among populations in West Africa (Durham 1991). Here, the frequency of the sickle cell genotype depends

Figure 12.4 A rounded forehead and short face are characteristic of the young of many vertebrates. After Thornhill & Thornhill (1983).

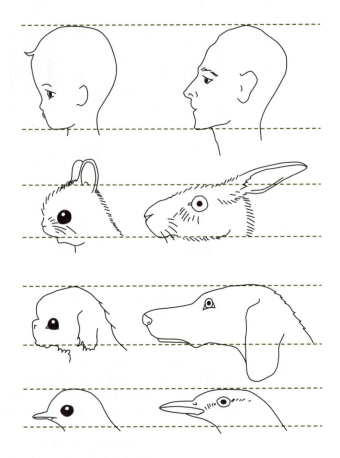

on a population's means of subsistence (cultural phenotype). Populations that cut down trees to cultivate yams create the conditions for rain to leave pools of standing water in which mosquitos carrying malaria can thrive. This leads to increased selection for the sickle cell trait because heterozygotes for the trait are resistant to malaria. Among yam cultivators, there is a correlation between the amount of standing water and the frequency of the sickle cell trait that is not present in comparable non-yam-producing populations. The selection pressure on the sickle cell gene thus depends on the prevalence of a cultural trait – yam cultivation – within the population and the two traits are selected together.

An important property of phenogenotype models is that they can lead to circumstances in which selection for cultural traits acts in direct conflict with selection on genotypes. Laland (1994), for example, modelled the evolution of mating preferences where females expressed one of two culturally inherited preferences for males of two different genotypes. One set of females mated indiscriminately with males of both genotypes, while the other showed a culturally inherited preference for just one of them. Laland's model showed that, if a cultural female preference for a particular male genotype reached a critical frequency, e.g. through social learning or drift, it could drive a male trait to fixation even if the trait imposes a high reproductive cost on the male. A well-known example of such conflict in a real population is the spread of the neurodegenerative disease 'kuru' in the Foré tribe of New Guinea. As part of their ritual to honour the dead, the Foré practised cannibalism and smeared brain tissue from the deceased over their bodies (Durham 1991). As a result, they suffered an epidemic of a degenerative disease of the central nervous system somewhat similar to Creutzfeld–Jakob disease. While this is a terrible price to pay, Cavalli-Sforza & Feldman (1981) have shown that a highly maladaptive, culturally driven trait such as this could in fact dispatch up to 50% of its carriers and still spread through the population. Thus a cultural trait can spread despite an enormous cost in genetic fitness.

The dual inheritance model

In the mid 1980s, the ecologists Robert Boyd and Peter Richerson attempted to dispense with Wilson's 'leash' altogether and develop a model in which genes and cultural traits evolved independently: the so-called **dual inheritance model** (Boyd & Richerson 1985). Their model centres around (a) the mechanisms of social learning through which cultural traits can spread, in particular the degree of bias in what is copied as a result of the role models available, and (b) the role of individual experience in shaping copied responses (Box 12.2). This can lead to both direct and indirect cultural selection for traits. In indirect bias (Box 12.2), for example, choice of role model, say a particularly productive herder, may lead to the spread of both productive agricultural practice and other traits that happen to be associated with the model but are not functionally related to productive practice, such as a lavish dress style or rich diet (Borgerhoff Mulder 1991). In principle, this kind of association could drive a runaway cultural selection for extreme traits in a manner akin to the Fisher effect in mate choice (10.1.3.2) (Boyd & Richerson 1985). Cultural transmission may also influence levels of selection within populations. For instance, frequency-dependent bias that encourages individuals to copy the commonest role model (Box 12.2) could, in principle, lead to group selection through discrimination against rare exemplars (Richerson & Boyd 1991). Even though mobility between human populations is often extensive (a factor normally militating against group selection [2.4.4.1]), discrimination against rare traits could theoretically reduce its impact on the potential for group selection (Borgerhoff Mulder 1991). Empirical studies, however, do not lend much support to the idea. Data on group extinctions through intertribal warfare in New Guinea, for example, suggest that, even where there is relatively little 'leakage' of cultural information from one group to another – conditions superficially conducive to group selection – cultural group selection would still be too slow to account for the rate of cultural change across tribes (Soltis *et al.* 1995).

Instead, Boyd & Richerson (1985) use evolutionarily stable strategy theory (2.4.4.2) to model the relative importance of social and individual learning in different environments. Outcomes suggest that the stable strategy for acquiring information depends on two factors: how difficult it is to learn accurately through individual experience, and the degree of similarity in environments over time. Cultural transmission is favoured when environmental conditions are reasonably stable in the short term, but genetic inheritance is favoured when environmental stability is high and learning carries an unnecessary risk of inaccuracy (Boyd & Richerson 1988, 1989).

While Boyd & Richerson's dual inheritance model gives far greater play to cultural evolution than the models of Lumsden & Wilson and Cavalli-Sforza & Feldman, it is not based on a system of independently replicating cultural units, and to that extent is still ultimately a reflection of underlying genetic evolution (Blackmore 1999). The idea that cultural traits are driven by an independent evolutionary process based on a non-genetic replicator, however, *is* central to the more recent concept of memes (Dawkins 1976; Brodie 1996; Blackmore 1999).

Underlying theory

Box 12.2 Mechanisms of cultural transmission

Boyd & Richerson (1985) suggest a number of mechanisms that might underlie cultural transmission.

Guided variation

Guided variation occurs when an individual is given some indication as to what to do by others around it, but relies on its own trial-and-error experience to modify its behaviour in response. Guided variation may take place through rational calculation or procedural rules of thumb (see 4.2), and proceeds by collecting information about the (here social) environment, evaluating the desirability of alternative outcomes for various courses of action, and using the outcome to guide changes in phenotype that are then transmitted culturally to the next generation.

Direct bias

In **direct bias**, individuals copy some cultural traits rather than others on the basis of a key property of the traits available. Cultural food preferences are a good example. As children we tend to prefer sweet foods and dislike those that are spicy. Such preferences may have a genetic basis (Lumsden & Wilson 1981). By the time we are adults, however, we have acquired a range of culturally influenced preferences that override some of our genetic predispositions but depend in their specifics on what is available within our particular culture: thus British and American palates like roast beef and pork but dislike dog and fermented fish, while the opposite is true in some Asiatic cultures (Boyd & Richerson 1985). Choosing which of the range of traits is best may take place through trial-and-error learning.

Indirect bias

Indirect bias occurs when individuals choose one particular role model, perhaps the most successful or noticeable, and copy that. A whole suite of traits possessed by the model may be copied without the imitator necessarily knowing which is responsible for the model's success (Borgerhoff Mulder 1991). Indirect bias can take different forms. In **frequency-dependent bias**, for example, it is the commonest exemplar that is copied, perhaps on the assumption that the commonest is also likely to be the most successful.

From discussion in Boyd & Richerson (1985) and Borgerhoff Mulder (1991).

12.1.1.2 Memes and memetics

As we saw in 2.5, memes can be thought of as units of information (relating to ideas, beliefs, fashions and so on) that are capable of being stored in one brain and transmitted to others by social learning. Whereas genes are replicators that dwell in cells, issue instructions for making proteins, and are passed on during reproduction, memes are replicators that dwell in brains (or artefacts such as books, pictures or buildings), inform behaviour, and are passed on by imitation (Cloak 1975; Dawkins 1976; Blackmore 1999).

Both genes and memes succeed by being copied. Good copying potential in genes arises from their effects on their individual bearer's reproductive performance, or by manipulating their representation in the bearer's cells (2.4.3.2). In memes, copying potential arises from gaining the attention of prospective imitators. Explaining exactly how and why particular memes take off, and the implications of these alternative replicators for cultural evolution, is the province of the emerging field of **memetics** (see Blackmore 1999). At first sight, it might seem that memes are similar to Lumsden & Wilson's culturgens (12.1.1.1). However, this is far from the case. The key point about culturgens is that they are not replicators in their own right, but cultural phenotypes of genes. Memes in contrast are true replicators, a parallel system of replicator-driven evolution running alongside its genetic counterpart.

It is not easy to predict in advance which memes will triumph in the copying stakes, but it is clear that certain kinds of meme are, or have been, exceedingly effective. It is not difficult to think of 'fads' in clothing style, painting or architectural design, for example, that have leapt to prominence and spread around the world. The notion of a god is an ancient and conspicuously successful meme that has arisen independently on many occasions and in a variety of cultural forms. So-called 'urban myths', such as the much-repeated story of the woman who tried to dry her poodle in a microwave oven, are more recent versions of successful memes (Blackmore 1999). Some of these memes succeed because they emanate from influential or attractive role models (indirect bias [Box 12.2]), the spread of Beatle haircuts in the early 1960s being a good example (see also Fig. 12.2). Others, like the god meme, may feed on deep psychological needs and be reinforced by powerful ritual (Box 12.3). The fact that memes seem to exist simply because of their capacity for replication tempts comparison with viruses (Dawkins 1976; Brodie 1996, but see Blackmore 1999). Just like their biological and computer counterparts, meme viruses effectively take up residence in one host for little purpose other than to use it as a springboard to another.

In some cases, memes improve their potential for replication by teaming up with other memes. Dawkins (1976) referred to such teams as **'coadapted meme complexes'** (by analogy with their genetic equivalent [see Table 10.1]), a term sometimes shortened to **'memeplexes'** (Blackmore 1999). So-called 'viral' sentences illustrate the idea of memeplexes. These are sentences designed solely to replicate, 'Say me!', 'Copy me!' and 'Repeat me!' being simple, but probably not very effective, examples. However, if we add a few extra ingredients, such as 'If you copy me, I'll grant you three wishes!', or 'Say me or I'll put a curse on you!', the incentive to repeat the sentence is likely to increase, at least in some individuals. Add the further phrase 'in the afterlife' and the persuasive power of the sentence is likely to increase more substantially (Hofstadter 1985; Blackmore 1999); the exhortation to repeat has become stronger because it has been reinforced by association with other memes, in this case suggesting mythical rewards or punishments according to whether the recipient obeys. Just these ploys are used by the sophisticated memeplexes of religions and cults that spread through populations with an armoury of copy-enhancing tricks (Box 12.3), sometimes with uplifting effects on their recipients, sometimes with disastrous ones.

While memes and memeplexes are independent replicators, and can be compared with functionally useless (from the host's point of view) viruses, that is not to say they are without impact on the genetic evolution of their hosts. As Blackmore (1999) in particular has argued, memes may be responsible for some of the key attributes that we think set us apart from our fellow inhabitants on the planet. We shall look briefly at three of them.

Difficulties and debates

Box 12.3 Religion as a memeplex

Religions collectively constitute one of the most successful meme complexes in the history of human culture. A vast number of people's lives are utterly governed by them, and much of what they encourage – celibacy, selflessness, frugality, aceticism and so on – is likely to be inimical to reproductive interests. At a rational level, it is easy to dismiss them as nonsense. So why do they have such a grip? One answer is that they reflect an adaptive willingness to conform to a collective 'will', and thus maintain group stability (Dunbar 2003). In doing so, they deploy a battery of powerful tricks to ensure their memes are copied. The following are some examples.

Threat and appeasement

A good trick is to frighten people with some dire consequence, but at the same time offer them a way out of their predicament. The Roman Catholic notion of a god is a good example. The Catholic god is watching you all the time, like a cosmic Big Brother. If you step out of line ('sin'), he will be very displeased and wreak an appalling punishment on you; not now necessarily, but certainly later, when you enter the afterlife. Short of committing suicide and seeing if it is true, it is not possible to check this out in any way. However, a tantalising bolt-hole is offered if you 'repent of your sins', bring up your offspring as Catholics and regularly go to mass. Salvation can be found by promoting the meme, and 'good' Catholics work very hard indeed to promote it.

The altruism 'trick'

Altruism – kindness, generosity and general selflessness – is a potent source of memes because it attracts lots of imitators (see text). Religions cash in on this in a big way. Many believers are indeed altruistic ('good') people; they give up their time to charity work, donate money to 'good' causes, and live frugal and impeccably honest lives. Their attractive qualities encourage mimicry. Crucially, however, it is not so much the generosity element that is mimicked as the associated trappings of devoutness – going to church on Sunday, putting in an appearance at the charity fair and so on. In many other respects 'goodness' mimics can be veritable spiritual hooligans. Hypocrisy or no, however, the memes are perpetuated. Even true generosity can be subverted into promoting the meme, as when, say, church donations go not to the poor and needy, but to building a more splendid steeple or paying priests.

Mysticism

A key ingredient in religion memes is the untestability of their claims. Central to this is the requirement to have 'faith', which means believing in something despite a lack of evidence that it is right, and often in the face of evidence that it is wrong. Belief in miracles, resurrection, divine judgement, the existence of 'good' and 'evil' are all acts of faith, as is accepting that the Bible is the word of God. Many required beliefs are also steeped in ambiguity and contradiction, the Bible being a paradigm of the art. However, they are often powerful ideas, or, in the Bible's case, powerful exhortations about how life should be lived. Coupled with the undercurrent of threat should you not comply, faith and obfuscation help reduce serious challenges to the tenets of the belief. All this is further reinforced by the stylised, repetitive rituals, and often grandiose settings, that accompany religious observances and help create an illusion of substance.

Based partly on discussion in Dawkins (1989) and Blackmore (1999).

Memes and brains

Intelligence and consciousness have long been central to the debate about humanness (see 4.2, 11.3). Arguments about them often go hand in hand with those about brain size. The human brain is disproportionately large compared with those of other animals (Fig. 3.7c). It is also extremely costly to run, comprising some 2% of body mass but demanding 20% of its metabolic energy, and makes life difficult when it comes to giving birth. Not surprisingly, therefore, people have sought naturally selected advantages to account for its extraordinary development. There has been no shortage of suggestions. Brains became bigger as tool use and technology advanced, driven initially by hunting and the need to outsmart prey, is a frequently proposed explanation. The need for cognitive maps and navigational skills (see 4.2.1.2 and 7.3) to find food in spatially and temporally unpredictable environments is another. Yet others include living in an increasingly complex social environment (4.2.1.5) and the effects of escalating signaller–receiver arms races (11.3), ideas that connect explanations of brain size with those of consciousness and theory of mind (4.2.1). Cartwright (2000) provides a readable introductory review.

All these explanations, on their own or in various combinations, continue to have their adherents. Blackmore (1999), however, argues that a quite different selection pressure has ultimately been responsible for the dramatic increase in brain size. In her view, the crucial step was the evolution of true imitation (6.2.1.4). As we saw in Chapter 6, true imitation seems to be rare in species other than our own. The reason for this, argues Blackmore, is that it requires not only a capacity to classify objects and actions and decide which are worth copying (Box 12.1), but an ability to imagine oneself in the position of another so that points of view can be transformed into self-perspective and matching behaviour produced. We certainly have these abilities, but they may also be present to some degree in some other primates (4.2.1). Given this, it is unlikely that they are responsible for our large brain in themselves. Instead they may provide the essential precursors for true imitation, and it is then imitation that has fuelled the last stage: the elaboration of our brain from its more modest primate ancestry (Blackmore 1999).

But why should imitation have done this? The answer, according to Blackmore, is that it opens up the boundless evolutionary potential of memes. Not just the vast scope for different ideas, but the capacity for resulting cultural evolution to escalate and proliferate the resulting demands for information processing and decision-making. Her argument hinges on the assumption that the ability to imitate will itself become the focus of imitation. Good imitators will acquire important skills first, so will be favoured by natural selection *and* become role models for further imitation. But, as we saw earlier (Box 12.1), many different characteristics of role models may be copied, not just the ones that initially merited attention. The snowball of memetic evolution thus starts to roll. As it goes, new attributes worth copying emerge, and become, along with the capacity to imitate them, the next focus for selection. Initially, the memes that are copied are likely to relate to reproductive potential: hunting skills, successful warmongering, warm clothing and so on, and will therefore reinforce naturally selected attributes. But the rapid proliferation of memes and the wherewithall to copy them will eventually lead to novel traits and a demand for new neural machinery to cope. Memes will therefore begin to exert their own selection pressure on genes to build bigger brains (Blackmore 1999). Thus our extraordinary brain, with its sophisticated higher-order properties, becomes a biological consequence of meme-driven cultural evolution.

Memes and language

Another defining human attribute is, of course, language (see 11.3). Indeed it is arguably the single most important thing that distinguishes us from other animals (Barrett *et al.* 2002). Whether language itself influences the way we think (11.3.2) is debated, but it certainly facilitates the exchange of information between different individuals. While other species communicate, and may well make reference to the world around them (11.3.2.1), none remotely approaches the sophistication of human language, not even when efforts are made to teach it to them. So why have we evolved this unique capability?

As yet, there is no generally agreed explanation, though, as for brain size, there have been several suggestions. These have tended to focus on the way language is designed to facilitate communication *per se* (see Pinker & Bloom 1990), the argument being that language would have allowed our ancestors to acquire and transmit information far more quickly than would be possible under natural selection, and so gain a crucial advantage in competition with other species. Exactly what information is involved, however, is not clear. Details of planned hunts, or the whereabouts of foodstuffs or danger, are suggestions from early studies, but these do not really explain why humans alone should have developed language, since many other social species have complex hunting or foraging problems to solve. Another suggestion is that language is a product of our complex social environment. Advanced social organisation imposes many demands on individuals as they deal with family relationships, aggressive alliances, dominance hierarchies, cooperative trust bonds and the like (see 9.1.2.1 and 9.3). However, non-human primates deal with these perfectly adequately with a range of gestures, facial expressions, vocalisations and other behaviours, and do not seem to require a sophisticated language. It is unlikely, therefore, that these kinds of demands by themselves have provided the necessary driving force for language. So what might have done? Surprisingly enough, the answer may come down to social grooming, or at least its verbal equivalent.

Grooming and gossip   In 1996, the evolutionary psychologist Robin Dunbar pointed out that an awful lot of what we say to each other is relatively inconsequential (Dunbar 1996). Parodying Churchill's famous 'Battle of Britain' line he asks . . . 'why on earth is so much time devoted by so many to the discussion of so little?' Analysis shows that we spend very little of our time talking about weighty matters that are intended to instruct or educate in some important way. Instead, most of our exchanges are seemingly trivial chit-chat. We talk endlessly about the weather, other people's sexual relationships, local goings on in the village and so on. None of this usually has much impact on anything, so why do we do it?

Dunbar's argument is that gossip is a social 'cement', a means of exchanging social information and maintaining cohesion within our particular group. A similar function is fulfilled in non-human primate groups by social grooming. So why don't we too just sit around scratching each other's backs instead of indulging in endless gabble? The answer it seems is that there are simply too many of us. It turns out that non-human primates spend a maximum of 20% of their total daily time in social interaction, mainly grooming. This is a hefty proportion of the day given the competing demands on the animals' time. But how much time is spent grooming depends on group size, increasing steadily as groups become bigger (Fig. 12.5). The apparent ceiling of 20% imposes a limit on the maximum size of group (around 70 individuals) that can be sustained as a coherent unit. Stable effective group sizes in humans average around 150 individuals, much bigger than our non-human counterparts. If we were to maintain our social fabric by grooming,

Figure 12.5 The percentage time spent in social grooming by monkeys and apes (●) increases with group size, but levels off at around 20%. If the relationship between time spent grooming and group size was linear, extrapolation would predict a commitment of 43% of time to grooming in humans (□), with their groups of around 150 individuals (solid line). In fact, the time spent in the (suggested) functional equivalent of social grooming – gossiping – averages around 20% (○), much the same as for grooming in other primates. See text. After Barrett *et al.* (2002).

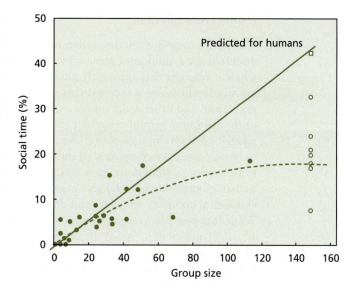

extrapolation of the relationship in Fig. 12.5 suggests we should end up committing nearly 45% of our day to it, more than twice the maximum commitment of our monkey and ape relatives (Dunbar 1993). Clearly this is unsustainable. What we need is some means of doing the whole thing more economically, and this is where gossip comes in. In Dunbar's view, gossip is a high-efficiency substitute for grooming. It is more efficient because it (a) allows us to interact with more individuals simultaneously (in the language of Chapter 11, it increases the communication network), (b) increases the amount of information flow between group members, for instance about who is trustworthy and worth cooperating with, and (c) provides a greater opportunity for detecting cheats freeloading on the benefits of group cooperation (see 9.3; Dunbar 1993). Interestingly, when the appropriate calculations are made, gossip takes up an average of 20% of our waking time (Fig. 12.5), the same as the apparent maximum time spent grooming in other primates! This seems to provide strong support for Dunbar's interpretation.

But what does all this have to do with memes? Well, there is still a problem to be overcome in the gossip argument above, and that is the sheer *extent* of the difference in language skills between humans and other primates. The effective group size hypothesis is compelling, but some other primates form very large groups and have not found it necessary to develop human-like language. We are unique by several orders of magnitude, and no orthodox neoDarwinian hypothesis seems quite up to explaining it. One thing that might account for the gulf, however, is a different replicator system that has a greater potential for rapid, explosive evolution. Memes fit the bill perfectly. Information flow is vital to the spread of memes, but the nature of the information changes as prolifically as the memes it represents. What is needed is a vastly flexible, open-ended and frequently operating system of exchange that registers with as many potential copiers as possible and keeps track of the continuously changing information. Language, and gossipy spoken language in particular, is tailor-made for the role. In this view, it is memetic selection that drives the elaboration of language, and, by the same argument as before, drags genetic selection for underlying neural mechanisms (big brains and language 'centres') along in its wake (Blackmore 1999). Once again, therefore, an apparent discontinuity between ourselves and other species finds a plausible account in the alternative evolutionary world of memes.

Memes and altruism

Our third example is another recognisably human attribute: altruism. We have already discussed the evolution of phenotypic altruism in animals generally in 2.4.4.5 and 9.3, where it emerged that apparently altruistic acts were really nothing of the sort. Instead, individual selflessness was revealed as genetic selfishness, as, via kin selection, reciprocity, mutualism and other mechanisms it helped to secure the transmission of more copies of the alleles for helpful behaviour into subsequent generations. The problem with this when it comes to humans is that our altruism seems too widespread and indiscriminate to be accounted for completely by these traditional mechanisms. Much of it is dispensed to recipients we shall never meet and are highly unlikely to be closely related to us. We donate blood anonymously, give money to charities for mistreated animals, adopt the children of complete strangers and do arduous, low-paid jobs caring for other people. While it is possible to explain some of this in terms of kin selection or reciprocity (see 12.2.1.3), not all of it, by any means, fits comfortably with their predictions. True, many of the examples might be construed as cultural *by-products* of kinship or reciprocity, reflections of historical biological pressures translated into a cultural context, but this rather begs the question rather than answering it: are these cultural permutations really just novel contexts for biology, or do they reflect an independent evolution of their own? Meme theory suggests the latter (Blackmore 1999).

An important feature of altruists is that they are likely to be nice to know and have a good reputation within their community (see Wedekind & Braithwaite 2002). Helpful and cooperative people are often recognisably so, even if some of their altruism is performed anonymously. This makes them attractive to imitate, because they are popular and accumulate friends. Indeed, it has even been suggested that individuals may *compete* to demonstrate their superior altruism because of its strong social (and possibly sexual [see 12.2.1.3]) currency (Alexander 1979; see also Zahavi & Zahavi 1997; Roberts 1998). Altruistic traits are thus excellent meme fodder and are likely to spread easily. It is important to realise, however, that the spreading potential of altruistic memes does not (necessarily anyway) reflect a desire by imitators to be altruistic themselves. It is merely a consequence of the attractiveness altruism imparts to its exemplars and the greater exposure to new imitators they therefore enjoy. We can therefore explain why helpfulness, kindness and cooperation come to dominate populations without having to torture kinship or reciprocal altruism theory to account for them.

But there is another twist. In its early stages at least, the memetic spread of altruism may complement its spread via natural selection, since gathering friends and admirers is likely to yield the payback for altruism that is central to the reciprocity argument (Wedekind & Milinski 2000; Wedekind & Braithwaite 2002). The initial act of altruism, however, is costly (otherwise it would not be altruism), so, just as in genetic evolution, there is a temptation to cheat. Copying the *traits* of a truly generous individual, but not actually being generous oneself, would reap the benefits of both worlds. This highlights an important point. It is not just the generosity of altruists that is likely to be copied, but any traits that identify or are strongly associated with them, such as clothing styles, general demeanour or tastes in music. Cultural traits may thus become biased in various, seemingly arbitrary, directions that have in fact been selected memetically through association with entirely different traits. Fundamentally attractive traits such as altruism may therefore become drivers for much else in cultural evolution (Box 12.2 and see Blackmore 1999 for an excellent discussion).

12.2 Natural selection and human behaviour

It is clear from our discussion so far that the forces of genetic and cultural evolution are likely to have some degree of impact on each other, even when they are based on independent replicator systems. It also stands to reason that our cultural life must have its roots somewhere in our genetic evolutionary past. We might therefore expect to find evidence of this in our present behaviour. But what exactly should we expect: that present-day behaviour is adaptive here and now, or that it is a reflection of some dim and distant set of selection pressures that shaped us in the past? This quandary effectively distinguishes two different schools of thought: **human behavioural ecology** and **evolutionary psychology**.

The modern evolutionary approach to behaviour is enshrined in behavioural ecology, as we saw in Chapters 1 and 2. While concerned with the ecology and evolution of behaviour in organisms generally, behavioural ecology has had various things to say about human behaviour over the last three decades, sometimes with almost incendiary effects (1.3.3). Just as with non-human organisms, behavioural ecologists (now sometimes called **human behavioural ecologists** or **Darwinian anthropologists**) are concerned with the effects of differences in behaviour between people on reproductive success; that is with the behaviour's *current* adaptive value (Table 12.1). Of course, behavioural

Table 12.1 Some differences between human behavioural ecology and evolutionary psychology as approaches to the evolution of human behaviour. Modified after Cartwright (2000).

Human behavioural ecology	Evolutionary psychology
Based on behavioural ecology	Based on cognitive psychology
Culture viewed in terms of current fitness maximisation, so optimality and games theory approaches are useful frameworks for investigating decision-making and behaviour	Current fitness maximisation is an unreliable guide to the human mind because most present environments differ substantially from ancestral 'environments of evolutionary adaptation'
Focus on behavioural outcomes	Focus on beliefs, values, emotions and other mental attributes
Views lifetime reproductive success as the yardstick of adaptiveness of current behaviour	Focuses on putative ancestral selection pressures rather than current reproductive success
Ancestral adaptations have given rise to 'domain-general' (see 4.2.1.6) mechanisms for solving problems	Ancestral adaptations have given rise to 'domain-specific' modules designed to solve particular problems (4.2.1.6)
Genetic variability in fitness-related characters remains and is currently acted upon by selection	Genetic variability in fitness-related characters (mental mechanisms) is low, and points to a universal 'human nature'

ecologists acknowledge the obvious influence of culture, but their approach has two things to recommend it. First, it has been enormously successful in its wider application to the living world, as previous chapters have demonstrated; second, it is conservative, given the likely powerful effects of cultural evolution. If a trait is shown to bring reproductive advantages despite potentially conflicting cultural influences, then it is indeed likely to have evolved by natural selection (Caro & Borgerhoff Mulder 1987). Unlike evolutionary psychologists (see below), behavioural ecologists are generally disinclined to speculate on influences in the past, because they are difficult to test unequivocally (Barrett *et al.* 2002).

A somewhat different view is taken by evolutionary psychologists. Coming as they do from the world of cognitive psychology rather than ecology (1.3.4), evolutionary psychologists are concerned with the adaptive design of higher brain function and psychology. But they do not seek answers in terms of current selection pressures. Instead, they assume we are living with a mental apparatus that was shaped in its essentials at some point in the past, when our ancestors inhabited our so-called **environment of evolutionary adaptation** (EEA) (Table 12.1; see also 1.3.4). Just when this was, and whether in fact the EEA is reflected in any one particular environment, is a matter of debate (see Crawford 1998). Tooby & Cosmides (1990) suggest the EEA is a weighted average of several pre-agricultural environments, spread over a period of some two million years from the early Pleistocene. Whenever it was, however, the EEA is assumed to have shaped the mind and body plan we have inherited, and what we see today are the effects of this ancestral template interacting with present-day culture.

Although we can contrast the approaches of behavioural ecologists and evolutionary psychologists in this way, there is, not surprisingly, considerable overlap in interests, and the distinction is beginning to blur (Daly & Wilson 1999; 1.3.4). Just as behavioural ecology has expanded into human behaviour, so evolutionary psychology has much to say about non-human species, so much so in fact that some prefer to identify **human evolutionary psychology** as just a particular branch of the subject. This two-way traffic can bring different perspectives to bear on a given behaviour. The familiar debate about intentionality and goal-directedness (see Chapter 4) is a case in point: evolutionary psychologists are interested in the adaptiveness of cognitive function, and interpret behaviour in that light, while behavioural ecologists remain more agnostic about the cognitive nature of underlying mechanisms. Nevertheless, the two approaches are jointly illuminating a number of problems in the evolution of human behaviour. We look at three here: sex differences in mating behaviour and parental care, and our general propensity for cooperation.

12.2.1 Sex, parenting and cooperation

The anthropologist Edmund Leach noted that 'human beings, wherever we meet them, display an almost obsessional interest in sex and kinship' (Leach 1966). We have also seen in Chapters 2, 9 and 10 that both factors are enormously important in the evolution of behaviour in other species. Indeed, they may be interwoven in their effects, as when relatedness between potential mates affects mating preferences (10.1.3.2). Kinship in nonhuman species is also important in determining other aspects of behaviour, including parental investment and care (10.2) and social relationships within groups (9.3). Is there any evidence that our 'obsessional interest' in sex and kinship has been shaped by the same kind of selection pressures?

12.2.1.1 Sexual behaviour

As in other mammals, females of our own species are burdened with the greater parental investment in terms of gamete size and nurturing developing embryos and infants. While men can also make considerable investments in parenting, their reproductive decisions as a whole are taken against the backdrop of this basic disparity and its consequences for female physiology and psychology. Even when not pregnant or nursing, women have eggs available for fertilisation on only a few days a month, and, before the widespread use of contraception, most of these were lost to long periods without ovulatory cycling (amenorrhoea) as women lactated and nursed offspring. So, like the majority of other species with anisogamous sex (see 10.1.2), our operational sex ratio (the ratio of fertilis-able females to reproductively active males) has been heavily skewed towards males for most of our evolutionary history. As a result, men generally compete for fertilisations, and women choose among them for mates.

However, we are a long-lived species with a long period of parental care, during which men are likely to contribute substantially to the survival and development of offspring. We might thus expect female choice to be influenced by a male's ability to provide. But since provisioning is likely to be costly to males, we might expect men to be fussy too, in this case choosing females that are fecund and able to produce robust, healthy offspring. They should also seek extra-pair copulations with females in other partnerships, gaining additional offspring at the expense of the other male's paternal care while guarding against the same thing happening to them. Thus human repro-ductive behaviour (at least in the heterosexual population – see Box 12.4) should be characterised by both cooperation and conflict between the sexes, something we have seen typifies reproductive relationships in many other species (10.1.2.2). So is this the case?

Mating strategies in men and women

To begin with, is there any evidence that women go for providers, while men go for fertility? One of the great conveniences of working on human beings is that you do not have to chase around trying to spot behaviour in action, but can instead simply ask questions of your subjects (there is, of course, a risk they will lie, but this can be taken into account in well-designed studies). Many studies have thus been conducted using questionnaires. Information is also sometimes made available spontaneously, as in the 'Personal Ads' section of magazines and newspapers. Analyses of these kinds of data provide considerable support for the predicted sex difference in mating preference. Figure 12.6(a), for example, shows that men seeking women in newspaper advertise-ments request good looks rather than resources (wealth, income, status, etc.). Women, on the other hand, give a much higher weighting to resources. This difference is apparent across a wide range of nationalities and cultures (e.g. Buss & Schmidt 1993). When it comes to advertising their *own* qualities, however, the opposite is true. Women stress their good looks, while men emphasise their wealth and status (Fig. 12.6b) (Waynforth & Dunbar 1995). That these differences reflect strategic preferences is supported by questionnaire studies in which the importance of income (a measure of provisioning capacity) was assessed in relation to the level of involvement with a partner. Both men and women valued a partner's income more in long-term relationships and marriage, but women attached far greater importance to it at all levels of involvement (Kenrick *et al.* 1990; Fig. 12.6c).

Difficulties and debates

Box 12.4 Homosexuality

While human sexual behaviour conforms closely to the predictions of evolutionary theory, there are some aspects of it that challenge explanation in evolutionary terms. One of these is the widespread existence of **homosexuality**.

Mating partnerships between people of the same sex are necessarily redundant reproductively, yet homosexual behaviour has been recorded across the animal kingdom, from acanthocephalan worms and fruit flies to gulls, cheetahs and humans (see Bagemihl 1999). There is also evidence for a genetic basis to homosexuality in various species, including ourselves (e.g. Gill 1963; Hamer *et al.* 1993; Kendler *et al.* 2000), and an association with certain aspects of brain anatomy and physiology (e.g. Swaab & Hofman 1990; LeVay 1991; Resko *et al.* 1996; see 3.1.3.1). But what maintains it in the population?

Suggestions relating to other species have ranged from enhancing intermale competition (e.g. through sealing up the reproductive tracts of rivals in acanthocephalan worms [Abele & Gilchrist 1977; see 10.1.3.1]) to making the best of a bad job when sex ratios are highly skewed (Conver *et al.* 1970) or epistatic benefits of homosexuality genes when in combination with certain other genes (see below). Explanations for its occurrence in humans have fallen into a number of categories, including the following:

Homosexuality is maladaptive

The 'cop out' explanation is that homosexuality is simply maladaptive, representing some form of developmental instability in the tails of the distribution of sexual proclivities in humans. Alternatively, it may be maladaptive from the individual's point of view, but reflect adaptive parental manipulation to reduce resource stress within family units, an argument that is loosely supported by a tendency for homosexuality (and other forms of reproductive withdrawal, such as religious celibacy) to be commoner among younger sons within families (E.M. Miller 2000). A completely different suggestion is that homosexuality may reflect the interests of selfish genetic elements (see 2.4.3.2) inherited maternally on mitochondrial DNA for which males are effectively a dead end. Some insects have been shown to carry 'male killing' genes on their mitochondrial DNA, so perhaps homosexuality is a similar, if more benign, route to sidelining males (Hurst 1991).

Homosexuality is an adaptive early phase of development

It may be that homosexuality performs an adaptive function during an individual's development by providing practice sexual experience or facilitating the formation of same sex alliances that will be useful later, arguments that are consistent with (a) the vast predominance of bisexuality over exclusive homosexuality, and (b) fertility among bisexual individuals being at least as high as among exclusively heterosexual people (Baker & Bellis 1995).

Genes for homosexuality confer reproductive advantages in other ways

Several people have suggested that putative genes for homosexuality may confer some kind of reproductive advantage in other ways. If sexual orientation is a polygenic trait (see 2.3.1.2), for example, homosexuality may be an occasional side effect of chance segregation among genes selected for feminised attributes in males, such as sensitivity, empathy and kindness, that increase the likelihood of long-term pair bonding and paternal care (E.M. Miller 2000).

Based partly on discussion in Barrett *et al.* (2002).

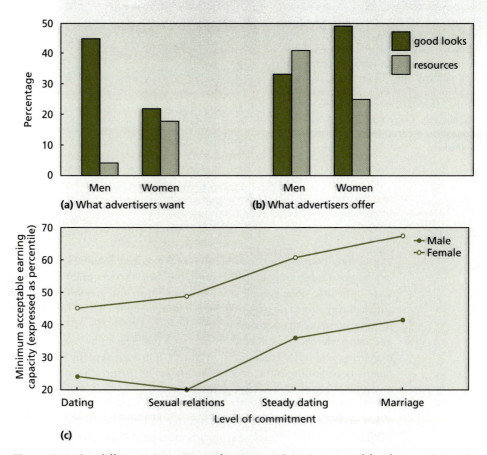

Figure 12.6 Sex differences in mating preferences. (a) Criteria requested for the opposite sex in the personal advertisement sections of newspapers and magazines. Women rate a healthy income more highly than do men, while men emphasise good looks. (b) Conversely, when advertising their *own* characteristics, men tend to stress their wealth and women their looks. After Waynforth & Dunbar (1995). (c) The importance of a partner's income increases with potential commitment in both sexes, but is always much higher among women. After Kenrick *et al.* (1990b). See text.

While it easy to see why women should be swayed by wealth and status in choosing a partner, it is less clear why men should go for looks. From the relative parental investment argument above, we might suspect attractive looks indicate something about fertility, or more precisely, **reproductive value** (see 2.4.5). Reproductive value is a measure of how many offspring a woman encountered now is likely have in the future. This is what matters if a male chooses a partner so as to maximise his reproductive success. Reproductive value, however, is not declared in billboard capitals on a woman's forehead. It has to be inferred from various physical cues. An obvious starting point is age.

Figure 12.7 shows the average curve of reproductive value for women plotted against age. As might seem obvious, the future reproductive prospects for younger women are much greater than those for older women. Thus men should find features relating to youth attractive. An accumulation of studies, over several different cultures, suggest they do. Preferred cues fall into two categories: the first relating to physical appearance, such as full lips, clear, smooth skin, clear eyes, healthy, shiny hair, good muscle tone and the right (see below) distribution of body fat; the second to behaviour, with animated facial

Figure 12.7 The curve of reproductive value for women (in traditional societies), showing how many children a woman is likely to have in the future as a function of her current age. From Buss, David M. *Evolutionary Psychology: The New Science of the Mind* © 1999. Published by Allyn and Bacon, MA. Copyright © 1999 by Pearson Education. Reprinted by permission of the publisher.

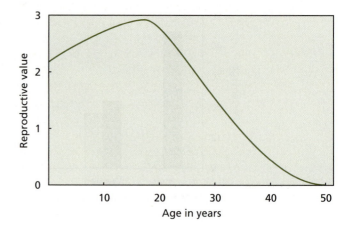

expression, 'bouncy' gait and overall vivacity scoring highly (Symons 1979, 1995). These characteristics certainly indicate youth, but they also say much about a woman's general state of health, since complexion, hair condition and general vigour are often quick to suffer during illness or stress. As the list above implies, facial features figure prominently in judgements of attractiveness (see e.g. Buss 1999). We all recognise beauty in faces, and our preferences are predictable enough for faces to be powerful marketing tools for magazines and other commercial products. But what actually makes a face attractive? A neat way of finding out is to manipulate facial features in computer-generated images and look at people's responses to them.

What makes a face attractive?   Johnstone & Franklin (1993) allowed male and female subjects to 'evolve' attractive women's faces on a computer screen, stopping when they deemed a face to be maximally attractive. The 40 maximally attractive faces so generated were then synthesised into a single 'beautiful composite' whose features were compared statistically with those of a composite formed from the faces of the female subjects themselves. The 'beautiful composite' and 'subject composite' images did not differ significantly in most measures, but they did differ in two. The 'beautiful composite' had a shorter lower face, with a smaller distance between the lips and the bottom of the chin; it also had a slightly smaller mouth with fuller lips. These are both features associated with youth, and, along with larger eyes relative to the face, have repeatedly emerged as key, and cross-cultural, attractiveness traits in facial analyses (Perrett *et al.* 1994; Jones 1996).

Composite images are also consistently rated as more attractive than images of individual faces, one possibility being that they blend out irregularities and make faces more symmetrical. We saw earlier that morphological asymmetry might be linked to underlying developmental problems (5.3.3), and that it may have a negative effect on mate quality and influence mating preferences in other species (10.1.3.2). Studies of morphological asymmetry and reproductive performance in humans also suggest that asymmetric people fare less well. Greater asymmetry is associated with a reduction in physical competitiveness among males (Manning & Pickup 1998; Manning & Wood 1998), attractiveness in both sexes (Gangestad *et al.* 1994), the likelihood of women achieving orgasm during copulation (and so possibly fertilisation [Baker & Bellis 1995]), and the number of children produced by women (Barrett *et al.* 2002). These effects may be well founded, since facial asymmetry correlates negatively with psychological and physical health measures (Shackelford & Larsen 1997), and increases with age (so providing yet another index of youthfulness). Interestingly, facial asymmetry may not lie solely in skeletal or other phenotypic characters that are stable in the long term. John Manning and coworkers have shown that soft-tissue asymmetry (for example, in ear size) in both men and women may be subject to short-term changes influenced by a range of hormones, including luteinising hormone, thyroxine, parathyroid hormone and follicle-stimulating hormone (Scutt & Manning 1996; Manning *et al.* 2002). These soft-tissue changes may thus reflect the reproductive or competitive state of an individual and affect their likelihood of mating. However, symmetry cannot be the whole story behind the attractiveness of composite faces, because composites remain more attractive even when their effects on symmetry are taken into account (Langlois *et al.* 1994). Moreover, symmetrical faces are rated more attractive when information about symmetry is removed (by presenting half faces) (Penton-Voak *et al.* 2001). Perhaps instead, it is being close to the average for the population, so within more people's range of what constitutes attractive, that matters.

Fat and attractiveness   Faces are a crucial element in assessing the desirability of potential mates, but they are not the only one. Overall body form, largely determined by the distribution of fat, plays an important role in the male view of female attractiveness. However, the relationship is not a simple one. Preference for a slim versus plump body form differs between cultures according to whether plumpness correlates with wealth or status. Where food is scarce, as in the Bushmen of Australia, plumpness indicates a lack of nutritional constraint, suggesting command of good resources. Where it is plentiful, as in western European and north American cultures, the relationship between body form and status is reversed, with slimness being valued more (Buss 1999) (although, comparisons with images of beauty from past centuries shows this is a relatively recent shift). As with facial features, there is some evidence that body forms around the population average are more attractive (Rozin & Fallon 1988).

Making people fat or thin is only one way fat deposition can affect attractiveness. The pattern of fat distribution also influences body *shape*, and the effect of this on attractiveness, especially in women, is much more consistent. The distribution of fat around the body in the two sexes changes markedly once puberty is reached. From a more or less similar distribution in prepubescent boys and girls, there is a sudden divergence, with men losing fat from their pelvic region (buttocks and thighs), but women depositing it there (hips and upper thighs). The upshot is that women end up with some 40% more fat in this area than men. This causes a reduction in the ratio of waist-to-hip size in healthy women from 0.85–0.95 before puberty to less than 0.80 after. The change in fat distribution in women is driven by the sex hormone oestrogen, and a reduced waist-to-hip ratio in the region of 0.67–0.80 turns out to be a good index of reproductive capability. Perhaps not surprisingly, therefore, it also emerges as a reliable predictor of mate preference in men (Singh 1993; Singh & Young 1995). The lower the ratio, the more attractive a woman is judged to be. Once again, this is consistent across cultures (Singh & Luis 1995).

Ovulation and attractiveness   The limited period for which women have eggs available for fertilisation each month might be expected to favour a preference among men for those that are ovulating. In most other primates males do seem to be able to tell when a female is ovulating, often with help from explicit visual and olfactory signals. Humans, however, are generally regarded as **concealed ovulators**, with women being sexually receptive throughout the ovulatory cycle and men unable to determine when eggs are available. For a long time, concealed ovulation in humans was regarded as needing special explanation, since it seemed to be unique to our species. Several, sometimes mutually contradictory, hypotheses have been put forward to explain it, many centring on its effects on certainty of paternity. For example, it has been argued that concealed ovulation increases paternal certainty, and thus the likelihood of paternal investment, by encouraging a particular male to mate with the female throughout the ovulatory cycle (Alexander & Noonan 1979). Conversely, if a number of males mate with a female over the cycle, it might lead to uncertainty of paternity, but reduce the subsequent risk of infanticide as a result (because males cannot be sure an offspring is not theirs and so help care for it [Hrdy 1981]).

Evidence in support of these various hypotheses has been equivocal (see e.g. Barrett *et al.* 2002). One reason for this may be that the starting assumption behind them is in fact wrong. More recent phylogenetic analyses suggest that the situation in humans may be closer to the ancestral state for anthropoid primates (great apes and humans), and not a recently evolved quirk at all (Sillén-Tullberg & Møller 1993; Dixson 1998; Nunn 1999). Unlike other mammals, in which reproductive behaviour often tracks hormonal changes within the oestrous cycle, anthropoid primates have **menstrual cycles**, in which reproductive behaviour reflects underlying hormonal events less closely. The continuous mating activity of humans is thus arguably only a slight elaboration of this hormonally more emancipated pattern of reproduction (Martin 1990; Barrett *et al.* 2002).

These arguments apart, however, there is in fact some evidence that sexual behaviour is influenced by the timing of ovulation. Careful studies of interactions between men and women have shown that men are more likely to touch and be more attentive to women at the time the latter are ovulating (Grammer 1996, cited in Buss 1999). But these effects are confounded by changes in the behaviour of the women themselves. Analyses have revealed that women increase the amount of sexual signalling around the time of ovulation, for instance by wearing tighter, shorter skirts and showing more skin. This fits with the evidence that soft-tissue asymmetry decreases at this point in the ovulatory cycle, perhaps increasing the attractiveness of the face (see above; Manning *et al.* 1996; Scutt & Manning 1996). The suggestion that heightened interaction at the time of ovulation is driven by female solicitation is supported by studies of reported 'sexual desire' by women in relation to their time of cycle (Stanislaw & Rice 1988; Fig. 12.8), and an increased

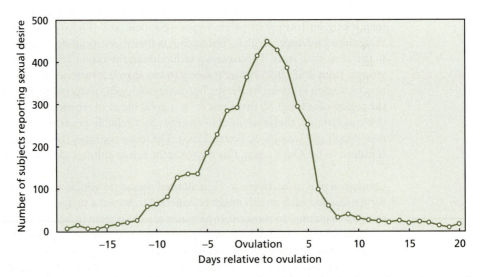

Figure 12.8 Reported sexual desire in women as a function of stage in the ovulatory (menstrual) cycle. Desire peaks strongly around the time of ovulation. After Stanislaw & Rice (1988).

Table 12.2 Female mating preferences and their functional significance. Modified from Grafen (1984).

Preferred trait	Functional significance *Selecting a partner who:*
Good financial prospects High social status Maturity Ambitiousness/industriousness Size/strength/athletic ability	Is able to invest in partner and offspring
Dependability/stability Love/commitment Positive interactions with children	Is willing to invest in partner and offspring
Size Strength Bravery Athletic ability	Is able to protect partner and offspring
Dependability Emotional stability Kindness Positive interactions with children	Will have good parenting skills
Similar values to self Similar age to self Similar personality to self	Is compatible with self

tendency for women in long-term partnerships to seek or fantasise about extra-pair copulations at this time (Baker & Bellis 1995; Gangestad *et al.* 2002).

Attractiveness in men Much attention has focused on what makes women attractive to men, but what about the other way round? We have already argued that prospective parental quality in a male is likely to be a key criterion for women, and seen some evidence that this might be a basis for preference (Fig. 12.6). But 'parental quality' is complex and multifaceted, both in how it might contribute to the reproductive success of offspring, and how it can be assessed in advance (Table 12.2). So what, specifically, do women go for?

   (i) **Provisioning capacity.** Various things might indicate a man's ability to provide for partner and offspring. Emphasis on financial wealth in present-day industrial societies (e.g. Fig. 12.6) arguably has its origins in the acquisition, defence and monopolisation of material resources in ancestral territorial and tool-using contexts (Buss 1999). Although men vary enormously in wealth or material resources, and their willingness to invest them in a given mate and her progeny, women have generally stood a good chance of getting more for their children from a single long-term partner than from several short-term ones (Buss 1999). This is because, in contrast to other primates, where males do not share resources with their mates, men provide food, shelter and protection, and tutor offspring in hunting, fighting, sport, social influence and other life skills (Smuts 1995; Buss 1999).

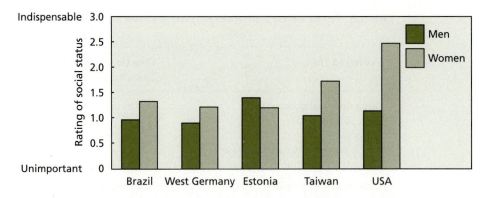

Figure 12.9 In most cultures, women rate social status in men more highly than men rate status in women. See text. After Buss (1999) from Buss, D.M. and Schmidt, D.P. (1993) Sexual strategies theory: an evolutionary perspective on human mating, *Psychological Review*, 100, 204–32. Copyright © 1993 by the American Psychological Association. Adapted with permission.

Fathers may also pass on social status to their offspring. In traditional hunter-gatherer societies, often deemed to reflect the ancestral social environment shaping our behaviour, men usually have clearly defined social hierarchies, with individuals at the top enjoying considerable power and prestige. Resources are readily available to these privileged few, but dwindle rapidly as the hierarchy is descended. Status-dependent resource acquisition and the possibility of their sons inheriting their father's status mean that social status is likely to become a criterion in female mate choice in itself. In keeping with this, women across many different cultures show a preference for men of high social status, whether it is measured as political or professional standing, or totemic tribal status (e.g. Buss 1999; Fig. 12.9).

The use by women of status, wealth and other cues relating to prowess and perform-ance in men is likely to motivate men to do well in these respects. Although people use a variety of means to get ahead in the world, including deception, sexual favours, social networking and education, the best predictor of achievement (boringly enough) turns out to be sheer hard graft (Kyl-Heku & Buss 1996). In line with this (but perhaps a little surprisingly, given the familiar adage 'all work and no play makes Jack a dull boy'), women seem to rate ambitiousness and hard work very highly in a potential partner (Buss 1989). Not only that, they are very likely to discontinue a long-term partnership if their partner becomes lazy and loses his career aspirations, and certainly if he loses his job (Betzig 1989).

(ii) **Age.** Power and privilege among men are often functions of age, just as they are in males of some other primate species. This can lead to a veritable gerontocracy (government by the old) in some societies. In the island-dwelling Tiwi of northern Australia, for example, power is vested almost entirely in a ruling group of old men who are able to control the mating system of the tribe through a network of alliances (Hart & Pilling 1960). The relationship between age and status can be based on a variety of things. In western industrialised societies it is often income levels and accumulated wealth, both of which tend to be greater in older men. Alternatively it may be holding important positions of responsibility or authority, though this is often confounded with income. In more traditional societies, it often reflects increasing physical strength and/or experience and skill. Whatever it is based on, however, it is clear that women are inclined choose older men (Fig. 12.10a). Thus there is a divergence in age preference between the sexes, with men preferring younger mates, and women preferring older ones. At least in western societies,

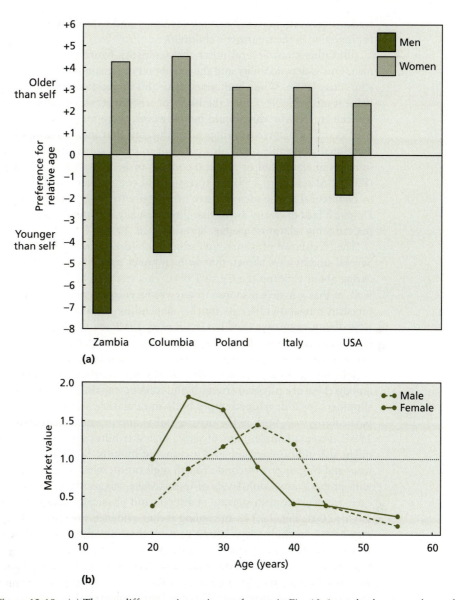

Figure 12.10 (a) The sex differences in mating preference in Fig. 12.6 are also borne out by preferences for relative age in partners: men prefer women younger than themselves, while women prefer men that are older. After Buss, D.M. and Schmidt, D.P. (1993) Sexual strategies theory: an evolutionary perspective on human mating, *Psychological Review*, 100, 204–32. Copyright © 1993 by the American Psychological Association. Adapted with permission. (b) The preferences in (a) are reflected in the relative 'market value' of men and women at different ages (deduced from criteria requested and advertised in personal advertisements – see text). After Pawlowski & Dunbar (1999).

this is reflected in estimates of the 'market value' of men and women of different ages (calculated as the number of individuals requesting a particular age group of partner in Personal Ads divided by the number in that group advertising their availability [Pawlowski & Dunbar 1999]). As Fig. 12.10(b) shows, men achieve their highest market value in their late thirties, whereas women are more marketable in their mid twenties. That male market value peaks in the thirties rather than continuing to increase, probably reflects a tradeoff between increased earning capability and reduced future longevity (i.e. potential for investing in their partner's children).

(iii) **Other cues.** Several other characteristics figure regularly in women's ratings of men. One is dependability and the likelihood of remaining in a stable long-term relationship (Buss 1999). While both sexes value this in questionnaire surveys, women generally value it more highly. Given the likely benefits in terms of future provisioning, it is not difficult to see why this should be. However, there may be compelling reasons besides predictability of provisioning, most notably that men coming over as unreliable or emotionally unstable often turn out later to inflict substantial costs on women, perhaps in terms of emotional or physical conflict, or being unreasonably dependent or jealous (Buss & Shackelford 1997). Expressions of commitment and long-term planning seem to be integral to the idea of 'love', the desire to bond with one particular individual. Despite a lack of clarity as to what 'love' actually is, women rate it more highly than men in evaluating relationships (e.g. Sprecher *et al.* 1994).

The likelihood of a man investing in children can also be assessed more directly. Several studies have shown that women prefer men who are seen to be attentive to and caring about children (La Cerra 1994). This seems to be true in some other primates as well, in that a male's response to any young encountered may lead to sexual reward or physical punishment by its mother depending on whether its response is affiliative (friendly) or aggressive (Keddy-Hector *et al.* 1989; see 11.2.1.1). Interestingly, in humans, it is specifically a man's response to children that matters, not indications of domesticity *per se*. Pictures of men doing housework, for instance, consistently rate poorly.

Like men, women use a number of physical attributes in judging mates. Prominent among these are physical strength and athletic capability. Men who are tall with wide shoulders, well-developed upper body musculature and relatively small buttocks fare particularly well, both in the mating stakes and in intrasexual competition (e.g. Barber 1995). From a female point of view, such attributes suggest an ability to protect, but might also indicate strength and stamina for hunting. Studies of the ratio in length of the fore- and ring fingers (second : fourth digit ratio), which appears to correlate negatively with prenatal and adult levels of testosterone, suggest this may have a basis in greater early exposure to testosterone in athletic and physically competitive men (Manning & Taylor 2001; Fig. 12.11). Intriguing recent evidence suggests that vocal characteristics may have something to say about some of these attributes, but while women show consistent preferences for particular kinds of male voices, their assumptions about associated physical characteristics of the males concerned tend to be wide of the mark, suggesting any correlations may be complex (Collins 2000). Finally, as we saw earlier, physical asymmetry is also important. Women find symmetrical faces more attractive than less symmetrical ones, though it is not clear that this is due to symmetry *per se* (see above). Nevertheless, symmetry, or some correlate of it, may have a role to play in screening for physical and mental health. Once again, this has obvious implications for the protection and provisioning a man is likely to be able to offer.

Health factors may underlie recent findings that women prefer certain odour characteristics in men. Claus Wedekind and coworkers at the University of Bern in Switzerland have shown that women prefer men with MHC (major histocompatibility complex; see 10.1.3.2) genotypes that differ from their own (Wedekind *et al.* 1995; Wedekind & Furi 1997). As in other species (see 10.1.3.2), differences in MHC genotype are reflected in various body odours, and women appear to be sensitive to differences in such odours in men. Since the MHC is involved in important aspects of immune response, preference for an MHC genotype differing from self may enhance immune capability of offspring by increasing heterozygosity at critical loci. Indeed, careful comparisons strongly suggest

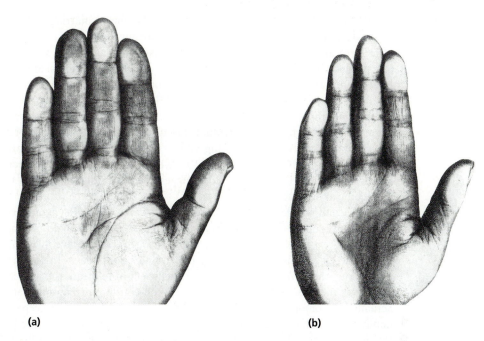

(a)   **(b)**

Figure 12.11 Photocopied right hands showing (a) an extreme masculinised second : fourth digit ratio (forefinger much shorter than ring finger) in a male, and (b) a slightly masculinised ratio (forefinger and ring finger very similar in length) in a female. See text. Photocopies courtesy of John Manning.

that choice does increase MHC heterozygosity rather than, say, biasing outcomes towards genotypes that are complementary to that of the female (Wedekind & Furi 1997; see 10.1.3.2). Interestingly, preferences for outbreeding with respect to MHC genotype are not evident in women regularly taking a contraceptive pill (Wedekind *et al.* 1995), suggesting that MHC-related odour judgements depend on underlying reproductive state. That odour preferences and the MHC are linked has been intriguingly confirmed in tests with commercial perfumes. Perfumes have been used for thousands of years, and are closely linked with sexual attraction. Tests with women of known MHC genotype showed a remarkable consistency in perfume preferences within genotypes, but no consistency in what women would prefer on a putative partner (Milinski & Wedekind 2001). Thus perfumes seem to be selected by women to somehow amplify odour cues that betray their individual immunogenetics.

Incest taboos   The MHC-driven outbreeding reported by Wedekind and coworkers leads to a broader consideration of outbreeding behaviour in humans. Many societies have culturally imposed incest taboos, sometimes carrying severe penalties if they are broken. At first sight this might seem to fit neatly with the MHC results above, since it would provide a nice synergy between cultural imperative and biological pragmatism. But two problems undermine this comfortable position.

First, it is clear that the widespread reluctance to mate with immediate family members in human populations is influenced by early social experience and thus familiarity – the so-called **Westermarck effect** (Westermarck 1891). A striking example comes from Shepher's (1971) study of Israeli *kibbutznik* (people who live in the communal communities known as *kibbutzim*). Unrelated children brought up together in the same *kibbutz* later show a marked reluctance to marry each other, preferring instead to marry partners from outside. Where the issue is forced, as in the 'minor marriage' tradition in Taiwan, where daughters are given to the family of the arranged eventual marriage partner shortly after birth and reared with their future spouse, marriages have a high failure rate and low productivity (Wolf & Huang 1980). The converse also holds. Siblings separated for long periods are much more likely to enter into incestuous matings if they meet later on than those that have been reared together (Bevc & Silverman 2000). From a biological perspective, the Westermarck effect probably reflects a developmental rule of thumb ('don't mate with familiar individuals because they're probably close relatives') that is driven by the kinds of outbreeding advantages/inbreeding disadvantages discussed in relation to the MHC and Table 10.1. However, the fact that more specific genetic cues are also used in outbreeding decisions suggests that considerations other than generalised inbreeding avoidance are at work.

The second problem with incest taboos reflecting inbreeding avoidance is that, rather like cultural concepts of 'kinship' generally (see below), they do not map on to blood relatedness very neatly. For example, many marriage systems have rules governing partnerships between cousins. But these often allow some kinds of between-cousin marriages while precluding others with exactly the same coefficient of relatedness (van den Berghe 1980). Moreover the effect of allowing some between-cousin marriages is that inbreeding can actually be promoted, often encouraged by cultural incentives for keeping wealth within family groups. That incest taboos are commoner, and more strictly enforced, within highly stratified societies also fits with the view that they are culturally, rather than genetically, driven, in this case by ruling elites attempting to prevent the concentration of wealth and power within rival family lineages (Barrett *et al.* 2002).

The overall message, therefore, is that, even with something such as incest avoidance, where the reproductive advantages would seem to be self-evident, care is needed before concluding that cultural traits are driven by naturally selected benefits.

Sex and memes The sex differences in mating preference above are as we might expect in an anisogamous species, and are reminiscent of the differences in other species we discussed in Chapter 10. However, we also saw in Chapter 10 that sexual selection is a powerful force for exaggeration and diversity in the qualities used to choose mates. Females may well choose males on the basis of social status, apparent parental qualities or 'good' genes, but what exactly indicates these things in any particular case is as variable as species themselves. This is because the mechanisms of sexual selection, such as the runaway effect and the handicap principle, can generate a huge variety of outcomes depending on starting conditions, conflicting selection pressures and many other factors. In humans, we can add to this list the most powerful agent of diversity of all: culture!

In a series of papers, the evolutionary psychologist Geoff Miller has linked the extraordinary evolutionary potential of memetics (see 12.1.1.2) to sexual selection (e.g. Miller 1997, 1998, 1999). Miller's thesis is that much (if not all) of the impetus behind cultural traits (memes) such as clothing fashions, story-telling and espousing particular beliefs and philosophies is sexually selected. The basis for this is that such traits are costly (in time, energy and/or lost opportunity to do other things), mostly have no survival value, are self-expressive with marked individual differences in quality, reflect creativity, intelligence and sometimes health, and play to the perceptual and cognitive preferences of onlookers – all classic qualities of sexually selected signals (Miller 1999; see Chapters 10 and 11). The elaborate, yet seemingly arbitrary, nature of human creativity thus becomes a cultural equivalent of the peacock's tail. In its wake came the development of the necessary language, cognitive and perceptual capabilities to transmit

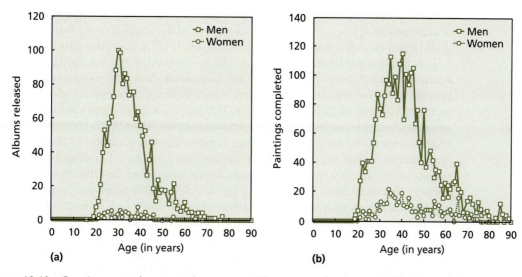

Figure 12.12 Creative output by men and women at different ages – here measured in terms of the number of (a) jazz records and (b) modern paintings produced. Output is generally higher among men, and dramatically so when they are around 30–40 years of age. See text. After Miller (1999).

and process the traits. Thus memetic sexual selection may well have provided the explosive power behind the proliferation of memes (indeed, the nature of memes themselves) and the dramatic evolution of their supporting mechanisms (brain size and specialisation) discussed in 12.1.1.2.

If Miller is right, and cultural creativity is all about courtship, then we might expect strong sexual dimorphism in creativity, with males usually displaying more of it. Given the age distributions of sexual 'market value' in Fig. 12.10(b), we might also expect the peak of male creativity to occur around the third decade of life. Using published reference works, such as museum catalogues, music discographies and writers' directories, as source material, Miller analysed the relationship between age and creative output in men and women for a variety of different types of cultural work. The results for just two of them, the numbers of jazz albums released and modern paintings produced, are shown in Fig. 12.12. In both cases, it is clear that vastly more output emanated from men, and, as predicted, their output peaked around 30–40 years of age. Very similar results emerged for older paintings, rock and classical music, books, philosophical tracts and plays (Miller 1999). These trends accord well with the suggestion that they reflect cultural forms of mate attraction by men. It is also interesting that murder rates among males peak just a bit earlier, in the mid to late twenties (Daly & Wilson 1988). This suggests a degree of coincidence between intra- and intersexual competition, but that the latter may require more time for income, status and other differentials to develop sufficiently to be reliable in mate choice. Not all sexually selected cultural traits may be as (apparently) arbitrary as artistic or academic achievement, however. Miller (2000) also suggests that altruism may have its roots in sexual selection, as males exploited its attractive characteristics to interest mates. Various lines of evidence lend some support to the idea. For example, studies of alms-giving to beggars in the street have shown that single men are more likely to give than those accompanied by a woman (Goldberg 1995), and accompanied men at an early stage in their pair bond were more likely to give than those in more established partnerships (Mulcahy 1999).

Sexual conflict Just as humans conform to general predictions about mating preferences in anisogamous species, so they also display some of the conflicts of interest between the sexes typical of such species (see 5.3.1 and Chapter 10). This can take various forms.

Extra-pair copulations and jealousy: As in many other species, the social mating system (see 10.1.2.2) in human societies may disguise the true pattern of mating behaviour between males and females. Both partners in an apparently monogamous relationship, for example, may seek extra-pair copulations. That this is common enough to have exerted a selection pressure on mating strategies is evidenced by the relatively (to body size) large size of human testes and the tendency for men to adjust the sperm content of their ejaculates in relation to the perceived risk of sperm competition from a rival (see Fig. 10.9b). It also seems that women seeking extra-pair copulations do so when the chances of conceiving are high (Baker & Bellis 1995); so at a reproductive, if not necessarily cognitive, level it is clandestine sex with intent. The reasons why men might seek extra-pair matings are straightforward and relate to the fact that their lifetime reproductive output is limited by opportunities to mate rather than parental effort (see 10.1.2). In women, it may reflect attempts to have their cake and eat it, in this case by entering a long-term relationship with a reliable, fatherly male, but duping him into raising offspring from a highly attractive (but paternally unreliable) male. In keeping with this, women's preferences for faces showing strongly masculine characters (i.e. those associated with high testosterone levels, such as a strong jaw line and high forehead) peak around the time of ovulation. At other times women prefer men with more feminine facial features (suggestive of caring and nurturing) (Penton-Voak & Perrett 2000; Johnston *et al.* 2001). While women may go for 'hunks' in their extra-pair matings, there is also evidence that they use extra-pair relationships to gain additional resources, or to judge the relative qualities of other available males so they can switch partners if they discover someone better (Buss 1999).

An obvious upshot of all this potential infidelity is **jealousy**. Sexual jealousy, and its sometimes extreme behavioural consequences, can be seen as a psychological mechanism encouraging mate guarding (10.1.3.1), for instance by increasing vigilance for mating opportunities by partners, or preventing a partner from contacting other members of the opposite sex. But, given the different criteria by which men and women seem to be driven in their sexual behaviour, we might expect jealousy to take different forms in the two sexes. This turns out to be the case. Across a wide range of cultures, female sexual jealousy tends to focus on the potential loss of a provider and accompanying emotional commitment, while male jealousy focuses on uncertainty of paternity and the risk of investing in another male's offspring (Daly *et al.* 1982; Buss *et al.* 1992; Fig. 12.13).

Sexual coercion: Mating usually seems to be achieved with a measure of agreement between male and female, albeit after a sometimes long and fraught preliminary interaction. Occasionally, however, it is forced on an unwilling partner. Forced copulation occurs in several species (e.g. Berry & Shine 1980; Thornhill 1980; Le Boeuf & Mesnick 1990), and, for obvious physical reasons, is almost always perpetrated by males. However, sexual aggression by females can sometimes coerce copulations from males, and is certainly known to do so in our own species.

Rape by men often carries severe penalties. Despite this, it occurs quite frequently in all cultures that have so far been studied (Thornhill & Palmer 2000). Understandably, the reasons why men sometimes resort to rape have generated a good deal of debate. Feminists have viewed it as an act of intimidation against women in order to maintain the social dominance of men (Brownmiller 1975), and frown on attempts to 'dignify' it with an adaptive evolutionary explanation (Brownmiller & Merhof 1992). But, from the evidence, it is difficult to conclude that rape reflects anything other than an adaptive,

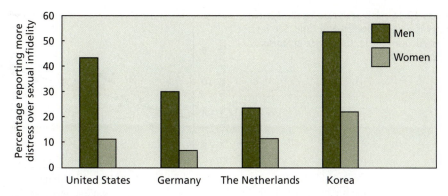

Figure 12.13 Men from different cultures tend to report greater distress over sexual infidelity by their partners than women. After Buunk *et al.* (1996).

Figure 12.14 Age profiles for the female population of North American cities, female victims of murder and female victims of rape. See text. After Thornhill & Thornhill (1983).

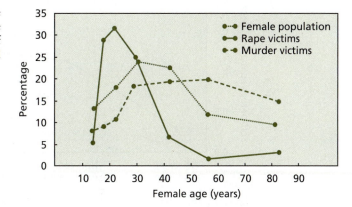

if extreme and distressing, reproductive response (Alcock 2001). This does not in any way lessen its offensiveness in a social context; it simply explains why it happens (the argument that explanation somehow provides justification crassly misunderstands both processes). But what *is* the evidence that rape is adaptive?

If rape is part of an adaptive reproductive strategy, some fairly obvious predictions follow. To start with, rape should be targeted at fertile young women who, at least sometimes, become pregnant as a result. Analysis of the age distribution of rape victims (Fig. 12.14) produces a remarkable match to the distribution of reproductive 'market value' of women in Fig. 12.10(b). The distribution is very different from those for the female population as a whole, and for female murder victims (Fig. 12.14), implying that the motivation for rape is not simply aggressive. Furthermore, even in societies where the use of birth control pills is widespread, rape can result in pregnancy. Taken together, these trends strongly suggest that rape is motivated by reproduction and can be successful in that regard. But why do some men turn to rape instead of competing for matings in the usual way?

Intuitively the most likely explanation is that rape is a conditional tactic (see 2.4.4.2) in response to a failure to gain access to mates through the more usual channels – the **mate deprivation hypothesis** (Thornhill *et al.* 1986; Lalumiere & Quinsey 1996). There is some evidence to support this view. For example, in societies where divorce rates are high, serial marriages by sexually successful men may make a large proportion of the desirable

Figure 12.15 Divorce rate in the United States is a good predictor of rape. See text. After Alcock (2001), from data in Starks & Blackie (2000).

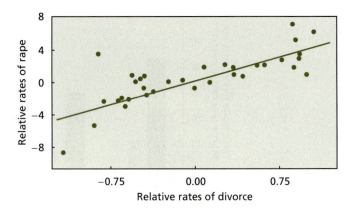

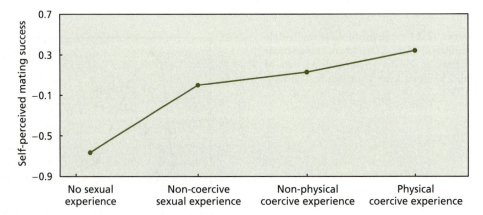

Figure 12.16 Men who rate their own mating success highly are more likely to indulge in coercive sex. See text. After Lalumiere *et al.* (1996).

female population unavailable to others, so increasing the variance in male mating success and creating an incentive for losers to rape. When Starks & Blackie (2000) looked at the relationship between divorce rate and the incidence of rape in the USA, they indeed found a strong positive correlation (Fig. 12.15). However, things are clearly not as simple as that. A troublesome finding for the mate deprivation hypothesis is that men who rate their own mating success very highly are *more*, rather than less, likely to rape (Lalumiere *et al.* 1996; Fig. 12.16). A possible explanation here is that such highly competitive men indulge in many brief, impersonal matings in which coercion may be more likely (Malamuth 1996). Moreover, precisely because of their high demand, it may be that these men rape in response to temporary shortfalls in the mate supply (so supporting the mate deprivation hypothesis in a shorter-term context) (Buss 1999).

12.2.1.2 Parental behaviour

Parental care is of paramount importance in humans. We typically invest an enormous amount of time and resources in nurturing our offspring to maturity and independent reproduction. In purely monetary terms, it has been estimated that rearing a child to age 17 in the United States or western Europe currently costs some $225 000 (€215 000), a vast sum in relation to the earning capacity of many people. Important or not, however, we have only to pick up a newspaper to be reminded that parental care is not a universal given in human societies, and that it can be fraught with conflict, often with distressing outcomes. Can evolutionary theory help us understand this?

Parental certainty and parental care

We saw in Chapter 10 that conflict between parents and offspring is a predictable consequence of the drain on parental resources by offspring. One factor in the equation was the relatedness between parents and offspring, since this determines the reproductive payoff for all that investment. In the majority of species, the two sexes are likely to differ in their certainty of parentage. A female giving birth to her young is in no doubt that they are hers, but a male partner looking on may not be quite so sure (10.1.3.1, 10.2). This applies in humans just as it does in many other species (though it is interesting that mothers nevertheless have mechanisms for discriminating the odours of their newborn babies after periods out of contact with them [Porter *et al.* 1983; Porter 1998]). On balance, therefore, we might expect male partners to be more circumspect than mothers in investing resources in the latters' offspring.

Relatedness and parental care   One scenario in which we might expect a clear difference in care between partners is where one is a stepparent and thus wholly unrelated to the offspring. Questionnaire surveys certainly bear this out, with only around half of all stepfathers interviewed and a quarter of all stepmothers claiming to have any 'parental feelings' towards their stepchildren (Duberman 1975). The expectation is further supported by studies of interactions between stepfathers and their stepchildren, which show that these tend to be less frequent and more aggressive than interactions involving genetic fathers (Flinn 1988). Stepparents are also less likely to invest material resources, such as college fees, in stepchildren (Buss 1999). The tendency towards a less caring attitude among stepparents sometimes becomes extreme and spills over into aggressive child abuse and even murder, as a catalogue of extensively reported court cases testifies (Lenington 1981; Daly & Wilson 1988). The risk of a pre-school-age child being killed by its adult carers in North America, for example, is 40–100 times greater for stepchildren than for children living with both genetic parents (Fig. 12.17). The ubiquitous 'wicked' stepparent of childhood fairy stories thus appears to have some basis in fact.

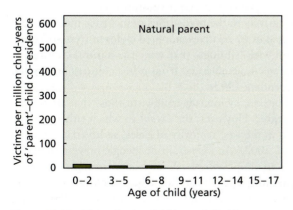 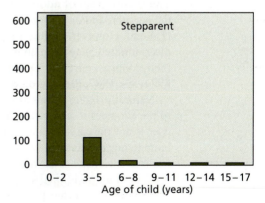

Figure 12.17   The risk of a child being killed by a genetic parent or a stepparent as a function of its age. See text. After Daly & Wilson (1988).

The risk of stepparental discrimination may lead to selection on mothers to convince fathers that offspring are indeed theirs. One way to do this might be to point out physical resemblances between babies and fathers early in the bonding process. Daly & Wilson (1982) studied this by video-recording births and the remarks made by parents at the time. All other things being equal, we should expect comments that the baby resembles the father to be just as frequent as comments that it resembles the mother. In fact, Daly & Wilson revealed, mothers were four times more likely to point out resemblances to the father. This was also the case later on when relatives of couples were asked to say which parent a child resembled (Daly & Wilson 1982; see also Regalski & Gaulin 1993). Intriguingly, a later study in which subjects were asked to compare photographs of 1-year-old children with those of three male and three female adults, one of which in each set was the child's father or mother (unknown to the subjects), found that children were matched to fathers, but not mothers, more often than expected by chance (Christenfeld & Hill 1995). If these results are repeatable, they might suggest young children benefit from resembling their fathers by reducing the risk of discrimination or rejection, perhaps through suppression of maternal genes during development (see 5.3.2). The importance of relatedness between parents and offspring for investment and avoidance of abuse may also account for the apparent ability of newborn babies to discriminate genetic parental cues at an early stage. Extensive experiments with facial images, vocal cues and odours have shown that babies as young as a day can discriminate in favour of maternal cues (e.g. Murry *et al.* 1975; Bushnell 1982; Porter 1998).

Of course, we are talking statistical tendencies in comparisons between step- and genetic parents. Many stepparents care wholeheartedly for their charges, supporting them as much as any genetic parents. And, surely, one might ask, if a lack of relatedness spelled lovelessness and neglect, why is there a widespread tendency for people to adopt unrelated children? **Adoption**, like celibacy, seems to be a cultural trait that defies explanation in traditional neoDarwinian terms, requiring as it does investment without prospect of reproductive benefit. But is adoption really the undiluted altruism it appears? Careful studies of adoption data suggest we should be cautious about this. Silk (1980), for example, found that adoption in Polynesian societies, where up to a quarter of children are raised in other families, occurred most often in infertile or post-reproductive couples. It also tended to take place between individuals that were at least distantly related to each other (e.g. adoptees were nieces or nephews of their adopters). Thus many cases of adoption did not involve lost opportunities for reproduction by the adopter, and/or conformed qualitatively to the predictions of inclusive fitness theory (2.4.4.5). However, despite being taken in by relatives, adopted children were still often discriminated against in comparison with 'siblings' that were the natural children of their adoptive parents. Similar findings have emerged from other cultures (e.g. Stack 1974; Silk 1987; Pennington & Harpending 1993).

Naturally, by no means all adoptions are by infertile couples or those that are related to the adoptees to any measurable degree. However, the *extent* to which either of these is or is not the case remains untested in the vast majority of cases, so it is dangerous to jump to any conclusions (Barrett *et al.* 2002). As a general rule, people in western industrialised societies are far more likely to adopt completely unrelated strangers than are those in more traditional societies, and the latter in turn are a quantum leap away from what happens in other primate societies, where adoption, even of orphaned young, is very rare (Silk 1990). One possibility is that our readiness to adopt reflects the high value that societies place on parenthood and children, and behaving parentally is a powerful meme that copies readily (12.1.1.2). Like altruism more generally (see above), adoption may thus be driven both by a biological incentive (to reproduce) and a cultural one (to be seen to be parental).

Reproductive value of offspring

Investment of parental resources might be expected to reflect the changing reproductive value of offspring as they develop. In traditional societies, reproductive value rises to a peak from birth to reproductive age (teens) due to early infant mortality (Fig. 12.7), though this part of the curve is flattened in western industrialised societies because of their dramatic reduction in infant mortality. Historically, therefore, and still in many developing countries, young infants have relatively low reproductive value. Perhaps not surprisingly then, it is these individuals that are most at risk from parental abuse and murder. That this trend is not due simply to older children being able to defend themselves is shown by studies of murder by non-parental relatives, which peaks during the teenage years of their victims, not early on (Daly & Wilson 1988).

Underlying this general scenario are some sex and birth order (first born, second born, etc.) effects, with male and last born infants being more vulnerable to abuse, probably because their resource demands are greater at times of stress for the parents (see 10.1.2.1; Lenington 1981). Sickness and general physical condition are other factors in the equation. Children with congenital defects, such as spina bifida or Down's syndrome, are generally of low reproductive value, and their rate of abandonment is correspondingly high. Of those that are institutionalised, some 22% receive only one parental visit a year or none at all (Buss 1999). The incidence of neglect among non-institutionalised cases is also high, ranging between 7% and 60% of individuals, compared with a base rate of 1.5% for children that are not congenitally afflicted (Daly & Wilson 1981; Buss 1999). Other health problems can spell trouble too. Parents faced with offspring that differ in general sickliness might be expected to invest more in the healthier ones, since they are probably a safer bet in terms of future reproductive success. A study of mothers' responses to twins of independently assessed health status showed that, even early on (four months of age), sick infants were likely to receive less attention than their healthier twin (50% of mothers directed more positive attention towards the healthy child). By the time twins were eight months old, *all* mothers biased their responses in favour of the healthier child (Mann 1992).

Reproductive value also appears to be reflected in the grief we feel for dead or seriously afflicted loved ones (Crawford *et al.* 1989). The devastating impact of the loss of a child is recognised everywhere, but the level of grief over an aged parent or grandparent is usually much lower (Sanders 1980). Lower too, very often, is the sense of loss when a young but chronically sick family member dies. Interestingly, the level of grief reported in any given scenario depends on the age (and thus reproductive value) of the victim and not on that of the reporting subject (Crawford *et al.* 1989).

Parent–offspring conflict and sibling rivalry Relative reproductive value underlies the evolutionary conflict between parents and offspring, and between offspring, that we discussed in 10.2.1.1. Humans are no exception to this. Conflict between mother and offspring begins before birth as the developing foetus attempts to secure more provisioning than the mother may be prepared to give (see Forbes 2002, also 5.3.2), and continues later in life in aggressive conflict with one or both parents. Heightened parent–offspring conflict would be expected around the time of puberty and early adulthood, as newly

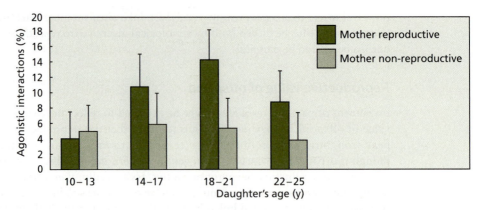

Figure 12.18 Mother–daughter conflict is more likely if the mother is still reproductively active, and when the daughter is of an age to reproduce independently instead of staying to help. See text. After Flinn (1989).

reproductively capable offspring attempt to maintain access to parental resources, are seen as reproductive competitors, or conflict with a parent over leaving home versus remaining to help rear other siblings. This is borne out by within-family murder statistics, which reveal a strong bias towards father–son conflict at this stage (Daly & Wilson 1990), and data on aggressive relationships between mothers and daughters (Flinn 1989). In mother–daughter relationships, conflict is especially likely where daughters are capable of leaving home to reproduce independently but are still living with a reproductively active mother (Fig. 12.18). Under these circumstances the inclusive fitness of the mother may be best served by retaining the daughter as a helper, while that of the daughter is best served by leaving. Thus aggressive conflict peaks when daughters are around 18–21 years of age (Fig. 12.18). If mothers successfully retain daughters, they stand to benefit not only through having 'more pairs of hands' to do things, but also because this reduces stress on the mother, which can otherwise depress her fertility (Jasienska & Ellison 1998). Sibling rivalry also follows predictable patterns. Jealousies over perceived parental favouritism are familiar at almost every stage of postnatal development among siblings, while parents encourage their children to value each other more highly than they are valued by unrelated peers, punish conflict and selfishness among brothers and sisters, and reward cooperation (Daly & Wilson 1988; Buss 1999).

The 'demographic transition'

One further aspect of parental behaviour is worth highlighting, and that is the somewhat puzzling phenomenon known as the **demographic transition**. The demographic transition is a profound cultural change in life history strategy in which both mortality *and* fertility have decreased to the point where the number of children produced per family is at or below that required to replace the parents (Coale & Treadway 1986; Mace 2000; Barrett *et al.* 2002). The change has come about recently, beginning about two centuries ago in Europe and North America, rather later in Asia and South America, and later still in Africa (Cleland 1995). This decline in reproductive output when resources are plentiful and mortality low again seems to fly in the face of an adaptive view of human behaviour.

While debate continues as to the cause of the transition, one widely held view is that it reflects an increased demand for investment in individual offspring. Evolutionary theory predicts that reproductive success will correlate positively with resource availability (of which wealth is an index in most human populations). This is broadly the case in the traditional populations studied by most anthropologists. In the larger, more heterogeneous, populations of developed countries, however, the opposite is true. Why should this be? Using a sophisticated optimisation model, Mace (1998) examined a range of life history and cultural factors, including mortality risk, environmental constraints on resource productivity and the cost of rearing children, that might affect optimal family size and are associated with demographic transitions. In particular, the model looked at the optimal amount of wealth allocated to a child at the end of the parents' reproductive lives (where optimal was defined as maximising the number of grandchildren). Of the variables entering the model, the only one predicting the drop in family size characteristic of a demographic transition was the cost of rearing children. As this rose, family size fell but more wealth passed to individual children from their parents. Thus, there appeared to be a direct tradeoff between maximising resources for some children and rearing fewer children overall. This outcome accords well with trends in populations where costs of child-rearing have increased measurably, as, for instance, in the wake of free medical care giving way to charging in many societies (Mace 2000). Social costs such as these have precipitated virtually immediate drops in population fertility (Hoem 1992; Conrad et al. 1996). Other modelling approaches have supported this general conclusion by focusing on the wealth investment necessary to ensure a child does not become trapped in a 'destitute underclass' (one consequence of the stratified social structure characteristic of many industrial populations showing the demographic transition) (Harpending & Rogers 1990). Another contributing factor may be the weakened, indeed often non-existent, kinship networks in many modern families, which results in the cost of child-rearing falling more heavily on unaided parents (Draper 1989; Turke 1989).

12.2.1.3 Cooperation and coalitions

Human beings are an intensely social species, a legacy of our social primate past. Our sociality, like that of other primates, stems from the advantages of group living for protection from predators and acquiring and defending limited resources (see Chapter 9). Unlike the aggregations of many other species, however, our sociality is highly structured, and characterised by extensive cooperation. We help each other at a local level via family and friends, and on a wider scale within community, national and even global social networks. However, all this cooperativeness comes at a price in that we often have to subordinate our own immediate interests to those of others in our social network. This can create a hefty incentive to cheat by accepting the benefits of a cooperative society but not paying the dues (e.g. reciprocating a kindly act, or paying taxes): the so-called **freeloader** or **freerider** problem (see Barrett et al. 2002).

One of the things that makes cheating more likely is an increase in the size of the group, with its consequences for reduced identity and cohesion within social networks. We saw in 12.1.1.2 that stable effective group sizes for social networking in humans peak at around 150 individuals. This appears to reflect a limit on the number of individuals with whom one can effectively exchange information and favours, and thus maintain friendships. Above this, some kind of higher-order structuring is necessary to ensure effective information flow and police relationships. In traditional societies this might take the form of occupational specialisation and dominance hierarchies; in an industrial business context it leads to various kinds of management system. Although modern industrial societies are socially diffuse, it seems people still hang on to their individual networks of roughly 150, but the networks are more widely dispersed and overlap much less with those of other individuals in the locality than in smaller societies. Thus people in large industrial communities have lost the old tribal village sense of everybody knowing everybody else. This, and the memetic potential for the spread of altruism outlined in 12.1.1.2, would seem to undermine the traditional roles of kinship and reciprocity as drivers for altruism within such populations. But what does the evidence suggest?

Kinship and cooperation

In fact there are several lines of evidence suggesting that kinship is important in cooperation within human societies, even large industrial ones. However, it is essential to point out first that 'kinship' has very different meanings when used by evolutionary biologists and anthropologists. To the former it implies blood relatedness, as it does when we have applied to other species in this book (2.4.4.5, 9.3). To the latter, kinship alliances may include blood kin, but they often include many individuals that are not related to each other. Moreover, the spectrum of individuals included within kinship alliances by anthropologists varies widely from society to society (e.g. Keesing 1975). While great play has been made of this by anthropologists hostile to adaptive evolutionary accounts of human behaviour (see particularly Sahlins 1976), careful analysis suggests the objection is misleading.

To begin with, there is good evidence that blood kin form the core of people's social networks. The ballpark figure of 150 above appears to represent one end of an hierarchy of network sizes each level of which is limited by practical and perhaps cognitive constraints (e.g. Dunbar & Spoors 1995). So-called 'sympathy groups', for instance, which reflect the cluster of people whose demise might cause an individual deep sadness, seem to hover around the 12 mark (Buys & Larsen 1979). In fact 12–15 people appears to be the general size range for networks of regular contact (e.g. Rands 1988; Dunbar & Spoors 1995). These smaller networks usually include both blood kin and unrelated individuals, but kin appear to provide the foundation, to which unrelated individuals may or may not be added, depending on the size of the network (Fig. 12.19).

Studies of traditional societies, such as those of the South American Yanomamö Indians, illustrate the importance of blood kinship quite nicely. Here, despite the obfuscation of blood kinship with cultural 'kinship' groupings, it is blood kinship that matters when the chips are down. When a village divides, for example, Yanomamö preferentially associate with close blood relatives rather than with less closely related or unrelated individuals

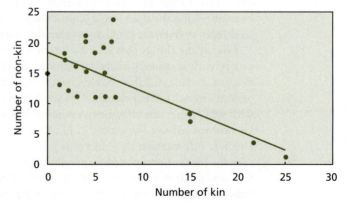

Figure 12.19 The number of blood kin versus non-kin in the social networks of individuals who were in regular (at least once a month) contact with ten or more people. The negative relationship suggests kin associations have greater weight than those with non-kin. See text. After Dunbar & Spoors (1995).

(Chagnon 1981). Close kin also tend to provide most of the help to combatants during fights, although other alliance relationships can play a role too (Chagnon & Bugos 1979). In some other South American tribes, such as the Ye'kwana, the cultural 'kinship' classification dictating the tribe's internal political relationships confounds different degrees of blood kinship, so that, for example, individuals with coefficients of relatedness of 0.5, 0.375, 0.25 and 0.125 are all called 'brother' or 'sister'. But despite their equivalent cultural 'relatedness', it is the degree of blood relatedness that predicts patterns of social interaction between them (Hames 1979). In fact, Hames found that the amount of social interaction *per se* increased in a steady linear fashion with blood relatedness.

These data reflect broad associations between kinship and cooperation. But kinship can influence interactions at much more refined levels. We have seen already that it is important in social relationships within families, where it influences parent–offspring relationships. But it also appears to affect cooperation between siblings (see also 10.2). A particularly intriguing study by Segal (1984) looked at competition and cooperation between pairs of monozygotic (derived from the same fertilised egg) and dizygotic (derived from different eggs) twins. The 47 pairs of twins used were aged between 6 and 11 years and of comparable intelligence (IQ). Each pair was asked to assemble a simple puzzle within three minutes by working together. Their behaviour during the enterprise was video-recorded and later scored by an experimenter who was unaware of the aim of the exercise. The results showed that 94% of the monozygotic twins completed the puzzle successfully, but only 46% of the dizygotic twins managed to do so. Analysis of the behaviour of two types of twin revealed a much greater tendency to cooperate and work for the benefit of the other individual among monozygotic twins. Segal's conclusion, not unreasonably, was that the greater degree of cooperation among monozygotic twins reflected the fact they were more closely related. While this may be the case, however, it could simply be that greater similarity in behavioural characteristics in monozygotic twins led to better coordination incidentally, not through any attempt to cooperate especially well with close kin (see Box 9.7). There is some evidence that greater genetic similarity between twins results in greater similarity in morphology and behaviour (and so could have such an incidental effect) but results vary across studies and the evidence is far from unequivocal (Wells 1987). The greater level of interaction and support evident between kin may account for a general association across cultures between living in a large (blood) kinship group and well-being and survivorship, particularly at times of stress (e.g. McCullough & York Barton 1991; Shavits *et al.* 1994).

Reciprocity and cooperation

Some examples of phenotypic altruism and cooperation, then, fit comfortably with kin selection theory. But many occur in situations where kinship is highly unlikely to play a role. The theoretical framework to which it is tempting to turn in these cases is reciprocal altruism (9.3.2). Superficially, many interactions in our lives seem to be based on reciprocal exchange of some kind, whether taking it in turns to buy each other lunch, exchanging goods for money or finding mutual support in friendships. But we have already noted that the extent and apparently indiscriminate nature of much human altruism seems beyond the explanatory power of reciprocity (12.1.1.2), so how convincing an account can it provide?

There are two central problems with reciprocity: variation in the currencies of reciprocation, and variation in the time taken to reciprocate – e.g. help with obtaining food now may be paid back as help with access to mates later, where 'later' may be anything from a few minutes to several days or months. This introduces uncertainty into the process, and thus ample opportunity for cheating. This in turn selects for caution in initiating helpful deeds so that they tend to be dispensed to trustworthy individuals, for instance by choosing partners that have already been seen, or have a reputation for, acting altruistically (the basis of so-called **indirect reciprocity** [e.g. Mohtashemi & Mui 2003]; see also 12.1.1.2). Forming trusted relationships in this way is the essence of **social contract theory** (Cosmides & Tooby 1992), where reciprocal exchange in variable currencies and over variable periods is made more likely through carefully selected partnerships. But if social contracts in the sense of Cosmides & Tooby are to explain cooperation on the scale seen in human societies, they must rely on some important assumptions. For example, people must be able to recognise and remember a large number of other individuals, otherwise they will be open to exploitation by cheats. They must also remember something of the history of interactions with different individuals if they are to keep track of who 'owes' what to whom. There must be some capacity for communicating one's needs to others and evaluating their needs in return. And there must be an ability to judge costs and benefits across a wide range of different currencies of exchange (food, sex, protection, money, information, etc.). Direct evidence exists for only some of these assumptions, so a robust test must await further studies. However, we saw in 4.2.1.6 that humans are particularly well tuned to recognise social cheating in different cognitive experimental contexts, so there is at least indirect evidence that our cognitive skills are geared to discriminating within the kind of social contract envisaged by Cosmides & Tooby. Whether this necessarily leads to the establishment of reciprocating networks, however, requires careful evaluation. It may also be that social contracts between particular, reciprocating individuals are not necessary. If altruism spreads memetically, because it is attractive to emulate (12.1.1.2), then 'reciprocal' payback may come from society at large as people other than the direct beneficiaries of altruism become more inclined to help altruists as a whole (Barrett *et al.* 2002).

Social contracts and friendship Social contracts and selective trusted partnerships would seem at first sight to be a good description of friendships. Indeed, our capacity for friendship is usually attributed to selective reciprocity. However, Tooby & Cosmides (1996) point to some problems with this assumption.

One of these is the so-called '**banker's paradox**'. Faced with more people wishing to borrow money than his bank can sensibly lend, a banker has to evaluate the chances of getting the money back should he lend to particular individuals. His best bet is to go for those that seem to have a good credit record. But precisely because they manage their money well, such people are likely to be the ones who need the current loan least. The people in dire straits will be the ones who need it most, but these are the very ones to whom the prudent banker should not lend. By analogy, the times at which we most need friends – when we are in deep trouble or seriously ill, say – are likely to be the times when we are least able to reciprocate. All other things being equal, we should expect selection to favour abandonment rather than assistance under these conditions. The widespread and familiar phenomenon of 'fair weather' friends may be a reflection of just such rational decision-making. One way out of the problem might be to cultivate an air of indispensability, for example by providing a service such as protection, technical assistance, medical care or mentoring, so that one's value is raised above that of other community members. Maintaining this position may require strategic decisions, such as avoiding groups where others possess your indispensable skills, or driving off newcomers that appear to possess them, while at the same time ensuring you are in a group where people are likely to benefit from, and so desire, what you have to offer (Tooby & Cosmides 1996). The ubiquitous tendency for large groupings – religious, political, philosophical, literary and so on – to split into smaller interest groups is one possible result of such manoeuvering (Buss 1999).

Another problem with friendships is that, as with many other things in life, their nature depends on sex. Same-sex friendships are likely to incur very different costs and benefits from those between the sexes. Intrasexual rivalry is a potential hazard in the former, while the latter can lead to opportunities for mating. In keeping with this, surveys show that men and women value different things in friends. Men are more likely than women to see sexual access as a benefit of intersexual friendships (and consequently report a greater lack of reciprocation in such relationships), while women are more likely to see benefits in terms of protection (Bleske & Buss 2000). However, both sexes report advantages from intersexual friendships in terms of gaining insights and information about the opposite sex that might later be put to good use with other partners (Bleske & Buss 2000). Both sexes also report problems with sexual rivalry in same-sex friendships, but men perceive more of a problem than women, a difference that fits with the expected greater motivation among men for opportunistic mating (Buss 1999).

In short, therefore, friendships are subject to a variety of pressures that bias the responses and perceptions of participants in different ways depending on the nature of the relationship, pressures that undermine simple interpretations in terms of reciprocity.

Social contracts and alliances Whatever their underlying motivation, personal friendships can be one form of **cooperative alliance** between individuals. But alliances extend well beyond local personal relationships, and are a familiar feature of human behaviour at community levels and beyond.

Paraphrasing Buss (1999), an alliance can broadly be defined as 'a group of people who identify with each other because they pursue common goals'. Not surprisingly, surveys reveal the important characteristics valued in fellow alliance partners to be those relating to industriousness, dependability, intelligence, a good sense of humour and an ability to motivate. These criteria are common to both sexes, but the sexes also show some differences in more specific requirements. Desirable qualities for men related more strongly to aggression and defence, such as bravery in the face of danger, physical strength, toleration of pain and ability to dominate socially (Buss 1999). This underlines the important role of alliance formation in competing for, and defending, resources, an importance that is emphasised further by the fact that mutual dependence between alliance partners may actively be engineered. A good example comes from the Yanomamö indians (Chagnon 1983; Ridley 1996). The Yanomamö exist in a state of almost continual warfare between villages. The key to maintaining the upper hand is forming alliances with other villages. As a result, complex networks of intervillage alliances spring up that are carefully nurtured through reciprocal trade. But one problem here is that most villages have the potential to produce all their own material requirements, so do not have to depend on trade with other communities. It thus seems that dependence is created artificially to cement bonds between alliance partners (Ridley 1996). Chagnon (1983) gives the example of one village that relied on another for its clay pots, its people claiming that they could not make, or had forgotten how to make, their own pots. When the pot-providing alliance eventually came unstuck, however, the village quickly recovered its independent skills in the art.

Alliances and warfare Depending on alliances for warfare is not unique to the Yanomamö. Neither is it unique to humans, though it is almost so. The only other species to show intergroup warfare is our closest living relative, the chimpanzee, which similarly relies on alliances between males to carry out attacks on its own kind (Wrangham & Peterson 1996). In many ways, war is as much an act of cooperation as it is of aggression, and specifically cooperation between males. This does not necessarily mean men are more aggressive than women, however. Women can be just as aggressive, but tend to show it verbally rather than physically (Campbell 1995). Whether between local gangs, villages, tribes, political factions, countries or multinational coalitions, war is effected by strategic alliances, and at whatever level it is conducted, it is about acquiring resources, especially (though it is rarely explicitly recognised) reproductive resources in the form of matings. Gang warfare in North American cities, for example, reaps measurable reproductive benefits. Members of successful gangs mate with a significantly greater number of partners in a given period than culturally equivalent non-gang members, with gang leaders securing the largest number (Palmer & Tilley 1995). Participation in warfare also seems to increase access to mates in the Yanomamö. In Chagnon's (1988) study, men that had previously killed on raids of other villages had almost four times as many wives as those that had not. Even among women, triggers for aggression often relate to competition for males or accusations of promiscuity or infidelity (Campbell 1995). But what about warfare on the national and international scale? Can this be reduced to a scramble for mates?

Civil and international wars are, in almost every case, to do with alliances of one kind or another attempting to gain economic and political control of a region (Barrett *et al.* 2002). The proximate cause is often economic hardship, political oppression or other constraint on resource or opportunity. However, an almost invariable consequence is the sexual abuse of women in the opposing camp by invading or victorious forces. This is as true of the recent conflicts in Kosovo, Bosnia and Kuwait as it was of the Roman and Viking invasions of the past (see Ridley 1993b). Historically, and in many different cultures, captives taken in war have tended to be women rather than men, and the systematic rape of women victims continues to be a feature of conflicts even now. Indeed, the prospect of free-for-all sex is privately acknowledged as a significant motivator among some combatants in military campaigns (Ridley 1993b). But even if there is the possibility of a substantial sexual reward, the cost of going into battle is potentially catastrophic, so can it be worth it? Clearly, if an individual went into battle alone, with the certain knowledge that he would die, the answer would be no, and we should be surprised if a psychology predisposed to war had evolved. But in an alliance, the risk to any given individual is diluted by the number of companions (see 9.1.1.1) and is unknown. If the risk is shared and no one knows who, if anyone, will die, the balance of risk to possible benefit tips substantially in favour of cooperative warfare (Buss 1999). So alliances provide both the physical might to mount a plausible attack, and adjust the likely benefit : cost ratio of the outcome to tempt attack in the first place.

Other causes of altruism

Kinship and reciprocity theories may provide *some* answers to the problem of altruism, but they are each heavily qualified in terms of the circumstances to which they are appropriate. Memetics provides another route by which altruism might spread through populations. However, apparent altruism can have a number of other causes that do not rely on any of these explanations. Following Barrett *et al.* (2002), we shall briefly mention two.

Tolerated theft and producer–scrounger relationships   Apparent altruism in the form of sharing hard-won resources may come about not because of egalitarian motives on the part of the sharers, but because sharing is forced on them. In hunter-gatherer societies, for example, the spoils of successful hunts tend to flow from certain individuals (good hunters) to the rest of the community, not in a reciprocating network where everyone takes turns. One reason for this may be that the cost of defending spoils against hungry usurpers exceeds their value to the hunter. If hunters bring in food in large units, as when a sizeable animal has been killed, they will be able to satisfy their needs with the first portions, after which the remaining meat is of less and less value to them. For onlookers, however, the remainder has considerable value because they have yet to eat at all. As a result, they will be strongly motivated to grab some of the food for themselves. The combination of their determination and the dwindling interest of the hunter means that it is likely to pay the hunter simply to let them have it rather than mount a costly defence. Thus the law of diminishing returns to the hunter makes passive sharing (or **tolerated theft**, as Blurton-Jones [1984] called it) a sensible economic decision. This argument should be familiar, because it follows the principle of the marginal value theorem in Chapter 8 (see Box 8.2), where we used it to predict the effect of the law of diminishing returns on patch residence times in predators.

Tolerated theft may develop into established producer–scrounger relationships (9.1.1.2 and Box 9.2) if some individuals (scroungers) give up gathering resources for themselves altogether and concentrate on taking them from others (producers). As we saw in 9.1.1.2, the benefits of scrounging may be both frequency- and density-dependent, so not everyone can do it or there would not be any producers to support them. Blurton-Jones (1984) has suggested that scroungers may actively encourage individuals to become (or remain) producers by giving them high social status. Thus, the traditionally high status of skilled hunters may come about not because of their skill *per se*, but because it is in the interests of community spongers to toady to them (Barrett *et al.* 2002). That such conferred status might translate into reproductive success is suggested by the higher survivorship of offspring born to the wives of lauded hunters in some traditional societies, and the frequency with which the latter sire progeny through extra-pair copulations (Kaplan & Hill 1985a). While there is evidence to support the tolerated theft and producer–scrounger scenarios, however, it can be difficult conclusively to rule out other causes of sharing, such as reciprocal benefits in different currencies (Kaplan & Hill 1985b). Barrett *et al.* (2002) provide a good summary of the problems.

Showing off   Another reason for sharing might be to **show off** prowess and so gain status and reproductive opportunities. The goal here would be to focus on those kinds of spoils that are most susceptible to being acquired by others back in the community, the key features being large size and unpredictability of availability. These qualities turn out to be sought by men rather than women (Fig. 12.20), suggesting the sexes have different objectives when harvesting resources (Hawkes 1991). Turtle hunting by islanders in the Torres Strait of northern Australia provides a good example (Bliege-Bird & Bird 1997).

The risks involved in hunting turtles varies with the season, but turtle meat is an important component of the islanders' diet so hunting is carried out through the year. Hunting is riskiest during the breeding season, when turtles are far out at sea. The dangers are greater and there is more chance of a hunt failing. During this period, it is only young, unmarried men that go out after the turtles, and any catches they bring back are shared in communal feasts. In the nesting season, the turtles come ashore to lay their eggs, and the costs of hunting are low. Now, men of all ages, married and unmarried, hunt turtles

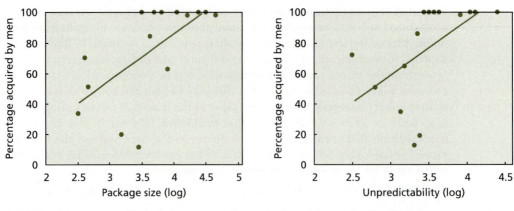

Figure 12.20 Men are more likely than women to forage for large (a) or unpredictable (b) resources, perhaps as a strategy to show off their prowess. See text. After Hawkes (1991).

and any that are caught are usually consumed within family households. The close association between risky hunting and both the age and mating status of the men who do it, coupled with the collective display of sharing afterwards, is consistent with the behaviour being a demonstration of hunting prowess for mating purposes – in other words, showing off for sex.

Summary

1. Behaviour in humans has been shaped both by natural selection (genetic evolution) and by selection within a rich diversity of cultural environments (cultural evolution). Understanding the interaction between genetic and cultural evolution is a difficult challenge, and models have considered a range of possibilities from mutual coevolution to more or less independent evolution based on separate replicator systems.

2. Cultural replicators have been dubbed 'memes', and can be thought of as 'units' of information capable of being stored in one brain and transmitted to others by social learning. Both genes and memes succeed by being copied. Understanding how and why particular memes succeed, and their consequences for cultural evolution, is the province of memetics. Memetic evolution, perhaps in synergy with sexual selection, may have been responsible for some of the defining features of human behaviour, including intelligence, language and altruism.

3. Genetic and cultural evolution are likely to have some impact on each other even when they are based on independent replicator systems. Cultural life must also have its roots somewhere in our genetic evolutionary past. So has natural selection shaped our present behaviour? Attempts to find out have taken two different approaches. The behavioural ecology approach assumes behaviour is adaptive in its present environment, whereas evolutionary psychology assumes current behaviour reflects adaptation to environmental conditions in the past. While the debate continues, several lines of

evidence suggest that human behaviour can be predicted from evolutionary principles that apply to other species.

4. Studies of sexual behaviour have shown that humans have much in common with other anisogamous species when it comes to differences between the sexes. Differences in mate choice between men and women are as expected if men compete for access to mates, and women rely on them for help rearing offspring. Like many other socially monogamous species, humans also show a propensity for extra-pair copulation, and several aspects of reproductive behaviour can be interpreted in terms of sperm competition, mate guarding and sneaky choice for 'good genes'.

5. Effects of certainty of parentage and variation in the reproductive value of offspring on parental care are in keeping with evolutionary parental investment theory, as are sibling rivalry and the nature and timing of conflict between parents and offspring.

6. Human beings show unusual levels of phenotypic altruism. Some of this can be accounted for by kinship theory, though care must be taken to distinguish between blood kin and cultural 'kinship' alliances. Other examples may be due to reciprocity, but it is important to note that apparent altruism can arise in several other ways that have nothing to do with reciprocity, even though they can sometimes look like it on the surface.

Further reading

For the brave, Cavalli-Sforza & Feldman (1981), Lumsden & Wilson (1981) and Boyd & Richerson (1985) discuss cultural evolution through the medium of mathematical models, though Lumsden & Wilson (1983) provide a synthesis of some of their ideas in a shorter, more accessible format. Dawkins (1989) and Blackmore (1999) develop the idea of memes as the underlying replicators in cultural evolution while G.F. Miller (2000) discusses their role in sexual selection. Buss (1999) and Barrett *et al.* (2002) are excellent, complementary introductions to evolutionary psychology, the former from a psychology background, the latter adopting more of a behavioural ecology approach. Diamond (1992, 1997), Ridley (1993b, 1996) and Zuk (2002) provide good popular accounts of human behaviour, sex, evolution and culture.

References

Abele, L.G. & Gilchrist, S. (1977) Homosexual rape and sexual selection in acanthocephalan worms. *Science* **197**: 81–3.

Able, K.P. (1982) Field studies of avian nocturnal migration orientation. I. Interaction of sun, wind and stars as directional cues. *Animal Behaviour* **30**: 761–7.

Able, K.P. (1993) Orientation cues used by migratory birds: a review of cue-conflict experiments. *Trends in Ecology & Evolution* **8**: 367–71.

Able, K.P. & Able, M.A. (1990) Ontogeny of migratory orientation in the savannah sparrow (*Passerculus sandwichensis*): mechanisms of sunset orientation. *Animal Behaviour* **39**: 905–13.

Abrams, P.A. (2000) The evolution of predator–prey interactions: theory and evidence. *Annual Review of Ecology & Systematics* **31**: 79–105.

Abu-Ghidieri, Y.B. (1966) The behaviour and neuro-anatomy of some developing teleost fishes. *Journal of Zoology* **149**: 215–41.

Adams, E.S. (2001) Approaches to the study of territory size and shape. *Annual Review of Ecology and Systematics* **32**: 277–303.

Adams, E.S. & Caldwell, R.L. (1990) Deceptive communication in asymmetric fights of the stomatopod crustacean *Gonodactylus bredini*. *Animal Behaviour* **39**: 706–16.

Ader, R.L., Felten, D.L. & Cohen, N. eds (2001) *Psychoneuroimmunology*, 3rd Edition. New York, Academic Press.

Adler, K. (1976) Extraocular photoreception in amphibians. *Photochemistry and Photobiology* **23**: 275–98.

Adler, K. & Phillips, J.B. (1985) Orientation in a desert lizard (*Uma notata*): time-compensated compass movement and polarotaxis. *Journal of Comparative Physiology A* **156**: 547–52.

Adrian, R.H. (1974) *The Nerve Impulse*. Oxford Biology Readers, Oxford, Oxford University Press.

Agrawal, A.A., Laforsch, C. & Tollrian, R. (1999) Transgenerational induction of defences in animals and plants. *Nature* **401**: 60–63.

Agrawal, A.F. (2001) The evolutionary consequences of mate copying on male traits. *Behavioral Ecology & Sociobiology* **51**: 33–40.

Alatalo, R.V. & Mappes, J. (1996) Tracking the evolution of warning signals. *Nature* **382**: 708–10.

Alatalo, R.V., Carlson, A. Lundberg, A. & Ulfstrand, S. (1981) The conflict between male polygamy and female monogamy: the case of the pied flycatcher, *Ficedula hypoleuca*. *The American Naturalist* **117**: 738–53.

Alberts, J.R. & Gubernick, D.J. (1983) Reciprocity and resource exchange: a symbiotic model of parent–offspring relations. In (L.A. Rosenblum & H. Moltz eds) *Symbiosis in Parent–offspring Interactions*, pp. 7–44. New York, Plenum Press.

Alcock, J. (1975) *Animal Behavior: an Evolutionary Approach*. Sunderland, MA, Sinauer.

Alcock, J. (1984) *Animal Behavior: An Evolutionary Approach*, 3rd Edition. Sunderland, MA, Sinauer.

Alcock, J. (1998) *Animal Behavior*, 6th Edition. Sunderland, MA, Sinauer.

Alcock, J. (2001) *Animal Behavior*, 7th Edition. Sunderland, MA, Sinauer.

Alcock, J., Jones, C.E. & Buchmann, S.L. (1977) Male mating strategies in the bee *Centris pallida*, Fox (Anthophoridae: Hymenoptera). *The American Naturalist* **111**: 145–55.

Alerstam, T. & Hedenström, A. (1998) Optimal migration. *Journal of Avian Biology* **29**: 337–636.

Alexander, J. & Stimson, W.H. (1988) Sex hormones and the course of parasitic infection. *Parasitology Today* **4**: 189–93.

Alexander, R.D. (1974) The evolution of social behavior. *Annual Review of Ecology & Systematics* 5: 325–83.

Alexander, R.D. (1979) *Darwinism and Human Affairs*. Seattle, University of Washington Press.

Alexander, R.D. & Borgia, G. (1978) Group selection, altruism, and the levels of organization of life. *Annual Review of Ecology & Systematics* 9: 449–74.

Alexander, R.D. & Noonan, K.M. (1979) Concealment of ovulation, parental care, and human social evolution. In (N.A. Chagnon & W. Irons eds) *Evolutionary Biology and Human Social Behavior*, pp. 402–35. North Scituate, MA, Duxbury Press.

Alexander, R.D. & Sherman, P.W. (1977) Local mate competition and parental investment in social insects. *Science* 196: 494–500.

Allen, J.A. (1972) Evidence for stabilizing and apostatic selection by wild blackbirds. *Nature* 237: 348–9.

Allen, N.D., Logan, K., Lally, G., Drage, D.J., Norris, M.L. & Keverne, E.B. (1995) Distribution of parthenogenetic cells in the mouse brain and their influence on brain development and behavior. *Proceedings of the National Academy of Sciences of the USA* 92: 10782–6.

Altmann, S.A. (1962) Sociobiology of rhesus monkeys. II. Stochastics of social communication. *Journal of Theoretical Biology* 5: 490–522.

Altmann, S.A. (1967) The structure of primate social communication. In (L.R. Aronson, E. Tobach, D.S. Lehrman & J.S. Rosenblatt eds) *Development and Evolution of Behavior*, pp. 53–74. New York, Freeman.

Altmann, S.A. (1979) Altruistic behaviour: the fallacy of kin deployment. *Animal Behaviour* 27: 958–9.

Andersen, D.C. (1996) A spatially-explicit model of search path and soil disturbance by a fossorial herbivore. *Ecological Modelling* 89: 99–108.

Andersson, M. (1980) Why are there so many threat displays? *Journal of Theoretical Biology* 86: 773–81.

Andersson, M. (1982) Female choice selects for extreme tail length in a widowbird. *Nature* 299: 818–20.

Andersson, M. (1984a) The evolution of eusociality. *Annual Review of Ecology & Systematics* 15: 165–89.

Andersson, M. (1984b) Brood parasitism within species. In (C.J. Barnard ed.) *Producers and Scroungers: Strategies of Exploitation and Parasitism*, pp. 195–228. London, Croom Helm.

Andersson, M. (1994) *Sexual Selection*. Princeton, NJ, Princeton University Press.

Andersson, M. & Eriksson, M.O.G. (1982) Nest parasitism in goldeneyes, *Bucephala clangula*: some evolutionary aspects. *American Naturalist* 120: 1–16.

Andersson, M. & Wiklund, C.C. (1978) Clumping versus spacing out: experiments on nest predation in fieldfares (*Turdus pilaris*). *Animal Behaviour* 26: 1207–12.

Annau, Z. & Kamin, L.J. (1961) The conditioned emotional response as a function of intensity of the US. *Journal of Comparative and Physiological Psychology* 54: 428–32.

Aoki, S. (1977) *Colophina clematis* (Homoptera, Pemphigidae), an aphid species with 'soldiers'. *Kontyû* 45: 276–82.

Apanius, V., Penn, D., Slev, P., Ruff, L.R. & Potts, W.K. (1997) The nature of selection on the major histocompatibility complex. *Critical Reviews in Immunology* 17: 179–224.

Applebaum, S.L. & Cruz, A. (2000) The role of mate-choice copying and disruption effects in mate preference determination of *Limia perugiae* (Cyprinodontiformes: Poeciliidae). *Ethology* 106: 933–44.

Arak, A. (1988) Female mate selection in the natterjack toad: active choice or passive attraction? *Behavioral Ecology & Sociobiology* 22: 317–27.

Arak, A. & Enquist, M. (1993) Hidden preferences and the evolution of signals. *Philosophical Transactions of the Royal Society of London Series B* 340: 207–14.

Arnold, E.N. (1988) Caudal autotomy as defense. In (C. Gans & R.B. Huey eds) *Biology of the Reptiles*, Vol. 16, pp. 235–73. New York, Allan Liss Inc.

Arnold, S.J. (1980) The microevolution of feeding behavior. In (A. Kamil & T. Sargent eds) *Foraging Behavior: Ecology, Ethological and Psychological Approaches*, pp. 98–124. New York, Garland STPM Press.

Arnold, S.J. (1981) Behavioural variation in natural populations: II. The inheritance of a feeding response in crosses between geographical races of the garter snake, *Thamnophis elegans*. *Evolution* 35: 510–15.

Arnqvist, G. & Nilsson, T. (2000) The evolution of polyandry: multiple mating and female fitness in insects. *Animal Behaviour* 60: 145–64.

Aschoff, J. (1965) Circadian rhythms in man. *Science* 148: 1427–32.

Aschoff, J. (1979) Circadian rhythms: influences of internal and external factors on the period measured in constant conditions. *Zeitschrift für Tierpsychologie* **49**: 225–49.

Astur, R.S., Ortiz, M.L. & Sutherland, R.J. (1998) A characterisation of performance by men and women in a virtual Morris water task: a large and reliable sex difference. *Behavioural Brain Research* **93**: 185–90.

Atchley, W.R., Herring, S.W., Riska, B. & Plummer, A.A. (1984) Effects of the muscular dysgenesis gene on developmental stability in the mouse mandible. *Journal of Craniofacial Genetics and Developmental Biology* **4**: 179–89.

Aubin, T. (1994) Adaptation for acoustic information through the environment: importance of duration parameters in the distress calls of the starling (*Sturnus vulgaris*). *Revue d'Ecologie* **49**: 405–15.

Aubin, T. & Jouventin, P. (1998) Cocktail-party effect in king penguin colonies. *Proceedings of the Royal Society of London Series B* **265**: 1665–73.

Aubin, T. & Jouventin, P. (2002) How to vocally identity kin in a crowd: the penguin model. *Advances in the Study of Behaviour* **31**: 243–77.

Austad, S.N. (1982) First male sperm priority in the bowl and doily spider, *Frontinella pyramitela*. *Evolution* **36**: 777–85.

Austad, S.N. (1983) A game theoretical interpretation of male combat in the bowl and doily spider, *Frontinella pyramitela*. *Animal Behaviour* **31**: 59–73.

Axelrod, R. (1984) *The Evolution of Cooperation*. New York, Basic Books.

Axelrod, R. & Hamilton, W.D. (1981) The evolution of cooperation. *Science* **211**: 1390–6.

Baalen, M. van & Sabelis, M.W. (1993) Coevolution of patch selection strategies of predator and prey and the consequences for ecological stability. *American Naturalist* **142**: 646–70.

Baalen, M. van & Sabelis, M.W. (1999) Non-equilibrium population dynamics of 'ideal and free' prey and predators. *American Naturalist* **154**: 69–88.

Baars, B. (1988) *A Cognitive Theory of Consciousness*. New York, Cambridge University Press.

Bachmann, G.C. & Chappell, M.A. (1998) The energetic cost of begging behaviour in nestling house wrens. *Animal Behaviour* **55**: 1607–18.

Baerends, G.P. (1941) Fortpflansungvershalten und Orientierung der Grabwespe *Ammophila campestris* Jur. *Tidjschrift voor Entomologie* **84**: 68–275.

Baerends, G.P. (1970) A model of the functional organization of incubation behaviour. In (Baerends, G.P. & Drent, R.G. eds) *The Herring Gull and its Egg. Behaviour Supplement* **XVIII**: 265–310.

Baerends, G.P., Brouwer, R. & Waterbolk, H.Tj (1955) Ethological studies of *Lebistes reticulatus* (Peters): I. An analysis of the male courtship pattern. *Behaviour* **8**: 249–334.

Bagemihl, B. (1999) *Biological Exuberance: Animal Homosexuality and Natural Diversity*. London, Profile Books.

Baggerman, B. (1962) Some endocrine aspects of fish migration. *General and Comparative Endocinology, Supplement* **1**: 188–205.

Baker, J.M.R. (1992) Body condition and tail height in great crested newts (*Triturus cristatus*). *Animal Behaviour* **43**: 157–9.

Baker, R.R. (1972) Territorial behaviour of the nymphalid butterflies *Aglais urticae* (L.) and *Inachis io* (L.). *Journal of Animal Ecology* **41**: 453–69.

Baker, R.R. (1978) *The Evolutionary Ecology of Animal Migration*. London, Hodder & Stoughton.

Baker, R.R. (1981) *Human Navigation and the Sixth Sense*. London, Hodder & Stoughton.

Baker, R.R. (1982) *Migration: Paths Through Time and Space*. London, Hodder & Stoughton.

Baker, R.R. (1984) *Bird Navigation: The Solution of a Mystery?* London, Hodder & Stoughton.

Baker, R.R. & Bellis, M.A. (1988) 'Kamikaze' sperm in mammals. *Animal Behaviour* **36**: 937–80.

Baker, R.R. & Bellis, M.A. (1989) Number of sperm in human ejaculates varies in accordance with sperm competition theory. *Animal Behaviour* **37**: 867–9.

Baker, R.R. & Bellis, M.A. (1995) *Human Sperm Competition: Copulation, Masturbation and Infidelity*. London, Chapman and Hall.

Balda, R.P. & Kamil, A.C. (1992) Long-term spatial memory in Clark's nutcracker, *Nucifraga columbiana*. *Animal Behaviour* **44**: 761–9.

Baldaccini, N.E., Benvenuti, S., Fiaschi, V. & Papi, F. (1975) Pigeon navigation: effects of wind deflection at home cage on homing behaviour. *Journal of Comparative Physiology* **99**: 177–86.

Balleine, B.W. & Dickinson, A. (1998) The role of incentive learning in instrumental outcome revaluation by sensory-specific satiety. *Animal Learning & Behavior* **26**: 46–59.

Baltz, A.P. & Clark, A.B. (1997) Extra-pair courtship behaviour of male budgerigars and the effect of an audience. *Animal Behaviour* **53**: 1017–24.

Barber, N. (1995) The evolutionary psychology of physical attractiveness: sexual selection and human morphology. *Ethology & Sociobiology* **16**: 395–424.

Barinaga, M. (1994) From fruit flies, rats, mice: evidence of genetic influence. *Science* **264**: 1690–93.

Barker, D.J.P. (1995) The foetal and infant origins of disease. *European Journal of Clinical Investigation* **25**: 457–63.

Barlow, D.P. (1995) Genetic imprinting in mammals. *Science* **270**: 1610–13.

Barlow, G.W. (1974) Hexagonal territories. *Animal Behaviour* **22**: 876–8.

Barnard, C.J. (1978) Aspects of winter flocking and food fighting in the house sparrow (*Passer domesticus domesticus* L.). Unpublished D. Phil. Thesis, University of Oxford.

Barnard, C.J. (1980a) Flock feeding and time budgets in the house sparrow (*Passer domesticus* L.) *Animal Behaviour* **28**: 295–309.

Barnard, C.J. (1980b) Factors affecting flock size mean and variance in a winter population of house sparrows (*Passer domesticus* L.) *Behaviour* **74**: 114–27.

Barnard, C.J. (1980c) Equilibrium flock size and factors affecting arrival and departure in feeding sparrows. *Animal Behaviour* **28**: 503–11.

Barnard, C.J. (1983) *Animal Behaviour: Ecology and Evolution*. London, Croom Helm.

Barnard, C.J. ed. (1984a) *Producers and Scroungers: Strategies of Exploitation and Parasitism*. London, Croom Helm.

Barnard, C.J. (1984b) When cheats may prosper. In (C.J. Barnard ed.) *Producers and Scroungers: Strategies of Exploitation and Parasitism*, pp. 6–33. London, Croom Helm.

Barnard, C.J. (1990) Kin recognition: problems, prospects and the evolution of discrimination systems. *Advances in the Study of Behavior* **19**: 29–82.

Barnard, C.J. (1991) Kinship and social behaviour: the trouble with relatives. *Trends in Ecology & Evolution* **6**: 310–11.

Barnard, C.J. (1999) Still having trouble with relatives. *Trends in Ecology & Evolution* **14**: 448.

Barnard, C.J. & Behnke, J.M. eds (1990) *Parasites and Host Behaviour*. London, Taylor & Francis.

Barnard, C.J. & Behnke, J.M. (2001) From psychoneuroimmunology to ecological immunology: life history strategies and immunity trade-offs. In (R.L. Ader, D.L. Felten & N. Cohen eds) *Psychoneuroimmunology*, 3rd Edition, Vol. 2, pp. 35–47. New York, Academic Press.

Barnard, C.J. and Brown, C.A.J. (1981) Prey size selection and competition in the common shrew (*Sorex araneus* L.). *Behavioral Ecology and Sociobiology* **8**: 239–43.

Barnard, C.J. and Brown, C.A.J. (1982) The effects of prior residence, competitive ability and food availability on the outcome of interactions between shrews (*Sorex araneus* L.) *Behavioral Ecology & Sociobiology* **10**: 307–11.

Barnard, C.J. & Brown, C.A.J. (1984) A payoff asymmetry in resident–resident disputes between shrews. *Animal Behaviour* **32**: 302–4.

Barnard, C.J. and Brown, C.A.J. (1985a) Risk-sensitive foraging in common shrews (*Sorex araneus* L.). *Behavioral Ecology and Sociobiology* **16**: 162–4.

Barnard, C.J. and Brown, C.A.J. (1985b) Competition affects risk-sensitivity in foraging shrews. *Behavioral Ecology and Sociobiology* **16**: 379–82.

Barnard, C.J. & Burk, T.E. (1979) Dominance hierarchies and the evolution of 'individual recognition'. *Journal of Theoretical Biology* **81**: 65–72.

Barnard, C.J. & Fitzsimons, J. (1988) Kin recognition and mate choice in mice: the effects of kinship, familiarity and social interference on intersexual interaction. *Animal Behaviour* **36**: 1078–90.

Barnard, C.J. & Fitzsimons, J. (1989) Kin recognition and mate choice in mice: fitness consequences of mating with kin. *Animal Behaviour* **38**: 35–40.

Barnard, C.J. & Hurst, J.L. (1996) Welfare by design: the natural selection of welfare criteria. *Animal Welfare* **5**: 405–33.

Barnard, C.J. & Sibly, R.M. (1981) Producers and scroungers: a general model and its application to feeding flocks of house sparrows. *Animal Behaviour* **29**: 543–50.

Barnard, C.J. & Thompson, D.B.A. (1985) *Gulls and Plovers: The Ecology and Behaviour of Mixed-species Feeding Groups*. London, Croom Helm.

Barnard, C.J., Behnke, J.M., Gage, A.R., Brown, H. & Smithurst, P.R. (1997a) Modulation of behaviour and testosterone concentration in immunodepressed male laboratory mice (*Mus musculus*). *Physiology & Behaviour* **61**: 907–17.

Barnard, C.J., Behnke, J.M., Gage, A.R., Brown, H. & Smithurst, P.R. (1997b) Immunity costs and behavioural modulation in male laboratory mice (*Mus musculus*) exposed to the odour of females. *Physiology & Behaviour* **62**: 857–66.

Barnard, C.J., Behnke, J.M., Gage, A.R., Brown, H. & Smithurst, P.R. (1998) Maternal effects on the development of social rank and immunity trade-offs in male laboratory mice (*Mus musculus*). *Proceedings of the Royal Society of London Series B* **265**: 2087–93.

Barnard, C.J., Behnke, J.M. & Sewell, J. (1996a) Social status and resistance to disease in house mice (*Mus musculus*): status-related modulation of hormonal responses in relation to immunity costs in different social and physical environments. *Ethology* **102**: 63–84.

Barnard, C.J., Behnke, J.M. & Sewell, J. (1996b) Environmental enrichment, immunocompetence and resistance to *Babesia microti* in male laboratory mice. *Physiology and Behaviour* **60**: 1223–31.

Barnard, C.J., Hurst, J.L. & Aldhous, P. (1991) Of mice and kin: the functional significance of kin bias in social behaviour. *Biological Reviews* **66**: 379–430.

Barnard, C.J., Sayed, E., Barnard, L.E., Behnke, J.M., Abdel Nabi, I., Sherif, N, Shutt, A. & Zalat, S. (2003) Local variation in helminth burdens of Egyptian spiny mice (*Acomys cahirinus dimidiatus*) from ecologically similar sites: relationships with hormone concentrations and social behaviour. *Journal of Helminthology* **77**: 197–208.

Barnes, R.D. (1968) *Invertebrate Zoology*, 2nd Edition. London, Saunders.

Barrett, L., Dunbar, R.I.M. & Lycett, J. (2002) *Human Evolutionary Psychology*. Basingstoke, Palgrave.

Bart, J. & Tornes, A. (1989) Importance of monogamous male birds in determining reproductive success: evidence for house wrens and a review of male removal studies. *Behavioral Ecology & Sociobiology* **24**: 109–16.

Barta, Z., Flynn, R. & Giraldeau, L.-A. (1997) Geometry for a selfish foraging group: a genetic algorithm approach. *Proceedings of the Royal Society of London Series B* **264**: 1233–8.

Barton, R.A. & Harvey, P.H. (2000) Mosaic evolution of brain structure in mammals. *Nature* **405**: 1055–8.

Basolo, A.L. (1990) Female preference predates the evolution of the sword in swordtail fish. *Science* **250**: 808–10.

Basolo, A.L. (1995) Phylogenetic evidence for the role of a pre-existing bias in sexual selection. *Proceedings of the Royal Society of London Series B* **259**: 307–11.

Basolo, A.L. (1998) Evolutionary change in a receiver bias: a comparison of female preference functions. *Proceedings of the Royal Society of London Series B* **265**: 2223–8.

Bastock, M. (1956) A gene mutation which changes a behaviour pattern. *Evolution* **10**: 421–39.

Bastock, M. (1967) *Courtship: An Ethological Study*. Chicago, Aldine.

Bateman, A.J. (1948) Intrasexual selection in *Drosophila*. *Heredity* **2**: 349–68.

Bateson, P. (1976) Specificity and the origins of behaviour. *Advances in the Study of Behavior* **6**: 1–20.

Bateson, P.P.G. (1979) How do sensitive periods arise and what are they for? *Animal Behaviour* **27**: 470–86.

Bateson, P. (1982) Preferences for cousins in Japanese quail. *Nature* **295**: 236–7.

Bateson, P. (1983) Optimal outbreeding. In (P. Bateson ed.) *Mate Choice*. Cambridge, Cambridge University Press.

Bateson, P. (1987) Imprinting as a process of competitive exclusion. In (J.P. Rauschecker & P. Marler eds) *Imprinting and Cortical Plasticity*, pp. 151–68. New York, John Wiley & Sons.

Bateson, P. & Martin, P. (1999) *Design for a Life: How Behaviour Develops*. London, Jonathan Cape.

Bateson, P.P.G., Mendl, M. & Fearer, J. (1990) Play in the domestic cat is enhanced by rationing the mother during lactation. *Animal Behviour* **40**: 514–25.

Batteau, D.W. (1968) The world as a source; the world as a sink. In (S.J. Freedman ed.) *The Neuropsychology of Spatially Oriented Behavior*, pp. 197–203. Homewood, IL, Dorsey Press.

Baum, W.M. (1974) On two types of deviation from the matching law: bias and under-matching. *Journal of the Experimental Analysis of Behavior* **22**: 231–42.

Baum, W.M., Schwendiman, J.W. & Bell, K.E. (1999) Choice, contingency discrimination and foraging theory. *Journal of the Experimental Analysis of Behavior* **71**: 355–73.

Baxter, M. (1983) Ethology in environmental design for animal production. *Applied Animal Ethology* **9**: 207–20.

Baylies, M.K., Bargiello, T.K., Jackson, F.R. & Young, M.W. (1987) Changes in abundance or structure of the *per* gene product can alter periodicity of the *Drosophila* clock. Nature **326**: 390–2.

Beach, F.A. (1975) Behavioral endocrinology: an emerging discipline. *American Scientist* **63**: 178–87.

Beani, L. & Turillazzi, S. (1999) Stripes display in hover-wasps (Vespidae: Stenogastrinae): a socially costly status badge. *Animal Behaviour* **57**: 1233–9.

Beason, R.C. & Semm, P. (1996) Does the avian ophthalmic nerve carry magnetic navigational information? *Journal of Experimental Biology* **199**: 1241–4.

Beason, R.C., Dussourd, N. & Deutschlander, M.E. (1995) Behavioural evidence for the use of magnetic material in magnetoreception by a migratory bird. *Journal of Experimental Biology* **198**: 141–6.

Beason, R.C. Wiltschko, R. & Wiltschko, W. (1997) Pigeon homing: effects of magnetic pulses on initial orientation. *Auk* **114**: 405–15.

Beauchamp, G. (2001) Should vigilance always decrease with group size? *Behavioral Ecology & Sociobiology* **51**: 47–52.

Beauchamp, G. & Kacelnik, A. (1991) Effects of the knowledge of partners on learning rates in zebra finches, *Taeniopygia guttata*. *Animal Behaviour* **41**: 247–53.

Beckage, N.E. ed. (1997) *Parasites and Pathogens: Effects on Host Hormones and Behavior*. New York, Chapman and Hall.

Bednarz, J.C. (1988) Cooperative hunting in Harris' hawks (*Parabuteo unicinctus*). *Science* **239**: 1525–7.

Bee, M.A. & Gerhardt, H.C. (2002) Individual voice recognition in a territorial frog (*Rana catesbeiana*). *Proceedings of the Royal Society of London Series B* **269**: 1443–8.

Beecher, M.D. (1982) Signature systems and kin recognition. *American Zoologist* **22**: 477–90.

Beehler, B.M. & Foster, M.S. (1988) Hotshots, hotspots and female preferences in the organization of mating systems. *The American Naturalist* **131**: 203–19.

Beissinger, S.R. & Snyder, N.F.R. (1987) Mate desertion in the snail kite. *Animal Behaviour* **35**: 477–87.

Bekoff, A. & Kauer, J.A. (1984) Neural control of hatching: fate of the generator for the leg movements of hatching in post-hatching chicks. *Journal of Neuroscience* **4**: 2659–66.

Bekoff, M. (1995a) Vigilance, flock size, and flock geometry: information gathering by western evening grosbeaks (Aves: Fringillidae). *Ethology* **99**: 150–61.

Bekoff, M. (1995b) Play signals as punctuation: the structure of social play in canids. *Behaviour* **132**: 419–29.

Bekoff, M. & Byers, J.A. (1981) A critical re-analysis of the ontogeny of mammalian social and locomotor play: an ethological hornets' nest. In (K. Immelmann, G.W. Barlow, L. Petrinovich & M. Main eds) *Behavioral Development: the Bielefeld Interdisciplinary Project,* pp. 296–337. New York, Cambridge University Press.

Bekoff, M. & Byers, J.A. eds (1998) *Animal Play: Evolutionary, Comparative and Ecological Perspectives*. Cambridge, Cambridge University Press.

Bekoff, M. & Dugatkin, L.A. (2000) Winner and loser effects and the development of dominance hierarchies in young coyotes: an integration of data and theory. *Evolutionary Ecology Research* **2**: 871–83.

Beletsky, L.D., Orians, G.H. & Wingfield, J.C. (1992) Year-to-year patterns of circulating levels of testosterone and corticosterone in relation to breeding density, experience, and reproductive success of the polygynous red-winged blackbird. *Hormones and Behavior* **26**: 420–32.

Bell, G. (1982) *The Masterpiece of Nature: The Evolution and Genetics of Sexuality*. Berkeley, University of California Press.

Bellrose, F.C. (1958) Celestial orientation in wild mallards. *Bird Banding* **29**: 75–90.

Belovsky, G.E. (1978) Diet optimization in a generalist herbivore: the moose. *Theoretical Population Biology* **14**: 105–34.

Belovsky, G.E. (1990) How important are nutrient constraints in optimal foraging models, or are spatial and temporal factors more important? In (R.N. Hughes ed.) *Behavioural Mechanisms of Food Selection*, pp. 255–80. NATO ASI Series G, Ecological Sciences Vol. 20. Heidelberg, Springer.

Belyaev, D.K. (1979) Destabilizing selection as a factor in domestication. *Journal of Heredity* **70**: 301–8.

Benhamou, S. & Bovet, P. (1989) How animals use their environment: a new look at kinesis. *Animal Behaviour* **38**: 375–83.

Benhamou, S. & Bovet, P. (1992) Distinguishing between elementary orientation mechanisms by means of path analysis. *Animal Behaviour* **43**: 371–7.

Bensch, S. & Hasselquist, D. (1992) Evidence for active female choice in a polygynous warbler. *Animal Behaviour* **44**: 301–11.

Bentley, D. (1975) Single gene cricket mutations: effects on behavior, sensillae, sensory neurons, and identified interneurons. *Science* **187**: 760–4.

Bentley, D. & Hoy, R. (1972) Genetic control of the neuronal network generating cricket (*Teleogryllus*) song patterns. *Animal Behaviour* **20**: 478–92.

Bentley, D. & Kishishian, H. (1982) Pathfinding by peripheral pioneer neurons in grasshoppers. *Science* **218**: 1082–8.

Benton, T.G. & Foster, W.A. (1992) Altruistic housekeeping in a social aphid. *Proceedings of the Royal Society of London Series B* **247**: 199–202.

Benvenuti, S. Gagliardo, A., Guilford, T. & Luschi, O. (1996) Overcast tests with clock-shifted pigeons. *Experientia* **52**: 608–12.

Benzer, S. (1973) Genetic dissection of behavior. *Scientific American* **229**: 24–37.

Berghe, P.L. van den (1980) Incest and exogamy: a sociobiological reconsideration. *Ethology and Sociobiology* **1**: 151–62.

Berry, J.F. & Shine, R. (1980) Sexual dimorphism and sexual selection in turtles (Order Testudines). *Oecologia* **44**: 185–91.

Berthold, P. (1998) Bird migration: genetic programs with high adaptability. *Zoology – Analysis of Complex Systems* **101**: 235–45.

Berthold, P. (1999) *Bird Migration: a General Survey*, 2nd Edition. Oxford, Oxford University Press.

Berthold, P. (2001) Bird migration: a novel theory for the evolution, the control and the adaptability of bird migration. *Journal für Ornithologie* **142**: 148–59.

Berthold, P. & Querner, U. (1981) Genetic basis of migratory behavior in European warblers. *Science* **212**: 77–9.

Berthold, P., Helbig, A.J., Mohr, G. & Querner, U. (1992) Rapid microevolution of migratory behaviour in a wild bird species. *Nature* **360**: 668–70.

Berthold, P., Wiltschko, W., Mildenberger, H. & Querner, U. (1990) Genetic transmission of migratory behavior into a nonmigratory bird population. *Experientia* **46**: 107–8.

Bertram, B.C.R. (1978) Living in groups: predators and prey. In (J.R. Krebs & N.B. Davies eds) *Behavioural Ecology: An Evolutionary Approach*, 1st Edition, pp. 64–96. Oxford, Blackwell.

Bertram, B.C.R. (1980) Vigilance and group size in ostriches. *Animal Behaviour* **28**: 278–86.

Bertram, B.C.R. (1992) *The Ostrich Communal Nesting System*. Princeton, Princeton University Press.

Bethel, W.M. & Holmes, J.C. (1974) Correlation of development of altered evasive behavior in *Gammarus lacustris* (Amphipoda) harboring cystacanths of *Polymorphus paradoxus* (Acanthocephala) with the infectivity to the definitive host. *Journal of Parasitology* **60**: 272–4.

Betzig, L.L. (1989) Causes of conjugal dissolution. *Current Anthropology* **30**: 654–76.

Bevc, I. & Silverman, I. (2000) Early separation and sibling incest: a test of the revised Westermarck theory. *Evolution and Human Behavior* **21**: 151–62.

Beynon, R.J. & Hurst, J.L. (2003) Multiple roles of major urinary proteins in the house mouse, *Mus domesticus*. *Biochemical Society Transactions* **31**: 142–6.

Beynon, R.J., Veggerby, C., Payne, C.E., Robertson, D.H.L., Gaskell, S.J., Humphries, R.E. & Hurst, J.L. (2002) Polymorphism in major urinary proteins: molecular heterogeneity in a wild mouse population. *Journal of Chemical Ecology* **28**: 1429–46.

Biben, M. (1998) Squirrel monkey playfighting: making the case for a cognitive training function for play. In (M. Bekoff & J.A. Byers eds) *Animal Play: Evolutionary, Comparative and Ecological Perspectives*, pp. 161–82. Cambridge, Cambridge University Press.

Biebach, H. (1983) Genetic determination of partial migration in the European robin (*Erithacus rubecula*). *Auk* **100**: 601–6.

Bindra, D. (1978) How adaptive behaviour is produced: a perceptual-motivational alternative to response reinforcement. *Behavioural and Brain Sciences* **1**: 41–91.

Bingman, V.P. (1981) Savannah sparrows have a magnetic compass. *Animal Behaviour* **29**: 962–3.

Bingman, V.P. (1998) Spatial representations and homing pigeon navigation. In (S.D. Healy ed.) *Spatial Representation in Animals*, pp. 69–85. Oxford, Oxford University Press.

Bingman, V.P. & Benvenuti, (1996) Olfaction and the homing ability of pigeons in the southeastern United States. *Journal of Experimental Biology* **276**: 186–92.

Birkhead, T.R. & Clarkson, K. (1980) Mate selection and precopulatory guarding in *Gammarus pulex*. *Zeitschrift für Tierspychologie* **52**: 365–80.

Birkhead, T.R. & Møller, A.P. (1998) *Sperm Competition and Sexual Selection*. London, Academic Press.

Birkhead, T.R., Fletcher, F. & Pellatt, E.J. (1998) Sexual selection in the zebra finch, *Taeniopygia guttata*: condition, sex traits and immune capacity. *Behavioral Ecology & Sociobiology* **44**: 179–91.

Bjorksten, T, David, P., Pomiankowski, A. & Fowler, K. (2000) Fluctuating asymmetry of sexual and non-sexual traits in stalk-eyed flies: a poor indicator of developmental stress and genetic quality. *Journal of Evolutionary Biology* **13**: 89–97.

Blackmore, S. (1999) *The Meme Machine*. Oxford, Oxford University Press.

Blaizot, X., Landeau, B., Baron, J.C. & Chavoix, C. (2000) Mapping the visual recognition memory network with PET in the behaving baboon. *Journal of Cerebral Blood Flow and Metabolism* **20**: 213–19.

Blakemore, C. & Cooper, G.F. (1970) Development of the brain depends on the visual environment. *Nature* **228**: 477–8.

Blakemore, R.P. (1975) Magnetotactic bacteria. *Science* **190**: 377–9.

Bleske, A.L. & Buss, D.M. (2000) A comprehensive theory of mating must explain between-sex and within-sex differences in mating strategies. *Behavioral and Brain Sciences* **23**: 593–617.

Bliege Bird, R. & Bird, D.W. (1997) Delayed reciprocity and tolerated theft: the behavioural ecology of food-sharing strategies. *Current Anthropology* **38**: 49–78.

Blurton-Jones, N. (1984) A selfish origin for human food sharing: tolerated theft. *Ethology and Sociobiology* **5**: 1–3.

Boakes, R. (1984) *From Darwin to Behaviourism: Psychology and the Minds of Animals*. Cambridge, Cambridge University Press.

Boas, S.B. (1966) *The Earthly Venus* by Pierre-Louis Moreau de Maupertuis; translated from *Venus Physique* (1753) by S.B. Boas. New York, Johnson Reprint Corporation.

Boesch, C. (1991) Teaching among wild chimpanzees. *Animal Behaviour* **41**: 530–2.

Bolhuis, J.J. (1999) Early learning and the development of filial preferences in the chick. *Behavioural Brain Research* **98**: 245–52.

Bolhuis, J.J. ed. (2000) *Brain, Perception and Memory: Advances in Cognitive Neuroscience*. Oxford, Oxford University Press.

Bolles, R.C. (1979) *Learning Theory*, 2nd Edition. New York, Holt, Rinehart & Winston.

Bond, A.B. & Kamil, A.C. (1999) Searching image in blue jays: facilitation and interference in sequential priming. *Animal Learning and Behavior* **27**: 461–71.

Bonner, J.T. (1980) *The Evolution of Culture in Animals*. Princeton, NJ, Princeton University Press.

Bono, M. de (2003) Molecular approaches to aggregation behavior and social attachment. *Journal of Neurobiology* **54**: 78–92.

Bono, M. de, Tobin, D.M., Davis, M.W., Avery, L. & Bargmann, C.I. (2002) Social feeding in *Caenorhabditis elegans* is induced by neurons that detect aversive stimuli. *Nature* **419**: 899–903.

Borgerhoff Mulder, M. (1991) Human behavioural ecology. In (J.R. Krebs & N.B. Davies eds) *Behavioural Ecology: An Evolutionary Approach*, 3rd Edition, pp. 69–98. Oxford, Blackwell.

Boriss, H. & Gabriel, W. (1998) Vertical migration in *Daphnia*: the role of phenotypic plasticity in the migration pattern for competing clones or species. *Oikos* **83**: 129–38.

Bourke, A.F.G. (1997) Sociality and kin selection in insects. In (J.R. Krebs & N.B. Davies eds) *Behavioural Ecology: An Evolutionary Approach*, 4th Edition, pp. 203–27. Oxford, Blackwell.

Bourke, A.F.G. & Franks, N.R. (1995) *Social Evolution in Ants*. Princeton, NJ, Princeton University Press.

Bourne, G.R. (1993) Proximate costs and benefits of mate acquisition at leks of the frog *Ololygon rubra*. *Animal Behaviour* **45**: 1051–9.

Bovet, P. & Benhamou, S. (1985) La clinocinèse: un mécanisme élémentaire de direction. In (J. Paillard ed.) *La Lecture Sensorimotrice et Cognitive de l'Expérience Spatiale*, pp. 171–8. Paris, CNRS.

Bowlby, J. (1969) *Attachment: Attachment and Loss*, Vol. I. New York, Basic Books.

Boyd, H. & Fabricius, E. (1965) Observations on the incidence of following of visual and auditory stimuli in naïve mallard ducklings (*Anas platyrhynchos*). *Behaviour* **25**: 1–15.

Boyd, R. & Richerson, P.J. (1985) *Culture and the Evolutionary Process*. Chicago, Chicago University Press.

Boyd, R. & Richerson, P.J. (1988) The evolution of reciprocity in sizeable groups. *Journal of Theoretical Biology* **132**: 337–56.

Boyd, R. & Richerson, P.J. (1989) The evolution of indirect reciprocity. *Social Networks* **11**: 213–36.

Boyd, R. & Richerson, P.J. (1994) Why does culture increase human adaptability? *Ethology & Sociobiology* **16**: 125–43.

Bradbury, J.W. & Gibson, R.M. (1983) Leks and mate choice. In (P.P.G. Bateson ed.) *Mate Choice*, pp. 109–38. Cambridge, Cambridge University Press.

Bradbury, J.W. & Vehrencamp, S.L. (1984) Mating systems and ecology. In (J.R. Krebs & N.B. Davies eds) *Behavioural Ecology: An Evolutionary Approach*, 2nd Edition, pp. 251–78. Oxford, Blackwell.

Braithwaite, V.A. (1998) Spatial memory, landmark use and orientation in fish. In (S.D. Healy ed.) *Spatial Representation in Animals*, pp. 86–102. Oxford, Oxford University Press.

Brannon, E.L. (1982) Orientation mechanisms of homing salmonids. In (E.L. Brannon & E.O. Salo eds) *Salmon and Trout Migratory Behavior Symposium*, pp. 219–27. Seattle, School of Fisheries, University of Washington,

Brannon, E.L. & Quinn, T.P. (1990) A field test of the pheromone hypothesis for homing in Pacific salmon. *Journal of Chemical Ecology* **16**: 603–9.

Braude, S., Tang-Martinez, Z. & Taylor, G.T. (1999) Stress, testosterone and the immunredistribution hypothesis. *Behavioral Ecology* **10**: 345–50.

Bray, O.E., Kennelly, J.J. & Guarino, J.L. (1975) Fertility of eggs produced on territories of vasectomized red-winged blackbirds. *Wilson Bulletin* **87**: 187–95.

Breder, C.M. (1976) Fish schools as operational structures. *Fishery Bulletin of the Fisheries and Wildlife Service of the United States* **74**: 471–502.

Breland, K. & Breland, M. (1961) The misbehavior of organisms. *American Psychologist* **16**: 681–4.

Brenowitz, E.A., Lent, K. & Kroodsma, D.E. (1995) Brain space for learned song in birds develops independently of song learning. *Journal of Neuroscience* **15**: 6281–6.

Breuker, C.J. & Brakefield, P.M. (2002) Female choice depends on size but not symmetry of dorsal eyespots in the butterfly *Bicyclus anyana*. *Proceedings of the Royal Society of London Series B* **269**: 1233–9.

Briffa, M. & Elwood, R.W. (2001) Decision rules, energy metabolism and vigour of hermit crab fights. *Proceedings of the Royal Society of London Series B* **268**: 1841–8.

Briskie, J.V., Naugler, C. & Leech, S.M. (1994) Begging intensity of nestling birds varies with sibling relatedness. *Proceedings of the Royal society of London Series B* **258**: 73–8.

Brncic, D. & Koref-Santibañez, S. (1964) Mating activity of homo- and heterokaryotypes in *Drosophila pavani*. *Genetics* **49**: 585–91.

Brockmann, H.J. (1980) The control of nest depth in a digger wasp (*Sphex ichneumoneus* L.). *Animal Behaviour* **28**: 426–45.

Brockmann, H.J. (2001) The evolution of alternative strategies and tactics. *Advances in the Study of Behaviour* **30**: 1–51.

Brockmann, H.J., Grafen, A. & Dawkins, R. (1979) Evolutionarily stable nesting strategy in a digger wasp. *Journal of Theoretical Biology* **77**: 473–96.

Brodie, E.D. & Janzen, F.J. (1995) Experimental studies of coral snake mimicry: generalized avoidance of ringed snake patterns by free-ranging avian predators. *Functional Ecology* **9**: 186–90.

Brodie, R. (1996) *Virus of the Mind: the New Science of the Meme*. Seattle, Integral Press.

Bro Larsen, E. (1952) On subsocial beetles from the saltmarsh, their care of progeny and adaptation to the salt and tide. *Transactions of the Eleventh International Congress of Entomology*, pp. 502–6. Beltsville, MD, US Department of Agriculture.

Brooke, M. de L. (1998) Ecological factors influencing the occurrence of 'flash' marks in wading birds. *Functional Ecology* **12**: 339–46.

Broom, D.M. (1991) Assessing welfare and suffering. *Behavioural Processes* **25**: 117–23.

Broom, D.M. & Johnson, K. (1993) *Stress and Animal Welfare*. London, Chapman & Hall.

Brower, L.P. & Brower, J.V. (1964) Birds, butterflies and plant poisons: a study in ecological chemistry. *Zoologica* **49**: 137–59.

Brown, C.H. (1982) Ventriloquial and locatable vocalizations in birds. *Zeitschrift für Tierpsychologie* **59**: 338–50.

Brown, C.H. (1988) Enhanced foraging efficiency through information centres: a benefit of coloniality in cliff swallows. *Ecology* **69**: 602–13.

Brown, C.H. & Brown, M.B. (1986) Ectoparasitism as a cost of coloniality in cliff swallows (*Hirundo pyrrhonota*). *Ecology* **67**: 1206–18.

Brown, J.L. (1964) The evolution of diversity in avian territorial systems. *Wilson Bulletin* **76**: 160–9.

Brown, J.L. (1983) Cooperation – a biologist's dilemma. *Advances in the Study of Behavior* **4**: 1–37.

Brown, K.M. (1998) Proximate and ultimate causes of adoption in ring-billed gulls. *Animal Behaviour* **56**: 1529–43.

Brown, R.E. & McFarland, D.J. (1979) Interaction of hunger and sexual motivation in the male rat: a time-sharing approach. *Animal Behaviour* **27**: 887–96.

Brown, R.E., Singh, P.B. & Roser, B. (1987) The major histocompatibility complex and the chemosensory recognition of individuality in rats. *Physiology & Behavior* **40**: 65–73.

Brownmiller, S. (1975) *Against Our Will: Men, Women and Rape.* New York, Simon Shuster.

Brownmiller, S. & Merhof, B. (1992) A feminist response to rape as an adaptation in men. *Behavioral & Brain Sciences* **15**: 381–2.

Bruderer, B. (1982) Do migrating birds fly along straight lines? In (F. Papi & H.G. Wallraff eds) *Avian Navigation*, pp. 3–14. Heidelberg, Springer.

Bruning, D.F. (1974) Social structure and reproductive behaviour in the greater rhea. *Living Bird* **13**: 251–94.

Brunner, D., Kacelnik, A. & Gibbon, J. (1992) Optimal foraging and timing processes in the starling, *Sturnus vulgaris*: effect of intercatch interval. *Animal Behaviour* **44**: 597–613.

Brunner, D., Kacelnik, A. & Gibbon, J. (1996) Memory for inter-reinforcement interval variability and patch departure decisions in the starling, *Sturnus Vulgaris*. *Animal Behaviour* **51**: 1025–45.

Brunton, D.H. (1990) The effects of nesting stage, sex and type of predator on parental defense by killdeer (*Charadrius vociferus*): testing models of parental defense. *Behavioural Ecology & Sociobiology* **26**: 181–90.

Buchanan, K.L., Catchpole, C.K., Lewis, J.W. & Lodge, A. (1999) Song as an indicator of parasitism in the sedge warbler. *Animal Behaviour* **57**: 307–14.

Buchanan, K.L., Evans, M.R., Goldsmith, A.R., Bryant, D.M. & Rowe, L.V. (2001) Testosterone influences basal metabolic rate in male house sparrows: a new cost of dominance signalling. *Proceedings of the Royal Society of London Series B* **268**: 1337–44.

Budgell, P. (1970) Modulation of drinking by ambient temperature changes. *Animal Behaviour* **18**: 753–7.

Bull, J.J. (1981) Sex ratio evolution when fitness varies. *Heredity* **46**: 9–26.

Bullock, S.P. & Rogers, L.J. (1992) Hemispheric specialization for the control of copulation in the young chick and effects of 5-alpha-dihydrotestosterone and 17-beta-oestradiol. *Behavioural Brain Research* **48**: 9–14.

Bulmer, M.G. & Parker, G.A. (2002) The evolution of anisogamy: a game-theoretic approach. *Proceedings of the Royal Society of London Series B* **269**: 2381–8.

Burger, J. & Gochfeld, M. (1994) Vigilance in mammals: differences among mothers, other females, and males. *Behaviour* **131**: 153–69.

Burgess, J.W. (1976) Social spiders. *Scientific American* **234**: 100–6.

Burghardt, G.M. (1998) The evolutionary origins of play revisited: lessons from turtles. In (M. Bekoff & J.A. Byers eds) *Animal Play: Evolutionary, Comparative and Ecological Perspectives*, pp. 1–26. Cambridge, Cambridge University Press.

Burghardt, G.M. & Greene, H.W. (1988) Predator simulation and duration of death feigning in neonate hognose snakes. *Animal Behaviour* **36**: 1842–4.

Burke, G.H., Mook, D.G. & Blass, E.M. (1972) Hyperreactivity to quinine associated with osmotic thirst in the rat. *Journal of Comparative & Physiological Psychology* **78**: 32–9.

Burke, T., Davies, N.B., Bruford, M.W. & Hatchwell, B.J. (1989) Parental care and mating behaviour of polyandrous dunnocks, *Prunella modularis*, related to paternity by DNA fingerprinting. *Nature* **338**: 249–51.

Burkhardt, D. and Motte, I. de la (1987) Physiological, behavioural and morphometric data elucidate the evolutive significance of stalked eyes in Diopsidae (Diptera). *Entomologie Generalis* **12**: 221–33.

Burley, N. (1986) Sex ratio manipulation in color-banded populations of zebra finches. *Evolution* **40**: 1191–206.

Burnell, K. (1998) Cultural variation in savannah sparrow, *Passerculus sandwichensis*, songs: an analysis using the meme concept. *Animal Behaviour* **56**: 995–1003.

Bushnell, I.R.W. (1982) Discrimination of faces by young infants. *Journal of Experimental Child Psychology* **33**: 298–308.

Buskirk, R.E., Frohlich, C. & Ross, K.G. (1984) The natural selection of sexual cannibalism. *The American Naturalist* **123**: 612–25.

Buss, D.M. (1989) Sex differences in human mate preferences: evolutionary hypotheses testing in 37 cultures. *Behavioral & Brain Sciences* **12**: 1–49.

Buss, D.M. (1999) *Evolutionary Psychology: The New Science of the Mind*. London, Allyn & Bacon.

Buss, D.M. & Schmidt, D.P. (1993) Sexual strategies theory: an evolutionary perspective on human mating. *Psychological Review* **100**: 204–32.

Buss, D.M. & Shackelford, T.K. (1997) Susceptibility to infidelity in the first year of marriage. *Journal of Research in Personality* **31**: 1–29.

Buss, D.M., Larsen, R.J., Westen, D., & Semmelroth, J. (1992) Sex differences in jealousy: evolution, physiology and psychology. *Psychological Science* **3**: 251–5.

Buunk, B.P., Angleitner, A., Oubaid, V. & Buss, D.M. (1996) Sex differences in jealousy in evolutionary and cultural perspective: tests from the Netherlands, Germany and the United States. *Psychological Science* **7**: 359–63.

Buys, C.J. & Larsen, K.L. (1979) Human sympathy groups. *Psychological Report* **45**: 547–53.

Byers, J.A. (1992) Dirichlet tessellation of bark beetle spatial attack points. *Journal of Animal Ecology* **61**: 759–68.

Byers, J.A. & Bekoff, M. (1986) What does kin recognition mean? *Ethology* **72**: 342–5.

Byers, J.A. & Walker, C. (1995) Refining the motor training hypothesis for the evolution of play. *The American Naturalist* **146**: 25–40.

Byrne, R.W. (1995) *The Thinking Ape: Evolutionary Origins of Intelligence*. Oxford, Oxford University Press.

Byrne, R. & Whiten, A. eds (1988) *Machiavellian Intelligence: Social Expertise and the Evolution of Intelligence in Monkeys, Apes and Humans*. Oxford, Oxford Science Publications.

Cabib, S. (1993) Neurobiological basis of stereotypies. In (A.B. Lawrence & J. Rushen eds) *Stereotypic Animal Behavior: Fundamentals and Applications to Welfare*. Toronto, CAB International.

Cabib, S., Orsini, C., Moal, M. le & Piazza, P.V. (2000) Abolition and reversal of strain differences in behavioral responses drugs of abuse after a brief exposure. *Science* **289**: 463–5.

Cade, W.H. (1979) The evolution of alternative male reproductive strategies in field crickets. In (M. Blum & N.A. Blum eds) *Sexual Selection and Reproductive Competition in Insects*, pp. 343–79. London, Academic Press.

Caldwell, R.L. & Dingle, H. (1976) Stomatopods. *Scientific American* **234**: 80–9.

Call, J. & Tomasello, M. (1995) Use of social information in the problem solving of orangutans (*Pongo pygmaeus*) and human children (*Homo sapiens*). *Journal of Comparative Psychology* **109**: 308–20.

Calvert, W.H., Hedrick, L.E. & Brower, L.P. (1979) Mortality of the monarch butterfly, *Danaus plexippus*: avian predation at five over-wintering sites in Mexico. *Science* **204**: 847–51.

Camacho, J.P.M., Bakkali, M., Corral, J.M., Cabrero, J., López-León, M.D., Aranda, I., Martín-Alganza, A. and Perfectti, F. (2002) Host recombination is dependent on the degree of parasitism. *Proceedings of the Royal Society of London Series B* **269**: 2173–7.

Camara, M.D. (1997) Physiological mechanisms underlying the costs of chemical defence in *Junonia coenia* Hubner (Nymphalidae): a gravimetric and quantitative genetic analysis. *Evolutionary Ecology* **11**: 451–69.

Cameron, E.Z., Linklater, W.L., Stafford, K.J. & Veltman, C.J. (1999) Birth sex ratios relate to mare condition at conception in Kainanawa horses. *Behavioral Ecology* **10**: 472–5.

Camhi, J.M. (1984) *Neuroethology*. Sunderland, MA, Sinauer.

Camhi, J.M., Tom, W. & Volman, S. (1978) The escape behavior of the cockroach *Periplaneta americana*: II. Detection of natural predators by air displacement. *Journal of Comparative Physiology* **128**: 203–12.

Campbell, A. (1995) A few good men: evolutionary psychology and female adolescent aggression. *Ethology & Sociobiology* **16**: 99–123.

Campbell, B.A. & Mabry, P.D. (1972) Ontogeny of behavioral arousal: a comparative study. *Journal of Comparative and Physiological Psychology* **81**: 371–9.

Campbell, N.A., Reece, J.B. & Mitchell, L.G. (1999) *Biology*, 5th Edition. New York, Addison-Wesley.

Candolin, U. & Reynolds, J.D. (2002) Sexual signaling in the European bitterling: females learn the truth by direct inspection of the resource. *Behavioral Ecology* **12**: 407–11.

Cant, M.A. & Johnstone, R.A. (2000) Power struggles, dominance testing, and reproductive skew. *American Naturalist* **155**: 406–17.

Caraco, T. (1979) Time budgeting and group size: a test of theory. *Ecology* **60**: 618–27.

Caraco, T. (1981) Risk-sensitivity and foraging groups. *Ecology* **62**: 527–31.

Caraco, T. & Wolf, L.L. (1975) Ecological determinants of group sizes of foraging lions. *American Naturalist* **109**: 343–52.

Caraco, T., Martindale, S. & Pulliam, H.R. (1980b) Avian flocking in the presence of a predator. *Nature* **285**: 400–1.

Caraco, T., Martindale, S. & Whitham, T. (1980a) An empirical demonstration of risk-sensitive foraging preferences. *Animal Behaviour* **28**: 820–30.

Carayon, J. (1974) Insemination traumatique hétérosexuelle et homosexuelle chez *Xylocoris maculipennis* (Hem. Anthocoridae) *Comptes Rendus de l'Academie des Sciences Paris D* **278**: 2803–6.

Carlin, N.F. (1989) Discrimination between and within colonies of social insects: two null hypotheses. *Netherlands Journal of Zoology* **39**: 86–100.

Caro, T.M. (1980) The effects of experience on the predatory patterns of cats. *Behavioral and Neural Biology* **29**: 1–28.

Caro, T.M. (1986a) The functions of stotting: a review of the hypotheses. *Animal Behaviour* **34**: 649–62.

Caro, T.M. (1986b) The functions of stotting in Thomson's gazelles: some tests of the predictions. *Animal Behaviour* **34**: 663–84.

Caro, T.M. (1995) Short-term costs and correlates of play in cheetahs. *Animal Behaviour* **49**: 333–45.

Caro, T.M. & Borgerhoff Mulder, M. (1987) Problem of adaptation in the study of human behaviour. *Ethology & Sociobiology* **8**: 61–72.

Caro, T.M. & Hauser, M.D. (1992) Is there teaching in nonhuman animals? *Quarterly Review of Biology* **67**: 151–74.

Carpenter, F.L. & MacMillen, R.E. (1976) Threshold model of feeding territoriality and test with a Hawaiian honeycreeper. *Science* **194**: 639–42.

Carpenter, F.L., Paton, D.C. & Hixon, M.A. (1983) Weight gain and adjustment of feeding territory size in migrant hummingbirds. *Proceedings of the National Academy of Sciences of the USA* **80**: 7259–63.

Carter, R. (1998) *Mapping the Mind*. Berkeley, CA, University of California Press.

Cartwright, J. (2000) *Evolution and Human Behaviour*. Basingstoke, Macmillan Press.

Caryl, P.G. (1979) Communication by agonistic displays: what can games theory contribute to ethology? *Behaviour* **68**: 136–69.

Caryl, P.G. (1982) Animal signals: a reply to Hinde. *Animal Behaviour* **30**: 240–4.

Catania, A.C. (1963) Concurrent performances: a baseline for the study of reinforcement magnitude. *Journal of the Experimental Analysis of Behavior* **6**: 299–300.

Catchpole, C.K. (1979) *Vocal Communication in Birds*. London, Edward Arnold.

Catchpole, C.K., Leisler, B. & Winkler, H. (1985) The evolution of polygyny in the great reed warbler, *Acrocephalus arundinaceus*: a possible case of deception. *Behavioral Ecology & Sociobiology* **16**: 285–91.

Catterall, C.P., Elgar, M.A. & Kikkawa, J. (1992) Vigilance does not covary with group size in an island population of silvereyes (*Zosterops lateralis*). *Behavioral Ecology* **3**: 207–10.

Cavalli-Sforza, L.L. & Feldman, M.W. (1973) Cultural versus biological inheritance: phenotypic transmission from parent to children (a theory of the effect of parental phenotypes on children's phenotype). *American Journal of Human Genetics* **25**: 618–37.

Cavalli-Sforza, L.L. & Feldman, M.W. (1981) *Cultural Transmission and Evolution: A Quantitative Approach*. Princeton, NJ, Princeton University Press.

Cawthorn, J.M., Morris, D.L., Ketterson, E.D. & Nolan, V. Jr (1998) Influence of experimentally elevated testosterone on nest defence in dark-eyed juncos. *Animal Behaviour* **56**: 617–21.

Chadab, R. & Rettenmeyer, C.W. (1975) Mass recruitment by army ants. *Science* **188**: 1124–5.

Chagnon, N.A. (1981) Terminological kinship, genealogical relatedness and village fission among the Yanomamö Indians. In (R.D. Alexander & D.W. Tinkle eds) *Natural Selection and Social Behavior*, pp. 490–508. New York, Chiron Press.

Chagnon, N.A. (1983) *Yanomamö, the Fierce People*, 3rd Edition. New York, Holt Rhinehart and Winston.

Chagnon, N.A. (1988) Life histories, blood revenge, and warfare in a tribal population. *Science* **239**: 985–92.

Chagnon, N.A. & Bugos, P.E. (1979) Kin selection and conflict: an analysis of a Yanomamö ax fight. In (N.A. Chagnon & W. Irons eds) *Evolutionary Biology and Human Social Behavior: An Anthropological Perspective*, pp. 213–49. North Scituate, MA, Duxbury Press.

Chalfie, M. & Sulston, J. (1981) Developmental genetics of the mechanosensory cells of *Caenorhabditis elegans*. *Developmental Biology* **82**: 358–70.

Champoux, M., Bennett, A., Lesch, K.-P., Heils, A., Neilsen, D.A., Higley, J.D. & Suomi, S.J. (1999) Serotonin transporter gene polymorphism and neuro-behavioural development in rhesus monkey neonates. *Society for Neuroscience Abstracts* 25: 32.12.

Chance, M.R.A. & Russell, W.M.S. (1959) Protean displays: a form of allaesthenic behaviour. *Proceedings of the Zoological Society of London* 132: 65–70.

Chang, S. & Kung, C. (1973) Temperature-sensitive pawns: conditional behavioural mutants of *Paramecium aurelia*. *Science* 180: 1197–9.

Chapais, B. (1992) The role of alliances in social inheritance of rank among female primates. In (A.H. Harcourt & F.B.M. de Waal eds) *Coalitions and Alliances in Humans and Other Animals*, pp. 29–59. Oxford, Oxford University Press.

Chapman, G. & Barker, W.B. (1966) *Zoology for Intermediate Students*. London, Longman.

Chapman, R.F. (1971) *The Insects: Structure and Function*. London, The English Universities Press.

Chapman, T., Arnqvist, G., Bangham, J. & Rowe, L. (2003) Sexual conflict. *Trends in Ecology & Evolution* 18: 41–7.

Chappell, J. & Kacelnik, A. (2002) Tool selectivity in a non-primate, the New Caledonian crow (*Corvus moneduloides*). *Animal Cognition* 5: 71–8.

Chapuis, N. & Scardigli, P. (1993) Shortcut ability in hamsters (*Mesocricetus auratus*): the role of environmental and kinaesthetic information. *Animal Learning and Behavior* 21: 255–65.

Charnov, E.L. (1976a) Optimal foraging: attack strategy of a mantid. *American Naturalist* 110: 141–51.

Charnov, E.L. (1976b) Optimal foraging: the marginal value theorem. *Theoretical Population Biology* 9: 129–36.

Charnov, E.L. (1979) The genetical evolution of patterns of sexuality: Darwinian fitness. *American Naturalist* 113: 465–80.

Charnov, E.L. (1982) *The Theory of Sex Allocation*. Princeton, Princeton University Press.

Charnov, E.L. & Krebs, J.R. (1975) The evolution of alarm calls: altruism or manipulation? *American Naturalist* 110: 247–59.

Charnov, E.L., Orians, G. & Hyatt, K. (1976) The ecological implications of resource depression. *American Naturalist* 110: 247–59.

Chater, N. & Heyes, C. (1994) Animal concepts: content and discontent. *Mind and Language* 9: 209–47.

Cheney, D.L. (1977) The acquisition of rank and the development of reciprocal alliances among free-ranging immature baboons. *Behavioral Ecology & Sociobiology* 2: 303–18.

Cheney, D.L. & Seyfarth, R.M. (1990a) Attending to behaviour versus attending to knowledge: examining monkeys' attribution of mental states. *Animal Behaviour* 40: 742–53.

Cheney, D.L. & Seyfarth, R.M. (1990b) *How Monkeys See the World*. Chicago, University of Chicago Press.

Cheng, K. (1986) A purely geometric module in the rat's spatial representation. *Cognition* 23: 149–78.

Cheng, K. (1989) The vector sum model of pigeon landmark use. *Journal of Comparative Psychology: Animal Behavior Processes* 15: 366–75.

Cheng, K. & Spetch, M.L. (1998) Mechanisms of landmark use in mammals and birds. In (S.D. Healy ed.) *Spatial Representation in Animals*, pp. 1–17. Oxford, Oxford University Press.

Chittka, L., Kunze, J., Shipman, C. & Buchmann, S.L. (1995) The significance of landmarks for path integration in homing honeybee foragers. *Naturwissenschaften* 82: 635–45.

Chivers, D.P. & Smith, R.J.F. (1998) Chemical alarm signalling in aquatic predator–prey systems: a review and prospectus. *Ecoscience* 5: 338–52.

Christenfeld, N.J.S. & Hill, E.A. (1995) Whose baby are you? *Nature* 378: 699.

Christian, J.J. & Davis, D.E. (1964) Endocrines, behavior and population. *Science* 146: 1550–60.

Clark, A.B. (1978) Sex ratio and local resource competition in a prosimian primate. *Science* 201: 163–5.

Clark, C.W. & Mangel, M. (1986) The evolutionary advantages of group foraging. *Theoretical Population Biology* 30: 45–79.

Clarke, B.C. (1962) Balanced polymorphism and the diversity of sympatric species. *Systematics Association Publications* 4: 47–70.

Clarke, F.M., Miethe, G.H. & Bennett, N.C. (2001) Reproductive suppression in female Damaraland mole-rats, *Cryptomys damarensis*: dominant control or self-restraint? *Proceedings of the Royal Society of London Series B* 268: 899–909.

Clarke, G.M. & McKenzie, J.A. (1987) Developmental stability of insecticide resistant phenotypes in a blowfly: a result of canalizing natural selection. *Nature* 325: 345–6.

Clarke, G.M., Brand, G.W. & Whitten, M.J. (1986) Fluctuating asymmetry: a technique for measuring developmental stress caused by inbreeding. *Journal of Australian Biological Science* **39**: 145–53.

Clayton, D. & Moore, J. (1997) *Host–Parasite Evolution: General Principles and Avian Models*. Oxford, Oxford University Press.

Clayton, D.F. (2000) The neural basis of avian song learning and perception. In (J.J. Bolhuis ed.) *Brain, Perception, Memory: Advances in Cognitive Neuroscience*, pp. 113–25. Oxford, Oxford University Press.

Clayton, N.S. & Dickinson, A. (1998) Episodic-like memory during cache recovery by scrub jays. *Nature* **395**: 272–4.

Clayton, N.S. & Krebs, J.R. (1994) Hippocampal growth and attrition in birds affected by experience. *Proceedings of the National Academy of Sciences of the USA* **91**: 7410–14.

Clayton, N.S., Reboreda, J.C. & Kacelnik, A. (1997) Seasonal changes in hippocampus volume in parasitic cowbirds. *Behavioural Processes* **41**: 237–43.

Cleland, J. (1995) Obstacles to fertility decline in developing countries. In (R.I.M. Dunbar ed.) *Human Reproductive Decisions*, pp. 207–29. London, MacMillan.

Clemens, L.G. (1974) Neurohormonal control of male sexual behaviour. In (W. Montagna & S. Sadler eds) *Reproductive Behavior*. New York, Plenum.

Cloak, F.T. (1975) Is a cultural ethology possible? *Human Ecology* **3**: 161–82.

Clutton-Brock, T.H. (1989) Mammalian mating systems. *Proceedings of the Royal Society of London Series B* **236**: 339–72.

Clutton-Brock, T.H. (1991) *The Evolution of Parental Care*. Princeton, NJ, Princeton University Press.

Clutton-Brock, T.H. (1998) Reproductive skew, concessions and limited control. *Trends in Ecology and Evolution* **13**: 288–92.

Clutton-Brock, T.H. & Godfray, H.C.J. (1991) Parental investment. In (J.R. Krebs & N.B. Davies eds) *Behavioural Ecology: An Evolutionary Approach*, 3rd Edition, pp. 234–62. Oxford, Blackwell.

Clutton-Brock, T.H. & Harvey, P.H. (1977) Primate ecology and social organization. *Journal of Zoology* **183**: 1–39.

Clutton-Brock, T.H. & Iason, G.R. (1986) Sex ratio variation in mammals. *Quarterly Review of Biology* **61**: 339–74.

Clutton-Brock, T.H. & Parker, G.A. (1995) Sexual coercion in animal societies. *Animal Behaviour* **49**: 1345–65.

Clutton-Brock, T.H. & Vincent, A.C.J. (1991) Sexual selection and the potential reproductive rates of males and females. *Nature* **351**: 58–60.

Clutton-Brock, T.H., Albon, S.D., Gibson, R.M. & Guinness, F.E. (1979) The logical stag: adaptive aspects of fighting in red deer (*Cervus elephus* L.). *Animal Behaviour* **27**: 211–25.

Clutton-Brock, T.H., Albon, S.D., Gibson, R.M. & Guinness, F.E. (1981) Parental investment in male and female offspring in polygynous mammals. *Nature* **289**: 487–9.

Clutton-Brock, T.H., Albon, S.D. & Guinness, F.E. (1986) Great expectations: maternal dominance sex ratios and offspring reproductive success in red deer. *Animal Behaviour* **34**: 460–71.

Clutton-Brock, T.H., Guinness, F.E. & Albon, S.D. (1982) *Red Deer: Behaviour and Ecology of Two Sexes*. Chicago, University of Chicago Press.

Clutton-Brock, T.H., Hiraiwa-Hasegawa, M. & Robertson, A. (1989) Mate choice on fallow deer leks. *Nature* **340**: 463–5.

Coale, A.J. & Treadway, R. (1986) A summary of the changing distribution of overall fertility, marital fertility and the proportion married in the provinces of Europe. In (A.J. Coale & C. Watkins eds) *The Decline of Fertility in Europe*, pp. 31–181. Princeton, Princeton University Press.

Cody, M.L. (1971) Finch flocks in the Mohave desert. *Theoretical Population Biology* **2**: 142–8.

Cohen, S. & McFarland, D.J. (1979) Time-sharing as a mechanism for the control of behaviour sequences during the courtship of the male three-spined stickleback (*Gasterosteus aculeatus*). *Animal Behaviour* **27**: 270–83.

Cohen, T.E., Henzi, V., Kandel, E.R. & Hawkins, R.D. (1991) Further behavioral and cellular studies of dishabituation and sensitisation in *Aplysia*. *Society of Neuroscience Abstracts* **17**: 1302.

Cohen, T.E., Kaplan, S.W., Kandel, E.R. & Hawkins, R.D. (1997) A simplified preparation for relating cellular events to behavior: mechanisms contributing to habituation, dishabituation, and sensitization of the *Aplysia* gill-withdrawal reflex. *Journal of Neuroscience* **17**: 2886–99.

Colagross, A.M.L. & Cockburn, A. (1993) Vigilance and grouping in the eastern gray kangaroo, *Macropus giganteus*. *Australian Journal of Zoology* **41**: 325–34.

Colgan, P.W. (1989) *Animal Motivation*. New York, Routledge, Chapman and Hall.

Collett, T.S. (1983) Sensory guidance of motor behaviour. In (T.R. Halliday & P.J.B. Slater eds) *Animal Behaviour 1: Causes and Effects*, pp. 40–74. Oxford, Blackwell.

Collett, T.S. & Zeil, J. (1998) Places and landmarks: an arthropod perspective. In (S.D. Healy ed.) *Spatial Representation in Animals*, pp. 18–53. Oxford, Oxford University Press.

Collins, S.A. (1993) Is there only one type of male handicap? *Proceedings of the Royal Society of London Series B* **252**: 193–7.

Collins, S.A. (2000) Men's voices and women's choices. *Animal Behaviour* **60**: 773–80.

Colwill, R.M. (1994) Associative representations of instrumental contingencies. *Psychology of Learning and Motivation* **31**: 1–72.

Colwill, R.M. & Rescorla, R.A. (1985) Post-conditioning devaluation of a reinforcer affects instrumental responding. *Journal of Experimental Psychology: Animal Behavior Processes* **11**: 120–32.

Congdon, J.D., Dunham, A.E. & Tinkle, D.W. (1982) Energy budgets and life histories of reptiles. In (C. Gans & F. Billett eds) *Biology of the Reptilia*, pp. 131–46. New York, Academic Press.

Connor, R.C. (1986) Pseudoreciprocity: investing in mutualism. *Animal Behaviour* **34**: 1652–4.

Conrad, C., Lechner, M. & Werner, W. (1996) East German fertility after unification: crisis or adaptation. *Population Development Review* **22**: 331–58.

Conver, M.R., Miller, D.E. & Hunt, G.L. (1970) Female–female pairs and other unusual reproductive associations in ring-billed and California gulls. *Auk* **96**: 6–10.

Cook, M.N., Bolivar, V.J., McFadyen, M.P. & Flaherty, L. (2002) Behavioral differences among 129 substrains: implications for knockout and transgenic mice. *Behavioral Neuroscience* **116**: 600–11.

Cook, R.M. & Cockrell, B.J. (1978) Predator ingestion rate and its bearing on feeding time and the theory of optimal diets. *Journal of Animal Ecology* **47**: 529–47.

Cooley, J.R. & Marshall, D.C. (2001) Sexual signalling in periodical cicadas, *Magicicada* spp. (Hemiptera: Cicadidae). *Behaviour* **138**: 827–55.

Cooper, J.J., Odberg, F. & Nicol, C.J. (1996) Limitations on the effectiveness of environmental improvement in reducing stereotypic behaviour in bank voles (*Clethrionomys glare-olus*). *Applied Animal Behaviour Science* **48**: 237–48.

Cordero, C. (2001) A different look at the false head of butterflies. *Ecological Entomology* **26**: 106–8.

Cosmides, L. & Tooby, J. (1992) Cognitive adaptations for social exchange. In (J. Barkow, L. Cosmides & J. Tooby eds) *The Adapted Mind: Evolutionary Psychology and the Generation of Culture*, pp. 163–228. New York, Oxford University Press.

Cosmides, L. & Tooby, J. (1994) Origins of domain specificity: the evolution of functional organization. In (L.A. Hirschfeld & S.A. Gelman eds) *Mapping the Mind*, pp. 85–116. Cambridge, Cambridge University Press.

Cosmides, L. & Tooby, J. (1995) From function to structure: the role of evolutionary biology and computational theories in cognitive neuroscience. In (M. Gazzaniga ed.) *The Cognitive Neurosciences*, pp. 1199–210. Cambridge, MA, MIT Press.

Costa, P.T.J. & McCrae, R.R. (1988) Personality in adulthood: a six-year longitudinal study of self-reports and spouse ratings on the NEO personality inventory. *Journal of Personality and Social Psychology* **54**: 853–63.

Costa, R. & Kyriacou, C.P. (1998) Functional and evolutionary implications of natural variation in clock genes. *Current Opinion in Neurobiology* **8**: 659–64.

Costa, R., Peixoto, A.R., Barbujani, G. & Kyriacou, C.P. (1992) A latitudinal cline in a *Drosophila* clock gene. *Proceedings of the Royal Society of London Series B* **250**: 43–9.

Coulson, J.C. (1968) Differences in the quality of birds nesting in the centre and on the edges of a colony. *Nature* **217**: 478–9.

Courchamp, F., Rasmussen, G.S.A. & Macdonald, D.W. (2002) Small pack size imposes a trade-off between hunting and pup-guarding in the painted hunting dog *Lycaeon pictus*. *Behavioral Ecology* **13**: 20–7.

Coussi-Korbel, S. (1994) Learning to outwit a competitor in mangabeys (*Cercocebus torquatus torquatus*). *Journal of Comparative Psychology* **108**: 164–71.

Cowey, A. & Stoerig, P. (1995) Blindsight in monkeys. *Nature* **373**: 247–9.

Cowie, R.J. (1977) Optimal foraging in great tits (*Parus major*). *Nature* **268**: 137–9.

Cowie, R.J. & Krebs, J.R. (1979) Optimal foraging in patchy environments. In (R.M. Anderson, B.D. Turner & L.R. Taylor eds) *The British Ecological Society Symposium 20*.

Population Dynamics, pp. 183–205. Oxford, Blackwell Scientific Publications.

Crabbe, J.C., Wahlsten, D. & Dudek, B.C. (1999) Genetics of mouse behavior: interactions with laboratory environment. *Science* 284: 1670–2.

Craig, J.L. (1984) Are communal pukeko caught in the prisoner's dilemma? *Behavioral Ecology & Sociobiology* 14: 147–50.

Crawford, C.B. (1998) Environments and adaptations: then and now. In (C.B. Crawford & D.L. Krebs eds) *Handbook of Evolutionary Psychology*. Mahwah, NJ, Lawrence Erlbaum.

Crawford, C.B., Salter, B.E. & Jang, K.L. (1989) Human grief: is its intensity related to the reproductive value of the deceased? *Ethology & Sociobiology* 10: 297–307.

Crean, C.S., Dunn, D.W., Day, T.H. & Gilburn, A.S. (2000) Female mate choice for large males in several species of seaweed fly (Diptera: Coelopidae). *Animal Behaviour* 59: 121–6.

Creel, S.R. (2001) Social dominance and stress hormones. *Trends in Ecology & Evolution* 16: 491–7.

Crespi, B.J. (1992) Eusociality in Australian gall thrips. *Nature* 359: 724–6.

Crespi, B.J. (1994) Three conditions for the evolution of eusociality: are they sufficient? *Insectes Sociaux* 41: 395–400.

Crespi, B.J. (2001) The evolution of social behavior in microoganisms. *Trends in Ecology & Evolution* 16: 178–83.

Cresswell, J.L., Egger, P., Fall, C.H.D., Osmond, C., Fraser, R.B. & Barker, D.J.P. (1997) Is the age of menopause determined *in utero*? *Early Human Development* 49: 143–8.

Crisp, D.J. (1961) Territorial behaviour in barnacle settlement. *Journal of Experimental Biology* 38: 429–46.

Crowley, P.H., Travers, S.E., Linton, M.C., Cohn, S.L., Sih, A. & Sargent, R.C. (1991) Mate density, predation risk, and the seasonal sequence of mate choice: a dynamic game. *American Naturalist* 137: 567–96.

Croze, H.J. (1970) Searching image in carrion crows. *Zeitschrift für Tierpsychologie Beiheft* 5: 1–86.

Crudgington, H.S. & Siva-Jothy, M.T. (2000) Genital damage, kicking and early death: the battle of the sexes takes a sinister turn in the bean weevil. *Nature* 407: 855–6.

Cullen, J.M. (1966) Ritualization of animal activities in relation to phylogeny, speciation and ecology: reduction in ambiguity through ritualization. *Philosophical Transactions of the Royal Society of London Series B* 241: 363–74.

Cullen, J.M. (1972) Some principles of animal communication. In (R.A. Hinde ed.) *Nonverbal Communication*, pp. 101–22. Cambridge, Cambridge University Press.

Curio, E. (1976) *The Ethology of Predation*. Berlin, Springer.

Curio, E. (1978) The adaptive significance of avian mobbing: I. Teleonomic hypotheses and predictions. *Zeitschrift für Tierpsychologie* 48: 175–83.

Curio, E., Ernst, U. & Vieth, W. (1978) The adaptive significance of avian mobbing. *Zeitschrift für Tierpsychologie* 48: 184–202.

Curtis, C. (1968) *Biology*. New York, Worth.

Cuthill, I.C. (1985) Experimental studies of optimal foraging theory. Unpubl. D. Phil. thesis, University of Oxford.

Cuthill, I.C., Kacelnik, A., Krebs, J.R., Haccou, P. & Iwasa, Y. (1990) Patch use by starlings: the effect of recent experience on foraging decisions. *Animal Behaviour* 40: 625–40.

Cuthill, I.C., Haccou, P. & Kacelnik, A. (1994) Starlings (*Sturnus vulgaris*) exploiting patches: response to long term changes in travel time. *Behavioural Ecology* 5: 81–90.

Daan, S. & Tinbergen, J. (1997) Adaptation and life history strategies. In (J.R. Krebs & N.B. Davies eds) *Behavioural Ecology: An Evolutionary Approach*, 4th Edition, pp. 311–33. Oxford, Blackwell.

Daly, M. & Wilson, M.I. (1981) Abuse and neglect of children: an evolutionary perspective. In (R.D. Alexander & D.W. Tinkle eds) *Natural Selection and Social Behavior*, pp. 405–16. New York, Chiron.

Daly, M. & Wilson, M.I. (1982) Whom are newborn babies said to resemble? *Ethology & Sociobiology* 3: 69–78.

Daly, M. & Wilson, M.I. (1988) *Homicide*. New York, Aldine.

Daly, M. & Wilson, M.I. (1990) Is parent–offspring conflict sex-linked? Freudian and Darwinian models. *Journal of Personality* 58: 163–89.

Daly, M. & Wilson, M.I. (1999) Human evolutionary psychology and animal behaviour. *Animal Behaviour* 57: 509–19.

Daly, M., Wilson, M.I. & Weghorst, S.J. (1982) Male sexual jealousy. *Ethology & Sociobiology* 3: 11–27.

D'Amato, M.R. & Sant, S. van (1988) The person concept in monkeys. *Journal of Experimental Psychology: Animal Behavior Processes* 11: 35–51.

Danhardt, J. & Lindström, A. (2001) Optimal departure decisions of songbirds from an experimental stopover site and the significance of weather. *Animal Behaviour* 62: 235–43.

Darwin, C.R. (1859) *On the Origin of Species*. London, John Murray.

Darwin, C.R. (1871) *The Descent of Man and Selection in Relation to Sex*. London, John Murray.

Darwin, C.R. (1872) *The Expressions of Emotions in Man and Other Animals*. London, John Murray.

Datta, S.B. (1983) Relative power and the acquisition of rank. In (R.A. Hinde ed.) *Primate Social Relationships*, pp. 93–103. Sunderland, MA, Sinauer.

Davies, N.B. (1977) Prey selection and the search strategy of the spotted flycatcher (*Muscicapa striata*): a field study on optimal foraging. *Animal Behaviour* **25**: 1016–33.

Davies, N.B. (1978a) Ecological questions about territorial behaviour. In (J.R. Krebs & N.B. Davies eds) *Behavioural Ecology: An Evolutionary Approach*, 1st Edition, pp. 317–50. Oxford, Blackwell.

Davies, N.B. (1978b) Territorial defence in the speckled wood butterfly (*Pararge aegeria*): the resident always wins. *Animal Behaviour* **26**: 138–47.

Davies, N.B. (1981) Calling as an ownership convention on pied wagtail territories. *Animal Behaviour* **29**: 529–34.

Davies, N.B. (1985) Cooperation and conflict among dunnocks, *Prunella modularis*, in a variable mating system. *Animal Behaviour* **33**: 628–48.

Davies, N.B. (1991) Mating systems. In (J.R. Krebs & N.B. Davies eds) *Behavioural Ecology: An Evolutionary Approach*, 3rd Edition, pp. 263–94. Oxford, Blackwell.

Davies, N.B. (1992) *Dunnock Behaviour and Social Evolution*. Oxford, Oxford University Press.

Davies, N.B. (1999) Cuckoos and cowbirds versus hosts: coevolutionary lag and equilibrium. *Ostrich* **70**: 71–9.

Davies, N.B. & Halliday, T.R. (1978) Deep croaks and fighting assessment in toads, *Bufo bufo*. *Nature* **274**: 683–5.

Davies, N.B. & Halliday, T.R. (1979) Competitive mate searching in common toads, *Bufo bufo*. *Animal Behaviour* **27**: 1253–67.

Davies, N.B. & Hartley, I.R. (1996) Food patchiness, territory overlap and social systems: an experiment with dunnocks (*Prunella modularis*). *Journal of Animal Ecology* **65**: 837–46.

Davies, N.B. & Houston, A.I. (1981) Owners and satellites: the economics of territory defence in the pied wagtail, *Motacilla alba*. *Journal of Animal Ecology* **50**: 157–80.

Davies, N.B. & Houston, A.I. (1983) Time allocation between territories and flocks and owner–satellite conflict in foraging pied wagtails, *Motacilla alba*. *Journal of Animal Ecology* **52**: 621–34.

Davies, N.B. & Houston, A.I. (1984) Territory economics. In (J.R. Krebs & N.B. Davies eds) *Behavioural Ecology: An Evolutionary Approach*, 2nd Edition, pp. 148–69. Oxford, Blackwell.

Davies, N.B., Kilner, R.M. & Noble, D.G. (1998) Nestling cuckoos, *Cuculus canorus*, exploit hosts with begging calls that mimic a brood. *Proceedings of the Royal Society of London Series B* **265**: 673–8.

Davis, R.L. & Dauwalder, B. (1991) The *Drosophila* dunce locus. *Trends in Genetics* **7**: 224–9.

Davis, W.J., Mpitsos, G.J. & Pinneo, J.M. (1974a) The behavioral hierarchy of the mollusk Pleurobranchaea: I. The dominant position of feeding behavior. *Journal of Comparative Physiology* **90**: 207–24.

Davis, W.J., Mpitsos, G.J. & Pinneo, J.M. (1974b) The behavioral hierarchy of the mollusk Pleurobranchaea: II. Hormond suppression if feeding associated with egg-laying. *Journal of Comparative Physiology* **90**: 225–43.

Dawkins, M.S. (1971) Perceptual changes in chicks: another look at the 'search image' concept. *Animal Behaviour* **19**: 566–74.

Dawkins, M.S. (1980) *Animal Suffering*. London, Chapman & Hall.

Dawkins, M.S. (1983) The organisation of motor patterns. In (T.R. Halliday & P.J.B. Slater eds) *Animal Behaviour. 1. Causes and Effects*, pp. 75–99. Oxford, Blackwell.

Dawkins, M.S. (1986) *Unravelling Animal Behaviour*, 1st Edition. Harlow, Longman.

Dawkins, M.S. (1990) From an animal's point of view: motivation, fitness and animal welfare. *Behavioral and Brain Science* **13**: 1–61.

Dawkins, M.S. (1993a) *Through Our Eyes Only*. Oxford, Freeman.

Dawkins, M.S. (1993b) Are there general principles of signal design? *Philosophical Transactions of the Royal Society of London Series B* **340**: 251–5.

Dawkins, M.S. (1995) *Unravelling Animal Behaviour*, 2nd Edition. Harlow, Longman.

Dawkins, M.S. & Guilford, T. (1991) The corruption of honest signalling. *Animal Behaviour* **41**: 865–73.

Dawkins, R. (1976) Hierarchical organization: a candidate principle for ethology. In (P.P.G. Bateson & R.A. Hinde eds) *Growing Points in Ethology*, pp. 7–54. Cambridge, Cambridge University Press.

Dawkins, R. (1978) Replicator selection and the extended phenotype. *Zeitschrift für Tierpsychologie* 47: 61–76.

Dawkins, R. (1979) Twelve misunderstandings of kin selection. *Zeitschrift für Tierpsychologie* 51: 184–200.

Dawkins, R. (1989) *The Selfish Gene*, 2nd Edition. Oxford, Oxford University Press.

Dawkins, R. (1998) *Unweaving the Rainbow*. London, Allen Lane.

Dawkins, R. (1999) *The Extended Phenotype: The Long Reach of the Gene*, 2nd Edition. Oxford, Oxford University Press.

Dawkins, R. & Brockmann, H.J. (1980) Do digger wasps commit the Concorde fallacy? *Animal Behaviour* 28: 892–6.

Dawkins, R. & Carlisle, T.R. (1976) Parental investment, mate desertion and a fallacy. *Nature* 262: 131–3.

Dawkins, R. & Krebs, J.R. (1978) Animal signals: information or manipulation? In (J.R. Krebs & N.B. Davies eds) *Behavioural Ecology: An Evolutionary Approach*, 1st Edition, pp. 282–309. Oxford, Blackwell.

Dawkins, R. & Krebs, J.R. (1979) Arms races between and within species. *Proceedings of the Royal Society of London Series B* 205: 489–511.

Dean, J., Anehansley, J.J., Edgerton, H.E. & Eisner, T. (1990) Defensive spray of the bombardier beetle: a biological pulse jet. *Science* 248: 1219–21.

DeBach, P. & Smith, H.S. (1941) The effect of host density on the rate of reproduction of entomophagous parasites. *Journal of Economic Entomology* 34: 741–5.

DeCoursey, P.J. (1960) Phase control of activity in a rodent. *Cold Spring Harbor Symposium on Quantitative Biology* 25: 49–55.

DeFries, J.C., Thomas, E.A., Hegmann, J.P. & Weir, M.W. (1967) Open-field behaviour in mice: analysis of maternal effects by means of ovarian transplantation. *Psychonomic Science* 8: 207–8.

DeFries, J.C., Hegmann, J.P. & Halcomb, R.A. (1974) Response to 20 generations of selection for open-field activity in mice. *Behavioral Biology* 11: 481–95.

Degos, L., Colombani, J., Chavantr, A., Bengtson, B. & Jacqard, A. (1974) A selective pressure on HLA polymorphism. *Nature* 249: 62–3.

Delius, J.D. (1992) Categorical discrimination of objects and pictures by pigeons. *Animal Learning and Behavior* 20: 301–11.

Denisenko-Nehrbass, N.I., Jarvis, E., Scharff, C., Nottebohm, F. & Mello, C.V. (2000) Site-specific retinoic acid production in the brain of adult songbirds. *Neuron* 27: 359–70.

Dennett, D.C. (1991) *Consciousness Explained*. Boston, Little, Brown.

Dennett, D.C. (1995) *Darwin's Dangerous Idea: Evolution and the Meanings of Life*. New York, Simon & Shuster.

Dethier, V.G. (1982) The contribution of insects to the study of motivation. In (A.R. Morrison & P.L. Strick eds) *Changing Concepts in the Nervous System*, pp. 445–55. New York, Academic Press.

Deutsch, J.A. (1960) *The Structural Basis of Behaviour*. Cambridge, Cambridge University Press.

DeVoogd, T.J., Krebs, J.R. & Healy, S.D. (1993) Relations between repertoire size and the volume of brain nuclei related to song: comparative evolutionary analyses among oscine birds. *Proceedings of the Royal Society of London Series B* 254: 75–82.

DeVries, P.J. (1997) *The Butterflies of Costa Rica and their Natural History. Volume II: Riodinidae*. Princeton, NJ, Princeton University Press.

Dewsbury, D.A. (1984) *Comparative Psychology in the Twentieth Century*. New York, Hutchinson Ross.

Dewsbury, D.A. (1978) *Comparative Animal Behavior*. New York, McGraw-Hill.

Diamond, J. (1992) *The Rise and Fall of the Third Chimpanzee*. London, Vintage.

Diamond, J. (1997) *Guns, Germs and Steel: A Short History of Everybody for the Last 13 000 Years*. London, Jonathan Cape.

Dickinson, A. (1980) *Contemporary Animal Learning Theory*. Cambridge, Cambridge University Press.

Dickinson, A., Hall, G. & Mackintosh, N.J. (1976) Surprise and the attenuation of blocking. *Journal of Experimental Psychology: Animal Behavior Processes* 2: 313–22.

Dickinson, J.L. & Weathers, W.W. (1999) Replacement males in the western bluebird: opportunity for paternity, chick-feeding rules, and fitness consequences of male parental care. *Behavioral Ecology & Sociobiology* 25: 201–9.

Dilger, W. (1962) The behavior of lovebirds. *Scientific American* **206**: 88–98.

Dingle, H. (1972) Aggressive behavior in stomatopods and the use of information theory in the analysis of animal communication. In (H.E. Winn & B. Olla eds) *Behavior of Marine Animals*, Vol. 1, pp. 126–56. New York, Plenum Press.

Dingle, H. (1996) *Migration: The Biology of Life on the Move*. New York, Oxford University Press.

Dittus, W.P.J. (1977) The social regulation of population density and age–sex distribution in the toque monkey. *Behaviour* **63**: 281–322.

Dixson, A.F. (1998) *Primate Sexuality*. Oxford, Oxford University Press.

Dobzhansky, T. (1973) Nothing in biology makes sense except in the light of evolution. *American Biology Teacher* **35**: 125–9.

Dominey, W.J. (1981) Anti-predator function of bluegill sunfish nesting colonies. *Nature* **290**: 586–8.

Domjan, M. (1983) Biological constraints on instrumental and classical conditioning: implications for general process theory. *Psychology of Learning and Motivation* **17**: 215–77.

Dongen, S. van (2001) Modelling developmental instability in relation to individual fitness: a fully Bayesian latent variable model approach. *Journal of Evolutionary Biology* **14**: 552–63.

Dorn, N.J., Cronin, G. & Lodge, D.M. (2001) Feeding preferences and performance of an aquatic lepidopteran on macrophytes: plant hosts as food and habitat. *Oecologia* **128**: 406–15.

Dornfeldt, K. (1982) Dependence of the homing pigeon's initial orientation on topographical and meteorological variables: a multivariate study. In (F. Papi & H.G. Wallraff eds) *Avian Navigation*, pp. 253–64. Heidelberg, Springer.

Doutrelant, C. & McGregor, P.K. (2000) Eavesdropping and mate choice in female fighting fish. *Behaviour* **137**: 1655–69.

Draper, P. (1989) African marriage systems: perspectives from evolutionary ecology. *Ethology & Sociobiology* **10**: 145–69.

Drent, R.H. & Daan, S. (1980) The prudent parent: energetic adjustment in avian breeding. In (H. Klomp & J.W. Woldendorp eds) *The Integrated Study of Bird Populations*, pp. 225–52. Amsterdam, North Holland.

Drewry, G.E. & Rand, A.S. (1983) Characteristics of an acoustic community: Puerto Rican frogs of the genus *Eleutherodactylus*. *Copeia* **4**: 941–53.

Drickamer, L.C. & Lenington, S. (1987) T-locus effects on the male urinary chemosignal that accelerates puberty in female mice. *Animal Behaviour* **35**: 1581–3.

Drickamer, L.C., Vessey, S.H. & Meikle, D. (1996) *Animal Behavior: Mechanisms, Ecology, Evolution*, 4th Edition. Chicago, W.C. Brown.

Driver, P.M. & Humphries, D.A. (1988) *Protean Behaviour: the Biology of Unpredictability*. Oxford, Oxford Science Publications.

Duberman, L. (1975) *The Reconstituted Family: A Study of Remarried Couples and Their Children*. Chicago, IL, Nelson-Hall.

Duffy, D.L. & Ball, G.F. (2002) Song predicts immunocompetence in male European starlings (*Sturnus vulgaris*). *Proceedings of the Royal Society of London Series B* **269**: 847–52.

Duffy, S. (1996) Eusociality in a coral-reef shrimp. *Nature* **381**: 512–14.

Dugatkin, L.A. (1992a) Sexual selection and imitation: females copy the mate choice of others. *The American Naturalist* **139**: 1384–9.

Dugatkin, L.A. (1992b) Tendency to inspect predators predicts mortality risk in the guppy *Poecilia reticulata*. *Behavioral Ecology* **3**: 124–8.

Dugatkin, L.A. (1997) *Cooperation Among Animals: An Evolutionary Perspective*. Oxford, Oxford University Press.

Dugatkin, L.A. (2002) Winning streak. *New Scientist* **173**, No. 2333: 32–5.

Dugatkin, L.A. & Godin, J.-G.J. (1992) Predator inspection: shoaling and foraging under predation hazard in the Trinidadian guppy. *Environmental Fish Biology* **34**: 265–75.

Dugatkin, L.A. & Reeve, H.K. eds (1998) *Advances in Game Theory and the Study of Animal Behavior*. New York, Oxford University Press.

Dugatkin, L.A., Lucas, J.S. & Godin, J.-G.J. (2002) Serial effects of mate-choice copying in the guppy (*Poecilia reticulata*). *Ethology, Ecology & Evolution* **14**: 45–52.

Dukas, R. (1999) Ecological relevance of associative learning in fruit fly larvae. *Behavioral Ecology & Sociobiology* **45**: 195–200.

Dunbar, R.I.M. (1984) *Reproductive Decisions: An Economic Analysis of Gelada Baboon Social Strategies*. Princeton, NJ, Princeton University Press.

Dunbar, R.I.M. (1993) The coevolution of neocortical size, group size and language in humans. *Behavioral and Brian Sciences* **16**: 681–735.

Dunbar, R.I.M. (1996) *Grooming, Gossip and the Evolution of Language*. London, Faber & Faber.

Dunbar, R.I.M. (2003) Evolution: five big questions. *New Scientist* **178**, no. 2399: 32.

Dunbar, R.I.M. & Spoors, M. (1995) Social networks, support cliques, and kinship. *Human Nature* **6**: 273–90.

Dunbrack, R.L., Coffin, C. & Howe, R. (1995) The cost of males and the paradox of sex: an experimental investigation of the short-term competitive advantages of evolution in sexual populations. *Proceedings of the Royal Society of London Series B* **262**: 45–9.

Duncan, I.J.H. (1993) Welfare is to do with what animals feel. *Journal of Agricultural and Environmental Ethics* **6**: 8–14.

Duncan, J., Seitz, R.J., Kolodny, J., Bor, D., Herzog, H., Ahmed, A., Newell, F.N. & Emslie, H. (2000) A neural basis for general intelligence. *Science* **289**: 457–60.

Duncan, P. & Vigne, N. (1979) The effect of group size in horses on the rate of attacks by blood-sucking flies. *Animal Behaviour* **27**: 623–5.

Durham, W.H. (1991) *Coevolution: Genes, Culture and Human Diversity*. Stanford, Stanford University Press.

Dyer, F.C. (1991) Bees acquire route-based memories but not cognitive maps in a familiar landscape. *Animal Behaviour* **41**: 239–46.

Dyer, F.C. (2002) The biology of the dance language. *Annual Review of Entomology* **47**: 917–49.

Dyer, F.C. & Seeley, T.D. (1989) On the evolution of the dance language. *The American Naturalist* **133**: 580–90.

Dyer, F.C. & Seeley, T.D. (1991) Dance dialects and foraging range in three Asian honey bee species. *Behavioral Ecology & Sociobiology* **28**: 227–33.

Earn, D.J.D. & Johnstone, R.A. (1997) A systematic error in tests of ideal free theory. *Proceedings of the Royal Society of London Series B* **264**: 1671–5.

Eberhard, W.G. (1985) *Sexual Selection and Animal Genitalia*. Cambridge, MA, Harvard University Press.

Eberhard, W.G. & Guttiérres, E.E. (1991) Male dimorphisms in beetles and earwigs and the question of developmental constraints. *Evolution* **45**: 18–28.

Edmunds, M. (1972) Defensive behaviour in Ghanaian praying mantids. *Zoological Journal of the Linnean Society* **51**: 1–12.

Edmunds, M. (1974) *Defence in Animals*. Harlow, Longman.

Eggert, A.K. & Sakaluk, S.K. (1995) Female-coerced monogamy in burying beetles. *Behavioral Ecology & Sociobiology* **37**: 147–53.

Ehrman, L. & Parsons, P.A. (1981) *Behavior Genetics and Evolution*. New York, McGraw-Hill.

Eibl-Eibesfeldt, I. (1963) Angoborenes und Erworbenes im Verhalten einiger Säuger. *Zeitschrift für Tierpsychologie* **20**: 705–45.

Eibl-Eibesfeldt, I. (1975) *Ethology: The Biology of Behaviour*. New York, Holt Rinehart and Winston.

Eichenbaum, H., Fagan, A. & Cohen, N.J. (1986) Normal olfactory discrimination learning set and facilitation of reversal learning after medial-temporal damage in rats: implications for an account of preserved learning abilities in amnesia. *Journal of Neuroscience* **6**: 1876–84.

Eisner, T. & Meinwald, J. (1966) Defensive secretions of arthropods. *Science* **153**: 1341–50.

Eisner, T. & Meinwald, J. (1995) The chemistry of sexual selection. *Proceedings of the National Academy of Sciences of the USA* **92**: 50–5.

Eisner, T., Kluge, A.F., Ikeda, M.I. & Meinwald, Y.C. (1971) Defence of phalangid: liquid repellent administered by leg dabbing. *Science* **173**: 650–2.

Ekman, J. & Askenmo, C. (1986) Reproductive cost, age specific survival and a comparison of the reproductive strategy in two European tits (Genus *Parus*). *Evolution* **40**: 159–68.

Elgar, M.A. (1986) House sparrows establish foraging flocks by giving chirrup calls if the resources are divisible. *Animal Behaviour* **34**: 169–74.

Elgar, M.A. (1989) Predator vigilance and group size in mammals and birds: a critical review of the empirical evidence. *Biological Reviews* **64**: 13–33.

Elgar, M.A. & Catterall, C.P. (1981) Flocking and predator surveillance in house sparrows: test of an hypothesis. *Animal Behaviour* **29**: 868–72.

Elgar, M.A., Burren, P.J. & Posen, M. (1984) Vigilance and perception of flock size in foraging house sparrows (*Passer domesticus* L.). *Behaviour* **90**: 215–23.

Ellegren, H., Gustafsson, L. & Sheldon, B.C. (1996) Sex ratio adjustment in relation to paternal attractiveness in a wild bird population. *Proceedings of the National Academy of Sciences of the USA* **93**: 11723–8.

Elner, R.W. & Hughes, R.N. (1978) Energy maximization in the diet of the shore crab, *Carcinus maenas*. *Journal of Animal Ecology* **47**: 103–16.

Elwood, R.W. & Briffa, M. (2001) Information gathering and communication during agonistic encounters: a case study of hermit crabs. *Advances in the Study of Behaviour* **30**: 53–97.

Embleton, T.F.W., Piercy, J.E. & Olson, N. (1976) Outdoor sound propagation over ground of finite impedance. *Journal of the Acoustical Society of America* **59**: 267–77.

Emery, N.J., Lorincz, E.N., Perrett, D.I., Oram, M.W. & Baker, C.I. (1997) Gaze following and joint attention in rhesus monkeys (*Macaca mulatta*). *Journal of Comparative Psychology* **111**: 286–93.

Emlen, S.T. (1967) Migratory orientation in the indigo bunting, *Passerina cyanea*. Part I: the evidence for use of celestial cues. *Auk* **84**: 309–42.

Emlen, S.T. (1970) Celestial rotation: its importance in the development of migratory orientation. *Science* **170**: 1198–201.

Emlen, S.T. (1972) The ontogenetic development of orientation capabilities. In (S.R. Galler, K. Schmidt-Koenig, G.J. Jacobs & R.E. Belleville eds) *Animal Orientation and Navigation*, pp. 191–210. Washington, DC, NASA SP-262 US Government Printing Office.

Emlen, S.T. (1975) Migration: orientation and navigation. In (D.S. Farner & J.R. King eds) *Avian Biology*, Vol. V, pp. 129–219. London, Academic Press.

Emlen, S.T. (1978) Cooperative breeding. In (J.R. Krebs & N.B. Davies eds) *Behavioural Ecology: An Evolutionary Approach*, 1st Edition, pp. 245–82. Oxford, Blackwell.

Emlen, S.T. (1984) Cooperative breeding in birds and mammals. In (J.R. Krebs & N.B. Davies eds) *Behavioural Ecology: An Evolutionary Approach*, 2nd Edition, pp. 305–39. Oxford, Blackwell.

Emlen, S.T. (1991) Evolution of cooperative breeding in birds and mammals. In (J.R. Krebs & N.B. Davies eds) *Behavioural Ecology: An Evolutionary Approach*, 3rd Edition, pp. 301–37. Oxford, Blackwell.

Emlem, S.T. (1995) An evolutionary theory of the family. *Proceedings of the National Academy of Sciences of the USA* **92**: 8092–9.

Emlen, S.T. (1997a) Predicting family dynamics in social vertebrates. In (J.R. Krebs & N.B. Davies eds) *Behavioural Ecology: An Evolutionary Approach*, 4th Edition, pp. 228–53. Oxford, Blackwell.

Emlen, S.T. (1997b) When mothers prefer daughters over sons. *Trends in Ecology & Evolution* **12**: 291–2.

Emlen, S.T. & Demong, N.J. (1978) Orientation strategies used by free-flying bird migrants: a radar tracking study. In (K. Schmidt-Koenig & W.T. Keeton eds) *Animal Migration: Navigation and Homing*, pp. 283–93. Heidelberg, Springer.

Emlen, S.T. & Emlen, J.T. (1966) A technique for recording migratory orientation of captive birds. *Auk* **83**: 361–7.

Emlen, S.T. & Oring, L.W. (1977) Ecology, sexual selection and the evolution of mating systems. *Science* **197**: 215–23.

Emlen, S.T. & Wrege, P.H. (1988) The role of kinship in helping decisions among white-fronted beeaters. *Behavioral Ecology & Sociobiology* **23**: 305–15.

Endler, J.A. (1983) Natural and sexual selection on color patterns in poeciliid fishes. *Environmental Biology of Fishes* **9**: 173–90.

Endler, J.A. (1984) Progressive background matching in moths and a quantitative measure of crypsis. *Biological Journal of the Linnean Society of London* **22**: 187–231.

Endler, J.A. (1987) Predation, light intensity, and courtship behaviour in *Poecilia reticulata* (Pisces: Poeciliidae). *Animal Behaviour* **35**: 1376–85.

Endler, J.A. (1991) Interactions between predators and prey. In (J.R. Krebs & N.B. Davies eds) *Behavioural Ecology: An Evolutionary Approach*, 3rd Edition, pp. 169–96. Oxford, Blackwell.

Enquist, M. (1985) Communication during aggressive interactions with particular reference to variation in choice of behaviour. *Animal Behaviour* **33**: 1152–61.

Enquist, M. & Arak, A. (1994) Symmetry, beauty and evolution. *Nature*, **372**: 169–72.

Enquist, M. & Leimar, O. (1990) The evolution of fatal fighting. *Animal Behaviour* **39**: 1–9.

Enquist, M., Leimar, O., Ljungberg, T., Mallner, Y. & Segerdahl, M. (1990) A test of the sequential assessment game: fighting in the cichlid fish *Nannacara anomala*. *Animal Behaviour* **40**: 1–14.

Enquist, M., Plane, E. & Roed, J. (1985) Aggressive communication in fulmars (*Fulmaris glacialis*) competing for food. *Animal Behaviour* **33**: 1107–20.

Enright, J.T. (1972) When the beachhopper looks at the moon: the moon-compass hypothesis. In (S.R. Galler, K. Schmidt-Koenig, G.J. Jacobs & R.E. Belleville eds) *Animal Orientation and Navigation*, pp. 523–55. Washington, DC, NASA SP-262 US Government Printing Office.

Epstein, R., Kirshnit, C.E., Lanza, R.P. & Rubin, L.C. (1984) 'Insight' in the pigeon: antecedents and determinants of an intelligent performance. *Nature* **308**: 61–2.

Epstein, R., Lanza, R.P. & Skinner, B.F. (1980) Symbolic communication between two pigeons (*Columba livia domestica*). *Science* **207**: 543–5.

Epstein, R., Lanza, R.P. & Skinner, B.F. (1981) 'Self-awareness' in the pigeon. *Science* **212**: 695–6.

Erichsen, J.T., Krebs, J.R. & Houston, A.I. (1980) Optimal foraging and cryptic prey. *Journal of Animal Ecology* **49**: 271–6.

Escós, J.M., Alados, C.L. & Emlen, J.M. (1995) Fractal structures and fractal functions as disease indicators. *Oikos* **74**: 310–14.

Eshel, I. & Feldman, M.W. (1984) Initial increase of mutants and some continuity properties of ESS in two-locus systems. *The American Naturalist* **124**: 631–40.

Esterling, L.E., Yoshikawa, T., Turner, G., Bengel, D., Gershon, E.S., Berrettini, W.H. & Detera-Wadleigh, S.D. (1997) Serotonin transporter (5-HTT) gene and bipolar affective disorder. *American Journal of Medical Genetics* **81**: 37–40.

Estes, R.D. (1976) The significance of breeding synchrony in the wildebeeste. *East African Wildlife Journal* **14**: 135–52.

Estes, R.D. & Goddard, J. (1967) Prey selection and hunting behaviour of the African wild dog. *Journal of Wildlife Management* **31**: 52–70.

Etienne, A.S., Berlie, J., Georgakopoulos, J. & Maurer, R. (1998) Role of dead reckoning in navigation. In (S.D. Healy ed.) *Spatial Representation in Animals*, pp. 54–68. Oxford, Oxford University Press.

Evans, C.S. & Marler, P. (1994) Food calling and audience effects in male chickens, *Gallus gallus*: their relationships to food availability, courtship and social facilitation. *Animal Behaviour* **47**: 1159–70.

Evans, C.S. & Marler, P. (1995) Language and animal communication: parallels and contrasts. In (H.L. Roitblat & J.-A. Meyer eds) *Comparative Approaches to Cognitive Science*, pp. 342–82. Cambridge, MA, MIT Press.

Evans, H.E. (1977) Extrinsic versus intrinsic factors in the evolution of insect sociality. *Bioscience* **27**: 613–17.

Evans, M.R., Goldsmith, A.R. & Norris, S.R.A. (2000) The effects of testosterone on antibody production and plumage coloration in male house sparrows (*Passer domesticus*). *Behavioral Ecology & Sociobiology* **47**: 156–63.

Evans, R.M. (1982) Foraging flock recruitment at a black-billed gull (*Larus bulleri*) colony: implications for the information center hypothesis. *Auk* **99**: 24–30.

Evans, S.M. (1966) Non-associative behavioural modifications in the polychaete *Nereis diversicolor*. *Animal Behaviour* **14**: 107–19.

Ewert, J.-P. (1974) The neural basis of visually guided behavior. *Scientific American* **230**: 34–42.

Ewert, J.-P. (1980) *Neuroethology*. New York, Springer Verlag.

Ewert, J.-P. (1997) Neural correlates of key stimulus and releasing mechanism: a case study and two concepts. *Trends in Neurosciences* **20**: 332–9.

Ewert, J.P. & Trand, R. (1979) Releasing stimuli for antipredator behaviour in the common toad, *Bufo bufo* CL. *Behaviour* **68**: 170–80.

Ezaki, Y. (1990) Female choice and the causes and adaptiveness of polygyny in great reed warblers. *Journal of Animal Ecology* **59**: 103–19.

Faaborg, J.F., deVries, T., Patterson, C.B. & Griffin, C.R. (1980) Preliminary observations on the occurrence and evolution of polyandry in the Galapagos hawk (*Buteo galapagoensis*). *Auk* **97**: 581–90.

Fagen, R. (1976) Three-generation family conflict. *Animal Behaviour* **24**: 874–9.

Fagen, R. (1977) Selection for optimal age-dependent schedules of play behavior. *The American Naturalist* **112**: 395–414.

Fagen, R. (1981) *Animal Play Behavior*. New York, Oxford University Press.

Fantino, E. (2001) Context: a central concept. *Behavioural Processes* **54**: 95–110.

Fantino, E. & Logan, C.A. (1979) *The Experimental Analysis of Behavior*. San Francisco, Freeman.

Farmer D.C. & Barnard C.J. (2000) Fluctuating asymmetry and sperm transfer in male decorated field crickets (*Gryllodes sigillatus*). *Behavioral Ecology & Sociobiology* **47**: 287–92.

Farner, D.S. & Lewis, R.A. (1971) Photoperiodism and reproductive cycles in birds. *Photophysiology* **6**: 325–70.

Faulkes, C. & Bennett, N. (2001) Family values: group dynamics and social control of reproduction in African mole-rats. *Trends in Ecology & Evolution* **16**: 181–90.

Feldman, M.W. & Laland, K.N. (1996) Gene–culture coevolutionary theory. *Trends in Ecology & Evolution* **11**: 453–7.

Feltmate, B.W. & Williams, D.D. (1989) A test of crypsis and predator avoidance in the stonefly *Paragnetina media* (Plecoptera: Perlidae). *Animal Behaviour* **37**: 992–9.

Fenton, M.B. (1984) The case of vespertilionid and rhinolophid bats. In (R.L. Smith ed.) *Sperm Competition and Animal Mating Systems*, pp. 573–87. London, Academic Press.

Fernald, R.D. & Hirata, N.R (1977) Field study of *Haplochromis burtoni*: quantitative behavioural observations. *Animal Behaviour* 25: 964–75.

Festa-Bianchet, M. (1989) Individual differences, parasites and the costs of reproduction for bighorn ewes (*Ovis canadensis*). *Journal of Animal Ecology* 58: 785–95.

Fischer, J.H., Freake, M.J., Borland, S.C. & Phillips, J.B. (2001) Evidence for the use of magnetic map information by an amphibian. *Animal Behaviour* 62: 1–10.

Fisher, J. & Hinde, R.A. (1949) The opening of milk bottles by birds. *British Birds* 42: 347–57.

Fisher, R.A. (1930) *The Genetical Theory of Natural Selection*. Oxford, Clarendon Press.

Fitzgibbon, C.D. & Fanshawe, J.H. (1988) Stotting in Thomson's gazelles: an honest signal of condition. *Behavioural Ecology & Sociobiology* 23: 69–74.

Fitzsimons, J.T. & LeMagnen, J. (1969) Eating as a regulatory control of drinking in the rat. *Journal of Comparative Physiology and Psychology* 67: 273–83.

Fletcher, D.J.C. & Michener, C.D. eds (1987) *Kin Recognition in Animals*. New York, Wiley.

Flinn, M. (1988) Parent–offspring interactions in a Caribbean village: daughter guarding. In (L. Betzig, M. Borgerhoff Mulder & P. Turke eds) *Human Reproductive Behavior: A Darwinian Perspective*, pp. 189–200. Cambridge, Cambridge University Press.

Flinn, M. (1989) Household compositions and female strategies in a Trinidadian village. In (A.E. Rasa, C. Vogel & E. Voland eds) *The Sociobiology of Sexual and Reproductive Strategies*, pp. 206–33. New York, Chapman and Hall.

Flynn, R.E. & Giraldeau, L.-A. (2001) Producer–scrounger games in a spatially explicit world: tactic use influences flock geometry of spice finches. *Ethology* 107: 249–57.

Fodor, J. (1983) *The Modularity of Mind*. Cambridge, MIT Press.

Folstad, I. & Karter, A.J. (1992) Parasites, bright males and the immunocompetence handicap. *The American Naturalist* 139: 603–22.

Forbes, S. (2002) Pregnancy sickness and embryo quality. *Trends in Ecology and Evolution* 17: 115–19.

Fordyce, J.A. (2001) The lethal plant defense paradox: inducible host–plant aristolochic acids and the growth and defense of the pipevine swallowtail. *Entomologia Experimentalis et Applicata* 100: 339–46.

Forrester, G.E. (1991) Social rank, individual size and group composition as determinants of food consumption by humbug damselfish, *Dascyllus aruanus*. *Animal Behaviour* 42: 701–11.

Foster, W.A. & Treherne, J.E. (1981) Evidence for the dilution effect in the selfish herd from fish predation of a marine insect. *Nature* 293: 466–7.

Fraenkel, G.S. & Gunn, D.L. (1961) *The Orientation of Animals*. New York, Dover.

Fragaszy, D.M. & Visalberghi, E. (1989) Social influences on the acquisition of tool-using behaviors in tufted capuchin monkeys (*Cebus apella*). *Journal of Comparative Psychology* 103: 159–70.

Fraley, N.B. & Fernald, R.D. (1982) Social control of developmental rate in the African cichlid *Haplochromis burtoni*. *Zeitschrift für Tierpsychologie* 60: 66–82.

Francis, R.C., Soma, K. & Fernald, R.D. (1993) Social regulation of the brain–pituitary–gonadal axis. *Proceedings of the National Academy of Sciences of the USA* 90: 7794–8.

Frank, L.G., Weldele, M.L. & Glickman, S.E. (1995) Masculinization costs in hyenas. *Nature* 377: 584–5.

Frankham, R., Ballou, J.D. & Briscoe, D.A. (2002) *Introduction to Conservation Genetics*. Cambridge, Cambridge University Press.

Fraser, D. (1995) Science, values and animal welfare: exploring the 'inextricable connection'. *Animal Welfare* 4: 103–17.

Freeman, S. & Herron, J.C. (2001) *Evolutionary Analysis*, 2nd Edition. Upper Saddle River, NJ, Prentice Hall.

Fretwell, S.D. & Lucas, H.L. (1970) On territorial behaviour and other factors influencing habitat distribution in birds. *Acta Biotheoretica* 19: 16–36.

Friedhoff, A.J., Miller, J.C., Armour, M., Schweitzer, J.W. & Mohan, S. (2000) Role of maternal biochemistry in fetal brain development: effect of maternal thyroidectomy on behavior and biogenic amine metabolism in rat progeny. *International Journal of Neuropsychopharmacology* 3: 89–97.

Frisch, K. von (1941) Uber einen Schreckstoff der Fischhaut und seine biologische Bedeutung. *Zeitschrift für Vergleischende Physiologie* 29: 46–145.

Frisch, K. von (1953) *The Dancing Bees*. New York, Harcourt Brace.

Frisch, K. von (1976) *The Dance Language and Orientation of Bees*. Cambridge, MA, University of Harvard Press.

Fugle, G.N. & Rothstein, S.I. (1987) Experiments on the control of deceptive signals of status in white-crowned sparrows. *Auk* **104**: 188–97.

Fuller, J.L. & Thompson, W.R. (1978) *Foundations of Behavior Genetics*. St Louis, Mosby.

Gage, A.R. & Barnard, C.J. (1996) Male crickets increase sperm number in relation to competition and female size. *Behavioral Ecology & Sociobiology* **38**: 349–53.

Gage, M.J.G. (1991) Risk of sperm competition directly affects ejaculate size in the Mediterranean fruit fly. *Animal Behaviour* **44**: 587–9.

Gagliardi, J.L., Kirkpatrick-Steger, K.K., Thomas, J., Allen, G.L. & Blumberg, M.S. (1995) Seeing and knowing: knowledge attribution versus stimulus control in adult humans (*Homo sapiens*). *Journal of Comparative Psychology* **109**: 107–14.

Gagliardo, A., Ioale, P., Odetti, F. & Bingman, V.P. (2001) The ontogeny of the homing pigeon navigational map: evidence for a sensitive learning period. *Proceedings of the Royal Society of London Series B* **268**: 197–202.

Gahr, M. (2001) Distribution of sex steroid hormone receptors in the avian brain: functional implications for neural sex differences and sexual behaviors. *Microscopy Research and Technique* **55**: 1–11.

Gahr, M., Sonnenschien, E. & Wickler, W. (1998) Sex difference in the size of the neural song control regions in a duetting songbird with similar song repertoire sizes of males and females. *Journal of Neuroscience* **18**: 1124–31.

Galef, B.G., Jr (1996a) Social enhancement of food preferences in Norway rats: a brief review. In (C.M. Heyes & B.G. Galef Jr eds) *Social Learning and Imitation in Animals: The Roots of Culture*, pp. 49–64. New York, Academic Press.

Galef, B.G., Jr (1996b) Tradition in animals: field observations and laboratory analyses. In (M. Bekoff & D. Jamieson eds) *Readings in Animal Cognition*, pp. 91–105. Cambridge, MA, MIT Press.

Galef, B.G., Jr & Allen, C. (1995) A new model system for studying animal tradition. *Animal Behaviour* **50**: 705–17.

Galef, B.G., Jr & Giraldeau, L.-A. (2001) Social influences on foraging in vertebrates: causal mechanisms and adaptive functions. *Animal Behaviour* **61**: 3–15.

Galef, B.G., Jr & Sherry, D. (1973) Mother's milk: a medium for the transmission of cues reflecting the flavor of mother's diet. *Journal of Comparative and Physiological Psychology* **83**: 374–8.

Galef, B.G., Jr & Wigmore, S.W. (1983) Transfer of information concerning distant foods: a laboratory investigation of the 'information centre' hypothesis. *Animal Behaviour* **31**: 748–58.

Galef, B.G., Jr, Manzig, L.A. & Field, R.M. (1986) Imitation learning in budgerigars: Dawson & Foss (1965) revisited. *Behavioural Processes* **13**: 191–202.

Gallese, V. & Goldman, A. (1998) Mirror neurons and the simulation theory of mind. *Trends in Cognitive Science* **2**: 493.

Gallup, G.G., Jr (1970) Chimpanzees: self-recognition. *Science* **167**: 86–7.

Gallup, G.G., Jr, Burch, R.L. Zappieri, M.L., Parrez, R.A., Stockwell, M.L. & Davis, J.A. (2003) The human penis as a sperm displacement device. *Evolution and Human Behaviour* **24**: 277–89.

Gangestad, S.W., Thornhill, R. & Garver, C.E. (2002) Changes in women's sexual interests and their partners' mate-retention tactics across the menstrual cycle: evidence for shifting conflicts of interest. *Proceedings of the Royal Society of London Series B* **269**: 975–82.

Gangestad, S.W., Thornhill, R. & Yeo, R.A. (1994) Facial attractiveness, developmental stability and fluctuating asymmetry. *Ethology & Sociobiology* **15**: 73–85.

Garcia, J. & Koelling, R.A. (1966) Relation of cue to consequence in avoidance learning. *Psychonomic Science* **4**: 123–4.

Garcia, J., Ervin, F.R. & Koelling, R.A. (1966) Learning with prolonged delay of reinforcement. *Psychonomic Science* **5**: 121–2.

Garcia, J.F., McGowan, B.K. & Green, K.F. (1972) Biological constraints on conditioning. In (A.H. Black & W.K. Prokasy eds) *Classical Conditioning: Current Research and Theory* pp. 3–27. New York, Appleton-Century-Crofts.

Gardner, R.A. & Gardner, B.T. (1969) Teaching sign language to a chimpanzee. *Science* **165**: 664–72.

Garner, J.P. & Mason, G.J. (2002) Evidence for a relationship between cage stereotypies and behavioural disinhibition in laboratory rats. *Behavioural Brain Research* **136**: 83–92.

Gatti, S., Ferveur, J.F. & Martin, J.R. (2000) Genetic identification of neurons controlling a sexually dimorphic behaviour. *Current Biology* **10**: 667–70.

Gaulin, S.J.C. & FitzGerald, R.W. (1986) Sex differences in spatial ability: an evolutionary hypothesis and test. *American Naturalist* **127**: 74–88.

Gendron, R.P. & Staddon, J.E. (1983) Searching for cryptic prey: the effects of search rate. *American Naturalist* **121**: 172–86.

Gerhardt, H.C. (1978) Temperature coupling in the vocal communication system of the gray treefrog, *Hyla versicolor. Science* **199**: 992–4.

Gerhardt, H.C. (1982) Sound pattern recognition in some North American treefrogs (Anura: Hylidae): implications for male choice. *American Zoologist* **22**: 581–95.

Gerhardt, H.C. (1983) Communication and the environment. In (T.R. Halliday & P.J.B. Slater eds) *Animal Behaviour, Volume 2: Communication*, pp. 82–133. Oxford, Blackwell.

Getty, T. (1998a) Handicap signalling: when fecundity and viability do not add up. *Animal Behaviour* **56**: 127–30.

Getty, T. (1998b) Reliable signals need not be a handicap. *Animal Behaviour* **56**: 253–5.

Getty, T. (2002) Signalling health versus parasites. *American Naturalist* **159**: 363–71.

Gibb, J.A. (1957) Food requirements and other observations on captive tits. *Bird Study* **4**: 207–15.

Gibb, J.A. (1960) Populations of tits and goldcrests and their food supply in pine plantations. *Ibis* **102**: 163–208.

Gibbon, J. (1977) Scalar expectancy theory and Weber's Law in animal timing. *Psychological Reviews* **84**: 279–325.

Gibbs, C.M., Latham, S.B. & Gormezano, I. (1978) Classical conditioning of the rabbit nictitating membrane response: effects of reinforcement schedule on response maintenance and resistance to extinction. *Animal Learning and Behavior* **6**: 209–15.

Gibo, D.L. (1986) Flight strategies of migratory monarch butterflies (*Danaus plexippus* L.) in southern Ontario. In (W. Danthanarayana ed.) *Insect Flight: Dispersal and Migration*, pp. 172–84. Berlin, Springer.

Gibson, R.M., Bradbury, J.W. & Vehrencamp, S. (1991) Mate choice in lekking sage grouse revisited: the roles of vocal display, female site fidelity, and copying. *Behavioral Ecology* **2**: 165–80.

Giebel, L.B., Tripathi, R.K., King, R.A. & Spritz, R.A. (1991) A tyrosinase gene missense mutation in temperature-sensitive type-1 oculocutaneous albinism: a human homolog to the Siamese cat and the Himalayan mouse. *Journal of Clinical Medicine* **87**: 1119–22.

Gigerenzer, G. & Hug, K. (1992) Domain-specific reasoning: social contracts, cheating and perspective change. *Cognition* **43**: 127–71.

Gil, D. & Gahr, M. (2002) The honesty of bird song: multiple constraints for multiple traits. *Trends in Ecology & Evolution* **17**: 133–41.

Gil, D., Graves, J., Hazon, N. & Wells, A. (1999) Male attractiveness and differential testosterone investment in zebra finches. *Science* **286**: 126–8.

Gilburn, A.S., Foster, S.P. & Day, T.H. (1993) Genetic correlation between a female mating preference and the male preferred character in the seaweed fly *Coelopa frigida. Evolution* **47**: 1788–95.

Gill, K.S. (1963) A mutation causing abnormal courtship and mating behavior in *Drosophila melanogaster. American Zoologist* **3**: 507.

Gillette, R., Huang, R.-C., Hatcher, N. & Moroz, L.L. (2000) Cost–benefit analysis potential in feeding behavior of a predatory snail by integration of hunger, taste and pain. *Proceedings of the National Academy of Sciences of the USA* **97**: 3585–90.

Gimenez-Llort, L., Fernandez-Teruel, A., Escorihuela, R.M., Fredholm, B.B., Tobena, A., Pekny, M. & Johansson, B. (2002) Mice lacking the adenosine A(1) receptor are anxious and aggressive, but are normal learners with reduced muscle strength and survival rate. *European Journal of Neuroscience* **16**: 547–50.

Giraldeau, L.-A. (1997) Ecology of information use. In (J.R. Krebs & N.B. Davies eds) *Behavioural Ecology: An Evolutionary Approach*, 4th Edition, pp. 42–68. Oxford, Blackwell.

Giraldeau, L.-A. & Gillis, D. (1985) Optimal group size can be stable: reply to Sibly. *Animal Behaviour* **33**: 666–7.

Giraldeau, L.-A. & Lefebvre, L. (1986) Exchangeable producer–scrounger roles in a captive flock of feral pigeons. *Animal Behaviour* **34**: 797–803.

Giraldeau, L.-A. & Lefebvre, L. (1987) Scrounging prevents the cultural transmission of food-finding behaviour in pigeons. *Animal Behaviour* **35**: 387–94.

Giraldeau, L.-A., Soos, C. & Beauchamp, G. (1994) A test of the producer–scrounger foraging game in captive flocks of spice finches, *Lonchura punculata. Behavioral Ecology & Sociobiology* **34**: 251–6.

Giurfa, M., Eichmann, B. & Menzel, R. (1996) Symmetry perception in an insect. *Nature* **382**: 458–61.

Gleason, J.M., Nuzhdin, S.V. & Ritchie, M.G. (2002) Quantitative trait loci affecting a courtship signal in *Drosophila melanogaster. Heredity* **89**: 1–6.

Godin, J.-G.J. & Keenleyside, M.H.A. (1984) Foraging on patchily distributed prey by a cichlid fish (Teleostii: Cichlidae): a test of the ideal free distribution theory. *Animal Behaviour* **32**: 120–31.

Godin, J.-G.J. & Smith, S. (1989) A fitness cost of foraging in the guppy. *Nature* **333**: 69–71.

Goldberg, T.A. (1995) Altruism towards panhandlers: who gives? *Human Nature* **6**: 79–90.

Golding, Y.C. & Edmunds, M. (2000) Behavioural mimicry of honeybees (*Apis mellifera*) by droneflies (Diptera: Syrphidae: *Eristalis* spp.). *Proceedings of the Royal Society of London Series B* **267**: 903–9.

Golding, Y.C., Ennos, A.R. & Edmunds, M. (2001) Similarity in flight behaviour between the honeybee *Apis mellifera* (Hymenoptera: Apidae) and its presumed mimic, the dronefly *Eristalis tenax* (Diptera: Syrphidae). *Journal of Experimental Biology* **204**: 139–45.

Goldman, S.A. & Nottebahm, F. (1983) Neuronal production, migration and differentiation in a vocal control nucleus of the adult female canary brain. *Proceedings of the National Academy of Sciences of the USA* **80**: 2390–4.

Goncalves, D.M., Simoes, P.C., Chumbinho, A.C., Correia, M.J., Fagundes, T. & Oliveira, R.F. (2002) Fluctuating asymmetries and reproductive success in the peacock blenny. *Journal of Fish Biology* **60**: 810–20.

Gonzalez, G., Sorci, G., Smith, L.C. & Lope, F. de (2002) Social control and physiological cost of cheating in status signalling male house sparrows (*Passer domesticus*). *Ethology* **108**: 289–302.

Goodall, J. van Lawick (1968) The behaviour of free-living chimpanzees in the Gombe Stream Reserve. *Animal Behaviour Monographs* **1**: 161–311.

Goodenough, J., McGuire, B. & Wallace, R. (1993) *Perspectives on Animal Behavior*. New York, Wiley.

Gosling, L.M. (1982) A reassessment of the function of scent marking in territories. *Zeitschrift für Tierpsychologie* **60**: 89–118.

Gosling, L.M. (1986) The evolution of mating strategies in male antelope. In (D.I. Rubenstein & R.W. Wrangham eds) *Ecological Aspects of Social Evolution*, pp. 244–81. Princeton, NJ, Princeton University Press.

Gosling, L.M., Roberts, S.C., Thornton, E.A. & Andrew, M.J. (2000) Life history costs of olfactory status signalling in mice. *Behavioral Ecology & Sociobiology* **48**: 328–32.

Goss, R.J. (1983) *Deer Antlers: Regeneration, Function and Evolution*. New York, Academic Press.

Gottlieb, G. (1961) Developmental age as a baseline for determination of the critical period in imprinting. *Journal of Comparative Physiology and Psychology* **54**: 422–7.

Gould, J.L. (1976) The dance-language controversy. *Quarterly Review of Biology* **51**: 211–44.

Gould, J.L. (1986) The locale map of honey bees: do insects have a cognitive map? *Science* **232**: 861–3.

Gould, J.L. (1998) Sensory bases of navigation. *Current Biology* **8**: R731–8.

Gould, J.L. & Gould, C.G. (1988) *The Honey Bee*. New York, Scientific American Library.

Gould, J.L. & Gould, C.G. (1994) *The Animal Mind*. New York, Scientific American Library.

Gould, J.L. & Gould, C.G. (1989) *Sexual Selection*. New York, Scientific American Library.

Gould, J.L. & Towne, W.F. (1987) Evolution of the dance language. *American Naturalist* **130**: 317–38.

Gould, S.J. (1980) *The Panda's Thumb: More Reflections on Natural History*. New York, Norton.

Gould, S.J. & Lewontin, R.C. (1979) The spandrels of San Marco and the Panglossian paradigm: a critique of the adaptationist programme. *Proceedings of the Royal Society of London Series B* **205**: 581–98.

Grafe, T.U. (1999) A function of synchronous chorusing and a novel female preference shift in an anuran. *Proceedings of the Royal Society of London Series B* **266**: 2331–6.

Grafen, A. (1984) Natural, kin and group selection. In (J.R. Krebs & N.B. Davies eds) *Behavioural Ecology: An Evolutionary Approach*, 2nd Edition, pp. 62–84. Oxford: Blackwell.

Grafen, A. (1990a) Do animals really recognize kin? *Animal Behaviour* **39**: 42–54.

Grafen, A. (1990b) Sexual selection unhandicapped by the Fisher process. *Journal of Theoretical Biology* **144**: 473–516.

Grafen, A. (1990c) Biological signals as handicaps. *Journal of Theoretical Biology* **144**: 517–46.

Grafen, A. (1991) Kin vision? A reply to Stuart. *Animal Behaviour* **41**: 1095–6.

Graham, J.H. & Felley, J.D. (1985) Genomic coadaptation and developmental stability within introgressed populations of *Enneacanthus gloriosus* and *E. obesus* (Pisces: Centrachidae). *Evolution* **39**: 104–14.

Graham, J.H., Freeman, D.C. & Emlen, J.M. (1993) Developmental stability: a sensitive indicator of populations under stress. In (Landis, W.G., Hughes, J.S. & Lewis, M.A. eds) *Environmental Toxicology and Risk Assessment*, ASTM STP Vol. 1179, pp. 136–58. Philadelphia, American Society for Testing Materials.

Graham-Rowe, D. (1998) March of the biobots. *New Scientist* 160, No. 2163: 26–32.

Grand, T.C. & Dill, L.M. (1999) The effect of group size on the foraging behaviour of juvenile coho salmon: reduction of predation risk or increased competition? *Animal Behaviour* 58: 443–51.

Grant, J.W.A. & Green, L.D. (1996) Mate copying versus preference for actively courting males by female Japanese medaka (*Oryzias latipes*). *Behavioral Ecology* 7: 165–7.

Gray, A. (1972) *Mammalian Hybrids*. Farnham Royal, Slough, Commonwealth Agricultural Bureaux.

Gray, R.D. (1994) Sparrows, matching and the ideal free distribution: can biological and psychological approaches be synthesized? *Animal Behaviour* 48: 411–23.

Gray, S. & Hurst, J.L. (1995) The effects of cage cleaning on aggression within groups of male laboratory mice. *Animal Behaviour* 49: 821–6.

Green, A.J. (1991) Large male crests, an honest indicator of condition, are preferred by female smooth newts, *Triturus vulgaris* (Salamandridae) at the spermatophore transfer stage. *Animal Behaviour* 41: 367–9.

Green, R.F. (1984) Stopping rules for optimal foragers. *American Naturalist* 123: 30–40.

Greene, E. & Meagher, T. (1998) Red squirrels, *Tamiasciurus hudsonicus*, produce predator-class specific alarm calls. *Animal Behaviour* 55: 511–18.

Greenfield, M.D. & Weber, T. (2000) Evolution of ultrasonic signalling in wax moths: discrimination of ultrasonic mating calls from bat echolocation signals and the exploitation of anti-predator receiver bias by sexual advertisement. *Ethology, Ecology & Evolution* 12: 259–79.

Greenfield, M.D., Tourtellot, M.K. & Snedden, W.A. (1997) Precedence effects and the evolution of chorusing. *Proceedings of the Royal Society of London Series B* 264: 1355–61.

Greenfield, S. (2000) *The Private Life of the Brain*. London, Allen Lane.

Greenberg, B.D., Li, Q., Lucas, F.R., Hu, S., Sirota, L.A., Bejamin, J., Lesch, K.-P., Hamer, D. & Murphy, D.L. (2000) Association between the serotonin transporter promotor polymorphism and personality traits in a primarily female population. *American Journal of Medical Genetics* 96: 202–16.

Greenberg, L. (1979) Genetic component of bee odor in kin recognition. *Science* 206: 1095–7.

Gregory, R.L. (1981) *Mind in Science*. London, Weidenfeld and Nicolson.

Gregory, R.L. (1998) *Eye and Brain: the Psychology of Seeing*, 5th Edition. Oxford, Oxford University Press.

Greig-Smith, P.W. (1981) Responses to disturbance in relation to flock size in foraging groups of barred ground doves (*Geopelia striata*). *Ibis* 123: 103–6.

Griffin, A.S. & West, S.A. (2002) Kin selection: fact and fiction. *Trends in Ecology & Evolution* 17: 15–21.

Griffin, D.R. (1981) *The Question of Animal Awareness*. New York, Rockefeller University Press.

Griffin, D.R. (1984) *Animal Thinking*. Cambridge, MA, Harvard University Press.

Griffin, D.R. (1991) Progress towards a cognitive ethology. In (C.S. Ristau ed.) *Cognitive Ethology: The Minds of Other Animals*, pp. 3–17. Hillsdale, NJ, Lawrence Erlbaum.

Griffith, S.C., Owens, I.P.F. & Thuman, K.A. (2002) Extra pair paternity in birds: a review of interspecific variation and adaptive function. *Molecular Ecology* 11: 2195–212.

Grillner, S.P. & Wallen, P. (1985) Central pattern generators for locomotion, with special reference to vertebrates. *Annual Review of Neuroscience* 8: 233–61.

Grindstaff, J.L., Buerkle, C.A., Casto, J.M., Nolan, V. Jr & Ketterson, E.D. (2001). Offspring sex ratio is unrelated to male attractiveness in dark-eyed juncos (*Junco hyemalis*). *Behavioral Ecology & Sociobiology* 50: 312–16.

Grohmann, J. (1939) Modifikation oder Funktionsrei Fung? *Zeitschrift für Tierpsychologie* 2: 132–44.

Groot, P. de (1980) Information transfer in a socially roosting weaverbird (*Quelea quelea*: Ploceinae): an experimental study. *Animal Behaviour* 28: 1249–54.

Grosberg, R.K. & Quinn, J.F. (1986) The genetic control and consequences of kin recognition by the larvae of a colonial marine invertebrate. *Nature* 332: 456–9.

Gross, M.R. (1982) Sneakers, satellites and parentals: polymorphic mating strategies in North American sunfishes. *Zeitschrift für Tierpsychologie* 60: 1–26.

Gross, M.R. (1996) Alternative reproductive strategies and tactics: diversity within sexes. *Trends in Ecology & Evolution* 11: 92–7.

Groves, P.M. & Thompson, R.F. (1970) Habituation: a dual-process theory. *Psychological Review* 77: 419–50.

Grunt, J.A. & Young, W.C. (1952) Differential reactivity of individuals and the response of the male guinea pig to testosterone proprionate. *Endocrinology* 51: 237–48.

Gudmundsson, G.A., Benvenuti, S., Alerstam, T., Papi, F., Lillendahl, K & Åkesson, S. (1995) Examining the limits of flight and orientation performance: satellite tracking of Brent geese migrating across the Greenland ice-cap. *Proceedings of the Royal Society of London Series B* 261: 73–9.

Guilford, T. (1990) The evolution of aposematism. In (D.L. Evans & J.O. Schmidt eds) *Insect Defences: Adaptive Mechanisms and Strategies of Prey and Predators*, pp. 23–61. New York, State University of New York Press.

Guilford, T. (1994) Go-slow signalling and the problem of automimicry. *Journal of Theoretical Biology* 170: 311–16.

Guilford, T. & Dawkins, M.S. (1987) Search images not proven: a reappraisal of recent evidence. *Animal Behaviour* 35: 1838–45.

Guilford, T. & Dawkins, M.S. (1991) Receiver psychology and the evolution of animal signals. *Animal Behaviour* 42: 1–14.

Guilford, T., Nicol, C., Rothschild, M. & Moore, B.P. (1987) The biological roles of pyrazines: evidence for a warning odour function. *Biological Journal of the Linnaean Society* 31: 113–28.

Guilford, T., Gagliardo, A., Chappell, J., Bonadonna, F., de Perera, T.B. & Holland, R. (1998) Homing pigeons use olfactory cues for navigation in England. *Journal of Experimental Biology* 201: 895–900.

Guillery, R.W., Casagrande, V.A. & Oberdorfer, M.D. (1974) Congenitally abnormal vision in Siamese cats. *Nature* 252: 195–9.

Guillery, R.W., Jeffery, G. & Saunders, N. (1999) Visual abnormalities in albino wallabies: a brief note. *Journal of Comparative Neurology* 403: 33–8.

Guiton, P. (1959) Socialization and imprinting in brown leghorn chicks. *Animal Behaviour* 7: 26–34.

Günther, J. & Walther, J.B. (1971) Funktionelle Anatomie der dorsalen Riesenfaser-System von *Lumbricus terrestris* L. (Annelida, Oligocaeta).

Zeitschrift für Morphologie der Tiere 70: 253–80.

Guthrie, D.M. (1980) *Neuroethology: An Introduction*. Oxford, Blackwell.

Gvozdev, V.A. (1986) Mobile genetic elements in *Drosophila melanogaster*: a study of distribution and saltatory transpositions coupled with fitness changes. *Soviet Science Review D. Physiochemical Biology* 6: 107–38.

Gwadz, R. (1970) Monofactorial inheritance of early sexual receptivity in the mosquito *Aedes atropalus. Animal Behaviour* 18: 358–61.

Gwinner, E. & Dittami, J. (1990) Endogenous reproductive rhythms in a tropical bird. *Science* 249: 906–8.

Gwynne, D.T. (1981) Sexual difference theory: Mormon crickets show role reversal in mate choice. *Science* 213: 779–80.

Gygax, L. (1995) Hiding behaviour of long-tailed macaques (*Macaca fascicularis*): I. Theoretical background and data on mating. *Ethology* 101: 10–24.

Gyger, M. (1990) Audience effects on alarm calling. *Ethology, Ecology & Evolution* 2: 227–32.

Hagstrum, J.T. (2001) Infrasound and the avian navigational map. *Journal of Navigation* 54: 377–91.

Haig, D. (1993) Genetic conflicts in human pregnancy. *Quarterly Review of Biology* 68: 495–532.

Haig, D. (1996) Gestational drive and the green-bearded placenta. *Proceedings of the National Academy of Sciences of the USA* 93: 6547–51.

Haigh, J. & Rose, M.R. (1980) Evolutionary game auctions. *Journal of Theoretical Biology* 85: 381–97.

Hales, C.N. (1997) Metabolic consequences of intrauterine growth retardation. *Acta Paediatrica* 86: 184–7.

Hailman, J.P. (1967) How an instinct is learned. *Scientific American* 221: 98–108.

Hailman, J.P. & Sustare, R.D. (1973) What a stuffed toy tells a stuffed shirt. *BioScience* 23: 644–51.

Halbach, U. (1971) Zum Adaptivwert der zyklomorphen Dornenbildung von *Brachionus calyciflorus* Pallas (Rotatoria): 1. Räuber-Beute-Beziehung in Kurzzeit-Versuchen. *Oecologia* 6: 267–88.

Hall, C.L. & Fedigan, L.M. (1997) Spatial benefits afforded by high rank in white-faced capuchins. *Animal Behaviour* 53: 1069–82.

Hall, J.C. (1977) Portions of the central nervous system controlling reproductive behavior in *Drosophila melanogaster. Behavior Genetics* 7: 291–312.

Hall, J.C. (1979) Control of male reproductive behavior by the central nervous system of *Drosophila*: dissection of a courtship pathway by genetic mosaics. *Genetics* **92**: 437–57.

Halliday, T.R. (1975) An observational and experimental study of sexual behaviour in the smooth newt, *Triturus vulgaris. Animal Behaviour* **23**: 291–322.

Halliday, T.R. (1977) The effects of experimental manipulation of breathing behaviour on the sexual behaviour of the smooth newt, *Triturus vulgaris. Animal Behaviour* **25**: 39–45.

Halliday, T.R. (1983) Information and communication. In (T.R. Halliday & P.J.B. Slater eds) *Animal Behaviour 2: Communication*, pp. 43–81. Oxford, Blackwell.

Halliday, T.R. & Slater, P.J.B. eds (1983) *Animal Behaviour*, Vols 1–3. Oxford, Blackwell.

Halliday, T.R. & Sweatman, H.P.A. (1976) To breathe or not to breathe: the newt's problem. *Animal Behaviour* **24**: 551–61.

Halpern, D.F. ed. (2000) *Sex Differences in Cognitive Abilities*, 3rd Edition. Mahwah, NJ, Lawrence Erlbaum.

Hamer, D. & Copeland, P. (1994) *The Science of Desire*. New York, Simon & Shuster.

Hamer, D. & Copeland, P. (1998) *Living With Our Genes: Why They Matter More Than You Think*. New York, Doubleday.

Hamer, D., Hu, S., Magnusson, V.L., Hu, N. & Pattatucci, A.M.L. (1993) A linkage between DNA markers on the X chromosome and male sexual orientation. *Science* **261**: 321–7.

Hames, R.B. (1979) Relatedness and interaction among the Ye'kwana: a preliminary analysis. In (N.A. Chagnon & W. Irons eds) *Evolutionary Biology and Human Social Behavior: An Anthropological Perspective*, pp. 239–48. North Scituate, MA, Duxbury Press.

Hamilton, W.D. (1964a) The genetical evolution of social behaviour: I. *Journal of Theoretical Biology* **7**: 1–16.

Hamilton, W.D. (1964b) The genetical evolution of social behaviour: II. *Journal of Theoretical Biology* **7**: 17–52.

Hamilton, W.D (1967) Extraordinary sex ratios. *Science* **156**: 477–88.

Hamilton, W.D. (1971) Geometry for the selfish herd. *Journal of Theoretical Biology* **31**: 295–311.

Hamilton, W.D. (1980) Sex versus non-sex versus parasites. *Oikos* **35**: 282–90.

Hamilton, W.D. (1987) Discriminating nepotism: expectable, common, overlooked. In (D.J.C. Fletcher & C.D. Michener eds) *Kin Recognition in Animals*, pp. 417–38. New York, Wiley.

Hamilton, W.D. (1996) *Narrow Roads of Gene Land: The Collected Papers of W.D. Hamilton Volume I: Evolution of Social Behaviour*. Oxford, Freeman.

Hamilton, W.D. (2001) *Narrow Roads of Gene Land: The Collected Papers of W.D. Hamilton Volume II: The Evolution of Sex*. Oxford, Freeman.

Hamilton, W.D. & Zuk, M. (1982) Heritable true fitness and bright birds. A role for parasites? *Science* **218**: 384–7.

Hamm, S.L. & Shettleworth, S.J. (1987) Risk aversion in pigeons. *Journal of Experimental Psychology: Animal Behavior Processes* **13**: 376–83.

Hampton, N.G., Bolhuis, J.J. & Horn, G. (1995) Induction and development of a filial predisposition in the chick. *Behaviour* **132**: 451–77.

Hanlon, R.T., Forsythe, J.W. & Joneschild, D.E. (1999) Crypsis, conspicuousness, mimicry and polyphenism as antipredator defences of foraging octopuses on Indo-Pacific coral reefs, with a method of quantifying crypsis from video tapes. *Biological Journal of the Linnean Society of London* **66**: 1–22.

Hanstrom, B. (1928) Vergleischende Anatomie des Nervensystems der wirbellosan. *Tiere unter Berucksichtigung seiner Funktion*, Berlin.

Harcourt, A.H., Harvey, P.H. Larson, S.G. & Short, R.V. (1980) Testis weight, body weight and breeding system in primates. *Nature* **293**: 55–7.

Harcourt, R. (1991) Survivorship costs of play in the south American fur seal. *Animal Behaviour* **42**: 509–11.

Harlow, H.F. (1949) The formation of learning sets. *Psychological Review* **56**: 51–65.

Harlow, H.F. & Zimmerman (1959) Affectional responses in the infant monkey. *Science* **130**: 421–32.

Harpending, H.C. & Rogers, A. (1990) Fitness in stratified societies. *Ethology & Sociobiology* **11**: 497–509.

Harper, D.G.C. (1982) Competitive foraging in mallards: ideal free ducks. *Animal Behaviour* **30**: 575–84.

Harris, G.W. & Levine, S. (1965) Sexual differentiation of the brain and its experimental control. *Journal of Physiology* **181**: 379–400.

Hart, A. & Lendrem, D.W. (1984) Vigilance and scanning patterns in birds. *Animal Behaviour* **32**: 1216–24.

Hart, C.W. & Pilling, A.R. (1960) *The Tiwi of North Australia*. New York, Holt, Hart, Rhinehart & Winston.

Hartmann, G. & Wehner, R. (1995) The ant's path integration system: a neural architecture. *Biological Cybernetics* **73**: 483–97.

Harvey, P.H. & Bradbury, J. (1991) Sexual selection. In (J.R. Krebs & N.B. Davies eds) *Behavioural Ecology: An Evolutionary Approach*, 3rd Edition, pp. 203–33. Oxford, Blackwell.

Harvey, P.H. & Greenwood, P.J. (1978) Antipredator defence strategies: some evolutionary problems. In (J.R. Krebs & N.B. Davies eds) *Behavioural Ecology: An Evolutionary Approach*, 1st Edition, pp. 129–51. Oxford, Blackwell.

Hassell, M.P. (1976) *The Dynamics of Competition and Predation*. London, Edward Arnold.

Hassell, M.P. & May, R.M. (1974) Aggregation of predators and insect parasitoids and its effects on stability. *Journal of Animal Ecology* **43**: 567–94.

Hasselquist, D., Marsh, J.A., Sherman, P.W. & Wingfield, J.C. (1999) Is avian humoral immunocompetence suppressed by testosterone? *Behavioral Ecology & Sociobiology* **45**: 167–75.

Hastings, J.W., Rusak, B. & Boulos, Z. (1991) The physiology of biological timing. In (C.L. Prosser ed.) *Neural and Integrative Animal Physiology*, pp. 435–546. New York, Wiley-Liss.

Hatcher, M.J. (2000) Persistence of selfish genetic elements: population structure and conflict. *Trends in Ecology and Evolution* **15**: 271–7.

Hatchwell, B.J. & Davies, N.B. (1990) Provisioning of nestlings by dunnocks, *Prunella modularis*, in pairs and trios: compensation reactions by males and females. *Behavioral Ecology & Sociobiology* **27**: 199–209.

Hauber, M.E., Russo, S.A. & Sherman, P.W. (2001) A password for species recognition in a brood-parasitic bird. *Proceedings of the Royal Society of London Series B* **268**: 1041–8.

Hauber, W. (1998) Involvement of basal ganglia transmitter systems in motor initiation. *Progress in Neurobiology* **56**: 507–40.

Hauser, M.D. (1988) How infant vervet monkeys learn to recognize starling alarm calls: the role of experience. *Behaviour* **195**: 187–201.

Hauser, M.D. (1996) *The Evolution of Communication*. Cambridge, MA, MIT Press.

Hauser, M.D. (2000) *Wild Minds: What Animals Really Think*. New York, Henry Holt.

Hauser, M.D., Kralik, J., Botto-Mahan, C., Garrett, M. & Oser, J. (1995) Self-recognition in primates: phylogeny and the salience of species-typical features. *Proceedings of the National Academy of Sciences of the USA* **92**: 10811–14.

Hausfater, G. (1975) Dominance and reproduction in baboons (*Papio cynocephalus*). *Contributions to Primatology* **7**.

Hawkes, K. (1991) Showing off: tests of another hypothesis about men's foraging goals. *Ethology & Sociobiology* **11**: 29–54.

Hawkins, R.D., Greene, W. & Kandel, E.R. (1993) Classical conditioning, differential conditioning, and second-order conditioning of the *Aplysia* gill-withdrawal reflex in an isolated mantle organ preparation. *Society for Neuroscience Abstracts* **19**: 17.

Hawryshyn, C.W. (1992) Polarization vision in fish. *American Scientist* **80**: 164–75.

Hay, D.A. (1985) *Essentials of Behaviour Genetics*, Oxford, Blackwell.

Hayes, K.J. & Hayes, C. (1951) The intellectual development of a home-raised chimpanzee. *Proceedings of the American Philosophical Society* **95**: 105.

Hazlett, B.A. & Bossert, W.H. (1965) A statistical analysis of the aggressive communications of some hermit crabs. *Animal Behaviour* **13**: 357–73.

Healy, S.D. ed. (1998) *Spatial Representation in Animals*. Oxford, Oxford University Press.

Healy, S.D. & Krebs, J.R. (1992) Food storing and the hippocampus in corvids: amount and volume are correlated. *Proceedings of the Royal Society of London Series B* **248**: 241–5.

Hearst, E. & Franklin, S.R. (1977) Positive and negative relations between a signal and food: approach–withrawal behaviour. *Journal of Experimental Psychology: Animal Behavior Processes* **3**: 37–52.

Heiligenberg, W. (1966) The stimulation of territorial singing in house crickets (*Acheta domesticus*). *Zeitschrift für Vergleichende Physiologie* **53**: 114–29.

Heinrich, B. (1978) The economics of insect sociality. In (J.R. Krebs & N.B. Davies eds) *Behavioural Ecology: An Evolutionary Approach*, 1st Edition, pp. 97–128. Oxford, Blackwell.

Heinrich, B. & Smolker, R. (1998) Play in common ravens (*Corvus corax*). In (M. Bekoff & J.A. Byers eds) *Animal Play: Evolutionary, Comparative and Ecological Perspectives*, pp. 27–44. Cambridge, Cambridge University Press.

Helbig, A.J. (1991) Inheritance of migratory direction in a bird species: a cross-breeding experiment with SE- and SW-migratory blackcaps (*Sylvia atricapilla*). *Behavioural Ecology & Sociobiology* **28**: 9–12.

Held, S., Mendl, M., Devereux, C. & Byrne, R.W. (2000) Social tactics of pigs in a competitive foraging task: the 'informed forager' paradigm. *Animal Behaviour* **59**: 569–76.

Henderson, N.D. (1970) Genetic influences on the behaviour of mice can be obscured by laboratory rearing. *Journal of Comparative and Physiological Psychology* **72**: 505–11.

Hendry, A.P., Berg, O.K. & Quinn, T.P. (2001) Breeding location choice in salmon: causes (habitat competition, body size, energy stores) and consequences (life span, energy stores). *Oikos* **93**: 407–18.

Hepper, P.G. (1986) Kin recognition: function and mechanisms. A review. *Biological Reviews* **61**: 63–93.

Hepper, P.G. ed. (1990) *Kin Recognition*. Cambridge, Cambridge University Press.

Herbers, J.M. (1981) Time resources and laziness in animals. *Oecologia* **49**: 252–62.

Herrnkind, W.F. (1972) Orientation in shore-living arthropods, especially the sand fiddler crab. In (H.E. Winn & B. Olla eds) *Behavior of Marine Animals*, Vol. I, pp. 1–59. New York, Plenum Press.

Herrnstein, R.J. (1961) Relative and absolute strength of response as a function of frequency of reinforcement. *Journal of the Experimental Analysis of Behavior* **4**: 267–72.

Herrnstein, R.J. (1964) Aperiodicity as a factor in choice. *Journal of the Experimental Analysis of Behavior* **7**: 178–82.

Herrnstein, R.J. (1970) On the law of effect. *Journal of the Experimental Analysis of Behavior* **13**: 243–66.

Herrnstein, R.J. & Loveland, D.H. (1975) Maximizing and matching on concurrent ratio schedules. *Journal of the Experimental Analysis of Behavior* **24**: 107–16.

Herrnstein, R.J., Loveland, D.H. & Cable, C. (1976) Natural concepts in pigeons. *Journal of Experimental Psychology: Animal Behavior Processes* **2**: 285–311.

Herter, K. (1927) Reizphysiologische Untersuchungen an der Karpfenlaus (*Argulus foliaceus* L.). *Zeitschrift für vergleichende Physiologie* **5**: 283–370.

Hess, E.H. (1973) *Imprinting*. New York, van Nostrand.

Hews, D.K. (1988) Alarm response in larval western toads, *Bufo boreas*: release of larval chemicals by a natural predator and its effect on predator capture efficiency. *Animal Behaviour* **36**: 125–33.

Heyes, C.M. (1993) Imitation, culture and cognition. *Animal Behaviour* **46**: 999–1010.

Heyes, C.M. (1994) Reflections on self-recognition in primates. *Animal Behaviour* **47**: 909–19.

Heyes, C.M. (1996) Introduction: identifying and defining imitation. In (C.M. Heyes & B.G. Galef Jr eds) *Social Learning in Animals: The Roots of Culture*, pp. 211–20. San Diego, Academic Press.

Heyes, C.M. & Dawson, G.R. (1990) A demonstration of observational learning in rats using a bidirectional control. *Quarterly Journal of Experimental Psychology* **45B**: 229–40.

Heyes, C.M., Ray, E.D., Mitchell, C.J. & Nokes, T. (2000) Stimulus enhancement: controls for social facilitation and local enhancement. *Learning and Motivation* **31**: 83–98.

Hilden, O. & Vuolanto, S. (1972) Breeding biology of the red-necked phalarope, *Phalaropus lobatus*, in Finland. *Ornis Fennica* **49**: 57–85.

Hill, S., Burroughs, M.T. & Hughes, R.N. (2000) Increased turning per unit distance as an area-restricted search mechanism in a pause-travel predator, juvenile plaice, foraging for buried bivalves. *Journal of Fish Biology* **56**: 1497–508.

Hillgarth, N. & Wingfield, J.C. (1997) Parasite-mediated sexual selection: endocrine aspects. In (D. Clayton & J. Moore eds) *Host–Parasite Evolution: General Principles and Avian Models*, pp. 78–104. Oxford, Oxford University Press.

Hillgarth, N., Ramenofsky, M. & Wingfield, J.C. (1997) Testosterone and sexual selection. *Behavioral Ecology* **8**: 108–9.

Hilton, G.M., Cresswell, W. & Ruxton, G.D. (1999) Intraflock variation in the speed of escape-flight response on attack by an avian predator. *Behavioral Ecology* **10**: 391–5.

Hinde, R.A. (1965) Interaction of internal and external factors in integration of canary reproduction. In (F.A. Beach ed.) *Sex and Behavior*, pp. 381–415. New York, Wiley.

Hinde, R.A. (1970) *Animal Behaviour: A Synthesis of Ethology and Comparative Psychology*, 2nd Edition. New York: McGraw-Hill.

Hinde, R.A. (1981) Animal signals: ethological and games-theory approaches are not incompatible. *Animal Behaviour* **29**: 535–42.

Hinde, R.A. (1982) *Ethology*. London: Fontana.

Hinde, R.A. & Barden, L.A. (1985) The evolution of the teddy bear. *Animal Behaviour* **33**: 1371–3.

Hinde, R.A. & Fisher, J. (1951) Further observations on the opening of milk bottles by birds. *British Birds* **44**: 392–6.

Hinde R.A. & Rowell, T.E. (1962) Communication by postures and facial expressions in the rhesus monkey (*Macaca mulatta*). *Proceedings of the Zoological Society of London* **138**: 1–21.

Hino, T. (1998) Mutualistic and commensal organization of avian mixed species foraging flocks in a forest of western Madagascar. *Journal of Avian Biology* **29**: 17–24.

Hitchcock, C.L. & Sherry, D.F. (1990) Long-term memory for cache sites in the black-capped chickadee. *Animal Behaviour* **40**: 701–12.

Hock, B.J. & Bunsey, M.D. (1998) Differential effects of dorsal and ventral hippocampal lesions. *Journal of Neuroscience* **18**: 7027–32.

Hockett, C.F. (1960) The origin of speech. *Scientific American* **203**: 89–96.

Hodge, M.A. & Uetz, G.W. (1996) Foraging advantages of mixed species association between solitary and colonial orb-weaving spiders. *Oecologia* **107**: 578–87.

Hoem, J. (1992) Social policy and recent fertility change in Sweden. *Population Development Review* **16**: 735–48.

Hoffman, S.G. (1983) Sex-related foraging behaviour in sequentially hermaphroditic hogfishes (*Bodianus* spp.). *Ecology* **64**: 798–808.

Hoffmann, A.A. (1994) Behaviour genetics and evolution. In (P.J.B. Slater & T.R. Halliday eds) *Behaviour and Evolution*, pp. 7–42. Cambridge, Cambridge University Press.

Hoffmann, K. (1954) Versuche zu der im Richtungfinden der Vögel enthaltenen Zeitschätzung. *Zeitschrift für Tierpsychologie* **11**: 453–75.

Hofstadter, D.R. (1985) *Metamagical Themas: Questing for the Essence of Mind and Pattern*. New York, Basic Books.

Hogan, J.A. (1997) Energy models of motivation: a reconsideration. *Applied Animal Behaviour Science* **53**: 89–105.

Högstedt, G. (1980) Evolution of clutch size in birds: adaptive variation in relation to territory quality. *Science* **210**: 1148–50.

Holland, O. & McFarland, D.J. (2001) *Artificial Ethology*. Oxford, Oxford University Press.

Hollard, V.D. & Delius, J.D. (1982) Rotational invariance in visual pattern recognition by pigeons and humans. *Science* **218**: 804–6.

Hölldobler, B. (1977) Communication in social Hymenoptera. In (T.A. Sebeok ed.) *How Animals Communicate*. pp. 418–71. Bloomington, Indiana University Press.

Hölldobler, B. & Lumsden, C.J. (1980) Territorial strategies in ants. *Science* **210**: 732–9.

Hölldobler, B. & Wilson, E.O. (1990) *The Ants*. Berlin, Springer.

Holling, C.S. (1959) Some characteristics of simple types of predation and parasitism. *Canadian Entomologist* **91**: 385–98.

Holling, C.S. (1965) The functional response of predators to prey density and its role in mimicry and population regulation. *Memoirs of the Entomological Society of Canada* **45**: 43–60.

Hollis, K.L. (1982) Pavlovian conditioning of signal-centered action patterns and autonomic behavior: a biological analysis of function. *Advances in the Study of Behavior* **12**: 1–64.

Hollis, K.L. (1999) The role of learning in the aggressive and reproductive behavior of blue gouramis, *Trichogaster trichopterus*. *Environmental Biology of Fishes* **54**: 355–69.

Hollis, K.L., Cate, C. ten & Bateson, P. (1991) Stimulus representation: a subprocess of imprinting and conditioning. *Journal of Comparative Psychology* **105**: 307–17.

Holmes, R.T. (1970) Differences in population density, territoriality, and food supply of dunlin on arctic and subarctic tundra. In (A. Watson ed.) *Animal Populations in Relation to their Food Resources*, pp. 303–19. Oxford, Blackwell.

Holmes, W.G. & Sherman, P.W. (1982) The ontogeny of kin recognition in two species of ground squirrels. *American Zoologist* **22**: 491–517.

Holmes, W.G. & Sherman, P.W. (1983) Kin recognition in animals. *American Scientist* **71**: 46–55.

Holtby, L.B. & Healey, M.C. (1986) Selection for adult size in Coho salmon. *Canadian Journal of Fisheries and Aquatic Science* **43**: 949–1057.

Hoogland, J.L. (1979) Aggression, ectoparasitism and other possible costs of prairie dog (Sciuridae: *Cynomys* spp.) coloniality. *Behaviour* **69**: 1–35.

Hoogland, J.L. (1983) Nepotism and alarm calling in the black-tailed prairie dog, *Cynomys ludovicianus*. *Animal Behaviour* **31**: 472–9.

Hoogland, R., Morris, D. & Tinbergen, N. (1957) The spines of sticklebacks (*Gasterosteus* and *Pygosteus*) as a means of defence against predators (*Perca* and *Esox*). *Behaviour* **10**: 205–36.

Hopkins, C.D. (1977) Electric communication. In (T.A. Sebeok ed.) *How Animals Communicate*, pp. 263–89. Bloomington, Indiana University Press.

Hori, M. (1993) Frequency-dependent natural selection in the handedness of scale-eating cichlid fish. *Science* **260**: 216–19.

Horn, H.S. (1968) The adaptive significance of colonial nesting in the Brewer's blackbird *Euphagus cyanocephalus*. *Ecology* **49**: 682–94.

Horn, H.S. & Rubenstein, D.I. (1984) Behavioural adaptations and life history. In (J.R. Krebs & N.B. Davies eds) *Behavioural Ecology: An Evolutionary Approach*, 2nd Edition, pp. 279–98. Oxford, Blackwell.

Horridge, G.A. (1996) The honeybee (*Apis mellifera*) detects bilateral symmetry and discriminates its axis. *Journal of Insect Physiology* **42**: 755–64.

Hotta, Y. & Benzer, S. (1969) Abnormal electroretinograms in visual mutants of *Drosophila*. *Nature* **222**: 354–6.

Hotta, Y. & Benzer, S. (1970) Genetic dissection of *Drosophila* nervous systems by means of mosaics. *Proceedings of the National Academy of Sciences of the USA* **67**: 1156–63.

Houde, A.E. & Endler, J.A. (1990) Correlated evolution of female mating preferences and male color patterns in the guppy *Poecilia reticulata*. *Science* **248**: 1405–8.

Houle, D. (1997) A meta-analysis of the heritability of developmental stability: comment. *Journal of Evolutionary Biology* **10**: 17–20.

Houle, D. & Kondrashev, A.S. (2002) Coevolution of costly mate choice and condition-dependent display of good genes. *Proceedings of the Royal Society of London Series B* **269**: 97–104.

Houston, A.I. (1982) Transitions and time-sharing. *Animal Behaviour* **30**: 615–25.

Houston, A.I. (1998) Models of avian migration: state, time and predation. *Journal of Avian Biology* **29**: 395–404.

Houston, A.I. & McFarland, D.J. (1980) Behavioural resilience and its relation to demand functions. In (J.E.R. Staddon ed.) *Limits to Action*, pp. 177–203. New York, Academic Press.

Houston, A.I. & McNamara, J.M. (1988) The ideal free distribution when competitive abilities differ: an approach based on statistical mechanics. *Animal Behaviour* **36**: 166–74.

Houston, A.I. & McNamara, J. (1999) *Models of Adaptive Behaviour*. Cambridge, Cambridge University Press.

Houston, A.I. & Sumida, B.H. (1987) Learning rules, matching and frequency-dependence. *Journal of Theoretical Biology* **126**: 289–308.

Houston, A.I., Halliday, T.R. & McFarland, D.J. (1977) Towards a model of the courtship of the smooth newt, *Triturus vulgaris*, with special emphasis on problems of observability in the simulation of behaviour. *Medical and Biological Engineering and Computing* **15**: 49–61.

Houston, A.I., Krebs, J.R. & Erichsen, J.T. (1980) Optimal prey choice and discrimination time in the great tit (*Parus major* L.). *Behavioural Ecology & Sociobiology* **6**: 169–75.

Howard, R. (1978) The evolution of mating strategies in bullfrogs, *Rana catesbiana*. *Evolution* **32**: 850–71.

Howse, P.E. (1974) Design and function in the insect brain. In (L.B. Browne ed.) *Experimental Analysis of Insect Behaviour*, pp. 180–95. Berlin, Springer.

Hrdy, S.B. (1981) *The Woman that Never Evolved*. Cambridge, MA, Harvard University Press.

Hsia, Y.-Y. (1970) Phenylketonuria and its variants. In (A.G. Steinberg & A.G. Bearn eds) *Progress in Medical Genetics*, Vol. 7, pp. 29–68. New York, Grune & Stratton.

Huber, F. (1990) Nerve cells and insect behavior: studies on crickets. *American Zoologist* **30**: 609–27.

Huck, U.W., Soltis, R.L. & Coopersmith, C.B. (1982) Infanticide in male laboratory mice: effects of social status, prior social experience, and the basis for discrimination between related and unrelated young. *Animal Behaviour* **30**: 1158–65.

Hudson, R., Schaal, B. & Bilko, A. (1999) Transmission of information from mother to young in the European rabbit. In (H.O. Box & K.R. Gibson eds) *Mammalian Social Learning*, pp. 141–57. Cambridge, Cambridge University Press.

Hughes, B.O., Duncan, I.J.H. & Brown, M.F. (1989) The performance of nest building by domestic hens: is it more important than the construction of a nest? *Animal Behaviour* **37**: 210–14.

Hugie, D.M. & Dill, L.M. (1994) Fish and game: a game theoretic approach to habitat selection by predators and prey. *Journal of Fish Biology* 45: 151–69.

Humphrey, N. (1983) *Consciousness Regained*. Oxford, Oxford University Press.

Humphries, R.E., Robertson, D.H.L., Beynon, R.J. & Hurst, J.L. (1999) Unravelling the chemical basis of competitive scent marking in house mice. *Animal Behaviour* 58: 1177–90.

Hunt, G.J., Page, R.E. Fondrk, M.K. & Dullum, C.J. (1995) Major quantitative trait loci affecting honey bee foraging. behaviour. *Genetics* 141: 1537–45.

Hunt, G.R. (2000) Tool use by the New Caledonian crow, *Corvus moneduloides*, to obtain Cerambycidae from dead wood. *Emu* 100: 109–14.

Hunt, J. & Simmons, L.W. (1998) Patterns of fluctuating asymmetry in beetle horns: no evidence for reliable signalling. *Behavioral Ecology* 9: 465–70.

Hunter, M.L. & Krebs, J.R. (1979) Geographical variation in the song of the great tit (*Parus major*) in relation to ecological factors. *Journal of Animal Ecology* 48: 759–85.

Huntingford, F.A. (1986) Development of behaviour in fish. In (T.J. Pitcher ed.) *The Behaviour of Teleost Fishes*. Baltimore, Johns Hopkins University Press.

Hurst, J.L. (1989) The complex network of olfactory communication in populations of wild house mice, *Mus domesticus* Rutty: markings and investigation within family groups. *Animal Behaviour* 37: 705–25.

Hurst, J.L. (1990a) Urine marking in populations of wild house mice (*Mus domesticus* Rutty): I. Communication between males. *Animal Behaviour* 40: 209–22.

Hurst, J.L. (1990b) Urine marking in populations of wild house mice (*Mus domesticus* Rutty): II. Communication between females. *Animal Behaviour* 40: 223–32.

Hurst, J.L. (1990c) Urine marking in populations of wild house mice (*Mus domesticus* Rutty): III. Communication between the sexes. *Animal Behaviour* 40: 233–43.

Hurst, J.L. (1993) The priming effects of urine substrate marks on interactions between male house mice, *Mus musculus domesticus* Schwarz and Schwarz. *Animal Behaviour* 45: 55–81.

Hurst, J.L. & Barnard, C.J. (1992) Kinship and social behaviour in wild house mice: effects of social group membership and relatedness on the responses of dominant males towards juveniles. *Behavioral Ecology* 3: 196–206.

Hurst, J.L. & Barnard, C.J. (1995) Kinship and social tolerance among female and juvenile wild house mice: kin bias but not kin discrimination. *Behavioral Ecology & Sociogiology* 5: 333–42.

Hurst, J.L., Barnard, C.J., Hare, R., Wheeldon, E.B. & West, C.D. (1996) Housing and welfare in laboratory rats: status-dependent time-budgeting and pathophysiology in single sex groups maintained in open rooms. *Animal Behaviour* 52: 335–60.

Hurst, J.L., Barnard, C.J., Nevison, C.M. & West, C.D. (1997) Housing and welfare in laboratory rats: welfare implications of isolation and social contact among caged males. *Animal Welfare* 6: 329–47.

Hurst, J.L., Barnard, C.J., Tolladay, U., Nevison, C.M. & West, C.D. (1999) Housing and welfare in laboratory rats: effects of cage stocking density and behavioural predictors of welfare. *Animal Behaviour* 58: 563–86.

Hurst, J.L., Payne, C.E., Nevison, C.M., Marie, A.D., Humphries, R.E., Robertson, D.H.L., Cavaggioni, A. & Beynon, R.J. (2001) Individual recognition in mice mediated by major urinary proteins. *Nature* 414: 631–4.

Hurst, L.D. (1991) The incidence and evolution of cytoplasmic male killers. *Proceedings of the Royal Society of London Series B* 244: 91–9.

Hurst, L.D. & Hamilton, W.D. (1992) Cytoplasmic fusion and the nature of sexes. *Proceedings of the Royal Society of London Series B* 247: 189–94.

Huxley, J.S. (1923) Courtship activities in the red-throated diver *Colymbus stellatus pontopp*: together with a discussion on the evolution of courtship in birds. *Journal of the Linnaean Society of London* 25: 253–92.

Huxley, J.S. (1934) A natural experiment on the territorial instinct. *British Birds* 27: 270–7.

Huxley, J.S. (1942) *Evolution: the Modern Synthesis*. London, Allen and Unwin.

Huxley, J.S. (1966) A discussion of ritualisation of behaviour in animals and man: introduction. *Philosophical Transactions of the Royal Society of London Series B* 251: 247–71.

Iersel, J.J. van & Bol, A.C.A. (1958) Preening of two tern species: a study of displacement activities. *Behaviour* 13: 1–88.

Inglis, J.M. (1976) Wet season movements of individual wildebeests of the Serengeti migratory herd. *East African Wildlife Journal* 14: 17–34.

Ioalé, P., Papi, F., Fiaschi, V. & Baldaccini, N.E. (1978) Pigeon navigation: effects upon homing behaviour by reversing wind direction at the loft. *Journal of Comparative Physiology* 128: 285–95.

Irwin, D.E. & Price, T. (1999) Sexual imprinting, learning and speciation. *Heredity* **82**: 347–54.

Irwin, R. (1988) The evolutionary importance of behavioural development: the ontogeny and phylogeny of bird song. *Animal Behaviour* **36**: 814–24.

Ishida, H., Kobayashi, A., Nakomoto, T. & Moriisumi, T. (1999) Three dimensional odour compass. *IEEE Transactions on Robotics and Automation* **15**: 251–7.

Isles, A.R. & Wilkinson, L.S. (2000) Imprinted genes, cognition and behaviour. *Trends in Cognitive Sciences* **4**: 309–18.

Isles, A.R., Baum, M.J., Ma, D., Keverne, E.B. & Allen, N.D. (2001) Urinary odour preferences in mice. *Nature* **409**: 783–4.

Itani, J. & Nishimura, A. (1973) The study of infra-human culture in Japan. In (E. Menzel ed.) *Precultural Primate Behavior*, pp. 127–41. Basel, Karger.

Iwasa, Y. (1982) Vertical migration of zooplankton: a game between predator and prey. *American Naturalist* **120**: 171–80.

Iyengar, V.K., Reeve, H.K. & Eisner, T. (2002) Paternal inheritance of a female moth's mating preference. *Nature* **419**: 830–32.

Jacques, A.R. & Dill, L.M. (1980) Zebra spiders may use uncorrelated asymmetries to settle contests. *American Naturalist* **116**: 899–901.

Jakobsen, P.J., Birkeland, K. & Johnsen, G.H. (1994) Swarm location in zooplankton as an antipredator defense mechanism. *Animal Behaviour* **47**: 175–8.

Jakobsson, S., Radesäter, T. & Järvi, T. (1979) On the fighting behaviour of *Nannacara anomala* (Pisces: Cichlidae) males. *Zeitschrift für Tierpsychologie* **49**: 210–20.

James, W. (1890) *Principles of Psychology*. New York, Holt.

Janowitz, H.D. & Grossman, M.I. (1949) Some factors affecting the food intake of normal dogs and dogs with esophagotomy and gastric fistulas. *American Journal of Physiology* **159**: 143–8.

Jarman, P.J. (1974) The social organization of antelope in relation to their ecology. *Behaviour* **48**: 215–67.

Jarvis, J.U.M. (1981) Eusociality in a mammal: cooperative breeding in naked mole rat colonies. *Science* **212**: 571–3.

Jarvis, J.U.M. & Bennett, N.C. (1993) Eusociality has evolved independently in two genera of bathyergid mole-rats – but occurs in no other subterranean mammal. *Behavioral Ecology & Sociobiology* **33**: 353–60.

Jasienska, G. & Ellison, P. (1998) Physical work causes suppression of ovarian function in women. *Proceedings of the Royal Society of London Series B* **265**: 1847–51.

Jeffery, G. (1997) The albino retina: an abnormality that provides insight into normal retinal development. *Trends in Neurosciences* **20**: 165–9.

Jenni, D.A. & Collier (1972) Polyandry in the American jacana, *Jacana spinosa. Auk* **89**: 743–65.

Jennings, T. & Evans, S.M. (1980) Influence of position in the flock and flock size on vigilance in the starling, *Sturnus vulgaris. Animal Behaviour* **30**: 634–5.

Jerison, H. (1973) *Evolution of the Brain and Intelligence*. New York, Academic Press.

Jerison, H.J. (2001) The evolution of neural and behavioural complexity. In (Roth, G. & Williman, M.F. eds.) *Brain, Evolution and Cognition*, pp. 523–53, New York, Wiley.

Jiang, Y.H., Armstrong, D., Albrect, V., Atkins, C.M., Noebels, J.L., Eichele, G., Swealt, J.D. & Beaudet, A.L. (1998) Mutation of the Angelman ubiquitin ligase in mice causes increased cytoplasmic p53 and deficits of contextual learning and long-term potentiation. *Neuron* **21**: 799–811.

Jing, J. & Gillette, R. (2000) Escape swim network interneurons have diverse roles in behavioral switching and putative arousal in *Pleurobranchaea. Journal of Neurophysiology* **83**: 1346–55.

John, E.R., Tang, Y., Brill, A.B., Young, R. & Ono, K. (1986) Double-labelled metabolic maps of memory. *Science* **212**: 571–3.

Johnson, C.R. & Boerlijst, M.C. (2002) Selection at the level of the community: the importance of spatial structure. *Trends in Ecology & Evolution* **17**: 83–90.

Johnson, D.D.P., Kays, R., Blackwell, P.G. & Macdonald, D.W. (2002) Does the resource dispersion hypothesis explain group living? *Trends in Ecology and Evolution* **17**: 563–70.

Johnson, P.B. & Hasler, A.D. (1980) The use of chemical cues in upstream migration of coho salmon, *Oncorhynchus kisutch. Waldbaum Journal of Fish Biology* **17**: 67–73.

Johnson-Laird, P.N. (1988) A computational analysis of consciousness. In (A.J. Marcel & E. Bisiach eds) *Consciousness in Contemporary Science*, pp. 357–68. Oxford, Clarendon Press.

Johnston, V.S. & Franklin, M. (1993) Is beauty in the eyes of the beholder? *Ethology & Sociobiology* **14**: 183–99.

Johnston, V.S., Hagel, R., Franklin, M., Fink, B. & Grammer, K. (2001) Male facial attractiveness: evidence for hormone-mediated adaptive design. *Evolution & Human Behavior* **22**: 251–67.

Johnstone, R.A. (1994) Female preference for symmetrical males as a by-product of selection for mate recognition. *Nature* **372**: 172–5.

Johnstone, R.A. (1996) Multiple displays in animal communication: 'backup signals' and 'multiple messages'. *Philosophical Transactions of the Royal Society of London Series B* **351**: 329–38.

Johnstone, R.A., Reynolds, J.D. & Deutsch, J.C. (1996) Mutual mate choice and sex differences in choosiness. *Evolution* **50**: 1382–91.

Jones, D. (1996) *Physical Attractiveness and the Theory of Sexual Selection*. Ann Arbor, University of Michigan Press.

Jones, F.R.H. (1955) Photo-kinesis in the ammocoete larva of the brook lamprey. *Journal of Experimental Biology* **34**: 492–503.

Jones, K.J. & Hill, W.L. (2001) Auditory perception of hawks and owls for passerine alarm calls. *Ethology* **107**: 717–26.

Kacelnik, A. & Bateson, M. (1996) Risky theories – the effects of variance on foraging decisions. *American Zoologist* **36**: 402–34.

Kacelnik, A. & Cuthill, I.C. (1987) Starlings and optimal foraging theory: modelling in a fractal world. In (A.C. Kamil, J.R. Krebs & H.R. Pulliam eds) *Foraging Behaviour*, pp. 303–33. New York, Plenum Press.

Kacelnik, A. & Todd, I.A. (1992) Psychological mechanisms and the marginal value theorem: effect of variability in travel time on patch exploitation. *Animal Behaviour* **43**: 313–22.

Kacelnik, A., Brunner, D. & Gibbon, J. (1990) Timing mechanisms in optimal foraging: some applications of scalar expectancy theory. In (R.N. Hughes ed.) *Behavioural Mechanisms of Food Selection*, pp. 61–82. NATO ASI Series G, Ecological Sciences Vol. 20. Heidelberg, Springer.

Kacelnik, A., Krebs, J.R. & Ens, B. (1987) Foraging in a changing environment: an experiment with starlings. In (M.L. Commons, A. Kacelnik & S.J. Shettleworth eds) *Harvard Symposium on the Quantitative Analysis of Behavior* Volume 6: *Foraging*, pp. 63–87. Hillsdale, NJ, Lawrence-Erlbaum Associates.

Kahneman, D., Slovic, P. & Tversky, A. eds (1982) *Judgment Under Uncertainty: Heuristics and Biases*. New York, Cambridge University Press.

Kaitala, A., Kaitala, V. & Lundberg, P. (1993) A theory of partial migration. *American Naturalist* **142**: 59–81.

Kako, E. (1999) Elements of syntax in the systems of three language-trained animals. *Animal Learning & Behavior* **27**: 1–14.

Kamil, A.C. (1985) The evolution of higher learning abilities in birds. Paper presented at the XVIII International Ornithological Congress, Moscow, 1982.

Kamil, A.C., Misthal, R.L. & Stephens, D.W. (1993) Failure of simple optimal foraging models to predict residence time when patch quality is uncertain. *Behavioral Ecology* **4**: 350–63.

Kamin, L.J. & Schaub, R.E. (1963) Effects of conditioned stimulus intensity on the conditioned emotional response. *Journal of Comparative and Physiological Psychology* **56**: 502–7.

Kandel, E.R. (1976) *Cellular Basis of Behavior*. San Francisco, W.H. Freeman.

Kankel, D.R. & Hall, J.C. (1976) Fate mapping of nervous system and other internal tissues in genetic mosaics of *Drosophila melanogaster*. *Developmental Biology* **48**: 1–24.

Kanz, J.E. (1977) The orientation of migrant and non-migrant monarch butterflies. *Danaus plexippus* (L.). *Psyche* **84**: 120–41.

Kanzaki, R. (1996) Behavioral and neural basis of instinctive behavior in insects: odor-searching strategies without memory and learning. *Robotics and Autonomous Systems* **18**: 33–43.

Kaplan, H. & Hill, K. (1985a) Hunting ability and foraging success among Ache foragers. *Current Anthropology* **26**: 131–3.

Kaplan, H. & Hill, K. (1985b) Food sharing among Ache foragers: tests of explanatory hypotheses. *Current Anthropology* **26**: 223–45.

Karakashian, S.J., Gyger, M. & Marler, P. (1988) Audience effects on alarm calling in chickens (*Gallus gallus*). *Journal of Comparative Psychology* **102**: 129–35.

Kareem, A. & Barnard, C.J. (1982) The importance of kinship and familiarity in social interactions between mice. *Animal Behaviour* **34**: 1814–24.

Karlsson, B. & Wiklund, C. (1984) Egg weight variation and lack of correlation between egg weight and offspring fitness in the small brown butterfly *Lasiommata megera*. *Oikos* **43**: 376–85.

Katz, P.S. & Frost, W.N. (1995) Intrinsic neuromodulation in the *Tritonia* swim CPG: the serotonergic dorsal swim interneurons act presynaptically to enhance transmitter release from interneuron C2. *Journal of Neuroscience* **15**: 6035–45.

Kavaliers, M. & Colwell, D.D. (1995) Discrimination by female mice between the odours of parasitized and non-parasitized males. *Proceedings of the Royal Society of London Series B* **261**: 31–5.

Keddy-Hector, A.C., Seyfarth, R.M. & Raleigh, M.J. (1989) Male parental care, female choice and the effect of an audience in vervet monkeys. *Animal Behaviour* 38: 262–71.

Keesing, R.M. (1975) *Kinship Groups and Social Structure*. New York, Holt, Rhinehart & Winston.

Keeton, W.T. (1971) Magnets interfere with pigeon homing. *Proceedings of the National Academy of Sciences USA* 68: 102–6.

Keeton, W.T. (1981) Orientation and navigation in birds. In (D.J. Aidley ed.) *Animal Migration*, pp. 81–104. Cambridge, Cambridge University Press.

Keeton, W.T. & Gobert, A. (1970) Orientation by untrained pigeons requires the sun. *Proceedings of the National Academy of Sciences of the USA* 65: 853–6.

Keller, L. (ed.) (1999) *Levels of Selection in Evolution*. Princeton, NJ, Princeton University Press.

Keller, L. & Reeve, H.K. (1995) Why do females mate with multiple males: the sexual selected sperm hypothesis. *Advances in the Study of Behaviour* 24: 291–315.

Keller, L. & Ross, K.G. (1998) Selfish genes: a green beard in the red fire ant. *Nature* 394: 573–5.

Keller, M.J. & Gerhardt, H.C. (2001) Polyploidy alters advertisement call structure in gray treefrogs. *Proceedings of the Royal Society of London Series B* 268: 341–5.

Kelley, D.B. & Gorlick, D.L. (1990) Sexual selection and the nervous system. *BioScience* 40: 275–83.

Kemp, D.J. & Wiklund, C. (2001) Fighting without weaponry: a review of male–male contest competition in butterflies. *Behavioral Ecology & Sociobiology* 49: 429–42.

Kempenaers, B.G., Verheyen, M., Broeck M. van den, Burke, T.C., Broekhoven, C. van & Dhondt, A.A. (1992) Extra-pair paternity results from female preference for high-quality male in the blue tit. *Nature* 357: 494–6.

Kendler, K.S., Thornton, L.M., Gilman, S.E. & Kessler, R.C. (2000) Sexual orientation in a U.S. national sample of twin and non-twin sibling pairs. *American Journal of Psychiatry* 157: 1843–9.

Kennedy, C.E.J., Endler, J.A., Poynton, S.L. & McMinn, H. (1987) Parasite load predicts mate choice in guppies. *Behavioral Ecology & Sociobiology* 21: 291–5.

Kennedy, J.S. (1985) Migration: behavioral and ecological. In (M.A. Rankin ed.) *Migration: Mechanisms and Adaptive Significance*. Contributions to Marine Science 27 (Supplement): 5–26.

Kennedy, J.S. (1992) *The New Anthropomorphism*. Cambridge, Cambridge University Press.

Kennedy, M. & Gray, R.D. (1993) Can ecological theory explain the distribution of foraging animals? A critical analysis of experiments on the Ideal Free Distribution. *Oikos* 68: 158–66.

Kennedy, M., Spencer, H.G. & Gray, R.D. (1996) Hop, step and gape: do social displays of the Pelecaniformes reflect phylogeny? *Animal Behaviour* 51: 273–91.

Kenrick, D.T., Groth, G.E., Trost, M.R. & Sadalla, E.K. (1990a) Integrating evolutionary and social exchange perspectives on relationships: effects of gender, self-appraisal, and involvement level on mate choice. *Journal of Personality and Social Psychology* 25: 159–67.

Kenrick, D.T., Sadalla, E.K., Groth, G. and Trost, M.R. (1990b) Evolution, traits, and the stages of human courtship: qualifying the parental investment model, *Journal of Personality*, 58: 97–116.

Kent, D.S. & Simpson, J.A. (1992) Eusociality in the beetle *Austroplatypus incompertus* (Coleoptera: Curculionidae). *Naturwissenschaften* 79: 86–7.

Kenward, R.E. (1978) Hawks and doves: factors affecting success and selection in goshawk attacks on woodpigeons. *Journal of Animal Ecology* 47: 449–60.

Kenyon, K.W. (1969) The sea otter (*Enhydra lutris*) in the eastern Pacific Ocean. *North American Fauna* 68: 1–352.

Keverne, E.B., Fundele, R., Narashima, M., Barton, S.C. & Surani, M.A. (1996a) Genomic imprinting and the differential role of parental genomes in brain development. *Developmental Brain Research* 92: 91–100.

Keverne, E.B., Martel, F.L. & Nevison, C.M. (1996b) Primate brain evolution: genetic and functional considerations. *Proceedings of the Royal Society of London Series B* 262: 689–96.

Kieser, J.A. & Groeneveld, H. (1991) Fluctuating odontometric asymmetry, morphological variability, and genetic monomorphism in the cheetah *Acinonyx jubatus*. *Evolution* 45: 1175–83.

Kilner, R. & Johnstone, R.A. (1997) Begging the question: are offspring solicitation behaviours signals of needs? *Trends in Ecology & Evolution* 12: 11–15.

Kimball, R.T. & Ligon, J.D. (1999) Evolution of avian plumage dichromatism from a proximate perspective. *American Naturalist* 154: 182–93.

Kimchi, T. & Terkel, J. (2001) Magnetic compass orientation in the blind mole rat, *Spalax ehrenbergi*. *Journal of Experimental Biology* **204**: 751–8.

King, J.A. (1955) Social behavior, social organization and population dynamics in a black-tailed prairie dog town in the Black Hills of South Dakota. *Contributions of the Laboratory of Vertebrate Biology, University of Michigan* **67**: 1–123.

Kippert, F. (1996) An ultradian clock controls locomotor behaviour and cell division in isolated cells of *Paramecium tetraurelia*. *Journal of Cell Science* **109**: 867–73.

Kirkpatrick, M. & Ryan, M.J. (1991) The evolution of mating preferences and the paradox of the lek. *Nature* **350**: 33–8.

Kirschvink, J.L., Dizon, A.E. & Westphal, J.A. (1986) Evidence from strandings for geomagnetic sensitivity in cetaceans. *Journal of Experimental Biology* **120**: 1–24.

Kirschvink, J.L., Walker, M.M. & Diebel, C.E. (2001) Magnetite-based magnetoreception. *Current Opinion in Neurobiology* **11**: 462–7.

Kislaliogu, M. & Gibson, R.N. (1976) Prey 'handling time' and its importance in food selection by the 15-spined stickleback *Spinachia spinachia*. *Journal of Experimental Marine Biology and Ecology* **25**: 151–8.

Kiss, J.P. & Vizi, E.S. (2001) Nitric oxide: a novel link between synaptic and nonsynaptic transmission. *Trends in Neurosciences* **24**: 211–15.

Klein, S.L. (2000) The effects of hormones on sex differences in infection: from genes to behavior. *Neuroscience and Biobehavioral Reviews* **24**: 627–38.

Knight-Jones, E.W. (1953) Decreased discrimination during settling after prolonged planktonic life in larvae of *Spirorbis borealis* (Serpulidae). *Journal of the Marine Biological Association of the UK* **32**: 337–45.

Knowlton, N. (1974) A note on the evolution of gamete dimorphism. *Journal of Theoretical Biology* **46**: 283–5.

Knudsen, E.I. & Konishi, M. (1979) Mechanisms of sound location in the barn owl (*Tyto alba*). *Journal of Comparative Physiology* **133**: 13–21.

Kobayashi, A. & Kirschvink, J.L. (1995) Magnetoreception and electromagnetic field effects: sensory perception of the geomagnetic field in animals and humans. *Electromagnetic Fields: Advances in Chemistry Series* **250**: 367–94.

Kodric-Brown, A. & Brown, J.H. (1978) Influence of economics, interspecific competition and

sexual dimorphism on territoriality of migrant rufous hummingbirds. *Ecology* **59**: 285–96.

Köhler, K.L. (1978) Do pigeons use their eyes for navigation? A new technique! In (K. Schmidt-Koenig & W.T. Keeton eds) *Animal Migration: Navigation & Homing*, pp. 57–64. Heidelberg, Springer.

Köhler, W. (1927) *The Mentality of Apes*. New York, Harcourt Brace.

Kokko, H. & Jennions, M. (2003) It takes two to tango. *Trends in Ecology & Evolution* **18**: 103–4.

Kokko, H. & Lundberg, P. (2001) Dispersal, migration and offspring retention in saturated habitats. *American Naturalist* **157**: 188–202.

Komdeur, J., Daan, S., Tinbergen, J. & Matemann, C. (1997) Extreme adaptive modification in the sex ratio of Seychelles warbler's eggs. *Nature* **285**: 522–5.

Komisaruk, B.R., Adler, N.T. & Hutchison, J. (1972) Genital sensory field: enlargement by oestrogen treatment in female rats. *Science* **178**: 1295–8.

Konopka, R.J. & Benzer, S. (1971) Clock mutants of *Drosophila melanogaster*. *Proceedings of the National Academy of Sciences of the USA* **68**: 2112–16.

Konorski, J. (1967) *Integrative Activity of the Brain*. Chicago, University of Chicago Press.

Koolhaas, J.M., Korte, S.M., Boer S.F. de, Vegt, B.J. van der, Reenen, C.G. van, Hopster, H., Jong, I.C. de, Ruis, M.A.W. & Blokhuis, H.J. (1999) Coping styles in animals: current status in behavior and stress physiology. *Neuroscience and Biobehavioral Reviews* **23**: 925–35.

Kotiaho, J.S., Simmons, L.W. & Tomkins, J.L. (2001) Towards a resolution of the lek paradox. *Nature* **410**: 684–6.

Kraak, S.B.M. (1996) 'Copying mate choice': which phenomena deserve this term? *Behavioural Processes* **36**: 99–102.

Krakauer, D.C. (1995) Groups confuse predators by exploiting perceptual bottlenecks: a connectionist model of the confusion effect. *Behavioural Ecology & Sociobiology* **36**: 421–9.

Kramer, D.L. (1985) Are colonies superoptimal groups? *Animal Behaviour* **33**: 1031–2.

Kramer, G. (1957) Experiments on bird orientation and their interpretation. *Ibis* **99**: 196–227.

Krause, J. & Ruxton, G. (2002) *Living in Groups*. Oxford, Oxford University Press.

Kravitz, E.A. (1988) Hormonal control of behavior: amines and the biasing of behavioral output in lobsters. *Science* **241**: 1775–81.

Krebs, J.R. (1971) Territory and breeding density in the great tit, *Parus major* L. *Ecology* **52**: 2–22.

Krebs, J.R. (1973) Social learning and the significance of mixed species flocks of chickadees (*Parus* spp.). *Canadian Journal of Zoology* **51**: 1275–88.

Krebs, J.R. (1974) Colonial nesting and social feeding as strategies for exploiting food resources in the great blue heron (*Ardea herodias*). *Behaviour* **51**: 99–134.

Krebs, J.R. (1978) Optimal foraging: decision rules for predators. In (J.R. Krebs & N.B. Davies eds) *Behavioural Ecology: An Evolutionary Approach*, 1st Edition, pp. 23–63. Oxford, Blackwell.

Krebs, J.R. & Davies, N.B. (1987) *An Introduction to Behavioural Ecology*, 2nd Edition. Oxford, Blackwell.

Krebs, J.R. & Davies, N.B. (1993) *An Introduction to Behavioural Ecology*, 3rd Edition. Oxford, Blackwell.

Krebs, J.R. & Dawkins, R. (1984) Animal signals: mind-reading and manipulation. In (J.R. Krebs & N.B. Davies eds) *Behavioural Ecology: An Evolutionary Approach*, 2nd Edition, pp. 380–402. Oxford, Blackwell.

Krebs, J.R. & Kacelnik, A. (1991) Decision-making. In (J.R. Krebs & N.B. Davies eds) *Behavioural Ecology: An Evolutionary Approach*, 3rd Edition, pp. 105–36. Oxford, Blackwell.

Krebs, J.R. & McCleery, R.H. (1984) Optimization in behavioural ecology. In (J.R. Krebs & N.B. Davies eds) *Behavioural Ecology: An Evolutionary Approach*, 2nd Edition, pp. 91–121. Oxford, Blackwell.

Krebs, J.R., Ashcroft, R. & Webber, M. (1978a) Song repertoires and territory defence in the great tit (*Parus major*). *Nature* **271**: 539–42.

Krebs, J.R., Erichsen, J.T., Webber, M.I. & Charnov, E.L. (1977) Optimal prey-selection by the great tit (*Parus major*). *Animal Behaviour* **25**: 30–8.

Krebs, J.R., Kacelnik, A. & Taylor, P.J. (1978b) Test of optimal sampling by foraging great tits. *Nature* **275**: 27–31.

Krebs, J.R., MacRoberts, M.H. & Cullen, J.M. (1972) Flocking and feeding in the great tit, *Parus major*: an experimental study. *Ibis* **114**: 507–30.

Krebs, J.R., Ryan, J.C. & Charnov, E.L. (1974) Hunting by expectation or optimal foraging? A study of patch use by chickadees. *Animal Behaviour* **22**: 953–64.

Krebs, J.R., Stephens, D.W. & Sutherland, W.J. (1983) Perspectives in optimal foraging. In (A.H. Brush & G.A. Clarke Jr eds) *Perspectives in Ornithology*, pp. 165–216. New York, Cambridge University Press.

Krebs, J.R., Sherry, D.F., Healy, S.D., Perry, V.H. & Vaccarino, A.L. (1989) Hippocampal specialization of food-storing birds. *Proceedings of the National Academy of Sciences of the United States of America* **86**: 1388–92.

Kreithen, M.L. & Keeton, W.T. (1974) Detection of changes in atmospheric pressure by the homing pigeon, *Columba livia*. *Journal of Comparative Physiology* **89**: 83–92.

Kreithen, M.L. & Quine, D.B. (1979) Infrasound detection by the homing pigeon: a behavioral audiogram. *Journal of Comparative Physiology* **129**: 1–4.

Kruuk, H. (1964) Predators and anti-predator behaviour of the black-headed gull, *Larus ridibundus*. *Behaviour Supplement* **11**: 1–129.

Kruuk, H. (1972) *The Spotted Hyena*. Chicago, Chicago University Press.

Kummer, H., Anzenberger, G. & Hemelrijk, C.K. (1996) Hiding and perspective taking in long-tailed macaques (*Macaca fascicularis*). *Journal of Comparative Psychology* **110**: 97–102.

Kyl-Heku, L.M. & Buss, D.M. (1996) Tactics as units of analysis in personality psychology: an illustration using tactics of hierarchy negotiation. *Personality and Individual Differences* **21**: 497–517.

La Cerra, M.M. (1994) Evolved mate preferences in women: psychological adaptations for assessing a man's willingness to invest in offspring. Unpublished Doctoral Dissertation, University of California, Santa Barbara.

Lachlan, R.F. & Slater, P.J.B. (2000) The maintenance of vocal learning by gene–culture interaction: the cultural trap hypothesis. *Proceedings of the Royal Society of London Series B* **266**: 701–6.

Lafleur, D.L., Lozano, G.A. & Sclafani, M. (1997) Female mate-choice copying in guppies, *Poecilia reticulata*: a re-evaluation. *Animal Behaviour* **54**: 579–86.

Lakin-Thomas, P.L. (2000) Circadian rhythms: new functions for old clock genes? *Trends in Genetics* **16**: 106–14.

Laland, K.N. (1994) Sexual selection with a culturally transmitted mating preference. *Theoretical Population Biology* **45**: 1–15.

Lalumiere, M.L. & Quinsey, V.L. (1996) Sexual deviance, antisociality, mating effort, and the use of sexually coercive behaviors. *Personality and Individual Differences* **21**: 33–48.

Lalumiere, M.L., Chalmers, L.J., Quinsey, V.L. & Seto, M.C. (1996) A test of the mate deprivation hypothesis of sexual coercion. *Ethology & Sociobiology* **17**: 299–318.

Lambrinos, D., Maris, M., Kobayashi, H. Labhart, T., Pfiefer, R. & Wehner, R. (1997) An autonomous agent navigating with a polarized light compass. *Adaptive Behaviour* **6**: 175–206.

Lambrinos, D., Moller, R., Labhart, T., Pfiefer, R. & Wehner, R. (2000) A mobile robot employing insect strategies for navigation. *Robotics and Autonomous Systems* **30**: 39–64.

Lande, R. (1981) Models of speciation by sexual selection of polygenic traits. *Proceedings of the National Academy of Sciences of the USA* **78**: 3721–5.

Landsberg, J.W. (1976) Posthatch age and developmental age as a baseline for determination of the sensitive period for imprinting. *Journal of Comparative Physiology and Psychology* **90**: 47–52.

Lane, A., Kinsella, A., Waddington, J.L., Larkin, C. & O'Callagham, E. (1996) Quantitative evidence of developmental instability in schizophrenia an measured by fluctuating asymmetry. *Schizophrenia Research* **18**: IXF3.

Langbauer, W.R. (2000) Elephant communication. *Zoo Biology* **19**: 425–45.

Langley-Evans, S. (1997) Fetal programming of immune function and respiratory disease. *Clinical and Experimental Allergy* **27**: 1377–9.

Langlois, J.H., Roggman, L.A. & Musselman, L. (1994) What is average and what is not average about attractive faces? *Psychological Science* **5**: 214–19.

Larkin, S. & McFarland, D.J. (1978) The cost of changing from one activity to another. *Animal Behaviour* **26**: 1237–46.

Larkin, T.S. & Keeton, W.T. (1978) An apparent lunar rhythm in the day-to-day variations in the initial bearings of homing pigeons. In (K. Schmidt-Koenig & W.T. Keeton eds) *Animal Migration: Navigation and Homing*, pp. 92–106. Heidelberg, Springer.

Lauder, J.M. (1993) Neurotransmitters as growth regulatory signals: role of receptors and second messengers. *Trends in Neuroscience* **16**: 233–40.

Lawrence, P.A. (1981) A general cell marker for clonal analysis of *Drosophila* development. *Journal of Embryology and Experimental Morphology* **64**: 321–32.

Lazarus, J. (1978) Vigilance, flock size and domain of danger size in the white-fronted goose. *Wildfowl* **29**: 135–45.

Lazarus, J. (1979) The early warning function of flocking in birds: an experimental study with captive quelea. *Animal Behaviour* **27**: 855–65.

Lazarus, J. & Inglis, I.R. (1978) The breeding behaviour of the pink-footed goose: parental care and vigilant behaviour during the fledging period. *Behaviour* **65**: 62–8.

Leach, E. (1966) Virgin birth. *Proceedings of the Royal Anthropological Institute of Great Britain and Ireland* 39–49.

Leamy, L. & Atchley, W. (1985) Directional selection and developmental stability: evidence from fluctuating asymmetry of morphometric characters in rats. *Growth* **49**: 8–18.

Leary, R.F., Allendorf, F.W. & Knudsen, R.L. (1985) Developmental instability as an indicator of reduced genetic variation in hatchery trout. *Transactions of the American Fisheries Society* **114**: 230–5.

Le Boeuf, B.J. (1974) Male–male competition and reproductive success in elephant seals. *American Zoologist* **14**: 173–6.

Le Boeuf, B.J. & Mesnick, S. (1990) Sexual behaviour of male Northern elephant seals: I. Lethal injuries to adult females. *Behaviour* **116**: 143–62.

LeCroy, M. (1981) The genus *Paradisaea* – display and evolution. *American Museum Novitates* **2714**: 1–52.

Lefebvre, L. (1995) Culturally-transmitted feeding behaviour in primates: evidence for accelerating learning rates. *Primates* **36**: 227–39.

Lefebvre, L., Viville, S., Barton, S.C., Ishino, F., Keverne, E.B. & Surani, M.A. (1998) Abnormal maternal behaviour and growth retardation associated with loss of the imprinted gene *Mest*. *Nature Genetics* **20**: 163–9.

Leggett, W.C. (1977) The ecology of fish migrations. *Annual Review of Ecology and Systematics* **8**: 285–308.

Lehman, M.N., Silver, R., Gladstone, W.R., Kahn, R.M., Gibson, M. & Bittman, E. (1987) Circadian rhythmicity restored by neural transplant. Immunocytochemical characterization of the graft and its integration with the host brain. *Journal of Neuroscience* **7**: 1626–39.

Lehrer, M. & Bianco, G. (2000) The turn-back-and-look behaviour: bee versus robot. *Biological Cybernetics* **83**: 211–29.

Leising, A.W. (2001) Copepod foraging in patchy habitats and thin layers using a 2-D individual-based model. *Marine Ecology: Progress Series* **216**: 167–79.

Lendrem, D.W. (1985) Kinship affects puberty acceleration in mice (*Mus musculus*). *Behavioral Ecology & Sociobiology* **17**: 397–9.

Lendrem, D.W. (1986) *Modelling in Behavioural Ecology*. London, Croom Helm.

Lenington, S. (1981) Child abuse: the limits of sociobiology. *Ethology & Sociobiology* **2**: 17–29.

Lenington, S. (1991) The *t* complex: a story of genes, behavior and populations. *Advances in the Study of Behavior* **20**: 51–86.

Leonard, M.L. & Horn, A.G. (1998) Need and nestmates affect begging in tree swallows. *Behavioral Ecology & Sociobiology* **42**: 431–6.

Leonard, M.L. & Picman, J. (1987) Female settlement in marsh wrens: is it affected by other females? *Behavioral Ecology & Sociobiology* **21**: 135–40.

Lesch, K.-P., Bengel, D., Heils, A., Sabol, S.Z., Greenberg, B.D., Petri, S., Benjamin, J., Muller, C.R., Hamer, D.H. & Murphy, D.L. (1996) Association of anxiety-related traits with a polymorphism in the serotonin transporter gene regulatory region. *Science* **274**: 1527–31.

Lettvin, J.Y., Maturana, H.R., McCullough, W.S. & Pitts, W.H. (1959) What the frog's eye tells the frog's brain. *Proceedings of the Institute of Radio Engineering* **47**: 1940–51.

Leuthold, W. (1966) Variations in territorial behaviour of the Uganda kob, *Adenota kob thomasi*. *Behaviour* **27**: 215–58.

LeVay, S. (1991) A difference in hypothalamic structure between heterosexual and homosexual men. *Science* **253**: 1034–7.

LeVay, S. (1994) *The Sexual Brain*. Cambridge, MA, MIT Press.

Levine, R.B. & Truman, J.W. (1985) Dendritic reorganisation of abdominal motoneurons during metamorphosis of the moth *Manduca sexta*. *Journal of Neuroscience* **5**: 2424–31.

Lewontin, R.C. (1978) Adaptation. *Scientific American* **239**: 212–30.

Lewontin, R.C. (1991) *Biology as Ideology*. Concord, Anansi Press.

Lewontin, R.C., Rose, S. & Kamin, L.J. (1984) *Not in Our Genes*. New York, Pantheon.

Li, L.L., Keverne, E.B., Aparicio, S.A., Ishino, F., Barton, S.C. & Surani, M.A. (1999) Regulation of maternal behavior and offspring growth by paternally expressed *Peg3*. *Science* **284**: 330–3.

Lichtenstein, G. & Sealy, S.G. (1998) Nesting competition, rather than supernormal stimulus, explains the success of parasitic brown-headed cowbird chicks in yellow warbler nests. *Proceedings of the Royal Society of London Series B* **265**: 249–54.

Lima, S.L. (1995) Back to the basics of antipredatory vigilance: the group size effect. *Animal Behaviour* **49**: 11–20.

Lima, S.L. (2002) Putting predators back into behavioral predator–prey interactions. *Trends in Ecology & Evolution* **17**: 70–5.

Lima, S.L. & Bednekoff, P.A. (1999) Back to the basics of antipredatory behaviour: can non-vigilant animals detect attack? *Animal Behaviour* **58**: 537–43.

Lima, S.L., Valone, T.J. & Caraco, T. (1985) Foraging efficiency–predation risk tradeoff in the grey squirrel. *Animal Behaviour* **33**: 155–65.

Lincoln, G.A., Guinnen, F. & Short, R.V. (1972) The way in which testosterone controls the social and sexual behaviour of the red deer stag (*Cervus elephus*). *Hormones and behaviour* **3**: 375–96.

Lindauer, M. (1961) *Communication Among Social Bees*. Cambridge, MA, Harvard University Press.

Lindauer, M. & Martin, H. (1968) Die Schwere-orientierung der Bienen unter dem Einfluss des Erdmagnetfeldes. *Zeitschrift für vergleichende Physiologie* **60**: 219–43.

Lindström, A. & Alerstam, T. (1992) Optimal fut loads in migrating birds – a test of the time-minimization hypothesis. *American Naturalist* **140**: 477–91.

Lively, C.M. (1992) Parthenogenesis in a freshwater snail: reproductive assurance versus parasite release. *Evolution* **46**: 907–13.

Lloyd, D.G. (1987) Allocations to pollen, seeds and pollination mechanisms in self-fertilizing plants. *Functional Ecology* **1**: 83–9.

Lloyd, D.P.C. (1955) Synaptic mechanisms. In (J.F. Fulton ed.) *A Textbook of Physiology*, pp. 35–47. Philadelphia, Saunders.

Lockery, S.R. & Kristan, W.B. Jr (1990) Distributed processing of sensory information in the leech: II. Identification of interneurons contributing to the local bending reflex. *Journal of Neuroscience* **10**: 1816–29.

Lockery, S.R. & Sejnowski, T.J. (1992) Distributed processing of sensory information in the leech: III. A dynamical neural network model of the local bending reflex. *Journal of Neuroscience* **12**: 3877–95.

Loehlin, J.C. (1992) *Genes and Environment in Personality Development*. Newburg Park CA, Sage Publications.

Loesche, P., Stoddard, P.K., Higgins, B.J. & Beecher, M.D. (1991) Signature versus perceptual adaptations for individual vocal recognition in swallows. *Behaviour* **118**: 15–25.

Loftus-Hills, J.J. & Johnstone, B.M. (1970) Auditory function, communication and the brain-evoked response in anuran amphibians. *Journal of the Acoustical Society of America* **47**: 1131–8.

Logan, F.A. (1965) Decision making by rats. *Journal of Comparative Physiology and Psychology* **59**: 1–12.

Loher, W. (1972) Circadian control of stridulation in the cricket *Teleogryllus commodus* Walker. *Journal of Comparative Physiology* **79**: 173–90.

Lohmann, K.J. & Lohmann, C.M.F. (1996) Detection of magnetic field intensity by sea turtles. *Nature* **380**: 59–61.

Lolordo, V.M. (1979) Selective associations. In (A. Dickinson & R.A. Boakes eds) *Mechanisms of Learning and Motivation*, pp. 367–98. Hillsdale, NJ, Erlbaum.

Lorenz, K. (1950) The comparative method in studying innate behaviour patterns. *Symposia of the Society for Experimental Biology* **4**: 221–68.

Lorenz, K. (1971) *Studies in Animal and Human Behaviour*, Vol. 2. London, Methuen.

Lougheed, L.W. & Anderson, D.J. (1999) Parent blue-footed boobies suppress siblicidal behavior of offspring. *Behavioral Ecology & Sociobiology* **45**: 11–18.

Lovegrove, B.G. (1991) The evolution of eusociality in mole rats (Bathyergidae): a question of risks, numbers and costs. *Behavioral Ecology & Sociobiology* **28**: 37–45.

Lovibond, P.F. (1983) Facilitation of instrumental behavior by a Pavlovian appetitive conditioned stimulus. *Journal of Experimental Psychology: Animal Behavior Processes* **9**: 225–47.

Luca, D.W. de & Ginsberg, J.R. (2001) Dominance, reproduction and survival in banded mongooses: towards an egalitarian social system. *Animal Behaviour* **61**: 17–30.

Lucas, J.R. (1985) Partial prey consumption by antlion larvae. *Animal Behaviour* **33**: 945–59.

Lummaa, V. & Clutton-Brock, T.H. (2002) Early development, survival and reproduction in humans. *Trends in Ecology & Evolution* **17**: 141–7.

Lumsden, C.J. & Wilson, E.O. (1981) *Genes, Mind & Culture*. Cambridge, MA, Harvard University Press.

Lumsden, C.J. & Wilson, E.O. (1983) *Promethean Fire: Reflections on the Origin of Mind*. Cambridge, MA, Harvard University Press.

Lunau, K. (1996) Signalling functions of floral colour patterns for insect flower visitors. *Zoologischer Anzeiger* **235**: 11–30.

Lund, H.H., Webb, B. & Hallam, J. (1998) Physical and temporal scaling considerations in a robot model of cricket calling song preference. *Artificial Life* **4**: 95–107.

Lundberg, P. (1985) Dominance behaviour, body weight and fat variations, and partial migration in European blackbirds, *Turdus merula*. *Behavioral Ecology and Sociobiology* **17**: 185–9.

Lundberg, P. (1987) Partial bird migration and evolutionarily stable strategies. *Journal of Theoretical Biology* **125**: 351–60.

Lüters, W. & Birukow, G. (1963) Solar compass orientation of *Apodemus agrarius*. *Naturwissenschaften* **50**: 757–8.

Lynch, A. (1996) *Thought Contagion: How Belief Spreads Through Society*. New York, Basic Books.

Lyon, B.E., Eadie, J.M. & Hamilton, L.D. (1994) Parental choice selects for ornamental plumage in American coot chicks. *Nature* **371**: 240–3.

MacArthur, R.H. & Pianka, E.R. (1966) On optimal use of a patchy environment. *The American Naturalist* **100**: 603–9.

Mace, R. (1998) The coevolution of human fertility and wealth inheritance. *Philosophical Transactions of the Royal Society of London Series B* **353**: 389–97.

Mace, R. (2000) Evolutionary ecology of human life history. *Animal Behaviour* **59**: 1–10.

MacKaye, T.F.C. & Doyle, R.W. (1978) An ecological genetic analysis of the settling behaviour of marine polychaete: I. Probability of settlement and gregarious behaviour. *Heredity* **40**: 1–12.

Mackintosh, N.J. (1975) A theory of attention: variations in the associability of stimuli with reinforcement. *Psychological Review* **82**: 276–98.

Mackintosh, N.J. (1983) General principles of learning. In (T.R. Halliday & P.J.B. Slater eds) *Animal Behaviour 3: Genes, Development and Learning*, pp. 149–177. Oxford, Blackwell.

Mackney, P.A. & Hughes, R.N. (1995) Foraging behaviour and memory window in sticklebacks. *Behaviour* **132**: 1241–53.

Macphail, E.M. (1982) *Brain and Intelligence in Vertebrates*. Oxford, Clarendon Press.

Macphail, E.M. (1985) Vertebrate intelligence: the null hypothesis. *Philosophical Transactions of the Royal Society of London B* **308**: 37–52.

Macphail, E.M. (1987) The comparative psychology of intelligence. *Behavioral and Brain Sciences* **10**: 645–95.

Macphail, E.M. & Bolhuis, J. (2001) The evolution of intelligence: adaptive specialization versus general process. *Biological Reviews* **76**: 341–64.

Madsen, T.R., Shine, R., Loman, J. & Håkansson, T. (1992) Why do female adders copulate so frequently? *Nature* **355**: 440–1.

Maguire, E.A., Gadian, D.G., Johnsrude, I.S., Good, C.D., Ashburner, J., Frackowiak, R.S.J. & Frith, C.D. (2000) Navigation-related structural changes in the hippocampi of taxi drivers. *Proceedings of the National Academy of Sciences USA* **97**: 4398–403.

Magurran, A.E. (1990) The adaptive significance of schooling as an anti-predator defence in fish. *Annales Zoologici Fennici* **27**: 51–66.

Magurran, A.E. & Girling, S. (1986) Predator recognition and response habituation in shoaling minnows. *Animal Behaviour* **34**: 510–18.

Magurran, A.E. & Pitcher, T. (1987) Provenance, shoal size and the sociobiology of predator-evasion in minnow shoals. *Proceedings of the Royal Society of London Series B* **246**: 31–8.

Magurran, A.E., Irving, P.W. & Henderson, P.A. (1996) Is there a fish alarm pheromone? A wild study and critique. *Proceedings of the Royal Society of London Series B* **263**: 1551–6.

Maier, S.F. & Watkins, L.R. (1999) Bidirectional communications between the brain and the immune system: implications for behaviour. *Animal Behaviour* **57**: 741–51.

Major, P. (1978) Predator–prey interactions in two schooling fishes, *Caranx ignobilis* and *Stolephorus purpureus*. *Animal Behaviour* **26**: 760–77.

Malamuth, N.M. (1996) The confluence model of sexual aggression: feminist and evolutionary perspectives. In (D.M. Buss & N.M. Malamuth eds) *Sex, Power, Conflict: Evolutionary and Feminist Perspectives*, pp. 269–95. New York, Oxford University Press.

Mann, J. (1992) Nurturance or negligence: maternal psychology and behavioral preference among preterm twins. In (J. Barkow, L. Cosmides & J. Tooby eds) *The Adapted Mind*, pp. 367–90. New York, Oxford University Press.

Manning, A. & Dawkins, M.S. (1998) *An Introduction to Animal Behaviour*, 5th Edition. Cambridge, Cambridge University Press.

Manning, J.T. (1989) Age-advertisement and the age-dependency model of female choice. *Journal of Evolutionary Biology* **2**: 379–84.

Manning, J.T. & Ockenden, L. (1994) Fluctuating asymmetry in racehorses. *Nature* **370**: 185–6.

Manning, J.T. & Pickup, L.J. (1998) Symmetry and performance in middle distance runners. *International Journal of Sports Medicine* **19**: 205–9.

Manning, J.T. & Taylor, R.P. (2001) Second to fourth digit ratio and male ability in sport: implications for sexual selection in humans. *Evolution and Human Behavior* **22**: 61–9.

Manning, J.T. & Wood, D. (1998) Fluctuating asymmetry and aggression in boys. *Human Nature* **9**: 53–66.

Manning, J.T., Barley, L., Walton, J., Lewis-Jones, D.I., Trivers, R.L., Singh, D., Thornhill, R., Rohde, P., Bereczkei, T., Henzi, P., Soler, M. & Szwed, A. (2000) The 2nd : 4th digit ratio, sexual dimorphism, population differences and reproductive success: evidence for sexually antagonistic genes? *Evolution & Human Behavior* **21**: 163–83.

Manning, J.T., Gage, A.R., Diver, M.J., Scutt, D. & Fraser, W.D. (2002) Short-term changes in asymmetry and hormones in men. *Evolution & Human Behavior* **23**: 95–102.

Manning, J.T., Scutt, D., Whitehouse, G.H., Leinster, S.J. & Walton, J.M. (1996) Asymmetry and the menstrual cycle in women. *Ethology & Sociobiology* **17**: 129–43.

Manning, J.T., Trivers, R.L., Thornhill, R., Singh, D., Denman, J., Eklo, M.H. & Anderton, R.H. (1997) Ear asymmetry and left-side cradling. *Evolution & Human Behavior* **18**: 327–40.

Mappes, J. & Alatalo, R.V. (1997) Effects of novelty and gregariousness in survival of aposematic prey. *Behavioral Ecology* **8**: 174–7.

March, J., Chamizo, V.D. & Mackintosh, N.J. (1992) Reciprocal overshadowing between intra-maze and extra-maze cues. *Quarterly Journal of Experimental Psychology* **45B**: 49–63.

Marchetti, C. & Drent, P.J. (2000) Individual differences in the use of social information in foraging by captive great tits. *Animal Behaviour* **60**: 131–40.

Marconato, A. & Bisazza, A. (1986) Males whose nests contain eggs are preferred by female *Cottus gobio* L. (Pisces: Cottidae). *Animal Behaviour* **34**: 1580–2.

Mark, T.A. & Gallistel, C.R. (1994) Kinetics of matching. *Journal of Experimental Psychology: Animal Behavior Processes* **20**: 79–95.

Markl, L. (1977) Adaptive radiation of mechano-reception. In (M.A. Ali ed.) *Sensory Ecology*, pp. 49–69. New York, Plenum.

Markow, T.A. ed. (1994) *Developmental Instability: Its Origins and Evolutionary Implications.* Amsterdam, Kluwer Academic.

Markow, T.A. & Martin, J.F. (1993) Inbreeding and developmental stability in a small human population. *Annals of Human Biology* **20**: 389–94.

Marler, P. (1955) Characteristics of some animal calls. *Nature* **176**: 6–8.

Marler, P. (1956) The voice of the chaffinch and its function as a language. *Ibis* **98**: 231–61.

Marler, P. (1957) Specific distinctiveness in the communication signals of birds. *Behaviour* **11**: 13–39.

Marler, P. (1970) A comparative approach to vocal learning: song development in white-crowned sparrows. *Journal of Comparative and Physiological Psychology* **71**: 1–25.

Marler, P. & Peters, S. (1982) Subsong and plastic song: their role in the vocal learning process. In (D.E. Kroodsma & E.H. Miller eds) *Acoustic Communication in Birds*, Vol. 2. pp. 25–50. New York, Academic Press.

Marler, P. & Peters, S. (1989) Species differences in auditory responsiveness in early vocal learning. In (R.J. Dooling & S.H. Hulse eds) *The Comparative Psychology of Audition: Perceiving Complex Sounds*, pp. 243–73. Hillsdale, NJ, Lawrence Erlbaum Associates.

Marler, P. & Tamura, M (1962) Song 'dialects' in three populations of white-crowned sparrows. *Condor* **64**: 368–77.

Marler, P., Evans, C.S. & Hauser, M.D. (1992) Animal signals: motivational, referential or both? In (H. Papousek, U. Jurgens & M. Papousek eds) *Nonverbal Vocal Communication*, pp. 66–86. Cambridge, Cambridge University Press.

Marr, D. (1982) *Vision.* San Francisco, Freeman.

Marrow, P. & Johnston, R.A. (1996) Riding the evolutionary streetcar: where population genetics and game theory meet. *Trends in Ecology & Evolution* **11**: 445–6.

Martin, G.M. & Lett, B.T. (1985) Formation of associations of colored and flavored food with induced sickness in five main species. *Behavioral and Neural Biology* **43**: 223–37.

Martin, H. & Lindauer, M. (1977) Der Einfluss der Erdmagnetfeldes auf die Schwereorientierung der Honigbiene (*Apis mellifera*). *Journal of Comparative Physiology* **122**: 145–87.

Martin, P. & Bateson, P. (1993) *Measuring Behaviour: An Introductory Guide*, 2nd Edition. Cambridge, Cambridge University Press.

Martin, P. & Caro, T. (1985) On the functions of play and its role in behavioural development. *Advances in the Study of Behavior* **15**: 59–103.

Martin, R.D. (1990) *Primate Origins and Evolution.* London, Chapman and Hall.

Martin, S.M., Manning, J.T. & Dowrick, C.F. (1999) Fluctuating asymmetry, relative digit length and depression in men. *Evolution & Human Behavior* **20**: 203–14.

Mason, G.J. (1991) Stereotypies: a critical review. *Animal Behaviour* **41**: 1015–37.

Mason, G.J. & Mendl, M. (1993) Why is there no simple way of measuring animal welfare? *Animal Welfare* **2**: 301–19.

Mather, J.G. (1981) Wheel-running activity: a new interpretation. *Mammal Review* **11**: 41–51.

Mather, J.G. & Baker, R.R. (1980) A demonstration of navigation by rodents using an orientation cage. *Nature* **284**: 259–62.

Mather, J.G. & Baker, R.R. (1982) Magnetic compass sense in the large yellow underwing moth, *Noctua pronuba*. *Animal Behaviour* **30**: 543–8.

Mather, K. (1953) Genetic control of stability in development. *Heredity* **7**: 297–336.

Matthews, G.V.T. (1951) The sensory basis of bird navigation. *Journal of the Institute of Navigation* **4**: 260–75.

Matthews, G.V.T. (1955) *Bird Navigation.* Cambridge, Cambridge University Press.

May, M. (1991) Aerial defense tactics of flying insects. *American Scientist* **79**: 316–29.

Maynard Smith, J. (1964) Group selection and kin selection. *Nature* **201**: 1145–7.

Maynard Smith, J. (1972) *On Evolution.* Edinburgh, Edinburgh University Press.

Maynard Smith, J. (1974) The theory of games and the evolution of animal conflicts. *Journal of Theoretical Biology* **47**: 209–21.

Maynard Smith, J. (1977) Parental investment – a prospective analysis. *Animal Behaviour* **25**: 1–9.

Maynard Smith, J. (1978a) *The Evolution of Sex.* Cambridge, Cambridge University Press.

Maynard Smith, J. (1978b) The evolution of behavior. *Scientific American* **239**: 136–45.

Maynard Smith, J. (1982) *Evolution and the Theory of Games.* Cambridge, Cambridge University Press.

Maynard Smith, J. (1984) The ecology of sex. In (J.R. Krebs & N.B. Davies eds) *Behavioural Ecology: An Evolutionary Approach*, 2nd Edition, pp. 201–21. Oxford, Blackwell.

Maynard Smith, J. & Parker, G.A. (1976) The logic of asymmetric contests. *Animal Behaviour* 24: 159–75.

Maynard Smith, J. & Parker, G.A. (1990) Optimality theory in evolutionary biology. *Nature* 348: 27–33.

Maynard Smith, J. & Szathmáry, E. (1999) *The Origins of Life: From the Birth of Life to the Origins of Language*. Oxford, Oxford University Press.

Mazur, J.E. (1981) Optimization theory fails to predict performance of pigeons in a two-response situation. *Science* 214: 823–5.

Mazur, J.E. (1984) Tests of an equivalence rule for fixed and variable reinforcer delays. *Journal of Experimental Psychology: Animal Behavior Processes* 10: 426–36.

McCann, T.S. (1982) Aggressive and maternal activities of female southern elephant seals (*Mirounga leonina*). *Animal Behaviour* 30: 268–76.

McClean, I.G. & Rhodes, G. (1991) Enemy recognition and response in birds. In (D.M. Power ed.) *Current Ornithology*, Vol. 8, pp. 173–211. New York, Plenum Press.

McCleery, R.H. (1977) On satiation curves. *Animal Behaviour* 25: 1005–15.

McCleery, R.H. (1978) Optimal behaviour sequences and decision-making. In (J.R. Krebs & N.B. Davies eds) *Behavioural Ecology: An Evolutionary Approach*, 1st Edition, pp. 377–410. Oxford, Blackwell.

McCollom, R.E., Siegel, P.B. & VanKrey, H.P. (1971) Responses to androgen in lines of chickens selected for mating behavior. *Hormones & Behavior* 2: 31–42.

McComb, K.E. (1987) Roaring by red deer stags advances date of oestrus in hinds. *Nature* 330: 648–9.

McCracken, G.F. (1984) Communal nursing in Mexican free-tailed bat maternity colonies. *Science* 223: 1090–1.

McCracken, G.F. & Gustin, M.K. (1991) Nursing behavior in Mexican free-tailed bat maternity colonies. *Ethology* 89: 305–21.

McCullough, J.M. & York Barton, E. (1991) Relatedness and mortality risk during a crisis year: Plymouth colony. *Ethology & Sociobiology* 12: 195–209.

McFarland, D.J. (1970) Behavioural aspects of homeostasis. In *Advances in the Study of Behaviour* (Lehrman *et al.* ed), pp. 1–26, Academic Press.

McFarland, D.J. (1971) *Feedback Mechanisms in Animal Behaviour*. London, Academic Press.

McFarland, D.J. (1974) Time-sharing as a behavioural phenomenon. *Advances in the Study of Behaviour* 5: 201–25.

McFarland, D.J. (1989) *Problems of Animal Behaviour*. Harlow, Longman.

McFarland, D.J. (1991) Defining motivation and cognition in animals. *International Studies in the Philosophy of Science* 5: 153–70.

McFarland, D.J. (1993) *Animal Behaviour: Psychobiology, Ethology and Evolution*, 2nd Edition. Harlow, Longman.

McFarland, D.J. (1999) *Animal Behaviour*, 3rd Edition. Harlow, Longman.

McFarland, D.J. & Houston, A.I. (1981) *Quantitative Ethology: The State Space* Approach. London, Pitman.

McFarland, D.J. & Sibly, R.M. (1972) 'Unitary drives' revisited. *Animal Behaviour* 20: 548–63.

McGill, T.E. (1970) Genetic analysis of male sexual behavior. In (G. Lindsey & D.D. Thiessen eds) *Contributions to Behavior–Genetic Analysis: the Mouse as a Prototype*, pp. 57–88. New York, Appleton Century Crofts.

McGregor, P.K. (1993) Signalling in territorial systems: a context for individual identification, ranging and eavesdropping. *Philosophical Transactions of the Royal Society of London Series B* 340: 237–44.

McGregor, P.K. & Thompson, D.B.A. (1988) Constancy and change in local dialects of the corn bunting. *Ornis Scandinavica* 19: 153–9.

McKenzie, J.A., Batterham, P. & Baker, L. (1990) Fitness and asymmetry modification as an evolutionary process. A study in the Australian sheep blowfly, *Lucilia cuprina* and *Drosophila melanogaster*. In (J.S.F. Barker, W.T. Starmer & R.J. MacIntyre eds) *Ecological Genetics of Drosophila*, pp. 57–73. New York, Plenum Press.

McKinney, F., Derrickson, S.R. & Mineau, P. (1983) Forced copulation in waterfowl. *Behaviour* 86: 250–94.

McNair, J.N. (1982) Optimal giving-up times and the marginal value theorem. *American Naturalist* 119: 511–29.

McNamara, J.M. (1985) An optimal sequencing policy for controlling a Markov neural process. *Journal of Applied Probability* 22: 324–35.

McNamara, J.M. & Houston, A.I. (1985) Optimal foraging and learning. *Journal of Theoretical Biology* 117: 231–49.

McNeill Alexander, R. (1996) *Optima for Animals*, 2nd Edition. Princeton, NJ, Princeton University Press.

Mech, L.D. (1970) *The Wolf: The Ecology and Behavior of an Endangered Species.* New York, Natural History Press.

Medvin, M.B., Stoddard, P.K. & Beecher, M.D. (1993) Signals for parent–offspring recognition: a comparative analysis of the begging calls of cliff swallows and barn swallows. *Animal Behaviour* 45: 841–50.

Mehdiabadi, N.J., Reeve, H.K. & Mueller, U.G. (2003) Queens versus workers: sex-ratio conflict in eusocial Hymenoptera. *Trends in Ecology & Evolution* 18: 88–93.

Meikle, D.B., Kruper, J.H. & Browning, C.R. (1995) Adult male house mice born to undernourished mothers are unattractive to oestrous females. *Animal Behaviour* 50: 753–8.

Menaker, M. (1968) Light reception by the extraretinal receptors in the brain of the sparrow. *Proceedings of the 76th Annual Convention of the American Psychological Association*, pp. 299–300. Washington, DC, American Psychological Association.

Mendl, M. & Deag, J. (1995) How useful are concepts of alternative strategy and coping strategy in applied studies of social behaviour? *Applied Animal Behaviour Science* 44: 119–37.

Mendoza, S.P., Coe, C.L., Lowe, E.L. & Levine, S. (1979) The physiological response to group formation in adult squirrel monkeys. *Psychoneuroendocrinology* 3: 221–9.

Menzel, R. (1990) Learning, memory, and 'cognition' in honeybees. In (R.P. Kessler & D.S. Olton eds) *Neurobiology of Comparative Cognition*, pp. 237–92. Hillsdale, NJ, Lawrence Erlbaum.

Menzel, R. (2001) Searching for the memory trace in a mini-brain, the honeybee. *Learning and Memory* 8: 53–62.

Menzel, R., Greggers, U. & Hammer, M. (1993) Functional organization of appetitive learning and memory in a generalist pollinator, the honey bee. In (D. Papaj & A. Lewis eds) *Insect Learning: Ecological and Evolutionary Perspectives*, pp. 79–125. London, Chapman and Hall.

Messina, F.J. (1987) Genetic contribution to the dispersal polymorphism of the cowpea weevil (Coleoptrea: Bruchidae). *Annals of the Entomological Society of America* 80: 12–16.

Mesterton-Gibbons, M. (1992) Ecotypic variation in the asymmetric Hawk–Dove game: when is Bourgeois an evolutionarily stable strategy? *Evolutionary Ecology* 6: 198–222.

Metcalfe, R.A. (1980) Sex ratios, parent offspring conflict, and local competition for mates in the social wasps *Polistes metricus* and *Polistes variatus*. *American Naturalist* 116: 642–54.

Metzgar, L.H. (1967) An experimental comparison of screech owl predation on resident and transient white-footed mice (*Peromyscus leucopus*). *Journal of Mammalogy* 48: 387–91.

Michael, R.P. & Keverne, E.B. (1968) Pheromones in the communication of sexual status in primates. *Nature* 218: 746–9.

Michael, R.P. & Saayman, G.S. (1968) Differential effects on behaviour of the subcutaneous and intravaginal administration of oestrogen in the rhesus monkey (*Macaca mulatta*). *Journal of Endocrinology* 41: 231–46.

Michelsen, A. (1978) Sound reception in different environments. In (A.B. Ali ed.) *Perspectives in Sensory Ecology*, pp. 54–72. New York, Plenum Press.

Middendorf, A. von (1855) Die Isepipetsen Russlands: Grundlagen zur Erforschung der Zugzeiten und Zugrichtungen der Vögel Russlands. *Memoires of the Academy of Science, St Petersburg* 8: 1–143.

Miles, C.G. & Jenkins, H.M. (1973) Overshadowing in operant conditioning as a function of discriminability. *Learning and Motivation* 4: 11–27.

Milinski, M. (1977) Experiments on the selection by predators against spatial oddity of their prey. *Zeitschrift für Tierpsychologie* 43: 311–25.

Milinski, M. (1979) An evolutionarily stable feeding strategy in sticklebacks. *Zeitschrift für Tierpsychologie* 51: 36–40.

Milinski, M. (1987) Tit for tat and the evolution of cooperation in sticklebacks. *Nature* 325: 433–5.

Milinski, M. (1988) Games fish play: making decisions as a social forager. *Trends in Ecology & Evolution* 3: 325–30.

Milinski, M. & Bakker, T.C.M. (1991) Female sticklebacks use male coloration in mate choice and hence avoid parasitized males. *Nature* 344: 330–2.

Milinski, M. & Heller, R. (1978) Influence of a predator on the optimal foraging behaviour of sticklebacks (*Gasterosteus aculeatus*). *Nature* 275: 642–4.

Milinski, M. & Parker, G.A. (1991) Competition for resources. In (J.R. Krebs & N.B. Davies eds) *Behavioural Ecology: An Evolutionary Approach*, 3rd Edition, pp. 137–68. Oxford: Blackwell.

Milinski, M. & Wedekind, C. (2001) Evidence for MHC-correlated perfume preferences in humans. *Behavioral Ecology* 12: 140–9.

Miller, E.M. (2000) Homosexuality, birth order and evolution: toward an equilibrium reproductive economics of homosexuality. *Archives of Sexual Behavior* 29: 1–34.

Miller, G.F. (1997) Protean primates: the evolution of adaptive unpredictability in competition and courtship. In (A. Whiten & R.W. Byrne eds) *Machiavellian Intelligence II: Extensions and Evaluations*, pp. 312–40. Cambridge, Cambridge University Press.

Miller, G.F. (1998) How mate choice shaped human nature: a review of sexual selection and human evolution. In (C. Crawford & D. Krebs eds) *Handbook of Evolutionary Psychology: Ideas, Issues and Applications*, pp. 87–129. Mahwah, NJ, Lawrence Erlbaum.

Miller, G.F. (1999) Sexual selection for cultural displays. In (R.I.M. Dunbar, C. Knight & C. Power eds) *The Evolution of Culture*, pp. 71–91. Edinburgh, Edinburgh University Press.

Miller, G.F. (2000) *The Mating Mind*. London, Heinemann.

Miller, N.E. (1956) Effects of drugs on motivation: the value of using a variety of measures. *Annals of the New York Academy of Sciences* 65: 318–33.

Moberg, G.P. (1983) Using risk assessment to define animal welfare. *Journal of Agricultural and Environmental Ethics* 6, Supplement 2: 1–17.

Mock, D.W. & Parker, G.A. (1998) *The Evolution of Sibling Rivalry*. Oxford, Oxford University Press.

Mock, D.W. & Ploger, B.J. (1987) Parental manipulation of optimal hatch asynchrony in cattle egrets: an experimental study. *Animal Behaviour* 35: 150–60.

Moehlman, P.D. (1979) Jackal helpers and pup survival. *Nature* 277: 382–3.

Mohtashemi, M. & Mui, L. (2003) Evolution of indirect reciprocity by social information: the role of trust and reputation in the evolution of altruism. *Journal of Theoretical Biology* 223: 523–31.

Moiseff, A., Pollack, G.S. & Hoy, R.R. (1978) Steering responses of flying crickets to sound and ultrasound: mate attraction and predator avoidance. *Proceedings of the National Academy of Sciences of the USA* 75: 4052–6.

Møller, A.P. (1987a) Behavioural aspects of sperm competition in swallows, *Hirundo rustica*. *Behaviour* 100: 92–104.

Møller, A.P. (1987b) Social control of deception among status signalling house sparrows. *Behavioral Ecology & Sociobiology* 20: 307–11.

Møller, A.P. (1987c) Variation in badge size in male house sparrows (*Passer domesticus*): evidence for status signalling. *Animal Behaviour* 35: 1637–44.

Møller, A.P. (1988a) Paternity and paternal care in the swallow, *Hirundo rustica*. *Animal Behaviour* 36: 996–1005.

Møller, A.P. (1988b) Female choice selects for male sexual tail ornaments in a swallow. *Nature* 332: 640–2.

Møller, A.P. (1989) Viability costs of male tail ornaments in a swallow. *Nature* 339: 132–5.

Møller, A.P. (1990a) Effects of a haematophagous mite on the barn swallow *Hirundo rustica*: a test of the Hamilton and Zuk hypothesis. *Evolution* 44: 771–84.

Møller, A.P. (1990b) Deceptive use of alarm calls by male swallows, *Hirundo rustica*: a new paternity guard. *Behavioral Ecology* 1: 1–6.

Møller, A.P. (1992) Female swallow preference for symmetrical male sexual ornaments. *Nature* 357: 238–40.

Møller, A.P. & Swaddle, J.P. (1997) *Asymmetry, Developmental Stability and Evolution*. Oxford, Oxford University Press.

Molles, L.E. & Vehrencamp, S.L. (2001) Neighbour recognition by resident males in the banded wren, *Thryothorus pleurostictus*, a tropical songbird with high song type sharing. *Animal Behaviour* 61: 119–27.

Monaghan, P. & Metcalfe, N.B. (1985) Group foraging in brown hares: effects of resource distribution and social status. *Animal Behaviour* 33: 993–9.

Montgomery, G.G. & Sunquist, M.E. (1978) Habitat selection and use by two-toed and three-toed sloths. In (G.G. Montgomery ed.) *The Ecology of Arboreal Folivores*, pp. 329–60. Washington, DC, Smithsonian Institution Press.

Moodie, G.E.E. (1972) Predation, natural selection and adaptation in an unusual three-spined stickleback. *Heredity* 28: 155–67.

Moody, A.L., Houston, A.I. & McNamara, J.M. (1996) Ideal free distributions under predation risk. *Behavioural Ecology & Sociobiology* 38: 131–43.

Moore, A.J. & Boake, C.R.B. (1994) Optimality and evolutionary genetics: complementary procedures for evolutionary analysis in behavioural ecology. *Trends in Ecology & Evolution* **9**: 69–72.

Moore, B.R. (1973) The role of directed Pavlovian reactions in simple instrumental learning in the pigeon. In (R.A. Hinde & J. Stevenson-Hinde eds) *Constraints on Learning: Limitations and Predispositions*, pp. 159–88. London, Academic Press.

Moore, F.R. & Phillips, R.B. (1988) Sunset, skylight polarization and the migratory orientation of yellow-rumped warblers, *Dendroica coronata*. *Animal Behaviour* **36**: 1770–8.

Moore, J. (1984) Parasites and altered behavior. *Scientific American* **250**: 108–15.

Moore, J. (2002) *Parasites and the Behavior of Animals*. Oxford, Oxford University Press.

Moore, T. & Haig, D. (1991) Genomic imprinting in mammalian development: a parental tug-of-war. *Trends in Genetics* **7**: 45–9.

Mooring, M.S. & Hart, B.L. (1992) Animal grouping for protection from parasites: selfish herd and encounter-dilution effects. *Behaviour* **123**: 173–93.

Moranz, R. & Brower, L.P. (1998) Geographic and temporal variation of cardenolide-based chemical defenses of queen butterfly (*Danaus gilippus*) in northern Florida. *Journal of Chemical Ecology* **24**: 905–32.

Moreau, R.E. (1972) *The Palaearctic–African Bird Migration Systems*. London, Academic Press.

Morgan, M.J. & Nicholas, D.J. (1979) Discrimination between reinforced action patterns in the rat. *Learning & Motivation* **10**: 1–22.

Morris, D. (1957) 'Typical intensity' and its relation to the problem of ritualisation. *Behaviour* **11**: 1–12.

Morris, M.R., Wagner, W.E. Jr & Ryan, M.J. (1996) A negative correlation between trait and mate preference in *Xiphophorus pygmaeus*. *Animal Behaviour* **52**: 1193–203.

Morris, P.H., Reddy, V. & Bunting, R.C. (1995) The survival of the cutest: who's responsible for the evolution of the teddy bear? *Animal Behaviour* **50**: 1697–700.

Morse, D.W. (1970) Ecological aspects of some mixed species foraging flocks. *Ecological Monographs* **40**: 119–68.

Morton, E.S. (1975) Ecological sources of selection on avian sounds. *American Naturalist* **109**: 17–34.

Moser, E.I., Moser, M.B. & Andersen, P. (1993) Spatial learning impairment parallels the magnitude of dorsal hippocampal lesions, but is hardly present following ventral lesions. *Journal of Neuroscience* **13**: 3916–25.

Mouritsen, H. (1998a) Modelling migration: the clock-and-compass model can explain the distribution of ringing recoveries. *Animal Behaviour* **56**: 899–907.

Mouritsen, H. (1998b) Redstarts, *Phoenicurus phoenicurus*, can orient in a true-zero magnetic field. *Animal Behaviour* **55**: 1311–24.

Mouritsen, H. & Larsen, O.N. (1998) Migrating young pied flycatchers, *Ficedula hypoleuca*, do not compensate for geographical displacements. *Journal of Experimental Biology* **201**: 2927–34.

Mouritsen, H. & Larsen, O.N. (2001) Migrating songbirds tested in computer-controlled Emlen funnels use stellar cues for a time-independent compass. *Journal of Experimental Biology* **204**: 3855–65.

Mowrer, O.H. (1960) *Learning Theory and Behavior*. New York, Wiley.

Moynihan, M.H. (1970) The control, suppression, decay and replacement of displays. *Journal of Theoretical Biology* **29**: 85–112.

Moynihan, M.H. & Rodaniche, A.F. (1977) Communication, crypsis and mimicry in cephalopods. In (T.A. Sebeok ed.) *How Animals Communicate*, pp. 293–302. Bloomington, Indiana University Press.

Mpitsos, G.S. & Pinneo, J.M. (1974a) The behavioral hierarchy of the mollusk *Pleurobranchaea*: I. The dominant position of feeding behavior. *Journal of Comparative Psychology* **90**: 207–24.

Mpitsos, G.S. & Pinneo, J.M. (1974b) The behavioral hierarchy of the mollusk *Pleurobranchaea*. II. Hormonal suppression if feeding associated with egg-laying. *Journal of Comparative Psychology* **90**: 225–43.

Mueller, H.C. (1971) Oddity and specific searching image more important than conspicuousness in prey selection. *Nature* **233**: 345–6.

Mulcahy, N.J. (1999) Altruism towards beggars as a human mating strategy. Unpublished M.Sc. thesis, University of Liverpool.

Mullen, R.J. & Herrup, K. (1979) Chimeric analysis of mouse cerebellar mutants. In (D.R. Kankel & A. Ferrus eds) *Neurogenetics: Genetic Approaches to the Nervous System*, pp. 173–96. New York, Elsevier-North Holland.

Muller, H.J. (1964) The relation of recombination to mutational advance. *Mutation Research* **1**: 2–9.

Murdie, G. & Hassell, M.P. (1973) Food distribution, searching success and predator–prey models. In (M.S. Bartlett & R.W. Hiorns eds) *The Mathematical Theory of the Dynamics of Biological Populations*, pp. 67–86. London, Academic Press.

Murphey, R.K, Mendenhåll, B., Palka, J. & Edwards, J.S. (1975) Deafferentation slows the growth of specific dendrites of identified giant interneurones. *Journal of Comparative Neurology* 159: 407–18.

Murry, T., Hollien, H. & Müller, E. (1975) Perceptual responses to infant crying: maternal discrimination and sex judgments. *Journal of Child Language* 2: 199–204.

Naef-Danzer, B. (2000) Patch time allocation and patch sampling by foraging great and blue tits. *Animal Behaviour* 59: 989–99.

Naguib, M. (1999) Effects of song overlapping and alternating on nocturnally singing nightingales. *Animal Behaviour* 58: 1061–7.

Naguib, M. & Todt, D. (1997) Effects of dyadic vocal interaction on other conspecific receivers in nightingales. *Animal Behaviour* 54: 1535–43.

Näslund, I. (1991) Partial migration and the development of seasonal habitat shifts in a landlocked arctic char (*Salvelinus alpinus*) population. Ph.D. Dissertation, Swedish University of Agricultural Sciences, Umeå.

Navins, P.M. & Capranica, R.R. (1976) Sexual differences in the auditory system of the tree frog *Eleutherodactylus coqui*. *Science* 192: 378–80.

Naylor, E. (1958) Tidal and diurnal rhythms of locomotor activity in *Carcinus maenas*. *Journal of Experimental Biology* 35: 602–10.

Neill, S.R. St J. & Cullen, J.M. (1974) Experiments on whether schooling by their prey affects the hunting behaviour of cephalopods and fish predators. *Journal of Zoology* 172: 549–69.

Nelson, D.A. (1999) Ecological influences on vocal development in the white-crowned sparrow. *Animal Behaviour* 58: 21–36.

Neumann, J. von & Morgenstern, O. (1953) *Theory of Games and Economic Behavior*. Princeton, NJ, Princeton University Press.

Nicol, A.C. (1948) Annelid giant fibres. *Quarterly Review of Biology* 23: 291–324.

Nicol, C.J. & Pope, S.J. (1994) Social learning in small flocks of laying hens. *Animal Behaviour* 47: 1289–96.

Nishida, T. (1987) Local traditions and cultural transmission. In (B.B. Smuts, D.L. Cheney, R.M. Seyfarth, R.W. Wrangham & T.T. Struhsaker eds) *Primate Societies*, pp. 462–74. Chicago, University of Chicago Press.

Noë, R. & Hammerstein, P. (1995) Biological markets. *Trends in Ecology & Evolution* 10: 336–9.

Nolan, T.G. & Høy, R.R. (1984) Phonotaxis in flying crickets: neural correlates. *Science* 226: 992–4.

Nonacs, P. (2001) State dependent behavior and the marginal value theorem. *Behavioral Ecology* 12: 71–83.

Nordenskiöld, E. (1928) *The History of Biology: A Survey*. New York, Tudor Publishing.

Norman, R.F., Taylor, P.D. & Robertson, R.J. (1977) Stable equilibrium strategies and penalty functions in a game of attrition. *Journal of Theoretical Biology* 69: 571–8.

Norris, K.S. & Lowe, C.H. (1964) An analysis of background color-matching in amphibians and reptiles. *Ecology* 45: 565–80.

Nottebohm, F. (1967) The role of sensory feedback in the development of avian vocalizations. *Proceedings of the 14th International Ornithological Congress*, Oxford, pp. 265–80. Oxford, Blackwell.

Nottebohm, F. (1981) A brain for all seasons: cyclical anatomical changes in song-control nuclei of the canary. *Science* 214: 1368–70.

Nottebohm, F., Alvarez-Buylla, A., Cynx, J.J., Kirn, J., Long, C.Y. & Nottebohm, M. (1990) Song learning in birds: the relation between perception and production. *Philosophical Transactions of the Royal Society of London B* 329: 115–24.

Nottebohm, F., Stokes, T.M. & Leonard, C.M. (1976) Central control of song in the canary. *Journal of Comparative Neurology* 165: 457–68.

Nudds, T. (1978) Convergence of group size strategies by mammalian social carnivores. *American Naturalist* 112: 957–60.

Nunn, C.L. (1999) The evolution of exaggerated sexual swellings in primates and the graded signal hypothesis. *Animal Behaviour* 58: 229–46.

Nyby, J., Matochik, J.A. & Barfield, R.J. (1992) Intracranial androgenic and estrogenic stimulation of male-typical behaviors in house mice (*Mus domesticus*). *Hormones and Behaviour* 26: 24–45.

Oatley, K. (1988) On changing one's mind: a possible function of consciousness. In (A.J. Marcel & E. Bisiach eds) *Consciousness in Contemporary Science*, pp. 369–89. Oxford, Clarendon Press.

Ober, C., Hyslop, T., Elias, S., Weitkamp, L.R. & Hauck, W.W. (1998) Human leucocyte antigen matching and fetal loss: a result of a 10-year prospective study. *Human Reproduction* **13**: 33–8.

Ober, C., Weitkamp, L.R., Cox, N., Dytch, H., Kostyu, D. & Elias, S. (1997) HLA and mate choice in humans. *American Journal of Human Genetics* **61**: 497–504.

O'Connor, R.J. (1981) Comparisons between migrant and non-migrant birds in Britain. In (D.J. Aidley ed.) *Animal Migration.* Cambridge, Cambridge University Press.

O'Donald, P. (1983) Sexual selection by female choice. In (P. Bateson ed.) *Mate Choice*, pp. 53–65. Cambridge, Cambridge University Press.

Ohara, Y., Nagasaka, K. & Ohsaki, I. (1993) Warning coloration in sawfly, *Athalia rosae*, larva and concealing coloration in butterfly, *Pieris rapae*, larva feeding on similar plants evolved through individual selection. *Researches on Population Ecology* **35**: 223–30.

Ohyama, T. & Mauk, M.D. (2001) Latent acquisition of timed responses in cerebellar cortex. *Journal of Neuroscience* **21**: 682–90.

O'Keefe, J. & Nadel, L. (1978) *The Hippocampus as a Cognitive Map.* Oxford, Oxford University Press.

Olds, J. (1961) Differential effects of drives and drugs on self-stimulation at different brain sites. In (D.E. Sheer ed.) *Electrical Stimulation of the Brain.* University of Texas, Institute of Mental Health.

Oleskin, A.V. (1994) Social behavior of microbial populations. *Journal of Basic Microbiology* **34**: 425–39.

Oliveira, R.F., McGregor, P.K. & Latruffe, C. (1998) Know thine enemy: fighting fish gather information from observing conspecific interactions. *Proceedings of the Royal Society of London Series B* **265**: 1045–9.

Olson, V.A. & Owens, I.P.F. (1998) Costly sexual signals: are carotenoids rare, risky or required? *Trends in Ecology & Evolution* **13**: 510–14.

Olsson, O., Brown, J.S. & Smith, H.G. (2001) Gain curves in depletable food patches: a test of five models with European starlings. *Evolutionary Ecology Research* **3**: 285–310.

Orians, G.H. (1969) On the evolution of mating systems in birds and mammals. *American Naturalist* **103**: 589–603.

Oring, L.W. & Maxson, S.S. (1978) Instances of simultaneous polyandry in the spotted sandpiper. *Living Bird* **11**: 59–73.

Osher, Y., Hamer, D. & Benjamin, J. (2000) Association and linkage of anxiety-related traits with a functional polymorphism of the serotonin transporter gene regulatory region in Israeli sibling pairs. *Molecular Psychiatry* **5**: 216–19.

Ottosson, U., Báckman, J. & Smith, H.G. (1997) Begging affects parental effort in the pied flycatcher, *Ficedula hypoleuca. Behavioral Ecology & Sociobiology* **41**: 318–84.

Owen, E.H., Christensen, S.C., Paylor, R. & Wehner, J.M. (1997) Identification of quantitative trait loci involved in contextual and auditory-cued fear conditioning in BXD recombinant inbred strains. *Behavioral Neuroscience* **111**: 292–300.

Owens, I.P.F. & Short, R.V. (1995) Hormonal basis of sexual dimorphism in birds: implications for new theories of sexual selection. *Trends in Ecology & Evolution* **10**: 44–7.

Packer, C. & Pusey, A.E. (1983) Male takeovers and female reproductive parameters: a simulation of oestrous synchrony in lions (*Panthera leo*). *Animal Behaviour* **31**: 334–40.

Packer, C. & Pusey, A.E. (1985) Asymmetric contests in social mammals: respect, manipulation and age-specific aspects. In (P.J. Greenwood and M. Slatkin eds) *Evolution: Essays in Honour of John Maynard Smith*, pp. 173–86. Cambridge, Cambridge University Press.

Packer, C., Gilbert, D., Pusey, A.E. & O'Brien, S. (1991) A molecular genetic analysis of kinship and cooperation in African lions. *Nature* **351**: 562–5.

Packer, C., Scheel, D. & Pusey, A.E. (1990) Why lions form groups: food is not enough. *The American Naturalist* **136**: 1–19.

Page, R.E., Robinson, G.E. & Fondrk, M.K. (1989) Genetic specialists, kin recognition and nepotism in honey bee colonies. *Nature* **338**: 576–9.

Page, T.L. (1985) Circadian organization in cockroaches: effects of temperature cycles on locomotor activity. *Journal of Insect Physiology* **31**: 235–42.

Palmer, A.R. (1999) Detecting publication bias in meta-analyses: a case study of fluctuating asymmetry and sexual selection. *The American Naturalist* **154**: 220–33.

Palmer, A.R. & Strobeck, C. (1986) Fluctuating asymmetry: measurement, analysis, patterns. *Annual Review of Ecology & Systematics* **17**: 391–421.

Palmer, C.T. & Tilley, C.F. (1995) Sexual access to females as a motivation for joining gangs: an evolutionary approach. *Journal of Sex Research* **32**: 213–17.

Pamilo, P. (1990) Sex allocation and the queen–worker conflict in polygynous ants. *Behavioral Ecology & Sociobiology* **27**: 31–6.

Panuis, T.M. & Wilkinson, G.S. (1999) Exaggerated male eye span influences contest outcome in stalk-eyed flies (Diopsidae). *Behavioral Ecology & Sociobiology* **46**: 221–7.

Papi, F. (2001) Animal navigation at the end of the century: a retrospect and a look forward. *Italian Journal of Zoology* **68**: 171–80.

Papi, F. & Pardi, L. (1963) On the lunar orientation of sand hoppers. *Biological Bulletin* **124**: 97–105.

Papi, F., Fiore, L., Fiaschi, V. & Benvenuti, S. (1972) Olfaction and homing in pigeons. *Monitore Zoologico Italiano* **6**: 85–95.

Papi, F., Ioalé, P., Fiaschi, V., Benvenuti, S. & Baldaccini, N.E. (1974) Olfactory navigation of pigeons: the effect of treatment with odorous air currents. *Journal of Comparative Physiology* **94**: 187–93.

Parker, G.A. (1970) Sperm competition and its evolutionary consequences in the insects. *Biological Reviews* **45**: 525–67.

Parker, G.A. (1974) The reproductive behaviour and the nature of sexual selection in *Scatophaga stercoraria* L: IX. Spatial distribution of fertilization rates and evolution of male search strategy within the reproductive area. *Evolution* **28**: 93–108.

Parker, G.A. (1979) Sexual selection and sexual conflict. In (M.S. Blum & N.A. Blum eds) *Sexual Selection and Reproductive Competition in Insects*, pp. 123–66. New York, Academic Press.

Parker, G.A. (1983a) Arms races in evolution: an ESS to the opponent-independent costs game. *Journal of Theoretical Biology* **101**: 619–48.

Parker, G.A. (1983b) Mate quality and mating decisions. In (P. Bateson ed.) *Mate Choice*, pp. 141–66. Cambridge, Cambridge University Press.

Parker, G.A. (1984) Courtship persistence and female-guarding as male time investment strategies. *Behaviour* **48**: 157–84.

Parker, G.A. (1985) Models of parent–offspring conflict: V. Effects of the behaviour of two parents. *Animal Behaviour* **33**: 519–33.

Parker, G.A. & Macnair, M.R. (1978) Models of parent–offspring conflict: I. Monogamy. *Animal Behaviour* **26**: 97–111.

Parker, G.A. & Macnair, M.R. (1979) Models of parent–offspring conflict: IV. Suppression: evolutionary retaliation of the parent. *Animal Behaviour* **27**: 1210–35

Parker, G.A., Baker, R.R. & Smith, V.G.F. (1972) The origin and evolution of gamete dimorphism and the male–female phenomenon. *Journal of Theoretical Biology* **36**: 529–53.

Parsons, J. (1970) Relationship between egg size and post-hatching chick mortality in the herring gull (*Larus argentatus*). *Nature* **228**: 1221–2.

Partridge, B.L. & Pitcher, T. (1979) Evidence against a hydrodynamic function of fish schools. *Nature* **279**: 418–19.

Partridge, L. (1974) Habitat selection in titmice. *Nature* **247**: 573–4.

Partridge, L. (1976) Field and laboratory observations on the foraging and feeding techniques of bluetits.(*Parus caeruleus*) and coaltits (*Parus ater*) in relation to their habitats. *Animal Behaviour* **24**: 534–44.

Partridge, L. (1978) Habitat selection. In (J.R. Krebs & N.B. Davies eds) *Behavioural Ecology: An Evolutionary Approach*, 1st Edition, pp. 351–76. Oxford: Blackwell.

Partridge, L. (1983) Genetics and behaviour. In (T.R. Halliday & P.J.B. Slater eds) *Animal Behaviour Vol. 3, Genes, Development and Learning*, pp. 11–51. Oxford, Blackwell.

Partridge, L. & Green, P. (1987) An advantage for specialist feeding in jackdaws, *Corvus monedula*. *Animal Behaviour* **35**: 982–90.

Partridge, L. & Nunney, L. (1977) Three-generation family conflict. *Animal Behaviour* **25**: 785–6.

Paton, D.C. (1986) Communication by agonistic displays: II. Perceived information and the definition of agonistic displays. *Behaviour* **99**: 157–75.

Paton, D.C. & Caryl, P.G. (1986) Communication by agonistic displays: I. Variation in information content between samples. *Behaviour* **98**: 213–39.

Paul, R.C. & Walker, T.J. (1979) Arboreal singing in a burrowing cricket, *Anurogryllus arboreus*. *Journal of Comparative Physiology* **132**: 217–23.

Paulissen, M.A., Walker, J.M., Cordes, J.E. & Taylor, H.L. (1993) Diet of diploid and triploid populations of parthenogenetic whiptail lizards of the *Cnemidophorus tesselatus* complex (Teiidae) in southeastern Colorado. *Southwestern Naturalist* **38**: 377–81.

Pawlowski, B. & Dunbar, R.I.M. (1999) Impact of market value on human mate choice decisions. *Proceedings of the Royal Society of London Series B* **266**: 281–5.

Payne, R. (1971) Acoustic location of prey by barn owls (*Tyto alba*). *Journal of Experimental Biology* **54**: 535–73.

Payne, R.B. (1986) Bird song and avian systematics. *Current Ornithology* **3**: 87–126

Patterson, I.J. (1965) Timing and spacing of broods of the black-headed gull *Larus ridibundus*. *Ibis* 107: 433–59.

Payne, R.J.H. & Pagel, M. (1996) Escalation and time costs in displays of endurance. *Journal of Theoretical Biology* 183: 185–93.

Peake, T.M., Terry, A.M.R., McGregor, P.K. & Dabelsteen, T. (2002) Do great tits assess rivals by combining direct experience with information gathered by eavesdropping? *Proceedings of the Royal Society of London Series B* 269: 1925–9.

Pearce, J.M. (1997) *Animal Learning and Cognition: An Introduction*, 2nd Edition. London, Taylor & Francis.

Pearce, J.M. & Hall, G. (1980) A model for Pavlovian learning: variations in the effectiveness of conditioned but not of unconditioned stimuli. *Psychological Review* 87: 532–52.

Pearce, J.M., Colwill, R.M. & Hall, G. (1978) Instrumental conditioning of scratching in the rat. *Learning & Motivation*, 9: 225–71.

Peeke, H.V.S. & Veno, A. (1973) Stimulus specificity of habituated aggression in three-spined sticklebacks (*Gasterosteus aculeatus*). *Behavioral Biology* 8: 427–32.

Peisl, P. (1997) Flowers as signal emitters. *Botanica Helvitica* 107: 3–28.

Peng, W.D., Xu, S.B. & Peng, X. (1996) Effects of suberogorgin and its derivatives on learning and memory in mice. *Acta Pharmacologica Sinica* 17: 215–18.

Penn, D. & Potts, W.K. (1998a) Chemical signals and parasite-mediated sexual signals. *Trends in Ecology & Evolution* 13: 391–6.

Penn, D. & Potts, W.K. (1998b) How do major histocompatibility complex genes influence odor and mating preferences? *Advances in Immunology* 69: 411–36.

Penn, D. & Potts, W.K. (1999) The evolution of mating preferences and major histocompatibility complex genes. *American Naturalist* 153: 145–64.

Pennington, R. & Harpending, H. (1993) *The Structure of an African Pastoralist Community: Demography, History and Ecology of the Ngamiland Herero*. Oxford, Oxford University Press.

Pennycuick, C.J. (1975) Movements of the migratory wildebeeste population in the Serengeti area between 1960 and 1973. *East African Wildlife Journal* 13: 65–87.

Penton-Voak, I.S. & Perrett, D.I. (2000) Female preference for male faces changes cyclically: further evidence. *Evolution and Human Behavior* 21: 39–48.

Penton-Voak, I.S., Jones, B.C., Little, A.C., Baker, S., Tiddeman, B., Burt, D.M. & Perrett, D.I. (2001) Symmetry, sexual dimorphism in facial proportions and male facial attractiveness. *Proceedings of the Royal Society of London Series B* 268: 1617–23.

Pepperberg, I.M. (1991) A communicative approach to animal cognition: a study of the conceptual abilities of an African grey parrot. In (C.A. Ristau ed.) *Cognitive Ethology: The Minds of Other Animals*, pp. 153–86. Hillsdale, NJ, Lawrence Erlbaum Associates.

Pepperberg, I.M., Garcia, S.E., Jackson, E.C. & Marconi, S. (1995) Mirror use by African grey parrots (*Psittacus erithacus*). *Journal of Comparative Psychology* 108: 36–44.

Perdeck, A.C. (1958) Two types of orientation in migrating starlings, *Sturnus vulgaris* L. and chaffinches, *Fringilla coelebs*, as revealed by displacement experiments. *Ardea* 46: 1–37.

Perrett, D.I., May, K.A. & Yoshikawa, S. (1994) Facial shape and judgments of female attractiveness. *Nature* 368: 239–42.

Perrins, C.M. (1968) The purpose of high-intensity alarm calls in small passerines. *Ibis* 110: 200–1.

Perrins, C.M. (1979) *British Tits*. London, Collins.

Perrone, M. & Zaret, T.M. (1979) Parental care patterns of fishes. *The American Naturalist* 113: 351–61.

Persson, O. & Öhström, P. (1989) A new avian mating system: ambisexual polygamy in the penduline tit *Remiz pendulinus*. *Ornis Scandinavica* 20: 105–11.

Petit, L.J. (1991) Adaptive tolerance of cowbird parasitism by prothonotary warblers: a consequence of site limitation? *Animal Behaviour* 41: 425–32.

Petrie, M., Krupa, A. & Burke, T. (1999) Peacocks lek with relatives even in the absence of social and environmental cues. *Nature* 401: 155–7.

Petrinovich, L., Patterson, T.L. & Baptista, L.F. (1981) Song dialects as barriers to dispersal: a re-evaluation. *Evolution* 35: 180–8.

Pettigrew, J.D. (1999) Electroreception in monotremes. *Journal of Experimental Biology* 202: 1447–54.

Peuhkuri, N., Ranta, E. & Seppa, P. (1997) Size-assortative schooling in free-ranging sticklebacks. *Ethology* 103: 318–24.

Pfennig, D. (1990) 'Kin recognition' among spadefoot road tadpoles: a side-effect of habitat selection? *Evolution* 44: 785–98.

Phelps, S.M. & Ryan, M.J. (2000) History influences signal recognition: neural network models of tungara frogs. *Proceedings of the Royal Society of London Series B* 267: 1633–9.

Phillips, D.I.W. (1996) Insulin-resistance as a programmed response to fetal undernutrition. *Diabetologia* 39: 1119–22.

Phillips, J.B. (1996) Magnetic navigation. *Journal of Theoretical Biology* 180: 309–19.

Phillips, J.B. & Moore, F.R. (1992) Calibration of the sun compass by sunset polarized light patterns in a migratory bird. *Behavioural Ecology & Sociobiology* 31: 189–93.

Piattelli-Palmarini, M. (1994) *Inevitable Illusions: How Mistakes of Reason Rule Our Minds.* New York, Wiley.

Pickard, G.E. & Turek, F.W. (1982) Splitting of the circadian rhythm of activity is abolished by unilateral lesions of the suprachiasmatic nuclei. *Science* 215: 119–21.

Pinheiro, C.E.G. (1996) Palatability and escaping ability in neotropical butterflies: tests with wild kingbirds (*Tyrannus melancholicus*, Tyrrandiae). *Biological Journal of the Linnean Society of London* 59: 351–65.

Pinker, S. (1994) *The Language Instinct.* New York, William Morrow.

Pinker, S. & Bloom, P. (1990) Natural language and natural selection. *Behavioral and Brain Sciences* 13: 707–84.

Pitcher, T. (1993) Who dares wins: the function and evolution of predator inspection behaviour in shoaling fish. *Netherlands Journal of Zoology* 42: 371–91.

Plaisted, K.C. & Mackintosh, N.J. (1995) Visual search for cryptic prey stimuli in pigeons: implications for the search image and search rate hypotheses. *Animal Behaviour* 50: 1219–32.

Plomin, R., Owen, M.J. & McGuffin, P. (1994) The genetic basis of complex human behaviors. *Science* 264: 1733–9.

Plotkin, H. (1999) *Evolution in Mind: An Introduction to Evolutionary Psychology.* London, Penguin.

Podos, J. (1996) Motor constraints on vocal developments in a songbird. *Animal Behaviour* 51: 1061–70.

Pomiankowski, A. (1987) Sexual selection: the handicap principle does work – sometimes. *Proceedings of the Royal Society of London Series B* 231: 123–45.

Pomiankowski, A. & Møller, A.P. (1995) A resolution of the lek paradox. *Proceedings of the Royal Society of London Series B* 260: 21–9.

Pomiankowski, A., Iwasa, Y. & Nee, S. (1991) The evolution of costly mate preferences: I. Fisher and biased mutation. *Evolution* 45: 1422–30.

Porter, R.H. (1998) Olfaction and human kin recognition. *Genetica* 104: 259–63.

Porter, R.H., Cernoch, J.M. & McLaughlin, F.J. (1983) Maternal recognition of neonates through olfactory cues. *Physiology & Behavior* 30: 151–4.

Potti, J. & Montalvo, S. (1991) Male arrival and female mate choice in pied flycatchers, *Ficedula hypoleuca*, in Central Spain. *Ornis Scandinavica* 22: 45–54.

Potts, W.K., Manning, C.J. & Wakeland, E.K. (1991) Mating patterns in seminatural populations of mice influenced by MHC genotype. *Nature* 352: 619–21.

Pough, F.H. (1976) Multiple cryptic effects of cross-banded and ringed patterns of snakes. *Copeia* 176: 619–21.

Poulin, R. (2000) Manipulation of host behaviour by parasites: a weakening paradigm? *Proceedings of the Royal Society of London Series B* 267: 787–92.

Povinelli, D.J. (1989) Failure to find self-recognition in Asian elephants (*Elephus maximus*) in contrast to their use of mirror cues to discover hidden food. *Journal of Comparative Psychology* 103: 122–31.

Povinelli, D.J. & Cant, J.G.H. (1995) Arboreal clambering and the evolution of self-conception. *Quarterly Review of Biology* 70: 393–421.

Povinelli, D.J. & Eddy, T.J. (1996) Factors affecting young chimpanzees' (*Pan troglodytes*) recognition of attention. *Journal of Comparative Psychology* 110: 336–45.

Povinelli, D.J., Dunphy-Lelii, S., Reaux, J.E. & Mazza, M.P. (2002) Psychological diversity in chimpanzees and humans: new longitudinal assessments of chimpanzees' understanding of attention. *Brain, Behavior and Evolution* 59: 33–53.

Povinelli, D.J., Gallup, G.G., Jr, Eddy, T.J., Bierschwale, D.T., Engstrom, M.C., Perilloux, H.K. & Toxopeus, I.B. (1997) Chimpanzees recognize themselves in mirrors. *Animal Behaviour* 53: 1083–8.

0Povinelli, D.J., Nelson, K.E. & Boysen, S.T. (1990) Inferences about guessing and knowing by chimpanzees (*Pan troglodytes*). *Journal of Comparative Psychology* 104: 203–10.

Povinelli, D.J., Parks, K.A. & Novak, M.A. (1991) Do rhesus monkeys (*Macaca mulatta*) attribute knowledge and ignorance to others? *Journal of Comparative Psychology* 105: 318–25.

Povinelli, D.J., Rulf, A.B., Landau, K.R. & Bierschwale, D.T. (1993) Self-recognition in chimpanzees (*Pan troglodytes*): distribution, ontogeny and patterns of emergence. *Journal of Comparative Psychology* **107**: 347–72.

Powell, G.V.N. (1974) Experimental analysis of the social value of flocking by starlings (*Sturnus vulgaris*) in relation to predation and foraging. *Animal Behaviour* **22**: 501–5.

Praag, H. van, Kempermann, G. & Gage, F.H. (2000) Neural consequences of environmental enrichment. *National Review of Neuroscience* **1**: 191–8.

Praw, J.C. & Grant, J.W.A. (1999) Optimal territory size in the convict cichlid. *Behaviour* **136**: 1347–63.

Premack, D. (1971) Language in chimpanzee? *Science* **172**: 808–22.

Premack, D. (1983) The codes and man and beasts. *The Behavioral and Brain Sciences* **6**: 125–67.

Premack, D. & Premack, A.J. (1983) *The Mind of an Ape*. New York, W.W. Norton & Company.

Premack, D. & Woodruff, G. (1978) Does the chimpanzee have a theory of mind? *Behavioral and Brain Sciences* **4**: 515–26.

Price, E.O. (2002) *Animal Domestication and Behavior*. New York, CABI Publishing.

Proske, U., Gregory, J.E. & Iggo, A. (1998) Sensory receptors in monotremes. *Philosophical Transactions of the Royal Society of London, Series B* **353**: 1187–98.

Pruett-Jones, S.G. & Pruett-Jones, M.A. (1990) Sexual selection through female choice in Lawes' Parotia, a lek mating bird of paradise. *Evolution* **44**: 486–501.

Pulido, F., Berthold, P. & Van Noordwijk, A.J. (1996) Frequency of migrants and migratory activity are genetically correlated in a bird population: evolutionary implications. *Proceedings of the National Academy of Sciences of the USA* **93**: 14642–7.

Pulliam, H.R. (1983) On the advantages of flocking. *Journal of Theoretical Biology* **38**: 419–22.

Pulliam, H.R. & Caraco, T. (1984) Living in groups. In (J.R. Krebs & N.B. Davies eds) *Behavioural Ecology: An Evolutionary Approach*, 2nd Edition, pp. 122–47. Oxford, Blackwell.

Pulliam, H.R. & Millikan, G.C. (1982) Social organization in the non-reproductive season. In (D.S. Farner & J.R. King eds) *Avian Biology*, pp. 132–45. New York, Academic Press.

Pulliam, H.R., Pyke, G.H. & Caraco, T. (1982) The scanning behavior of juncos: a game theoretical approach. *Journal of Theoretical Biology* **95**: 89–104.

Pusey, A. & Packer, C. (1983) Once and future Kings. *Natural History* **92**: 54–8.

Pyke, G.H. (1979) The economics of territory size and time budget in the golden-winged sunbird. *The American Naturalist* **114**: 131–45.

Pytte, C.L. & Suthers, R.A. (2000) Sensitive period for sensorimotor integration during vocal motor learning. *Journal of Neurobiology* **42**: 172–89.

Quiatt, D. & Reynolds, V. (1993) *Primate Behaviour*. Cambridge, Cambridge University Press.

Quinn, T.P. & Brannon, E.L. (1982) The use of celestial and magnetic cues by orientating sockeye salmon smolts. *Journal of Comparative Physiology A* **147**: 547–52.

Radinsky, L.B. (1968) Evolution of somatic sensory specialization in otter brains. *Journal of Comparative Neurology* **134**: 495–506.

Raisman, G. & Field, P.M. (1973) Sexual dimorphism in the neuropile of the preoptic area of the rat and its dependence on neonatal androgen. *Brain Research* **54**: 1–29.

Ramsay, A.O. & Hess, E.H. (1954) A laboratory approach to the study of imprinting. *Wilson Bulletin* **66**: 196–206.

Rand, A.S. & Rand, W.M. (1976) Agonistic behaviour in nesting iguanas: a stochastic analysis of dispute settlement dominated by minimisation of energy cost. *Zeitschrift für Tierpsychologie* **40**: 279–99.

Randall, D., Burggren, W. & French, K. (1997) *Eckert Animal Physiology: Mechanisms and Adaptations*, 4th Edition. New York, Freeman.

Randall, J.A. (1995) Modification of footdrumming signatures by kangaroo rats: changing territories and gaining new neighbours. *Animal Behaviour* **49**: 1227–37.

Randall, P.K. & Campbell, B.A. (1976) Ontogeny of behavioral arousal in rats: effect of maternal and sibling presence. *Journal of Comparative and Physiological Psychology* **90**: 453–9.

Rands, M. (1988) Changes in social networks following marital separation and divorce. In (R.M. Milardo ed.) *Families and Social Networks*, pp. 127–46. Newbury Park, CA, Sage.

Ranta, E. (1993) There is no optimal foraging group size. *Animal Behaviour* **46**: 1032–5.

Ranta, E. & Lindstrom, K. (1993) Body size and shelter possession in mature signal crayfish, *Pacifastacus leniusculus*. *Annales Zoologici Fennici* **30**: 125–32.

Ranta, E., Peukhuri, N., Hirvonen, H. & Barnard, C.J. (1998) Producers, scroungers and the price of a free meal. *Animal Behaviour* **55**: 737–44.

Ranta, E., Peukhuri, N., Laurila, A., Rita, H. & Metcalfe, N.B. (1996) Producers, scroungers and foraging group structure. *Animal Behaviour* **51**: 171–5.

Rao, R.H. (1996) Experimental evidence for the thrifty phenotype hypothesis in rats. *Diabetes* **45**: 911.

Razran, G. (1971) *Mind in Evolution: An East/West Synthesis of Learned Behavior and Cognition.* Boston, MA, Houghton Miflin.

Read, A.F. (1990) Parasites and the evolution of host sexual behaviour. In (C.J. Barnard & J.M. Behnke eds) *Parasitism and Host Behaviour*, pp. 117–57. London, Taylor & Francis.

Reboreda, J.C. & Kacelnik, A. (1992) Risk sensitivity in starlings: variability in food amount and food delay. *Behavioral Ecology* **2**: 301–8.

Rechten, C., Avery, M.I. & Stevens, T.A. (1983) Optimal prey selection: why do great tits show partial preferences? *Animal Behaviour* **31**: 576–84.

Reese, E.S. (1989) Orientation behavior of butterfly fishes (family Chaetodontidae) on coral reefs: spatial learning of route-specific landmarks and cognitive maps. *Environmental Biology of Fishes* **25**: 79–86.

Reeve, H.K. (1993) Haplodiploidy, eusociality and absence of male parental and alloparental care in Hymenoptera: a unifying genetic hypothesis distinct from kin selection theory. *Philosophical Transactions of the Royal Society of London Series B* **342**: 335–52.

Reeve, H.K., Emlen, S.T. & Keller, L. (1998) Reproductive sharing in animal societies: reproductive incentives or incomplete control by dominant breeders? *Behavioral Ecology* **9**: 267–78.

Regalski, J.M. & Gaulin, S.J.C. (1993) Whom are Mexican infants said to resemble? Monitoring and fostering paternal confidence in the Yucatan. *Ethology & Sociobioology* **14**: 97–113.

Regan, T. (1984) *The Case for Animal Rights.* Berkeley, University of California Press.

Reid, P.J. & Shettleworth, S.J. (1992) Detection of cryptic prey: search image or search rate? *Journal of Experimental Psychology: Animal Behavior Processes* **18**: 273–86.

Reilly, T. (2000) The menstrual cycle and human performance: an overview. *Biological Rhythm Research* **31**: 29–40.

Rescorla, R.A. & Solomon, R.L. (1967) Two-process learning theory: relationship between Pavlovian conditioning and instrumental learning. *Psychological Review* **88**: 151–82.

Rescorla, R.A. & Wagner, A.R. (1972) A theory of Pavlovian conditioning: variations in the effectiveness of reinforcement and nonreinforcement. In (A.H. Black & W.F. Prokasy eds) *Classical Conditioning II: Current Research and Theory*, pp. 64–99. New York, Appleton-Century-Crofts.

Resko, J.A., Perkins, A., Roselli, C.E., Fitzgerald, J.A., Choate, J.V.A. & Stormshak, F. (1996) Endocrine correlates of partner preference behavior in rams. *Biology of Reproduction* **55**: 120–6.

Rettenmeyer, C.W. (1963) Behavioral studies of army ants. *University of Kansas Scientific Bulletin* **44**: 281–465.

Reynolds, J.D. & Gross, M.R. (1990) Costs and benefits of female choice: is there a lek paradox? *The American Naturalist* **136**: 230–43.

Reznick, D. & Endler, J.A. (1982) The impact of predation on life history evolution in Trinidadian guppies (*Poecilia reticulata*). *Evolution* **36**: 160–77.

Rice, W.R. (2002) Experimental tests of the adaptive significance of sexual recombination. *Nature Reviews Genetics* **3**: 241–51.

Rice, W.R. & Holland, B. (1997) The enemies within: intergenomic conflict, interlocus contest evolution (ICE), and the intraspecific Red Queen. *Behavioural Ecology and Sociobiology* **41**: 1–10.

Richardson, W.J. (1978) Timing and amount of bird migration in relation to the weather: a review. *Oikos* **30**: 224–72.

Richerson, P.J. & Boyd, R. (1991) Cultural inheritance and evolutionary ecology. In (E.A. Smith & B. Winterhalder eds) *Ecology, Evolution and Human Behavior*, pp. 65–89. New York, Aldine de Gruyter.

Ricklefs, R. & Wikelski, M. (2002) The physiology/life history nexus. *Trends in Ecology & Evolution* **17**: 462–8.

Ridley, M. (1993a) Clutch size and mating frequency in parasitic Hymenoptera. *The American Naturalist* **142**: 893–910.

Ridley, M (1993b) *The Red Queen: Sex and the Evolution of Human Nature.* London, Viking.

Ridley, M. (1995) *An Introduction to Animal Behaviour*, 2nd Edition. Oxford, Blackwell.

Ridley, M. (1996) *The Origins of Virtue.* London, Viking.

Ridley, R.M. (1994) The psychology of perseverative and stereotyped behaviour. *Progress in Neurobiology* **44**: 221–31.

Ridpath, M.G. (1972) The Tasmanian native hen, *Tribonyx mortieri*. *C.S.I.R.O. Wildlife Research* **17**: 1–118.

Riechert, S. (1993) The evolution of behavioral phenotypes: lessons learned from divergent spider populations. *Advances in the Study of Behaviour* 22: 103–34.

Riley, D.A. & Langley, C.M. (1993) The logic of species comparisons. *Psychological Science* 4: 185–9.

Ritland, D.B. (1994) Variation in palatability of queen butterflies (*Danaus gilippus*) and implications regarding mimicry. *Ecology* 75: 732–46.

Riva, D. (2000) Cerebellar contribution to behaviour and cognition in children. *Journal of Neurolinguistics* 13: 215–25.

Rizzolatti, G. & Arbib, M. (1998) Language within our grasp. *Trends in Neuroscience* 21: 188.

Robbins, C.T. (1983) *Wildlife Feeding and Nutrition*. New York, Academic Press.

Robbins, R.K. (1981) The 'false head' hypothesis: predation and wing pattern variation in lycaenid butterflies. *American Naturalist* 118: 770–5.

Robertoux, P.L. & Carlier, M. (1988) Differences between CBA/H and NZB mice in intermale aggression: II. Maternal effects. *Behavior Genetics* 18: 175–84.

Roberts, G. (1996) Why individual vigilance declines as group size increases. *Animal Behaviour* 51: 1077–86.

Roberts, G. (1998) Competitive altruism: from reciprocity to the handicap principle. *Proceedings of the Royal Society of London Series B* 265: 427–31.

Roberts, G. & Sherratt, T.N. (1998) Development of cooperative relationships through increasing investment. *Nature* 394: 175–9.

Robertson, J.G.M. (1990) Female choice increases fertilization success in the Australian frog *Uperolia laevigata*. *Animal Behaviour* 39: 639–45.

Robertson, R.M. & Pearson, K.G. (1982) Interneuronal organisation in the flight system of the locust. *Journal of Insect Physiology* 30: 95–101.

Rodgers, R.J., Boullier, E., Chatzimichalaki, P., Cooper, G.D. & Shorten, A. (2002) Contrasting phenotypes of C57BL/6JOIaHsd, 129S2/SvHsd and 129/SvEv mice in two exploration-based tests of anxiety-related behaviour. *Physiology and Behavior* 77: 301–10.

Roeder, K.D. (1970) Episodes in insect brains. *American Scientist* 58: 378–89.

Roff, D.A. (1988) The evolution of migration and some life history parameters in marine fishes. *Environmental Biology of Fishes* 22: 133–46.

Rogers, A.R. (1989) Does biology constrain culture? *American Anthropologist* 90: 819–31.

Rohwer, S. & Ewald, P.W (1981) The cost of dominance and advantage of subordination in a badge signalling system. *Evolution* 35: 441–54.

Rohwer, S. & Rohwer, F.C. (1978) Status signalling in Harris sparrows: experimental deceptions achieved. *Animal Behaviour* 26: 1012–22.

Rohwer, S., Herron, J.C. & Daly, M. (1999) Stepparental behavior as mating effort in birds and other animals. *Evolution & Human Behavior* 20: 367–90.

Roitberg, L.I.C. & Mackauer, M. (1993) Patch residence time and parasitism of *Aphelinus asychis*: a simulation model. *Ecological Modelling* 69: 227–41.

Rolff, J. (2002) Bateman's principle and immunity. *Proceedings of the Royal Society of London Series B* 269: 867–72.

Rollenhagen, A. & Bischoff, H.J. (2000) Evidence for the involvement of two areas of the zebra finch forebrain in sexual imprinting. *Neurobiology of Learning and Memory* 73: 101–13.

Rollin, B.E. (1989) *The Unheeded Cry: Animal Consciousness, Animal Pain and Science*. Oxford, Oxford University Press.

Romey, W.L. (1995) Position preferences within groups: do whirligigs select positions which balance feeding opportunities with predator avoidance? *Behavioural Ecology & Sociobiology* 37: 195–200.

Roof, R.L. & Havens, M.D. (1992) Testosterone improves maze performance and induces development of a male hippocampus in females. *Brain Research* 572: 310–13.

Roper, T.J. (1983) Learning as a biological phenomenon. In (T.R. Halliday & P.J.B. Slater eds) *Animal Behaviour 3: Genes, Development and Learning*, pp. 178–212. Oxford, Blackwell.

Roper, T.J. (1997) Olfaction in birds. *Advances in the Study of Behavior* 28: 247–332.

Rose, G.J. & Canfield, J.G. (1991) Discrimination of the sine of frequency differences by *Sternopyrgus*, an electric fish without a jamming avoidance response. *Journal of Comparative Physiology* 168: 461–7.

Rose, S. (1998) Molecules of memory. In (R. Carter) *Mapping the Mind*, pp. 178–9. London, Weidenfeld & Nicolson.

Rosen, R.A. & Hales, D.C. (1981) Feeding of paddlefish (*Polyodon spathula*). *Copeia* 1981: 441–55.

Rosenblum, L.A. & Moltz, H. eds (1983) *Symbiosis in Parent–Offspring Relations*. New York, Plenum Press.

Rosenzweig, M.R. & Bennett, E.L. (1996) Psycho-biology of plasticity: effects of training and experience on brain and behavior. *Behavioral and Brain Research* **78**: 57–65.

Ross, K.G. & Robertson, J.L. (1990) Develop-mental stability, heterozygosity, and fitness in two introduced fire ants (*Solenopsis invicta* and *S. richteri*) and their hybrid. *Heredity* **64**: 93–103.

Rothenbuhler, N. (1964a) Behaviour genetics of nest cleaning honey bees: I. Responses of four inbred lines to disease-killed brood. *Animal Behaviour* **12**: 578–83.

Rothenbuhler, N. (1964b) Behaviour genetics of nest cleaning honey bees: IV. Responses of F1 and backcross generations to disease-killed brood. *American Zoologist* **4**: 111–23.

Rothschild, M. & Moore, B.P. (1987) Pyrazines as alerting signals in toxic plants and insects. In (V. Labeyrie, G. Fabres & D. Fachaise eds) *Insects–Plants*. Dordrecht, W. Junk.

Rothstein, S.I. (1982) Mechanisms of avian egg recognition: which egg parameters elicit responses by rejector species? *Behavioral Ecology & Sociobiology* **11**: 229–39.

Rousse, I., Beaulieu, S., Rowe, W., Meaney, M.J., Barden, N. & Rochford, J. (1997) Spatial memory in transgenic mice with impaired glucocorticoid receptor function. *Neuroreport* **8**: 841–5.

Rowe, C. (1999) Receiver psychology and the evolution of multicomponent signals. *Animal Behaviour* **58**: 921–31.

Rowe, C. (2002) Sound improves visual discrimina-tion learning in avian predators. *Proceedings of the Royal Society of London Series B* **269**: 1353–7.

Rowe, C. & Guilford, T. (1996) Hidden colour aversions in domestic chicks triggered by pyrazine odours of insect warning displays. *Nature* **383**: 520–22.

Rowe, C. & Guilford, T. (1999) The evolution of multimodal warning displays. *Evolutionary Ecology* **13**: 655–71.

Rowell, T. (1974) The concept of social domin-ance. *Behavioural Biology* **11**: 131–54.

Royle, N.J., Hartley, I.R. & Parker, G.A. (2002) Begging for control: when are offspring solicita-tion behaviours honest? *Trends in Ecology & Evolution* **17**: 434–40.

Rozin, P. & Fallon, A. (1988) Body image, atti-tudes to weight, and misperceptions of figure preferences of the opposite sex: a comparison of men and women in two generations. *Journal of Abnormal Psychology* **97**: 342–5.

Rozin, P. & Schull, J. (1988) The adaptive-evolutionary point of view in experimental psychology. In (R. Atkinson, R.J. Herrnstein, G. Lindzey & R.D. Luce eds) *S.S. Stevens' Hand-book of Experimental Psychology*, pp. 503–46. New York, Wiley.

Rubenstein, D.I. (1981) Individual variation and competition in the Everglades pygmy sunfish. *Animal Behaviour* **29**: 155–72.

Rubenstein, D.I. (1986) Ecology and sociality in horses and zebras. In (D.I. Rubenstein & R.W. Wrangham eds) *Ecological Aspects of Social Evolution*, pp. 282–302. Princeton, NJ, Princeton University Press.

Rulicke, T., Chapuisat, M., Homberger, F.R., Macas, E. & Wedekind, C. (1998) MHC-genotype of progeny influenced by parental infection. *Proceedings of the Royal Society of London Series B* **265**: 711–16.

Rumbaugh, D.M. ed. (1977) *Language Learning by a Chimpanzee*. New York, Academic Press.

Rumbaugh, D.M. & Savage-Rumbaugh, E.S. (1994) Language in comparative perspective. In (N.J. Mackintosh ed.) *Animal Learning and Cognition*, pp. 307–33. San Diego, Academic Press.

Rushen, J. & de Passillé, A.M.B. (1992) The scientific assessment of the impact of housing on animal welfare. *Applied Animal Behaviour Science* **28**: 381–6.

Rutowski, R.L., Newton, M. & Schaefer, J. (1983) Interspecific variation in the size of the nutrient investment made by male butterflies during copulation. *Evolution* **37**: 708–13.

Ryan, M.J. (1985) *The Túngara Frog: a Study in Sexual Selection and Communication*. Chicago, Chicago University Press.

Ryan, M.J. (1990) Sensory systems, sexual selec-tion and sensory exploitation. *Oxford Surveys in Evolutionary Ecology* **7**: 157–95.

Ryan, M.J. (1997) Sexual selection and mate choice. In (J.R. Krebs & N.B. Davies eds) *Behavioural Ecology: An Evolutionary Approach*, 4th Edi-tion, pp. 179–202. Oxford, Blackwell.

Ryan, M.J., Fox, J.H., Wilczynski, W. & Rand, A.S. (1990) Sexual selection for sensory exploitation in the frog *Physalaemus pustulosus*. *Nature* **343**: 66–7.

Ryan, M.J., Tuttle, M.D. & Taft, L.K. (1981) The costs and benefits of chorusing behavior. *Behavioral Ecology & Sociobiology* **8**: 273–8.

Ryder, J.J. & Siva-Jothy, M.T. (2000) Male call-ing song provides a reliable signal of immune function in a cricket. *Proceedings of the Royal Society of London Series B* **269**: 847–52.

Ryder, J.J. & Siva-Jothy, M.T. (2001) Quantitative genetics of immune function and body size in the house cricket, *Acheta domesticus*. *Journal of Evolutionary Biology* **14**: 646–53.

Saal, F.S. vom (1989) Sexual differentiation in litter-bearing mammals: influence of sex of adjacent foetuses in utero. *Journal of Animal Science* **67**: 1824–40.

Sade, D., Cushing, K., Cushing, P., Dunaif, J., Figuerola, A., Kaplan, J., Lauer, C., Rhodes, D. & Schneider, J. (1977) Population dynamics related to social structure on Cayo Santiago. *Yearbook of Physiological Anthropology* **20**: 253–62.

Sahlins, M. (1976) *The Use and Abuse of Biology: An Anthropological Critique of Sociobiology*. Ann Arbor, University of Michigan Press.

Saino, N., Martinelli, R. & Møller, A.P. (2001) Immunoglobulin plasma concentration in relation to egg laying and mate ornamentation of female barn swallows (*Hirundo rustica*). *Journal of Evolutionary Biology* **14**: 95–109.

Saino, N., Stradi, R., Ninni, P., Pini, E. & Møller, A.P. (1999) Carotenoid plasma concentration, immune profile, and plumage ornamentation of male barn swallows (*Hirundo rustica*). *The American Naturalist* **154**: 441–8.

Saledin, S.R. & Elgar, M.A. (1998) The influence of flock size and geometry on the scanning behaviour of spotted turtle doves, *Streptopelia chinensis*. *Australian Journal of Ecology* **23**: 177–80.

Sandberg, R., Backman, J., Moore, F.R. & Lohmus, M. (2000) Magnetic information calibrates celestial cues during migration. *Animal Behaviour* **60**: 453–62.

Sanders, C.M. (1980) A comparison of adult bereavement in the death of a spouse, child and parent. *Omega* **10**: 303–22.

Santschi, F. (1911) Observations et remarques critiques sur le mécanisme de l'orientation chez les fourmis. *Revue Swiss Zoologique* **19**: 305–38.

Sapolsky, R.M. (1992) Cortisol concentrations and the social significance of rank instability among wild baboons. *Psychoneuroendocrinology* **17**: 701–9.

Sargent, R.C. (1989) Alloparental care in the fathead minnow, *Pimephales promelas*: stepfathers discriminate against their adopted eggs. *Behavioral Ecology & Sociobiology* **25**: 379–86.

Sargent, T.D. (1968) Cryptic moths: effects on background selection of painting the circumocular scales. *Science* **159**: 100–1.

Sargent, T.D. (1969) Behavioural adaptations of cryptic moths. III. Resting attitudes of two bark-like species, *Melanolophia canadaria* and *Catocala ultronia*. *Animal Behaviour* **17**: 670–2.

Sauer, E.G.F. (1957) Die Sternorientierung nächtlich ziehender Grasmücken (*Sylvia atricapilla*, *borin* und *curruca*). *Zeitschrift für Tierpsychologie* **14**: 29–70.

Savage-Rumbaugh, E.S. (1986) *Ape Language: From Conditioned Response to Symbol*. New York, Columbia University Press.

Savage-Rumbaugh, E.S. & Brakke, K.E. (1996) Animal language: methodological and interpretive issues. In (M. Bekoff & D. Jamieson eds) *Readings in Animal Cognition*, pp. 269–88. Cambridge, MA, MIT Press.

Savage-Rumbaugh, E.S., McDonald, K., Sevcik, R.A., Hopkins, W.D. & Rubert, E. (1986) Spontaneous symbol acquisition and communicative use by a pygmy chimpanzee (*Pan paniscus*). *Journal of Experimental Psychology: General* **115**: 211–35.

Savage-Rumbaugh, E.S., Murphy, J., Sevcik, R.A., Brakke, K.E., Williams, S.L. & Rumbaugh, D.M. (1993) Language comprehension in ape and child. *Monographs of the Society for Research in Child Development* **58**: 1–256.

Savage-Rumbaugh, E.S., Pate, J.L., Lawson, J., Smith, S.T. & Rosenbaum, S. (1983) Can a chimpanzee make a statement? *Journal of Experimental Psychology: General* **112**: 457–92.

Savage-Rumbaugh, E.S., Rumbaugh, D.M. & Boyson, S. (1978) Symbolic communication between two chimpanzees (*Pan troglodytes*). *Science* **201**: 641–4.

Sawyer, L.A., Hennessy, J.M., Peixoto, A.A., Rosato, E., Parkinson, H., Costa, R. & Kyriacou, C.P. (1997) Natural variation in a *Drosophila* clock gene and temperature compensation. *Science* **278**: 2117–20.

Scannell, J., Roberts, G. & Lazarus, J. (2001) Prey scan at random to evade observant predators. *Proceedings of the Royal Society of London B* **268**: 541–7.

Schaller, G.B. (1963) *The Mountain Gorilla: Ecology and Behavior*. Chicago, Chicago University Press.

Schamel, D. & Tracy, D. (1977) Polyandry, replacement clutches, and site tenacity in the red phalarope (*Phalaropus fulicarius*) at Barrow, Alaska. *Bird Banding* **48**: 314–24.

Schantz, T. von, Göransson, G., Andersson, I., Fröberg, M., Grahn, M., Helgée, A. &

Wittzell, H. (1989). Female choice selects for a viability-based male trait in pheasants. *Nature* **337**: 166–9.

Schenkel, R. (1956) Zur Deutung der Phasianiden-balz. *Ornithologische Beobachter* **53**: 182.

Schenkel, R. (1958) Zur Deutung der Balzleistungen einiger Phasianen und Tetraoniden. *Ornithologische Beobachter* **55**: 65–95.

Schjelderup-Ebbe, T. (1935) Social behaviour of birds: In (C. Murchison ed.) *Handbook of Social Psychology*, pp. 947–72. Worcester, MA, Clark University Press.

Schlenoff, D.H. (1985) The startle responses of blue jays to *Catocala* (Lepidoptera: Noctuidae) prey models. *Animal Behaviour* **33**: 1057–67.

Schlupp, I. & Ryan, M.J. (1997) Male sailfin mollies (*Poecilia latipinna*) copy the mate choice of other males. *Behavioral Ecology* **8**: 104–7.

Schmajuk, N.A. & Thieme, A.D. (1992) Purposive behavior and cognitive mapping: a neural network model. *Biological Cybernetics* **67**: 165–74.

Schmid, J. & Schlund, W. (1993) Anosmia in $ZnSO_4$-treated pigeons: loss of olfactory information during ontogeny and the role of site familiarity in homing experiments. *Journal of Experimental Biology* **185**: 33–49.

Schmidt-Koenig, K. (1960) Internal clocks and homing. *Cold Spring Harbor Symposia on Quantitative Biology* **25**: 389–93.

Schmidt-Koenig, K. (1985) Migration strategies of monarch butterflies. *Contributions to Marine Science (Supplement)* **27**: 786–98.

Schmidt-Koenig, K. & Keeton, W.T. (1977) Sun compass utilization by pigeons wearing frosted eye contact lenses. *Auk* **94**: 143–5.

Schneirla, T.C. & Rosenblatt, J.S. (1961) Behavioral organization and genesis of the social bond in insects and mammals. *American Journal of Orthopsychiatry* **31**: 223–53.

Schneirla, T.C., Rosenblatt, J.S. & Tobach, E. (1963) Maternal behavior of the cat. In (H.L. Rheingold ed.) *Maternal Behavior of Mammals*, pp. 122–68. New York, John Wiley.

Schoener, T. W. (1971) Theory of feeding strategies. *Annual Review of Ecology and Systematics* **2**: 369–404.

Scholz, A.T., Horrall, R.M., Cooper, J.C. & Hasler, D. (1976) Imprinting to chemical cues: the basis for home stream selection in salmon. *Science* **192**: 1247–9.

Schradin, C. (2000) Confusion effect in a reptilian and a primate predator. *Ethology* **106**: 691–700.

Schulman, S.R. & Chapais, B. (1980) Reproductive value and rank relations among macaque sisters. *American Naturalist* **115**: 580–93.

Schumacher, M., Hendrick, J.C. & Balthazart, J. (1989) Sexual differentiation in quail: critical period and hormonal specificity. *Hormones & Behavior* **23**: 130–49.

Schutz, F. (1965) Sexuelle Pragung bei Anatiden. *Zeitschrift für Tierpsychologie* **22**: 50–103.

Schwabl, H. (1983) Ausprögung und Bedeutung des Teilzugverhaltens einer südwest-deutschen Population der Amsel, *Turdus merula*. *Journal für Ornithologie* **124**: 101–15.

Schwabl, H., Mock, D.W. & Geig, J.A. (1997) A hormonal mechanism for parental favoritism. *Nature* **386**: 231.

Schwartz, J.J. (1993) Male calling behavior, female discrimination and acoustic interference in the neotropical treefrog *Hyla microcephala* under realistic acoustic conditions. *Behavioral Ecology & Sociobiology* **32**: 401–14.

Schwartz, J.J., Buchanan, B.W. & Gerhardt, H.C. (2001) Female mate choice in the gray treefrog (*Hyla diversicolor*) in three experimental environments. *Behavioral Ecology & Sociobiology* **49**: 443–55.

Schwartz, J.J., Ressel, S.J. & Bevier, C.R. (1995) Carbohydrate and calling: depletion of muscle glycogen and the chorusing dynamics of the neotropical treefrog *Hyla microcephala*. *Behavioral Ecology & Sociobiology* **37**: 125–35.

Schwegler, H. & Lipp, H.-P. (1995) Variations in the morphology of the septo-hippocampal complex and maze learning in rodents: correlations between morphology and behaviour. In (E. Alleva, A. Fasolo, H.-P. Lipp, L. Nadel & L. Ricceri eds) *Behavioural Brain Research in Naturalistic and Semi-naturalistic Settings. Proceedings of the NATO Advanced Science Institute*, Aquafredda di Maratea, Italy, 10–20 September 1994, pp. 259–276. Dordrecht, Kluwer.

Scordalakes, E.M., Imwalle, D.B. & Rissman, E.F. (2002) Oestrogen's masculine side: mediation of mating in mice. *Reproduction* **124**: 331–8.

Scudamore, R.E. (1995) Aspects of the electric sense of Gymnotis carapo. Unpublished Ph.D. thesis, University of Nottingham, UK.

Scutt, D. & Manning, J.T. (1996) Symmetry and ovulation in women. *Human Reproduction* **11**: 2477–80.

Sebeok, T.A. (1962) Coding in the evolution of signalling behaviour. *Behavioural Science* **7**: 430–42.

Seeley, T. (1989) The honey bee colony as a super-organism. *American Scientist* 77: 546–53.

Seeley, T.D. (1995) *The Wisdom of the Hive*. Cambridge, MA, Harvard University Press.

Segal, N.L. (1984) Cooperation, competition, and altruism within twin sets: a reappraisal. *Ethology & Sociobiology* 5: 163–77.

Seger, J. (1991) Cooperation and conflict in social insects. In (J.R. Krebs & N.B. Davies eds) *Behavioural Ecology: An Evolutionary Approach*, 3rd Edition, pp. 338–73. Oxford, Blackwell.

Selander, R.K. (1966) Sexual dimorphism and differential niche utilization in birds. *Condor* 68: 113–51.

Seyfarth, R.M. & Cheney, D.L. (1990) The assessment by vervet monkeys of their and another species' alarm calls. *Animal Behaviour* 40: 754–64.

Seyfarth, R.M., Cheney, D.L. & Marler, P. (1980) Vervet monkey alarm calls: semantic communication in a free-ranging primate. *Animal Behaviour* 28: 1070–94.

Shackelford, T.K. & Larsen, R.J. (1997) Facial asymmetry as indicator of psychological, emotional and physiological distress. *Journal of Personality and Social Psychology* 72: 456–66.

Shaffery, J.P., Ball, N.J. & Amlaner, C.J. Jr (1985) Manipulating daytime sleep in herring gulls (*Larus argentatus*). *Animal Behaviour* 33: 566–72.

Shalter, M. (1978a) Mobbing in the pied flycatcher: effect of experiencing a live owl on responses to a stuffed facsimile. *Zeitschrift für Tierpsychologie* 47: 173–9.

Shalter, M. (1978b) Localization of passerine seet and mobbing calls by goshawks and pygmy owls. *Zeitschrift für Tierpsychologie* 46: 260–7.

Shanker, C.G., Savage-Rumbaugh, E.S. & Taylor, T.J. (1999) Kanzi: a new beginning. *Animal Learning & Behavior* 27: 24–5.

Shannon, C.E. & Weaver, W. (1949) *The Mathematical Theory of Communication*. Urbana, University of Illinois Press.

Shavits, Y., Fischer, C.S. & Koresh, Y. (1994) Kin and nonkin under collective threat: Israeli networks during the Gulf War. *Social Forces* 72: 1197–215.

Sheldon, B.C. & Verhulst, S. (1996) Ecological immunology: costly parasite defences and trade-offs in evolutionary ecology. *Trends in Ecology & Evolution* 11: 317–21.

Sheldon, B.C., Andersson, S., Griffith, S.C., Örnborg, J. & Sendecka, J. (1999) Ultraviolet colour variation influences blue tit sex ratios. *Nature* 112: 381–8.

Shepher, J. (1971) Mate selection among second generation kibbutz adolescents and adults: incest avoidance and negative imprinting. *Archives of Sexual Behaviour* 1: 293–307.

Sherman, P.A. (1981) Reproductive competition and infanticide in Belding's ground squirrels and other animals. In (R.P. Alexander & D.W. Tinkle eds) *Natural Selection and Social Behavior: Recent Research and New Theory*, pp. 311–31. New York, Chiron Press.

Sherman, P.W. (1989) Mate guarding as paternity insurance in Idaho ground squirrels. *Nature* 338: 418–20.

Sherman, P.W., Lacey, E.A., Reeve, H.K. & Keller, L. (1995) The eusociality continuum. *Behavioral Ecology* 6: 102–8.

Sherman, P.W., Reeve, H.K. & Pfennig, D.W. (1997) Recognition systems. In (J.R. Krebs & N.B. Davies eds) *Behavioural Ecology: An Evolutionary Approach*, 3rd Edition, pp. 69–96. Oxford, Blackwell.

Sherry, D.F. (1977) Parental food-calling and the role of the young in the Burmese red junglefowl (*Gallus gallus spadiceus*). *Animal Behaviour* 25: 594–601.

Sherry, D. F. (1984) Food storage by black-capped chickadees: memory of the location and content of caches. *Animal Behaviour* 32: 451–64.

Sherry, D. & Galef, B.G. Jr (1984) Cultural transmission without imitation: milk bottle opening by birds. *Animal Behaviour* 32: 937–8.

Sherry, D. & Galef, B.G. Jr (1990) Social learning without imitation: more about milk bottle opening by birds. *Animal Behaviour* 40: 987–9.

Sherry, D.F. & Shacter, D.L. (1987) The evolution of multiple memory systems. *Psychological Review* 94: 439–54.

Sherry, D.F., Forbes, M.R.L., Khurgel, M. & Ivy, G.O. (1993) Females have a larger hippocampus than males in the brood-parasitic brown-headed cowbird. *Proceedings of the National Academy of Sciences of the USA* 90: 7839–43.

Shettleworth, S.J. (1975) Reinforcement and the organization of behavior in golden hamsters: hunger, environment and food reinforcement. *Journal of Experimental Psychology: Animal Behavior Processes* 1: 56–87.

Shettleworth, S.J. (1978) Reinforcement and the organization of behavior in golden hamsters: sunflower seed and nest paper reinforcers. *Animal Learning and Behavior* 6: 352–62.

Shettleworth, S. J. (1998) *Cognition, Evolution and Behavior*. Oxford, Oxford University Press.

Shettleworth, S.J. (2001) Animal cognition and animal behaviour. *Animal Behaviour* **61**: 277–86.

Shettleworth, S.J. & Plowright, C.M.S. (1992) How pigeons estimate rates of prey encounter. *Journal of Experimental Psychology – Animal Behaviour Processes* **18**: 219–35.

Shields, W.M. (1982) *Philopatry, Inbreeding and the Evolution of Sex*. New York, State University of New York Press, Albany.

Shine, R. (1980) 'Costs' of reproduction in reptiles. *Oecologia* **46**: 92–100.

Shorey, L., Piertney, S., Stone, J. & Höglund, J. (2000) Fine-scale genetic structuring on *Manacus manacus* leks. *Nature* **408**: 352–3.

Sibly, R.M. (1975) How incentive and deficit determine feeding tendency. *Animal Behaviour* **23**: 437–46.

Sibly, R.M. (1983) Optimal group size is unstable. *Animal Behaviour* **31**: 947–8.

Sibly, R.M. & McFarland, D.J. (1974) A state-space approach to motivation. In (D.J. McFarland, ed.) *Motivational Control Systems Analysis*, pp. 213–50. London, Academic Press.

Sibly, R.M. & McFarland, D.J. (1976) On the fitness of behaviour sequences. *American Naturalist* **110**: 601–17.

Sigg, H. & Stolba, A. (1981) Home range and daily march in a Hamadryas baboon troop. *Folia Primatologia* **36**: 40–75.

Sih, A. (1984) The behavioral response race between predators and prey. *American Naturalist* **123**: 143–50.

Sih, A. (1998) Game theory and predator–prey response races. In (L.A. Dugatkin & H.K. Reeve eds) *Advances in Game Theory and the Study of Animal Behaviour*, pp. 221–38. New York, Oxford University Press.

Silberglied, R.E., Sheperd, J.G. & Dickinson, J.L. (1984) Eunuchs: the role of apyrene sperm in Lepidoptera? *American Naturalist* **123**: 255–65.

Silk, J.B. (1980) Adoption and kinship in Oceania. *American Anthropologist* **82**: 799–820.

Silk, J.B. (1987) Adoption among the Inuit. *Ethos* **15**: 320–30.

Silk, J.B. (1990) Which humans adopt adaptively and why does it matter? *Ethology & Sociobiology* **11**: 425–6.

Sillén-Tullberg, B. (1988) Evolution of gregariousness in aposematic butterfly larvae: a phylogenetic analysis. *Evolution* **42**: 293–305.

Sillén-Tullberg, B. & Møller, A.P. (1993) The relationship between concealed ovulation and mating systems in anthropoid primates: a phylogenetic analysis. *American Naturalist* **141**: 1–25.

Simmons, J.A. & Stein, R.A. (1980) Acoustic imaging in bat sonar: echolocating signals and the evolution of echolocation. *Journal of Comparative Physiology A* **135**: 61–84.

Simmons, L.W. (1986) Female choice in the field cricket, *Gryllus bimaculatus* (De Geer). *Animal Behaviour* **34**: 1463–70.

Simmons, L.W. (2001) *Sperm Competition and its Evolutionary Consequences in the Insects*. Princeton, NJ, Princeton University Press.

Simmons, L.W., Llorens, T., Schinzig, M., Hosken, D. & Craig, M. (1994) Courtship role reversal in bush crickets: another role for parasites? *Animal Behaviour* **47**: 117–22.

Simmons, L.W., Tomkins, J.L., Kotiaho, J.S. & Hunt, J. (1999) Fluctuating paradigm. *Proceedings of the Royal Society of London Series B* **266**: 593–5.

Simpson, M.J.A. (1968) The display of the Siamese fighting fish, *Betta splendens*. *Animal Behaviour Monographs* **1**: 1–73.

Singh, D. (1993) Adaptive significance of waist-to-hip ratio and female physical attractiveness. *Journal of Personality and Social Psychology* **65**: 293–307.

Singh, D. & Luis, S. (1995) Ethnic and gender consensus for the effect of waist-to-hip ratio on judgments of women's attractiveness. *Human Nature* **6**: 51–65.

Singh, D. & Young, R.K. (1995) Body weight, waist-to-hip ratio, breasts and hips: role in judgments of female attractiveness and desirability for relationships. *Ethology & Sociobiology* **16**: 483–507.

Skinner, B.F. (1953) *Science and Human Behavior*. New York, Macmillan.

Slabbekoorn, H. & Smith, T.B. (2002) Habitat-dependent song divergence in the little greenbul: an analysis of environmental selection pressures on acoustic signals. *Evolution* **56**: 1849–58.

Slabbekoorn, H., Ellers, J. & Smith, T.B. (2002) Birdsong and sound transmission: the benefits of reverberations. *Condor* **194**: 564–73.

Slater, P.J.B. (1983) The study of communication. In (T.R. Halliday & P.J.B. Slater eds) *Animal Behaviour 2: Communication*, pp. 9–42. Oxford, Blackwell.

Slater, P.J.B. (1999) *Essentials of Animal Behaviour*. Cambridge: Cambridge University Press.

Slater, P.J.B. & Ince, S.A. (1979) Cultural evolution in chaffinch song. *Behaviour* **71**: 146–66.

Sluyter, F., Oortmerssen, G.A. van, Ruiter, A.J.H. de & Koolhaas, J.M. (1996) Aggression in wild house mice: current state of affairs. *Behavior Genetics* **26**: 489–96.

Smith, D.F., Balagura, S. & Lubran, M. (1970) 'Antidotal thirst': a response to intoxication. *Science* **167**: 297–8.

Smith, J.C. & Roll, D.L. (1967) Trace conditioning with X-rays as the aversive stimulus. *Psychonomic Science* **9**: 11–12.

Smith, J.E. (1957) The nervous anatomy of the body segments of nereid polychaetes. *Philosophical Transactions of the Royal Society Series B* **240**: 135–96.

Smith, J.N.M. (1974) The food searching behaviour of two European thrushes: II. The adaptiveness of the search patterns. *Behaviour* **49**: 1–61.

Smith, J.N.M. & Sweatman, H.P. (1974) Food searching behaviour of titmice in patchy environments. *Ecology* **55**: 1216–32.

Smith, L.H. (1965) Changes in the tail feathers of the adolescent lyrebird. *Science* **147**: 510–13.

Smith, R.J.F. (1997) Does one result trump all others? A response to Magurran, Irving and Henderson. *Proceedings of the Royal Society of London Series B* **264**: 445–50.

Smith, R.L. (1980) Evolution of exclusive postcopulatory parental care in the insect. *Florida Entomologist* **63**: 65–78.

Smith, S.M. (1977) Coral snake pattern rejection and stimulus generalization by naïve great kiskadees (Aves: Tyrannidae). *Nature* **265**: 535–6.

Smith, S.M. (1988) Extra-pair copulations in black-capped chickadees: the role of the female. *Behaviour* **107**: 15–23.

Smith, W.J. (1965) Message, meaning and context in ethology. *American Naturalist* **99**: 405–9.

Smuts, B.B. (1995) The evolutionary origins of patriarchy. *Human Nature* **6**: 1–32.

Snyder, R.J. (1991) Migration and life histories of the threespine stickleback: evidence for adaptive variation in growth rate between populations. *Environmental Biology of Fishes* **31**: 381–8.

Sockman, K.W., Gentner, T.Q. & Ball, G.F. (2002) Recent experience modulates forebrain gene expression in response to mate choice cues in European starlings. *Proceedings of the Royal Society of London Series B* **269**: 2479–85.

Soler, M., Soler, J.J., Martinez, J.G. & Møller, A.P. (1995) Magpie host manipulation by great spotted cuckoos. Evidence for an avian Mafia? *Evolution* **49**: 770–5.

Solomon, M.E. (1969) *Population Dynamics*. London, Edward Arnold.

Soltis, J., Boyd, R. & Richerson, P.J. (1995) Can group-functional behaviors evolve by cultural group selection? *Current Anthropology* **36**: 473–94.

Sonerud, G.A., Smedshaug, C.A. & Brathen, O. (2001) Ignorant hooded crows follow knowledgeable roost-mates to food: support for the information centre hypothesis. *Proceedings of the Royal Society of London Series B* **268**: 827–31.

Sotthibandhu, S. & Baker, R.R. (1979) Celestial orientation by the large yellow underwing moth, *Noctua pronuba* L. *Animal Behaviour* **27**: 786–800.

Southwick, C.H. (1968) Effects of maternal environment on aggressive behaviour of inbred mice. *Communications in Behavioral Biology* **1**: 129–32.

Southwood, T.R.E. (1981) Ecological aspects of insect migration. In (D.J. Aidley ed.) *Animal Migration*, pp. 196–208. Cambridge, Cambridge University Press.

Sparks, J. (1983) *The Discovery of Animal Behaviour*. London, Collins.

Speed, M.P. (2001) Can receiver psychology explain the evolution of aposematism? *Animal Behaviour* **61**: 205–16.

Speiss, E.B. & Langer, B. (1964) Mating speed control by gene arrangements in *Drosophila pseudoobscura* homokaryotypes. *Proceedings of the National Academy of Sciences of the USA* **51**: 1015–18.

Speiss, E.B. & Langer, B. (1966) Mating control by gene arrangements in *Drosophila pseudoobscura*. *Genetics* **54**: 1139–49.

Spetch, M.L. (1995) Overshadowing in landmark learning: touch-screen studies with pigeons and humans. *Journal of Experimental Psychology: Animal Behavior Processes* **21**: 166–81.

Spetch, M.L. & Wilkie, D.M. (1994) Pigeons' use of landmarks presented in digitized images. *Learning and Motivation* **25**: 245–75.

Spinka, M., Newberry, R.C. & Bekoff, M. (2001) Mammalian play: training for the unexpected. *Quarterly Review of Biology* **76**: 141–68.

Sprecher, S., Aron, A., Hatfield, E., Cortese, A., Potapova, E. & Levitskaya, A. (1994) Love: American style, Russian style, and Japanese style. *Personal Relationships* **1**: 349–69.

Srygley, R.B. (1999) Incorporating motion into investigations of mimicry. *Evolutionary Ecology* **13**: 691–708.

Srygley, R.B. (2001) Sexual differences in tailwind drift compensation in *Phoebis sennae* butterflies (Lepidoptera: Pieridae) migrating over seas. *Behavioral Ecology* **12**: 607–11.

Srygley, R.B. & Ellington, C.P. (1999) Discrimination of flying mimetic, passion-vine butterflies, *Heliconius. Proceedings of the Royal Society of London Series B* **266**: 2137–40.

Stacey, P. & Ligon, J.D. (1987) Territory quality and dispersal options in the acorn woodpecker, and a challenge to the habitat saturation model of cooperative breeding. *American Naturalist* **130**: 654–76.

Stack, C.B. (1974) *All Our Kin*. New York, Harper & Row.

Staddon, J.E.R. (1983) *Adaptive Behavior and Learning*. Cambridge, Cambridge University Press.

Stanislaw, H. & Rice, F.J. (1988) Correlation between sexual desire and menstrual cycle characteristics. *Archives of Sexual Behavior* **17**: 499–508.

Starks, P. & Blackie, C. (2000) The relationship between serial monogamy and rape in the United States (1960–1995). *Proceedings of the Royal Society of London Series B* **267**: 1259–63.

Stebbins, G.L. (1982) *Darwin to DNA, Molecules to Humanity*. San Francisco, Freeman.

Steen, W.J. van der (1998) Bias in behaviour genetics: an ecological perspective. *Acta Biotheoretica* **46**: 369–77.

Stein, B.E., Wallace, M.T., Stanford, T.R. & McHaffie, J.G. (2000) Integrating information from different senses in the superior colliculus. In (J.J. Bolhuis ed.) *Brain, Perception, Memory: Advances in Cognitive Neuroscience*, pp. 17–34. Oxford, Oxford University Press.

Steiner, H. & Gerfen, C.R. (1998) Role of dynorphin and enkephalin in the regulation of striatal output pathways and behavior. *Experimental Brain Research* **123**: 60–76.

Stephens, D.W. & Krebs, J.R. (1986) *Foraging Theory*. Princeton, NJ, Princeton University Press.

Stevenson-Hinde, J. (1973) Constraints on reinforcement. In (R.A. Hinde & J. Stevenson-Hinde eds) *Constraints on Learning: Limitations and Predispositions*, pp. 285–99. London, Academic Press.

Stockley, P., Gage, M.J.G., Parker, G.A. & Møller, A.P. (1997) Sperm competition in fishes: the evolution of testis size and ejaculate characteristics. *American Naturalist* **149**: 933–54.

Strauss, R., Hanesch, U., Kinkelin, M., Wolf, R. & Heisemberg, M. (1992) No-bridge of *Drosophila melanogaster*: portrait of a structural brain mutant of the central complex. *Journal of Neurogenetics* **8**: 125–55.

Studd, M., Montgomerie, R.D. & Robertson, R.J. (1983) Group size and predator surveillance in foraging house sparrows (*Passer domesticus*). *Canadian Journal of Zoology* **61**: 226–31.

Stutt, A.D. & Willmer, P. (1998) Territorial defence in speckled wood butterflies: do the hottest males always win? *Animal Behaviour* **55**: 1341–7.

Suboski, M.D. & Bartashunas, C. (1984) Mechanisms for social transmission of pecking preferences to neonatal chicks. *Journal of Experimental Psychology: Animal Behavior Processes* **10**: 101–12.

Sueur, J. (2002) Cicada acoustic communication: potential sound partitioning in a multispecies community from Mexico (Hemiptera: Cicadomorpha: Cicadidae). *Biological Journal of the Linnean Society* **75**: 379–94.

Sullivan, K.A. (1985) Vigilance patterns in downy woodpeckers. *Animal Behaviour* **33**: 328–9.

Surani, M.A., Barton, S.C. & Norris, M.L. (1987) Influence of parental chromosomes on spatial specific distribution of cells in androgenetic-parthenogenetic chimeras in the mouse. *Nature* **326**: 395–7.

Sutherland, W.J. (1985) Chance can produce a sex difference in variance in mating success and explain Bateman's data. *Animal Behaviour* **33**: 1349–52.

Sutherland, W.J. (1996) *From Individual Behaviour to Population Ecology*. New York, Oxford University Press.

Sutherland, W.J. & Parker, G.A. (1992) The relationship between continuous input and interference models of ideal free distributions with unequal competitors. *Animal Behaviour* **44**: 345–55.

Sutherland, W.J. & Parker, G.A. (1985) Distribution of unequal competitors. In (R.M. Sibly & R.H. Smith eds) *Behavioural Ecology: Ecological Consequences of Adaptive Behaviour*, pp. 255–73. Oxford, Blackwell.

Swaab, D.F. & Hofman, M.A. (1990) An enlarged suprachiasmatic nucleus in homosexual men. *Brain Research* **537**: 141–8.

Swaddle, J.P. & Cuthill, I.C. (1994) Preference for symmetric males by female zebra finches. *Nature* **367**: 165–6.

Swaddle, J.P. & Witter, M.S. (1994) Food, feathers and fluctuating asymmetry. *Proceedings of the Royal Society Series B* **255**: 147–52.

Swaddle, J.P., Witter, M.S. & Cuthill, I.C. (1994) The analysis of fluctuating asymmetry. *Animal Behaviour* **48**: 986–9.

Swaddle, J.P., Witter, M.S., Cuthill, I.C., Budden, A. & McCowan, P. (1996) Plumage condition affects flight performance in starlings: implications for developmental homeostasis, abrasion and moult. *Journal of Avian Biology* **27**: 103–11.

Swaisgood, R.R., Rowe, M.P. & Owings, D.H. (1999) Assessment of rattlesnake dangerousness by California ground squirrels: exploitation of cues from rattling sounds. *Animal Behaviour* **57**: 1301–10.

Swingland, I.R. (1983) Intraspecific differences in movement. In (I.R. Swingland & P.J. Greenwood eds) *The Ecology of Animal Movement*. Oxford, Clarendon Press.

Swingland, I.R. & Lessells, C.M. (1979) The natural regulation of giant tortoise populations on Aldabra Atoll. Movement polymorphism, reproductive success and mortality. *Journal of Animal Ecology* **48**: 639–54.

Symons, D. (1978) *Play and Aggression: a Study of Rhesus Monkeys*. New York, Columbia University Press.

Symons, D. (1979) *The Evolution of Human Sexuality*. New York, Oxford.

Symons, D. (1995) Beauty is in the adaptations of the beholder: the evolutionary psychology of human female sexual attractiveness. In (P.R. Abramson & S.D. Pinkerton eds) *Sexual Nature, Sexual Culture*, pp. 80–118. Chicago, University of Chicago Press.

Számadó, S. (2000) Cheating as a mixed strategy in a simple model of aggressive communication. *Animal Behaviour* **59**: 221–30.

Szekèly, T., Catchpole, C.K., DeVoogd, A., Marchl, Z. & DeVoogd, T.J. (1996) Evolutionary changes in a song control area of the brain (HVC) are associated with evolutionary changes in song repertoire among European warblers (Sylvidae). *Proceedings of the Royal Society of London Series B* **263**: 607–10.

Tafti, M., Franken, P., Kitahama, K., Malafosse, A., Jouvet, M. & Valatx, J.L. (1997) Localization of candidate genomic regions influencing paradoxical sleep in mice. *Neuroreport* **8**: 3755–8.

Takahashi, J.S., Hamm, H. & Menaker, M. (1980) Circadian rhythms of melatonin release from individual superfused chicken pineal glands *in vitro*. *Proceedings of the National Academy of Sciences of the USA* **77**: 2319–22.

Takahashi, J.S., Pinto, L.H. & Vitaterna, M.H. (1994) Forward and reverse genetic approaches to behavior in the mouse. *Science* **264**: 1724–32.

Taylor, E.B. (1990) Phenotypic correlates of life history variation in juvenile chinook salmon, *Onchorhynchus tshawytscha*. *Journal of Animal Ecology* **59**: 455–68.

Taylor, L.R. (1961) Aggregation, variance and the mean. *Nature* **189**: 732–5.

Taylor, P.W. & Elwood, R.W. (2003) The mismeasure of animal contests. *Animal Behaviour* **65**: 1195–1202.

Taylor, P.W., Hasson, O. & Clark, D.L. (1999) Body postures and patterns as amplifiers of physical condition. *Proceedings of the Royal Society of London Series B* **267**: 917–22.

Tebb, G. & Thoday, J.M. (1954) Stability in development and relational balance of X chromosomes in lab populations of *Drosophila melanogaster*. *Nature* **174**: 1109.

Teghtsoonian, R. & Campbell, B.A. (1960) Random activity of the rat during food deprivation as a function of environmental conditions. *Journal of Comparative and Physiological Psychology* **53**: 242–4.

Teitelbaum, P. (1955) Sensory control of hypothalamic hyperphagia. *Journal of Comparative and Physiological Psychology* **48**: 156–63.

Teitelbaum, P. & Epstein, A.N. (1962) The lateral hypothalamic syndrome: recovery of feeding and drinking after lateral hypothalamic lesions. *Psychological Review* **69**: 74–90.

Terrace, H.S., Pettito, L.A., Sanders, R.J. & Bever, T.G. (1979) Can an ape create a sentence? *Science* **206**: 891–902.

Tets, G.W. van (1965) A comparative study of some social communication patterns in the Pelecaniformes. *Ornithological Monographs* **1**: 1–18.

Thomas, A.L.R. (1997) On the tails of birds. *Bioscience* **47**: 215–25.

Thomas, J.H. (1990) Genetic analysis of defecation in *Caenorhabditis elegans*. *Genetics* **124**: 855–72.

Thomson, A.L. (1926) *Problems of Bird Migration*. London, Witherby.

Thompson, D.B.A. & Barnard, C.J. (1983) Antipredator responses in mixed species associations of lapwings, golden plovers and gulls. *Animal Behaviour* **32**: 554–63.

Thompson, D.B.A. & Barnard, C.J. (1984) Prey selection by plovers: optimal foraging in mixed-species groups. *Animal Behaviour* **32**: 534–63.

Thompson, T.E. (1960) Defensive acid-secretion in marine gastropods. *Journal of the Marine Biological Association of the UK* **39**: 115–22.

Thorndike, E.L. (1898) Animal intelligence: an experimental study of the associative process in animals. *Psychological Review Monograph Supplement* **8**: 68–72.

Thorndike, E.L. (1911) *Animal Intelligence*. Darien, CT, Hafner Publishing.

Thornhill, R. (1976) Sexual selection and nuptial feeding behavior in *Bittacus apicalis* (Insecta: Mecoptera). *American Naturalist* **110**: 529–48.

Thornhill, R. (1980) Rape in *Panorpa* scorpion-flies and a general rape hypothesis. *Animal Behaviour* **28**: 52–9.

Thornhill, R. & Alcock, J. (1983) *The Evolution of Insect Mating Systems*. Cambridge, MA, Harvard University Press.

Thornhill, R. & Palmer, C.T. (2000) *A Natural History of Rape: The Biological Bases of Sexual Coercion*. Cambridge, MA, MIT Press.

Thornhill, R. & Thornhill, N.W. (1983) Human rape: an evolutionary analysis. *Ethology & Sociobiology* **4**: 137–73.

Thornhill, R., Møller, A.P. & Gangestad, S.W. (1999) The biological significance of fluctuating asymmetry and sexual selection: a reply to Palmer. *American Naturalist* **154**: 234–41.

Thornhill, R., Thornhill, N. & Dizinno, G. (1986) The biology of rape. In (S. Tomaselli & R. Porter eds) *Rape*. London, Basic Blackwell.

Thorpe, W.H. (1961) *Bird Song*. Cambridge, Cambridge University Press.

Thorpe, W.H. (1963) *Learning and Instinct in Animals*. London, Methuen.

Thorpe, W.H. (1965) The assessment of pain and distress in animals in intensive livestock husbandry systems. In *Report of the Technical Committee to Enquire into the Welfare of Animals Kept Under Intensive Livestock Husbandry Systems*. Command Paper 2836, pp. 71–9. London, HMSO.

Thorpe, W.H. (1972) The comparison of vocal communication in animals and man. In (R.A. Hinde ed.) *Non-verbal Communication*, pp. 27–48. Cambridge, Cambridge University Press.

Thuijsman, F., Peleg, B., Amitai, M. & Shmida, A. (1995) Automata, matching and foraging behavior in bees. *Journal of Theoretical Biology* **175**: 305–16.

Tilson, R.L. (1981) Family formation strategies of Kloss's gibbons. *Folia Primatologia* **35**: 259–81.

Tinbergen, L. (1960) The natural control of insects in pine woods: I. Factors influencing the intensity of predation in song birds. *Archives Néerlandaises de Zoologie* **13**: 265–343.

Tinbergen, N. (1951) *The Study of Instinct*. Oxford, Oxford University Press.

Tinbergen, N. (1952) Derived activities: their causation, biological significance, origin and emancipation during evolution. *Quarterly Review of Biology* **27**: 1–32.

Tinbergen, N. (1963) On aims and methods in ethology. *Zeitschrift für Tierpsychologie* **20**: 410–33.

Tinbergen, N. & Perdeck, A.C. (1950) On the stimulus situation releasing the begging response in the newly hatched herring gull chick (*Larus argentatus argentatus* Pont.). *Behaviour* **3**: 1–39.

Toates, F.M. (1986) *Motivational Systems*. Cambridge, Cambridge University Press.

Tolman, E.C. & Honzik, C.H. (1930) Introduction and removal of reward and maze performance in rats. *University of California Publications in Psychology* **4**: 257–75.

Tomasello, M. & Call, J. (1997) *Primate Cognition*. New York, Oxford University Press.

Tomasello, M., Kruger, A. & Ratner, H. (1993) Cultural learning. *Behavioral and Brain Sciences* **16**: 450–88.

Tomkins, J.L. & Simmons, L.W. (1999) Heritability of size but not asymmetry in a sexually selected trait chosen by female earwigs. *Heredity* **82**: 151–7.

Tooby, J. & Cosmides, L. (1990) The past explains the present: adaptations and the structure of ancestral environments. *Ethology and Sociobiology* **11**: 375–424.

Tooby, J. & Cosmides, L. (1992) The psychological foundations of cultures. In (J.H. Barkow, L. Cosmides & J. Tooby eds) *The Adapted Mind: Evolutionary Psychology and the Generation of Culture*. New York, Oxford University Press.

Tooby, J. & Cosmides, L. (1996) Friendship and the banker's paradox: other pathways to the evolution of adaptations for altruism. *Proceedings of the British Academy* **88**: 119–43.

Townsend, D.S. (1986) The costs of male parental care and its evolution in a neotropical frog. *Behavioral Ecology & Sociobiology* **19**: 187–95.

Toyoda, F., Ito, M., Tanaka, S. & Kikuyama, S. (1992) Hormonal induction of male courtship behavior in the Japanese newt *Cynops pyrrhogaster*. *Hormones and Behaviour* **27**: 511–22.

Traniello, J.F.A. & Beshers, S.N. (1991) Maximization of foraging efficiency and resource defense by group retrieval in the ant *Formica schaufussi*. *Behavioural Ecology & Sociobiology* **29**: 283–9.

Tregenza, T. & Wedell, N. (1998) Benefits of multiple mates in the cricket *Gryllus bimaculatus*. *Evolution* **52**: 1726–30.

Tregenza, T. & Wedell, N. (2000) Genetic compatibility, mate choice and patterns of parentage: invited review. *Molecular Ecology* **9**: 1013–27.

Treherne, J.E. & Foster, W.A. (1981) Group transmission of predator avoidance in a marine insect: the Trafalgar effect. *Animal Behaviour* **29**: 911–7.

Treisman, M. (1975) Predation and the evolution of gregariousness: I. Models for concealment and evasion. *Animal Behaviour* **23**: 779–88.

Treisman, M. & Dawkins, R. (1976) The cost of meoisis: is there any? *Journal of Theoretical Biology* **63**: 479–84.

Trivers, R.L. (1971) The evolution of reciprocal altruism. *Quarterly Review of Biology* **46**: 35–7.

Trivers, R.L. (1972) Parental investment and sexual selection. In (B. Campbell ed.) *Sexual Selection and the Descent of Man*, pp. 139–79. Chicago, Aldine Press.

Trivers, R.L. (1974) Parent–offspring conflict. *The American Zoologist* **14**: 249–64.

Trivers, R.L. (1976) Sexual selection and resource-accruing abilities in *Anolis garmani*. *Evolution* **30**: 253–69.

Trivers, R.L. (1985) *Social Evolution*. Menlo Park, CA, Benjamin-Cummings.

Trivers, R.L. & Hare, H. (1976) Haplodiploidy and the evolution of the social insects. *Science* **191**: 249–63.

Trivers, R.L. & Willard, D.E. (1973) Natural selection of parental ability to vary the sex ratio of offspring. *Science* **179**: 90–2.

Truman, J.W & Reiss, S.E. (1995) Neuromuscular metamorphosis in the moth *Manduca sexta*: hormonal regulation of synapse loss and remodeling. *Journal of Neuroscience* **15**: 4815–26.

Trune, D.R. & Slobodchikoff, C.N. (1976) Social effects of roosting on the metabolism of the pallid bat, *Antrozous pallidus*. *Journal of Mammalogy* **57**: 656–63.

Tully, T. & Quinn, W.G. (1985) Classical conditioning and retention in normal and mutant *Drosophila melanogaster*. *Journal of Comparative Physiology* **157**: 263–77.

Turke, P.W. (1989) Evolution and the demand for children. *Population and Development Review* **15**: 61–90.

Turner, J.R.G. (1971) Studies of Mullerian mimicry and its evolution in burnet moths and heliconiid butterflies. In (R. Creed ed.) *Ecological Genetics and Evolution*, pp. 224–60. Oxford, Blackwell.

Uetz, G.W. & Hieber, C.S. (1994) Group size and predation risk in colonial web-building spiders: analysis of attack abatement mechanisms. *Behavioral Ecology* **5**: 326–33.

Underwood, H., Barrett, R.K. & Siopes, T. (1990) The quail's eye: a biological clock. *Journal of Biological Rhythms* **5**: 257–65.

Unger, L.M. & Sargent, R.C. (1988) Alloparental care in the fathead minnow, *Pimphales promelas*: females prefer males with eggs. *Behavioral Ecology & Sociobiology* **23**: 27–32.

Upchurch, M. & Wehner, J.M. (1988) Differences between inbred strains of mice in Morris water maze performance. *Behavior Genetics* **18**: 55–68.

Urcuioli, P.J. & DeMarse, T.B. (1997) Memory processes in delayed spatial discriminations: response intentions or response mediation? *Journal of the Experimental Analysis of Behavior* **67**: 323–36.

Vahed, K. & Gilbert, F.S. (1996) Differences across taxa in nuptial gift size correlate with differences in sperm number and ejaculate volume in bushcrickets (Orthoptera: Tettigoniidae). *Proceedings of the Royal Society of London Series B* **263**: 1257–65.

Vale, J.R., Ray, D. & Vale, C.A. (1972) The interaction of genotype and exogenous neonatal androgen: agonistic behavior in female mice. *Behavioral Biology* **7**: 321–34.

Valen, L. van (1962) A study of fluctuating asymmetry. *Evolution* **16**: 125–42.

Valen, L. van (1973) A new evolutionary law. *Evolutionary Theory* **1**: 1–30.

Vallortigara, G., Zanforlin, M. & Pasti, G. (1990) Geometric modules in animals' spatial representations: a test with chicks (*Gallus gallus domesticus*). *Journal of Comparative Psychology* **194**: 248–54.

Valone, T.J. (1989) Group foraging, public information and patch estimation. *Oikos* **56**: 357–63.

Vehrencamp, S.L. (1978) The adaptive significance of communal nesting in groove-billed anis, *Crotophaga sulcirostris*. *Behavioral Ecology & Sociobiology* **4**: 1–33.

Vehrencamp, S.L. (1983) A model for the evolution of despotic versus egalitarian societies. *Animal Behaviour* **31**: 667–82.

Vehrencamp, S.L. & Bradbury, J.W. (1984) Mating systems and ecology. In (J.R. Krebs & N.B. Davies eds) *Behavioural Ecology: An Evolutionary Approach*, 2nd Edition, pp. 251–78. Oxford, Blackwell.

Verhulst, S., Dieleman, S.J. & Parmentier, H.K. (1999) A tradeoff between immunocompetence and sexual ornamentation in domestic fowl. *Proceedings of the National Academy of Sciences of the USA* **96**: 4478–81.

Verner, J. & Willson, M.F. (1966) The influence of habitats on the mating systems of North American passerine birds. *Ecology* **47**: 143–7.

Vestergaard, K.S. (1980) The regulation of dustbathing and other patterns in the laying hen: a Lorenzian approach. In (R. Moss ed.) *The Laying Hen and its Environment*, pp. 101–20. The Hague, Martinus Nijhoff.

Vestergaard, K.S., Damm, B.I., Abbott, U.K. & Bildsoe, M. (1999) Regulation of dustbathing in feathered and featherless domestic chicks: the Lorenzian model revisited. *Animal Behaviour* **58**: 1017–25.

Viehmann, W. (1979) The magnetic compass of blackcaps (*Sylvia atricapilla*). *Behaviour* **68**: 24–30.

Viguier, C. (1882) Le sens d'orientation et ses organes chez les animaux et chez l'homme. *Revue Philosophique* **14**: 1–36.

Vince, M.A. (1964) Use of the feet in feeding by the great tit *Parus major*. *Ibis* **106**: 508–29.

Vine, I. (1971) Risk of visual detection and pursuit by a predator and the selective advantage of flocking behaviour. *Journal of Theoretical Biology* **30**: 405–22.

Visalberghi, E. (1994) Learning processes and feeding behavior in monkeys. In (B.G. Galef, Jr, M. Mainardi & P. Valsecchi eds) *Behavioral Aspects of Feeding: Basic and Applied Research on Mammals*, pp. 257–70. Chur, Switzerland, Harwood Academic Publisher.

Visalberghi, E. & Alleva, E. (1979) Magnetic influences on pigeon homing. *Biological Bulletin* **156**: 246–56.

Vleck, C.M. & Brown, J.L. (1999) Testosterone and social and reproductive behaviour in *Aphelocoma* jays. *Animal Behaviour* **59**: 943–51.

Volman, S.F., Grubb, T.C. & Schuett, K.C. (1997) Relative hippocampal volume in relation to food-storing behavior in four species of woodpeckers. *Brain, Behavior and Evolution* **49**: 110–20.

Vreys, C. & Michiels, N.K. (1997) Flatworms flatten to size each other up. *Proceedings of the Royal Society of London Series B* **264**: 1559–64.

Waage, J.K. (1979a) Foraging for patchily distributed hosts by the parasitoid *Nemeritis canescens*. *Journal of Animal Ecology* **48**: 353–71.

Waage, J.K. (1979b) Dual function of the damselfly penis: sperm removal and transfer. *Science* **203**: 916–18.

Waal, F. de (1996) Conflict as negotiation. In (W.C. McGrew, L.F. Marchant & M.T. Nishida eds) *Great Ape Societies*, pp. 159–72. Cambridge, Cambridge University Press.

Waal, F. de (1982) *Chimpanzee Politics*. New York, Harper & Row.

Waas, J.R. (1991) The risks and benefits of signaling aggressive motivation: a study of cave-dwelling little blue penguins. *Behavioral Ecology & Sociobiology* **29**: 139–46.

Wade, M.J. (1977) An experimental study of group selection. *Evolution* **31**: 134–53.

Wade, M.J. & Shuster, S.M. (2002) The evolution of parental care in the context of sexual selection: a critical reassessment of parental investment theory. *American Naturalist* **160**: 285–92.

Wagner, A.R. (1981) SOP: a model of automatic memory processing in animal behavior. In (N.E. Spear & R.R. Miller eds) *Information Processing in Animals: Memory Mechanisms*, pp. 5–47. Hillsdale, NJ, Lawrence Erlbaum Associates.

Wagner, G. (1978) Homing pigeons flight over and under low stratus. In (K. Schmidt-Koenig & W.T. Keeton eds) *Animal Migration: Navigation & Homing*, pp. 455–70. Heidelberg, Springer.

Wajnberg, E., Fauvergue, X. & Pons, O. (2000) Patch leaving decision rules and the marginal value theorem: an experimental analysis and a simulation model. *Behavioural Ecology* **11**: 577–86.

Walcott, C. & Green, R.P. (1974) Orientation of homing pigeons is altered by a change in the direction of an applied magnetic field. *Science* **184**: 180–2.

Waldman, B. (1987) Mechanisms of kin recognition. *Journal of Theoretical Biology* **128**: 159–85.

Wallraff, H.G. (1991) Conceptual approaches to avian navigation systems. In (P. Berthold ed.) *Orientation in Birds*, pp. 128–65. Basel, Birkhäuser.

Wallraff, H.G. (1999) The magnetic map of homing pigeons: an evergreen phantom. *Journal of Theoretical Biology* **197**: 265–9.

Wallraff, H.G. (2000) Simulated navigation based on observed gradients of atmospheric trace gases (models on pigeon homing 3). *Journal of Theoretical Biology* **205**: 133–45.

Wallraff, H.G. (2001) Navigation by homing pigeons: updated perspective. *Ethology, Ecology & Evolution* **13**: 1–48.

Wallraff, H.G. & Foà, A. (1981) Pigeon navigation: charcoal filter removes relevant information from environmental air. *Behavioral Ecology & Sociobiology* **9**: 67–77.

Wallraff, H.G., Kiepenheuer, J., Neumann, M.F. & Streng, A. (1995) Homing experiments with starlings deprived of the sense of smell. *Condor* **97**: 20–6.

Walters, E.T., Carew, T.J. & Kandel, E.R. (1981) Associative learning in *Aplysia*: evidence for conditioned fear in an invertebrate. *Science* **211**: 504–6.

Ward, J.A. & Barlow, G.W. (1967) The maturation and regulation of glancing off the parents by young orange chromides (*Etroplus maculatus*: Pisces–Cichlidae). *Behaviour* **29**: 1–56.

Ward, P. & Zahavi, A. (1973) The importance of certain assemblages of birds as 'information centres' for food finding. *Ibis* **115**: 517–34.

Warren, J.M. (1965) Primate learning in comparative perspective. In (A.M. Schrier, H.F. Harlow & F. Stollnitz eds) *Behavior of Non-human Primates*, Vol. I, pp. 249–81. New York, Academic Press.

Warren, J.M. (1973) Learning in vertebrates. In (D.A. Dewsbury & D.A. Rethlingshafer eds) *Comparative Psychology: A Modern Survey*, pp. 471–509. New York, Academic Press.

Wasserman, E.A. & Astley, S.L. (1994) A behavioral analysis of concepts: its application to pigeons and children. *Psychology, Motivation & Learning* **31**: 73–132.

Wasserman, E.A., DeVolder, C.L. & Coppage, D.J. (1992) Non-similarity-based conceptualization in pigeons via secondary or mediated generalization. *Psychological Science* **3**: 374–9.

Watanabe, S., Lea, S.E.G. & Dittrich, W.H. (1993) What can we learn from experiments on pigeon concept discrimination? In (H.P. Zeigler & H.-J. Bischof eds) *Vision, Brain and Behavior in Birds*, pp. 351–76. Cambridge, MA, MIT Press.

Watson, J.B. (1913) Psychology as the behaviorist views it. *Psychological Review* **20**: 158–77.

Watts, D.P. (1985) Relations between group size and composition and feeding competition in mountain gorilla groups. *Animal Behaviour* **33**: 72–85.

Waynforth, D. & Dunbar, R.I.M. (1995) Conditional mate choice strategies in humans: evidence from lonely hearts advertisements. *Behaviour* **132**: 755–79.

Webb, B. (2000) What does robotics offer animal behaviour? *Animal Behaviour* **60**: 545–58.

Webb, B. (2002) Robots in invertebrate neuroscience. *Nature* **417**: 359–63.

Webb, B. & Scutt, T. (2000) A simple latency dependent spiking neuron model of cricket phonotaxis. *Biological Cybernetics* **82**: 247–69.

Wecker, S.C. (1963) Habitat selection. *Scientific American* **211**: 109–16.

Wedekind, C. & Braithwaite, V.A. (2002) The long term benefits of human generosity in indirect reciprocity. *Current Biology* **12**: 1012–15.

Wedekind, C. & Folstad, I. (1994) Adaptive and non-adaptive immunosuppression by sex hormones. *American Naturalist* **143**: 936–8.

Wedekind, C. & Furi, S. (1997) Body odour preferences in men and women: do they aim for specific MHC combinations or simply heterozygosity? *Proceedings of the Royal Society of London Series B* **264**: 1471–9.

Wedekind, C. & Milinski, M. (2000) Cooperation through image scoring in humans. *Science* **288**: 850–2.

Wedekind, C., Chapuisat, M., Macas, E. & Rülicke, T. (1996) Nonrandom fertilization in mice correlates with MHC and something else. *Heredity* **77**: 400–9.

Wedekind, C., Seebeck, T., Bettens, F. & Paepke, A. (1995) MHC-dependent mate preferences in humans. *Proceedings of the Royal Society of London Series B* **260**: 245–9.

Wehner, J.M., Radcliffe, R.A., Rosmann, S.T., Christensen, S.C., Rasmussen, D.L., Fulker, D.W. & Wiles, M. (1997) Quantitative trait locus analysis of contextual fear conditioning in mice. *Nature Genetics* **17**: 331–4.

Wehner, R. (1989) Neurobiology of polarization vision. *Trends in Neuroscience* **12**: 353–9.

Wehner, R. (1992) Arthropods. In (F. Papi ed.) *Animal Homing*, pp. 45–144. London, Chapman and Hall.

Wehner, R. (1998) Navigation in context: grand theories and basic mechanisms. *Journal of Avian Biology* **29**: 370–86.

Weidenmayer, C. (1997) Causation of the ontogenetic stereotypic development of stereotypic digging in gerbils. *Animal Behaviour* **53**: 461–70.

Weih, A.S. (1951) Unterschungen über das Wechselsingen (Anaphonie) und über das angeborene Lautschema einiger Feldheuschrecken. *Zeitschrift für Tierpsychologie* **8**: 1–41.

Weihs, D. (1973) Hydrodynamics of fish schooling. *Nature* **241**: 290–1.

Weiner, J. (1999) *Time, Love, Memory: A Great Biologist and his Quest for the Origins of Behavior.* New York, Alfred A. Knopf.

Weir, A.A.S., Chappell, J. & Kacelnik, A. (2002) Shaping of hooks in New Caledonian crows. *Science* **297**: 981.

Weiskrantz, L. (1986) *Blindsight.* Oxford, Clarendon Press.

Wells, P.A. (1987) Kin recognition in humans. In (D.J.C. Fletcher & C.D. Michener eds) *Kin Recognition in Animals,* pp. 395–415. New York, John Wiley.

Wenner, A.M. & Wells, P.H. (1990) *Anatomy of a Controversy.* New York, Columbia University Press.

Wenner, A.M., Meade, D.E. & Friesen, L.J. (1991) Recruitment, search behavior, and flight ranges of honey bees. *American Zoologist* **31**: 768–82.

West, K.J. & Alexander, R.D. (1963) Sub-social behavior in a burrowing cricket, *Anurogryllus muticus* (De Geer). Orthoptera: Gryllidae. *Ohio Journal of Science* **63**: 19–24.

West, S.A. & Sheldon, B.C. (2002) Constraints in the evolution of sex ratio adjustment. *Science* **295**: 1685–8.

Westerberg, H.G., Murphy, D.L. & Boer, J.A. den (1996) *Advances in the Neurobiology of Anxiety Disorders.* New York, Wiley.

Westermarck, E.A. (1891) *The History of Human Marriage.* New York, Macmillan.

Westneat, D.F., Walters, A., McCarthy, T.M., Hatch, M.I. & Hein, W.K. (2000) Alternative mechanisms of nonindependent mate choice. *Animal Behaviour* **59**: 467–76.

Wheeler, D.A., Kyriacou, C.P., Greenacre, M.L., Yu, Q., Rutila, M., Rosbash, M. & Hall, J.C. (1991) Molecular transfer of a species-specific behavior from *Drosophila simulans* to *Drosophila melanogaster. Science* **251**: 1082–5.

Wheeler, W.M. (1910) *Ants: Their Structure, Development and Behavior.* New York, Columbia University Press.

Whiten, A. (1994) Grades of mind-reading. In (C. Lewis & P. Mitchell eds) *Children's Early Understanding of Mind: Origins and Development,* pp. 47–70. Hove, Lawrence Erlbaum Associates.

Whiten, A, & Byrne, R. eds (1997) *Machiavellian Intelligence II: Extensions and Evaluations.* Cambridge, Cambridge University Press.

Whiten, A., Custance, D.M., Gomez, J.-C., Teixidor, P. & Bard, K.A. (1996) Imitative learning of artificial fruit processing in children (*Homo sapiens*) and chimpanzees (*Pan troglodytes*). *Journal of Comparative Psychology* **110**: 3–14.

Whiten, A., Goodall, J., McGrew, W.C., Nishida, T., Reynolds, V., Sugiyuma, Y., Tutin, C.E.G., Wrangham, R.W. & Boesch, C. (1999) Cultures in chimpanzees. *Nature* **399**: 682–5.

Whitham, T.G. (1978) Habitat selection by *Pemphigus* aphids in response to resource limitation and competition. *Ecology* **59**: 1164–76.

Whitham, T.G. (1980) The theory of habitat selection examined and extended using *Pemphigus* aphids. *American Naturalist* **115**: 449–66.

Whyatt, T. (2003) *Pheromones and Animal Behaviour.* Cambridge, Cambridge University Press.

Wichems, C., Sora, I., Andrews, A.M., Bengel, D., Uhl, G. & Murphy, D.L. (1998) Altered responses to psychoactive drugs and spontaneous behavior differences in mice lacking the serotonin transporter. *Abstract Fourth IUPHAR Satellite Meeting on Serotonin,* Rotterdam.

Wickler, W. (1968) *Mimicry in Plants and Animals.* London, Weidenfeld & Nicholson.

Wicksten, M.K. (1980) Decorator crabs. *Scientific American* **242**: 146–54.

Wilcock, J. (1969) Gene action and behavior: an evaluation of a major gene pleiotropism. *Psychological Bulletin* **72**: 1–29.

Wiley, R.H. (1983) The evolution of communication: information and manipulation. In (T.R. Halliday & P.J.B. Slater eds) *Animal Behaviour. Volume 2: Communication,* pp. 156–189. Oxford, Blackwell.

Wiley, R.H. & Richards, D.G. (1978) Physical constraints on acoustic communication in the atmosphere: implications for the evolution of animal vocalizations. *Behavioral Ecology & Sociobiology* **3**: 69–94.

Wilkinson, G.S. (1984) Reciprocal food sharing in vampire bats. *Nature* **308**: 181–4.

Wilkinson, G.S. & Reillo, P.R. (1994) Female choice response to artificial selection on an exaggerated male trait in a stalk-eyed fly. *Proceedings of the Royal Society of London Series B* **255**: 1–6.

Wilkinson, G.S. & Taper, M. (1999) Evolution of genetic variation for condition-dependent traits in stalk-eyed flies. *Proceedings of the Royal Society of London Series B* **266**: 1685–90.

Wilkinson, G.S., Presgraves, D.C. & Crymes, L. (1998) Male eye span in stalk-eyed flies indicates genetic quality by meiotic drive suppression. *Nature* **391**: 276–9.

Will, B.E. (1977) Neurochemical correlates of individual differences in animal learning capacity. *Behavioral Biology* **19**: 143–59.

Williams, B.A. (1988) Reinforcement, choice and response strength. In (R.C. Atkinson ed.) *Stevens Handbook of Experimental Psychology*, pp. 81–108. New York, Wiley.

Willams, C.L., Barnett, A.M. & Meck, W.H. (1990) Organizational effects of early gonadal secretions on sexual differentiation in spatial memory. *Behavioral Neuroscience* **104**: 84–97.

Williams, D.R. & Williams, H. (1969) Auto-maintenance in the pigeon: sustained pecking despite contingent non-reinforcement. *Journal of the Experimental Analysis of Behavior* **12**: 511–20.

Williams, G.C. (1975) *Sex and Evolution*. Princeton, NJ, Princeton University Press.

Williams, G.C. (1979) The question of adaptive sex ratio in outcrossed vertebrates. *Proceedings of the Royal Society of London Series B* **205**: 567–80.

Williams, G.C. (1992) *Natural Selection: Domains, Levels, Challenges*. New York, Oxford University Press.

Willows, A.O.D. & Hoyle, G. (1969) Neuronal network triggering a fixed action pattern. *Science* **166**: 1549–51.

Wilson, D.S. (1980) *The Natural Selection of Populations and Communities*. Menlo Park, CA, Benjamin Cummings.

Wilson, D.S. & Sober, E. (1989) Reviving the superorganism. *Journal of Theoretical Biology* **136**: 337–56.

Wilson, E.O. (1971) *The Insect Societies*. Cambridge, MA, Harvard University Press.

Wilson, E.O. (1975a) *Sociobiology: The New Synthesis*. Cambridge, MA, Belknap.

Wilson, E.O. (1975b) Animal communication. In (T. Eisner & E.O. Wilson eds) *Animal Behavior: Readings from Scientific American*, pp. 265–72. San Francisco, Freeman.

Wilson, E.O. (1978) *On Human Nature*. Cambridge, MA, Harvard University Press.

Wilson, M.E., Gordon, T.P. & Bernstein, I.S. (1978) Timing of births and reproductive success in rhesus monkey social groups. *Journal of Medical Primatology* **7**: 202–12.

Wiltschko, R. (1996) The function of olfactory input in pigeon orientation: does it provide navigational information or play another role? *Journal of Experimental Biology* **199**: 113–19.

Wiltschko, R. & Wiltschko, W. (1998) Pigeon homing: effect of various wave lengths of light during displacement. *Naturwissenschaften* **85**: 164–7.

Wiltschko, R. & Wiltschko, W. (1999a) The orientation system of birds: III. Migratory orientations. *Journal für Ornithologie* **140**: 273–308.

Wiltschko, R. & Wiltschko, W. (1999b) The orientation system of birds: IV. Evolution. *Journal für Ornithologie* **140**: 393–417.

Wiltschko, R. & Wiltschko, W. (1999c) The orientation system of birds: I. Compass mechanisms. *Journal für Ornithologie* **140**: 1–40.

Wiltschko, W. (1968) Uber den Einfluss statischer Magnetfelder auf die Zugorientierung der Rotkelchen (*Erithacus rubecula*). *Zeitschrift für Tierpsychologie* **25**: 537–58.

Wiltschko, W. (1982) The migratory orientation of garden warblers, *Sylvia borin*. In (F. Papi & H.G. Wallraff eds) *Avian Navigation*, pp. 50–8. Heidelberg, Springer.

Wiltschko, W. & Balda, R.P. (1989) Sun compass orientation in seed-caching scrub jays (*Aphelocoma coerulescens*). *Journal of Comparative Physiology A* **164**: 717–21.

Wiltschko, W., Daum, P., Fergenbauer-Kimmel, A. & Wiltschko, R. (1987) The development of the star compass in garden warblers, *Sylvia borin*. *Ethology* **74**: 285–92.

Wiltschko, W., Wiltschko, R. & Keeton, W.T. (1976) Effects of a 'permanent' clock-shift on the orientation of young homing pigeons. *Behavioural Ecology & Sociobiology* **1**: 229–43.

Wiltschko, W., Weindler, P. & Wiltschko, R. (1998) Interaction of magnetic and celestial cues in the migratory orientation of birds. *Journal of Avian Biology* **29**: 606–17.

Wingfield, J.C. & Moore, M.C. (1987) Hormonal, social and environmental factors in the reproductive biology of free-living male birds. In (D. Crews ed.) *Psychobiology of Reproductive Behavior: An Evolutionary Perspective*. Englewood Cliffs, NJ, Prentice Hall.

Wingfield, J.C., Hegner, R.E., Dufty, A.M. Jr & Ball, G.F. (1990) The 'challenge hypothesis': theoretical implications for patterns of testosterone secretion, mating systems and breeding strategies. *American Naturalist* **136**: 829–46.

Winkler, H. & Kothbauer-Hellmann, R. (2001) The role of search area in the detection of cryptic prey by crested tits and coal tits. *Behaviour* **138**: 873–83.

Winklhofer, M., Holtkamp-Rotzler, E., Hanzlik, M., Fleissner, G. & Peterson, N. (2001) Clusters of supermagnetic magnetite particles in the upper beak skin of homing pigeons: evidence of a magnetoreceptor? *European Journal of Mineralogy* **13**: 659–69.

Winn, P. (1995) The lateral hypothalamus and motivated behavior: an old syndrome reassessed and a new perspective gained. *Current Directions in Psychological Science* **4**: 1182–7.

Winn, P., Tarbuck, A. & Dunnett, S.B. (1984) Ibotenic acid lesions of the lateral hypothalamus: comparisons with the electrolytic lesion syndrome. *Neuroscience* **12**: 225–40.

Wisenden, B.D. & Thiel, T.A. (2002) Field verification of predator attraction to minnow alarm substance. *Journal of Chemical Ecology* **28**: 433–8.

Witte, K. & Ryan, M.J. (2002) Mate choice copying in the sailfin molly, *Poecilia latipinna*, in the wild. *Animal Behaviour* **63**: 943–9.

Wolf, A.P. & Huang, C. (1980) *Marriage and Adoption in China*. Stanford, Stanford University Press.

Wolfensohn, S. & Lloyd, M. (1994) *Handbook of Laboratory Animal Management and Welfare*. Oxford, Oxford University Press.

Wollerman, L. (1999) Acoustic interference limits call detection in a neotropical frog, *Hyla ebraccata*. *Animal Behaviour* **57**: 529–36.

Woolfenden, G.E. & Fitzpatrick, J.W. (1978) The inheritance of territory in group breeding birds. *BioScience* **28**: 104–8.

Woolfenden, G.E. & Fitzpatrick, J.W. (1984) *The Florida Scrub Jay*. Princeton, NJ, Princeton University Press.

Woolum, J.C. (1991) A re-examination of the role of the nucleus in generating the circadian rhythm in *Acetabularia*. *Journal of Biological Rhythms* **6**: 129–36.

Wrangham, R. & Peterson, D. (1996) *Demonic Males*. Boston, Houghton Mifflin.

Wrona, F.J. & Dixon, R.W.J. (1991) Group size and predation risk: a field analysis of encounter and dilution effects. *American Naturalist* **137**: 186–201.

Wu, J., Whittier, J.M. & Crews, D. (1985) Role of progesterone in the control of female sexual receptivity in *Anolis carolinensis*. *General and Comparative Endocrinology* **58**: 402–6.

Würbel, H. (2001) Ideal homes? Housing effects on rodent brain and behaviour. *Trends in Neurosciences* **24**: 207–11.

Wyman, J. (1967) The jackals of the Serengeti. *Animals* **10**: 79–83.

Wyman, R.L. & Ward, J.A. (1973) The development of behaviour in the cichlid fish *Etroplus maculatus*. *Zeitschrift für Tierpsychologie* **33**: 461–91.

Wynne-Edwards, V.C. (1962) *Animal Dispersion in Relation to Social Behaviour*. Edinburgh, Oliver & Boyd.

Wynne-Edwards, V.C. (1986) *Evolution Through Group Selection*. Oxford, Blackwell Scientific Publications.

Yamazaki, K., Beauchamp, G.K., Kupniewski, D., Bard, J., Thomas, L. & Boyse, E.A. (1988) Familial imprinting determines H-2 selective mating preferences. *Science* **240**: 1331–2.

Yamazaki, K., Boyse, E.A., Miké, V., Thaler, H.T., Mathieson, B.J., Abbott, J., Zayas, Z.A. & Thomas, L. (1976) Control of mating preferences in mice by genes in the major histocompatibility complex. *Journal of Experimental Medicine* **144**: 1324–35.

Ydenberg, R.C., Prins, H.H.Th. & van Dijk, J. (1983) Post roost gatherings and information centres. *Ardea* **71**: 125–32.

Young, D. (1996) *Nerve Cells and Animal Behaviour*. Cambridge, Cambridge University Press.

Young, W.C. (1965) The organization of sexual behaviour by hormonal action during the prenatal and larval periods in vertebrates. In (F.A. Beach ed.) *Sex and Behavior*. New York, Wiley.

Youthed, G.J. & Moran, R.C. (1969) The lunar day activity rhythm of myrmeleontid larvae. *Journal of Insect Physiology* **15**: 1259–71.

Yu, A.C. & Margoliash, D. (1996) Temporal hierarchical control of singing in birds. *Science* **273**: 1871–5.

Yu, Q., Jacquier, A.C., Citri, Y., Hamblen, M., Hall, J.C. & Rosbash, M. (1987) Molecular mapping of point mutations in the *period* gene that stop or speed up biological clocks in *Drosophila melanogaster*. *Proceedings of the National Academy of Sciences of the USA* **84**: 784–8.

Zahavi, A. (1971) The social behaviour of the white wagtail, *Motacilla alba alba*, wintering in Israel. *Ibis* **113**: 203–11.

Zahavi, A. (1975) Mate selection – a selection for a handicap. *Journal of Theoretical Biology* **53**: 205–14.

Zahavi, A. (1977) The cost of honesty (further remarks on the handicap principle). *Journal of Theoretical Biology* **67**: 603–5.

Zahavi, A. (1987) The theory of signal selection and some of its implications. In (V.P. Delfino ed.) *International Symposium of Biological Evolution*, pp. 305–27. Bari, Adriatica Editrice.

Zahavi, A. & Zahavi, A. (1997) *The Handicap Principle*. Oxford, Oxford University Press.

Zehring, W.A., Wheeler, D.A., Reddy, P., Konopka, J., Kyriacou, C.P., Rosbash, M. & Hall, J.C. (1984) P-element transformation with *period* locus DNA restores rhythmicity to mutant, arrhythmic, *Drosophila melanogaster*. *Cell* 39: 369–76.

Zeldin, R.K. & Olton, D.S. (1986) Rats acquire spatial learning sets. *Journal of Experimental Psychology: Animal Behavior Processes* 12: 412–19.

Zentall, T.R. (1996) An analysis of imitative learning in animals. In (C.M. Heyes & B.G. Galef Jr eds) *Social Learning in Animals: the Roots of Culture*, pp. 221–43. San Diego, Academic Press.

Zhang, H.Y. & Hoffmann, K.P. (1993) Retinal projections to the pretectum, accessory optic system and superior colliculus in pigmented and albino ferrets. *European Journal of Neuroscience* 5: 486–500.

Zill, S.N. & Moran, D.T. (1981) The exoskeleton and insect proprioception: I. Responses of tibial campaniform sensillae to external and muscle-generated forces in the American cockroach, *Periplaneta americana*. *Journal of Experimental Biology* 91: 1–24.

Zuberbuhler, K., Jenny, D. & Bshary, R. (1999) The predator deterrence function of primate alarm calls. *Ethology* 105: 477–90.

Zuk, M. (2002) *Sexual Selections: What We Can and Can't Learn about Sex from Animals*. Los Angeles, CA, University of California Press.

Zuk, M., Johnsen, T. & Maclarty, T. (1995) Endocrine–immune interactions, ornaments and mate choice in red jungle fowl. *Proceedings of the Royal Society of London Series B* 260: 205–10.

Zuk, M., Ligon. J.D. & Thornhill, R. (1992) Effects of experimental manipulation of male secondary sexual characters on female mate preference in red jungle fowl. *Animal Behaviour* 44: 999–1006.

INDEX